Springer Series in Optical Sciences

Volume 194

Springer Series in Optical Sciences

The Springer Series in Optical Sciences, under the leadership of Editor-in-Chief William T. Rhodes, Georgia Institute of Technology, USA, provides an expanding selection of research monographs in all major areas of optics: lasers and quantum optics, ultrafast phenomena, optical spectroscopy techniques, optoelectronics, quantum information, information optics, applied laser technology, industrial applications, and other topics of contemporary interest.

With this broad coverage of topics, the series is of use to all research scientists and engineers who need up-to-date reference books.

The editors encourage prospective authors to correspond with them in advance of submitting a manuscript. Submission of manuscripts should be made to the Editor-in-Chief or one of the Editors. See also www.springer.com/series/624

More information about this series at http://www.springer.com/series/624

Stefan Wabnitz · Benjamin J. Eggleton
Editors

All-Optical Signal Processing

Data Communication and Storage Applications

Springer

Editors
Stefan Wabnitz
Dipartimento di Ingegneria dell'Informazione
Università degli Studi di Brescia
Brescia
Italy

Benjamin J. Eggleton
CUDOS, School of Physics
University of Sydney
Sydney, NSW
Australia

ISSN 0342-4111 ISSN 1556-1534 (electronic)
Springer Series in Optical Sciences
ISBN 978-3-319-14991-2 ISBN 978-3-319-14992-9 (eBook)
DOI 10.1007/978-3-319-14992-9

Library of Congress Control Number: 2015933154

Springer Cham Heidelberg New York Dordrecht London

Printed on acid-free paper

Springer International Publishing AG Switzerland is part of Springer Science+Business Media (www.springer.com)

Preface

Although it holds the promise for substantial processing speed improvements, in today's communication infrastructure optics remains largely confined to the signal transport layer, as it lags behind electronics as far as signal processing is concerned. This situation is bound to change in the near future as the tremendous growth of data traffic requires the development of new, energy efficient, and fully transparent all-optical networks for telecom and datacom applications. This book provides a comprehensive review of the state of the art of all-optical devices based on nonlinear optical materials for applications to optical signal processing. Contributors to this book present breakthrough solutions for enabling a pervasive use of optics in data communication and signal storage applications. The book content ranges from the development of innovative materials and devices, such as graphene and photonic crystal structures, to the use of nonlinear optical signal processing for secure quantum information processing, for increasing the transmission channel capacity, and for enhancing the performance of broadband radio frequency signal processing. The book is expected to benefit all researchers in the fields of optical communications, photonic devices for optical signal processing, nonlinear guided wave optics, quantum information processing, and microwave photonics, including senior undergraduate and postgraduate students and industry researchers.

Chapter 1 summarizes the recent progress in materials and structures for all-optical signal processing that employ either second- or third-order optical nonlinearities. The dominant choice for quadratic materials is provided by periodically poled lithium niobate waveguides. For cubic nonlinearities, the range of materials ranges from glasses to both active and passive semiconductor devices: a brief summary of the advantages and disadvantages of each class of materials and device structure is provided. In Chap. 2, recent advances in new nonsilicon CMOS-compatible platforms for nonlinear integrated optics are revised, focusing on Hydex glass and silicon nitride. The promising new platform of amorphous silicon is also briefly discussed. These material systems have opened up new functionalities such as on-chip optical frequency comb generation, ultrafast optical pulse generation, and measurement. Chapter 3 overviews the principles of optical switching devices, based on either optical or electrical control signals, which permit to avoid the necessity of electro-optic conversion. Discussed devices include nonlinear mode couplers and interferometers based on optical fibers, and integrated waveguides based on photonic crystal structures or surface wave interactions in novel materials such as graphene. Chapter 4 reviews the recent progress on using nonlinear optical fibers for optical pulse shaping in the temporal and spectral domains. Significant examples that are most relevant for applications include the synthesis of specialized temporal waveforms, the generation of ultrashort pulses, and optical supercontinuum.

Given the exponential growth rate of the total volume of data transported across the communication network, energy consumption, alongside increased information capacity, has become a critical driver in deploying new technologies. In addition to transponders at the end terminals of an optical network, certain signal processing functions, such as regeneration, format conversion, wavelength conversion, and arbitrary waveform generation, are often proposed. The following chapters of the book discuss how many of these intermediate functions may be performed all-optically, with the primary advantage of increased bandwidth and consequent resource sharing. In Chap. 5 the need, the general principles, and the approaches used for the all-optical regeneration of mainly phase encoded signals of differing levels of coding complexity are discussed. The key underpinning technology and the current state of the art of optical regeneration, including a historic perspective, are presented.

Chapter 6 presents the theory and experiments of photonic signal processing, logic operations, and computing. These functionalities take advantage of nonlinear processes with ultrafast response time to perform high speed operations either on analog or digital optical signals directly in the optical domain. Practical all-optical frequency generation and conversion requires highly efficient parametric interactions across a wide spectral band. Chapter 7 presents a new class of traveling wave parametric mixers for efficient, cavity-free frequency generation. Driven by continuous-wave seeds, these devices combine inherently more stable lasers with distributed noise inhibition in dispersion-managed parametric processes. The operating principles, the design methodology, and the performance limits of parametric mixers are discussed, together with applications to signal multicasting and ultrafast channel processing.

Optical pulse shaping techniques are an active area of research for increasing the spectral efficiency of optical modulation formats in dense wavelength division multiplexed (DWDM) transmissions, by avoiding interchannel and intersymbol interference. In Chap. 8 the main pulse shapes of interest are introduced, the different available techniques for their generation are presented, along with the associated signal multiplexing schemes, namely orthogonal frequency division multiplexing (OFDM) and Nyquist pulse modulation. The relative performances of electronic and optical signal processors for implementing Fourier transforms and Nyquist pulse generation are discussed.

As previously mentioned, energy saving will be a main driver for the transition from electronic to optical signal processing solutions. Chapter 9 describes advanced functionalities for optical signal processing with reduced energy consumption using optical time lenses. This approach permits broadband optical processing, which is also capable of handling many bits in a single operation. In this way, the processing energy is shared by the many bits, and the energy per bit is reduced. The basic functionality is serial-to-parallel conversion in a single time lens. Combining time lenses into telescopic arrangements allows for more advanced signal processing solutions, such as conversion of OFDM signals into DWDM-like signals, which can be separated passively, i.e., without additional energy consumption. The previously discussed signal processing functions may also be performed by using the very optoelectronic devices used in the transponders themselves, either including decision circuitry and/or forward error correction or as linear media converters. In Chap. 10, the performance and energy consumption of digital coherent transponders, linear coherent repeaters, and modulator-based pulse shaping/frequency conversion is analyzed, thus setting an important benchmark for the proposed all-optical implementations.

The exponential growth in demand for information transmission capacity requires a rethinking of the maximum or Shannon capacity of the fiber-optic communication channel, in the presence of fiber nonlinearity. Chapter 11 addresses the problem of estimating the Shannon capacity for nonlinear communication channels, and discusses the potential of different optical signal coding, transmission, and processing techniques to improve the information capacity and increase the system reach of fiber-optic links.

Information and communication technologies based on quantum optics principles may lead to greatly improved functionalities such as enhanced sensing, exponentially faster computing, and the secure transfer of information. Chapter 12 overviews techniques for the nonlinear optics-based encoding and fully secure transfer of information in a quantum communication optical network. Chapter 13 discusses how classical optical signal processing techniques can be extended to nonclassical entangled photon states, thus permitting unprecedented control of the time frequency correlations shared by these light quanta. Moreover, in Chap. 13 it is shown that quantum properties produce interesting effects that are not observable with classical fields. Examples include Fourier transform pulse shaping, which relies on programmable spectral filtering and electro-optic modulation, where the temporal phase or amplitude of the entagled photon state is manipulated by means of an electrical signal.

Chapter 14 discusses how nonlinear optical effects in photonic chip scale devices may be exploited for enhancing the performance of radio frequency (RF) signal processing in microwave photonics applications. Specific examples presented in Chap. 14 include frequency agile and high suppression microwave bandstop filters, general purpose programmable analog signal processors, and high performance active microwave filters. Finally, Chap. 15 presents recent advances in optical signal processing techniques for wireless RF signals. Specifically, Chap. 15 discusses photonic architectures for wideband analog signal processing, including RF beamforming, co-channel interference cancelation, and physical layer security.

Besides broadband operation, photonics offers reduced size, weight, and power, in addition to low transmission loss, rapid reconfigurability, and immunity to electromagnetic interference.

We would like to thank Dr. Claus E. Ascheron, Springer Executive Editor for Physics, for inviting us, during the 2013 Conference on lasers and electro-optics in Munich, Germany, to bring a volume on all-optical signal processing to a wider audience. We also acknowledge helpful comments and suggestions by Dr. Herbert Venghaus of the Fraunhofer Institute for Telecommunications, and Editor of the Springer Series in Optical Sciences. Last but not the least, we are most grateful to all colleagues who contributed to this book for their brilliant work and continued effort in bringing this project to reality.

Brescia, Italy Stefan Wabnitz
Sydney, Australia Benjamin J. Eggleton

Contents

Contributors

Antonella Bogoni CNIT, National Laboratory of Photonic Networks, Pisa, Italy

Sonia Boscolo School of Engineering and Applied Science, Aston Institute of Photonic Technologies, Aston University, Birmingham, UK

Camille-Sophie Brès Institute of Electrical Engineering, Ecole Polytechnique Fédérale de Lausanne (EPFL), Lausanne, Switzerland

John Chang Princeton University, Princeton, NJ, USA

Matthew P. Chang Princeton University, Princeton, NJ, USA

Duk-Yong Choi Centre for Ultrahigh-bandwidth Devices of Optical Systems, Laser Physics Centre, RSPE, The Australian National University, Canberra, ACT, Australia

Alex S. Clark School of Physics, Centre for Ultrahigh bandwidth Devices for Optical Systems (CUDOS), Institute of Photonics and Optical Science (IPOS), University of Sydney, Camperdown, NSW, Australia

Anders Clausen DTU Fotonik, Department of Photonics Engineering, Technical University of Denmark, Kongens Lyngby, Denmark

Matthew J. Collins School of Physics, Centre for Ultrahigh bandwidth Devices for Optical Systems (CUDOS), Institute of Photonics and Optical Science (IPOS), University of Sydney, Camperdown, NSW, Australia

Costantino De Angelis Dipartimento di Ingegneria dell'Informazione, Università degli Studi di Brescia, Brescia, Italy

Benjamin J. Eggleton School of Physics, Centre for Ultrahigh bandwidth Devices for Optical Systems (CUDOS), Institute of Photonics and Optical Science (IPOS), University of Sydney, Camperdown, NSW, Australia

Andrew Ellis School of Engineering and Applied Science, Aston Institute of Photonic Technologies, Aston University, Birmingham, UK

Simon Fabbri School of Engineering and Applied Science, Aston Institute of Photonic Technologies, Aston University, Birmingham, UK

Julien Fatome Laboratoire Interdisciplinaire CARNOT de Bourgogne, UMR 6303 CNRS-Université de Bourgogne, Dijon, France

Christophe Finot Laboratoire Interdisciplinaire CARNOT de Bourgogne, UMR 6303 CNRS-Université de Bourgogne, Dijon, France

Xin Gai Centre for Ultrahigh-bandwidth Devices of Optical Systems, Laser Physics Centre, RSPE, The Australian National University, Canberra, ACT, Australia

Michael Galili DTU Fotonik, Department of Photonics Engineering, Technical University of Denmark, Kongens Lyngby, Denmark

Pengyu Guan DTU Fotonik, Department of Photonics Engineering, Technical University of Denmark, Kongens Lyngby, Denmark

Lukas G. Helt MQ Photonics Research Centre and CUDOS, Department of Physics and Astronomy, Macquarie University, North Ryde, NSW, Australia

Hao Hu DTU Fotonik, Department of Photonics Engineering, Technical University of Denmark, Kongens Lyngby, Denmark

Joseph Kakande Bell Laboratories, Alcatel-Lucent, Holmdel, NJ, USA

Bill P.-P. Kuo Department of Computer and Electrical Engineering, University of California San Diego, La Jolla, CA, USA

Juerg Leuthold Institute of Electromagnetic Fields, ETH-Zurich, Zurich, Switzerland

Mads Lillieholm DTU Fotonik, Department of Photonics Engineering, Technical University of Denmark, Kongens Lyngby, Denmark

Andrea Locatelli Dipartimento di Ingegneria dell'Informazione, Università degli Studi di Brescia, Brescia, Italy

Joseph M. Lukens Purdue University, West Lafayette, IN, USA

Barry Luther-Davies Centre for Ultrahigh-bandwidth Devices of Optical Systems, Laser Physics Centre, RSPE, The Australian National University, Canberra, ACT, Australia

Steve Madden Centre for Ultrahigh-bandwidth Devices of Optical Systems, Laser Physics Centre, RSPE, The Australian National University, Canberra, ACT, Australia

David Marpaung School of Physics, Centre for Ultrahigh Bandwidth Devices for Optical Systems (CUDOS), Institute of Photonics and Optical Science (IPOS), University of Sydney, Camperdown, NSW, Australia

Mary McCarthy School of Engineering and Applied Science, Aston Institute of Photonic Technologies, Aston University, Birmingham, UK

Guy Millot Laboratoire Interdisciplinaire CARNOT de Bourgogne, UMR 6303 CNRS-Université de Bourgogne, Dijon, France

Daniele Modotto Dipartimento di Ingegneria dell'Informazione, Università degli Studi di Brescia, Brescia, Italy

Roberto Morandotti INRS-EMT, Varennes, QC, Canada

David J. Moss School of Electrical and Computer Engineering, RMIT University, Melbourne, VIC, Australia

Hans Christian Hansen Mulvad DTU Fotonik, Department of Photonics Engineering, Technical University of Denmark, Kongens Lyngby, Denmark

Mitchell A. Nahmias Princeton University, Princeton, NJ, USA

Leif Katsuo Oxenløwe DTU Fotonik, Department of Photonics Engineering, Technical University of Denmark, Kongens Lyngby, Denmark

Evarist Palushani DTU Fotonik, Department of Photonics Engineering, Technical University of Denmark, Kongens Lyngby, Denmark

Ravi Pant School of Physics, Centre for Ultrahigh Bandwidth Devices for Optical Systems (CUDOS), Institute of Photonics and Optical Science (IPOS), University of Sydney, Camperdown, NSW, Australia

Francesca Parmigiani Optoelectronics Research Centre, University of Southampton, Southampton, Hampshire, UK

Periklis Petropoulos Optoelectronics Research Centre, University of Southampton, Southampton, Hampshire, UK

Paul R. Prucnal Princeton University, Princeton, NJ, USA

Stojan Radic Department of Computer and Electrical Engineering, University of California San Diego, La Jolla, CA, USA

David Richardson Optoelectronics Research Centre, University of Southampton, Southampton, Hampshire, UK

Bhavin J. Shastri Princeton University, Princeton, NJ, USA

Radan Slavík Optoelectronics Research Centre, University of Southampton, Southampton, Hampshire, UK

Mariia Sorokina School of Engineering and Applied Science, Aston Institute of Photonic Technologies, Aston University, Birmingham, UK

Kartik Srinivasan Center for Nanoscale Science and Technology, National Institute of Standards and Technology (NIST), Gaithersburg, MD, USA

Michael J. Steel MQ Photonics Research Centre and CUDOS, Department of Physics and Astronomy, Macquarie University, North Ryde, NSW, Australia

Alexander N. Tait Princeton University, Princeton, NJ, USA

Sergei K. Turitsyn School of Engineering and Applied Science, Aston Institute of Photonic Technologies, Aston University, Birmingham, UK

Stefan Wabnitz Dipartimento di Ingegneria dell'Informazione, Università degli Studi di Brescia, Brescia, Italy

Andrew M. Weiner Purdue University, West Lafayette, IN, USA

Alan Willner Department of Electrical and Computer Engineering, University of Southern California, Los Angeles, CA, USA

Ben Wu Princeton University, Princeton, NJ, USA

Chunle Xiong School of Physics, Centre for Ultrahigh bandwidth Devices for Optical Systems (CUDOS), Institute of Photonics and Optical Science (IPOS), University of Sydney, Camperdown, NSW, Australia

Chapter 1
Materials and Structures for Nonlinear Photonics

Xin Gai, Duk-Yong Choi, Steve Madden and Barry Luther-Davies

Abstract In this chapter we summarize progress in materials and structures for all-optical signal processing that employ either the second or third order optical nonlinearity. Three-wave mixing and cascading in periodically-poled lithium niobate waveguides dominates signal processing in second order materials. In the case of third-order nonlinearities, four wave mixing is the dominant physical process but here the range of materials that have been employed is wider and ranges from glasses to both active and passive semiconductor devices. This chapter provides a brief summary of the advantages and disadvantages of each class of materials and device structure as well as the state-of-the-art for each case.

1.1 Introduction

Research into all-optical processing dates back to the 1980s when the so-called "electronic bottleneck", the limited speed of electronics, was seen as the major barrier that would limit the information capacity of fiber-optic networks. This was the era pre-dating the development of wavelength division multiplexing (WDM) and, hence, research was exclusively focused on time division multiplexed (TDM) signals on a single wavelength channel [1]. In this "first era" of optical signal processing the main idea was to use the ultra-fast third order nonlinearity of Kerr-like materials to create fast switches so that an interleaved signal could be de-multiplexed efficiently to different output ports of a device. Although there was a lot of work on self-switched devices, where routing was controlled by varying the signal power itself, the most practical implementation involved switching the state of the device using separate control pulses.

X. Gai · D.-Y. Choi · S. Madden · B. Luther-Davies (✉)
Centre for Ultrahigh-bandwidth Devices of Optical Systems,
Laser Physics Centre, RSPE, The Australian National University,
Canberra, ACT 2600, Australia
e-mail: Barry.Luther-Davies@anu.edu.au

S. Wabnitz and B.J. Eggleton (eds.), *All-Optical Signal Processing*,
Springer Series in Optical Sciences 194, DOI 10.1007/978-3-319-14992-9_1

A canonical optical switch consisted of a nonlinear Michelson interferometer in which a third order nonlinear element existed in only one of the arms. At the input port the beam was split in two and one half propagated through the linear arm and the other through the nonlinear arm to the output. At the output the signals interfered at a second beam splitter and the power was divided between the two output ports depending on the relative phase of the beams. At low power if the phase shift in the two arms was identical, the two beams interfered constructively to create an output at the "through" port. However, if the power was increased (or a control pulse was present) such that the nonlinear phase change that occurred in the nonlinear arm was $\Delta\phi_{nl} = k_o n_2 IL = \pi$ (L is the length of the nonlinear element; n_2 its nonlinear refractive index, I the light intensity and k_o the free space wavenumber), the power would be switched to the "cross" port. This type of nonlinear optical switch, therefore, relied on a nonlinear phase shift caused by the *real* part of the third order nonlinearity.

Of course, in this description, losses are ignored and no account is taken of the time dependence of the pulse that can lead to partial switching and pulse distortion. This can only be eliminated for the special case where the propagating pulses are temporal solitons [2] or when the control pulse is flat-topped and long compared with the signal pulse. Losses are, however, invariably present and can have a number of important consequences. Firstly, linear (and nonlinear) losses lead to power dissipation and this causes heating of the device. Via the thermo-optic effect this creates a slow phase change that can imbalance the interferometer. Thermal effects can be particularly serious in high average power applications, such as processing of telecommunications signals, or when micro-resonators with very small thermal mass are used as a switch [3]. Secondly, losses linear or nonlinear, reduce the effective path length in the nonlinear medium and this can limit the available nonlinear phase change because the power decays in the propagation direction. Finally, in a case of a simple Michelson interferometer, loss can imbalance the arms reducing the contrast of the switch. These matters were considered extensively in the early 1990s and a number of figures of merit were proposed to compare different switching devices and materials [1]. Whilst there is now little research into these true optical switches, except in the case of photonic crystal cavities, these remain relevant today.

One of the most widely quoted figures of merit (FOM) is associated with two-photon absorption [4]. The nonlinear refractive index, n_2, and the two-photon absorption (2PA) coefficient, β_{2PA}, reflect the real and imaginary components of the third order susceptibility. They are related via the Kramers-Kronig relation just as is the case of the linear refractive index and the linear absorption coefficient. As a consequence, two-photon absorption can be present in any third order nonlinear material operated at photon energies close to half the bandgap. Figures of merit (FOM) define the nonlinear phase change that can be achieved over an absorption length. For two-photon absorption, the nonlinear phase change and the nonlinear absorption coefficient are both linearly dependent on intensity and hence the resulting FOM_{2PA} depends only on material parameters and is written $FOM_{2PA} = n_2/(\beta_{2PA}\lambda)$. For efficient all-optical processing we ideally require

$FOM_{2PA} \gg 1$. A similar FOM can be written for linear absorption as reads $FOM_{1PA} = n_2 I_{max}/(\lambda\alpha)$ [1] where I_{max} is the maximum intensity that the material can sustain (e.g. limited by optical damage, or by the maximum power available for the device) and α is the linear absorption coefficient. Taking this further, in materials where three photon absorption may exist (3PA) we obtain $FOM_{3PA} = n_2/(\beta_{3PA} I\lambda)$ where I is the beam intensity. Clearly in the case of 3PA the FOM_{3PA} is no longer a constant but gets smaller with as the intensity is increased.

There have been a number of attempts to predict the dispersion of the real and imaginary components of the third order nonlinearity. One of the more useful results was published by Sheik-Bahae et al. [5–7], and was based on a simple two-band model for direct-gap semiconductors. Their analysis resulted in expressions for the dependence of n_2 and β_{2PA} on the ratio of the photon to bandgap energies, and had the form $n_2 = A/\left(n_0^2 E_g^4\right)\bar{G}_2\left(\hbar\varpi/E_g\right)$ and $\beta_{2PA} = B/\left(n_o^2 E_g^3\right)\bar{F}_2\left(\hbar\varpi/E_g\right)$ where A and B are constants; n_o is the linear refractive index; and E_g the bandgap energy. The functions $\bar{F}_2$ and $\bar{G}_2$ are reproduced in Fig. 1.1a. These expressions have been shown to provide a reasonable fit to experimental measurements for a wide range of materials [5, 6]. For the case of indirect-gap semiconductors the situation is somewhat more complex but was analyzed by Dinu [8] whose results are shown in Fig. 1.1b. In both these cases the dispersion of the nonlinear coefficients contain similar features. Firstly, as the frequency increases, the third order nonlinearity becomes resonantly enhanced above its value at DC frequency as the two-photon resonance close to half the band gap of the material is approached. Once this resonance is passed, two photon absorption ($\bar{F}_2$, $\beta(\omega)$) grows and maintains a high value all the way up to the band edge of the material. The influence of the single photon resonance has a marked effect on the real part of the nonlinearity and, in general, results in a reversal in the sign of the nonlinearity from positive to negative. In the case of the direct gap semiconductors this is predicted to occur at $\approx 0.7E_g$ whilst for indirect materials it is located within the bandgap of the material at around $1.25E_g$. In both these models $\beta_{2PA} = 0$ when $h\omega < E_g/2$, thus in

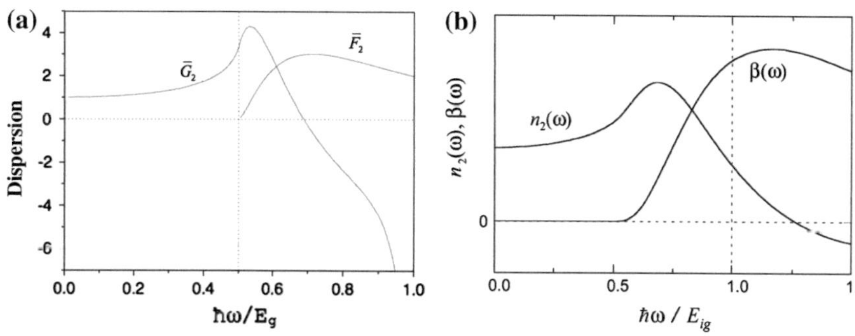

Fig. 1.1 **a** Dispersion functions for the nonlinear refraction ($\bar{G}_2$) and two photon absorption ($\bar{F}_2$) for direct gap materials (from [7]). **b** Similar relationships for $n_2(\omega)$ ($=n_2$) and $\beta(\omega)$ ($=\beta_{2PA}$) for indirect gap materials (from [8])

this regime the FOM_{2PA} becomes very large. Whilst this is true, one must remember that the nonlinear polarization of the material contains higher order terms. For example, the imaginary part of $\chi^{(5)}$ gives rise to 3PA and this is not necessarily small as, for example, has been demonstrated recently for silicon [9]. Thus, beyond half its band-gap silicon has the intensity dependent FOM_{3PA} mentioned above. In fact as the frequency is progressively reduced we expect that higher and higher orders of nonlinear absorption can appear. For the case of the direct gap materials, the dispersion of the high order multi-photon absorption was predicted by Wherrett [10] and consists of a series of similar curves to that shown for 2PA in Fig. 1.1a, but displaced progressively lower frequencies with a cut-off frequencies corresponding to $h\omega < E_g/3$, $<E_g/4$. etc. To our knowledge no similar analysis exists for indirect gap materials. Of course it is generally assumed that such higher order absorption can be neglected but this may not always be the case as has been illustrated by recent results in silicon, which shows that strong nonlinear absorption can exist at wavelengths well into the mid infrared [9].

So far we have concentrated on the ultrafast Kerr nonlinearity, but intensity-dependent changes in the phase of the light can also be due to other mechanisms. Apart from the thermal effects, mentioned above, the most common of these is due to the creation of free carriers that lead to a change of index as described by the Drude model. Free carrier effects in semiconductors can be strong and have proved very useful for all-optical switching. However, in passive materials such as silicon, free carriers created by single or multi-photon absorption lead to strong free carrier absorption (FCA) and this is generally detrimental. However, in active devices based on III–V semiconductors, the losses can be compensated using laser action and, hence, the semiconductor optical amplifier (SOA) has proved to be a very effective structure for nonlinear optics. As outlined in more detail below, SOAs main limitation arises because of the finite lifetime of the free carriers which leads to a speed limitation.

Finally in the late 1990s, the so-called "cascaded" second order nonlinearity was (re-)discovered and showed that a nonlinear phase shift could be generated using second order nonlinear materials [11]. Implemented in PPLN this has led to very efficient all-optical processors that are effective even at the single photon level. Their general characteristics are described below.

In what follows we summarize the properties of materials and devices using Kerr, cascaded and free carrier nonlinearities outlining the material and device properties and some of their capabilities and limitations for all-optical signal processing.

1.2 All-Optical Processing Using $\chi^{(2)}$ Nonlinearities

In the late 1990s it was realized that an intensity-dependent phase shift that was used for all-optical switching could also be obtained by cascading two second-order ($\chi^{(2)}$) frequency conversion processes, creating an effective third order nonlinearity without nonlinear loss [11]. In this early work, cascading involved imperfectly phase matched second harmonic generation and degenerate parametric down conversion operating simultaneously. Because of the phase mismatch, power flowed from the low frequency pump to its second harmonic and back again. In this process the down-converted signal originating from the second harmonic accumulated a small phase shift (positive or negative depending on the sign of the phase mismatch) relative to the unconverted pump so when these combined coherently, the phase of the pump was slightly modified. If the coherence length of the SHG/DFG processes was small compared with the device length (the case of large phase mismatch) the phase shift increased linearly with device length, and also linearly with pump intensity mimicking the behavior of a $\chi^{(3)}$ nonlinear material. Various schemes for all-optical switching using cascading were reported [12].

The advent of WDM significantly changed the emphasis of the research into all-optical signal processing away from switching devices. For example, the challenge in a WDM network was not, at least until recently, to de-multiplex signals at speeds beyond the limits of electronics, but to transfer data from one to another wavelength or to multi-cast data on many channels simultaneously. Conventionally such processes used direct detection and electronic modulation of a wavelength shifted laser or multiple lasers but this can become power-hungry and complex as channel counts rise. An alternative and potentially better solution is to use nonlinear optics to directly convert light at one wavelength to another using nonlinear mixing processes. Thus, optical signal processing using $\chi^{(3)}$ nonlinear optics became mostly based on four-wave mixing (4WM) whilst in the case of $\chi^{(2)}$ devices three-wave mixing (3WM) was employed. Many of the wave mixing devices are agnostic to the data format and can readily preserve the amplitude and phase of the incoming signal.

4WM and 3WM are shown schematically in Fig. 1.2. In the case of 4WM *two* pump photons at frequency ω_p mix with a signal photon at ω_s to create an idler photon at frequency ω_i for which energy conservation requires $2\omega_p = \omega_s + \omega_i$. In this case $\omega_p = \omega_{s,i} \pm \Delta\omega$ where $\Delta\omega \ll \omega_{s,i}$ which means that all three frequencies lie in the telecommunications band where there are readily available sources. Since two photons at the pump are destroyed and new photons created at ω_s, ω_i, the signal and idlers experience gain.

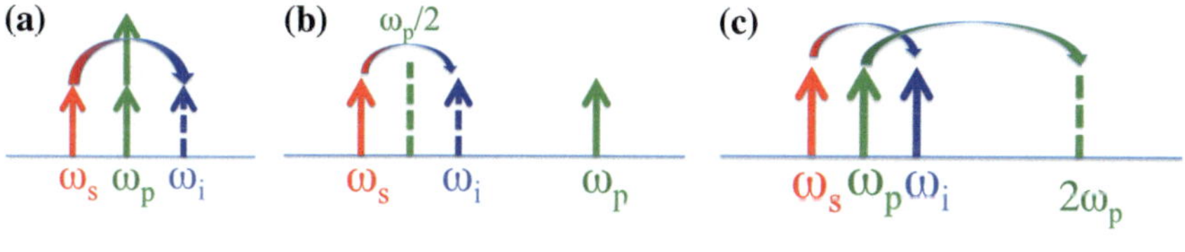

Fig. 1.2 **a** Schematic of 4WM in which two pump photons (ω_p) mix with a signal photon (ω_s) to create an idler (ω_i) such that $2\omega_p = \omega_s + \omega_i$; **b** schematic of 3WM in which difference frequency mixing between a single high frequency pump photon (ω_p) and a signal (ω_s) creates an idler (ω_i) symmetrically positioned around $\omega_p/2$ such that $\omega_p = \omega_s + \omega_i$; **c** cascaded 3WM where second harmonic generation converts a pump at ω_p to $2\omega_p$ which then undergoes 3WM with the signal to create an idler

The case of 3WM is quite similar, however, here a *single* pump photon decays into signal and idler photons so that the energy conservation condition now reads $\omega_p = \omega_s + \omega_i$. Again $\omega_p = \omega_{s,i} \pm \Delta\omega$ but now $\Delta\omega \approx \omega_{s,i}$ and hence a high frequency pump is required that lies well outside the telecommunication band where sources are generally less compatible with telecommunications technology. This difference between the pump and signal/idler frequencies in 3WM would be a large disadvantage were it not possible to eliminate it using cascading. To achieve this a strong pump at frequency $\omega_p \approx \omega_{s,i}$ in the telecommunications band is frequency doubled in the $\chi^{(2)}$ material to create a new second harmonic pump at $2\omega_p$. Difference frequency generation then allows this second harmonic pump to mix with the signal to create a new idler spaced symmetrically around the original low frequency pump. For this case the pump, signal and idler all exist in the telecommunications band although it must be remembered that the process inevitably generates a new high frequency pump at $2\omega_p$ which needs to optimally couple with the signal and idler waves to achieve high overall conversion efficiency. Just like in the 4WM process, two pump photons are thus destroyed and their energy transferred to the signal and idler exactly mimicking 4WM in a χ^3 material.

To achieve low operating powers, compatible with telecommunications systems, 3WM must take place in a non-centrosymmetric crystalline waveguide exhibiting very strong $\chi^{(2)}$ nonlinearity with perfect phase matching. The most common material of choice has been periodically-poled lithium niobate (PPLN) for which there are several well-established routes for waveguide fabrication. In addition, the technology for electric field poling of lithium niobate is well developed and complex grating patterns can be engineered, for example, to increase the bandwidth for phase matching. Of the approaches to waveguide fabrication that have been developed, the most widely used is the so-called reverse proton exchange process (RPE) [13, 14]. This involves, firstly, the preparation of a substrate using electric field poling to periodically invert the ferroelectric domains. This is then followed by proton exchange and annealing steps that define the waveguide channel creating as annealed proton exchange (APE) waveguides [15, 16]. The losses, nonlinearity and field overlap between second harmonic pump and signal and idler fields are not optimum in APE structures and so a second exchange process with a lithium-rich melt is used to replace proton by lithium near the channel surface thus creating a RPE waveguide. These waveguides have lower losses because the interacting fields are pushed further from the top surface, and better overlap between the interacting modes which also enhances the nonlinearity. RPE waveguides typically have losses are 0.1–0.2 dB/cm and modes well-matched to standard SMF-28 optical fiber which reduces coupling loss [17]. Whilst RPE waveguides have been used in many demonstrations of all-optical signal processing, they do suffer from some issues with photorefractive damage. This can be alleviated by operation at elevated temperatures or by using alternative materials and fabrication methods such as Ti in-diffusion. A fuller description of the various fabrication approaches can be found in the references [17].

An important parameter for any $\chi^{(2)}$ waveguide is its normalized efficiency η_{cor}. For the cascaded process the output power at the idler can be written in the non-depleted pump approximation as $P_{out} = (\eta_{cor}L^2P_p/2)^2P_s$ where P_p and P_s are the pump and signal powers respectively. Typically $\eta_{cor} \approx 1\ \mathrm{W}^{-1}\ \mathrm{cm}^2$ and this implies that if the losses are negligible, 100 % conversion can be obtained for a pump power of only 80 mW in a 5 cm long device at low signal levels. In fact one of the remarkable features of the devices based on cascaded nonlinearities is their high dynamic range and ability to up-convert single signal photons to access high detector sensitivity [18].

The second important parameter is the phase matching bandwidth for the interaction. For DFG the conversion scales with $\mathrm{sinc}^2(\Delta kL/2)$, where is the device length and $\Delta k = k_p - k_s - k_i - K_g$ is the phase mismatch with $k_{p,s,i}$ are the k-vectors of the pump, signal and idler at frequencies ω_p, ω_s and ω_i, respectively, and $K_g = 2\pi/\Gamma_g$ with Γ_g being the grating period. Similarly for the SHG process $\Delta k = k_p - 2k_{p/2} - K_g$. In either case the output falls to half it maximum when $\Delta k \approx 0.89\pi/L$. The DFG bandwidth substantially exceeds that for SHG and is typically 70 nm FWHM for a 50 mm long device, however, for SHG the bandwidth is only ≈0.2 nm for a similar device length. This rather narrow pump bandwidth can be overcome either by engineering multiple gratings into the crystal to allow several different pumps to be used simultaneously [19] or by using two tunable pump beams [17].

PPLN based all-optical processors have proven be highly capable in applications such as wavelength conversion [20]; dispersion compensation via phase conjugation [21]; digital signal processing including header recognition, time-slot interchange [22], etc. For an extensive review of their capabilities the reader is referred to recent reviews by Langrock et al. [17] and Willner et al. [23]. In terms of limitation there are three of consequence. The first arises from group velocity dispersion (GVD) because of the need to convert the pump in the telecom band near 1550 nm to its harmonic around 775 nm. This large frequency difference results in GVD of ≈0.3 ps/mm which means that 50 mm long devices are restricted to operate with pulses >15 ps in duration. Whilst techniques have been proposed to circumvent this issue, such as the introduction of delays lines to retime the fundamental and second harmonic waves [24], the large GVD makes it more difficult to process signals at very high rates. Nevertheless optical processing at 640 Gbit/s has been reported [25]. A second limitation arises from the single polarization nature of the interaction, and generally requires the use of polarization diversity techniques. Finally, in a photonic world increasingly dominated by silicon waveguide devices, PPLN remains a material that is difficult to integrate to create highly functional circuits and because of the interactions are phase matched require temperature stabilization. Countering these are the ability of PPLN devices to operate with low noise; minimal cross talk, high bandwidth and high efficiency and no chirp making them a powerful platform for all-optical signal processing.

1.3 All-Optical Processing Using $\chi^{(3)}$ Nonlinearities

3WM can achieve efficient all-optical processing but the use of the $\chi^{(2)}$ nonlinearity requiring a non-centro-symmetric material limits its applicability to only a few materials and these are generally incompatible with current CMOS processing technology. As a result 4WM using the $\chi^{(3)}$ nonlinearity that exists in all materials has started to dominate all-optical processing. In this case, the pump signal and idler all lie in the telecommunication band, group velocity dispersion can be small, and phase matching can be achieved by engineering the structure of the waveguide so that the dispersion is anomalous. A large number of materials can be used for FWM and combined with the simply device geometry and, in many cases, compatibility with CMOS processing, allow all-optical devices to be integrated at low cost. In this section we describe some of the materials dependent factors that influence the efficiency and bandwidth for FWM.

For waveguides that utilize the $\chi^{(3)}$ nonlinearity, the nonlinear phase change is generally written in the form $\Delta\phi_{nl} = \gamma PL$, where P is the laser power; L is the propagation length; and γ is the waveguide nonlinear parameter $\gamma = k_o n_2/A_{eff}$, with $k_o = 2\pi/\lambda$ is the free space wavenumber; and A_{eff} is the effective area of the waveguide mode. In a lossless waveguide, the 4WM conversion efficiency from signal to idler in the non-depleted pump approximation, can be written $\eta_{4WM} = (\Delta\phi_{nl})^2 = (\gamma PL)^2$, and is directly determined by nonlinear phase change. It is worth noting here that compared with switching devices where $\Delta\phi_{nl}$ needs to be large ($>\pi$), FWM can be quite efficient for small values of $\Delta\phi_{nl} \approx 0.3$. The nonlinear parameter γ is important for any $\chi^{(3)}$ waveguide because a high γ value minimizes P and L, and small, low power devices are essential in most applications. In practice P is often limited by energy constraints in the system, and L has a maximum value determined by the loss of the waveguide. In a lossy waveguide, L is replaced by an effective length written $L_{eff} = (1 - exp(-\alpha L))/\alpha$ where α is the linear loss coefficient. Because of these constraints, as high a value of γ as possible is always preferable and this motivates the use of materials with high n_2 and waveguides with small A_{eff}.

The phase matching condition for 4WM is generally written $-4\gamma P < \Delta k \approx \beta_2 \Delta\omega^2 < 0$, where $\Delta k = 2k_p - k_s - k_i$ is the phase mismatch with $k_{p,s,i}$ are the k-vectors of the pump, signal and idler at frequencies $\omega_{p,s,i}$ respectively, $\Delta\omega = |\omega_p - \omega_{s,i}|$ is the frequency difference between pump and signal (or idler), and β_2 is the second order dispersion of waveguide. The dispersion of waveguide $\beta(\omega) = k(\omega) = 2\pi n_{eff}/\lambda$ can be expanded as a Taylor series as $\beta(\omega_0 + \Delta\omega) = \beta_0 + \beta_1\Delta\omega + \beta_2\Delta\omega^2/2 + \beta_3\Delta\omega^3/6 + \ldots$, where $\beta_m = d^m\beta/d\omega^m$ $_{(m=1,\ 2,\ldots)}$ at ω_0; n_{eff} is the waveguide effective index. The first order of the dispersion $\beta_1 = 1/v_g = n_g/c_0$ where v_g is the group velocity, n_g is the group index and c_0 is the light speed in free space. The second orders of the dispersion $\beta_2 = d\beta_1/d\omega = -\lambda^2 D/2\pi c_0$, where D is the group velocity dispersion (GVD) of waveguide. β_1 and β_2 are the most critical parameters for all-optical processing because the pulse envelope propagates at v_g and pulse broadening is relating to GVD.

In order to achieve phase matching for FWM, the waveguide dispersion must be anomalous, that is $\beta_2 < 0$ and when this is achieved, power in the signal and idler waves grows exponentially at the expense of the pump. However, most materials exhibit normal dispersion, $\beta_2 > 0$, at telecommunication frequencies and it is, therefore, necessary to engineer the waveguide structure to produce sufficient anomalous waveguide dispersion so that the normal material dispersion is compensated. This means that for any particular material, only specific waveguide designs can be used and these determine the minimum mode area, A_{eff}, and hence the γ of the waveguide. Figure 1.3 is an example of such dispersion engineering for a 630 nm wide chalcogenide nanowire and shows the effect of changing the waveguide height on the TM mode dispersion [26]. In this example, the waveguide has to be >450 nm thick in order to achieve anomalous dispersion at 1550 nm. In other materials, like crystalline silicon, anomalous dispersion occurs in smaller structures potentially leading to higher γ. Thus material dispersion becomes the dominant parameter that ultimately determines the achievable value of γ at a particular wavelength.

In the phase matching relation, $\Delta\omega$ represents the bandwidth of 4WM and it is determined by β_2 and γP. β_2 close to zero and a large γP will support 4WM over a very wide bandwidth. This is important for many devices required for telecommunications and allows 4WM to span the whole S- C- and L-bands. As a result, the 'zero-dispersion' is a target for dispersion engineering as well as a large γ value.

The phase-matching condition of 4WM, in fact, demonstrates the relation between the nonlinear phase change and the dispersive phase change. However, this conventional phase matching relation ignores the influence of other nonlinearities, notably Raman scattering. In fact Raman scattering can cause an additional phase change that originates from the real part of the Fourier transform of the Raman response function $Re[h_R(\omega)]$ [27, 29]. Thus, in materials where the bandwidth for FWM extends as far as the Raman bands, the phase matching relation must be modified to include this additional phase term and thus reads $-4\gamma P - 4\gamma P f_R(Re[h_R(\omega) - 1]) < \Delta k \approx \beta_2 \Delta\omega^2 < 0$, where f_R is the fractional Raman factor which normally range from 0.1 to 0.2 depending on material. When the $Re[h_R(\omega)] \approx 1$, the

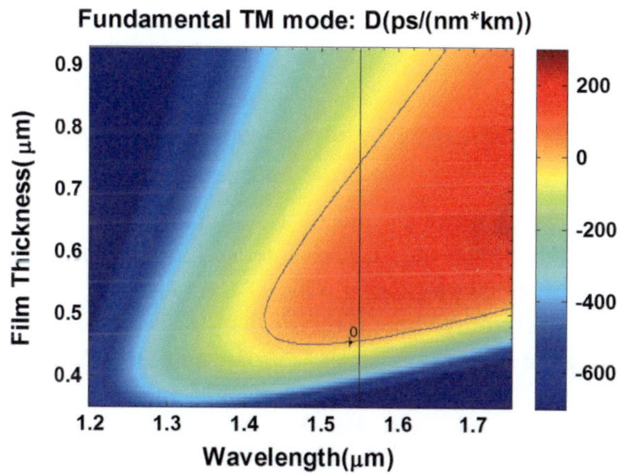

Fig. 1.3 The group velocity dispersion (GVD) of $Ge_{11.5}As_{24}Se_{64.5}$ nanowire as a function of film thickness and wavelength. The waveguide width is fixed at 630 nm [26]

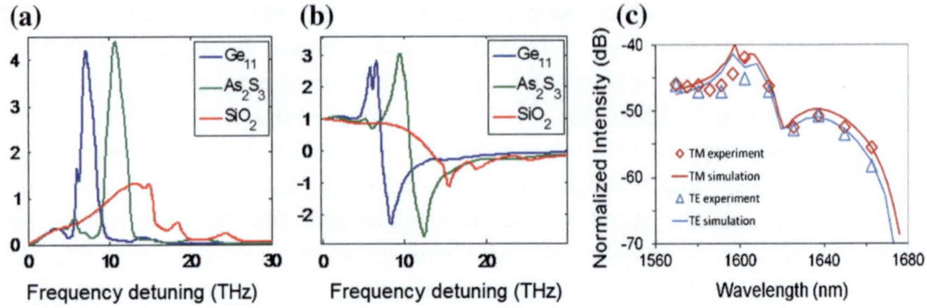

Fig. 1.4 **a** The imaginary part of Fourier transform of the Raman response function of $Ge_{11.5}As_{24}Se_{64.5}$, As_2S_3 and SiO_2 [27]. **b** The real part of Fourier transform of the Raman response function of $Ge_{11.5}As_{24}Se_{64.5}$, As_2S_3 and SiO_2 [27]. **c** The experimental result of Raman modulated 4WM in a $Ge_{11.5}As_{24}Se_{64.5}$ chalcogenide nanowire [28]

inequality becomes $-4\gamma P < \Delta k \approx \beta_2 \Delta\omega^2 < 0$, and Raman scattering does not contribute to the phase change. When $Re[h_R(\omega)] > 1$, it causes a positive phase change which broadens the phase-matching condition increases and the gain for 4WM. On the other hand, if $Re[h_R(\omega)] < 1$, then the Raman phase change is negative, the phase-matching condition narrows and the 4WM gain is reduced.

Figure 1.4a, b shows the imaginary and real parts of $h_R(\omega)$ for SiO_2, As_2S_3 and $Ge_{11.5}As_{24}Se_{64.5}$ glasses [27]. Typically $Re[h_R(\omega)]$ remains ≈ 1 for small detuning, and then exhibits a resonant behavior to increase up several times right after the Raman peak of $Im[h_R(\omega)]$ before dropping sharply. At large detuning, $Re[h_R(\omega)] \to 0$, but never returns to its original value of unity. As a result, the bandwidth of 4WM is restricted by the spectrum of Raman scattering even if β_2 is near zero and γ is large. Figure 1.4c shows the effect of $Re[h_R(\omega)]$ on 4WM for a $Ge_{11.5}As_{24}Se_{64.5}$ chalcogenide nanowire [28]. When the $Re[h_R(\omega)]$ recovers to zero at large detuning, the gain is less than 70 % of that $Re[h_R(\omega)] \approx 1$. The maximum bandwidth for high gain 4WM can be predicted using the Raman detuning frequency for the material, and is 130 nm for $Ge_{11.5}As_{24}Se_{64.5}$; 180 nm for As_2S_3; 190 nm for SiO_2; 220 nm for amorphous Si; and over 250 nm for crystalline Si assuming a pump at 1550 nm.

1.3.1 Properties of $\chi^{(3)}$ Materials and Devices

Very many materials and device structures have been considered for $\chi^{(3)}$ nonlinear optics although only a few of these have actually been used in demonstrations of optical signal processing. One of the most successful has been silica-based highly nonlinear optical fiber (HLNF) that has been used for parametric signal processing for more than two decades [30]. Despite its inherent compatibility with fiber optic systems, the small material nonlinearity implies large device lengths and additionally the fiber geometry makes integration of multiple structures for signal processing impractical. Several alternative media with much larger material nonlinearity have emerged including bismuth oxide glass; crystalline and amorphous silicon; silicon nitride; AlGaAs; chalcogenide glasses; and high-index doped silica-based glasses. In this section we summarize the characteristics of a few of these materials and discuss device properties relevant to a few on-chip platforms that use the nonlinear Kerr effect and 4WM based all-optical processing. Because of the large amount of work based on the following materials we focus on crystalline silicon (c–Si); hydrogenated amorphous silicon (a–Si:H); and silicon nitride that are all compatible to complementary metal oxide semiconductor (CMOS) processes and the somewhat less CMOS-compatible chalcogenide glasses.

1.3.1.1 Crystalline Silicon

In the past decade a new application for silicon has emerged as a material for photonics. The main driver for this has been the interconnect bottleneck now facing electronic processors. The major advantage of silicon photonic devices is they can leverage the precision of advanced CMOS processing technology. However, in order to create a small-scale communication network, various functional photonic components must be integrated onto the silicon platform. This has resulted in some remarkable developments including germanium light emitters [31] and germanium photo-detector [32] integrated onto silicon chips as well as high-speed optical modulators [33]. There has also been substantial interest in all-optical signal processing utilizing the $\chi^{(3)}$ nonlinearity of Si.

Si is transparent from 1100 to 7000 nm and linear absorption can, therefore, be low across the whole optical communications bands between 1260 and 1675 nm, but two-photon absorption is large. This results in a relatively low FOM_{2PA} of $\sim$0.4 at 1550 nm corresponding to a Kerr nonlinear coefficient $n_2 = 4.4 \times 10^{-14}$ cm^2/W and $\beta_{2PA} = 8.4 \times 10^{-10}$ cm/W [34]. This means that Si is, perhaps, not the best material for nonlinear photonics but since it can be used to produce complex circuits at low-cost in high-volume there has been a large incentive to explore its use in all-optical signal processing. Si also has excellent thermal conductivity and high damage resistance and because of its high refractive index can confine light into a very small mode and this can lead to exceptionally large values of γ. In addition, waveguide designs that lead to near optimal confinement also correspond to those which achieve anomalous dispersion.

The silicon-on-insulator (SOI) platform has become the foundation of silicon photonics and comprises of 220 nm thick top silicon layer sitting on thermally oxidized (1–2 μm SiO_2 layer) silicon substrate. Figure 1.1a shows a schematic of a typical silicon photonic nanowire, and the optical field distribution in the fundamental TE mode. The large refractive index contrast between Si ($n = 3.45$) and SiO_2 ($n = 1.45$) or air ($n = 1$) leads to a strong light confinement which makes it possible to scale down the size of the waveguides mode to approximately 0.1 μm^2. Combining high n_2 and small A_{eff} yields an extremely large nonlinear parameter γ of 300 W^{-1} m^{-1} [35]. On the other hand this rectangular geometry makes much lower effective index in TM so that the mode area is a few times larger than that of TE.

Silicon photonic wire waveguides are typically fabricated by plasma etching with electron-beam or deep-ultraviolet (DUV) lithography used to define the submicron pattern. Nowadays, propagation losses of 1–2 dB/cm [36] are routinely achieved mostly limited by light scattering from the sidewalls but this is low enough since the devices are short (a few cm or less). A more difficult issue has been light coupling into the photonic wire due to a large mode mismatch between nanoscale silicon waveguides and fibers. Two coupling schemes are commonly employed to overcome this issue are to use surface grating couplers [36] and Si inverse tapers [37] for which the coupling efficiency reaches around 40–70 %.

In spite of its many advantages over other media, silicon-based all-optical signal processing inevitably suffers from nonlinear loss and patterning effects resulting from large TPA and free carrier absorption (FCA). The lifetime of free carriers generated by two-photon absorption is in the range of several hundred ps to several hundred ns [38] and slow carrier dynamics can limit the speed of signal processing. Several approaches have been proposed to mitigate the issue as shown in Fig. 1.5b–d. The main idea is that direct light-silicon interaction can be avoided by engineering the Si waveguide structure. The first is to employ a thin strip waveguide and cover it with a material possessing both a large nonlinear Kerr coefficient (n_2) and a small nonlinear loss (β_{2PA}) [39]. In this case the structure makes TM mode extend far into the cladding material so that most of the nonlinear effect can be due to the upper cladding. Figure 1.5c shows a so-called "slotted waveguide" where trenches several tens to hundreds nanometer wide are formed along the middle of Si waveguide, and this slot is filled with another nonlinear medium. With an appropriate design the light intensity in the slot is enhanced a few times relative to the unstructured waveguide [39]. Photonic crystal waveguides have also been demonstrated to further increase nonlinearity by slowing-down the light speed [40]. This slow-light effect can boost the light-matter interaction proportional to the group index squared $(c/v_g)^2$.

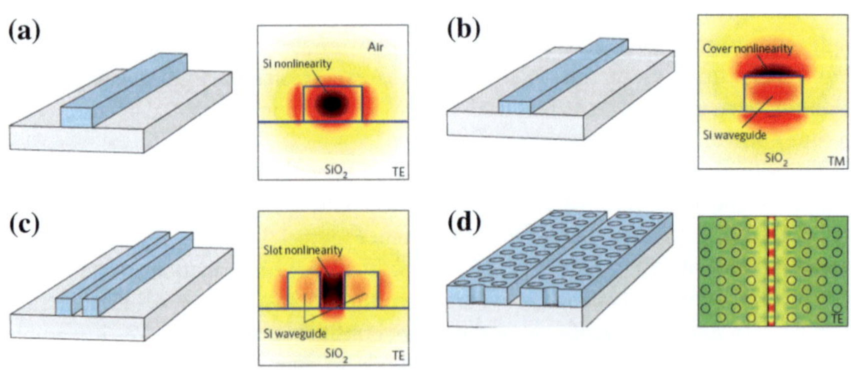

Fig. 1.5 Silicon waveguidesilicon waveguide structures and their respective electric field distributions. **a** Strip waveguide using silicon nonlinearity in core. **b** Strip waveguide using cover nonlinearities. **c** Slot waveguide using nonlinearities in the slot. **d** Slot slow-light waveguide [35]

To conclude, it is worthwhile noting that in spite of the limitations of silicon, two recent reports clearly show that these can be overcome with careful attention to device design. Forster et al. [41] achieved peak conversion efficiencies >10 % and conversion bandwidths >150 nm in carefully dispersion-engineered Si nanowire waveguides. For this they used rather fat and thick Si core (300 nm tall and 500 nm wide) and operated with the TM mode. The work of Denmark Technical University group is also remarkable in that ultra-high bit rate signals were processed in a nominal Si nanowire without incurring TPA and associated FCA problems [42]. They demonstrated the wavelength conversion of 640 Gbit/s signal by low-power 4WM in a 3.6 mm long silicon waveguide with a switching energy of ~110 fJ/bit, which is low enough to reduce nonlinear absorption.

1.3.1.2 Hydrogenated Amorphous Silicon (a–Si:H)

Over the years a–Si:H has been extensively deployed in applications from photovoltaics to flat-panel displays but its applications in lightwave technology is just evolving. As a nonlinear medium for all-optical signal processing, amorphous silicon outperforms crystalline silicon in some respects. First of all, this material retains CMOS-compatibility so that low-cost mass manufacturing is available. Another attractive feature of a–Si:H comes from the fact that it can be deposited at low temperature (below 400 °C) on almost any substrate including glass, metal and even plastic, whilst crystalline silicon (SOI) docs not offer this flexibility. Moreover, the low temperature deposition allows back-end integration of a–Si:H photonic components on pre-processed CMOS electronic chips without any damage to the underlying metal routing wires [43]. In addition it also proves to be a highly promising nonlinear material for power efficient photonic devices due to its ultra-high optical nonlinearity and low nonlinear loss relative to crystalline Si.

Plasma-enhanced chemical vapour deposition (PECVD) is a well-established technique for depositing high quality amorphous silicon-hydrogen alloy (~10 at.% H) at low temperature. Silane (SiH_4) is the dominant precursor gas, mixed with hydrogen, helium, or argon as a diluent. If amorphous silicon is deposited by sputtering it has mid-bandgap absorption in the near infrared due to dangling bonds. These Si dangling bonds, however, can be saturated by incorporating hydrogen during PECVD, resulting in the low optical attenuation below 1 dB/cm [44]. The fabrication of a–Si:H photonic nanowaveguides is identical to that of SOI; i.e. optical or electron-beam lithography followed by plasma etching and SiO_2 PECVD for cladding. a–Si:H nanowires have similar or slightly smaller dimension compared with crystalline Si because of its higher index. The linear propagation losses achieved in a–Si:H waveguides lies in the same range as c–Si; that is a few dB/cm, and even ~1 dB/cm has been achieved by adopting a delicate thickness control technique [45].

The nonlinear characteristics of the material are favourable with a Kerr nonlinearity (n_2) reported to be several or up to 10 times [46] higher than that of c–Si. It is debatable as to whether TPA in a–Si:H is a still the dominant nonlinear loss mechanism in the telecommunication band, since TPA should vanish since the photon energy is less than the half the bandgap ($E_g/2 \sim 0.85$ eV) of the medium. Nevertheless, it is generally considered to have similar level of nonlinear absorption to c–Si. Because of the high nonlinearity, γ can reach 3000 W^{-1} m^{-1} in a 500 nm wide, 200 nm high nanowire [47]. Utilizing this extremely high nonlinear parameter in a–Si:H, Wang et al. [47] demonstrated wavelength conversion through nonlinear parametric processes with maximum conversion efficiency of −13 dB using mere 15 mW of pump peak power. In addition on-chip parametric amplification with 26.5 dB gain was attained in a–Si:H photonic wire at telecommunication wavelengths [48].

A key concern of hydrogenated amorphous silicon devices has been its lack of stability. Kuyken et al. [49] observed a decrease of parametric amplification with the time of exposure to pump light. They asserted that this results from material degradation and has the same origin to the Staebler-Wronski effect [49], which is well known to the a–Si solar cell community. Most recent results, however, demonstrated that deterioration in amorphous silicon may not be an issue. The large scatter in n_2, β_{2PA} (or β_{3PA}), FOM_{2PA}, γ, and free carrier lifetime in the literature is another nagging problem. For instance, the measured β_{2PA} spans from 41×10^{-10} to 2.5×10^{-10} cm/W. The origin of this large variation remains an open question, but it is a general thought to be due to differences in the film deposition process. Even so, a–Si:H is quite a promising platform for ultra-compact all-optical signal processing chips operating at low power.

Silicon nitride (*SiN*): This is another CMOS-compatible material that has recently emerged as a candidate for nonlinear photonics. Since SiN can be deposited from gas precursors in a standard silicon processing environment, it benefits from flexibility in deposition parameters which produce different film characteristics. Despite n_2 of SiN ($n_2 = 2.5 \times 10^{-15}$ cm^2/W at 1.55 μm) being more than an order of magnitude lower than c–Si, the most significant advantage of SiN over crystalline and amorphous Si comes from the absence of 2PA or multi-photon absorptions and associated FCA due to its wide bandgap ($E_g \sim 5$ eV) [50].

Low-pressure chemical vapour deposition (LPCVD) at high temperature (~800 °C) is the preferred method to grow low loss SiN layer and leads to minimal hydrogen contamination in the film. Using this method, propagation losses of 0.5 dB/cm loss was obtained in a SiN waveguide with 1 μm^2 cross section [51]. In order to facilitate back-end integration with pre-existing electronic circuits, however, PECVD below 400 °C was also demonstrated to produce SiN waveguides with only ~1 dB/cm propagation losses. Here, the absorption in the near IR was controlled by reducing the number of hydrogen bonds (Si–H and N–H) in the grown SiN by employing techniques such as adding helium into the plasma [52] and replacing NH_3 with a N_2 precursor [53].

The group at Cornell University [51] demonstrated on-chip multiple wavelength source exploiting 4FM optical parametric oscillation in a SiN micro-ring (Fig. 1.6). They used a single pump laser tuned to the resonance of the ring, then ~1000 times higher light power circulating in the ring which leads to cascaded 4WM, and allowing narrow multi-wavelength lines.

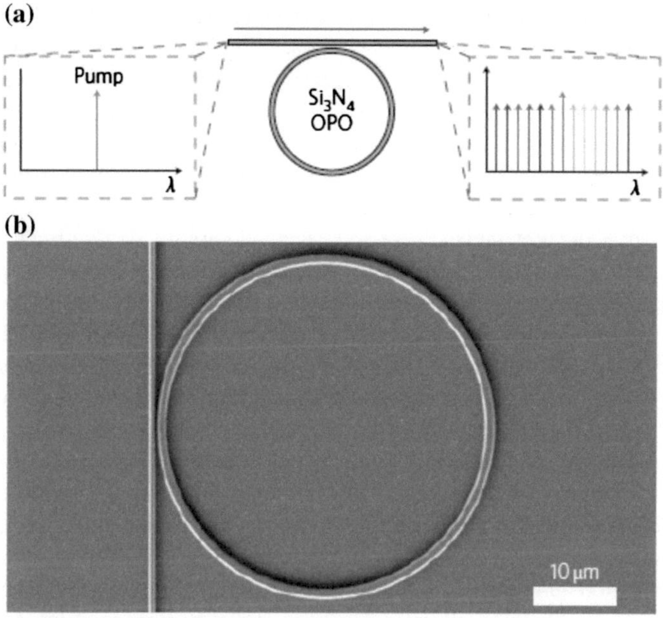

Fig. 1.6 On-chip optical parametric oscillator in SiN waveguide. **a** A single pump laser tuned to the resonance of an integrated SiN microring allows the generation of numerous narrow linewidth sources at precisely defined wavelengths. **b** A scanning electron micrograph of a SiN microring resonator coupled to a bus waveguide [51]

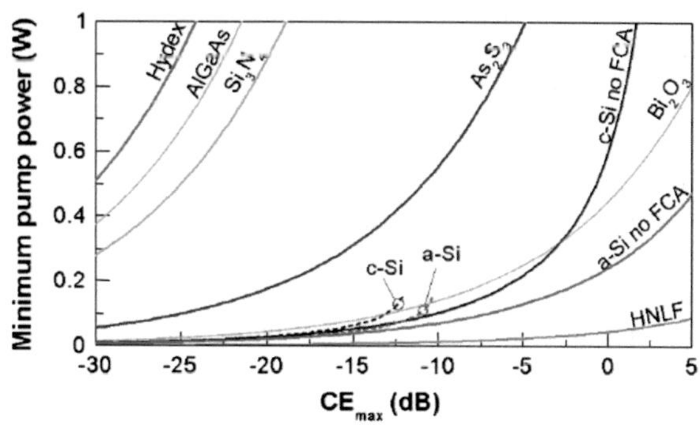

Fig. 1.7 Minimum pump power required to achieve a given CE for the different materials [54]

Diaz et al. [54] reported a systematic comparison of eight nonlinear media for parametric all-optical signal processing using degenerate 4WM wavelength conversion. Figure 1.7 represents the minimum pump power required to attain a given conversion efficiency (CE). Here CE is the ratio of the idler power at the waveguide output to the signal power at its input. Crystalline and amorphous silicon can operate at low pump power because of their very high γ, but their CE is limited to around 10 % due to TPA and FCA. Whilst the performance of SiN is inferior to c–Si and a–Si because of its smaller nonlinear parameter, it can handle much higher pump power without suffering nonlinear losses. An additional difficulty, however, is that the large mode area needed for dispersion engineering which is around 1 μm^2 results in the footprint of SiN being two orders higher than the equivalent devices made in c–Si and a–Si:H and this makes it difficult to achieve compact devices.

Chalcogenide glass: Chalcogenide glasses have attracted a lot of interest for all-optical signal processing. These are amorphous, highly nonlinear materials containing the chalcogen elements S, Se and Te covalently bonded with glass forming materials such as Ge, As, P, Sb or Si. Chalcogenide glasses provide low optical loss in the telecommunication bands (1310 and 1550 nm) and transmission out to beyond 8 μm in the infrared. These glasses have high linear refractive indices (2.0–3.0 at 1550 nm) which leads to small mode volumes in dispersion-engineered nanowires and a Kerr nonlinear index similar to c–Si (≈a few hundred × silica) [55, 56]. Most importantly they generally have negligible 2PA which leads to $FOM_{2PA} > 100$ [57–59]. In most chalcogenide glasses, there are no free carrier effects and this provides a distinct advantage compared with the semiconductors.

As_2S_3 is one of the most well known chalcogenide glasses with a linear index of 2.43 at 1550 nm, a nonlinear index over 120 times silica and FOM_{2PA} over 320 and a glass transition temperature over 170 °C. The first reports demonstrating the efficacy of As_2S_3 fiber for all-optical processing were produced by Asobe [60]. More recently a large number of ultra-fast nonlinear devices have been demonstrated using rib waveguides made from this glass including a signal regenerator [61]; RF spectrum analyser [62, 63]; dispersion compensator using mid-span spectral inversion [64]; TDM demultiplexers [65]; wavelength convertor [66]; parametric amplifier [29, 67]; as well as a supercontimuum source [68].

Although, As_2S_3 glass has, therefore, exhibited some good properties for all-optical processing, even better chalcogenide materials would be of advantage particularly if higher nonlinearity and an absence of photosensitivity can be obtained [69]. High nonlinear index and low TPA has been demonstrated in Ge–As–Se, Ge–As–S–Se and As–S–Se compositions [55, 56, 58]. Germanium rich compositions have reported nonlinearities between 200 and 900 times that of fused silica. It was found that the optical properties of the glasses can be improved by fine tuning the composition in the Ge–As–Se system.

The $Ge_{11.5}As_{24}Se_{64.5}$ glass is one of a small family of Ge–As–Se glass compositions that show high nonlinearity, particularly high thermal and optical stability when exposed to near band edge light and good film forming properties. It is known that the physical properties of ternary chalcogenides vary significantly as a function of their chemical composition and mean coordination number (MCN = the sum of the products of the valency times the atomic abundance of the constituent atoms) and that MCN can be used to categorize the basic properties of the glass network. In the case of Ge–As–Se glasses, it has been found that their linear and nonlinear refractive indices; optical losses; elastic properties, etc., all vary strongly with MCN. $Ge_{11.5}As_{24}Se_{64.5}$, for which MCN = 2.47, lies in the so-called "intermediate" phase (IP) which lies between the "floppy" and "stressed-rigid" glass networks. Of particular significance is that films produced by thermal evaporation in this region have similar properties to those of bulk glasses—an unusual behaviour for films made from ternary chalcogenides.

Compared with As_2S_3 glass, which has been the workhorse for all-optical devices to date, $Ge_{11.5}As_{24}Se_{64.5}$ has a higher linear refractive index (2.65 compared with 2.43) which leads to better mode confinement and, as would be expected from Miller's rule, a higher nonlinear index, n_2 (8.6×10^{-14} cm^2/W c.f. 3×10^{-14} cm^2/W for As_2S_3 at 1500 nm) [70]. No TPA could be detected during z-scan measurements made on bulk $Ge_{11.5}As_{24}Se_{64.5}$ samples, however, the *FOM* has been found to be ≈60 from measurements of the power-dependence of the transmission for $Ge_{11.5}As_{24}Se_{64.5}$ rib waveguides, as shown in Fig. 1.8a [71]. $Ge_{11.5}As_{24}Se_{64.5}$ nanowires with $\gamma \approx 135$ W^{-1} m^{-1} have been reported with losses around 1.5–2.5 dB/cm [26, 28]. 4WM and SC were demonstrated in such nanowires as well as polarization independent properties using the structure shown in Fig. 1.8b [28]. These nanowires were also used to demonstrate correlated quantum photon pair generation [27, 72]. High Q (>700,000) photonic crystal cavities are also made from this material for optical switching application, as shown in Fig. 1.9 [73].

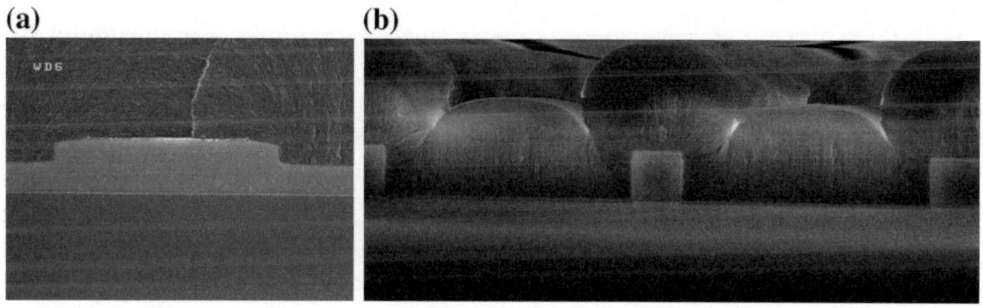

Fig. 1.8 **a** The As_2S_3 rib waveguide. **b** The polarization independent $Ge_{11.5}As_{24}Se_{64.5}$ nanowires

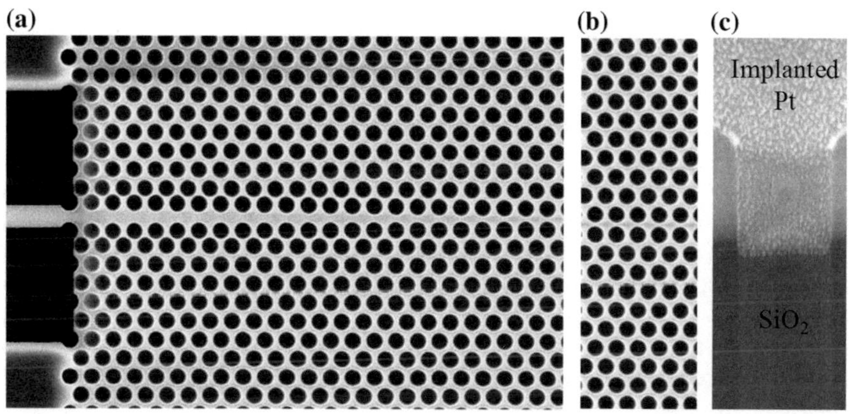

Fig. 1.9 SEM images of an end coupled $Ge_{11.5}$ photonic crystal. **a**, **b** The profile from *top surface*. **c** Cross section of holes cut by FIB and filled with Pt for imaging the side walls

1.4 All-Optical Processing in Semiconductor Optical Amplifiers

Semiconductor optical amplifiers (SOAs) have a long history in all-optical processing with their popularity stemming from the range and strength of nonlinear effects available; from the early commercial availability of fibre pigtailed devices; from the capacity for on chip integration of multiple functions/devices; and from the potential for low power operation with net conversion gain. Such is the diversity of architectures that it is beyond the scope of this summary to cover these in their entirety and the reader is instead referred to some of the excellent recent review articles and books on the topic for the wider view, e.g. [74–81]. Here the types of nonlinear effects will be briefly reviewed, the limitations facing SOAs examined along with some of the proposed remedies and the potential limits, and the best and latest achievements briefly summarized.

1.4.1 Types and Origins of Nonlinear Effects in Semiconductor Optical Amplifiers

As electrically pumped devices, SOAs are dominated by carrier density driven nonlinearities. These result in effects operating on the amplitude, phase and polarization of optical signals. The first and most obvious effect is due to the signal power dependence of the SOA gain which saturates as a result of carrier depletion as the input power is raised. Thus, cross gain modulation (XGM) occurs where an intense pulse modulates a co-propagating CW beam at a different wavelength. This process is commonly used in wavelength conversion. Gain saturation is also used in optical limiting where the SOA output power becomes essentially independent of the input power, thereby reducing amplitude noise on high signal levels. This is particularly useful for non-return to zero (NRZ) modulation.

The Kramers-Kronig relation means gain or absorption changes are always accompanied by changes in refractive index and also therefore phase change in the propagating signals. Refractive index power dependence of course manifests as a χ^3 effect. Ascribing a precise value to the nonlinearity is troublesome due to the differing contributing parts [82], but to give some comparative indication of the effective size of n_2, values from 2×10^{-12} up to $\sim 1 \times 10^{-9}$ cm^2/W have been reported, e.g. [83–85]. However the response time of the electron plasma is not as fast as the response of bound electron states in glasses, for example, leading to speed limitations as will be discussed shortly. Cross phase modulation (XPM) between a strong pulsed pump and a weaker CW probe can also be exploited in SOAs for wavelength conversion by placing the SOA inside a Mach-Zehnder or other interferometer.

Nonlinear polarisation rotation (NPR) in SOAs has also been increasingly harnessed for high speed nonlinear processing. This results from a combination of the different gain saturation behaviours of the TE and TM modes, and also from the intrinsic waveguide birefringence in the SOA (with both geometrical and stain components) which can also be modulated by changes in the carrier density. Thus a signal launched into both polarization states sees a "waveplate" with an effective input power dependent retardation, and this can be used for all optical processing.

1.4.2 Impairments in SOA Devices

1.4.2.1 Carrier Induced Switching Speed Limitations

The strengths of SOAs for nonlinear processing stem from the strong carrier-based nonlinearity in the devices, but this also leads to their main shortcoming. A major driver for all-optical signal processing is speed, but the relaxation times of carrier-based nonlinearities are long compared with those due to bound electron (Kerr) nonlinearities in dielectrics. This is the most significant impairment for SOA based devices and so will be explored in some detail.

The dynamics of the gain compression and its associated phase shift has been studied by quite a number of groups, e.g. [83, 86–101] with good summaries of the varying contributing processes and their distinct timescales being available e.g. [75, 76, 102]. It should also be noted that the SOA gain region structure (bulk vs. Quantum well (QW) vs. Quantum dot (QD) [103]) also strongly influences the relative contributions of these components. The recovery time is also affected by the input power, the polarization, the level of bias current applied to the SOA, the dimensions and design of the active region, and the pump wavelength, e.g. [92, 94, 97–101]. This makes it hard to compare recovery times in any absolute sense, but nonetheless some broad generalizations of structural impacts, and mitigating approaches is in order. It is also important to note that the phase recovery has different characteristics to the gain recovery.

Hall et al. [83], undertook one of the most comprehensive earlier studies at 1550 nm in bulk and strained multi-quantum well devices, showing that carrier injection and carrier heating effects dominated the dynamic recovery of the gain and phase. Under the conditions studied in their pump-probe experiment, the QW device exhibited faster gain recovery than the bulk albeit with a small but long lived tail, but broadly similar phase recovery to the bulk device, however at about one third the amplitude. Responses from the devices studied are reproduced from their paper below in Fig. 1.10.

QD SOAs exhibit a different response again with very rapid gain recovery but rather slower and smaller phase response as shown in Fig. 1.11. The reason for the more rapid gain recovery in QD devices is considered to be the action of the wetting layer used to grow the dots as a charge reservoir that feeds the dots locally and therefore quickly. The wetting layer itself can also deplete but this has to refill via normal interband mechanisms and so displays the characteristic long time constant (visible as a very small and long exponential recovery tail in Fig. 1.11).

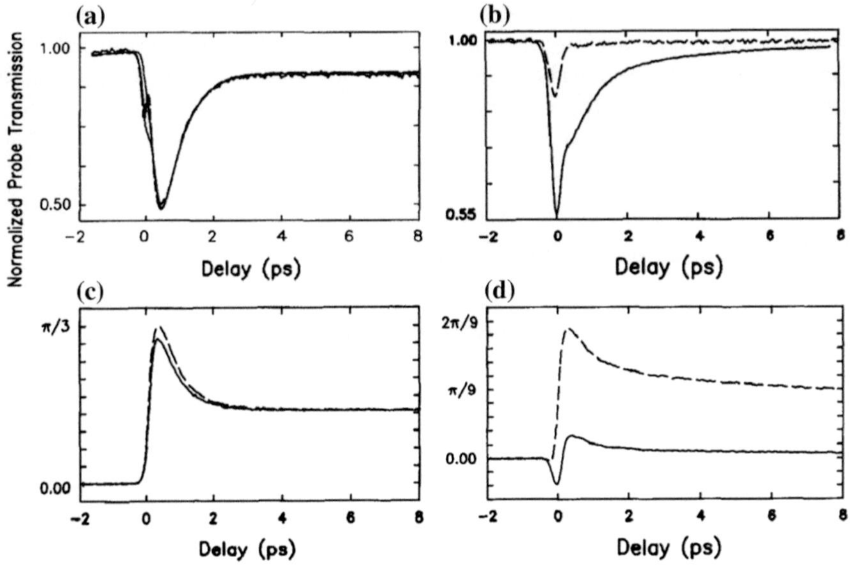

Fig. 1.10 Bulk gain left and phase underneath, MQW to the right from [83]

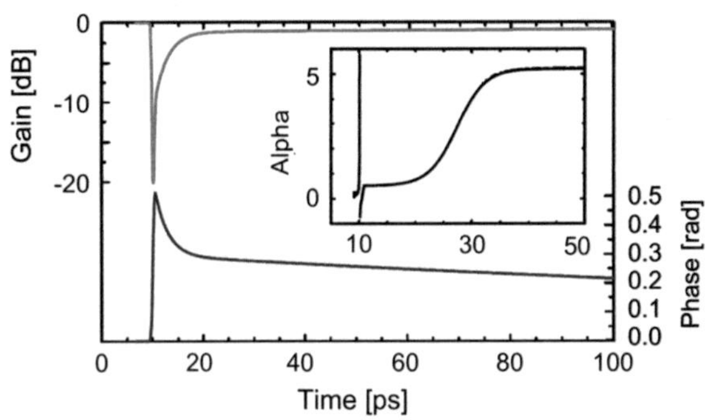

Fig. 1.11 Gain and phase recovery of typical QD SOA (from [104])

Taking the published results as a whole, the obvious question is how short can the recovery times go? As noted earlier operating parameters influence the recovery times as was recently graphically illustrated by Cleary et al. [75, 105]. By optimizing the pump and probe wavelengths relative to the SOA gain peak, they achieved 10–90 % gain recovery times from 3–5 ps in an MQW SOA, up to 7× improvement over some of the unoptimized conditions. The phase response was improved by factors up to 2×. QD SOAs have also demonstrated similar gain recovery times [104, 106]. These results suggest bit rates of 3–500 Gbit/s should be possible at least using XGM effects.

There are other effects that can also be harnessed in the quest for ultimate speed. To improve pure carrier recovery these include doping to reduce the carrier lifetime in the bulk/feeding zones, exploiting tunnelling effects in QW and QD structures

[91], the use of pulsed or CW optical holding/reset beams [86, 107, 108], and increasing the energy gap of QDs to the wetting layer [97] amongst others. Exactly how far the intrinsic recovery time can be pushed is not yet clear, but modelling has shown that speeds of 1 Tbit/s appear possible in some devices [93].

Sub-picosecond phase response times without long lived tails have also been reported both experimentally and theoretically in MQW SOAs and have been used in NPR based wavelength conversion experiments under carefully tailored low current injection regimes where two photon absorption dominates [109–111]. However the major issue with this mode of operation is the large pulse energy required for switching ($\sim$10 pJ meaning an average power at the SOA input of 5 W for a 1 Tbit/s PRBS RZ stream!) and the observation that the output pulses were broadened considerably (nonetheless at 500 fs these are still the shortest pulses ever wavelength translated in an SOA).

In addition to directly attacking the carrier recovery, architectural design can also aid in improving the overall system response time. To cite a single example from many, in XGM based wavelength conversion the nature of the phase response can also be exploited as it causes the long recovery tail to be red shifted on the probe and so a blue shifted bandpass filter can remove it, e.g. [112]. This effect was used to demonstrate XGM based wavelength conversion at 640 Gbit/s SOAs with a carrier recovery time of $\sim$60 ps [113]. A wide variety of other architectures have also been developed to enhance overall switching speed as are discussed elsewhere, e.g. [74–81].

The discussion above relates to nonlinear devices utilising the gain or XPM effects, but the Kerr effect in SOAs is large enough to enable Four Wave Mixing (FWM) based processing. There are two distinct scenarios for FWM operation, one where intense CW pumping is used and the pulsed input signal is relatively weak (often for optical phase conjugation), and the other where the input pulsed signal is strong and the CW probe used to generate the wavelength shifted idler is weak. In the latter case all the recovery processes and timescales discussed above clearly apply and this scenario has also been analysed by several authors for interactions between pulses, e.g. [88, 114, 115]. In the former case, the recovery processes are strongly suppressed due to the strong CW pump leading to the question of how fast a signal can be accurately phase conjugated or wavelength translated by this process?

Surprisingly there seem to be no analyses of the ultimate speeds achievable, though Kikcuhi et al. [116] Summerfield et al. [117] and Kim et al. [118] recognized that the ratio of CW pump power to signal power is critical and that a trade off exists with conversion efficiency (due to carrier lifetime effects) and SNR for bulk SOAs. In QD SOAs different and less stringent restrictions apply to the conversion efficiency e.g. [88, 103] but the trade off still exists. Experimentally, Kikuchi et al. [119] demonstrated phase conjugation of a 1.6 ps pulse train in an SOA cascade and successful transmission through 40 km of standard single mode fibre limited only by third-order fibre dispersion differences resulting from the wavelength translation. Taken together this suggests that the potential for Tbit/s CW pumped FWM wavelength translation and OPC in SOAs are good.

1.4.2.2 Other Limitations and Impairments of SOA Nonlinear Elements

Speed aside, there are other potential limitations in using SOAs as nonlinear elements. First and foremost all the nonlinear effects in an SOA except the Kerr effect require amplitude modulation. Phase, subcarrier, and polarization encoding are increasingly important in today's communication systems and amplitude modulation is no longer implicit. Where modulation formats such as NRZ DQPSK are employed, the signal has no amplitude modulation and XGM, XPM, and NPR cannot be harnessed for nonlinear signal processing. FWM however is a format transparent effect and remains highly useful for wavelength translation and phase conjugation.

Another obvious issue in using a device with gain is the linear addition of amplified spontaneous emission noise and its effect on optical signal to noise ratio, but also the less well studied nonlinear interactions between noise and the signal. OSNR is well understood and can be engineered and optimized with well proven approaches. Nonlinear noise based interactions in SOAs have been studied by numerous groups but most recently and with the least approximation by Connelly et al. [120] for phase encoded signals in the context of line amplifiers operating in the saturated regime (i.e. as nonlinear elements are). However further study is needed to clarify whether the effects of the various nonlinear noise mixing processes in operation present a limiting factor to high speed data processing.

A further limitation often considered to limit the applicability of SOAs is multichannel crosstalk. This relates to the imposition of the modulation of other channels onto the channel under consideration by XGM, XPM, and NPR as a channel ensemble passes through an SOA. In general this is an issue for devices relying on these effects as the modulated signal is usually the pump and so it is rather difficult to suppress the crosstalk as the nonlinear effect being exploited relies upon it. However it can sometimes be managed enabling multichannel operation with acceptable performance, e.g. [121].

There is however one potential exception to this which exploits the properties of QD SOAs. QD SOAs have a range of dot sizes, each size amplifying a reasonably narrow portion of the optical spectrum through the homogenous broadening of the theoretically atom like DOS. Thus if the signals are sufficiently separated in wavelength that they each address different QD cohorts, then they will not impact on each other's gain to first order [103] (at a more detailed level they are all fed by the same wetting layer which may deplete based on the total instantaneous power, this however as seen previously is a lower level effect in QD SOAs with appropriate design and operating conditions).

Multichannel Crosstalk is also an issue in devices employing FWM. Here all the same XGM mediated effects can apply if the powers are high enough, but in addition, there are also multiple interchannel FWM products that can fall inside the desired signal bandwidth [117]. Whilst certain optimisations can minimise this, the effects cannot be eliminated completely. Multichannel devices usually operate with a strong CW pump that suppresses the gain compression based effects, and early work demonstrated that even for purely intensity modulated data that the effects were acceptable for four channel conversion [122]. With NRZ phase modulated data the gain compression effects are absent leaving only the intermixing effects, and four channel NRZ DQPSK modulated signals were wavelength converted by FWM in a QD SOA with <4 dB penalties from intermixing [123]. Much further work however remains to optimise the systems architecture to attain higher channel counts.

1.4.3 Current State of the Art and Future Prospects for SOA Nonlinear Processing

Whilst the range of applications and therefore relevant results is too broad to summarise meaningfully here, there are a number of key achievements for SOA based nonlinear devices related to speed, conversion range, and conversion efficiency which will now be briefly discussed. Taking speed first, Table 1.1 gives an overview of some of the fastest demonstrations using both pulses and full data streams.

There are a number of pertinent points to consider from this summary. High bit rates, up to 640 Gbit/s, have been achieved with error free transmission and under special circumstances that even shorter pulses have been converted. However this has not been achieved to date will all nonlinear effects, nor with very low detection penalties. The detection penalties at best were 4 dB, this is not a small number compared to the essentially penalty free results achieved in for example passive waveguide demultiplexers using FWM at even 1.2 Tbit/s [65], although such schemes are much less power efficient than SOAs. The origins of the penalties seem to be connected in these two cases of 640 Gbit/s performance to the filtering required to speed up the SOA response, particularly for the XGM based method when used for demultiplexing.

Conversion efficiencies for the system experiments were also relatively low though the input powers were much lower than used in passive nonlinear waveguide devices. In particular, [126] shows performance that appears close to usable in a real system. However there was only a single demonstration at high speed of a device that could be format independent using FWM [128]. This would of course allow for higher bit rates through multilevel coding schemes, and the use of phase only modulation would also reduce XGM effects considered to be partly responsible for the detection penalty (the output OSNR of ~ -23 dB was though to make up the remnant part).

With further research and perhaps utilizing some of the techniques discussed earlier the detection penalties might be decreased in the future, conversion efficiencies increased and possibly operation at 1.2 Tbit/s demonstrated as some modelling has predicted [93].

Where speed is a lesser concern (e.g. for 40 Gbit/s RZ-OOK or 100 Gbit/s coherent schemes), then some impressive results have been achieved in terms of conversion efficiency, range, and regenerative performance. Table 1.2 summarizes a few of the more recent and impressive examples.

Table 1.1 Details of fastest and most recent nonlinear switching demonstrations in SOAs

References	Nonlinear effect	Data rate (Gbit/s)	Data format	Pulse width (ps)	Operation	Conversion efficiency (dB)	Pump power	Detection penalty (dB @10^{-9} BER)
[113]	XGM, blue shifted bandpass filter	640	RZ-OOK	0.8	TDM demux, 640 to 40 Gbit/s	~−22	8 dBm Ave.	13.5
[124]	XPM in symmetric Mach-Zehnder	640	RZ-DPSK	0.38	TDM demux, 640 to 40 Gbit/s	?	17 dBm Ave.	4
[125]	XGM, blue shifted bandpass filter	320	RZ-OOK	1.2	TDM demux, 320 to 40 Gbit/s	~−19	8 dBm Ave.	8
[126]	XGM, blue shifted bandpass filter	320	RZ-OOK	1.2	Full C band wavelength shift	−13	7 dBm Ave.	5
[127]	XGM, blue shifted bandpass filter	320	RZ-OOK	1.0	Wavelength shift	~−17	5.4 dBm Ave.	~10
[128]	FWM	320	RZ-OOK	1.2	Wavelength shift and phase conjugation	−18	9 dBm CW	4
[119]	FWM	–	Pulses	1.6	Wavelength shift and phase conjugation	−5	7.3 dBm CW	–
[110]	NPR	–	pulses	0.12	Wavelength shift	12	10 pJ per pulse	Output pulse 0.5 ps

Table 1.2 High conversion efficiency and multiple format NLO processing demonstrations in SOAs

References	Nonlinear effect	Operation	Conversion range (nm @ X dB eff.)	Pump power	Detection penalty	Bit rate	Data format	No of Channels
[130]	Single pump FWM	Wavelength shift/ OPC	11.2 @ >0 28 @ >−10	6 dBm CW	1	20 30 40	QPSK 8 PSK 16 QAM	1
[129]	Dual pump FWM	Wavelength shift/ OPC	100 @ >0 140 @ >−10	3 dBm CW per pump	? error free max error vector mag. 14 %	20 30 40 40	QPSK 8 PSK 16 QAM OFDM	1
[134]	Single pump FWM	Wavelength shift/ OPC	32 @ >−10	5 dBm CW per pump	~1.5 ~2	256 320	16 QAM 64 QAM	1
[123]	Single pump FWM	Multichannel OPC	–	10 dBm CW	4	50	NRZ-DQPSK	4
[135]	Single pump FWM	Regeneration via saturation of FWM eff.	10 @ >−20	13.5 dBm CW	Q raised from 6 to >12.5	10	NRZ-DPSK	1
[121]	XPM plus filter	Regeneration	N/A	1 dBm Ave/ channel	−2	20	PM-RZ-PSK	8

The work of Contestabile et al. [129, 130] has conclusively demonstrated that ultra-broadband wavelength conversion with gain is possible in a single device with minimal signal degradation, and that it can be format and bitrate independent at least to symbol rates <100 GBaud/s. Additionally the phase conjugation function can also be obtained if desired for distortion compensation. Perhaps the only shortcoming of their experiment was that the device was not polarization insensitive, but this can be addressed with dual pump schemes [131, 132] and an appropriately optimized QD SOA, or by using a polarization diversity approach to obtain the best performance [133].

Filion et al. [134] recently demonstrated wideband conversion with >10 % efficiency of even high order multiplexed signals for the first time. They also demonstrated data rates up to 320 Gbit/s with low penalties including the phase conjugation function.

Given the de facto nature of DWDM technology, the conversion of multiple channels in a single device is clearly also a priority. Matsuura et al. [123] demonstrated that acceptable penalties can be achieved for wavelength translation/OPC in a single SOA even with coherent transmission. The penalties resulted from intra-channel mixing products, meaning that to process a complete DWDM comb four or less channel banding would be required, as partly demonstrated by Andriolli et al. [136].

Taking this body of work, it is clear that the prognosis for system deployment of format transparent highly efficient wideband wavelength translation devices using in SOAs is very good. All optical regeneration of coherent signals in an SOA is becoming competitive. Unlike On-Off Keying where an electronic regenerator can be comparatively simple, coherent versions are relatively complex and expensive. Saturation effects in SOAs can also be used to limit amplitude noise without introducing significant phase noise (or even reducing it in some cases) to improve the signal Q factor or reduce the detection penalty. Porzi et al. [135] and Wu et al. [121] demonstrated significant improvements in the signal quality of coherently coded data this way, and in the case of Wu et al. did this with 8 channels of polarisation and wavelength division multiplexed data. In both cases these operations were performed in a simple integratable single SOA based device.

Finally there have also been developments to enable effects based on FWM to be mimicked in SOAs. Optical phase conjugation has recently been accomplished for the first time based on non-parametric processes [137], and it remains to be seen whether this enables new types of processing or superior performance to conventional approaches.

References

1. G.I. Stegeman, E.M. Wright, All-optical wave-guide switching. Opt. Quant. Electron. **22**, 95–122 (1990)
2. N.J. Doran, D. Wood, Nonlinear-optical loop mirror. Opt. Lett. **13**, 56–58 (1988)

3. L.D. Haret, T. Tanabe, E. Kuramochi, M. Notomi, Extremely low power optical bistability in silicon demonstrated using 1D photonic crystal nanocavity. Opt. Express **17**, 21108–21117 (2009)
4. V. Mizrahi, K.W. Delong, G.I. Stegeman, M.A. Saifi, M.J. Andrejco, 2-Photon absorption as a limitation to all-optical switching. Opt. Lett. **14**, 1140–1142 (1989)
5. M. Sheikbahae, D.J. Hagan, E.W. Vanstryland, Dispersion and band-gap scaling of the electronic kerr effect in solids associated with 2-photon absorption. Phys. Rev. Lett. **65**, 96–99 (1990)
6. M. Sheik-Bahae, D.C. Hutchings, D.J. Hagan, E.W. Van Stryland, Dispersion of bound electronic nonlinear refraction in solids. IEEE J. Quantum Electron. **27**, 1296–1309 (1991)
7. M.H.M. Sheik-Bahae, Third order optical nonlinearities, in *Handbook of Optics* (OSA, 2000)
8. M. Dinu, Dispersion of phonon-assisted nonresonant third-order nonlinearities. IEEE J. Quantum Electron. **39**, 1498–1503 (2003)
9. X. Gai, Y. Yu, B. Kuyken, P. Ma, S.J. Madden, J. Van Campenhout, P. Verheyen, G. Roelkens, R. Baets, B. Luther-Davies, Nonlinear absorption and refraction in crystalline silicon in the mid-infrared. Laser Photonics Rev. **7**, 1054–1064 (2013)
10. B.S. Wherrett, Scaling rules for multiphoton interband absorption in semiconductors. J. Opt. Soc. Am. B: **1**, 67–72 (1984)
11. G.I. Stegeman, D.J. Hagan, L. Torner, Chi((2)) cascading phenomena and their applications to all-optical signal processing, mode-locking, pulse compression and solitons. Opt. Quantum Electron. **28**, 1691–1740 (1996)
12. H. Kanbara, H. Itoh, M. Asobe, K. Noguchi, H. Miyazawa, T. Yanagawa, I. Yokohama, All-optical switching based on cascading of second-order nonlinearities in a periodically poled titanium-diffused lithium niobate waveguide. IEEE Photonic Tech. L. **11**, 328–330 (1999)
13. J.L. Jackel, J.J. Johnson, Reverse exchange method for burying proton exchanged waveguides. Electron. Lett. **27**, 1360–1361 (1991)
14. Y.N. Korkishko, V.A. Fedorov, T.M. Morozova, F. Caccavale, F. Gonella, F. Segato, Reverse proton exchange for buried waveguides in LiNbO3. J. Opt. Soc. Am. A: **15**, 1838–1842 (1998)
15. M.L. Bortz, M.M. Fejer, Measurement of the 2nd-order nonlinear susceptibility of proton-exchanged Linbo3. Opt. Lett. **17**, 704–706 (1992)
16. Y.N. Korkishko, V.A. Fedorov, F. Laurell, The SHG-response of different phases in proton exchanged lithium niobate waveguides. IEEE J. Sel. Top. Quantum **6**, 132–142 (2000)
17. C. Langrock, S. Kumar, J.E. McGeehan, A.E. Willner, M.M. Fejer, All-optical signal processing using chi((2)) nonlinearities in guided-wave devices. J. Lightwave Technol. **24**, 2579–2592 (2006)
18. R.V. Roussev, C. Langrock, J.R. Kurz, M.M. Fejer, Periodically poled lithium niobate waveguide sum-frequency generator for efficient single-photon detection at communication wavelengths. Opt. Lett. **29**, 1518–1520 (2004)
19. M.H. Chou, K.R. Parameswaran, M.M. Fejer, I. Brener, Multiple-channel wavelength conversion by use of engineered quasi-phase-matching structures in LiNbO3 waveguides. Opt. Lett. **24**, 1157–1159 (1999)
20. G.W. Lu, S. Shinada, H. Furukawa, N. Wada, T. Miyazaki, H. Ito, 160 Gb/s all-optical phase-transparent wavelength conversion through cascaded SFG-DFG in a broadband linear-chirped PPLN waveguide. Opt. Express **18**, 6064–6070 (2010)
21. M.H. Chou, I. Brener, G. Lenz, R. Scotti, E.E. Chaban, J. Shmulovich, D. Philen, S. Kosinski, K.R. Parameswaran, M.M. Fejer, Efficient wide-band and tunable midspan spectral inverter using cascaded nonlinearities in LiNbO3 waveguides. IEEE Photonics Technol. Lett. **12**, 82–84 (2000)
22. M.C. Cardakli, D. Gurkan, S.A. Havstad, A.E. Willner, K.R. Parameswaran, M.M. Fejer, I. Brener, Tunable all-optical time-slot-interchange and wavelength conversion using difference-frequency-generation and optical buffers. IEEE Photonics Technol. Lett. **14**, 200–202 (2002)

23. A.E. Willner, O.F. Yilmaz, J.A. Wang, X.X. Wu, A. Bogoni, L. Zhang, S.R. Nuccio, Optically efficient nonlinear signal processing. IEEE J. Sel. Top. Quantum **17**, 320–332 (2011)
24. J. Huang, J.R. Kurz, C. Langrock, A.M. Schober, M.M. Fejer, Quasi-group-velocity matching using integrated-optic structures. Opt. Lett. **29**, 2482–2484 (2004)
25. A. Bogoni, X.X. Wu, S.R. Nuccio, A.E. Willner, 640 Gb/s all-optical regenerator based on a periodically poled lithium niobate waveguide. J. Lightwave Technol. **30**, 1829–1834 (2012)
26. X. Gai, S. Madden, D.Y. Choi, D. Bulla, B. Luther-Davies, Dispersion engineered Ge(11.5) As(24)Se(64.5) nanowires with a nonlinear parameter of 136 W(−1)m(−1) at 1550 nm. Opt. Express **18**, 18866–18874 (2010)
27. X. Gai, R.P. Wang, C. Xiong, M.J. Steel, B.J. Eggleton, B. Luther-Davies, Near-zero anomalous dispersion Ge11.5As24Se64.5 glass nanowires for correlated photon pair generation: design and analysis. Opt. Express **20**, 776–786 (2012)
28. X. Gai, D.-Y. Choi, S. Madden, B. Luther-Davies, Polarization-independent chalcogenide glass nanowires with anomalous dispersion for all-optical processing. Opt. Express **20**, 13513–13521 (2012)
29. X. Gai, D.Y. Choi, S. Madden, B. Luther-Davies, Interplay between Raman scattering and four-wave mixing in As(2)S(3) chalcogenide glass waveguides. J. Opt. Soc. Am. B: **28**, 2777–2784 (2011)
30. M.H.T. Okuno, T. Nakanishi, M. Onishi, Highly-nonlinear optical fibers and their applications. SEI Tech. Rev. **62**, 34–40 (2006)
31. R.E. Camacho-Aguilera, Y. Cai, J.T. Bessette, L.C. Kimerling, J. Michel, High active carrier concentration in n-type, thin film Ge using delta-doping. Opt. Mater. Express **2**, 1462–1469 (2012)
32. L. Vivien, J. Osmond, J.M. Fedeli, D. Marris-Morini, P. Crozat, J.F. Damlencourt, E. Cassan, Y. Lecunff, S. Laval, 42 GHz p.i.n Germanium photodetector integrated in a silicon-on-insulator waveguide. Opt. Express **17**, 6252–6257 (2009)
33. A.S. Liu, R. Jones, L. Liao, D. Samara-Rubio, D. Rubin, O. Cohen, R. Nicolaescu, M. Paniccia, A high-speed silicon optical modulator based on a metal-oxide-semiconductor capacitor. Nature **427**, 615–618 (2004)
34. V.G. Ta'eed, N.J. Baker, L.B. Fu, K. Finsterbusch, M.R.E. Lamont, D.J. Moss, H.C. Nguyen, B.J. Eggleton, D.Y. Choi, S. Madden, B. Luther-Davies, Ultrafast all-optical chalcogenide glass photonic circuits. Opt. Express **15**, 9205–9221 (2007)
35. J. Leuthold, C. Koos, W. Freude, Nonlinear silicon photonics. Nat. Photonics **4**, 535–544 (2010)
36. X. Chen, C. Li, H.K. Tsang, Device engineering for silicon photonics. Npg Asia Mater. **3**, 34–40 (2011)
37. V.R. Almeida, R.R. Panepucci, M. Lipson, Nanotaper for compact mode conversion. Opt. Lett. **28**, 1302–1304 (2003)
38. A.R. Motamedi, A.H. Nejadmalayeri, A. Khilo, F.X. Kartner, E.P. Ippen, Ultrafast nonlinear optical studies of silicon nanowaveguides. Opt. Express **20**, 4085–4101 (2012)
39. C. Koos, L. Jacome, C. Poulton, J. Leuthold, W. Freude, Nonlinear silicon-on-insulator waveguides for all-optical signal processing. Opt. Express **15**, 5976–5990 (2007)
40. J.M. Brosi, C. Koos, L.C. Andreani, M. Waldow, J. Leuthold, W. Freude, High-speed low-voltage electro-optic modulator with a polymer-infiltrated silicon photonic crystal waveguide. Opt. Express **16**, 4177–4191 (2008)
41. M.A. Foster, A.C. Turner, R. Salem, M. Lipson, A.L. Gaeta, Broad-band continuous-wave parametric wavelength conversion in silicon nanowaveguides. Opt. Express **15**, 12949–12958 (2007)
42. H. Hu, J.D. Andersen, A. Rasmussen, B.M. Sorensen, K. Dalgaard, M. Galili, M. Pu, K. Yvind, K.J. Larsen, S. Forchhammer, L.K. Oxenlowe, Forward error correction supported 150 Gbit/s error-free wavelength conversion based on cross phase modulation in silicon. Opt. Express **21**, 3152–3160 (2013)

43. F.G. Della Corte, S. Rao, Use of amorphous silicon for active photonic devices. IEEE Trans. Electron Devices **60**, 1495–1505 (2013)
44. G. Cocorullo, F.G. Della Corte, R. De Rosa, I. Rendina, A. Rubino, E. Terzi, Amorphous silicon-based guided-wave passive and active devices for silicon integrated optoelectronics. IEEE J. Sel. Top. Quantum **4**, 997–1002 (1998)
45. K. Furuya, K. Nakanishi, R. Takei, E. Omoda, M. Suzuki, M. Okano, T. Kamei, M. Mori, Y. Sakakibara, Nanometer-scale thickness control of amorphous silicon using isotropic wet-etching and low loss wire waveguide fabrication with the etched material. Appl. Phys. Lett. **100** (2012)
46. C. Lacava, P. Minzioni, E. Baldini, L. Tartara, J.M. Fedeli, I. Cristiani, Nonlinear characterization of hydrogenated amorphous silicon waveguides and analysis of carrier dynamics. Appl. Phys. Lett. **103** (2013)
47. K.Y. Wang, A.C. Foster, Ultralow power continuous-wave frequency conversion in hydrogenated amorphous silicon waveguides. Opt. Lett. **37**, 1331–1333 (2012)
48. B. Kuyken, S. Clemmen, S.K. Selvaraja, W. Bogaerts, D. Van Thourhout, P. Emplit, S. Massar, G. Roelkens, R. Baets, On-chip parametric amplification with 26.5 dB gain at telecommunication wavelengths using CMOS-compatible hydrogenated amorphous silicon waveguides. Opt. Lett. **36**, 552–554 (2011)
49. B. Kuyken, H. Ji, S. Clemmen, S.K. Selvaraja, H. Hu, M. Pu, M. Galili, P. Jeppesen, G. Morthier, S. Massar, L.K. Oxenlowe, G. Roelkens, R. Baets, Nonlinear properties of and nonlinear processing in hydrogenated amorphous silicon waveguides. Opt. Express **19**, 146–153 (2011)
50. D.J. Moss, R. Morandotti, A.L. Gaeta, M. Lipson, New CMOS-compatible platforms based on silicon nitride and hydex for nonlinear optics. Nat. Photonics **7**, 597–607 (2013)
51. J.S. Levy, A. Gondarenko, M.A. Foster, A.C. Turner-Foster, A.L. Gaeta, M. Lipson, CMOS-compatible multiple-wavelength oscillator for on-chip optical interconnects. Nat. Photonics **4**, 37–40 (2010)
52. G.N. Parsons, J.H. Souk, J. Batey, Low hydrogen content stoichiometric silicon-nitride films deposited by plasma-enhanced chemical vapor-deposition. J. Appl. Phys. **70**, 1553–1560 (1991)
53. S.C. Mao, S.H. Tao, Y.L. Xu, X.W. Sun, M.B. Yu, G.Q. Lo, D.L. Kwong, Low propagation loss SiN optical waveguide prepared by optimal low-hydrogen module. Opt. Express **16**, 20809–20816 (2008)
54. J.B. Nadal, J.M. Díaz, D. Vukovic, F.Da Ros, E. Palushani, C. Peucheret, A comparison of nonlinear media for parametric all-optical signal processing. IEEE Photonic Conf. **ThG1.5** (2013)
55. F. Smektala, C. Quemard, L. Leneindre, J. Lucas, A. Barthelemy, C. De Angelis, Chalcogenide glasses with large non-linear refractive indices. J. Non-Cryst. Solids **239**, 139–142 (1998)
56. J.T. Gopinath, M. Soljacic, E.P. Ippen, V.N. Fuflyigin, W.A. King, M. Shurgalin, Third order nonlinearities in Ge–As–Se-based glasses for telecommunications applications. J. Appl. Phys. **96**, 6931–6933 (2004)
57. J.S. Sanghera, L.B. Shaw, P. Pureza, V.Q. Nguyen, D. Gibson, L. Busse, I.D. Aggarwal, C. M. Florea, F.H. Kung, Nonlinear properties of chalcogenide glass fibers. Int. J. Appl. Glass Sci. **1**, 296–308 (2010)
58. C. Quemard, F. Smektala, V. Couderc, A. Barthelemy, J. Lucas, Chalcogenide glasses with high non linear optical properties for telecommunications. J. Phys. Chem. Solids **62**, 1435–1440 (2001)
59. R.A. Ganeev, A.I. Ryasnyansky, M.K. Kodirov, T. Usmanov, Two-photon absorption and nonlinear refraction of amorphous chalcogenide films. J. Opt. Pure Appl. Opt. **4**, 446–451 (2002)
60. M. Asobe, Nonlinear optical properties of chalcogenide glass fibers and their application to all-optical switching. Opt. Fiber Technol. **3**, 142–148 (1997)

61. S.J. Madden, D.Y. Choi, D.A. Bulla, A.V. Rode, B. Luther-Davies, V.G. Ta'eed, M.D. Pelusi, B.J. Eggleton, Long, low loss etched As2S3 chalcogenide waveguides for all-optical signal regeneration. Opt. Express **15**, 14414–14421 (2007)
62. M. Pelusi, F. Luan, T.D. Vo, M.R.E. Lamont, S.J. Madden, D.A. Bulla, D.Y. Choi, B. Luther-Davies, B.J. Eggleton, Photonic-chip-based radio-frequency spectrum analyser with terahertz bandwidth. Nat. Photonics **3**, 139–143 (2009)
63. T.D. Vo, M.D. Pelusi, J. Schroder, F. Luan, S.J. Madden, D.Y. Choi, D.A.P. Bulla, B. Luther-Davies, B.J. Eggleton, Simultaneous multi-impairment monitoring of 640 Gb/s signals using photonic chip based RF spectrum analyzer. Opt. Express **18**, 3938–3945 (2010)
64. M.D. Pelusi, F. Luan, D.Y. Choi, S.J. Madden, D.A.P. Bulla, B. Luther-Davies, B. J. Eggleton, Optical phase conjugation by an As(2)S(3) glass planar waveguide for dispersion-free transmission of WDM-DPSK signals over fiber. Opt. Express **18**, 26686–26694 (2010)
65. T.D. Vo, H. Hu, M. Galili, E. Palushani, J. Xu, L.K. Oxenlowe, S.J. Madden, D.Y. Choi, D. A.P. Bulla, M.D. Pelusi, J. Schroder, B. Luther-Davies, B.J. Eggleton, Photonic chip based transmitter optimization and receiver demultiplexing of a 1.28 Tbit/s OTDM signal. Opt. Express **18**, 17252–17261 (2010)
66. M. Galili, J. Xu, H.C.H. Mulvad, L.K. Oxenlowe, A.T. Clausen, P. Jeppesen, B. Luther-Davies, S. Madden, A. Rode, D.Y. Choi, M. Pelusi, F. Luan, B.J. Eggleton, Breakthrough switching speed with an all-optical chalcogenide glass chip: 640 Gbit/s demultiplexing. Opt. Express **17**, 2182–2187 (2009)
67. M.R.E. Lamont, B. Luther-Davies, D.Y. Choi, S. Madden, X. Gai, B.J. Eggleton, Net-gain from a parametric amplifier on a chalcogenide optical chip. Opt. Express **16**, 20374–20381 (2008)
68. M.R.E. Lamont, B. Luther-Davies, D.Y. Choi, S. Madden, B.J. Eggleton, Supercontinuum generation in dispersion engineered highly nonlinear (gamma = 10/W/m) As(2)S(3) chalcogenide planar waveguide. Opt. Express **16**, 14938–14944 (2008)
69. A. Zakery, P.J.S. Ewen, A.E. Owen, Photodarkening in As-S films and its application in grating fabrication. J. Non-Cryst. Solids **198**, 769–773 (1996)
70. A. Prasad, C.J. Zha, R.P. Wang, A. Smith, S. Madden, B. Luther-Davies, Properties of GexAsySe1-x-y glasses for all-optical signal processing. Opt. Express **16**, 2804–2815 (2008)
71. A. Prasad, Ge–As–Se chalcogenide glasses for all-optical signal processing, in *Laser Physics Center* (Australian National University, Australia, 2010)
72. J. He, C. Xiong, A.S. Clark, M.J. Collins, X. Gai, D.Y. Choi, S.J. Madden, B. Luther-Davies, B.J. Eggleton, Effect of low-Raman window position on correlated photon-pair generation in a chalcogenide Ge11.5As24Se64.5 nanowire, J Appl. Phys. **112** (2012)
73. X. Gai, B. Luther-Davies, T.P. White, Photonic crystal nanocavities fabricated from chalcogenide glass fully embedded in an index-matched cladding with a high Q-factor (>750,000). Opt. Express **20**, 15503–15515 (2012)
74. E. Ciaramella, Wavelength conversion and all-optical regeneration: achievements and open issues. J. Lightwave Technol. **30**, 572–582 (2012)
75. C.S. Cleary, High speed nonlinear optical components for next-generation optical communications, *PhD. Thesis, Photonic Systems Group, Tyndall National Institute, Department of Physics,* University College Cork, 2013
76. R. Bonk, *Linear and Nonlinear Semiconductor Optical Amplifiers for Next-Generation Optical Networks* (KIT Scientific Publishing, 2013)
77. J. Leuthold, W. Freude, G. Boettger, J. Wang, A. Marculescu, P. Vorreau, R. Bonk, All-optical regeneration, in *ICTON 2006: 8th International Conference on Transparent Optical Networks, Vol 1, Proceedings: ICTON, MPM, INDUSTRIAL, PICAW, GOWN*, ed. by Marciniak, (2006), pp. 28–31
78. J. Leuthold, J. Wang, T. Vallaitis, C. Koos, R. Bonk, A. Marculescu, P. Vorreau, S. Sygletos, W. Freude, New approaches to perform all-optical signal regeneration, in *ICTON 2007: Proceedings of the 9th International Conference on Transparent Optical Networks, Vol 2*, ed. by Marciniak, (2007), pp. 222–225

79. R.J. Manning, R. Giller, X. Yang, R.P. Webb, D. Cotter, Faster switching with semiconductor optical amplifiers, in *2007 Photonics in Switching* (2007), pp. 145–146
80. S. Diez, C. Schmidt, R. Ludwig, H.G. Weber, K. Obermann, S. Kindt, I. Koltchanov, K. Petermann, Four-wave mixing in semiconductor optical amplifiers for frequency conversion and fast optical switching. IEEE J. Sel. Top. Quantum Electron. **3**, 1131–1145 (1997)
81. H. Ishikawa, *Ultrafast All-Optical Signal Processing Devices* (Wiley, Chichester, 2008)
82. N. Christodoulides, I. Khoo, G. Salamo, G. Stegeman, E. Van Stryland, Nonlinear refraction and absorption: mechanisms and magnitudes. Adv. Opt. Photonics **2**, 60–200 (2010)
83. K. Hall, G. Lenz, A. Darwish, E. Ippen, Subpicosecond gain and index nonlinearities in InGaAsP diode lasers. Opt. Commun. **111**, 589–612 (1994)
84. R.S. Grant, W. Sibbett, Observatiobns of ultrafast nonlinear refraction in an InGaAsP optical amplifier. Appl. Phys. Lett. **58**, 1119–1121 (1991)
85. G.P. Agrawal, N.A. Olsson, Self phase modulation and spectral braodening of optical pulses in semiconductor laser amplifiers. IEEE J. Quantum Electron. **25**, 2297–2306 (1989)
86. K. Abedi, H. Taleb, Phase recovery acceleration in quantum-dot semiconductor optical amplifiers. J. Lightwave Technol. **30**, 1924–1930 (2012)
87. L.G. Chen, R.G. Lu, S.J. Zhang, J.F. Li, Y. Liu, Time-resolved chirp properties of semiconductor optical amplifiers in high-speed all-optical switches, Opt. Eng. **52** (2013)
88. A.H. Flayyih, A.H. Al-Khursan, Theory of four-wave mixing in quantum dot semiconductor optical amplifiers, J. Phy. D-Appl. Phys. **46** (2013)
89. W. Freude, R. Bonk, T. Vallaitis, A. Marculescu, A. Kapoor, E.K. Sharma, C. Meuer, D. Bimberg, R. Brenot, F. Lelarge, G.H. Duan, C. Koos, J. Leuthold, IEEE linear and nonlinear semiconductor optical amplifiers, in *2010 12th International Conference on Transparent Optical Networks* (2011)
90. S.R. Hosseini, M. Razaghi, N.K. Das, Analysis of ultrafast nonlinear phenomena's influences on output optical pulses and four-wave mixing characteristics in semiconductor optical amplifiers. Opt. Quantum Electron. **42**, 729–737 (2011)
91. X. Huang, C. Qin, Y. Yu, X.L. Zhang, Acceleration of carrier recovery in a quantum well semiconductor optical amplifier due to the tunneling effect. J. Opt. Soc. Am. B-Opt. Phy. **29**, 2990–2994 (2012)
92. J.H. Kim, Influence of optical pumping wavelength on the ultrafast gain and phase recovery acceleration of quantum-dot semiconductor optical amplifiers, J. Opt. **15** (2013)
93. A. Kotb, 1 Tb/s high quality factor NOR gate based on quantum-dot semiconductor optical amplifier. Opt. Quantum Electron. **45**, 1259–1268 (2013)
94. T. Liu, K. Obermann, K. Petermann, F. Girardin, G. Guekos, Effect of saturation caused by amplified spontaneous emission on semiconductor optical amplifier performance. Electron. Lett. **33**, 2042–2043 (1997)
95. Y. Liu, L.G. Chen, X. Zheng, S.J. Zhang, R.G. Lu, Ultrafast nonlinear dynamics in semiconductor optical amplifiers for optical signal processing, in *Semiconductor Lasers and Applications V*, eds. by Zhu, Li, Peters, Yu (Spie-Int Soc Optical Engineering, Bellingham, (2012)
96. F.D. Mahad, A.S.M. Supa'at, S.M. Idrus, D. Forsyth, Analyses of semiconductor optical amplifier (SOA) four-wave mixing (FWM) for future all-optical wavelength conversion. Optik **124**, 1–3 (2013)
97. K. Solis-Trapala, H.J.S. Dorren, Dynamic and static gain characteristics of quantum-dot semiconductor optical amplifiers operating at 1.55 mu m. Opt. Commun. **298**, 106–113 (2013)
98. R. Giller, R.J. Manning, G. Talli, R.P. Webb, M.J. Adams, Analysis of the dimensional dependence of semiconductor optical amplifier recovery speeds. Opt. Express **15**, 1773–1782 (2007)
99. F. Ginovart, J. Simon, I. Valiente, Gain recovery dynamics in semiconductor optical amplifier. Opt. Commun. **199**, 111–115 (2001)
100. F. Girardin, G. Guekos, A. Houbavlis, Gain recovery of bulk semiconductor optical amplifiers. IEEE Photonics Technol. Lett. **10**, 784–786 (1998)

101. R. Manning, A. Ellis, A. Poustie, K. Blow, Semiconductor laser amplifiers for ultrafast all-optical signal processing. JOSA B **14**, 3204–3216 (1997)
102. J. Mørk, M. Nielsen, T. Berg, The dynamics of semiconductor optical amplifier: modeling and applications. Opt. Photonics News **14**, 42–48 (2003)
103. M. Sugawara, T. Akiyama, N. Hatori, Y. Nakata, K. Otsubo, H. Ebe, Quantum-dot semiconductor optical amplifiers, in *APOC 2002: Asia-Pacific Optical and Wireless Communications; Materials and Devices for Optical and Wireless Communications*, eds. by Chang Hasnain, Xia, Iga (2002), pp. 259–275
104. T. Vallaitis, C. Koos, R. Bonk, W. Freude, M. Laemmlin, C. Meuer, D. Bimberg, J. Leuthold, Slow and fast dynamics of gain and phase in a quantum dot semiconductor optical amplifier. Opt. Express **16**, 170–178 (2008)
105. C. Cleary, M. Power, S. Schneider, R. Webb, R. Manning, Fast gain recovery rates with strong wavelength dependence in a non-linear SOA. Opt. Express **18**, 25726–25737 (2010)
106. T. Akiyama, H. Kuwatsuka, T. Simoyama, Y. Nakata, K. Mukai, M. Sugawara, O. Wada, H. Ishikawa, Ultrafast nonlinear processes in quantum-dot optical amplifiers. Opt. Quantum Electron. **33**, 927–938 (2001)
107. A. Rostami, H. Nejad, R. Qartavol, H. Saghai, Tb/s optical logic gates based on quantum-dot semiconductor optical amplifiers. IEEE J. Quantum Electron. **46**, 354–360 (2010)
108. S. Alavizadeh, H. Baghban, A. Rostami, Quantum-dot semiconductor optical amplifier performance management under optical injection. J. Mod. Opt. **60**, 509–514 (2013)
109. H.J.S. Dorren, A.K. Mishra, Z.G. Li, H.K. Ju, H. de Waardt, G.D. Khoe, T. Simoyama, H. Ishikawa, H. Kawashima, T. Hasama, All-optical logic based on ultrafast gain and index dynamics in a semiconductor optical amplifier. IEEE J. Sel. Top. Quantum Electron. **10**, 1079–1092 (2004)
110. A.K. Mishra, X. Yang, D. Lenstra, G.D. Khoe, H.J.S. Dorren, Wavelength conversion employing 120-fs optical pulses in an SOA-based nonlinear polarization switch. IEEE J. Sel. Top. Quantum Electron. **10**, 1180–1186 (2004)
111. X. Yang, A.K. Mishra, D. Lenstra, F.M. Huijskens, H. de Waardt, G.D. Khoe, H.J.S. Dorren, Sub-picosecond all-optical switch using a multi-quantum-well semiconductor optical amplifier. Opt. Commun. **236**, 329–334 (2004)
112. J. Leuthold, R. Bonk, T. Vallaitis, A. Marculescu, W. Freude, C. Meuer, D. Bimberg, R. Brenot, F. Lelarge, G.H. Duan, Linear and nonlinear semiconductor optical amplifiers, paper OThI3," in *2010 Conference on Optical Fiber Communication* (*OFC*) *Collocated with National Fiber Optic Engineers Conference* (*NFOEC*) (IEEE, 2010)
113. E. Tangdiongga, Y. Liu, H. de Waardt, G.D. Khoe, A.M.J. Koonen, H.J.S. Dorren, X. Shu, I. Bennion, All-optical demultiplexing of 640–40 Gbits/s using filtered chirp of a semiconductor optical amplifier. Opt. Lett. **32**, 835–837 (2007)
114. A.H. Flayyih, A.H. Al-Khursan, Four-wave mixing in quantum dot semiconductor optical amplifiers. Appl. Optics **52**, 3156–3165 (2013)
115. J. Inoue, H. Kawaguchi, Highly nondegenerate four-wave mixing among subpicosecond optical pulses in a semiconductor optical amplifier. IEEE Photonics Technol. Lett. **10**, 349–351 (1998)
116. K. Kikuchi, C. Lorattanasane, K. Saito, Telenor, Phase-conjugation characteristics of semiconductor optical amplifiers, in *22nd European Conference on Optical Communications, Proceedings, Vols 1–6: Co-Located With: 2nd European Exhibition on Optical Communication—Eeoc '96*(1996), pp. C269-C272
117. M.A. Summerfield, R.S. Tucker, Frequency-domain model of multiwave mixing in bulk semiconductor optical amplifiers. IEEE J. Sel. Top. Quantum Electron. **5**, 839–850 (1999)
118. Y. Kim, H. Lee, J. Jeong, Analysis of four-wave mixing in semiconductor optical amplifiers for optical frequency conversion and optical phase conjugation. J. Korean Phys. Soc. **34**, S582–S586 (1999)
119. K. Kikuchi, K. Matsuura, Transmission of 2-ps optical pulses at 1550 nm over 40 km standard fiber using midspan optical phase conjugation in semiconductor optical amplifiers. IEEE Photonics Technol. Lett. **10**, 1410–1412 (1998)

120. M.J. Connelly, C.L. Janer, Modeling of semiconductor optical amplifier RIN and phase noise for optical PSK systems. Opt. Quantum Electron. **44**, 219–225 (2012)
121. W.H. Wu, Y. Yu, B.R. Zou, W.L. Yang, X.L. Zhang, Single SOA based simultaneous amplitude regeneration for WDM-PDM RZ-PSK signals. Opt. Express **21**, 6718–6723 (2013)
122. J.P.R. Lacey, S.J. Madden, M.A. Summerfield, R.S. Tucker, A.I. Faris, 4-Channel WDM optical-phase conjugator using 4-wave-mixing in a single semiconductor optical amplifier. Electron. Lett. **31**, 743–744 (1995)
123. M. Matsuura, N. Calabretta, O. Raz, H.J.S. Dorren, Multichannel wavelength conversion of 50 Gbit/s NRZ-DQPSK signals using a quantum-dot semiconductor optical amplifier. Opt. Express **19**, 560–566 (2011)
124. T. Hirooka, M. Okazaki, T. Hirano, P.Y. Guan, M. Nakazawa, S. Nakamura, All-optical demultiplexing of 640 Gb/s OTDM-DPSK signal using a semiconductor SMZ switch. IEEE Photonics Technol. Lett. **21**, 1574–1576 (2009)
125. M. Matsuura, O. Raz, F. Gomez-Agis, N. Calabretta, H.J.S. Dorren, Error-free 320–40 Gbit/s optical demultiplexing based on blueshift filtering in a quantum-dot semiconductor optical amplifier. Opt. Lett. **38**, 238–240 (2013)
126. M. Matsuura, O. Raz, F. Gomez-Agis, N. Calabretta, H.J.S. Dorren, Ultrahigh-speed and widely tunable wavelength conversion based on cross-gain modulation in a quantum-dot semiconductor optical amplifier. Opt. Express **19**, 551–559 (2011)
127. Y. Liu, E. Tangdiongga, Z. Li, H. de Waardt, A.M.J. Koonen, G.D. Khoe, X.W. Shu, I. Bennion, H.J.S. Dorren, Errior-free 320 Gb/s all-omtical wavelength conversion using a single semiconductor optical amplifier. J. Lightwave Technol. **25**, 103–108 (2007)
128. M. Matsuura, O. Raz, F. Gomez-Agis, N. Calabretta, H.J.S. Dorren, 320 Gbit/s wavelength conversion using four-wave mixing in quantum-dot semiconductor optical amplifiers. Opt. Lett. **36**, 2910–2912 (2011)
129. G. Contestabile, Y. Yoshida, A. Maruta, K. Kitayama, Ultra-broad band, low power, highly efficient coherent wavelength conversion in quantum dot SOA. Opt. Express **20**, 27902–27907 (2012)
130. G. Contestabile, Y. Yoshida, A. Maruta, K.I. Kitayama, Coherent wavelength conversion in a quantum dot SOA. IEEE Photonics Technol. Lett. **25**, 791–794 (2013)
131. J. Lu, J.J. Yu, H. Zhou, Y. Li, L. Chen, Polarization insensitive wavelength conversion based on dual-pump four-wave mixing for polarization multiplexing signal in SOA. Opt. Commun. **284**, 5364–5371 (2011)
132. R.M. Jopson, R.E. Tench, Polarization independent phase conjugation of lightwave signals. Electron. Lett. **29**, 2216–2217 (1993)
133. J.P.R. Lacey, M.A. Summerfield, S.J. Madden, Tunability of polarization-insensitive wavelength converters based on four-wave mixing in semiconductor optical amplifiers. J. Lightwave Technol. **16**, 2419–2427 (1998)
134. B. Filion, W.C. Ng, A.T. Nguyen, L.A. Rusch, S. LaRochelle, Wideband wavelength conversion of 16 Gbaud 16-QAM and 5 Gbaud 64-QAM signals in a semiconductor optical amplifier. Opt. Express **21**, 19825–19833 (2013)
135. C. Porzi, A. Bogoni, G. Contestabile, Regenerative wavelength conversion of DPSK signals through FWM in an SOA. IEEE Photonics Technol. Lett. **25**, 175–178 (2013)
136. N. Andriolli, S. Faralli, X.J.M. Leijtens, J. Bolk, G. Contestabile, Monolithically integrated all-optical regenerator for constant envelope WDM signals. J. Lightwave Technol. **31**, 322–327 (2013)
137. R.X. Yu, N.K. Fontaine, R. Proietti, B.B. Guan, S.J.B. Yoo, All-optical phase conjugation using a 90° optical hybrid and nested SOA-MZIs, in *2013 Optical Fiber Communication Conference and Exposition and the National Fiber Optic Engineers Conference* (2013)

Chapter 2
CMOS Compatible Platforms for Integrated Nonlinear Optics

David J. Moss and Roberto Morandotti

Abstract Nonlinear photonic chips are capable of generating and processing signals all-optically with performance far superior to that possible electronically—particularly with respect to speed. Although silicon-on-insulator has been the leading platform for nonlinear optics, its high two-photon absorption at telecommunications wavelengths poses a fundamental limitation that is an intrinsic property of silicon's bandstructure. We review recent progress in new non-silicon CMOS-compatible platforms for nonlinear optics, focusing on Hydex glass and silicon nitride, and briefly discuss the promising new platform of amorphous silicon. These material systems have opened up many new capabilities such as on-chip optical frequency comb generation, ultrafast optical pulse generation and measurement. We highlight their potential future impact as well as the challenges to achieving practical solutions for many key applications.

2.1 Introduction

All-optical signal generation and processing [1, 2] have been highly successful at enabling a vast array of capabilities, such as switching and de-multiplexing of signals at unprecedented speeds [3, 4], achieving parametric gain [5] on a chip, Raman lasing [6], wavelength conversion [7], optical logic [8], all-optical regeneration [9, 10], radio-frequency (RF) spectrometers operating at THz speeds [11, 12], as well as entirely new functions such as ultra-short pulse measurement [13, 14] and generation [15] on a chip, optical temporal cloaking [16], and many others. Phase sensitive functions [14, 17], in particular, will likely be critical for

D.J. Moss (✉)
School of Electrical and Computer Engineering,
RMIT University, Melbourne, VIC 3001, Australia
e-mail: david.moss@rmit.edu.au

R. Morandotti
INRS-EMT, 1650 Boulevard Lionel Boulet, Varennes, QC J3X 1S2, Canada

S. Wabnitz and B.J. Eggleton (eds.), *All-Optical Signal Processing*,
Springer Series in Optical Sciences 194, DOI 10.1007/978-3-319-14992-9_2

telecommunications systems that are already using phase encoding schemes [18, 19]. The ability to produce these devices in integrated form—photonic integrated circuits (PICs)—will reap the greatest dividends in terms of cost, footprint, energy consumption and performance, where demands for greater bandwidth, network flexibility, low energy consumption and cost must all be met.

The quest for high performance platforms for integrated nonlinear optics has naturally focused on materials with extremely high nonlinearity, and silicon-on-insulator (SOI) has led this field for several years [2]. Its high refractive index allows for tight confinement of light within nanowires that, combined with its high Kerr nonlinearity (n_2) has yielded extremely high nonlinear parameters of $\gamma = 200{,}000\ \mathrm{W}^{-1}\ \mathrm{km}^{-1}$ ($\gamma = \omega\ n_2/c\ A_{eff}$, where A_{eff} is the effective area of the waveguide, c is the speed of light and ω is the pump frequency).

As a platform for linear photonics, SOI has already gained a foothold as a foundation for the silicon photonic chip industry [20] aimed at replacing the large network interface cards in optical communications networks, with the ability to combine electronics and photonics on the same chip. These first-generation silicon photonic chips typically employ photonics for passive splitters, filters, and multiplexers and offer the substantial benefit of potentially exploiting the enormous global CMOS infrastructure, although not without challenges [21, 22]. What is clear, however, is that any progress made in this direction for these first generation SOI chips will have a direct benefit on future generation all-optical chips, whether in SOI directly or in other CMOS-compatible platforms. Many of the issues that make CMOS compatibility compelling for linear passive devices apply equally well to all-optical-processing chips.

Indeed, silicon is so attractive as a platform for both linear *and* nonlinear photonics that were it not for one single issue, the quest for the ideal all-optical platform might already be solved. In 2004 it was realized [23] that crystalline silicon (the basis of SOI) suffers from high nonlinear absorption due to (indirect) two-photon absorption (TPA) in telecommunications bands at wavelengths less than 2000 nm (Fig. 2.1). Furthermore, the problem is compounded by the fact that TPA generates free carriers which themselves produce significant linear absorption. This was first realized as being a fundamental limitation, at that time in terms of achieving net Raman gain on a chip [23]. Since then, many methods have been developed for reducing the effect of TPA generated free carriers, such as sweeping carriers out by the use of reverse-biased p-i-n junctions [24], reducing carrier lifetime by ion-implantation, and other approaches.

However, this only solves the secondary effect of free carriers—it has no effect on silicon's *intrinsic* nonlinear figure of merit (FOM = $n_2/\beta\ \lambda$, where β is the TPA coefficient and λ the wavelength), which is only 0.3 near 1550 nm [25–27] (Fig. 2.1). This represents a *fundamental* limitation—an intrinsic property of silicon's bandstructure. The fact that many impressive all-optical demonstrations have been made in silicon despite its low FOM is a testament to how exceptional its linear and nonlinear optical properties are. Nonetheless, the critical impact of silicon's large TPA was illustrated [28, 29] in 2010 by the demonstration of high parametric gain at wavelengths beyond 2 μm, where TPA vanishes. Indeed, it is

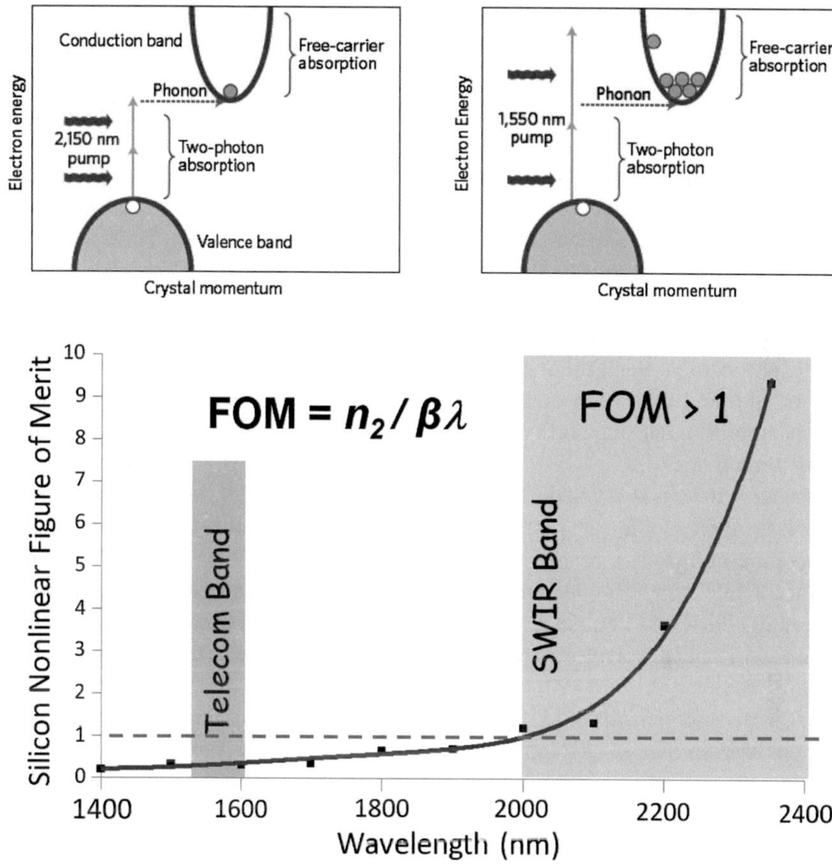

Fig. 2.1 Source of two-photon absorption in silicon in the telecom band near 1550 nm and nonlinear FOM of silicon from 1400 to 2400 nm obtained by taking an average of results reported in the literature [25–27], showing that the FOM is well below 1 in the telecom band due to the indirect TPA, and exceeds 1 beyond 2000 nm, making it an attractive nonlinear platform in this wavelength range

likely that in the mid-infrared wavelength range where silicon is transparent to both one photon and two photon transitions—between 2 and 6 μm—it will undoubtedly remain a highly attractive platform for nonlinear photonics, despite the influence of higher order absorption [30].

For the telecom band, however, the search has continued for the ideal nonlinear platform. Historically, two important platforms have been chalcogenide glasses [31] and AlGaAs [32]. Chalcogenide glasses have achieved very high performance as nonlinear devices, but pose some unique challenges. Realizing nanowires with ultrahigh nonlinearity (>10,000 W^{-1} km^{-1}) has proven elusive due to fabrication challenges, as has been achieving a material reliability on par with semiconductors. AlGaAs was the first platform proposed for nonlinear optics in the telecom band

[32] and offers the powerful ability to tune the nonlinearity and FOM by varying the alloy composition. A significant issue for AlGaAs, however, is that nanowires require very challenging fabrication [33] methods. Nonetheless, both platforms offer significant advantages and will undoubtedly play a role in future all-optical photonic chips.

With the motivation of overcoming this fundamental shortcoming of silicon, new platforms for nonlinear optics have recently been proposed [34–38] that have achieved considerable success and also offer CMOS compatibility. They are based on silicon nitride SiN and high-index doped silica (trade-named Hydex reference patents). Originally developed for linear optics [39–41], these platforms are particularly promising due to their low linear loss, relatively large nonlinearities and, most significantly, negligible nonlinear loss at telecommunication wavelengths [42]. In addition, their high quality CMOS-compatible fabrication processes, high material stability, and the ability to engineer dispersion [36] make these platforms highly attractive.

Indeed, within a short period of time significant progress has been made with respect to their nonlinear performance—particularly in the context of optical frequency comb (OFC) generation in microresonators [43–50]. Since the demonstration of OFCs in SiN [36] and Hydex [37] in 2010, this field has proliferated. Extremely wide-band frequency combs [42–44], sub 100-GHz combs [46], line-by-line arbitrary optical waveform generation [47], ultrashort pulse generation [15, 49], and dual frequency combs [50] have been reported. In addition to OFC generation, optical harmonic generation [51] has been observed. These achievements have not been possible in SOI at telecom wavelengths because of its low FOM.

Here, we review the substantial progress made towards nonlinear optical applications of these new CMOS-compatible platforms as well as the newly emerging promising platform of amorphous silicon. The high performance, reliability, manufacturability of all of these platforms combined with their intrinsic compatibility with electronic-chip manufacturing (CMOS) has raised the prospect of practical platforms for future low-cost nonlinear all-optical PICs.

2.2 Platforms

Silicon nitride (Si_3N_4), a CMOS-compatible material well known in the computer chip industry as a dielectric insulator, has been used as a platform for linear integrated optics [39] for some time. However, only recently [34] has it been proposed as a platform for nonlinear optics. Historically, the challenge for SiN optical devices has been to grow low loss layers thicker than 250 nm, due to tensile film stress. Achieving such thick layers is critical for nonlinear optics since both high mode confinement as well as dispersion engineering for phase matching [36] are needed. Thick (>500 nm) low loss SiN layers were recently grown (Fig. 2.2) by plasma-enhanced chemical vapor deposition (PECVD) [34] as well as by low-pressure chemical vapor deposition (LPCVD) [36] (Fig. 2.3). The latter approach employed

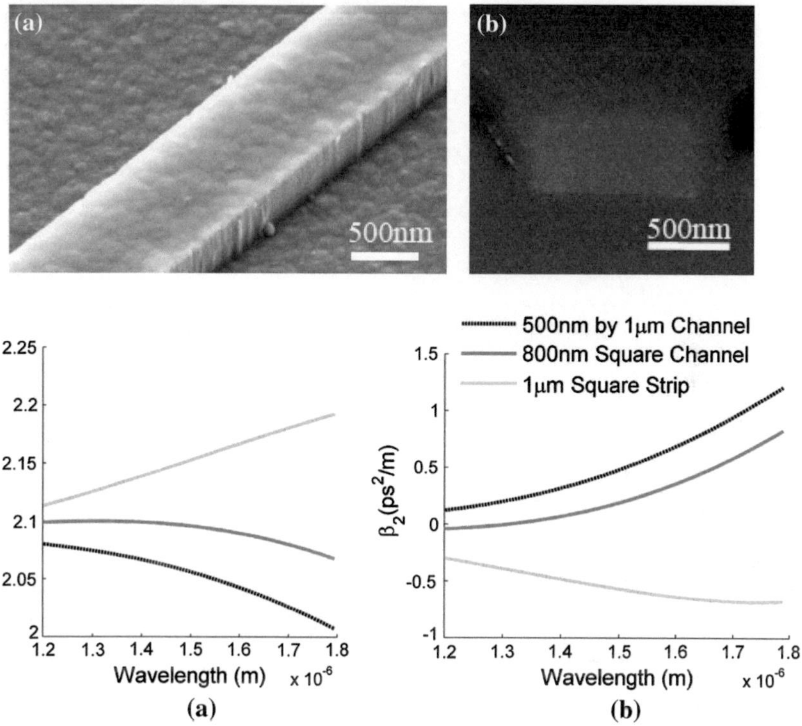

Fig. 2.2 Silicon nitride nanowires (**a**) SEM micrograph of the SiN/SiO_2 waveguides **a** before and **b** after SiO_2 upper-cladding deposition, respectively [34]; *Bottom* calculated group index (**a**) and group velocity dispersion (**b**) of SiN nanowires with different geometries from [35]

a thermal cycling process that resulted in very thick (700 nm) films that yielded nanowires with very low propagation loss (0.4 dB/cm).

The first nonlinear optical studies of SiN waveguides were reported in 2008 [34] showing nonlinear shifting of the resonances in 700 nm thick SiN ring resonators with 200 mW CW optical pump power. Time resolved measurements enabled thermal and Kerr contributions to be separated, resulting in an $n_2 \sim 10\times$ that of silica glass, which is consistent with Millers rule [52]. This value has been validated in subsequent reports of nonlinear optics in SiN nanowires and resonators. Figure 2.2 also shows the group refractive index and linear dispersion β_2 as a function of wavelength for different nanowire dimensions. This represents the first report of dispersion engineering in SiN waveguides.

Parametric gain in SiN was first demonstrated [36] in low loss nanowires by centering the pump for the FWM process in the anomalous group-velocity dispersion (GVD) regime near the zero-GVD point (Fig. 2.3). This allowed for broad-bandwidth phase matching, and hence signal amplification, over a wide range of wavelengths. Net gain was achieved in long (6 cm) SiN waveguides with a nonlinear γ parameter of 1200 W^{-1} km^{-1} and a zero-GVD point near 1560 nm.

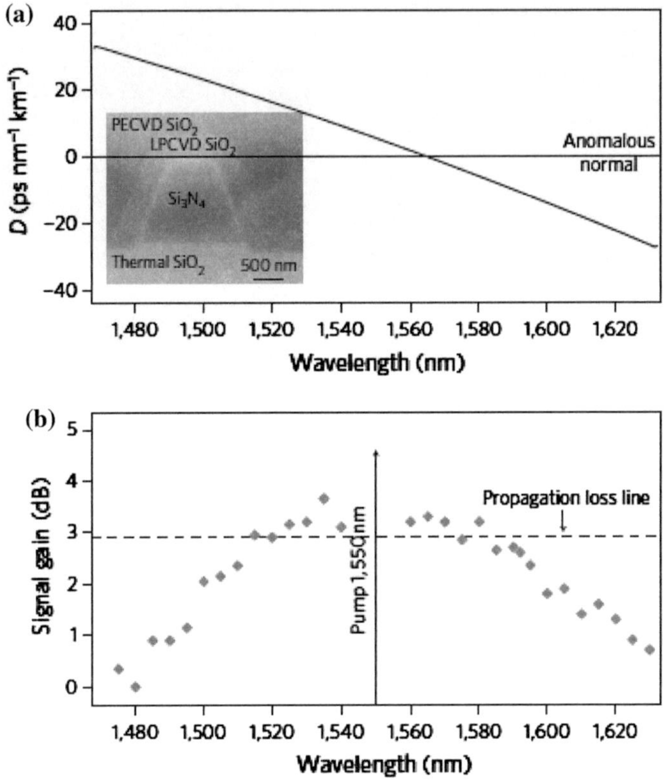

Fig. 2.3 Silicon nitride nanowires. **a** Theoretical dispersion [36] of a SiN nanowire grown by low pressure CVD, showing a zero-GVD point at 1560 nm with anomalous dispersion in the C-band. *Inset* shows a scanning electron micrograph (SEM) of the cross-section of the SiN nanowire highlighting the trapezoidal shape of the core and the cladding materials; (**b**) parametric gain achieved with a pump wavelength at 1550 nm, from [36]

An on/off signal gain as high as 3.6 dB was observed over a 150-nm bandwidth, and since the total propagation loss through the waveguides was 3 dB, this represented net parametric amplification.

Hydex glass was developed [41] as a low loss CMOS compatible optical platform primarily for advanced linear filters. Its refractive index range of n = 1.5–1.9 is slightly lower than SiN, being comparable to SiON, and so a buried waveguide geometry is typically used rather than nanowires. Nonetheless, the core-cladding contrast of 17 % still allows for relatively tight bend radii of 20 μm. Its proprietary composition is primarily aimed at reducing the need for annealing by reducing the effect of N–H bonds—the main source of absorption loss in the telecom band. This platform has resulted in extraordinarily low linear propagation losses of 5–7 dB/m, allowing for the use of extremely long waveguide spirals [14, 42, 53]. Figure 2.4 shows a schematic of a 45-cm-long spiral waveguide contained within a square area of 2.5 mm × 2.5 mm, pigtailed to single-mode fiber via low loss on-chip beam

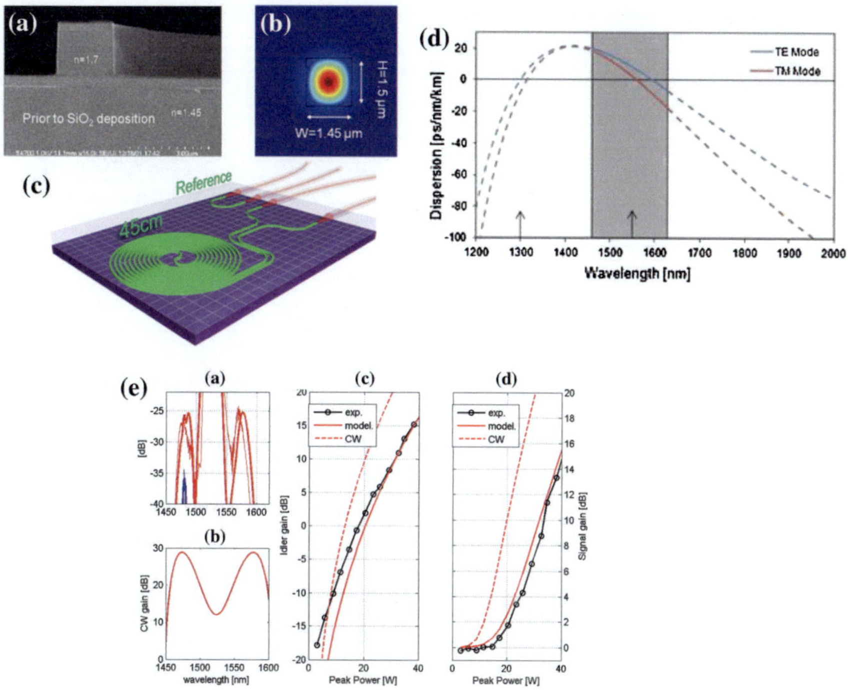

Fig. 2.4 Hydex waveguides. **a** SEM image of the cross-section of the Hydex waveguide [14] prior to the final deposition of the SiO_2 upper cladding; **b** theoretical mode profile and **c** *top down* schematic view of the 45 cm long spiral waveguide; **d** dispersion curves for the quasi-TE and quasi-TM modes of the Hydex waveguides [54]. The *shaded region* indicates the experimentally measured wavelength range (*solid lines*); the dispersion is extrapolated outside this region (*dashed lines*). The two zero dispersion points are illustrated by the *vertical arrows* **e** parametric gain curves from [53] when pumping with a pulsed source at peak powers up to 40 W with a wavelength at 1540 nm

expanders and a SEM picture of its cross section (before cladding deposition). The films were fabricated with CMOS compatible processes that yielded exceptionally low sidewall roughness in the core layer. Self-phase modulation experiments [42] yielded a Kerr nonlinearity of $n_2 = 1.15 \times 10^{-19}$ m^2/W, or $\sim$4.6× silica glass, and roughly half as large as that for SiN, with a nonlinearity parameter $\gamma \cong 233$ W^{-1} km^{-1} ($\sim$200× standard single-mode telecommunications fiber). This enhancement in n_2 is, like SiN, in agreement with Miller's rule [52], meaning that the proprietary chemistry of Hydex is not relevant to its nonlinear optical performance. The waveguides were engineered to yield anomalous dispersion [54] (Fig. 2.4), critical for efficient, wide bandwidth FWM over most of the C-band with zero-GVD points being 1600 nm for TE polarization and 1560 nm for TM. This resulted in a large FWM wavelength tuning range with efficient parametric gain of +15 dB and a signal to idler conversion efficiency of +16.5 dB (Fig. 2.4) [53].

Table 2.1 Nonlinear parameters for CMOS compatible optical platforms

	a-Si [65]	c-Si [2, 27]	SiN [34–36]	Hydex [42, 54]
n_2 (×fused silica[a])	700	175	10	5
γ [W^{-1} m^{-1}]	1200	300	1.4	0.25
β_{TPA} [cm/GW]	0.25	0.9	Negligible[b]	Negligible[c]
FOM	5	0.3	≫1	≫1

[a] n_2 for fused silica = 2.6×10^{-20} m^2 /W [1]
[b] No nonlinear absorption has been observed in SiN nanowires
[c] No nonlinear absorption has been observed in Hydex waveguides up to intensities of 25 GW/cm^2 [42]

Table 2.1 summarizes the linear and nonlinear properties of these CMOS compatible platforms. The empirical Millers rule [52] predicts that n_2 increases as the 4th power of n_0. The price of increasing n_0 in order to increase the Kerr nonlinearity is that this also tends to increase propagation losses. The relatively low refractive indices of Hydex (1.75) and Si_3N_4 (2.0) result in extremely low linear and nonlinear losses—their greatest strength—but also results in fairly modest values of n_2. The low linear and nonlinear losses have largely accounted for the success of these platforms for nonlinear optics in the telecommunications band, resulting in a combination of many factors that include negligible TPA, a moderately high nonlinearity, the ability to engineer dispersion, their high quality growth and fabrication processes, and their excellent material reliability. However, the *ultimate* platform would be one with a nonlinearity ideally larger than silicon *and* with a much larger FOM.

While amorphous silicon has been studied as a nonlinear material for some time [55], and developed as a platform for linear photonics in the telecom band for a number of years [56, 57], only recently has it been proposed [58] as an alternative to SOI for nonlinear optics in the telecom band, with the hope that a-Si could possibly offer lower TPA than silicon. Table 2.2 surveys the measured nonlinear properties for a-Si, which show a near universal improvement in both the nonlinearity and FOM over c-Si. Although initial measurements yielded a FOM no better than c-Si ($\sim$0.5) [58, 59], more recent results have yielded FOMs from 1 [60] to as high as 2 [61, 62]. This has enabled very high parametric gain of more than +26 dB over the C-band [63]. However, to date a key issue for this material has been a lack of stability resulting in a significant degradation in performance over relatively short timescales [64].

Table 2.2 Nonlinear parameters from the literature, for amorphous silicon

References	[65]	[63, 64]	[60]	[59]	[61]	[62]
n_2 (10^{-17} m^2/W)	2.1	1.3	4.2	0.05	0.3	7.43
γ (W^{-1} m^{-1})	1200	770	2000	35	N/A	N/A
β_{TPA} (cm/GW)	0.25	0.392	4.1	0.08	0.2	4 from [58, 60]
FOM	5 ± 0.3	2.2 ± 0.4	0.66 ± 0.3	0.4	0.97	1.1

Very recently, a-Si nanowires were demonstrated [65] that displayed a combination of high FOM, high nonlinearity, and good material stability at telecom wavelengths. Figure 2.5 shows a cross section SEM picture of a hydrogenated amorphous silicon nanowire (a-Si:H) waveguide fabricated in a 200 mm CMOS pilot line at CEA-LETI, France. The a-Si:H film was deposited by plasma enhanced chemical vapor deposition (PECVD) at 350C on 1.7 μm oxide deposited on a bulk wafer. After deposition of a silica hard mask, two steps of 193 nm DUV lithography and HBr silicon etching were used to define grating couplers that were well aligned with serpentine waveguides with varying lengths (1.22–11 cm). The fabricated waveguides are ~220 nm in thickness and ~500 nm in width. A 500 nm oxide was deposited to provide an upper cladding. The group velocity dispersion for the TE mode confined within a 500 nm × 220 nm nanowire was calculated with FEMSIM, yielding an anomalous second-order dispersion parameter $\beta_2 = -4.2 \times 10^{-25}$ s^2/m at $\lambda = 1550$ nm.

Figure 2.5 shows a schematic of the experimental setup used for the measurement of both the linear and nonlinear propagation characteristics of the a-Si:H nanowires. A modelocked fiber laser with near transform limited ~1.8 ps long pulses at a repetition rate of 20 MHz at 1550 nm was coupled into the TE mode of the nanowires via in-plane gratings. The fiber to waveguide coupling loss per coupler was ~10.6 and 12.4 dB per entry and exit, respectively—higher than expected due to the grating couplers not being optimized. The propagation loss of the TE mode was measured to be about 4.5 dB/cm, via a cut-back method on serpentine waveguides with lengths varying from 1.22 to 11 cm. A comparison with linear measurements performed on 1.3 mm long straight nanowires yielded a loss contribution due to the bends (10 μm radius) of about 4 dB, i.e. on the order of 0.04 dB/bend.

To determine the nonlinear parameters of the waveguides a series of self- phase modulation (SPM) measurements in a 1.22 cm long nanowire were performed with a coupled peak power up to ~3 W. The output spectrum was then measured as a function of input power.

Figure 2.5 shows the measured output spectra of the pulses for three coupled peak powers as well as the experimental contour plots of the output spectra as a function of coupled peak power ranging between 0.03 and 3 W. Strong spectral broadening is observed, which is the signature of self-phase modulation of the pulse propagating along the nonlinear nanowire. At higher coupled powers (>1.8 W), the spectral broadening of the pulse was limited by the spectral transfer function of the grating couplers, which had a 3 dB bandwidth of around ~35 nm centered near 1550 nm. Split-step-Fourier method simulations were used to solve the nonlinear Schrödinger equation governing the propagation of the picosecond optical pulse in the nonlinear waveguide, in the presence of second-order dispersion and TPA. The impact of dispersion was negligible, as expected for $\beta_2 = -4.2 \times 10^{-25}$ s^2/m, associated with a dispersion length exceeding 7 m for the 1.8 ps pulses—well over the physical length of the waveguide. Figure 2.5 shows the output spectra from the simulations, showing good agreement with measurement, when taking the TPA contribution stated above and a nonlinear waveguide parameter $\gamma = 1200$/W/m,

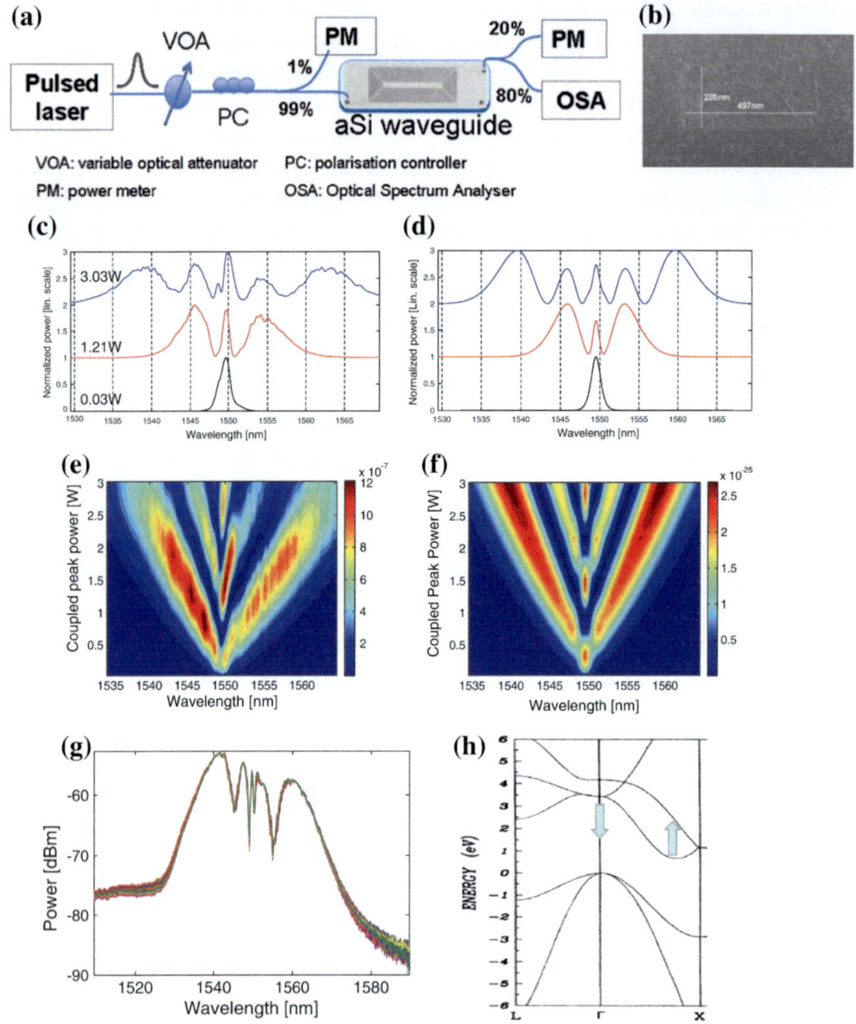

Fig. 2.5 Amorphous silicon nanowires from [65]. *Top* **a** Experimental setup for measuring self phase modulation and nonlinear transmission in order to extract the Kerr nonlinearity (n_2) and the two-photon absorption coefficient. **b** SEM image of nanowire cross-section. Output spectra for 0.03, 1.21 and 3.03 W coupled peak power **c** Experiment **d** Simulation. The *curves* are normalized and shifted upwards with increasing powers for clarity. **e** Experimental and **f** theory 2D plots showing the spectral broadening of the output pulse spectra versus coupled peak power. Note the linear intensity scale at the right is relative. **g** self phase modulation broadened spectrum taken over a period of an hour, showing negligible change. **h** Band diagram of silicon showing the principle of simultaneous increasing the nonlinearity and nonlinear FOM

associated with a Ker nonlinearity of $n_2 = 2.1 \times 10^{-17}$ m^2/W. The low TPA of a-Si:H was also reflected in the absence of any blue shift in the output spectra. Figure 2.5 also shows the output spectra taken over a period of several hours, showing no degradation in the nonlinear characteristics. Experimental measurements of self-phase modulation and nonlinear transmission yielded a record high nonlinear FOM of 5—over 10× that of SOI—together with an n_2 of 3–4 times c-Si.

The ability for a:Si to display both high nonlinearity and FOM has surprised many researchers, since in most materials, higher nonlinearity is accompanied by higher TPA. The key to the "anomalous" behavior of amorphous silicon may reside in the difference between *direct* and *indirect* transitions. In direct bandgap semiconductors (GaAs/AlGaAs [66]), below the bandgap both n_2 and TPA *always* increase with decreasing bandgap energy. For these simple direct bandgap semiconductors this is largely a result of Kramers–Kronig relations (adapted for nonlinear response functions [67]) that predict that increasing the bandgap to decrease nonlinear absorption also decreases the nonlinear response. In Si, however, the situation is more complex since TPA in the telecom band arises solely from *indirect* transitions, while n_2 (the real part of $\chi^{(3)}$) is dominated by *direct* transitions [67]. The direct bandgap in Si is >3.2 eV while the indirect bandgap is only 1.1 eV—well below the two-photon energy (1.6 eV) in the telecom band. For a:Si, it is likely that the effective *direct* "bandgap" (or mobility gap, being amorphous) is decreased compared to Si, thus increasing n_2, while the contribution of *indirect* transitions is greatly reduced, thus decreasing the TPA. This is entirely plausible since indirect transitions involve phonons which depend on long range order which, even for optical phonons, is clearly reduced in a:Si. Figure 2.5 shows the band structure of silicon and illustrates the principle of simultaneously increasing the nonlinearity and FOM.

A key goal for all-optical chips is to reduce device footprint and operating power, and the dramatic improvement in the FOM of a-Si raises the possibility of using slow-light structures [68, 69] to allow devices to operate at mW power levels with sub-millimeter lengths.

Finally, we note that the broad use of the description "CMOS compatible" in this context is intended to reflect a general compatibility in terms of growth temperatures (<400 C) and materials that are familiar and used in the CMOS process (silicon nitride, silicon oxynitride). It does not address the complexities and challenges of integrating optical and electronic devices with substantially different size scales, nor does it address the challenges of adapting CMOS production lines to optical device fabrication, both of which have been discussed at length [20–22]. A central issue in terms of integrating waveguides and nanowires with electronic components is the rather thick nature of both the core and cladding films. In this regard, SiN, with a higher refractive index contrast of about 0.5, offers a significant advantage over Hydex where the index contrast is only about 0.3. Both glasses, however, require noticeably thicker layers (both core and cladding) than SOI and this is probably a key area where SOI out-performs these platforms. Nonetheless, the concept of CMOS compatibility presented here, that these new platforms adhere to, is a powerful one that will go a long way towards enabling the broad application of CMOS techniques and manufacturing infrastructure to nonlinear photonic chips.

2.3 Low Power Nonlinear Optics in Ring Resonators

Optical microcavities, including ring resonators, can dramatically enhance nonlinear processes [70] and offer a powerful approach to greatly reducing operating powers for many applications such as frequency comb generation for spectroscopy and metrology [70, 71]. Microcavities are particularly conducive to enhancing FWM processes because FWM typically involves a pump, signal and idler beams with frequencies (ω_{Pump}, ω_{Idler}, ω_{Signal}) related by energy conservation: $\omega_{Idler} = 2\omega_{Pump} - \omega_{Signal}$. This process can occur with (classical) or without (spontaneous) a separate input signal at ω_{Signal}, the classical effect being much stronger and the basis for most all-optical signal processing. Very low CW operation was achieved in silica and single crystal micro-toroids and spheres with Q-factors from 10^7 to 10^{10} [70, 71]. For microcavities, achieving phase-matching is critical, which is equivalent to achieving equal resonance spacings (constant FSR) (with due allowance for the Kerr induced resonance shifts [36]) that results in pump, signal and idler all being in resonance—a triple resonance that greatly reduces the power requirement for the round-trip parametric gain to exceed the loss, thus producing oscillation. A comprehensive review of the use of microcavities for frequency comb generation is found in [48].

The first demonstration of low power nonlinear optics in these CMOS platforms was in Hydex [38] in 2008, where low power CW nonlinear optics was achieved via FWM in integrated ring resonators. Figure 2.6 shows a four port micro-ring resonator, radius ≅ 48 μm, along with a scanning electron microscope (SEM) cross section of the waveguide, having similar dimensions and composition to the spiral structure. Figure 2.6 also shows the transmission spectrum of the resonator showing a Q-factor of ≅65,000, a free-spectral range (FSR) of 575 GHz, and a full-width-half-maximum (FWHM) of 3 GHz. The bus waveguides, buried in SiO_2 beneath the ring, couple light in and out of the resonator.

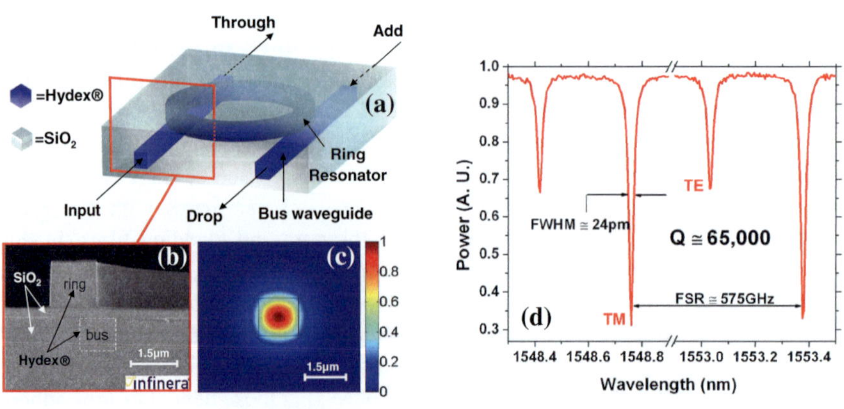

Fig. 2.6 Hydex micro-ring resonator [38] with moderate Q factor = 65,000, resulting in a bandwidth of 3 GHz, with an FSR of 575 GHz

Figure 2.7 shows the experimental setup as well as the basic physics behind FWM in ring resonators. The experiments were conducted with the ring resonator using only a few mW of CW pump power, with the signal tuned to an adjacent resonance to the pump. FWM was achieved near 1558 nm with 5 mW of CW pump power (in the waveguide) and a signal power of 570 μW tuned to an adjacent resonance. This yielded an idler that was almost exactly on resonance indicating

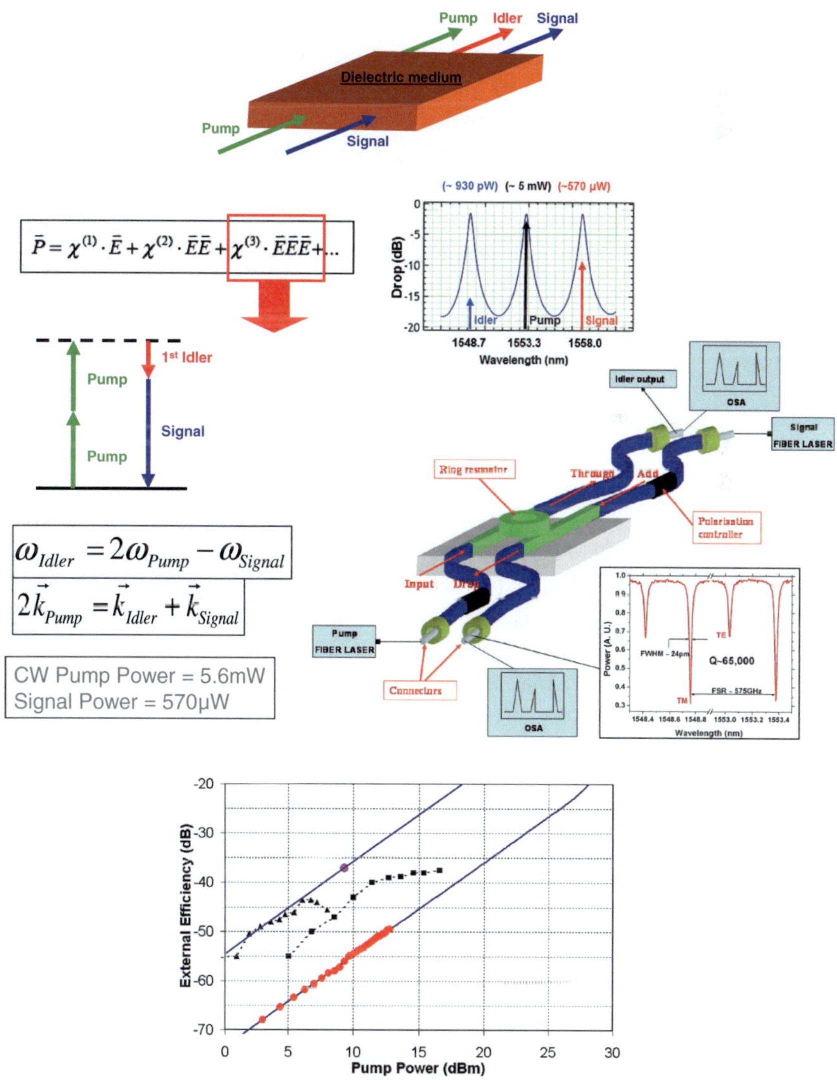

Fig. 2.7 Low power FWM in Hydex ring resonators with moderate Q factor = 65,000, resulting in a bandwidth of 3 GHz, with an FSR of 575 GHz [38]. *Bottom* External conversion efficiency along with estimated efficiency of silicon RR reported in [72]

that the dispersion in the system was indeed negligible and confirmed that the process was triply resonantly enhanced.

The experiments were performed with about 5 mW [38] of CW pump power tuned to a ring resonance at 1553.38 nm (TM polarization) and directed to the INPUT port, while a signal laser (TM polarization) with a power of 550 μW was tuned to an adjacent resonance at 1558.02 nm and directed to the ADD port. Figure 2.7 shows the resulting output power spectra (TM polarization) as recorded from both the THROUGH and the DROP ports after the pigtail; experiments carried out with TE polarized modes led to similar results. Two detectable idlers were generated by the FWM process—the first-idler power in the bus waveguide was ≅930 pW, whereas the second idler had a power of ≅100 pW. The ratio of the powers for the two idlers agrees remarkably well with the ratio of the pump to the signal power, $P_{idler(-1)}/P_{idler(-2)} \cong P_{pump}/P_{signal}$, as expected for FWM. The first idler was almost exactly on resonance at 1548.74 nm, indicating that the dispersion in the system is indeed negligible.

The theoretical conversion efficiency η takes into account the cavity enhancement factor due to the ring geometry:

$$\eta \equiv \frac{P_{idler}}{P_{signal}} = |2\pi R\gamma|^2 P_{pump}^2 \cdot \left(FE_p\right)^4 \cdot (FE_s)^2 \cdot (FE_i)^2 \tag{2.1}$$

$$FE_\mu = \frac{\sqrt{2(1-\sigma_\mu)}}{2(1-\sigma_\mu) + \alpha_\mu \pi R} \qquad \sigma_\mu = \left(1 - \frac{\pi}{2Finesse_\mu}\right) \exp\left(\frac{\alpha_\mu \pi R}{2}\right) \tag{2.2}$$

where R is the ring radius, FE_μ describes the field enhancement of the ring and α_μ is the mode linear loss coefficient for the mode μ. Finally, σ_μ is the self-coupling coefficient between ring and channels. The product $(FE_p)^4(FE_s)^2(FE_i)^2$ identifies an overall field enhancement factor.

Figure 2.7 shows the external conversion efficiency (including coupling losses), along with the external conversion efficiency of comparable experiments carried out in silicon ring resonators [72], estimating the efficiency from the coupling loss. Although the conversion efficiency of the Hydex device was predictably lower than the SOI ring resonators due to the much lower γ, the negligible nonlinear absorption potentially allows the scaling of pump powers to levels at which high bit rate all-optical signal processing in silicon nanowires is typically performed [73, 74].

2.4 Microresonator-Based Frequency Combs

The area where these platforms have arguably had the greatest impact is in integrated OPOs based on ring resonators. These devices have significant potential for many applications including spectroscopy and metrology as well as the ability to provide an on-chip link between the RF and optical domains [75]. As discussed above, micro-cavities enhance nonlinear optical processes, such as FWM involving a continuous-wave (CW) pump, signal and idler beams with frequencies (ω_{Pump}, ω_{Idler}, ω_{Signal}) related by energy conservation: $\omega_{Idler} = 2\omega_{Pump} - \omega_{Signal}$. Achieving phase-matching of the propagation constants for three interacting waves is essential for efficient FWM, which for the microcavities is equivalent to having near-equal resonance spacings, or a constant FSR, with due allowance for the Kerr-induced resonance shifts [36]. This results in the pump, signal and idler waves all being in resonance—a triple resonance that greatly reduces the power requirement for the round-trip parametric gain to exceed the loss, thus producing oscillation. Phase matching, or low and anomalous waveguide dispersion, can be achieved with a suitable design of the waveguide cross-section.

Reports in early 2010 of OPOs in both SiN [36] and Hydex [37] opened the door to achieving these devices in practical CMOS-compatible integration platforms with much lower Q factors than previous micro-cavity oscillators, and hence with much less sensitivity to environmental perturbations and without the need for delicate tapered fibre coupling. This will allow these devices to benefit not just the scientific community but telecommunications, computing, and precision spectroscopy and timekeeping.

Figure 2.8 shows a four port micro-ring resonator based in Hydex glass with a Q factor of 1.2 million [37] along with a scanning electron microscope (SEM) of the waveguide cross section and the corresponding optical transmission spectrum. Figure 2.9 shows the experimental setup [37] for demonstrating an on-chip OPO

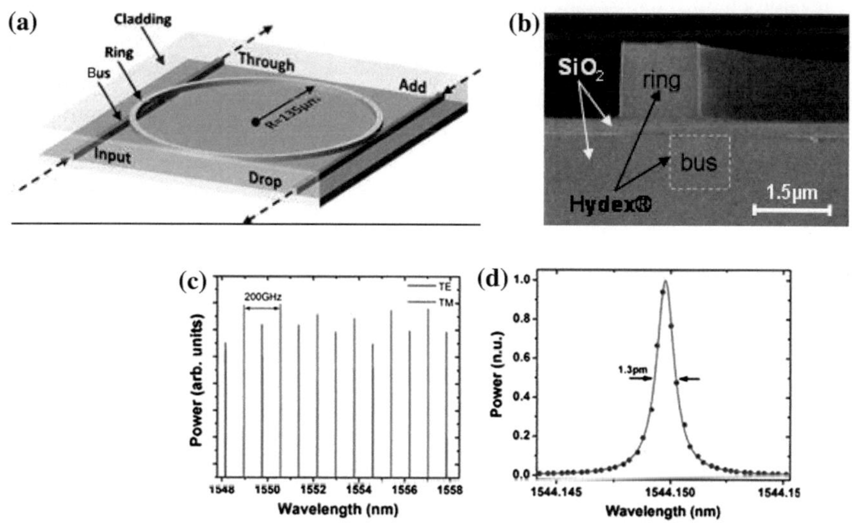

Fig. 2.8 Integrated ring resonators based in Hydex [37]. **a** Hydex four-port microring resonator (fibre pigtails not shown) with Q = 1.2 × 10^6. **b** SEM image of the Hydex ring resonator cross section before depositing the upper cladding of SiO_2; **c** *bottom left* linear transmission through ring resonator from the INPUT to THROUGH port for TM and TE polarizations and **d** *bottom right* high resolution transmission plot showing a resonance linewidth of 1.2 pm

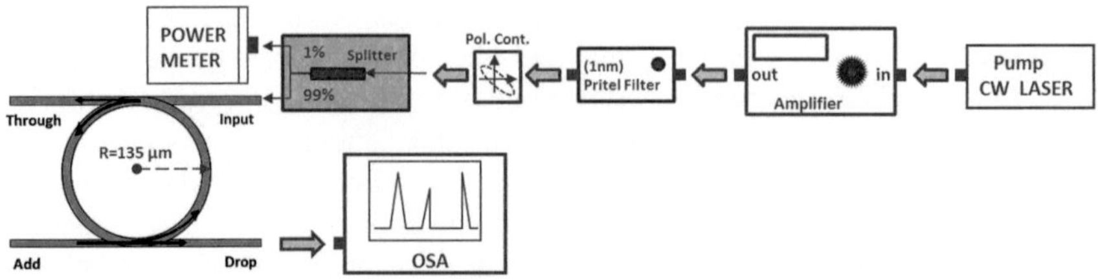

Fig. 2.9 Experimental setup for OPO via external pumping with soft thermal locking [37]

using an externally pumped laser, highlighting the use of soft thermal locking to achieve resonant coupling between the pump laser and the cavity resonances.

The optical frequency comb generated by the Hydex device (Fig. 2.10) exhibited a very wide spacing of almost 60 nm when pumped at 1544.15 nm, in the anomalous GVD regime (Fig. 2.3). The output power versus pump power shows a very high single line differential slope efficiency above threshold of 7.4 %, with a maximum power of 9 mW achieved in all oscillating modes out of both ports, representing a remarkable absolute total conversion efficiency of 9 %. When pumping at a slightly different wavelength closer to the zero-GVD wavelength (but still anomalous), the device oscillated with significantly a different spacing of 28.15 nm.

These observations are consistent with parametric gain based on a combination of FWM and MI described above, where the spacing depends on the waveguide dispersion characteristics and agrees well with calculations. Oscillation begins via modulational instability (MI) gain—essentially pure spontaneous (degenerate) FWM (with only a pump present). Once oscillation is achieved, "cascaded" FWM among different cavity modes takes over, resulting in the generation of a frequency comb of precisely spaced modes in the frequency domain. However, the enhancement of the fields in the cavity (all fields in the case of phase matching) that is responsible for lowering the oscillation threshold, also enhances the nonlinear losses, and it is primarily for this reason that oscillation in silicon ring resonators in the telecom band where the FOM <1, has not been achieved.

Figure 2.11 shows the calculated MI gain curve when pumping at 1544 nm with 54 mW of power, using the experimentally measured dispersion curves. This shows a gain peak at 1590 nm—close to the observed initial oscillation peaks. This confirms that oscillation begins via MI gain near the peak. Once oscillation is achieved, "cascaded" FWM can occur among many different cavity modes resulting in the generation of a frequency comb of precisely spaced modes in frequency. This illustrates the degree of freedom one can achieve in varying the frequency comb spacing—chiefly through dispersion engineering, and so one is not restricted by the FSR of the resonator itself. The trade-off is that MI generated combs can themselves further seed sub-combs that are poorly related in terms of coherence to the original comb, limiting the degree to which modelocking, or ultrashort pulse generation, can be achieved [47].

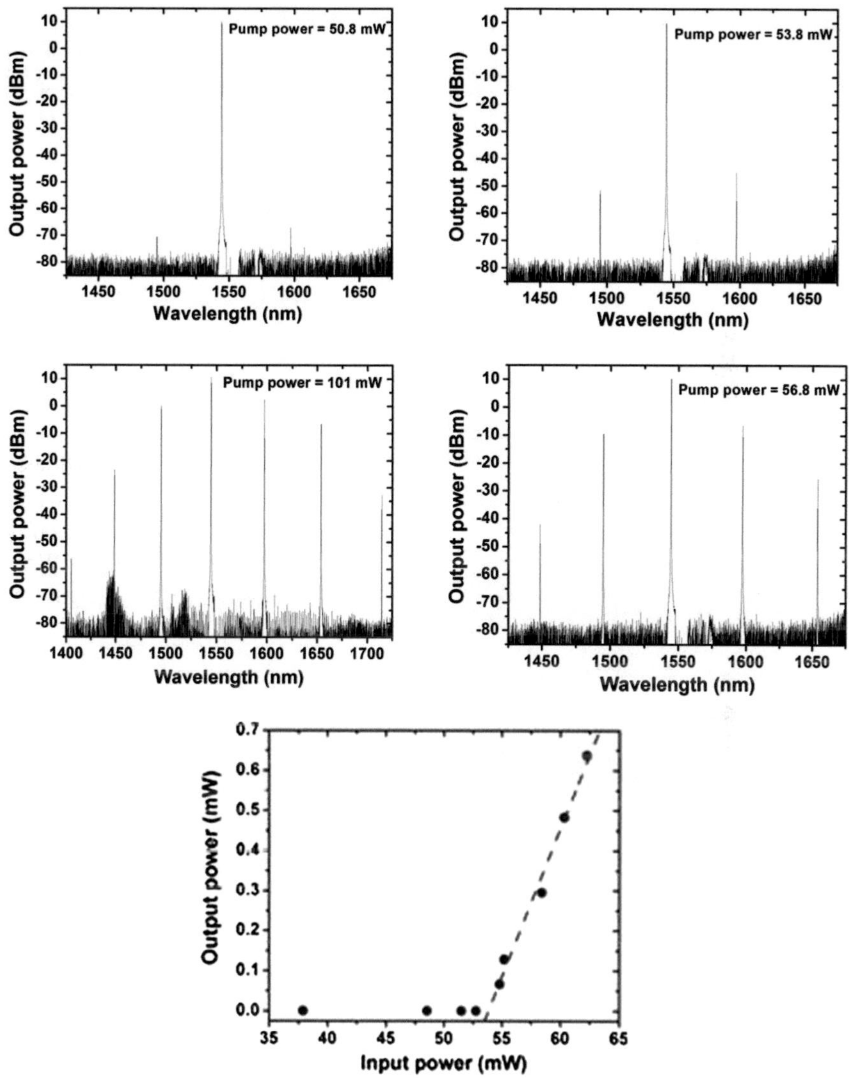

Fig. 2.10 (*Top*) Output spectra of Hydex hyper-parametric oscillator [37] with a pump power varying from 50.8 mW (just above threshold) tuned to a resonance at 1544.15 nm (TM polarization) to well above threshold at 101 mW. (*Bottom*) Output power vs pump power (at 1544.15 nm) in the drop port waveguide of the Hydex oscillator for a single oscillating mode at 1596.98 nm, showing a threshold of 54 mW with a differential slope efficiency of 7.4 % (linear fit —*blue dashed line*) above threshold

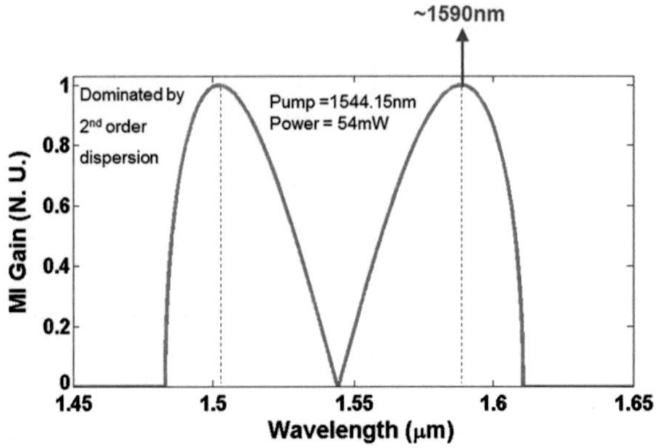

Fig. 2.11 Theoretical calculation of modulational instability (MI) gain [37] based on the measured dispersion, at a pump power of 54 mW and pump wavelength near the zero dispersion point at 1544 nm, showing a gain peak about 46 nm away from the pump wavelength

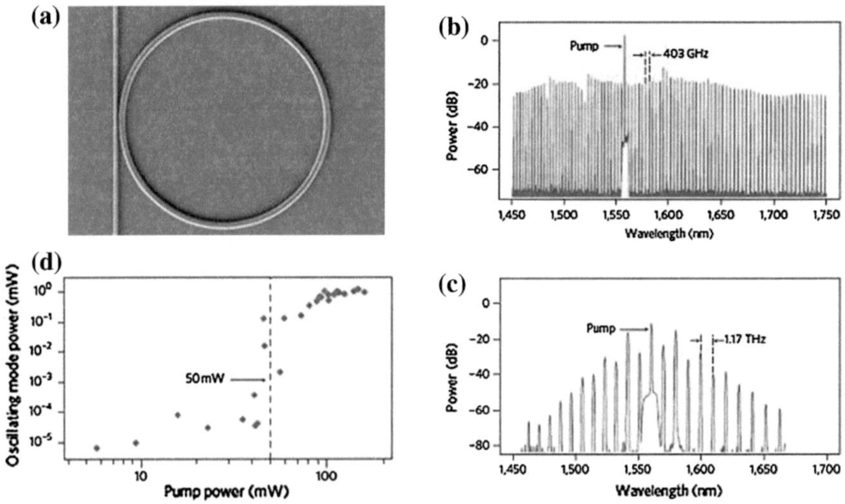

Fig. 2.12 Integrated OPO multiple wavelength sources in SiN [36] ring resonators [36]. **a** SEM image of SiN microring resonator (58 μm radius and Q = 500,000, FSR = 403 GHz) from [36] coupled to a bus waveguide, with cross section height of 711 nm, base width of 1700 nm and a sidewall angle of 20°, giving anomalous GVD in the C-band and a zero-GVD point at 1610 nm; **b** Output spectra of a 58-μm-radius SiN ring resonator OPO with a single pump wavelength tuned to a resonance at 1557.8 nm, numerous narrow linewidths at precisely defined wavelengths. The 87 generated wavelengths were equally spaced in frequency, with an FSR of 3.2 nm **c** spectra of a 20-μm radius SiN ring resonator generating a comb with a different frequency spacing, pumped at 1561 nm and producing 21 wavelengths over a 200-nm span, with a spacing of 9.5 nm **d** output power in the first generated mode of SiN OPO versus pump power, showing a threshold of 50 mW with a slope efficiency of 2 %

Optical parametric oscillation was simultaneously reported in SiN [36] by resonantly pumping the rings with CW light near 1550 nm using a soft 'thermal lock' process in which the cavity heating is counteracted by diffusive cooling. Figure 2.12 shows a schematics of 2-port SiN micro-ring resonators: [36] a 58 μm radius resonator (Q factor = 500,000, FSR = 403 GHz) with dimensions designed to yield anomalous GVD in the C-band with a zero GVD point at 1610 nm. Oscillation of multiple lines over a very broad (>200 nm) wavelength range was achieved (Fig. 2.12) at a pump threshold of 50 mW. Eighty-seven new frequencies were generated between 1450 and 1750 nm, corresponding to wavelengths covering the S, C, L and U communications bands. Several designs were employed with different ring radii, or FSR. A smaller ring with a Q factor of 100,000 generated oscillation in 20 resonator modes when pumped with modest input powers (150 mW) with THz mode spacing. These results represent a significant step toward a stabilized, robust integrated frequency comb source that can be scaled to other wavelengths.

2.5 Advanced Frequency Comb Generation

Since these initial demonstrations of multiple wavelength oscillation, significant progress has been made in advanced comb generation, including both very wide bandwidth octave spanning combs [45] and very low (sub 100 GHz) FSR spacing combs [46].

The development of microresonator-based frequency combs with a free spectral range (FSR) significantly less than 100 GHz is critical to provide a direct link between the optical and electrical domains in order to produce highly stable microwave signals detectable with photodiodes. The challenge is that simple ring geometries with sub-100 GHz FSR spacings do not fit on typical single e-beam fields and so novel ring geometries such as spirals need to be employed. Figure 2.13 shows spiral ring resonators with unique geometries for different FSRs below 100 GHz [46], all having a constant semicircular coupling region to enable critical coupling between the bus and resonator, independent of path length. Bends in the resonators had radii >100 μm to ensure that bend induced dispersion was negligible, a critical requirement for proper operation of the frequency comb. The experimental spectra for 80, 40, and 20 GHz combs are shown in Fig. 2.13, typically requiring about 2 W pump power to fill the entire comb spans.

Octave-spanning frequency combs are of great interest for spectroscopy, precision frequency metrology, and optical clocks and are highly desirable for comb self-stabilization using f to $2f$ interferometry for precision measurement of absolute optical frequencies [76]. Figure 2.14 shows an optical frequency comb in a SiN ring resonator spanning more than an octave [45] from 1170 to 2350 nm, corresponding to 128 THz, achieved by suitable dispersion engineering and employing higher pump powers of up to 400 mW inside the waveguide, detuned slightly from a

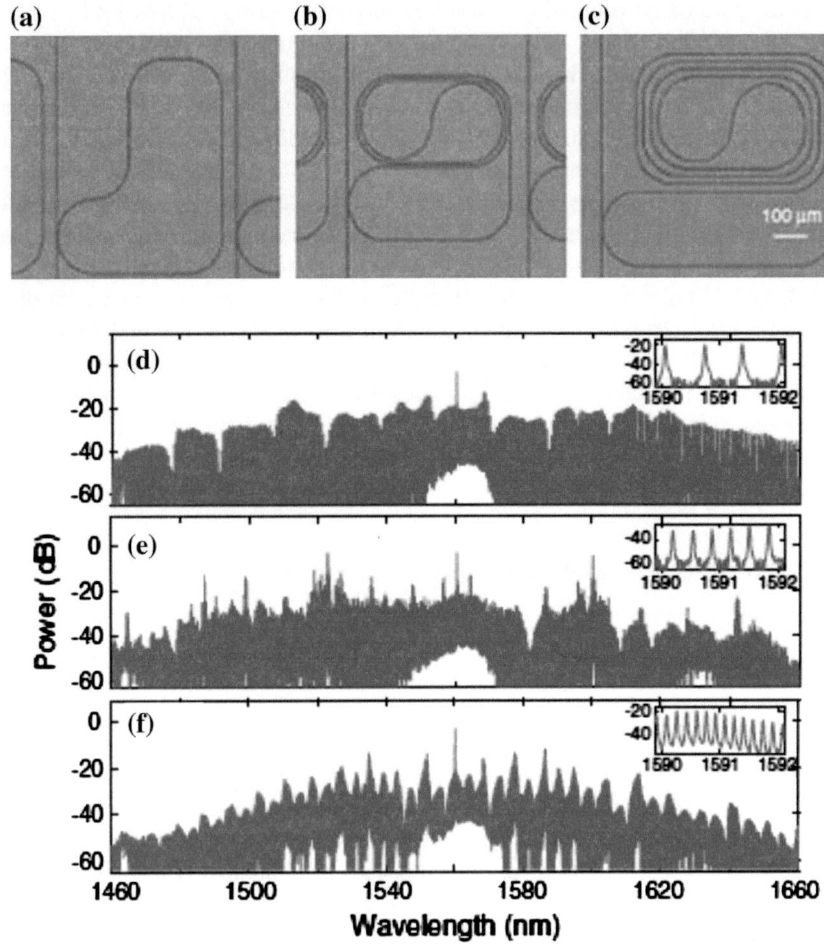

Fig. 2.13 Advanced frequency combs in SiN ring resonators [46]. Sub-100GHz spacing SiN ring resonators. Micrographs of the **a** 80 GHz, **b** 40 GHz, and **c** 20 GHz FSR resonators and the corresponding linear transmission spectra (*insets*). A 2 nm section of each comb is *inset* in each figure to illustrate the spacing of the *comb lines*. The nanowire cross section was 725 nm by 1650 nm, with a loaded Q of 100,000. Output spectra are **d** 300 nm wide for the **d** 80 GHz and **e** 40 GHz FSR rings and **f** 200 nm for the 20 GHz FSR ring

cavity resonance. Figure 2.14 shows the simulated dispersion for nanowires with varying widths (1200, 1650, and 2000 nm) indicating that large anomalous-GVD bandwidths spanning nearly an octave are possible with appropriate design. These results represent a significant step toward a stabilized, robust integrated frequency comb source that can be scaled to other wavelengths.

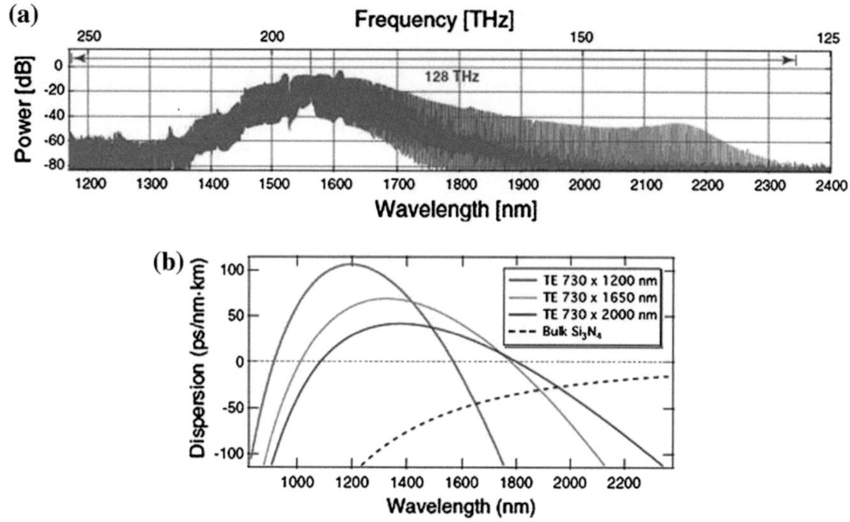

Fig. 2.14 Advanced frequency combs in SiN ring resonators [45]. **a** *Top* Optical spectrum of octave-spanning parametric frequency comb generated in a SiN ring resonator and **b** *Bottom* dispersion simulations for the fundamental TE mode of SiN waveguide with a height of 730 nm and widths of 1200, 1650, and 2000 nm. The *dashed curve* shows the dispersion for bulk silicon nitride

2.6 Supercontinuum Generation

By injecting ultrashort modelocked pulse trains into suitably designed waveguides very broadband spectra can be generated. This super-continuum (SCG) spectra also can result in octave-spanning frequency combs and has been of great interest for spectroscopy, precision frequency metrology, and optical clocks and is highly desirable for comb self-stabilization using *f* to 2*f* interferometry for precision measurement of absolute optical frequencies [76–80]. Wide bandwidth SCG has been demonstrated in microstructured fibers [81–83], ChG waveguides [84], periodically poled lithium niobate (PPLN) [85], and in Si [86–88]. Hydex and SiN offer advantages of much lower linear and nonlinear loss as well as transparency well into the visible.

Figure 2.15 compares the broadband frequency combs with SCG in Hydex waveguides [89] where a spectral width >350 nm was achieved (limited by experimental measurement capability), with that generated in 1100-nm wide SiN nanowires [90] that yielded a spectrum spanning 1.6 octaves, from 665 to 2025 nm, pumping at 1335 nm. The SiN results in particular represent the broadest recorded SCG to date in a CMOS compatible chip. Both of these results were enabled by a high effective nonlinearity, negligible TPA and most significantly very flexible dispersion engineering. SCG is significantly enhanced if the pulse is launched near a zero group-velocity dispersion (GVD) point or in the anomalous GVD regime. The former minimizes temporal pulse broadening, thereby preserving high peak

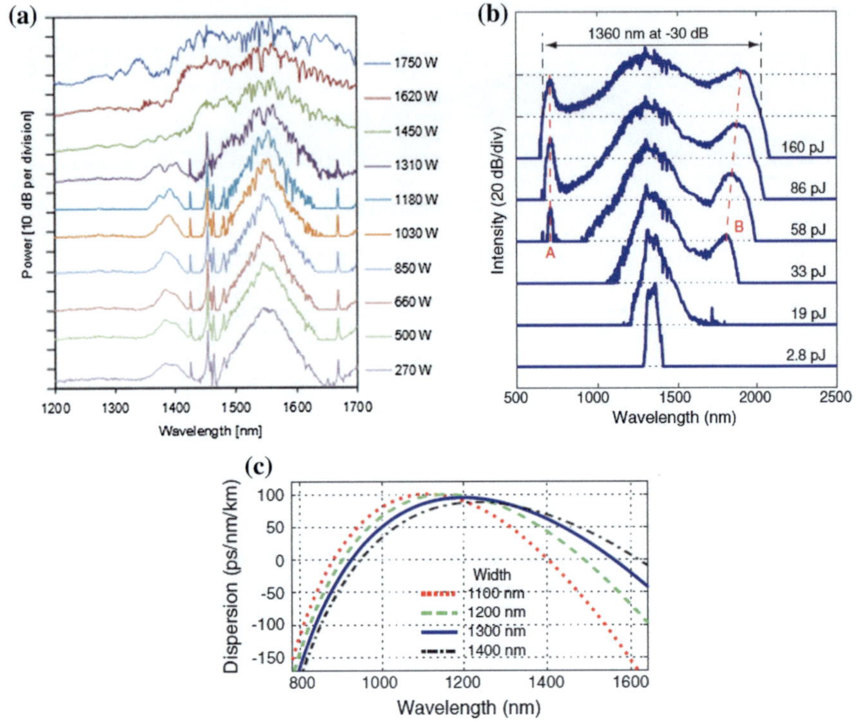

Fig. 2.15 Supercontinuum generation in **a** Hydex waveguides [89] and **b** silicon nitride nanowires [90] and **c** engineered dispersion in silicon nitride waveguides [90]

powers and thus maintaining a strong nonlinear interaction. The latter regime enables soliton propagation, whose dynamics can contribute to spectral broadening. The higher flexibility of dispersion engineering in SiN as compared to Hydex is partly enabled by its higher available core/cladding index contrast and is probably one of its most significant advantages.

2.7 Comb Coherence and Dynamic Properties

As with conventional modelocked lasers, parametric frequency combs can also potentially serve as sources of ultrashort laser pulses that, depending on the pump laser and material system, can produce ultrashort pulses from the visible to the mid-infrared at repetition rates in the GHz to THz regimes. The last few years have seen significant breakthroughs in understanding the dynamics and coherence behavior of frequency comb formation [44, 47, 48]. These reports have revealed complex and distinct paths to comb formation that can result in widely varying degrees of coherence, wavelength spacing, and RF stability of these sources. This field has recently been highlighted by the achievement of an ultrastable, ultrashort optical pulse source via modelocking [15] based on an integrated microcavity. Understanding and harnessing the coherence properties of these monolithic frequency comb sources is crucial for exploiting their full potential in the temporal domain, as well as for generally bringing this promising technology to practical fruition.

The first investigation of the coherence properties of these on-chip oscillators [47] focused on phase tuning of the comb via programmable optical pulse shaping (Fig. 2.16). This "line-by-line" pulse shaping of microresonator frequency combs was enabled by the relatively large mode spacings and represented a significant development in the field of optical arbitrary waveform generation [77–79]. In this approach, transform limited can be realized for any spectral phase signature of coherent combs by appropriately compensating the relative phase of the different comb lines (Fig. 2.16). The ability to achieve successful pulse compression also provides information on the passive stability of the frequency-dependent phase of the Kerr combs. These time domain experiments revealed different pathways to comb formation in terms of phase coherence properties, with the ability to effectively modelock the combs varying significantly depending on many factors such as pump conditions and waveguide dispersion.

Recent studies of the dynamics of comb formation [44] (Fig. 2.17) have shown that initial oscillation often begins at resonances near the parametric gain, or modulational instability, peak with spacings varying widely from the FSR to values as wide as 50–100 nm. Cascaded FWM then replicates a comb with this initial spacing. More complex dynamics arise when these comb lines themselves seed their own mini-comb bands based on the local dispersion and pumping conditions, often with a spacing at or near the cavity FSR in that wavelength region. While these "sub-combs" maintain coherence within themselves, they are not coherently related to the sub-combs generated by other lines of the initial, more widely spaced, comb, and so this results in broadband comb with complex and generally limited coherence. One approach to achieve high coherence is to design the device such that the initial comb intrinsically oscillates at the FSR, producing a wideband comb with high coherence properties. Figure 2.17 shows the output of a ring resonator under different pumping conditions [44], dramatically showing the transition to from an incoherent state where the frequency sub-combs are well separated and hence incoherent, to a coherent state resulting in a much improved ability to lock the modes to produce optical pulses, resulting in much lower RF noise in the output.

Further studies [49] of the temporal and optical and radio-frequency spectral properties of a parametric frequency combs generated in SiN microresonators demonstrate that the system undergoes a transition to a modelocked state and that ultrashort pulse generation occurs. From a 25-nm filtered section of the 300-nm comb spectrum, sub-200-fs pulses are observed and calculations indicate that the pulse generation process is consistent with soliton modelocking, which is consistent with very recent work involving comb generation in MgF_2 microresonators [91] and could be explained by the formation of temporal cavity solitons [91, 92], where contributions from dispersion and loss are compensated by nonlinearity and a coherent driving beam.

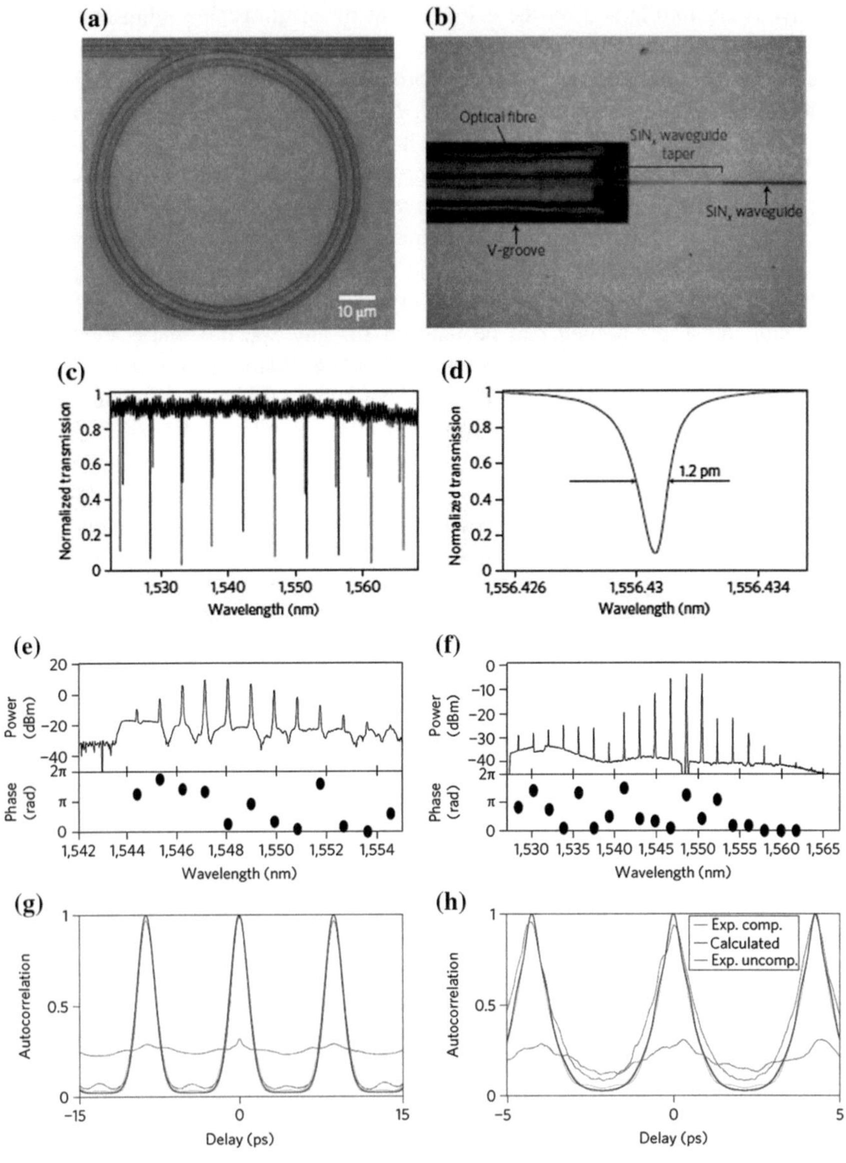
(a)
10 μm
(b)
Optical fibre
SiN$_x$ waveguide taper
SiN$_x$ waveguide
V-groove
(c)
Normalized transmission
Wavelength (nm)
(d)
1.2 pm
Normalized transmission
Wavelength (nm)
(e)
Power (dBm)
Phase (rad)
Wavelength (nm)
(f)
Power (dBm)
Phase (rad)
Wavelength (nm)
(g)
Autocorrelation
Delay (ps)
(h)
Exp. comp.
Calculated
Exp. uncomp.
Autocorrelation
Delay (ps)

◀**Fig. 2.16** **a** SEM image of SiN microring resonator from [47] (radius 40 μm) along with **b** fibre-pigtailed coupling waveguide to resonator; **c** Optical transmission spectrum for TM polarized light of ring resonator [46] including **d** high resolution spectrum for a mode at 1556.43 nm showing a linewidth of 1.2 pm, corresponding to a loaded Q factor of 1.3×10^6 with a FSR of 4.8 nm. *Bottom* Coherence and frequency comb formation dynamics. **e–h**, Frequency comb coherence properties from [47]. The output spectrum generated by a high-Q silicon nitride microring shows the ability of type-I Kerr combs to perform pulse compression. Spectrum of the combs after processing with a pulse shaper, together with the phase applied to the liquid-crystal modulator pixels required to achieve optimum SHG signals **e**, **f**. Autocorrelation traces **g**, **h** corresponding to **e**, **f** [47]. The *red lines* depict the compressed pulses after applying a phase correction, the *blue lines* are uncompressed pulses and the *black lines* are calculated from the spectra shown in **e** and **f** by assuming a flat spectral phase. The contrast ratios of the autocorrelations measured after phase compensation are 14:1 (**g**) and 12:1 (**h**). *Light grey* traces show the range of the simulated autocorrelation traces

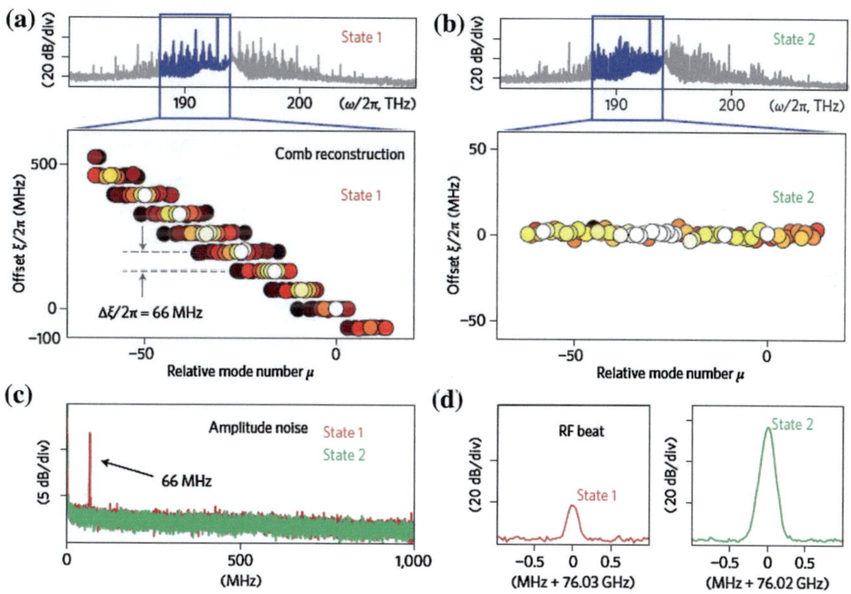

Fig. 2.17 Studies of coherence evolution in SiN microring resonators [44] showing the transition to a low phase noise Kerr comb. **a** Optical spectra of two microresonator comb states 1 (**a**) and 2 (**b**) (pump power 6W). State 2 evolves from state 1 when reducing the detuning of the pump laser. A transition is observed from multiple sub-combs to a single (sub) comb over the bandwidth of the Kerr comb reconstruction. In state 1, all sub-combs have the same mode spacing, but have different offsets which differ by a constant of 66 MHz. **c**, **d** In the transmission from state 1 to state 2, the amplitude noise peak resulting from the beating between overlapping offset sub-combs disappears

2.8 Ultrashort Pulsed Modelocked Lasers

The first demonstration of stable modelocking of a frequency comb based on a microresonator was recently achieved (Fig. 2.18) [15, 50, 93] by embedding the resonator in an active fiber loop. The unique aspect of this system is that the microresonator is used as both a linear filter and as the nonlinear element. This scheme, termed Filter-Driven Four-Wave-Mixing (FD-FWM), is an advancement inspired by dissipative FWM (DFWM) [94–96] where the nonlinear interaction occurs in-fibre and is then "filtered" separately by a linear FP filter. This new approach is intrinsically more efficient and so allows for substantially reduced main cavity lengths which in turn has enabled the achievement of highly stable operation —something that has so far eluded DFWM based lasers. A fundamental challenge with DFWM lasers is that the main cavity mode spacing is typically much finer than the microcavity, allowing many cavity modes to oscillate within each micro-resonator mode, giving rise to so-called "supermode instability". The FD-FWM approach allows a substantial reduction in the main cavity length to the point where only a very small number (<3) cavity modes—and even a single mode—exist within each microresonator resonance, allowing for the possibility of only one main cavity mode oscillating within each ring resonance. Hence, FD-FWM has achieved highly stable operation at high repetition rates over a large range of operating conditions, without a stabilization system, robust to external (i.e., thermal) perturbations. In addition, it can potentially produce much narrower linewidths than ultrashort cavity modelocked lasers because the long main cavity results in a much smaller Schawlow-Towns phase noise limit [97]. Figure 2.19 compares the optical spectra, time resolved optical waveforms (measured by autocorrelation) and radio-frequency (RF) spectra of short and long cavity length lasers, clearly showing that the RF spectra for the short cavity laser is highly stable whereas the long cavity length laser not only shows large, long timescale, instabilities in the RF output, but is also unstable on a short timescale where the optical autocorrelation traces show a very limited contrast ratio—a hallmark of unstable oscillation. Recently [50] stable modelocked laser operation with two modes oscillating within each ring cavity was demonstrated, which produced a highly pure RF tone over the modelocked train, enabling high resolution RF spectral measurement of the optical linewidth, which achieved the remarkably narrow width of about 10 kHz.

2.9 Ultrafast Phase Sensitive Pulse Measurement

Coherent optical communications [18, 19] have created a compelling need for ultrafast phase-sensitive measurement techniques operating at milliwatt peak power levels. Ultrafast optical signal measurements have been achieved using time-lens temporal imaging on a silicon chip [13, 98] and waveguide-based

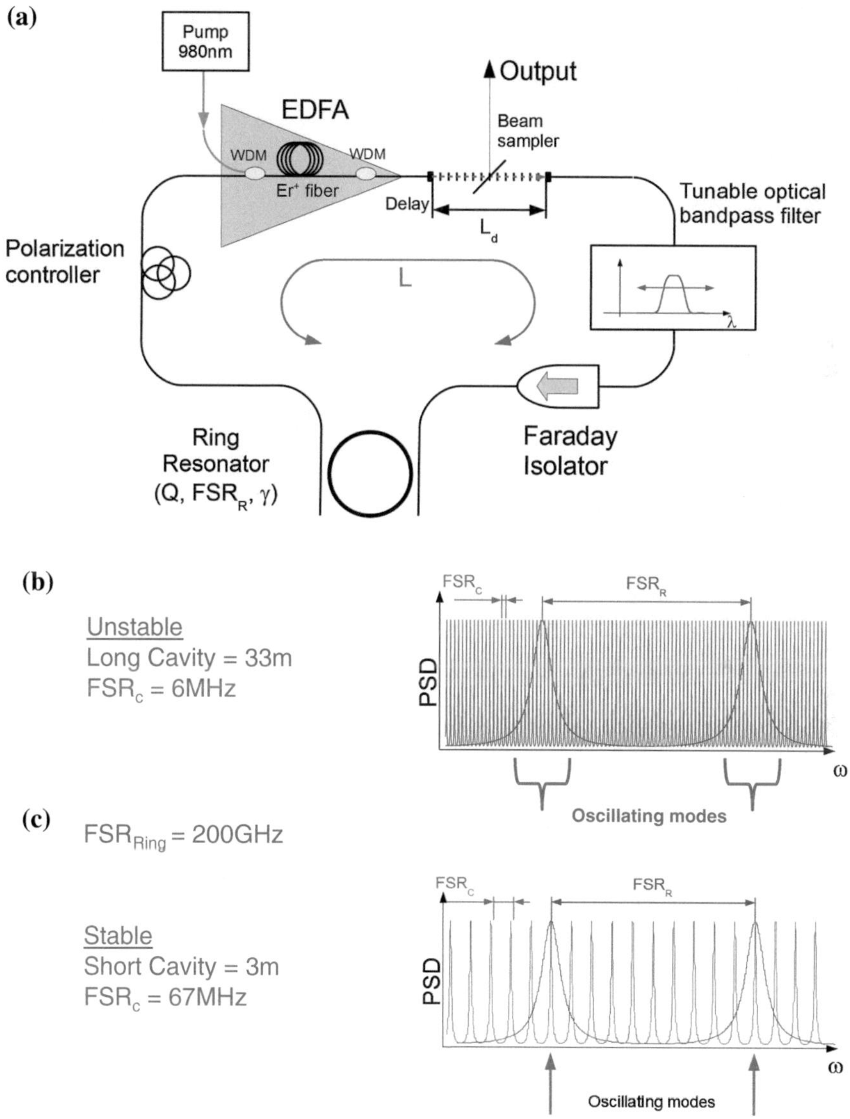

Fig. 2.18 Ultrafast modelocked laser based on filter driven four-wave mixing in a modelocked fiber loop laser [15]. **a** Experimental configuration for fiber loop modelocked laser based on filter driven four wave mixing (FDFWM) based on a microring resonator, where the resonator performs the dual function of linear filtering and nonlinear interaction. **b**, **c** oscillating modes of the laser superimposed on the filter profle of the microresonator structure. **b** long (33 m) cavity length showing 6 MHz spacing and **c** short (3 m) cavity showing much larger 67 MHz spacing

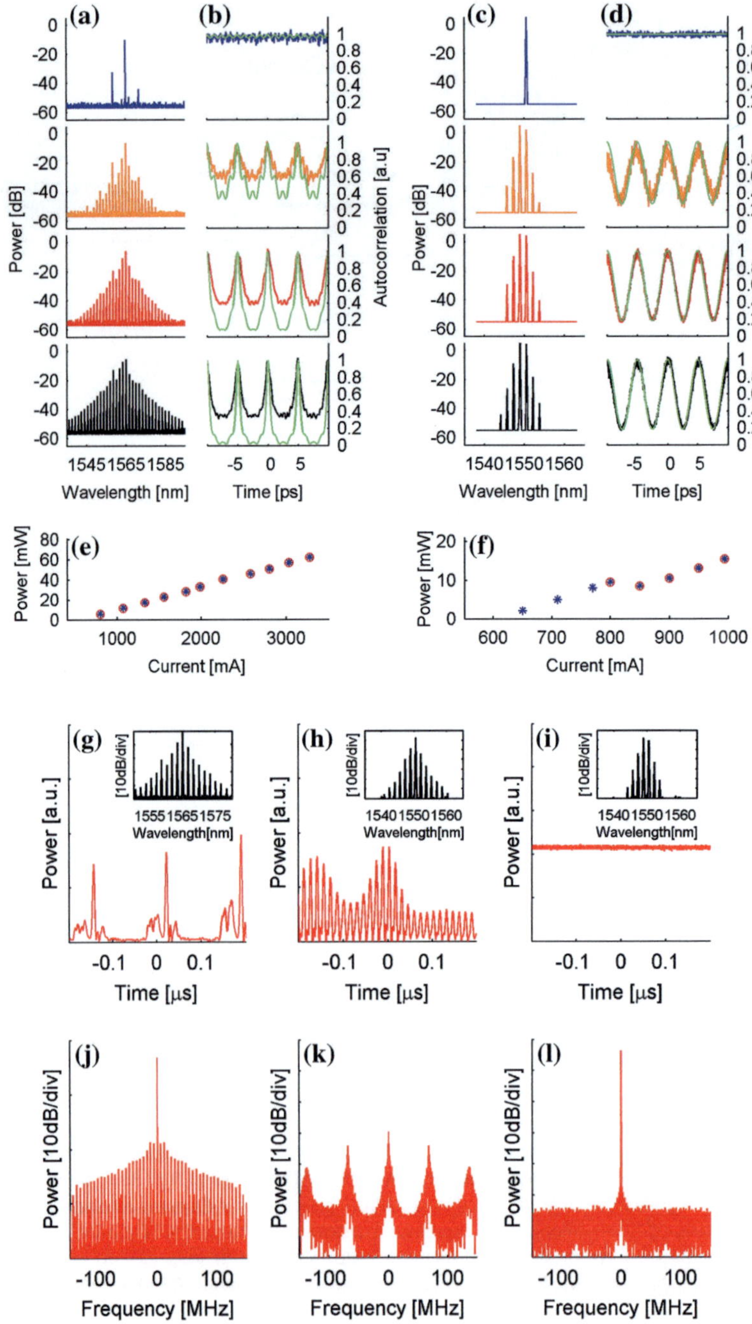
(a)
(b)
(c)
(d)
Power [dB]
Autocorrelation [a.u]
Wavelength [nm]
Time [ps]
(e)
(f)
Power [mW]
Current [mA]
(g)
(h)
(i)
Power [a.u.]
[10dB/div]
Wavelength[nm]
Time [μs]
(j)
(k)
(l)
Power [10dB/div]
Frequency [MHz]

◀**Fig. 2.19** Ultrafast modelocked laser based on filter driven four-wave mixing in a modelocked fiber loop laser [15]. Experimental optical output of unstable laser (**a**, **b**) and stable laser (**c**, **d**). **a** optical spectrum and **b** autocorrelation trace of unstable (long length) laser, and **c** optical spectrum and **d** autocorrelation trace of stable (short length) laser. Power output versus pump power for unstable (**e**) and stable (**f**) lasers. The autocorrelation plots show theoretical calculations (*green*) starting from the experimental optical spectra for a fully coherent and transform-limited system calculated by considering each line of the experimental optical spectra as being perfectly monochromatic and in-phase with the others, yielding an output pulse with a FWHM of 730 fs for the highest excitation power condition. The measured autocorrelation for the long cavity laser (**b**) shows a considerably higher background than the expected autocorrelation (50:1). Conversely, the calculated autocorrelation trace for the short-cavity laser (**d**) perfectly matches the measured trace, indicating stable modelocking, and corresponding to a transform-limited pulse width (FWHM) of 2.3 ps with a peak to background ratio of 5:1. Stable oscillation was obtained by adjusting the phase of the cavity modes for the short main cavity length laser relative to the ring resonator modes via a free space delay line. Temporal waveforms measured via radio frequency (RF) detectors (**g**–**i**) and RF spectral (**j**–**l**) output of the fiber loop laser for different main cavity lengths. (**g**, **j**) are for the long cavity (L = 33 m, FSR = 6.0 MHz) unstable laser, (**h**, **k**) are for the short cavity laser (L = 3 m, FSR = 68.5 MHz) laser but with the main cavity modes detuned from the microresonator mode, thus allowing more than one cavity mode to oscillate. (**i**, **l**) are for the short cavity laser where the main cavity is tuned (via free space delay) such that the cavity modes are aligned to the microcavity mode—thus restricting the laser to a single mode

Frequency-Resolved Optical Gating (FROG) [105], but these are either phase insensitive or require waveguide-based tunable delay lines—a difficult challenge. Recently [14], a device capable of both amplitude and phase characterization of ultrafast optical pulses was reported, operating via a novel variation of Spectral Phase Interferometry for Direct Electric-Field Reconstruction (SPIDER) [100–108] based on FWM in a Hydex waveguides. Here, pulse reconstruction was obtained with the aid of a synchronized incoherently related clock pulse.

SPIDER approaches are ultrafast, single-shot, and have a simple, direct and robust phase retrieval procedure. However they traditionally have employed either three-wave mixing (TWM) or linear electro-optic modulation, both of which require a second-order nonlinearity that is absent in CMOS compatible platforms. Also, typical SPIDER methods work best with optical pulses shorter than 100 fs (small time-bandwidth products (TBP)) and peak powers >10 kW, and hence are not ideally suited to telecommunications. The device reported in [14] measured pulses with peak powers below 100 mW, with a frequency bandwidth greater than a terahertz and on pulses as wide as 100 ps, yielding a record TBP >100 [105].

Figure 2.20 shows a schematic of the device. The pulse under test (PUT) is split into two replicas and nonlinearly mixed with a highly chirped pump inside the chip. The resulting output is then captured with a spectrometer and numerically processed to extract the complete information (amplitude and phase) of the incident pulse. Figure 2.21 compares the results from the X-SPIDER device using both a standard algorithm and a new extended (Fresnel) phase-recovery extraction

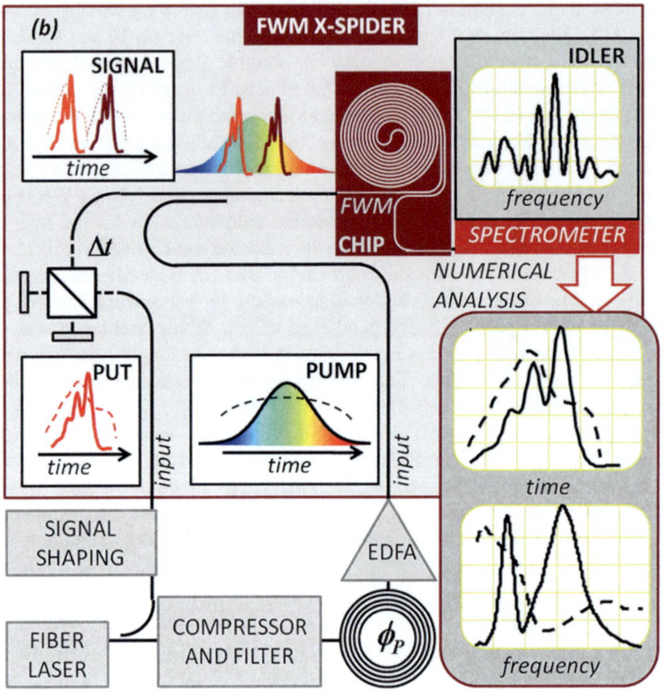

Fig. 2.20 Experimental setup for the phase measurement of ultrafast pulses. Phase and amplitude measurement of ultrafast optical pulses using spectral phase interferometry for direct electric-field reconstruction (X-SPIDER) device based on FWM in Hydex spiral waveguide [14]. The pulse under test (*PUT*) is split in two replicas and nonlinearly mixed with a highly chirped pump inside the chip (extracted from the same laser source in these experiments). The resulting output is captured by a spectrometer and numerically processed by a suitable algorithm to extract the complete information (amplitude and phase) of the incident pulse

algorithm [14, 108] designed for pulses with large TBP, with results from SHG-based FROG measurements. As expected, for low TBP pulses (short-pulse regime) the SPIDER device yielded identical results with both algorithms and agreed with the FROG spectrogram. For large TBP pulses (highly chirped, long pulsewidths) the X-SPIDER results obtained with the new algorithm agreed very well with the FROG trace, whereas results using the standard phase-recovery algorithm were unable to accurately reproduce the pulse.

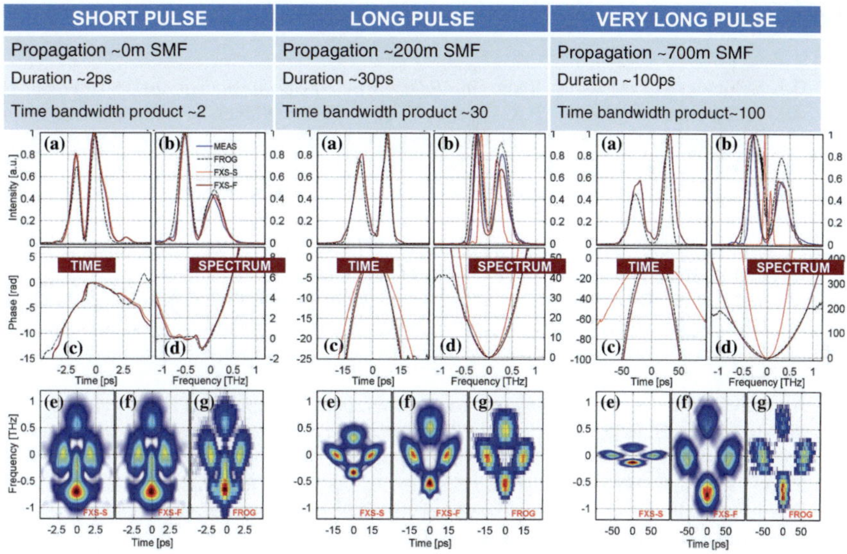

Fig. 2.21 Measurement of Ultrafast Pulses. X-SPIDER device based on FWM in Hydex spiral waveguide [14]. Retrieved phase and amplitude profiles for pulses with time bandwidth products (TBP) of 2 (*left*), 30 (*middle*) and 100 (*right*). For each pulse, plots **a** are the temporal amplitude output **b** the spectral amplitude **c** temporal phase and **d** spectral phase. The contour plots for each pulse (short, long, ultra-long) are shown in **e–g**. **e** was obtained by the X-SPIDER device using a standard algorithm, **f** was obtained from the X-spider using a new algorithm [14] for large time-bandwidth product (TBP) pulses, while **g** (*plots at the right*) were experimentally measured FROG SHG measurements

2.10 Conclusions

We have reviewed the recent progress in CMOS-compatible, non c-Si integrated optical platforms for nonlinear optics, focusing on SiN, Hydex glass, and amorphous Si. The combination negligible nonlinear (two-photon) absorption, low linear loss, the ability to engineer dispersion, and moderately high nonlinearities has enabled these new platforms to achieve new capabilities and novel devices not possible in c-Si because of its low FOM. These include on-chip multiple wavelength oscillators, optical frequency comb sources, ultrashort optical pulse generators, high-gain optical parametric amplifiers, phase-sensitive optical pulse measurement and many others. These platforms will likely have a significant impact on future all-optical devices, complementing the capabilities already offered by silicon nanophotonics. The high performance, reliability, manufacturability of all of these platforms combined with their intrinsic compatibility with electronic-chip manufacturing (CMOS) raises the prospect of practical platforms for future low-cost nonlinear all-optical PICs that offer a smaller footprint, lower energy consumption and lower cost than present solutions.

References

1. B.J. Eggleton, D.J. Moss, S. Radic, in *Optical Fiber Telecommunications V: Components and Sub-systems*, ed. by P.K. Ivan, T. Li, A.E. Willner. Nonlinear Optics in Communications: From Crippling Impairment to Ultrafast Tools, Chapter 20 (Academic Press, Oxford, 2008), pp. 759–828
2. J. Leuthold, C. Koos, W. Freude, Nonlinear silicon photonics. Nat. Photonics **4**, 535–544 (2010)
3. H. Ji et al., 1.28-Tb/s Demultiplexing of an OTDM DPSK data signal using a silicon waveguide. IEEE Photonics Technol. Lett. **22**, 1762 (2010)
4. M. Galili et al., Breakthrough switching speed with all-optical chalcogenide glass chips: 640 Gb/s Demux. Opt. Express **17**, 2182 (2009)
5. M. Foster et al., Broad-band optical parametric gain on a silicon chip. Nature **441**, 04932 (2006)
6. H.S. Rong et al., A cascaded silicon Raman laser. Nat. Photonics **2**, 170–174 (2008)
7. W. Mathlouthi, H. Rong, M. Paniccia, Characterization of efficient wavelength conversion by four-wave mixing in sub-micron silicon waveguides. Opt. Express **16**, 16735–16745 (2008)
8. F. Li, T.D. Vo, C. Husko, M. Pelusi, D.-X. Xu, A. Densmore, R. Ma, S. Janz, B.J. Eggleton, D.J. Moss, All-optical XOR logic gate for 40 Gb/s DPSK signals via FWM in a silicon nanowire. Opt. Express **19**, 20364–20371 (2011)
9. R. Salem et al., Signal regeneration using low-power FWM on a silicon chip. Nat. Photonics **2**, 35–38 (2008)
10. V.G. Ta'eed, M.Shokooh-Saremi, L.B.Fu, D.J. Moss, M.Rochette, I.C.M. Littler, B.J. Eggleton, Y. Ruan, B.Luther-Davies. Integrated all-optical pulse regeneration in chalcogenide waveguides. Opt. Lett. **30**, 2900 (2005)
11. M. Pelusi, F. Luan, T. Vo, M. Lamont, S. Madden, D. Choi, B. Luther-Davies, B. Eggleton, Photonic-chip-based radio-frequency spectrum analyser with terahertz bandwidth. Nat. Photonics **3**, 139 (2009)
12. B. Corcoran, T.D. Vo, M. Pelusi, C. Monat, D.-X. Xu, A. Densmore, R. Ma, S. Janz, D.J. Moss, B.J. Eggleton, Silicon nanowire based radio-frequency spectrum analyser. Opt. Express **18**, 20190 (2010)
13. M. Foster et al., Silicon-chip-based ultrafast optical oscilloscope. Nature **456**, 81 (2008)
14. A. Pasquazi, M. Peccianti, Y. Park, B.E. Little, S.T. Chu, R. Morandotti, J. Azaña, D.J. Moss, Sub-picosecond phase-sensitive optical pulse characterization on a chip. Nat. Photonics **5**, 618 (2011)
15. M. Peccianti et al., Demonstration of an ultrafast nonlinear microcavity modelocked laser. Nat. Commun. **3**, 765 (2012)
16. M. Fridman, A. Farsi, Y. Okawachi et al., Demonstration of temporal cloaking. Nature **481**, 62–65 (2012)
17. R. Slavik et al., All-optical phase and amplitude regenerator for next-generation telecom systems. Nat. Photonics **4**, 690–695 (2010)
18. P.J. Winzer, Beyond 100G ethernet. IEEE Commun. Mag. **48**, 26–30 (2010)
19. E. Rene-Jean et al., Capacity limits of optical fiber networks. J. Lightwave Technol. **28**, 662 (2010)
20. R. Won, M. Paniccia, Integrating silicon photonics. Nat. Photonics **4**, 498–499 (2010)
21. T. Baehr-Jones et al., Myths and rumours of silicon photonics. Nat. Photonics **6**, 206–208 (2012)
22. M. Hochberg, T. Baehr-Jones, Towards Fabless silicon photonics. Nat. Photonics **4**, 492 (2010)
23. T.K. Liang, H.K. Tsang, Nonlinear absorption and Raman scattering in silicon-on-insulator optical waveguides. IEEE J. Sel. Top. Quantum Electron. **10,** 1149–1153 (2004)
24. A.S. Liu, H.S. Rong, M. Paniccia, Net optical gain in a low loss silicon-on-insulator waveguide by stimulated Raman scattering. Opt. Express **12**, 4261–4268 (2004)

25. F. Gholami et al., Third-order nonlinearity in silicon beyond 2350 nm. Appl. Phys. Lett. **99,** 081102 (2011)
26. Q. Lin et al., Dispersion of silicon nonlinearities in the near infrared region. Appl. Phys. Lett. **91**, 021111 (2007)
27. M. Dinu, F. Quochi, H. Garcia, Third-order nonlinearities in silicon at telecom wavelengths. Appl. Phys. Lett. **82**, 2954–2956 (2003)
28. S. Zlatanovic et al., Mid-infrared wavelength conversion in silicon waveguides using ultracompact telecom-band-derived pump source. Nat. Photonics **4**, 561–564 (2010)
29. X. Liu et al., Mid-infrared optical parametric amplifier using silicon nanophotonic waveguides. Nat. Photonics **4**, 557–560 (2010)
30. X. Gai et.al., Nonlinear absorption and refraction in crystalline silicon in the midinfrared. Laser Photon. Rev. **7**, 1054–1064 (2013)
31. B.J. Eggleton, B. Luther-Davies, K. Richardson, Chalcogenide photonics. Nat. Photonics **5**, 141–148 (2011)
32. J.S. Aitchison et al., The nonlinear optical properties of AlGaAs at the half band gap. IEEE J. Quantum Electron. **33**, 341–348 (1997)
33. K. Dolgaleva et al., Compact highly-nonlinear AlGaAs waveguides for efficient wavelength conversion. Opt. Express **19**, 12440–12455 (2011)
34. K. Ikeda, R.E. Saperstein, N. Alic et al., Thermal and Kerr nonlinear properties of plasma-deposited silicon nitride/silicon dioxide waveguides. Opt. Express **16**, 12987–12994 (2008)
35. D.T.H. Tan, K. Ikeda, P.C. Sun, Y. Fainman, Group velocity dispersion and self phase modulation in silicon nitride waveguides. Appl. Phys. Lett. **96**, 061101 (2010)
36. J.S. Levy et al., CMOS-compatible multiple-wavelength oscillator for on-chip optical interconnects. Nat. Photonics **4**, 37–41 (2010)
37. L. Razzari et al., CMOS compatible integrated optical hyper-parametric oscillator. Nat. Photonics **4**, 41–44 (2010)
38. M. Ferrera et al., Low-power continuous-wave nonlinear optics in doped silica glass integrated waveguide structures. Nat. Photonics **2**, 737–740 (2008)
39. C.H. Henry, R.F. Kazarinov, H.J. Lee, K.J. Orlowsky, L.E. Katz, Low loss Si3N4-SiO2 optical waveguides on Si. Appl. Opt. **26**, 2621–2624 (1987)
40. N. Daldosso et al., Comparison among various Si3N4 waveguide geometries grown within a CMOS fabrication pilot line. IEEE J. Ligthtwave Technol. **22**, 1734–1740 (2004)
41. B.E. Little et al., Very high-order microring resonator filters for WDM applications. IEEE Photon. Technol. Lett. **16**, 2263–2265 (2004)
42. D. Duchesne et al., Efficient self-phase modulation in low loss, high index doped silica glass integrated waveguides. Opt. Express **17**, 1865 (2009)
43. J.S. Levy et al., High-performance silicon nitride based multiple wavelength source. IEEE Photonics Technol. Lett. **24,** 1375–1377 (2012)
44. T. Herr et al., Universal formation dynamics and noise of Kerr-frequency combs in microresonators. Nat. Photonics **6**, 480–487 (2012)
45. Y. Okawachi et al., Octave-spanning frequency comb generation in a silicon nitride chip. Opt. Lett. **36**, 3398–3400 (2011)
46. A. Johnson et al., Chip-based frequency combs with sub-100 GHz repetition rates. Opt. Lett. **37**, 875–877 (2012)
47. F. Ferdous et al., Spectral line-by-line pulse shaping of on-chip microresonator frequency combs. Nat. Photonics **5**, 770–776 (2011)
48. T.J. Kippenberg, R. Holzwarth, S.A. Diddams, Microresonator-based optical frequency combs. Science **332**, 555–559 (2011)
49. Y. Saha et al., Modelocking and femtosecond pulse generation in chip-based frequency combs. arXiv:1211.1096 (2012)
50. M. Peccianti et al., Dual frequency comb mode-locked laser based on an integrated nonlinear microring resonator. Opt. Express **20**, 27355–27362 (2012)
51. J.S. Levy et al., Harmonic generation in silicon nitride ring resonators. Opt. Express **19**, 11415–11421 (2011)

52. T.M. Monro et al., Progress in microstructured optical fibers. Annu. Rev. Mater. Res. **36**, 467–495 (2006)
53. A. Pasquazi et al., Efficient wavelength conversion and net parametric gain via four wave mixing in a high index doped silica waveguide. Opt. Express **18**, 7634–7641 (2010)
54. M. Ferrera et al., Low power four-wave mixing in an integrated, microring resonator with Q = 1.2 million. Opt. Express **17**, 14098–14103 (2009)
55. D.J. Moss, H. Driel, J. Sipe, Third harmonic generation as a structural diagnostic of ion implanted amorphous and crystalline silicon. Appl. Phys. Lett. **48**, 1150 (1986)
56. R. Orobtchouk, S. Jeannot, B. Han, T. Benyattou, J.M. Fedeli, P. Mur, in *Proceedings of SPIE 6183, conference on Integrated Optics, Silicon Photonics, and Photonic Integrated Circuits*. Ultra compact optical link made in amorphous silicon waveguide (Strasbourg, France, 2006) paper 618304
57. J.M. Fedeli, M. Migette, L. Di Cioccio, L. El Melhaoui, R. Orobtchouk, C. Seassal, P. Rojo-Romeo, F. Mandorlo, D. Marris-Morini, L. Vivien, in *Proceedings of 3rd IEEE International Conference on Group IV Photonics*. Incorporation of a photonic layer at the metallization levels of a CMOS circuit (2006) pp. 200–202
58. K. Ikeda, Y.M. Shen, Y. Fainman, Enhanced optical nonlinearity in amorphous silicon and its application to waveguide devices. Opt. Express **15**, 17761–17771 (2008)
59. Y. Shoji et al., Ultrafast nonlinear effects in hydrogenated amorphous silicon wire waveguide. Opt. Express **18**, 5668–5673 (2010)
60. K. Narayanan, S.F. Preble, Optical nonlinearities in hydrogenated amorphous silicon waveguides. Opt. Express **18**, 8998–9905 (2010)
61. S. Suda et al., Pattern-effect-free all-optical wavelength conversion using a hydrogenated amorphous silicon waveguide with ultra-fast carrier decay. Opt. Lett. **37**, 1382–1384 (2012)
62. K.-Y. Wang, A.C. Foster, Ultralow power continuous-wave frequency conversion in hydrogenated amorphous silicon waveguides. Opt. Lett. **37**, 1331–1333 (2012)
63. B. Kuyken et al., On-chip parametric amplification with 26.5 dB gain at telecommunication wavelengths using CMOS-compatible hydrogenated amorphous silicon waveguides. Opt. Lett. **36**, 552–554 (2011)
64. B. Kuyken et al., Nonlinear properties of and nonlinear processing in hydrogenated amorphous silicon waveguides. Opt. Express **19**, B146–B153 (2011)
65. C. Grillet et al., Amorphous silicon nanowires combining high nonlinearity. FOM and optical stability. Optics Express **20**, 22609 (2012)
66. L. Caspani et al., Optical frequency conversion in integrated devices. JOSA B **28**(12), A66–A81 (2011)
67. D. Moss, E. Ghahramani, J.E. Sipe, H.M. van Driel, Band structure calculation of dispersion and anisotropy in $\chi^{(3)}$ (3ω: ω,ω,ω) for third harmonic generation in Si, Ge, and GaAs. Phys. Rev. B **41**, 1542 (1990)
68. B. Corcoran et al., Optical signal processing on a silicon chip at 640 Gb/s using slow-light. Opt. Express **18**, 7770–7781 (2010)
69. Corcoran B et al., Slow light enhanced nonlinear optics in dispersion engineered slow-light silicon photonic crystal waveguides. IEEE J. Sel. Top. Quantum Electron. **16,** 344–356 (2010)
70. P. Del'Haye et al., Optical frequency comb generation from a monolithic microresonator. Nature **450**, 1214–1217 (2007)
71. I.S. Grudinin, N. Yu, L. Maleki, Generation of optical frequency combs with a CaF2 resonator. Opt. Lett. **34**, 878–880 (2009)
72. A.C. Turner, M.A. Foster, A.L. Gaeta, M. Lipson, Ultra-low power parametric frequency conversion in a silicon microring resonator. Opt. Express **16**, 4881–4887 (2008)
73. H. Fukuta et al., Four-wave mixing in silicon wire waveguides. Opt. Express **13**, 4629–4637 (2005)
74. H. Rong, Y. Kuo, A. Liu, M. Paniccia, High efficiency wavelength conversion of 10 Gb/s data in silicon waveguides. Opt. Express **14**, 1182–1188 (2006)

75. A. Bartels et al., Femtosecond-laser-based synthesis of ultrastable microwave signals from optical frequency references. Opt. Lett. **30**, 667–669 (2005)
76. Th. Udem, R. Holzwarth, T.W. Hänsch, Optical frequency metrology. Nature **416**, 233–237 (2002)
77. S.T. Cundiff, A.M. Weiner, Optical arbitrary waveform generation. Nat. Photonics **4**, 760–766 (2010)
78. Z. Jiang, C.B. Huang, D.E. Leaird, A.M. Weiner, Optical arbitrary waveform processing of more than 100 spectral comb lines. Nat. Photonics **1**, 463–467 (2007)
79. M.H. Khan et al., Ultrabroad-bandwidth arbitrary radiofrequency waveform generation with a silicon photonic chip-based spectral shaper. Nat. Photonics **4**, 117–122 (2010)
80. D. Jones et al., Carrier-envelope phase control of femtosecond mode-locked lasers and direct optical frequency synthesis. Science **288**, 635–639 (2000)
81. J. K. Ranka, R. S. Windeler, A. J. Stentz, Visible continuum generation in air-silica microstructure optical fibers with anomalous dispersion at 800 nm. Opt. Lett. **25**, 25–27 (2000)
82. J.M. Dudley, G. Genty, S. Coen, Supercontinuum generation in photonic crystal fiber Rev. Mod. Phys. **78**, 1135–1184 (2006)
83. J. Herrmann et al., Experimental evidence for supercontinuum generation by fission of higher-order solitons in photonic fibers. Phys. Rev. Lett. **88**, 173901 (2002)
84. M.R. Lamont, B. Luther-Davies, D.-Y. Choi, S. Madden, B.J. Eggleton, Supercontinuum generation in dispersion engineered highly nonlinear (γ = 10/W/m) As(2)S(3) chalcogenide planar waveguide. Opt. Express **16**, 14938 (2008)
85. C. Phillips, J. Jiang, C. Langrock, M.M. Fejer, M.E. Fermann, Supercontinuum generation in quasi-phase-matched LiNbO3 waveguide pumped by a Tm-doped fiber laser system. Opt. Lett. **36**, 3912–3914 (2011)
86. M.A. Foster, A.L. Gaeta, Ultra-low threshold supercontinuum generation in sub-wavelength waveguides. Opt. Express **12**, 3137–3143 (2004)
87. P. Koonath, D.R. Solli, B. Jalali, Limiting nature of continuum generation in silicon. Appl. Phys. Lett. **93**, 091114 (2008)
88. I.-W. Hsieh et al., Supercontinuum generation in silicon photonic wires. Opt. Express **15**, 15242–15249 (2007)
89. D. Duchesne et al., Supercontinuum generation in a high index doped silica glass spiral waveguide. Opt. Express **18**, 923–930 (2010)
90. R. Halir et al., Ultrabroadband supercontinuum generation in a CMOS-compatible platform. Opt. Lett. **37**, 1685 (2012)
91. T. Herr, V. Brasch, M.L. Gorodetsky, T.J. Kippenberg. *Soliton Mode-locking in Optical Microresonators*, (2012). arXiv:1211.0733
92. F. Leo, S. Coen, P. Kockaert, S.-P. Gorza, P. Emplit, M. Haelterman, Temporal cavity solitons in one-dimensional Kerr media as bits in an all-optical buffer. Nature Photon. **4**, 471–476 (2010)
93. A. Pasquazi, L. Caspani, M. Peccianti, M. Clerici, L. Razzari, M. Ferrera, D.Duchesne, B. Little, S.T. Chu, D.J. Moss, R. Morandotti, Self-locked optical parametric oscillation in a CMOS compatible microring resonator: a route to robust optical frequency comb generation on a chip. Opt. Express **21**(11), 13333–13341 (2013)
94. M. Quiroga-Teixeiro et al., Passive mode locking by dissipative four-wave mixing. J. Opt. Soc. Am. B **15**, 1315–1321 (1998)
95. T. Sylvestre et al., Self-induced modulational instability laser revisited: normal dispersion and dark-pulse train generation. Opt. Lett. **27**, 482–484 (2002)
96. M. Yoshida, A. Ono, M. Nakazawa, 10 GHz regeneratively modelocked semiconductor optical amplifier fibre ring laser and its linewidth characteristics. Opt. Lett. **32**, 3513–3515 (2007)
97. A.L. Schawlow, C.H. Townes, Infrared and optical masers. Phys. Rev. **112**, 1940–1949 (1958)

98. A. Pasquazi A et al., Time-lens measurement of subpicosecond optical pulses in CMOS compatible high-index glass waveguides. IEEE J. Sel. Top. Quantum Electron. **18**, 629–636 (2012)
99. R. Trebino, *Frequency Resolved Optical Gating: The Measurement of Ultrashort Laser Pulses* (Kluwer Academic, Boston, 2002)
100. C. Iaconis, I.A. Walmsley, Spectral phase interferometry for direct electric field reconstruction of ultrashort optical pulses. Opt. Lett. **23**, 792–794 (1998)
101. C. Iaconis, I.A. Walmsley, Self-referencing spectral interferometry for measuring ultrashort optical pulses. IEEE J. Quant. Electron. **35**, 501–509 (1999)
102. L. Gallmann et al., Characterization of sub-6-fs optical pulses with spectral phase interferometry for direct electric-field reconstruction. Opt. Lett. **24**, 1314–1316 (1999)
103. C. Dorrer et al., Single-shot real-time characterization of chirped-pulse amplification systems by spectral phase interferometry for direct electric-field reconstruction. Opt. Lett. **24**, 1644–1646 (1999)
104. J. Bromage, C. Dorrer, I.A. Begishev, N.G. Usechak, J.D. Zuegel, Highly sensitive, single-shot characterization for pulse widths from 0.4 to 85 ps using electro-optic shearing interferometry. Opt. Lett. **31**, 3523–3525 (2006)
105. C. Dorrer, J. Bromage, High-sensitivity optical pulse characterization using Sagnac electro-optic spectral shearing interferometry. Opt. Lett. **35**, 1353–1355 (2010)
106. I.A. Walmsley, C. Dorrer, Characterization of ultrashort electromagnetic pulses. Adv. Opt. Photon. **1**, 308–437 (2009)
107. M.E. Anderson, A. Monmayrant, S.P. Gorza, P. Wasylczyk, I.A. Walmsley, SPIDER: a decade of measuring ultrashort pulses. Laser Phys. Lett. **5**, 259–266 (2008)
108. A. Pasquazi, M. Peccianti, J. Azana, D.J. Moss, R. Morandotti, FLEA: Fresnel-limited extraction algorithm for SPIDER. Opt. Express **21**(5), 5743–5758 (2013)

Chapter 3
Optical Guided Wave Switching

Costantino De Angelis, Daniele Modotto, Andrea Locatelli and Stefan Wabnitz

Abstract Optical switching is a key functionality for enabling transparent all-optical networks. We present an overview of optical switching devices, based on either optical or electrical control signals, which permit to avoid the necessity of optics-electronics-optics conversion. We describe the basic principles of various guided wave optical switching devices, which exploit either relatively long interaction lengths in order to reduce the operating power requirements, or strong transverse confinement to reduce device dimensions. These devices include nonlinear mode couplers and interferometers based on optical fibers, as well as integrated waveguides based on photonic crystal structures or surface wave interactions in novel materials such as graphene.

3.1 Introduction: Optical Switching Using Guided-Waves

In recent years, telecommunication networks have witnessed a dramatic increase of capacity, mostly driven by the exponential growth of IP traffic. Researchers had to tackle and solve several problems, and certainly the challenge of realizing transparent all optical switching was among the most important issues which have been addressed. In this framework, the goal is to use optics not only at the transmission level, but also at the switching level; this in turn requires to conceive devices which can perform switching directly in the optical domain, thus overtaking the unavoidable limitations of switching at the electronic level. Remarkably the goal can be achieved using either an optical or an electrical control; the common denominator is not the physical nature of the control signal, but the possibility of

C. De Angelis · D. Modotto · A. Locatelli · S. Wabnitz (✉)
Dipartimento di Ingegneria dell'Informazione, Università degli Studi di Brescia,
Via Branze 38, Brescia 25123, Italy
e-mail: stefan.wabnitz@unibs.it

S. Wabnitz and B.J. Eggleton (eds.), *All-Optical Signal Processing*,
Springer Series in Optical Sciences 194, DOI 10.1007/978-3-319-14992-9_3

avoiding optics-electronics-optics (OEO) conversions which lead to inefficient and nontransparent switching.

In particular, in this paper we review some basic ideas and devices for all-optical switching in guided wave geometries, as this scenario offers important possibilities for tuning the light confinement, and thus exploring more efficient light-matter interactions. In Sect. 3.2 we will discuss some relevant examples of all-optical switching devices exploiting optical fibers; in Sect. 3.3 we will move into the field of integrated optics and we will discuss optical switching in photonic crystal waveguides, as well as using surface waves as a tool to enhance light matter-interactions in novel nonlinear materials such as graphene.

3.2 All-Optical Pulse Switching in Optical Fibers

We review the different implementations of switching in optical fibers, including polarization effects in both high and low-birefringence fibers, and nonlinear optical loop mirrors. Both self-switching and two-beam, phase or cross-phase modulation (XPM) controlled switching will be analyzed.

3.2.1 Nonlinear Mode Coupling

We start by considering the continuous wave (CW) case and focus our attention on the linear and nonlinear coupling between the two modes of a structure composed of two waveguides placed in close proximity. The basic model to describe the evolution of the modal amplitudes $A_{1,2}$ along the coupler [1] is given by the following coupled equations:

$$
\begin{aligned}
i\frac{dA_1}{dz} + CA_2 + \gamma\left[|A_1|^2 + \rho|A_2|^2\right]A_1 = 0 \\
i\frac{dA_2}{dz} + CA_1 + \gamma\left[|A_2|^2 + \rho|A_1|^2\right]A_2 = 0
\end{aligned}
\tag{3.1}
$$

where C is the linear coupling coefficient, γ is the nonlinear coefficient and ρ is the XPM factor. The previous equations were introduced for the nonlinear coherent directional coupler (NLDC) [2, 3], and they work well for dual-core fibers and for integrated couplers [4].

In the case of a birefringent optical fiber the XPM factor is $2/3$ and nonlinear four-wave mixing terms must be included, as well [5–7]:

$$i\frac{da_x}{d\zeta}+\frac{1}{2}a_x+\left[|a_x|^2+\frac{2}{3}|a_y|^2\right]a_x+\frac{1}{3}a_y^2a_x^*=0$$

$$i\frac{da_y}{d\zeta}-\frac{1}{2}a_y+\left[|a_y|^2+\frac{2}{3}|a_x|^2\right]a_y+\frac{1}{3}a_x^2a_y^*=0 \quad (3.2)$$

where $A_{x,y}$ are the complex envelopes of the fundamental modes polarized along the x, y principal axes of the fiber with propagation constants $\beta_{x,y}$, respectively. Moreover, $a_{x,y} \equiv \sqrt{\gamma/\Delta\beta}\exp\{-i(\beta_x+\beta_y)z/2\}A_{x,y}$, $\zeta = \Delta\beta z$ is a dimensionless distance, $\Delta\beta = \beta_x - \beta_y = 2\pi/L_b$ is the linear fiber birefringence and L_b the beat length. Let us express (3.2) in a more compact form in terms of the circularly polarized components $a_\pm = (a_x \pm ia_y)/\sqrt{2}$ as

$$i\frac{da_+}{d\zeta}+\frac{1}{2}a_-+\frac{2}{3}\left[|a_+|^2+2|a_-|^2\right]a_+=0$$

$$i\frac{da_-}{d\zeta}+\frac{1}{2}a_++\frac{2}{3}\left[|a_-|^2+2|a_+|^2\right]a_-=0 \quad (3.3)$$

At relatively low powers, fiber birefringence leads to the periodic exchange of power among the two circular polarizations: the spatial period of such linear coupling is equal to the linear beat length $L_b = 2L_c$ (L_c is the linear coupling length).

In order to better visualize the action of nonlinear coupling, it proves convenient to rewrite (3.2) in terms of the real Stokes parameters $s_i \equiv S_i/S_0$, where $S_0 \equiv P = |a_x|^2 + |a_y|^2$, $S_1 \equiv |a_x|^2 - |a_y|^2$, $S_2 \equiv a_xa_y^* + c.c.$, and $S_3 \equiv -ia_xa_y^* + c.c.$:

$$\frac{ds_1}{d\zeta} = 2ps_2s_3$$

$$\frac{ds_2}{d\zeta} = -s_3 - 2ps_1s_3$$

$$\frac{ds_3}{d\zeta} = s_2 \quad (3.4)$$

where we defined the dimensionless input power $p \equiv P/P_c$, with the critical power level $P_c \equiv 3\Delta\beta/2\gamma$. In vector notation, (3.4) can be written as

$$\frac{d\mathbf{s}}{d\zeta} = (\Omega_L + \Omega_{NL}(\mathbf{s})) \times \mathbf{s} \quad (3.5)$$

which describes polarization evolution on the Poincaré sphere as the motion of a rigid body subject to the sum of the fixed and the position-dependent angular velocities $\Omega_L \equiv (1, 0, 0)$ and $\Omega_{NL} = (0, 0, -ps_3)$, respectively [6, 8]. Figure 3.1a

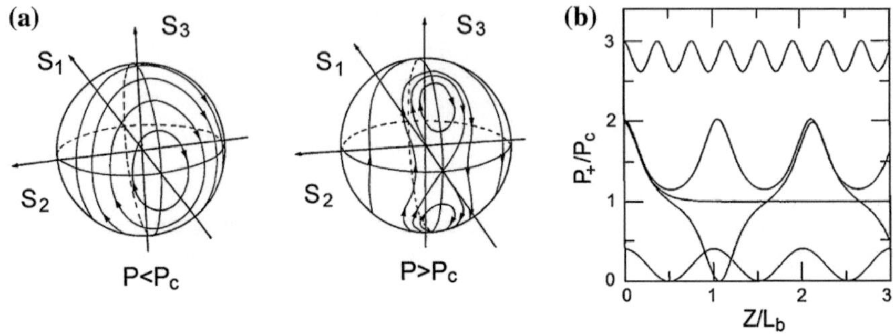

Fig. 3.1 **a** Poincaré sphere description of switching in a nonlinear coupler; **b** evolution with distance of power in the input mode for different input power levels (Figures adapted from [6, 8])

illustrates the radically different behavior of the polarization evolution trajectories of the tip of the Stokes vector **s** as the input power is below or above the critical power P_c. Below the critical power, all points on the Poincaré sphere rotate around the axis defined by Ω_L: as a result, the two circular polarizations periodically exchange their power along the fiber length. On the other hand, for $P > P_c$ the mode polarized along the fast axis of the fiber (i.e., whose Stokes vector is $\mathbf{s_F} = (-1, 0, 0)$) loses its spatial stability and becomes an unstable saddle (see right part of Fig. 3.1a) [6, 8–10]: as it is shown in Fig. 3.1b, an input right-handed circular polarization (whose power is $P_+ \equiv |A_+|^2$) no longer periodically couples into the orthogonal left-handed polarization whenever $P_+ > 2P_c$.

3.2.2 Nonlinear Fiber Couplers

The NLDC is a basic device which permits all-optical routing and switching operations [2, 3]. Experiments have demonstrated nonlinear transmission and self-switching in fiber-optic NLDCs, such as: dual-core fibers [11–15], low birefringence fibers [16], and periodically twisted birefringent filters [17, 18]. Even though fiber optic NLDCs exhibit ultrafast nonlinear response, whenever their length $L \simeq L_b$, the associated switching power $P \simeq P_c$ is relatively high, owing to the weak nonlinearity of silica. Indeed, because of imperfections in fiber fabrication (in the case of dual core couplers) or of random fiber birefringence (for polarization couplers), beat lengths exceeding one or a few meters are not possible in practice. Earlier experiments with fiber NLDCs involved just one or two linear coupling distances L_c. In this case, the associated nonlinear transmission exhibits a single switching power, that is inversely proportional to L_c.

Nevertheless, full switching in a fiber NLDC of any length L is still possible when operating in the multi-beatlength regime, i.e., whenever $L \gg L_b$ [19, 20]. The main advantage of the multi-beatlength NLDC is that the relevant switching power is inversely proportional to the total length of the fiber L. Therefore, the switching performance is comparable to that of other nonlinear fiber switches such as the nonlinear Mach-Zehnder interferometer (MZI) and the nonlinear optical loop mirror (NOLM). Using a circularly polarized beam of power P at the input of a birefringent fiber, the exact solution of (3.3) leads to the following expression for the output power P_+ in the same circular polarization

$$P_+ = \frac{P}{2}[1 + cn(\pi L/L_c|\tilde{p})] \tag{3.6}$$

where $\tilde{p} = p/2$ and cn is a Jacobi elliptic function. From (3.6), one obtains that the nonlinear period of polarization coupling between the two circular polarizations is

$$L_{nc}(\tilde{p}) = \frac{2L_c K(\tilde{p})}{\pi} \tag{3.7}$$

where K is the complete elliptic integral of the first kind. Whenever $\tilde{p} < 1$, one obtains that

$$K(\tilde{p}) = \frac{\pi}{2}\left[1 + \frac{\tilde{p}^2}{4} + O(\tilde{p})^2\right] \tag{3.8}$$

Therefore for $\tilde{p} \ll 1$ (3.6) reduces to

$$P_+ = P\cos^2(\pi L/2L_{nc}(\tilde{p})) \tag{3.9}$$

The associated ON–OFF switching power p_s corresponds to the power-induced increase of the coupling distance which leads to a $\pi/2$ shift in the argument of the cosine in (3.9). One obtains $p_s = 2\sqrt{2L_b/L}$, i.e., in real units

$$P_s = \frac{12\pi}{\gamma\sqrt{2L_b L}} \tag{3.10}$$

The switching power of a fiber-based NLDC operating in the multi-beatlength regime may be further reduced by injecting an input elliptically (as opposed to circularly) polarized beam [21, 22]. Let us set $s_1(\zeta = 0) = \varepsilon\cos(2\phi)$, $s_1(\zeta = 0) = \varepsilon\sin(2\phi)$, and $s_3(\zeta = 0) = \sqrt{1 - \varepsilon^2}$. From the exact solution of (3.3), one obtains $P_+ = P(1 + s_3(L, p))/2$ for the right handed circular polarization component at the output of a fibre of length L, where

$$s_3 = \sqrt{1 - \varepsilon^2\cos(2\phi)^2} f(p)\cos(2\pi r(p)L/L_b) \tag{3.11}$$

and $f(p) = (1 + p\varepsilon\cos(2\phi)/2)$. In this case, the nonlinear beat length $\tilde{L}_b(p) = L_b/f(p) \simeq L_b(1 - p\varepsilon\cos(2\phi)/2)$. By supposing $|\cos(2\phi)| = 1$, the switching power p_s is defined by $2\pi(r(p) - 1)L/L_b = \pi$, so that $p_s = L_b/\varepsilon L$. In real units, the switching power reads as

$$P_s = \frac{3\pi}{\gamma\varepsilon L} \tag{3.12}$$

In Fig. 3.2a we show an example of nonlinear CW transmissions computed from (3.9) at the output of a $L = 20L_b$ long birefringent fiber. Here we plot the transmission dependence on input power for different input polarization angles ϕ, and we have set $\varepsilon = 0.37$, $\gamma L = 1$. As can be seen from Fig. 3.2a, the switching power agrees well with the estimate $P_s \simeq 25$ W from (3.12). In Fig. 3.2 we also display the change of the nonlinear transmission as the input polarization angle ϕ is varied, for L = 10–20L_b and $p = 0.15$.

An experimental confirmation of all-optical power and polarization controlled switching in a multi-beatlength fiber NLDC was performed using 205 m of York ultralow-birefringence spun fiber [19, 23]. The fiber was wound on a drum with radius of 15 cm, in order to introduce a bending-induced linear birefringence with $L_b \simeq 20$ m, so that $L \simeq 10L_b$. By increasing the input power of a right-handed circularly polarized beam from low values up to the switching power, an accumulated nonlinear variation of the beat length equal to L_c or 10 m (or 5 % nonlinear variation per beat length L_b) was obtained.

Quasi-CW operating conditions were achieved using a mode-locked 100 ps input pulse train from an Nd-YAG laser operating at 1.06 μm. At the fiber output, the two circular polarization components of the emerging pulses were the two output channels of the birefringent fiber NLDC. These components were separated by means of a calibrated Babinet-Soleil compensator, followed by a Wollaston prism.

Figure 3.3a, b shows the streak camera pictures of the pulse profiles emerging from the bar (input) and the crossed circular polarization components at the fiber output, for an input average power of 320 mW [19]. As it can be seen, the center

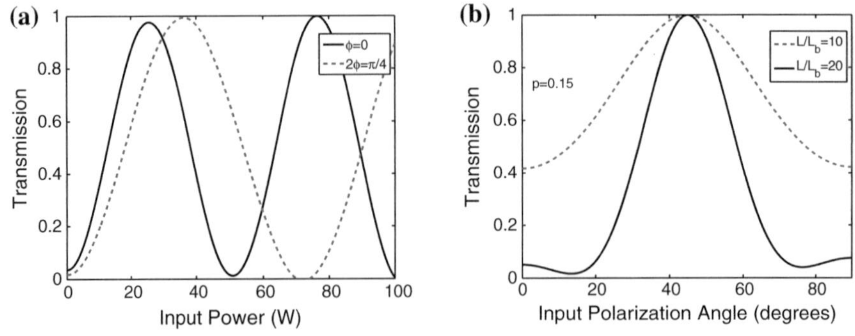

Fig. 3.2 Theoretical CW transmissions at the output of $L = 20L_b$ long birefringent fiber: **a** transmission dependence on input power for different input polarization angles; **b** transmission variation with input angle for two values of the total number of linear beat lengths L_b

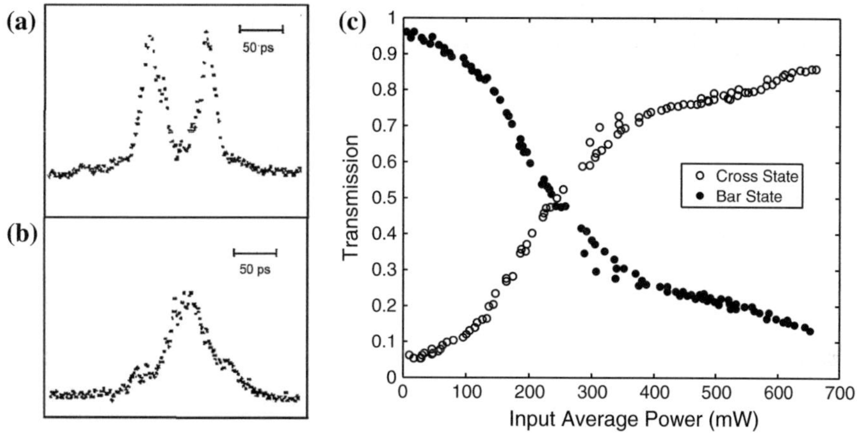

Fig. 3.3 Experimental pulse energy transmissions from orthogonal polarizations at the output of a 10 beat length long birefringent fiber: **a**, **b** streak-camera traces of pulses emerging from the input (*bar state*) and the orthogonal (*cross state*) polarizations; **c** average energy transmissions from cross and bar states versus input average power (Figures adapted from [19, 22])

and the wings of the pulse have opposite handedness, leading to a symmetric pulse breakup.

Figure 3.3c shows the measured circular polarization transmissions at the fiber output, as a function of the input average power into the fiber [22]. Because of the slow response time of the photodetectors, the transmission displays the average power in each output polarization. At low powers, the input circular polarization emerges from the fiber almost unchanged. About 50 % switching of the transmission is observed at about 250 mW: the corresponding peak pulse power is of about 20 W, in relatively good agreement with the estimate of (3.12), since in the experiment the condition $\gamma L \simeq 1$ is verified.

3.2.3 Nonlinear Mach-Zehnder Interferometers

A relatively simple nonlinear switch may be implemented using an all-fiber-based Mach-Zehnder interferometer (MZI): here two 3-dB linear directional couplers are used to split the input signal and then recombine the two arms of the interferometer of lengths L_1 and L_2, respectively (see Fig. 3.4). The components of the signal in the two arms of the MZI thus experience different overall linear and SPM induced phase shifts [24–28]. The transmission through the bar port of the MZI reads as

$$T(P) = \sin^2(\Delta\phi_L + \Delta\phi_{NL}) \tag{3.13}$$

where $\Delta\phi_{NL} = \gamma P(L_1 - L_2)$. Alternatively, a different nonlinear phase shift in the two arms of the MZI may be induced whenever $L_1 = L_2$ by using an unbalanced

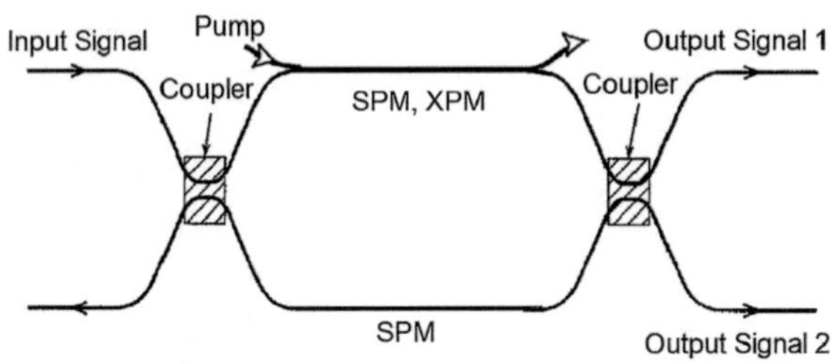

Fig. 3.4 Structure of a fiber based nonlinear MZ interferometer switch exploiting SPM or XPM (in the presence of a control pump)

linear directional coupler, so that the two signal components have different powers. A different implementation of a MZI type of switch consists of a highly birefringent fiber, where the two signal paths are represented by the two fiber modes which are orthogonally polarized along the principal axes of the fiber. For an input beam of power P that is initially linearly polarized with an orientation at the angle θ with respect to the x-axis of a birefringent fiber of length L, one has $P_x = P_0 \cos^2(\theta)$, $P_y = P_0 \sin^2(\theta)$, and $\Delta\phi_{NL} = \gamma PL \cos(2\theta)/3$.

Besides SPM-activated switching, a fiber MZI can be used to switch a signal among its two output ports through XPM by injecting a control pump at a different wavelength, that shifts the signal phase in the upper arm of the interferometer (see Fig. 3.4). In the presence of a control pump composed by a pulse train, the input CW signal may also be converted into a pulse train. The drawback of a fiber MZI is the presence of two separate paths for the signal, which make it sensitive to environmental perturbations, so that active stabilization of the fiber lengths may be required [25].

3.2.4 Nonlinear Loop Mirrors

The environmental stability of an interferometric fiber switch may be ensured by using a Sagnac loop configuration [28–33]: here the transmitted signal results from the coherent superposition at the directional coupler output port of two signals that have traveled in opposite directions around the same loop of fiber (see Fig. 3.5a). Clearly, whenever the directional coupler equally splits the input signal (i.e., a 3-dB coupler is used), there is no differential nonlinear phase shift among the two signal components that travel in opposite directions around the loop, and the Sagnac loop acts as a perfect mirror by reflecting all of the incoming signal: the transmitted signal vanishes. On the other hand, whenever an asymmetric coupler is used, there is a differential nonlinear phase shift which leads to a signal to be transmitted at high powers. Thus the Sagnac loop of Fig. 3.5a is also known as nonlinear optical loop mirror (NOLM).

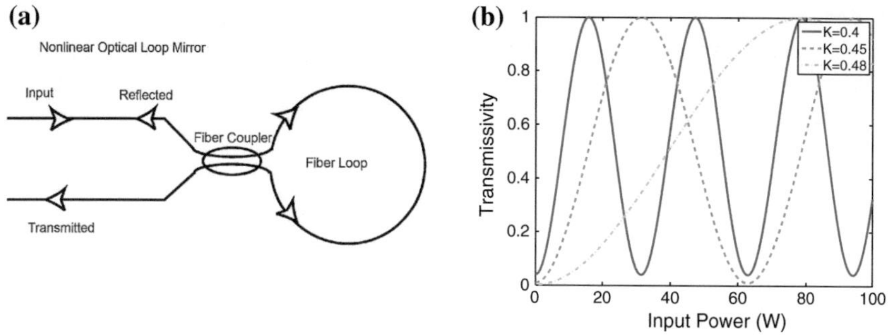

Fig. 3.5 **a** Structure of a fiber based NOLM and **b** associated power dependence of transmission for different power split coefficients K of the coupler

The nonlinear transmission of the NOLM can be simply obtained as follows. The input amplitude of the waves traveling clockwise and counter-clockwise in the loop reads as $A^i_{CW} = \sqrt{K}A_{in}$ and $A^i_{CCW} = i\sqrt{1-K}A_{in}$, respectively, where K is the power splitting ratio of the coupler and A_{in} is the input signal amplitude. After one round-trip through the loop, the two fields traveling in opposite directions acquire equal linear but different nonlinear (as determined by both SPM and XPM) phase shifts, so that at the input of the coupler one has the two fields

$$A^o_{CW} = A^i_{CW} \exp\left\{i\beta L + i\gamma\left(|A^i_{CW}|^2 + 2|A^i_{CCW}|^2\right)L\right\}$$

$$A^o_{CCW} = A^i_{CCW} \exp\left\{i\beta L + i\gamma\left(|A^i_{CCW}|^2 + 2|A^i_{CW}|^2\right)L\right\} \tag{3.14}$$

where β is the linear propagation constant in the fiber loop of length L. At the coupler output, one obtains $A_t = \sqrt{K}A^o_{CW} + i\sqrt{1-K}A^o_{CCW}$, so that the transmissivity of the NOLM reads as

$$T(P) \equiv |A_t|^2/|A_i|^2 = 1 - 4K(1-K)\cos^2[(K-0.5)\gamma P_{in}L] \tag{3.15}$$

where $P_{in} = |A_{in}|^2$. The resulting input signal power dependence of the NOLM transmissivity is illustrated in Fig. 3.5b for different values of the coupling ratio K, and for $\gamma L = 1$. The associated switching power (corresponding to the first unit peak of the transmissivity as P is increased from zero) reads as

$$P_s = \frac{\pi}{\gamma L(1-2K)} \tag{3.16}$$

which leads to $P_s \simeq= 16$ W and $P_s \simeq= 31$ W for $K = 0.4$ and $K = 0.45$, respectively (see Fig. 3.5b).

As in the nonlinear MZI, signal switching in the NOLM may also be induced via XPM by injecting in the loop, by means of a directional coupler, a pump pulse at a different wavelength (see Fig. 3.6a) [34–37]. The control pulse propagates in the CCW direction, thus it only induces a nonlinear phase shift into the corresponding component of the signal, so that propagation in the NOLM is un-balanced even when $K = 0.5$. In this case, the NOLM acts as a perfect reflecting mirror for a signal in the absence of the control pump, whereas in the presence of the pump the signal may be fully transmitted. For evaluating the nonlinear phase shift induced via XPM on the signal by a short pump pulse, one should take into account the temporal walk-off between pump and signal due to fiber GVD. The total XPM phase shift is

$$\Delta\phi(T) = 2\gamma \int_0^L |A_p(T - \delta x)dx, \tag{3.17}$$

where T is time in the reference frame that moves with the group velocity at the signal wavelength V_s, and $\delta = V_p^{-1} - V_s^{-1}$ is the group velocity mismatch between pump and signal. For a Gaussian pump pulse of the form $A_p(T) = P\exp\left(-T^2/T_0^2\right)$ (so that the pump pulse full width at half maximum is $T_{fwhm} \simeq 1.66T_0$), the resulting nonlinear phase shift reads as

$$\Delta\phi(T) = \frac{\gamma LP\sqrt{\pi}T_0}{T_w}\left[erf\left(\frac{T}{T_0}\right) - erf\left(\frac{T - T_w}{T_0}\right)\right], \tag{3.18}$$

where $erf(x)$ is the error function, and the total walk-off is $T_w = \delta L$. The associated nonlinear signal transmission is

$$T = 1 - 4K(1 - K)\cos^2(\Delta\phi/2), \tag{3.19}$$

Let us consider a balanced coupler with $K = 0.5$, a peak XPM-induced phase shift $2\gamma LP = \pi$, and set $T_0 = 10$ ps. Figure 3.6b shows examples of signal transmission windows for different values of the walk-off time T_w. As it can be seen,

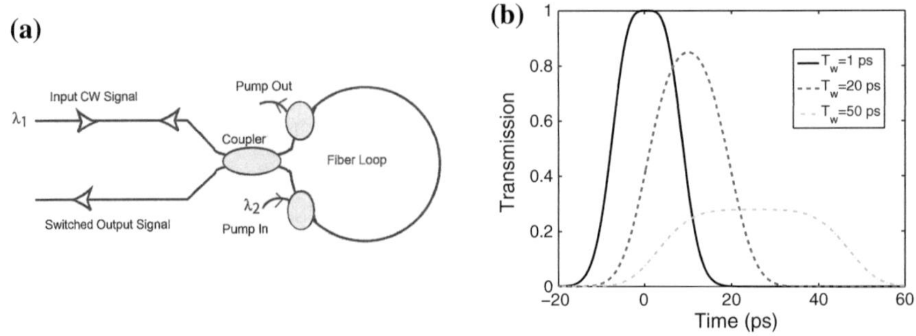

Fig. 3.6 **a** Schematic of XPM-controlled switching in a NOLM and **b** transmission function for $T_0 = 10$ ps and different values of the walk-off time T_w

whenever $T_w \ll T_0$ the transmission window is a replica of the control pulse; whereas as soon as $T_w > T_0$ the peak value of the transmission drops from unity and the window broadens in time.

3.2.5 Nonlinear Passive Loop Resonators

A variant of the nonlinear Sagnac interferometer is provided by the nonlinear fiber ring resonator [38–42]. As an example, consider the scheme of Fig. 3.7a: instead of connecting the two output ports of the coupler, a fiber loop now connects the first output port (port 3 in Fig. 3.7a) with the second input port (port 2) of the coupler. The signal E_1 at input port 1 enters the coupler and it continuously recirculates in the loop; the cavity field can be monitored at the second output port (port 4) of the coupler. The complex amplitude transmission $Y = A_4/A_1$ of the ring resonator is obtained from the relations $A_3 = \sqrt{1-\tau}(\sqrt{1-K}A_1 + i\sqrt{K}A_2)$ and $A_4 = \sqrt{1-\tau}$ $(i\sqrt{K}A_1 + \sqrt{1-K}A_2)$, where τ is the fractional power loss of the coupler and $A_2 = A_3 \exp(-\alpha L + i\beta L + i\Delta\phi_{NL})$ with $\Delta\phi_{NL} = \gamma|A_3|^2 L$. The result is provided by the implicit equation

$$Y = \frac{(1-\tau)\exp(i\delta - \alpha L) + i\sqrt{1-\tau}\sqrt{K}}{1 - i\sqrt{K}\sqrt{1-\tau}\exp(i\delta - \alpha L)}, \tag{3.20}$$

where

$$\delta = \beta L + \gamma|A_1|^2 L\left(\frac{1-\tau}{1-K} + \frac{K}{1-K}|Y|^2 + i\frac{\sqrt{K}\sqrt{1-\tau}}{1-K}(Y - Y^*)\right), \tag{3.21}$$

Clearly (3.20) is equivalent to two real equations, whose numerical solution yields the power transmissivity $T = |A_4|^2/|A_1|^2$ of the nonlinear resonator (see Fig. 3.7b,

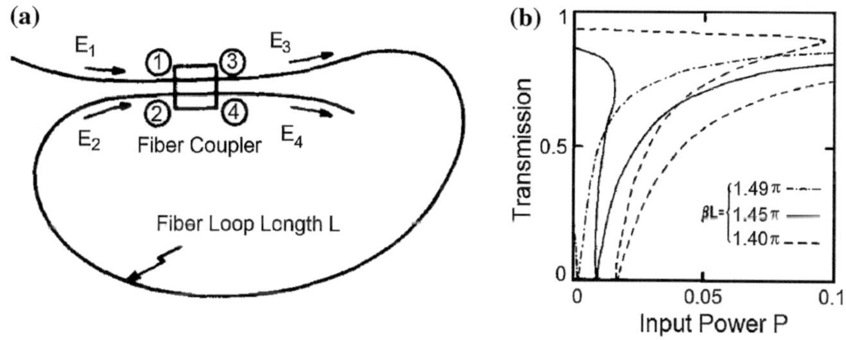

Fig. 3.7 **a** Schematic passive fiber loop resonator and **b** its transmissivity for different values of βL; here $K = 0.95$ (Figures adapted from [40])

where the resonant coupling coefficient is $M = (1 - \tau)\exp(-\alpha L) = 0.95$). As it can be seen, by varying the linear phase delay βL right below the resonance condition $\beta L = 3\pi/2 + 2\pi m$ (where m is an arbitrary integer), one obtains quite different nonlinear transmission behaviors. For small detunings from resonance (i.e., $\beta L = 1.49\pi$), it is only necessary to add a weak nonlinear phase shift (of the order of $\pi/100$) in order to bring the resonator back into resonance. At higher input powers, the resonator gets out-of-resonance and transmissivity rapidly increases. On the other hand, for larger linear cavity detunings (e.g., for $\beta L = 1.45\pi$), the transmissivity becomes multi-valued, and optical multistability results: again, the ON–OFF switching power is relatively low, i.e., it is obtained for nonlinear phase shifts $\gamma|A_3|^2 L \simeq \pi/100$, which is accessible at CW signal power levels.

3.2.6 Optical Soliton Switching

As we have seen in Sect. 3.2.1, the efficiency of all-optical switching in nonlinear couplers using optical pulses is severely limited unless square pulses (e.g., the non-return-to-zero (NRZ) data modulation format) are used [15]. In fact, the CW transmission curve is effectively averaged over the pulse profile, and pulse break-up occurs at the device output, since different portions of the pulse profile are independently switched according to their instantaneous power level (see Fig. 3.3a, b); the same occurs with nonlinear inteferometers. The pulse break-up effect may be avoided by operating in the soliton regime, that is whenever the signal pulses represent optical fiber solitons for the fibers used in the switching device [25, 32, 33, 43–45]. In the short pulse regime, the equations describing propagation in a nonlinear coupler should be extended to include group velocity dispersion terms, which leads to the coupled nonlinear Schrödinger equations

$$i\frac{\partial u}{\partial \xi} + \frac{\beta}{2}\frac{\partial^2 u}{\partial \tau^2} + \kappa v + \left(|u|^2 + \rho|v|^2\right)u = 0$$

$$i\frac{dv}{d\xi} + \frac{\beta}{2}\frac{\partial^2 v}{\partial \tau^2} + \kappa u + \left(|v|^2 + \rho|u|^2\right)v = 0 \tag{3.22}$$

where the dimensionless distance is $\xi = z|\beta_2|/t_s^2$, β_2 is the GVD coefficient, $\beta = \pm 1$ for anomalous or normal dispersion, respectively, $\tau = z/t_s$ where t_s is a reference pulse width. Moreover, the dimensionless coupling coefficient $\kappa = Ct_s^2/|\beta_2| = P_c/2P_{sol}$, where $P_c = 2C/\gamma$ is the NLDC critical power, $P_{sol} = |\beta_2|/\gamma t_s^2$ is the soliton power, and ρ is the XPM coefficient.

Figure 3.8a, b illustrates beam propagation solutions of (3.22) in the anomalous GVD regime. Here $\rho = 0$ (as in a linear dual-core fiber coupler) and $\kappa = 1/4$. As it can be seen, input soliton-like pulses of the form $u(0,\tau) = u_0 sech(\tau/\tau_0)$ (here $v(0,\tau) = 0$ and $\tau_0 = 1$) are entirely switched as a single entity from the cross to the

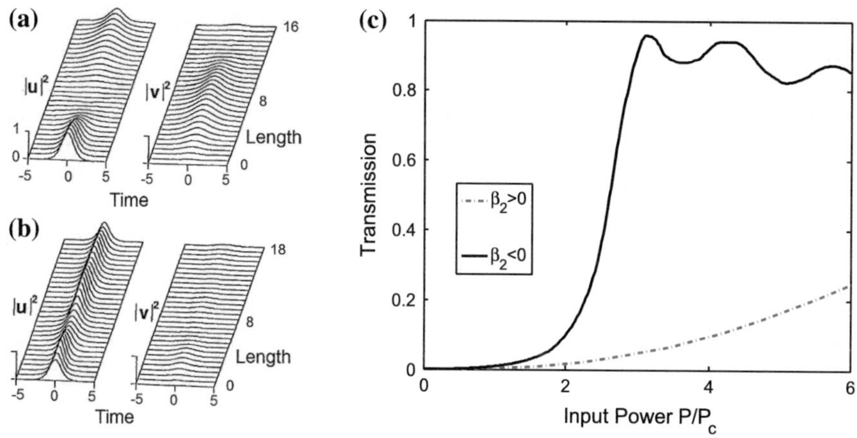

Fig. 3.8 Evolution of power with distance in the bar and cross state of soliton NLDC at **a** low and **b** high powers; **c** comparison of power-dependent transmission in the bar state for dispersive pulses in the anomalous (*solid curve*) and normal (*dashed curve*) dispersion regime (Figures adapted from [43, 45])

bar output state as the input soliton peak power $p = |u_0|^2/P_c$ is increased from $p = 2$ to $p = 3$. For $p = 2$, the input pulse periodically couples back and forth between the two channels, with a relatively small distortion of the pulse profile. Whereas for $p = 3$ the soliton transfer between the channels is inhibited. Therefore the optical soliton exhibits a particle-like switching behavior, and the pulse break-up which is observed in the absence of GVD (or for quasi-CW input signals) may thus be avoided.

Figure 3.8c compares the fraction of energy transmitted in the bar state as a function of the input peak power p, when using a NLDC in either the anomalous ($\beta > 0$) or normal ($\beta < 0$) GVD regime, respectively. Here we have chosen $\kappa = \pi/2$, and $\tau_0 = 1/\sqrt{2\pi}$, so that twice the CW critical power is equal to the fundamental soliton peak power. As it can be seen, with normal GVD the dispersive pulse broadening combined with SPM and linear coupling nearly inhibits the self-switching behavior. Conversely, in the anomalous GVD regime, the transmission is similar to that obtained for CWs or ideal square pulses (besides an increase of the effective switching power).

3.3 Optical Switching in Integrated Optical Waveguide Structures

3.3.1 All-Optical Switching in Photonic Crystal Couplers

In recent years Photonic Crystals (PC) have received increasing attention from the scientific community, especially for their ability to control the propagation of light

[46]. The basic building blocks for all-optical data processing such as waveguides with sharp bends, high-Q resonant cavities, perfect mirrors and so on could indeed be integrated on a single PC chip, in order to achieve complex functions with high performance and small size [47]. PC structures seem to be the ideal choice to get efficient nonlinear devices for optical switching, because of the strong confinement of the fields that permits to optimize the nonlinear interactions. In particular, the development of nonlinear PCs exploiting the ultrafast Kerr nonlinearity has become an important issue. The feasibility of bistable switching devices [48], optical diodes and nonlinear bends [49], and optical isolators [50] has been reported. Directional couplers are fundamental components for optical networks, and it has been demonstrated that PC couplers exhibit smaller size and better performance than the conventional ones [51–53]. Here we describe the properties of an all optical switch based on an ultrashort PC coupler. Switching is performed by exploiting the strong Kerr nonlinearity of AlGaAs, by controlling the intensity of the input signal. The reported two-dimensional finite-difference time-domain (2D FDTD, [54, 55]) analysis shows that the resulting ultra-compact device is characterized by a switching power comparable with the one reported in the literature for centimeter-long conventional nonlinear directional couplers [56].

The schematic view of the proposed structure is shown in Fig. 3.9. The PC is formed by a square lattice of AlGaAs rods in air. AlGaAs seems to be a proper material since it has a large nonlinear refractive index, with minimal linear and nonlinear absorption in the 1550 nm telecommunications window [57]. The PC lattice constant is $a = 400$ nm, whereas the radius of the rods for the bulk crystal is $r = 130$ nm. The resulting structure has a wide bandgap for TM polarization in a range of wavelengths between 1400 and 1750 nm. Two waveguides are formed by introducing linear defects reducing the radius of the rods to $r_{dif} = 70$ nm [58].

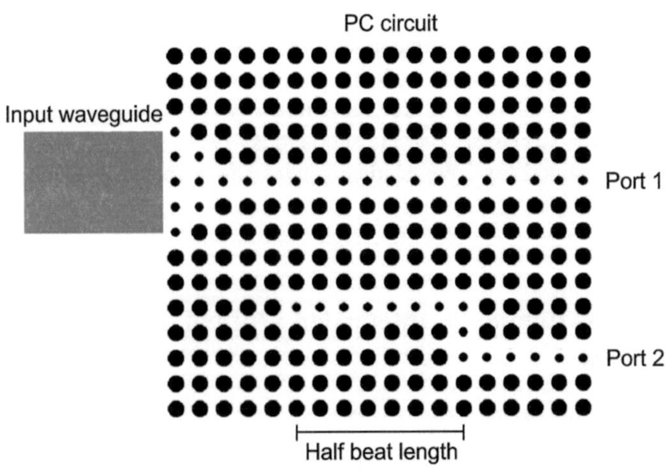

Fig. 3.9 Schematic view of the PC structure. $L_B = 140$ μm is the beat length of the coupler. It is worth to note that the real length is scaled to fit in the figure (Adapted from [56])

The reason for this choice is twofold. First, single-mode waveguides are essential to get a directional coupler, with only one even and one odd supermode. It is well known that reduced-index waveguides satisfy this requirement, whereas the increased-index ones tend to be multimode [58]. Second, the need to optimize the nonlinear interactions suggests to reduce only partially the radius of the defect rods, in order to maximize the semiconductor fill factor. We have calculated that in the previously described guiding structure over 50 % of the modal field energy is confined into the nonlinear dielectric defects [59]. The correct design of the coupler section is the key issue for our problem. We have found that the critical parameter to achieve switching with reasonable input intensity of the light is the distance between the two waveguides. Therefore fully vectorial eigenmodes of Maxwell's equations for a set of couplers were computed using a freely available mode solver [59], and 2D FDTD simulations [54, 55] were analyzed to study the device behavior when the waveguides separation is varied.

Moreover, a very simple and powerful coupled-mode theory was developed to model nonlinear propagation in PC couplers. As expected, there is a tradeoff between length of the structure and switching intensity. If the waveguides are close to each other the linear coupling is strong, thus it is very difficult to decouple them exploiting the effects of the ultrafast Kerr nonlinearity. If the waveguides are far away, the beat length of the coupler is very large, and the resulting device is not ultra-compact. We have then chosen a coupler composed of two waveguides separated by five lattice constants, as shown in Fig. 3.9. In Fig. 3.10 we report the projected band structure of the coupler, evaluated through the mode solver [59]. It is straightforward to calculate the beat length $L_B = 2\pi/(k_{even} - k_{odd})$ of the device, which is about 140 μm at the wavelength $\lambda = 1560\,\text{nm}$; k_{even} and k_{odd} are the effective wavenumbers of the even and the odd supermode. Light can be coupled from a dielectric slab waveguide (width W = 3 μm) to the coupler through a tapered input section [60], as shown in a schematic way in Fig. 3.9. More complex and

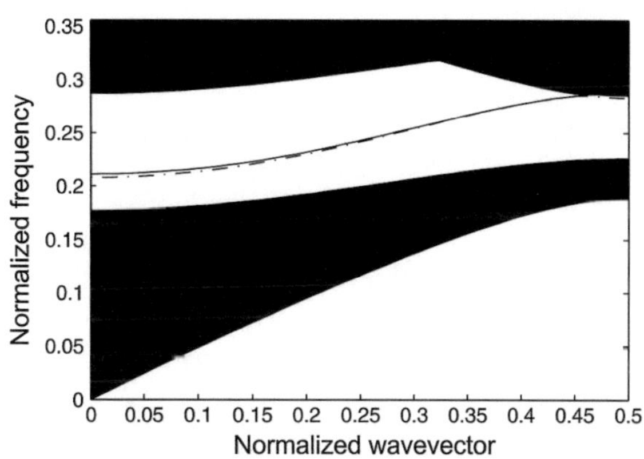

Fig. 3.10 Projected band structure of the coupler, with the normalized dispersion relations for the odd (*solid line*) and the even (*dash-dotted line*) supermode. Note that in our case we are working around $a/\lambda = 0.256$ (Adapted from [56])

efficient PC tapers have been proposed (e.g. in [61]), nevertheless a systematic analysis of the input section of the PC chip in order to optimize the coupling efficiency is well beyond the aim of this section. The switch layout is completed with the double sharp bend that decouples the two waveguides just in proximity of the half beat length of the coupler (see Fig. 3.9). The final structure, composed of the tapered input section, the coupler and the double sharp bend is about 75 μm long.

In order to show the basic operation principle of the proposed structure we exploit a coupled-mode theory [62–64], developed to study coupled defects in nonlinear PCs. In case of weak interactions between similar single-mode defects, the modal field of each defect can be considered unperturbed, so that only the field amplitudes vary in time. The evolution of the state of each PC waveguide, considered as a straight chain of resonators, is governed by a set of differential equations

$$i\frac{da_n}{dt} + C_n(a_{n+1} + a_{n-1}) + \gamma_n|a_n|^2 a_n = 0 \tag{3.23}$$

where a_n is the field amplitude in the n-th defect, C_n is the nearest-neighbor linear coupling coefficient and γ_n is the self-phase modulation strength [62–64]. This theory was proposed for the analysis of propagation in coupled-resonator optical waveguides (CROW). In our case, the individual defect rods composing the linear defect are strongly coupled, therefore the accuracy of the model could be questionable. Nevertheless, we show that this rough and simple theory can help to understand the behavior of the nonlinear PC coupler. We can introduce the coupling between the two waveguides by just defining two sets of equations,

$$i\frac{da_n}{dt} + C_n(a_{n+1} + a_{n-1}) + D_n \cdot b_n + \gamma_n|a_n|^2 a_n = 0 \tag{3.24}$$

$$i\frac{db_n}{dt} + C_n(b_{n+1} + b_{n-1}) + D_n \cdot a_n + \gamma_n|b_n|^2 b_n = 0 \tag{3.25}$$

where a_n and b_n are the field amplitudes in the n-th defect of the first and the second waveguide (called WG_1 and WG_2 respectively), and D_n is the linear coupling coefficient between the n-th defect in the first and in the second waveguide. We have solved numerically the system composed of (3.24) and (3.25) in the linear regime ($\gamma_n = 0$). The coupling coefficient C_n was fixed to 2 s^{-1} for every n, whereas $D_n = 0.035\ s^{-1}$ for $n < 170$ and $D_n = 0$ elsewhere, in order to simulate the introduction of the double sharp bend. Figure 3.11 shows that the coupled-defect theory is able to reproduce the behavior of the PC coupler. We report the calculated values of a_n and b_n for a CW excitation when a steady-state is reached. The input field is injected into WG_1, and the field flows toward WG_2 because of the linear coupling between the two waveguides. The sharp bend is in proximity of the half beat length of the coupler, thus at the output all the optical energy is in WG_2. In Fig. 3.12 we

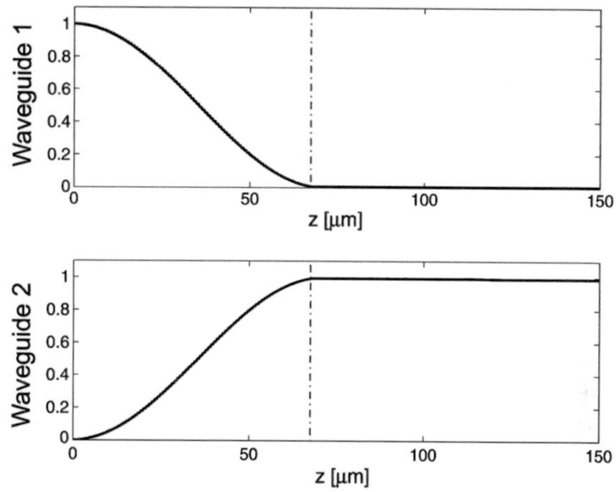

Fig. 3.11 Normalized intensity of the field in WG_1 (at the *top*) and WG_2 (on the *bottom*) in linear regime ($\gamma_n = 0$) calculated through the coupled defects model. The *sharp bend* is placed near the defect $n = 170$ (see the *dash-dotted line*) (Adapted from [56])

report the numerical solution of the system composed of (3.24) and (3.25) in the nonlinear regime ($\gamma_n |a_{in}|^2 = 0.75\ s^{-1}$). It is straightforward to note the effect of the nonlinearity: now at the output all the optical energy is in WG_2, which shows that it is possible to switch the output channel by varying the input field intensity.

Now we show a rigorous 2D FDTD analysis of the feasibility of all-optical switching in the previously described device. The problem of the implementation of the structure on a real PC slab is a big issue that would require huge 3D FDTD simulations, in order to estimate scattering losses in the third dimension. We highlight that the 2D numerical modeling of the real 3D device is commonly used in the literature for the study of phenomena due to Kerr effect in PC (see [48–50]), since this permits to focus the attention mainly on the nonlinear interactions. The nonlinear PC coupler is simulated by injecting into the taper section a CW Gaussian field that approximates the fundamental mode of the dielectric slab waveguide. It is worth to note that in this way we take into account the effects due to the coupling efficiency into the PC chip. By varying the maximum value of the Gaussian, and then the related maximum intensity of the input light, we may characterize the behavior of the power-controlled switch.

In Figs. 3.13, 3.14 and 3.15 we show the intensity of the electromagnetic field in the device for three different intensities of the input signal. Figure 3.13 shows the simulation result for a maximum input intensity $I_{IN_{\max}} = 1$ GW/cm^2. It is possible to see that all the optical energy flows through the output port 2, thus the coupler is in cross state. This behavior is in good agreement with the previously described design procedure, in fact the double sharp bend decouples the two waveguides just near the half beat length. In Fig. 3.14 we have increased the maximum input intensity to $I_{IN_{\max}} = 3.8$ GW/cm^2. Here it is clear the effect of the nonlinear phase shift: the input intensity is equally divided between the two output waveguides, and the structure behaves as a 50 % power splitter. Figure 3.15 shows the behavior of the device increasing $I_{IN_{\max}}$ to 5 GW/cm^2. In this case all the optical energy flows

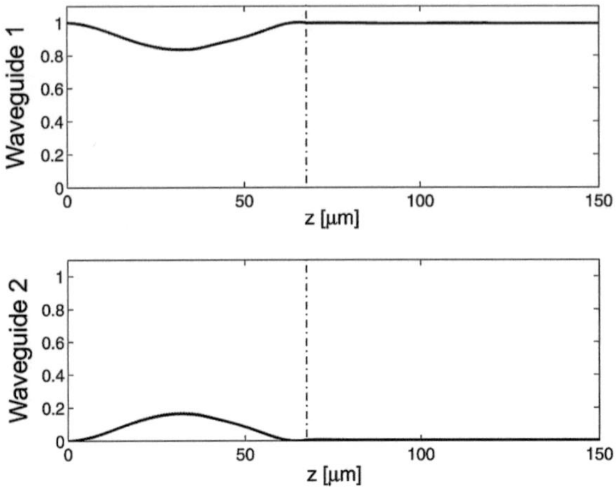

Fig. 3.12 Normalized intensity of the field in WG_1 (at the *top*) and WG_2 (on the *bottom*) in the nonlinear regime ($\gamma_n |a_{in}|^2 = 0.75\ s^{-1}$) calculated through the coupled defects model. The sharp bend is placed near the defect $n = 170$ (see the *dash-dotted line*) (Adapted from [56])

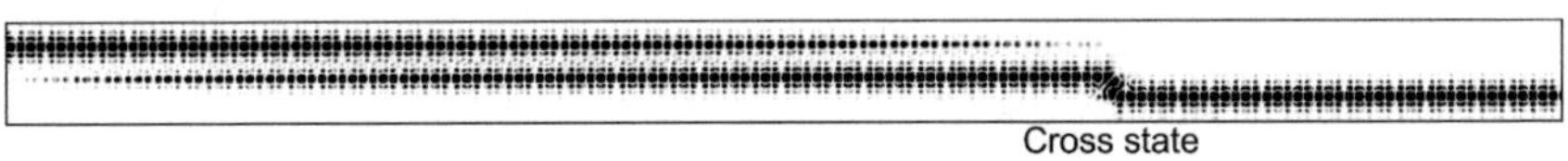

Fig. 3.13 Intensity of the field in the PC coupler with maximum input intensity 1 GW/cm^2: the coupler is in cross state, the power ratio is about 20 dB (Adapted from [56])

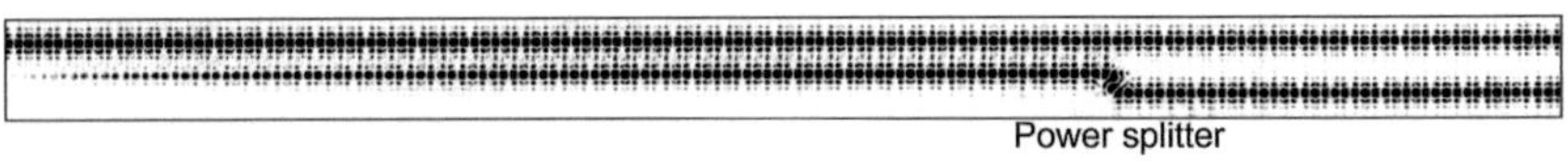

Fig. 3.14 Intensity of the field in the PC coupler with maximum input intensity 3.8 GW/cm^2: the coupler behaves as a 50 % power splitter, the power ratio is about 0 dB (Adapted from [56])

through the output port 1, thus the coupler is in a bar state induced by the nonlinearity. It is worth pointing out that the result of the FDTD simulation shown in Fig. 3.15 is in agreement with the solution of the coupled defects model (see Fig. 3.12). The field injected in WG_1 initially couples to WG_2 as in an asymmetric coupler, but at the double sharp bend position all the optical energy is in WG_1, as desired. Figure 3.16 summarizes the operation principle: a high-intensity signal propagates along the input waveguide and flows through the output port 1, whereas a low-intensity signal is switched toward output port 2.

In Fig. 3.17 we show the power ratio in decibel, i.e. the ratio between the optical energy in the output ports 2 and 1 of the structure, versus the maximum value of the input intensity at $\lambda = 1555$ nm, $\lambda = 1560$ nm and $\lambda = 1565$ nm, respectively. The simulations demonstrate that the device can be considered an optically-controlled

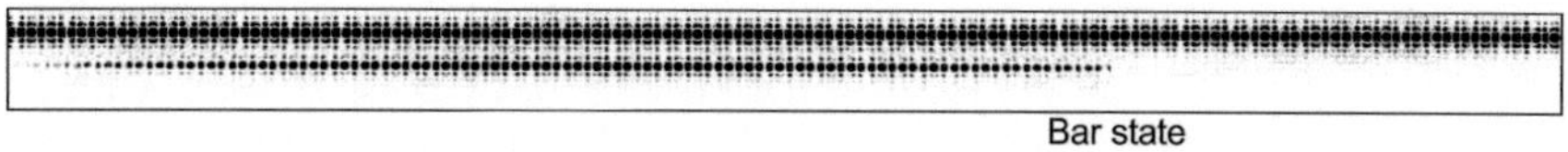

Fig. 3.15 Intensity of the field in the PC coupler with maximum input intensity 5 GW/cm^2: the coupler is in bar state, the power ratio is about -20 dB (Adapted from [56])

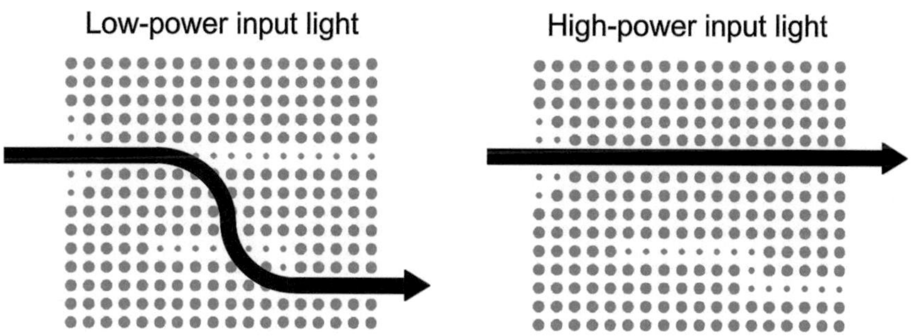

Fig. 3.16 The power-controlled switching function: a low-intensity signal is switched toward output port 2 (linear coupler), whereas a high-intensity signal propagates along the input waveguide because of the nonlinear phase shift (Adapted from [56])

Fig. 3.17 Power ratio versus maximum input intensity of the light at $\lambda = 1560$ nm (*solid line*), $\lambda = 1555$ nm (*dash-dotted line*) and $\lambda = 1565$ nm (*dashed line*) (Adapted from [56])

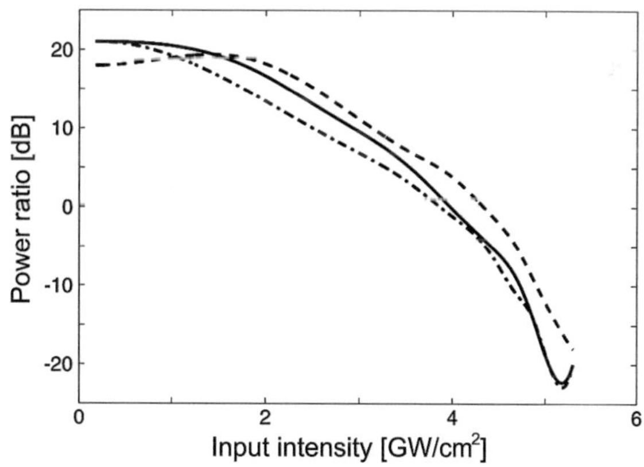

switch, with a power ratio larger than 20 dB in a wide range of wavelengths. In presence of pulsed excitations the performance of the power-controlled switch decreases with respect to the CW case. In particular, as has already been described in Sect. 3.2, with nonsquare input pulses the power ratio is reduced and output pulse break-up is observed [65, 66]. Nevertheless these phenomena are strongly dependent on the shape and the duration of the pulses, and they do not affect the validity of the proof-of-principle nonlinear PC coupler. It is interesting to compare the device behavior with what has already been reported in the literature for NLDCs made by AlGaAs semiconductor waveguides, and operating in the third-telecommunications window. In this second case, both simulations and experiments show switching powers from 50 to 90 W, for a few centimeters long devices [65, 66]. Assuming that the field at the input of the PC chip is a Gaussian beam, with the same spot-size $w_0 = 1$ μm we used in the 2D FDTD simulations, we can integrate the input intensity finding an approximated power for the total switching of about 70 W, which is comparable with the previously given values. Nevertheless, it is fundamental to note that the length of the nonlinear PC coupler is less than 80 μm, which is significantly shorter than a standard waveguide NLDC. The one-pulse scheme analyzed so far could be extended to a pump signal configuration, with a strong pump acting through cross-phase modulation (XPM) on a weak signal at different wavelength.

3.3.2 Graphene-Assisted Control of Coupling Between Surface Plasmon Polaritons

We discuss in this section the tuning of the coupling of surface plasmon polaritons between two graphene layers with nanometer spacing. We demonstrate that, by slightly changing the electrical doping and then shifting the chemical potential, a graphene coupler can switch from the bar to the cross state. As a consequence, the coupling coefficient in such structures can be easily controlled in an ultrafast fashion either by means of an applied electrical signal [67] or by changing the intensity of the signal at the device input. These findings open the way to fully exploit the huge nonlinearity of graphene for all optical signal processing: from one side giving more degrees of freedom to already proposed devices [68–72], from the other side paving the way to new devices.

Graphene can sustain surface plasmon polariton (SPP) having unique properties as compared to what we are used to with noble metals. In fact a single layer of graphene can support either TE or TM polarized plasmons without suffering from huge loss [73–75]; moreover, as far as the TM polarization is concerned, the extremely high confinement factor is particularly favourable to explore the huge $\chi^{(3)}$ nonlinearity of graphene [68, 76, 77]. Experimental endeavors have demonstrated the evidence of graphene plasmons by measuring the plasmon resonance of graphene nanoribbon arrays [78], and by acquiring their near field images [79, 80]. The coupling of SPP between separated graphene layers has been recently analyzed in [81]; however the very interesting properties arising from the easily tunable optical properties of graphene have not been exploited yet in this framework. Here in particular we show that by slightly changing the chemical potential, a graphene coupler can switch from the bar to the cross state.

In Fig. 3.18 we report the basic geometry that we are going to consider in this section; two graphene layers are embedded in a dielectric structure: region 1 (of width $2s$) is the dielectric in between the two graphene layers. At the graphene boundary we set the following conditions on the tangential components of the electromagnetic field:

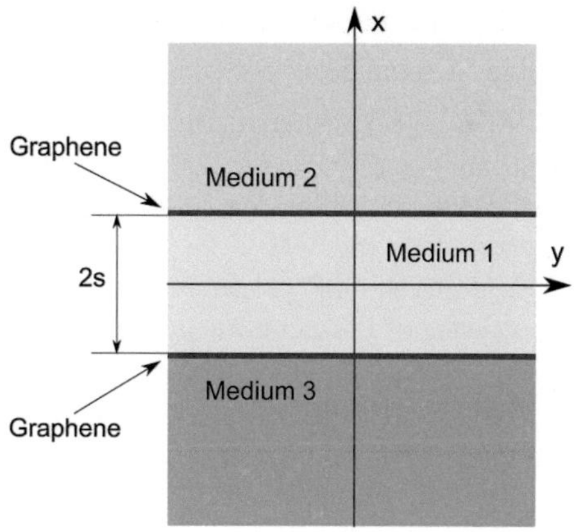

Fig. 3.18 Schematic of the graphene directional coupler: the separation between the layers is equal to $2s$ (Adapted from [67])

$$\left(\mathbf{E}_{2,3} - \mathbf{E}_1\right) \times \hat{x} = 0$$
$$\left(\mathbf{H}_{2,3} - \mathbf{H}_1\right) \times \hat{x} = \pm i\omega\varepsilon_0\varepsilon_{rS1-2,3}\mathbf{E}_{||}(x = \pm s) \quad (3.26)$$

where $E_{||}$ is the electric field tangent to the graphene layer and ε_{rS1-2} (ε_{rS1-3}) is the relative surface permittivity of the graphene layer between regions 1 and 2 (3). As far as the electromagnetic constants of graphene are concerned, we write the linear contribution to the relative complex permittivity as [82, 83]:

$$\varepsilon_{rC} = \frac{\varepsilon_{rS}}{d_g} = 1 + \frac{\sigma^{(1)}_{\Sigma,I}}{d_g\omega\varepsilon_0} - i\frac{\sigma^{(1)}_{\Sigma,R}}{d_g\omega\varepsilon_0} = \varepsilon_{rC,R} + i\varepsilon_{rC,I} \quad (3.27)$$

where d_g is the graphene thickness and the surface complex conductivity $\sigma^{(1)}_{\Sigma} = \sigma^{(1)}_{\Sigma,R} + i\sigma^{(1)}_{\Sigma,I}$ (in Siemens) is obtained from theoretical models now well established and experimentally validated [84, 85], which give the following dependence of the real and imaginary parts of the conductivity on frequency (ω), temperature (T) and chemical potential (μ):

$$\sigma^{(1)}_{\Sigma,\mathrm{R}}(\omega) \simeq \frac{\sigma_0}{2}\left(\tanh\frac{\hbar\omega + 2\mu}{4k_BT} + \tanh\frac{\hbar\omega - 2\mu}{4k_BT}\right)$$
$$\sigma^{(1)}_{\Sigma,\mathrm{I}}(\omega) \simeq \frac{\sigma_0}{\pi}\left[\frac{4}{\hbar\omega}\left(\mu - \frac{2\mu^3}{9t^2}\right) - \log\frac{\hbar\omega + 2\mu}{\hbar\omega - 2\mu}\right] \quad (3.28)$$

where $t = 2.7\mathrm{e}V$ is the hopping parameter, $\hbar$ and k_B are the reduced Planck's and Boltzmann's constants, respectively, and $\sigma_0 = e^2/(4\hbar) \simeq 6.0853 \times 10^{-5}$ S, with e the electron charge.

Note also that this model can be easily extended into the nonlinear regime by adding a nonlinear correction to the surface conductivity as: $\sigma_\Sigma = \sigma_\Sigma^{(1)} + \sigma_\Sigma^{(3)}|E|^2 E_{y,z}$ [74]. Moreover, thanks to the extremely small thickness of the graphene layer, nonlinearity can be analyzed by introducing a parameter embedded into the coefficients describing the continuity of the tangential components of the electromagnetic field [68, 77].

To describe SPP propagation along z, we first note that, at first order, the y dependence of the electromagnetic field can be neglected; we then look for guided modes with harmonic temporal dependence $\exp(i\omega t)$ and spatial variation $\mathbf{E}_{1,2,3}(x,z),\ \mathbf{H}_{1,2,3}(x,z) \sim \exp(-i\beta z \pm \Gamma_{1,2,3}x)$ with $= \Gamma_{1,2,3}^2 = \beta^2 - \varepsilon_{r1,2,3}k_0^2$. Obviously the complex wavenumber β, through its real and imaginary parts, describes the evolution of both the phase and the amplitude of the guided modes. We can apply the above modeling to derive the dispersion relation of both TE and TM modes. In the following we describe in details the TM polarization. We first consider a very general situation where the two graphene layers can be biased in a different way to give rise to an asymmetric coupler. After straightforward algebra we find that coupled SPP in the system are determined by setting to zero the determinant of the following matrix:

$$M = \begin{bmatrix} e^{\Gamma_1 s} & e^{-\Gamma_1 s} & -e^{-\Gamma_2 s} & 0 \\ e^{-\Gamma_1 s} & e^{\Gamma_1 s} & 0 & -e^{-\Gamma_3 s} \\ \frac{i\omega\varepsilon_1}{\Gamma_1}e^{\Gamma_1 s} & -\frac{i\omega\varepsilon_1}{\Gamma_1}e^{-\Gamma_1 s} & g_{1-2}e^{-\Gamma_2 s} & 0 \\ -\frac{i\omega\varepsilon_1}{\Gamma_1}e^{-\Gamma_1 s} & \frac{i\omega\varepsilon_1}{\Gamma_1}e^{\Gamma_1 s} & 0 & g_{1-3}e^{-\Gamma_3 s} \end{bmatrix} \tag{3.29}$$

where g_{1-2} and g_{1-3} take into account the contribution of the two graphene layers in the continuity conditions:

$$g_{1-2} = i\omega\varepsilon_0\varepsilon_{rS,1-2} + \frac{i\omega\varepsilon_2}{\Gamma_2}\ ,\ g_{1-3} = i\omega\varepsilon_0\varepsilon_{rS,1-3} + \frac{i\omega\varepsilon_3}{\Gamma_3}$$

where $\varepsilon_{rS,1-2}$ and $\varepsilon_{rS,1-3}$ refer to the relative dielectric constant of the two graphene layers, which in general may have different values due to different carriers concentrations. This asymmetric coupler offers a wide variety of possible settings which certainly deserve to be investigated both in the linear and in the nonlinear regime. Here we describe a prototype example illustrating the possibilities offered by the tunability of graphene parameters in this framework; we thus focus our attention on a very particular situation corresponding to a linear and symmetric case ($\varepsilon_2 = \varepsilon_3$ and $\varepsilon_{rS,1-2} = \varepsilon_{rS,1-3}$); moreover we use T = 300 K and $\lambda = 10$ μm. For the sake of simplicity, we also set $\varepsilon_{r1} = \varepsilon_{r2} = \varepsilon_{r3} = 2.25$. In this regime the graphene directional coupler has two different eigenstates: the even (odd) supermode corresponding to the out of phase (in phase) hybridization of the SPP guided by the single graphene layers. Note also that the even mode here has always the highest value of the propagation constant.

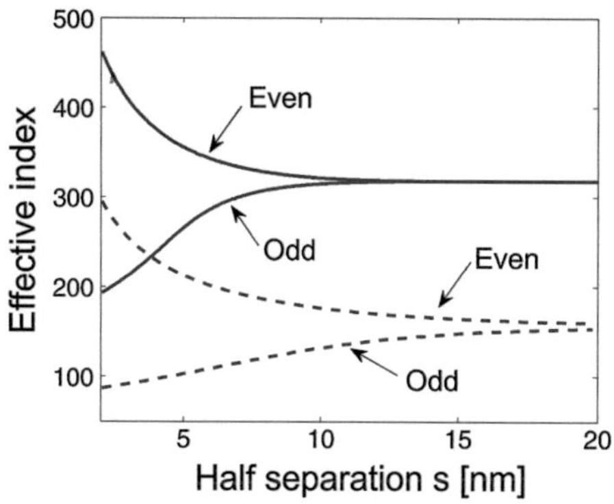

Fig. 3.19 Effective index $n_{eff} = \Re e(\beta)/k_0$ of even and odd supermodes of the coupled graphene layers as a function of the separation among the layers. Continuous (*dashed*) lines refer to a chemical potential of $\mu_1 = 0.1\,\text{eV}$ ($\mu_2 = 0.15\,\text{eV}$) (Adapted from [67])

In Fig. 3.19 we report the solution of the dispersion relation as a function of s for two different situations: continuous lines here refer to the even and odd supermodes corresponding to a chemical potential $\mu_1 = 0.1\,\text{eV}$ in (3.28), while the dashed lines refer to a choice of the chemical potential $\mu_2 = 0.15\,\text{eV}$ in (3.28). It is straightforward to note that, for large enough s, the two supermodes of the coupler tend to degeneracy, and their propagation constants approximate the propagation constant of the SPP of a single graphene layer. The main message we can read from Fig. 3.19 is that a very small change in the chemical potential can induce a very big change in the behaviour of the coupler. In the following, we focus our attention to the case with s = 5 nm. For this value of the separation between the layers, we computed the beat length of the directional coupler as a function of the chemical potential: the corresponding results are reported in Fig. 3.20. We can clearly see there that a very small change of the chemical potential can be used to induce huge changes of the beat length of the coupler. The two particular points (open square and open circle) enlightened in Fig. 3.20 are the initial conditions in Fig. 3.21, where we describe the propagation of the electromagnetic signal in the graphene coupler. In both panels in Fig. 3.21 total propagation length is set to $L \simeq 90$ nm. On the left panel in Fig. 3.21 the input condition corresponds to the square in Fig. 3.20 and the coupler is in the cross state; on the right panel the input condition corresponds to the circle in Fig. 3.20 and the coupler is in the bar state.

3.3.3 Graphene-Assisted Control of Coupling Between Optical Waveguides

In this subsection we demonstrate that, thanks to the ultrafast tunability of losses which are introduced by graphene layers deposited onto the structures, a careful design of silicon on insulator ridge waveguides can be used to explore the so-called

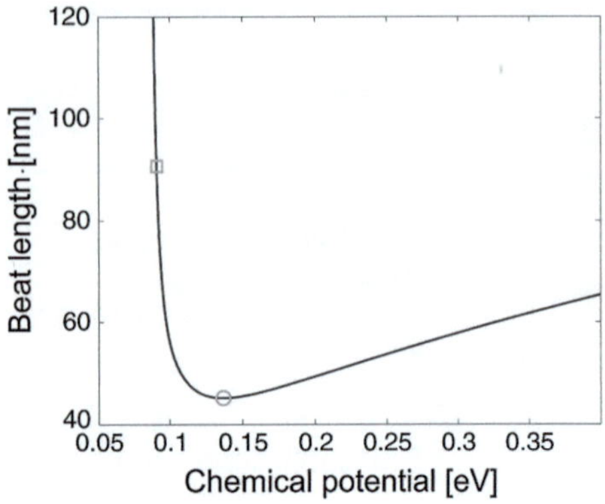

Fig. 3.20 Beat length versus chemical potential for a graphene plasmon coupler. Here 2*s* = 10 nm (Adapted from [67])

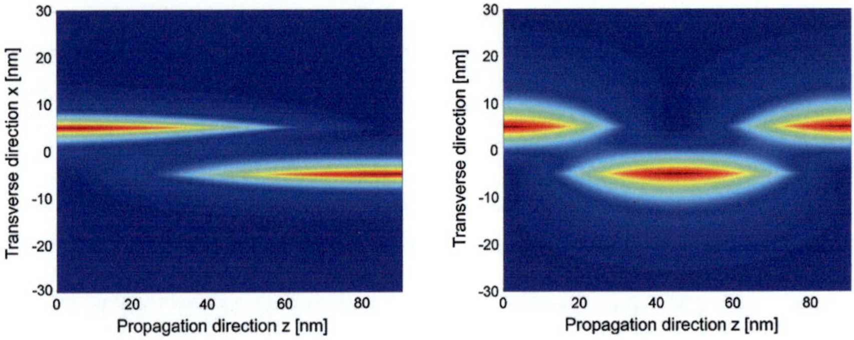

Fig. 3.21 Field evolution in a graphene plasmon directional coupler: *left* (*right*) refers to a chemical potential of $\mu_1 = 0.0908\,\text{eV}$ ($\mu_2 = 0.1367\,\text{eV}$). Here 2*s* = 10 nm (Adapted from [67])

passive parity-time (PT) symmetry breaking in directional couplers. A quantum system characterized by a Hamiltonian H is PT-symmetric if H commutes with the operator PT, where P is the parity operator and T is the temporal operator [84]. At an exceptional point, two or more eigenvalues are degenerate. We prove that the exceptional point of the coupler can be probed by varying the applied voltage, which may lead to very compact photonic structures for the control of coupling among waveguides, and for tailoring discrete diffraction in arrays [86].

In particular, in this section we present numerical results which demonstrate the huge potential of graphene as a means to control coupling between optical dielectric waveguides. The possibility of tuning losses in each waveguide by acting on a thin loss element permits to break the symmetry of the coupled waveguides without introducing a strong perturbation in each single waveguide. Remarkably, we demonstrate that tunable losses induced by graphene and a careful design of ridge waveguides allows to probe passive PT-symmetry breaking in directional couplers [87–89]. Moreover, the exceptional point of the coupler can be dynamically controlled by varying the applied voltage. We will thus explore these properties to mould energy exchange between waveguides and to finely tune discrete diffraction in waveguide arrays. These results, together with the strong saturable absorption of graphene [76, 90], suggest also the possible use of this configuration in nonlinear devices with a strong pump beam used as a probe for all optical switching of weak input signals from the bar to cross state.

In order to prove our statement we first need to consider the behavior of silicon waveguides on a silica substrate in a wavelength range between 1350 and 1600 nm. The structure has been inspired by the modulator proposed in [91]. In particular, a layer of silicon with thickness equal to 50 nm is deposited onto the substrate. The 400 nm wide ridge waveguide is composed of lower and higher layers made of silicon (both with thickness 200 nm) which sandwich a central region including three alternating layers of alumina (thickness 7 nm) and two absorption layers composed of three graphene monolayers with thickness 0.34 nm. Graphene can be electrically controlled in order to tune doping (and then conductivity), as suggested in [91–98]. The dielectric constants of silicon, silica and alumina were taken equal to 12.1, 2.1 and 3. Figure 3.22a displays a schematic view of the structure.

The behavior of graphene in the optical regime has been numerically modeled by following the approach suggested in [82, 83], as already described in the previous subsection. Indeed, we assigned to each graphene monolayer with thickness Δ a volume conductivity equal to $\sigma_{g,v} = \sigma_{\Sigma}^{(1)}/\Delta$, where $\sigma_{\Sigma}^{(1)}$ is the conductivity of the 2D sheet (see (3.27) and (3.28)). It was demonstrated that, as a first approximation, few-layer graphene is characterized by the same band structure (and then by the same excellent electronic properties) of the monolayer. Moreover, if N is small enough the conductivity of N-layer graphene ($N = 3$ in our design) can be evaluated as N times conductivity of the single layer [93].

We then performed a modal analysis of the waveguide in Fig. 3.22a by resorting to finite-element simulations. We focused the attention on the TE-like mode which is depicted in the inset of Fig. 3.22b, since the electric field is tangential with

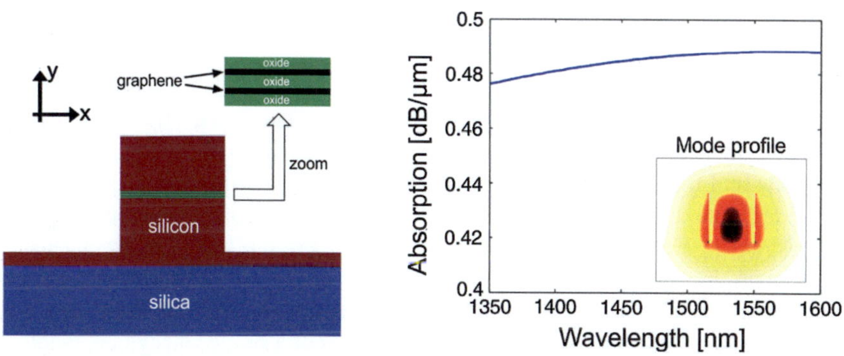

Fig. 3.22 *Left* Schematic view of the waveguide structure, with a detail of the central region with graphene layers. *Right* Losses of the single waveguide (in dB/μm) when graphene is in OFF state (null voltage), and x-component of the electric field of the TE-like mode (*inset*) (Adapted from [86])

respect to graphene layers. In the same figure we report the absorption of the TE-like mode when the graphene layers are in OFF state (null voltage). Of course, when graphene is in the ON state (a control voltage is applied), losses are close to zero. First, it is important to note that the real part of the effective index of the mode (not shown here) is barely affected by state of the graphene layers. Then, we emphasize that when graphene is in OFF state losses are quite large (0.48 dB/μm, which corresponds to $1100\,\mathrm{cm}^{-1}$) and almost constant over the entire bandwidth: indeed, a 6 dB modulation contrast between ON and OFF states can be achieved with a 12.5 μm long waveguide. In the next paragraphs we will study the properties of coupled waveguides wherein the described structure is the basic building block.

Full-wave simulations of photonic devices including graphene layers are characterized by huge computational burden [99]. This effect is obviously emphasized when structures composed of multiple waveguides must be analyzed. Therefore conventional coupled-mode theory (CMT) [100, 101] has been reformulated to study the present structures. Using full wave simulations, we first numerically proved that neither the profile of TE-like modes, nor the propagation constant β of each isolated waveguide are affected by the status of graphene layers. Whereas switching between ON and OFF states has the effect of turning off and on losses in the single waveguide, which are modeled by the attenuation constant α. Notice that modal evolution reads as $\exp(i\beta z)\exp(-\alpha z)$.

Under these conditions, it is possible to verify that the system of governing equations for $A_{1,2}$, which are the modal field amplitudes in the first and second waveguide of a directional coupler composed of two identical graphene-based waveguides, can be approximated as

$$\frac{d}{dz}\begin{bmatrix} A_1 \\ A_2 \end{bmatrix} = i\begin{bmatrix} \beta + i\alpha_1 & C \\ C^* & \beta + i\alpha_2 \end{bmatrix}\begin{bmatrix} A_1 \\ A_2 \end{bmatrix}, \tag{3.30}$$

where $\alpha_{1,2}$ can be tuned between zero (ON state) and α_{max} (OFF state) by controlling the voltage applied to the graphene layers, and C is a complex coupling coefficient [88].

When $\alpha_1 = 0$ and $\alpha_2 = \alpha$, the eigenvalues of the matrix in (3.29) read as $\lambda_{1,2} = \beta + i(\alpha/2) \pm \sqrt{|C|^2 - (\alpha/2)^2}$. Therefore the so-called exceptional point (EP) for the onset of PT-symmetry breaking is located at the critical loss value $\alpha_c = 2C$. Indeed, when $\alpha < \alpha_c$ the two supermodes have different propagation constants and the same attenuation constant $\alpha/2$. Beyond the critical loss the supermodes coalesce: they are characterized by the same propagation constant β and by different loss coefficients. In particular, one supermode experiences increasing losses with increasing α, whereas the other supermode is characterized by the opposite trend [87, 88]. Whenever α is much larger than C, one supermode is characterized by losses which are close to the losses of a single waveguide, whereas the other supermode is nearly loss-free.

At this point we may examine the behavior of the directional coupler which is obtained by placing two identical graphene-based waveguides close to each other,

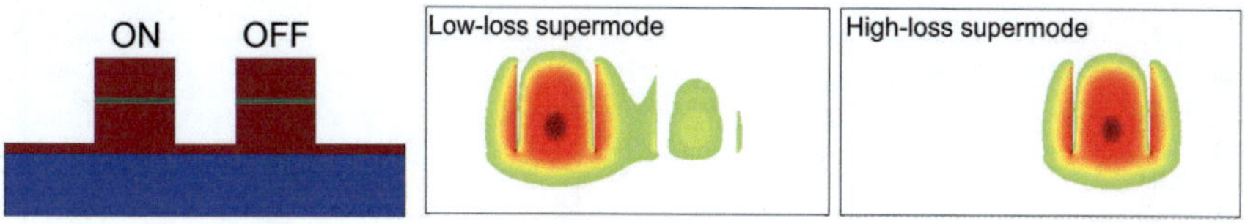

Fig. 3.23 Schematic view of the 300-nm-gap coupler (*left*), with the electric field of the low- (*center*) and high-loss mode (*right*) at 1530 nm. Graphene layers are in ON–OFF states (Adapted from [86])

separated by a 300 nm gap (see Fig. 3.23 on the left). We performed a modal analysis at 1530 nm focusing the attention on the supermodes which originate from the interaction between TE-like modes of the single waveguides, and we varied the state of the graphene layers. Numerical results in Fig. 3.23 illustrate in a qualitative way how the behavior of the structure can be controlled by exploiting the properties of graphene. In particular, when graphene layers in both waveguides are in the same state, the symmetry of the structure is preserved, and the two modes (not shown here) are even and odd. Viceversa, Fig. 3.23 shows that when symmetry is broken by switching to the ON–OFF state, the effect of losses on the modal properties is huge, and a trend toward decoupling between the waveguides is apparent.

The dispersive properties of the two modes have been characterized through full-wave and CMT simulations, and these quantitative results confirm the intuitive analysis we have reported above. Indeed, in Fig. 3.24a losses of the two supermodes are depicted when the graphene layers are in the ON–OFF state. In this case symmetry is broken: as a consequence, one mode is characterized by absorption which is close to zero, whereas the other mode experiences large losses, which are very close to those of a single lossy waveguide. It is worth noting that this effect tends to blur with increasing wavelength due to the dependence of coupling coefficient on frequency (C gets larger with increasing wavelength). A thorough treatment on phenomena arising from the wavelength dependence of the PT-symmetry condition is reported in [89]. The noticeable agreement in Fig. 3.24a between simulations performed by using a full-wave mode solver and the results evaluated by using CMT (in the latter case the imaginary part of $\lambda_{1,2}$ is reported) allows to confirm the accuracy of CMT.

These phenomena stem from the breaking of passive PT-symmetry in complex potentials. Indeed, in Fig. 3.24b we plot the attenuation constants of the two supermodes, evaluated by using CMT, as a function of the attenuation constant of the single waveguide α. Data are normalized with respect to twice the coupling coefficient, so that we have the exceptional point when the abscissa is equal to 1. The vertical dotted line indicates α_{max}, i.e. the value of α when our structure is in the OFF state: it is straightforward to see that one may operate beyond the exceptional point, in agreement with the results of Fig. 3.24a. It is worth to emphasize that graphene-based waveguides exhibit superior properties with respect to waveguides wherein losses are introduced by depositing metal layers [88]. Losses induced by sandwiching graphene layers inside silicon waveguides can be orders of magnitude larger (thousands of cm^{-1} with respect to tens of cm^{-1}), therefore it is possible to

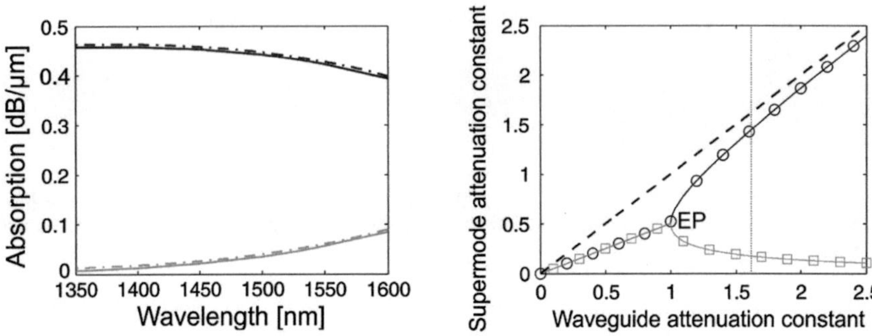

Fig. 3.24 *Left* Losses of low- (*red line*) and high-loss mode (*blue line*) from mode solver (*solid line*) and CMT (*dashed-dotted line*). *Right* Normalized attenuation constant of low- (*red line* with *squares*) and high-loss mode (*blue line* with *circles*) versus normalized attenuation constant of the single waveguide at $\lambda = 1530\,\text{nm}$ (Adapted from [86]). The *vertical thin line* indicates $\alpha = \alpha_{max}$

probe the exceptional point even in structures characterized by strong coupling. Last, but not least, it is important to note that graphene is electrically tunable, therefore losses in each single waveguide can be varied between zero (ON state) and a maximum value α_{max}, which is only determined by geometry (OFF state).

We envisage that switching of the state of one waveguide can be exploited to finely tune coupling between waveguides. In order to verify the effectiveness of this approach we applied CMT to our reference structure at the wavelength of 1530 nm, and we show the results in Fig. 3.25. When the coupler is in the ON–ON state losses are zero, and the predicted beat length $L_B = \pi/(\beta_{even} - \beta_{odd})$ is around 80 µm. Viceversa, when graphene layers are ON and OFF in the input and output channels, the two waveguides tend to decouple and the field intensity in the first waveguide is larger than in the second one. It is possible to justify this behavior by recalling that when we inject light into the waveguide in ON state the low-loss supermode is mainly excited.

These results have been validated by comparison with simulations of the 80 µm long coupler performed by using the commercial software CST Microwave Studio, which allows to solve Maxwell's equations in the time domain through the finite-integration technique. Indeed, the ratio between output and injected power evaluated by using CMT is -3 and $-12\,\text{dB}$ if the coupler is in the ON–OFF state and we consider as output port waveguides 1 and 2. CST simulations exhibit a good agreement, in fact the corresponding calculated values are about -5 and $-13\,\text{dB}$, respectively.

The unique properties we have described in the previous paragraph open the way to novel possibilities for controlling discrete diffraction in waveguide arrays [100–105]. Let us take for example an array composed of eleven identical waveguides, with the same geometrical and optical parameters that we have used throughout the chapter, and the same spacing (300 nm) that we considered for the coupler. The structure was simulated by using CMT in order to reduce the computational burden. The input excitation covers only the central waveguide, and the

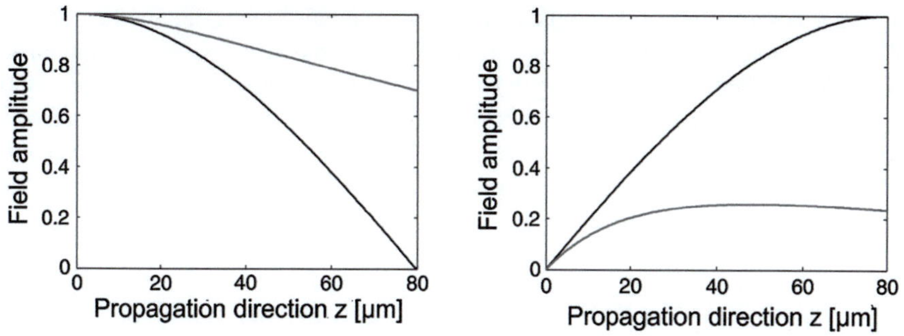

Fig. 3.25 Field amplitude in (*left*) first and (*right*) second waveguide of the coupler (Adapted from [86]). Graphene layers are in ON–ON (*black line*), and ON–OFF (*red line*) states

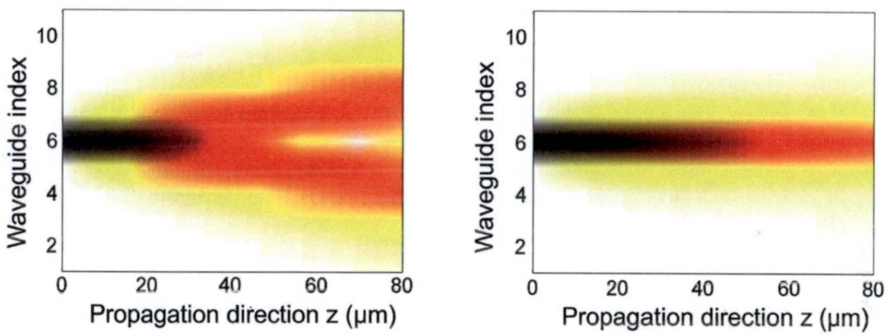

Fig. 3.26 Discrete diffraction along the array. *Left* All the graphene layers are in ON state. *Right* Only graphene layers inside the central waveguide are in ON state (Adapted from [86])

propagation length was taken equal to the beat length of the coupler (80 μm). Moreover, we assumed that the state of the graphene layers in each waveguide can be controlled independently from each other.

In Fig. 3.26a we show the field inside the structure when all the graphene layers are in the ON state: the typical pattern of discrete diffraction is clearly visible, with two pronounced outermost wings [100, 101]. In Fig. 3.26b all the waveguides except for the central one are switched to the OFF state, and two phenomena can be clearly noticed. First, beam broadening is reduced with respect to the previous case, so that most of the optical energy remains concentrated into the central waveguide. Second, losses are smaller with respect to the case of a single lossy waveguide.

3.4 Conclusions

Optical switching will be a key enabling functionality in future transparent all-optical networks. In this chapter we have provided an overview of several guided wave optical switching devices, where the input–output path of optical signals is controlled by either optical or electrical signals, thus avoiding the need for OEO conversion for optical signal processing. We have first presented the basic principles of fiber optics switching devices, whose long interaction lengths permit to significantly reduce the operating power requirements. Next we have discussed nonlinear couplers based on integrated waveguides with strong field confinement, hence reduced device dimensions, thanks to photonic crystal structures or surface wave interactions in graphene layers.

Acknowledgments This work was funded by Fondazione Cariplo (grants no. 2011-0395 and no. 2013-0736), the Italian Ministry of University and Research (grant no. 2012BFNWZ2), and the US Army (grants no. W911NF-12-1-0590 and no. W911NF-13-1-0466).

References

1. A. Yariv, Coupled-mode theory for guided-wave optics. IEEE J. Quantum Electron. **9**, 919–933 (1973)
2. S.M. Jensen, The nonlinear coherent coupler. IEEE J. Quantum Electron. **18**, 1580–1583 (1982)
3. A.A. Maier, Optical transistors and bistable devices utilizing nonlinear transmission of light in systems with undirectional coupled waves. Sov. J. Quantum Electron. **12**, 1490–1494 (1982)
4. G.I. Stegeman, E.M. Wright, All-optical waveguide switching. Opt. Quant. Electron. **22**, 95–122 (1990)
5. H.G. Winful, Self-induced polarization changes in birefringent optical fibers. Appl. Phys. Lett. **47**, 213–215 (1985)
6. B. Daino, G. Gregori, S. Wabnitz, New all-optical devices based on third-order nonlinearity of birefringent fibers. Opt. Lett. **11**, 42–44 (1986)
7. S. Trillo, S. Wabnitz, Nonlinear dynamics and instabilities of coupled waves and solitons in optical fibers, in *Anisotropic and Nonlinear Optical Waveguides*, ed. by C.S. Someda, G.I. Stegeman (Elsevier, Amsterdam, 1992), pp. 185–236
8. B. Daino, G. Gregori, S. Wabnitz, Stability analysis of nonlinear coherent coupling. J. Appl. Phys. **58**, 4512–4514 (1985)
9. H.G. Winful, Polarization instabilities in birefringent nonlinear media: application to fiber-optics devices. Opt. Lett. **11**, 33–35 (1986)
10. S. Wabnitz, E.M. Wright, C.T. Seaton, G.I. Stegeman, Instabilities and all-optical phase-controlled switching in a nonlinear directional coherent coupler. Appl. Phys. Lett. **49**, 838–840 (1986)
11. D.D. Gusovskii, E.M. Dianov, A.A. Maier, V.B. Neustruev, E.I. Shklovskii, I.A. Shcherbakov, Nonlinear light transfer in tunnel-coupled optical waveguides. Sov. J. Quantum Electron. **15**, 1523–1526 (1985)
12. D.D. Gusovskii, E.M. Dianov, A.A. Maier, V.B. Neustruev, V.V. Osiko, A.M. Prokhorov, KYu. Sitarskii, I.A. Shcherbakov, Experimental observation of the self-switching of radiation in tunnel-coupled optical waveguides. Sov. J. Quantum Electron. **17**, 724–727 (1987)
13. S.R. Friberg, Y. Silberberg, M.K. Oliver, M.J. Andrejco, M.A. Saifi, P.W. Smith, Ultrafast all-optical switching in a dual-core fiber nonlinear coupler. Appl. Phys. Lett. **51**, 1135–1137 (1987)
14. A.M. Weiner, Y. Silberberg, S.R. Friberg, B.G. Sfez, P.W. Smith, Femtosecond all-optical switching in a dual-core fiber nonlinear coupler. Opt. Lett. **13**, 904–906 (1988)

15. A.M. Weiner, Y. Silberberg, H. Fouckhardt, D.E. Leaird, M.A. Saifi, M.J. Andrejco, P.W. Smith, Use of femtosecond square pulses to avoid pulse breakup in all-optical switching. IEEE J. Quantum Electron. **25**, 2648–2655 (1989)
16. S. Trillo, S. Wabnitz, R.H. Stolen, G. Assanto, C.T. Seaton, G.I. Stegeman, Experimental observation of polarization instability in a birefringent optical fiber. Appl. Phys. Lett. **49**, 1224–1226 (1986)
17. S. Trillo, S. Wabnitz, W.C. Banyai, N. Finlayson, C.T. Seaton, G.I. Stegeman, R.H. Stolen, Picosecond nonlinear polarization switching with a fiber filter. Appl. Phys. Lett. **53**, 837–839 (1988)
18. S. Trillo, S. Wabnitz, W.C. Banyai, N. Finlayson, C.T. Seaton, G.I. Stegeman, R.H. Stolen, Observation of ultrafast nonlinear polarization switching induced by polarization instability in a birefringent fiber rocking filter. IEEE J. Quantum Electron. **25**, 104–112 (1989)
19. P. Ferro, M. Haelterman, S. Trillo, S. Wabnitz, B. Daino, Polarization switching in spun birefringent fiber. Appl. Phys. Lett. **59**, 2082–2084 (1991)
20. P. Ferro, M. Haelterman, S. Trillo, S. Wabnitz, B. Daino, All-optical polarization switch with a long low-birefringence fiber. Electron. Lett. **27**, 1407–1408 (1991)
21. P. Ferro, S. Trillo, S. Wabnitz, All-optical polarization differential amplification with a birefringent fiber. Electron. Lett. **30**, 1616–1617 (1994)
22. P. Ferro, S. Trillo, S. Wabnitz, Phase control of a nonlinear coherent coupler: the multibeatlength twisted birefringent fiber. Appl. Phys. Lett. **64**, 2872–2874 (1994)
23. P. Ferro, S. Trillo, S. Wabnitz, Demonstration of nonlinear nonreciprocity and logic operations with a twisted birefringent optical fiber. Opt. Lett. **19**, 263–265 (1994)
24. N.J. Doran, D. Wood, Soliton processing element for all-optical switching and logic. J. Opt. Soc. Am. B **4**, 1843–1846 (1987)
25. N. Imoto, S. Watkins, Y. Sasaki, A nonlinear optical-fiber interferometer for nondemolitional measurement of photon number. Opt. Commun. **61**, 159–163 (1987)
26. B.K. Nayar, N. Finlayson, N.J. Doran, S.T. Davey, W.L. Williams, J.W. Arkwright, All-optical switching in a 200-m twin-core fiber nonlinear Mach-Zehnder interferometer. Opt. Lett. **16**, 408–410 (1991)
27. M. Asobe, Effects of group-velocity dispersion in all-optical switching devices using highly nonlinear optical waveguides. J. Opt. Soc. Am. B **12**, 1287–1299 (1995)
28. N.J. Doran, D.S. Forrester, B.K. Nayar, Experimental investigation of all-optical switching in a fibre loop mirror device. Electron. Lett. **25**, 267–268 (1989)
29. K. Otsuka, Nonlinear antiresonant ring interferometer. Opt. Lett. **8**, 471–473 (1983)
30. D.B. Mortimore, Fiber loop reflectors. J. Lightwave Technol. **6**, 1217–1224 (1988)
31. N.J. Doran, D. Wood, Nonlinear optical loop mirror. Opt. Lett. **13**, 56–58 (1988)
32. K.J. Blow, N.J. Doran, B.K. Nayar, Experimental demonstration of optical soliton switching in an all-fiber nonlinear Sagnac interferometer. Opt. Lett. **14**, 754–756 (1989)
33. M.N. Islam, E.R. Sunderman, R.H. Stolen, W. Pleibel, J.R. Simpson, Soliton switching in a fiber nonlinear loop mirror. Opt. Lett. **14**, 811–813 (1989)
34. M.C. Farries, D.N. Payne, Optical fiber switch employing a Sagnac interferometer. Appl. Phys. Lett. **55**, 25–26 (1989)
35. K.J. Blow, N.J. Doran, B.K. Nayar, B.P. Nelson, Pulse shaping, compression, and pedestal suppression employing a nonlinear-optical loop mirror. Opt. Lett. **15**, 248–250 (1990)
36. M. Jinno, T. Matsumoto, Ultrafast, low power, and highly stable all-optical switching in an all polarization maintaining fiber Sagnac interferometer. IEEE Photon. Technol. Lett. **2**, 349–351 (1990)
37. M. Jinno, T. Matsumoto, Ultrafast all-optical logic operations in a nonlinear Sagnac interferometer with two control beams Opt. Lett. **16**, 220–222 (1991)
38. L.F. Stokes, M. Chodorow, H.J. Shaw, All-single-mode fiber resonator. Opt. Lett. **7**, 288–290 (1982)
39. H. Nakatsuka, S. Asaka, H. Itoh, K. Ikeda, M. Matsuoka, Observation of bifurcation to chaos in an all-optical bistable system. Phys. Rev. Lett. **50**, 109–112 (1983)

40. B. Crosignani, B. Daino, P. Di Porto, S. Wabnitz, Optical multistability in a fiber-optic passive-loop resonator. Opt. Commun. **59**, 309–312 (1986)
41. M. Haelterman, S. Trillo, S. Wabnitz, Dissipative modulation instability in a nonlinear dispersive ring cavity. Opt. Commun. **91**, 401–407 (1992)
42. S. Coen, M. Haelterman, P. Emplit, L. Delage, L.M. Simohamed, F. Reynaud, Experimental investigation of the dynamics of a stabilized nonlinear fiber ring resonator. J. Opt. Soc. Am. B **15**, 2283–2293 (1998)
43. S. Trillo, S. Wabnitz, E.M. Wright, G.I. Stegeman, Soliton switching in fiber nonlinear directional couplers. Opt. Lett. **13**, 672–674 (1988)
44. S. Trillo, S. Wabnitz, Weak-pulse-activated soliton switching in nonlinear couplers. Opt. Lett. **16**, 1–3 (1991)
45. M. Romagnoli, S. Trillo, S. Wabnitz, Soliton switching in nonlinear couplers. Opt. Quant. Electron. **24**, S1237–S1267 (1992)
46. J.D. Joannopoulos, R.D. Meade, J.N. Winn, *Photonic Crystals: Molding the Flow of Light* (Princeton University Press, Princeton, 1995)
47. J.D. Joannopoulos, P.R. Villeneuve, S. Fan, Photonic crystals: putting a new twist on light. Nature **386**, 143–149 (1997)
48. M. Soljacic, M. Ibanescu, S.G. Johnson, Y. Fink, J.D. Joannopoulos, Optimal bistable switching in nonlinear photonic crystals. Phys. Rev. E **66**, 055601 (2002)
49. S.F. Mingaleev, Y.S. Kivshar, Nonlinear transmission and light localization in photonic-crystal waveguides. J. Opt. Soc. Am. B **19**, 2241–2249 (2002)
50. M. Soljacic, C. Luo, J.D. Joannopoulos, S. Fan, Nonlinear photonic crystal microdevices for optical integration. Opt. Lett. **28**, 637–639 (2003)
51. S. Boscolo, M. Midrio, C.G. Someda, Coupling and decoupling of electromagnetic waves in parallel 2-D photonic crystal waveguides. IEEE J. Quantum Electron. **38**, 47–53 (2002)
52. A. Martinez, F. Cuesta, J. Marti, Ultrashort 2-D photonic crystal directional couplers. IEEE Photon. Technol. Lett. **15**, 694–696 (2003)
53. M. Thorhauge, L.H. Frandsen, P.I. Borel, Efficient photonic crystal directional couplers. Opt. Lett. **28**, 1525–1527 (2003)
54. A. Taflove, S.C. Hagness, *Computational Electrodynamics: The Finite-Difference Time-Domain Method* (Artech House, Norwood, 2000)
55. R.M. Joseph, A. Taflove, FDTD Maxwell's equations models for nonlinear electrodynamics and optics. IEEE Trans. Antennas Propag. **45**, 364–374 (1997)
56. A. Locatelli, D. Modotto, D. Paloschi, C. De Angelis, All optical switching in ultrashort photonic crystal couplers. Opt. Commun. **237**, 97–102 (2004)
57. J.S. Aitchison, D.C. Hutchings, J.U. Kang, G.I. Stegeman, A. Villeneuve, The nonlinear optical properties of AlGaAs at the half band gap. IEEE J. Quantum Electron. **33**, 341–348 (1997)
58. S.G. Johnson, P.R. Villeneuve, S. Fan, J.D. Joannopoulos, Linear waveguides in photonic-crystal slabs. Phys. Rev. B **62**, 8212–8222 (2000)
59. S.G. Johnson, J.D. Joannopoulos, Block-iterative frequency-domain methods for Maxwell's equations in a planewave basis. Opt. Express **8**, 173–190 (2001)
60. T.D. Happ, M. Kamp, A. Forchel, Photonic crystal tapers for ultracompact mode conversion. Opt. Lett. **26**, 1102–1104 (2001)
61. P. Bienstman, S. Assefa, S.G. Johnson, J.D. Joannopoulos, G.S. Petrich, L.A. Kolodziejski, Taper structures for coupling into photonic crystal slab waveguides. J. Opt. Soc. Am. B **20**, 1817–1821 (2003)
62. A.L. Reynolds, U. Peschel, F. Lederer, P.J. Roberts, T.F. Krauss, P.J.I. de Maagt, Coupled defects in photonic crystals. IEEE Trans. Microwave Theory Tech. **49**, 1860–1867 (2001)
63. U. Peschel, A.L. Reynolds, B. Arredondo, F. Lederer, P.J. Roberts, T.F. Krauss, P.J.I. de Maagt, Transmission and reflection analysis of functional coupled cavity components. IEEE J. Quantum Electron. **38**, 830–836 (2002)
64. D.N. Christodoulides, N.K. Efremidis, Discrete temporal solitons along a chain of nonlinear coupled microcavities embedded in photonic crystals. Opt. Lett. **27**, 568–570 (2002)

65. K. Al-hemyari, A. Villeneuve, J.U. Kang, J.S. Aitchison, C.N. Ironside, G.I. Stegeman, Ultrafast all-optical switching in GaAlAs directional couplers at 1.55 μm without multiphoton absorption. Appl. Phys. Lett. **63**, 3562–3564 (1993)
66. J.S. Aitchison, A. Villeneuve, G.I. Stegeman, All-optical switching in two cascaded nonlinear directional couplers. Opt. Lett. **20**, 698–700 (1995)
67. A. Auditore, C. De Angelis, A. Locatelli, A.B. Aceves, Tuning of surface plasmon polaritons beat length in graphene directional couplers. Opt. Lett. **38**, 4228–4231 (2013)
68. A. Auditore, C. De Angelis, A. Locatelli, S. Boscolo, M. Midrio, M. Romagnoli, A.-D. Capobianco, G. Nalesso, Graphene sustained nonlinear modes in dielectric waveguides. Opt. Lett. **38**, 631–633 (2013)
69. D.A. Smirnova, A.V. Gorbach, I.V. Iorsh, I.V. Shadrivov, Y.S. Kivshar, Nonlinear switching with a graphene coupler. Phys. Rev. B **88**, 045433 (2013)
70. P.I. Buslaev, I.V. Iorsh, I.V. Shadrivov, P.A. Belov, Y.S. Kivshar, Plasmons in waveguide structures formed by two graphene layers. JETP Lett. **97**, 535–539 (2013)
71. A. Locatelli, A.-D. Capobianco, G. Nalesso, S. Boscolo, M. Midrio, C. De Angelis, Graphene based electro-optical control of the beat length of dielectric couplers. Opt. Commun. **318**, 175–179 (2014)
72. Y.V. Bludov, D.A. Smirnova, YuS Kivshar, N.M.R. Peres, M.I. Vasilevskiy, Nonlinear TE-polarized surface polaritons on graphene. Phys. Rev. B **89**, 035406 (2014)
73. S.A. Mikhailov, K. Ziegler, New electromagnetic mode in graphene. Phys. Rev. Lett. **99**, 016803 (2007)
74. S.A. Mikhailov, K. Ziegler, Nonlinear electromagnetic response of graphene: frequency multiplication and the self-consistent-field effects. J. Phys.: Condens. Matter **20**, 1–10 (2008)
75. M. Jablan, H. Buljan, M. Soljacic, Plasmonics in graphene at infra-red frequencies. Phys. Rev. B **80**, 245435 (2009)
76. H. Zhang, S. Virally, Q. Bao, L.K. Ping, S. Massar, N. Godbout, P. Kockaert, Z-scan measurement of the nonlinear refractive index of graphene. Opt. Lett. **37**, 1856–1858 (2012)
77. A.V. Gorbach, Nonlinear graphene plasmonics: amplitude equation for surface plasmons. Phys. Rev. A **87**, 013830 (2013)
78. L. Ju, B. Geng, J. Horng, C. Girit, M. Martin, Z. Hao, H.A. Bechtel, X. Liang, A. Zettl, Y.R. Shen, F. Wang, Nat. Nanotechnol. **6**, 630 (2011)
79. J. Chen, M. Badioli, P. Alonso-Gonzalez, S. Thongrattanasiri, F. Huth, J. Osmond, M. Spasenovic, A. Centeno, A. Pesquera, P. Godignon, A. Zurutuza Elorza, N. Camara, F. Javier Garcia De Abajo, R. Hillenbrand, F.H.L. Koppens, Optical nano-imaging of gate-tunable graphene plasmons. Nature **486**, 77–81 (2012)
80. Z. Fei, A.S. Rodin, G.O. Andreev, W. Bao, A.S. McLeod, M. Wagner, L.M. Zhang, Z. Zhao, G. Dominguez, M. Thiemens, M.M. Fogler, A.H. Castro-Neto, C.N. Lau, F. Keilmann, D.N. Basov, Gate-tuning of graphene plasmons revealed by infrared nano-imaging. Nature **487**, 82–85 (2012)
81. B. Wang, X. Zhang, X. Yuan, J. Teng, Optical coupling of surface plasmons between graphene sheets. Appl. Phys. Lett. **100**, 131111 (2012)
82. A. Vakil, N. Engheta, Transformation optics using graphene. Science **332**, 1291–1294 (2011)
83. T. Stauber, N.M.R. Peres, A.K. Geim, Optical conductivity of graphene in the visible region of the spectrum. Phys. Rev. B **78**, 085432 (2008)
84. C.M. Bender, S. Boettcher, Real spectra in non-Hermitian Hamiltonians having PT symmetry. Phys. Rev. Lett. **80**, 5243–5246 (1998)
85. G.W. Hanson, Dyadic Green's functions for an anisotropic, non-local model of biased graphene. IEEE Trans. Antennas Propag. **56**, 747–757 (2008)
86. A. Locatelli, A.-D. Capobianco, M. Midrio, S. Boscolo, C. De Angelis, Graphene-assisted control of coupling between optical waveguides. Opt. Express **20**, 28479 (2012)
87. S. Klaiman, U. Gunther, N. Moiseyev, Visualization of branch points in PT-symmetric waveguides. Phys. Rev. Lett. **101**, 080402 (2008)

88. A. Guo, G.J. Salamo, D. Duchesne, R. Morandotti, M. Volatier-Ravat, V. Aimez, G.A. Siviloglou, D.N. Christodoulides, Observation of PT-symmetry breaking in complex optical potentials. Phys. Rev. Lett. **103**, 093902 (2009)
89. S. Yu, G.X. Piao, D.R. Mason, S. In, N. Park, Spatiospectral separation of exceptional points in PT-symmetric optical potentials. Phys. Rev. A **86**, 031802 (2012)
90. H. Yang, X. Feng, Q. Wang, H. Huang, W. Chen, A.T.S. Wee, W. Ji, Giant two-photon absorption in bilayer graphene. Nano Lett. **11**, 2622–2627 (2011)
91. K. Kim, J.Y. Choi, T. Kim, S.H. Cho, H.J. Chung, A role for graphene in silicon-based semiconductor devices. Nature **479**, 338–344 (2011)
92. F. Bonaccorso, Z. Sun, T. Hasan, A.C. Ferrari, Graphene photonics and optoelectronics. Nat. Photon. **4**, 611–622 (2010)
93. M. Liu, X. Yin, E. Ulin-Avila, B. Geng, T. Zentgraf, L. Ju, F. Wang, X. Zhang, A graphene-based broadband optical modulator. Nature **474**, 64–67 (2011)
94. M. Liu, X. Yin, X. Zhang, Double-layer graphene optical modulator. Nano Lett. **12**, 1482–1485 (2012)
95. M. Midrio, S. Boscolo, M. Moresco, M. Romagnoli, C. De Angelis, A. Locatelli, A.-D. Capobianco, Graphene-assisted critically-coupled optical ring modulator. Opt. Express **20**, 23144–23155 (2012)
96. Q. Bao, H. Zhang, B. Wang, Z. Ni, C. Lim, Y. Wang, D.Y. Tang, K.P. Loh, Broadband graphene polarizer. Nat. Photon. **5**, 411–415 (2011)
97. J.T. Kim, C.G. Choi, Graphene-based polymer waveguide polarizer. Opt. Express **20**, 3556–3562 (2012)
98. Z.Q. Li, E.A. Henriksen, Z. Jiang, Z. Hao, M.C. Martin, P. Kim, H.L. Stormer, D.N. Basov, Dirac charge dynamics in graphene by infrared spectroscopy. Nat. Phys. **4**, 532–535 (2008)
99. A.-D. Capobianco, A. Locatelli, C. De Angelis, M. Midrio, S. Boscolo, Finite-difference beam propagation method for graphene-based devices. IEEE Photon. Technol. Lett. **26**, 1007–1010 (2014)
100. D.N. Christodoulides, R.I. Joseph, Discrete self-focusing in nonlinear arrays of coupled waveguides. Opt. Lett. **13**, 794–796 (1988)
101. T. Pertsch, T. Zentgraf, U. Peschel, A. Brauer, F. Lederer, Anomalous refraction and diffraction in discrete optical systems. Phys. Rev. Lett. **88**, 093901 (2002)
102. A. Locatelli, M. Conforti, D. Modotto, C. De Angelis, Diffraction engineering in arrays of photonic crystal waveguides. Opt. Lett. **30**, 2894–2896 (2005)
103. A. Locatelli, M. Conforti, D. Modotto, C. De Angelis, Discrete negative refraction in photonic crystal waveguide arrays. Opt. Lett. **31**, 1343–1345 (2006)
104. M. Guasoni, A. Locatelli, C. De Angelis, Peculiar properties of photonic crystal binary waveguide arrays. J. Opt. Soc. Am. B **25**, 1515–1522 (2008)
105. M. Conforti, M. Guasoni, C. De Angelis, Subwavelength diffraction management. Opt. Lett. **33**, 2662–2664 (2008)

Chapter 4
Temporal and Spectral Nonlinear Pulse Shaping Methods in Optical Fibers

Sonia Boscolo, Julien Fatome, Sergei K. Turitsyn, Guy Millot and Christophe Finot

Abstract The combination of the third-order optical nonlinearity with chromatic dispersion in optical fibers offers an extremely rich variety of possibilities for tailoring the temporal and spectral content of a light signal, depending on the regime of dispersion that is used. Here, we review recent progress on the use of third-order nonlinear processes in optical fibers for pulse shaping in the temporal and spectral domains. Various examples of practical significance will be discussed, spanning fields from the generation of specialized temporal waveforms to the generation of ultrashort pulses, and to stable continuum generation.

4.1 Introduction

The third-order $\chi^{(3)}$ optical nonlinearity in silica-based single-mode (SM) fibers is responsible for a wide range of phenomena such as third-harmonic generation, nonlinear refraction (Kerr nonlinearity), and stimulated Raman and Brillouin scattering [1]. In high-speed optical communications nonlinear effects generally degrade the integrity of the transmitted signal, but the same effects, when properly managed, can be used to realize a variety of optical functions that have practical applications in the field of lightwave technology. Nonlinear processes that have been exploited in demonstrations and applications include stimulated Brillouin and Raman scattering, as well as aspects of the Kerr effect variously called self-phase

S. Boscolo (✉) · S.K. Turitsyn
Aston Institute of Photonic Technologies, School of Engineering and Applied Science, Aston University, Birmingham B4 7ET, UK
e-mail: s.a.boscolo@aston.ac.uk

J. Fatome · G. Millot · C. Finot
Laboratoire Interdisciplinaire CARNOT de Bourgogne, UMR 6303 CNRS-Université de Bourgogne, Dijon, France
e-mail: christophe.finot@u-bourgogne.fr

S. Wabnitz and B.J. Eggleton (eds.), *All-Optical Signal Processing*,
Springer Series in Optical Sciences 194, DOI 10.1007/978-3-319-14992-9_4

modulation (SPM), cross-phase modulation, four-photon (four-wave) mixing (FWM) and parametric gain.

Techniques for generating, controlling, manipulating, and measuring ultrashort optical pulses and specialized waveforms have become increasingly strategically important in many scientific areas including, amongst others, ultrahigh-speed optical communications, optical signal processing, and biophotonics. Picosecond and femtosecond pulse shaping in the optical domain have been extensively implemented using the Fourier-domain approach [2], which employs spectral manipulation of the intensity and phase components of a pulse in order to create the desired field distribution. Though powerful and flexible, this pulse shaping strategy has the drawback that the bandwidth of the output spectrum is determined by the bandwidth of the input spectrum. Indeed, a linear manipulation cannot increase the pulse bandwidth, and so to create shorter pulses nonlinear effects must be used. In addition, a linear pulse shaper can only subtract power from the frequency components of the signal while manipulating its intensity, thereby potentially making the whole process extremely power inefficient. The combination of third-order nonlinear processes and chromatic dispersion in optical fibers can provide efficient new solutions to overcome the aforementioned limitations. As we will see in this chapter, nonlinear fiber-based pulse shaping has become a remarkable tool to tailor both the temporal and spectral content of a light signal without the need for dedicated linear shapers. It is important to note that though we will use a silica optical fiber as a key example system to demonstrate various effects, these general nonlinear science concepts can be applied to a range of other nonlinear optical materials and media.

In this chapter, we review recent progress in the research on pulse shaping using nonlinear effects in optical fibers. We would like to note that the main attention in this chapter will be focused on results obtained in our groups, and that it is not our intention here to comprehensively cover all the possible examples of fiber-based pulse shaping. The chapter is organized as follows: Sect. 4.2 introduces the key model governing the optical pulse propagation in a fiber, and briefly discusses the resultant qualitative features of the pulse evolution. Section 4.3 emphasizes how the normal dispersion regime of a fiber is particularly well-suited for the generation of advanced temporal waveforms and stable continua with high power spectral density, as well as for spectral compression of the pulses. Section 4.4 highlights some of the benefits offered by the governing soliton dynamics in the anomalous dispersion region in terms of nonlinear shaping, namely, the possibility of generating ultrashort temporal structures and ultrabroad spectra, and of tuning the central frequency of the pulses. Finally, we conclude in Sect. 4.5.

4.2 Pulse Propagation in Optical Fibers

A variety of nonlinear phenomena associated with optical pulse propagation in SM fibers is well described by the nonlinear Schrödinger (NLS) equation [3], in spite of the fact that this equation only includes two physical effects, namely, linear group velocity dispersion (GVD) and nonlinear SPM. Here, a modified NLS equation for the pulse envelope, $\psi(z,t)$, is taken, which includes the effect of linear gain, namely,

$$\psi_z = -\frac{i\beta_2}{2}\psi_{tt} + i\gamma|\psi|^2\psi + \frac{1}{2}g\psi, \tag{4.1}$$

where z is the propagation distance, t is the reduced time, β_2 and γ are the respective GVD and cubic nonlinearity coefficients of the fiber, and g is the distributed gain coefficient (which is nonzero only for an active fiber). This equation neglects higher-order linear effects such as third-order dispersion as well as higher-order nonlinear effects such as self-steepening or intrapulse Raman scattering [3]. Although these effects can have noticeable impact on pulses shorter than 1 ps, here we neglect them as the leading-order behavior is well approximated by (4.1). Moreover, neglecting higher-order gain effects such as gain bandwidth and saturation is well-suited to describe experiments that use high-gain, broadband rare-earth fiber amplifiers [4].

It is useful to introduce here the dispersion length $L_D = \tau_0^2/|\beta_2|$ and the nonlinear length $L_{\mathrm{NL}} = 1/(\gamma P_0)$, which provide the length scales over which dispersive or nonlinear effects become important for pulse evolution. The parameters τ_0 and P_0 are some characteristic temporal value and the peak power of the input pulse, respectively. Linear propagation regime occurs when $N^2 = L_{\mathrm{D}}/L_{\mathrm{NL}} \ll 1$. In this case pulse evolution can be simply determined by using the Fourier-transform method. As a result of GVD, the phase of each spectral component of the pulse is changed by an amount $\beta_2\omega^2 z/2$, where ω is the angular frequency of the pulse envelope. Even though such phase changes do not affect the pulse spectrum, they lead to temporal broadening of the pulse and a time dependence of the pulse phase, which can be characterized by the instantaneous frequency shift or chirp $\delta\omega = -\phi_t$. In the opposite case when $N^2 \gg 1$, the nonlinearity-dominant regime is applicable. As a result of the nonlinear term in (4.1), upon propagation in the fiber the pulse acquires an intensity-dependent nonlinear phase shift as $\phi_{\mathrm{NL}}(z,t) = \gamma|\psi(0,t)|^2 \int_0^z \mathrm{d}z' \exp[G(z')]$ with $G(z') = \int_0^{z'} \mathrm{d}\zeta\,[g(\zeta)]$, namely the frequency chirp $\delta\omega = -(\phi_{\mathrm{NL}})_t$. In the absence of GVD and for an initially transform-limited pulse, the presence of a chirp causes a nonlinear broadening of the pulse spectrum.

When the effects of chromatic dispersion are considered in combination with the Kerr nonlinearity, rich pulse dynamics arise from the interplay between dispersive and nonlinear effects depending on the sign of the dispersion and the relative magnitudes of the associated length scales. A well-known and fascinating example is the formation of optical solitons [5, 6] in the anomalous dispersion ($\beta_2 < 0$) regime of a passive fiber as a result of a cooperation between GVD and SPM. In the anomalous dispersion region, if $N = 1$, the chirp induced by chromatic dispersion can be compensated by the SPM-induced chirp, which has opposite sign. The pulse shape adjusts itself during propagation to make such cancelation as complete as possible, hence leading to soliton evolution. On the contrary, in the regime of normal dispersion pulse dynamics are highly affected by the phenomenon of optical wave breaking [7–10]. In the nonlinearity-dominant regime of propagation, the combination of normal GVD and SPM makes a pulse broaden and change shape toward an almost rectangular pulse form with a linear frequency chirp variation over most of the pulse, as it is highlighted by Fig. 4.1 which shows the temporal and spectral evolution of an initially unchirped hyperbolic secant pulse in the normal and anomalous dispersion regimes of a fiber using $N \approx 11.4$ and $g = 0$. This evolution can be understood as arising from the frequency chirp, which makes the red- (blue-) shifted light near the leading (trailing) edge travel faster (slower) than the unshifted pulse parts, i.e., the pulse center and low-intensity wings. Ultimately, when the shifted light overruns the pulse tails, the wave breaks; oscillations appear in the wings of the pulse because of interference and, concomitantly, new frequencies are generated and side lobes appear in the pulse spectrum (Fig. 4.4a).

In the next sections we will discuss some newly emerged qualitative features of the optical pulse evolution in a fiber under the combined action of dispersion and nonlinearity.

4.3 Pulse Shaping in Normally Dispersive Fibers: From the Generation of Specialized Temporal Waveforms to Spectral Sculpturing

4.3.1 Generation of Specialized Temporal Waveforms

As a first illustration of nonlinear pulse shaping in normally dispersive (ND) fibers, we discuss here the possibility of generating advanced temporal waveforms, such as parabolic or triangular profiles, which are of interest for various applications in optical signal processing and manipulation [11].

4.3.1.1 Parabolic Pulse Generation

We have seen in the previous section that pulses that propagate in normal dispersion media are susceptible to distortion and breakup owing to optical wave breaking, which is a consequence of excessive nonlinear phase accumulation combined with dispersion. In 1993 Anderson et al. showed that a sufficient condition to avoid wave breaking is that a pulse acquires a monotonic frequency chirp as it propagates, and that wave breaking-free solutions of the NLS equation (4.1) exist when the GVD is normal [12]. These are asymptotic solutions of (4.1) in the quasi-classical or WKB limit (i.e., the limit of high amplitude or small dispersion such that $\beta_2|(|\psi|)_{tt}|/(2\gamma|\psi|^3)$ can be neglected), and take the form

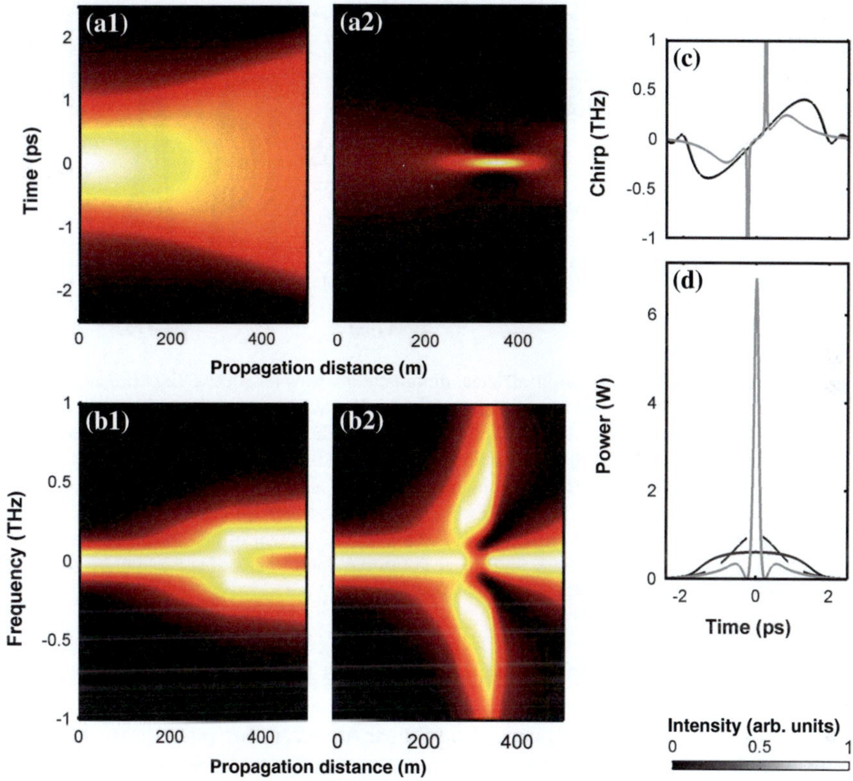

Fig. 4.1 Evolution of an initially unchirped hyperbolic secant pulse with $P_0 = 1$ W and $\tau_0 = 0.85$ ps propagating in an highly nonlinear (HNL) fiber with $|\beta_2| = 0.64\,\text{ps}^2/\text{km}$ and $\gamma = 10\ (\text{W km})^{-1}$, as obtained by numerical integration of the NLS equation. Longitudinal evolution of the **a** temporal and **b** spectral intensity profiles in the normal (*subplots 1*) and anomalous (*subplots 2*) dispersion regimes of the fiber. Temporal **c** chirp and **d** intensity profiles after 350 m of propagation, i.e. 3.5 L_{NL}, in the fiber in the case of normal (*black solid*) and anomalous (*gray solid*) GVD. Also shown is the initial temporal profile (*dashed black*)

$$|\psi(z,t)| = a(z)\sqrt{1-(t/\tau(z))^2}\theta(\tau(z)-|t|), \quad \arg\psi(z,t) = b(z)t^2 + \phi_0(z), \quad (4.2)$$

and thus have parabolic intensity profiles. Here, $\theta(x) = 1,\ x > 0;\ \theta(x) = 0,\ x<0$ is the Heaviside function. Such a pulse maintains its shape, and is always a scaled version of itself; i.e., it evolves self-similarly. In contrast to solitons, self-similar pulses, which have been dubbed similaritons, can tolerate strong nonlinearity ($\phi_{\text{NL},0} = \int \text{d}z\gamma(z)P_0(z) \gg \pi$, where $\phi_{\text{NL},0}$ is the peak nonlinear phase shift) without distortion or wave breaking (Fig. 4.4a). The normal GVD effectively linearizes the accumulated phase of the pulse allowing for the spectral bandwidth to increase without destabilizing the pulse. Similaritons were demonstrated theoretically and

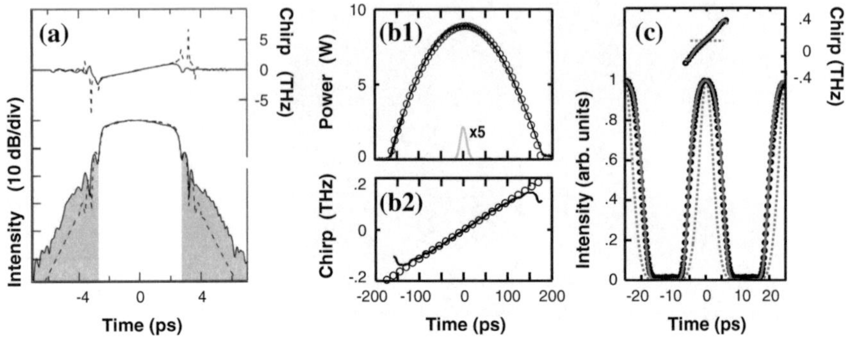

Fig. 4.2 Parabolic pulse generation at telecommunication wavelengths. **a** Experimental (*solid*) and numerical (*dashed*) temporal intensity and chirp profiles at the output of an erbium-doped fiber amplifier. Adapted from [17]. **b** Results obtained in a Raman fiber amplifier: experimental temporal intensity and chirp profiles at the fiber output (*solid black*) compared to a parabolic and a linear fit (*circles*), respectively. The initial intensity profile is also shown (*solid gray*). More details on the experiment can be found in [20]. **c** Experimentally measured 40-GHz pulse train and chirp profile at the input (*dashed*) and output (*solid*) of a passive segment of ND fiber followed by an HNL fiber. Also shown are parabolic and linear fits to the output pulse shape and chirp (*circles*). More details can be found in [73]

experimentally in fiber amplifiers in 2000 [4], and they continue to attract much attention [13]. Similaritons in fiber amplifiers are along with solitons in passive fiber the most well-known classes of nonlinear attractors for pulse propagation in optical fibers [4, 14, 15], so they take on major fundamental importance. Various experimental demonstrations relying on amplification from either rare-earth doping [4, 16–18] (Fig. 4.2a) or Raman scattering [19, 20] (Fig. 4.2b) have been achieved. Experimental interest has been strongly driven by the possibility of fully canceling the highly linear chirp developed during parabolic amplification and, thus, generating ultrashort high-power pulses with very low substructures [4, 16–18]. Furthermore, recent fiber lasers that support similariton evolution in the passive [21] or gain [22–25] fiber segment of the laser cavity have been demonstrated to achieve pulse energy and peak power performances much higher than those of prior approaches.

In addition to fiber amplifiers, similariton formation with nonlinear attraction can be achieved in passive fibers provided a suitable longitudinal variation of the dispersion is introduced [26–28]. This approach is based on the observation that a longitudinal decrease of the normal dispersion is formally equivalent to linear gain. Recently, a simple approach to the generation of parabolic pulses that uses progressive nonlinear pulse reshaping in a ND fiber with fixed dispersion has been also demonstrated [29] (Fig. 4.2c). However, in contrast with the asymptotic attracting solutions obtained in fiber amplifiers, the generated parabolic waveforms represent transient states of the nonlinear pulse evolution in the fiber medium [30] and can be associated with an intermediate asymptotic regime [31] of the pulse propagation. As such, they have a finite life distance (characteristic length scale of the self-similar pulse dynamics) that depends on the initial conditions. Nevertheless, stabilization of the parabolic features is possible by use of a second propagation stage in a fiber with specially adjusted nonlinear and dispersive characteristics relative to the first fiber [29, 30].

4.3.1.2 Triangular Pulse Generation

Parabolic shapes are not the only pulse waveforms that can be generated in a passive ND fiber. Indeed, it has been shown numerically that temporal triangular intensity profiles can result from the evolution of initially rectangular pulses in the highly nonlinear (quasi-classical) regime of propagation [32]. Another approach that has been recently demonstrated theoretically relies on the progressive reshaping of initially parabolic pulses driven by the fourth-order dispersion (FOD) of the fiber [33]. The overall temporal effect of FOD on parabolic pulse propagation is to stretch and enhance the power reduction in the pulse wings, leading to a triangular profile. In [30], the combination of pulse prechirping and nonlinear propagation in a section of ND fiber has been introduced as a method for passive nonlinear pulse shaping, which provides a simple way of generating various advanced field distributions, including flat-top- and triangular-profiled pulses with a linear frequency chirp. In this scheme, Kerr nonlinearity and GVD lead to various reshaping processes of an initial conventional field distribution (e.g., a Gaussian pulse) according to the chirping value and power level at the entrance of the fiber. In particular, triangular pulses can be generated for an anomalous (negative) initial chirp and sufficiently high energies. These theoretical results have been confirmed experimentally by intensity (Fig. 4.3b) and phase measurements of the generated pulses [34, 35]. In the experimental setup used (Fig. 4.3a), the control of the pulse prechirping value was realized by propagation through different lengths of standard SM fiber with anomalous GVD, which imposed a negative chirp on the pulse. The prechirped pulses were amplified to different power levels using an erbium-doped

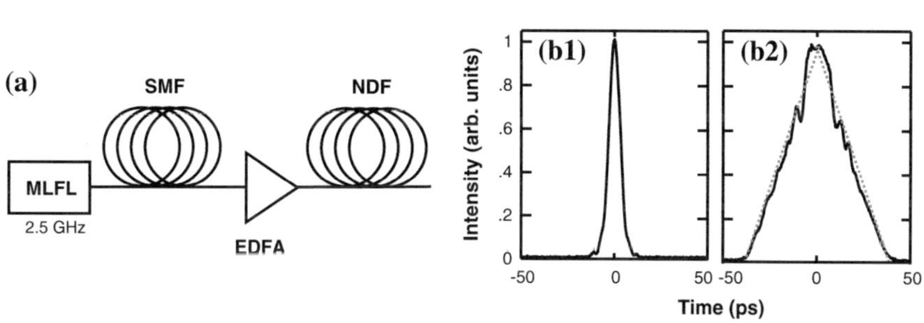

Fig. 4.3 **a** Experimental setup for triangular pulse generation. *MLFL* mode-locked fiber laser. **b** Measured (*solid black*) temporal intensity profiles at the entrance (*subplot b1*) and the exit (*subplot b2*) of the reshaping fiber. Also shown is the corresponding triangular fit (*dashed gray*). More details can be found in [35]

fiber amplifier (EDFA) and then propagated through a ND fiber to realize the pulse reshaping.

Furthermore, the possibility of triangular pulse shaping in mode-locked fiber lasers has been first reported in [36]. It has been numerically demonstrated that for normal net dispersion, formation of two distinct steady-state solutions of stable single pulses can be obtained in a laser cavity in different regions of the system parameter space: the previously known similariton [21] and a triangular-profiled pulse with a linear chirp.

4.3.2 Spectral Sculpturing

4.3.2.1 Generation of Highly Coherent Continua

Nonlinear pulse shaping represents the ideal way of generating spectral bandwidth that exceeds significantly that of the input seed pulse. This is obviously not possible with linear shaping techniques where one cannot shift energy outside the initial pulse bandwidth. Optical fiber-based supercontinuum (SC) sources have become a significant scientific and commercial success, with applications ranging from frequency comb production to advanced medical imaging and telecommunication applications [37, 38]. While typically the broadest spectra are generated in anomalously dispersive fibers, where the broadening mechanism is strongly influenced by soliton dynamics (Sect. 4.4), ND fibers are undoubtedly advantageous as far as the flatness and temporal coherence of the generated continuum is concerned. Specifically, nonlinear propagation in the normal dispersion regime inhibits modulation instability (MI) phenomena to occur, which reduces significantly the spontaneous amplification of noise and thereby shot-to-shot fluctuations [39]. Consequently, a high degree of coherence is maintained. However, the main limits to spectral pulse quality in the normal dispersion regime are the spectral ripple arising from SPM of conventional laser pulses, and the effects of optical wave breaking which may lead to significant changes in the temporal pulse shape and severe energy transfer into the wings of the spectrum. These effects can in principle be avoided by using preshaped input pulses with a parabolic temporal intensity profile, which would preserve their shape whilst propagating within the fiber and, thus, result in spectrally flat, highly coherent pulses (Fig. 4.4a). Continuum with low spectral ripple and high-energy density in the central part has been indeed demonstrated by use of parabolic pulses generated in erbium-doped fibers [40] or resulting from linear shaping by fiber Bragg gratings [41] (Fig. 4.4b) and reconfigurable liquid-crystal-on-silicon devices [42]. More advanced techniques for achieving even flatter spectra involve the use of a feedback loop relying on a genetic algorithm to obtain a nearly rectangular output spectrum [43], as Fig. 4.5 shows. This demonstrates the potential of nonlinear pulse shaping assisted by a linear shaping element as an additional sculpturing element for generating on-demand tailored spectra without the requirement of expensive laser sources.

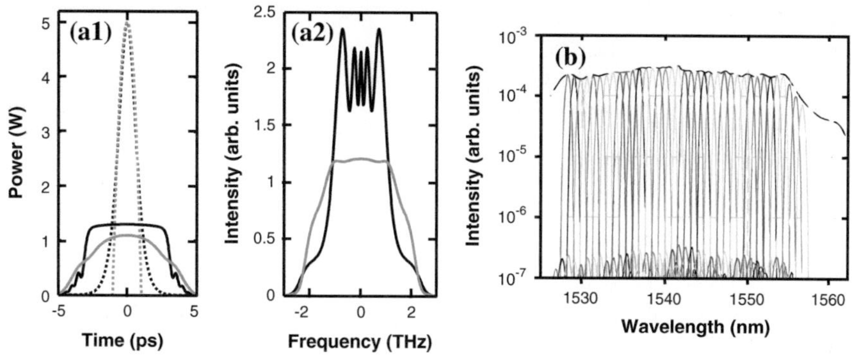

Fig. 4.4 **a** Temporal (*subplot a1*) and spectral (*subplot a2*) intensity profiles at the exit of a ND-HNL fiber with the same parameters as in Fig. 4.1 for initially unchirped hyperbolic secant (*solid black*) and parabolic (*solid gray*) pulses with $P_0 = 5$ W, as obtained by numerical integration of the NLS equation. Also shown are the initial pulse temporal profiles (*dashed*). **b** Experimental spectrum of a parabolic-shaped pulse after propagation in a ND-HNL fiber (*black dashed*) and superposition of the measured spectra of 38 sliced channels. Adapted from [41]

The high degree of coherence of the spectra generated in the normal dispersion regime is clearly attractive for all-optical pulse processing in HNL fibres [11], or the generation of SC covering the entire C-band of optical telecommunications. Indeed, several studies have reported that such continua can be sliced into several tens of high-quality, high-repetition-rate picosecond channels where isolated spectral lines of the 10 or 40-GHz resulting comb provide several hundreds of continuous waves [41, 44, 45] (Fig. 4.4b). Furthermore, high temporal coherence enables pulse compression with very high compression ratios as the SPM-induced nonlinear chirp can be compensated to a large extent [17, 46]. This technique has been efficiently exploited in photonic crystal (PC) fibers where the dispersion characteristics can be accurately controlled so as to obtain a fiber with normal dispersion at all wavelengths [33], which enables the generation of highly broadband coherent spectra spanning more than several hundreds of nanometers. Such spectra are particularly attractive from the perspective of pulse compression down to durations as short as a few optical cycles after compensation of the nonlinear phase [47, 48].

4.3.2.2 Spectral Compression

Spectral narrowing of optical pulses can be rather easily realized with standard optical bandpass filtering, but at the expenses of significant optical losses. Indeed, as we mentioned in the introduction, linear filtering introduces a power penalty that is at least proportional to the ratio of the target spectrum to the input one. The effect of SPM on a pulse propagating in an optical fiber is generally associated with spectral broadening of the pulse. However, this is not always the case with chirped

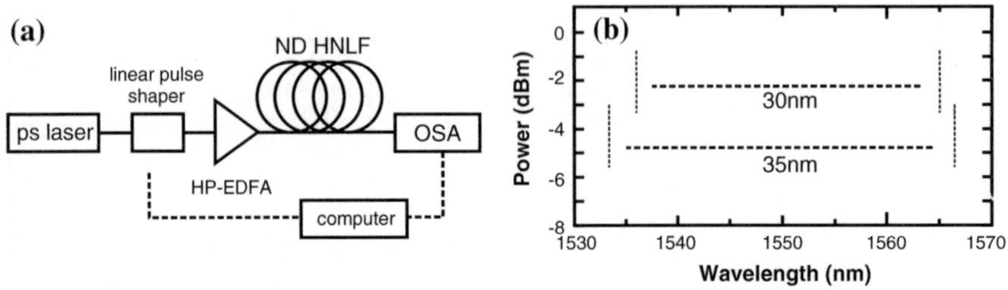

Fig. 4.5 **a** Experimental setup for the nonlinear generation of ultra-flat broadband frequency combs assisted by adaptive pulse shaping. **b** SPM spectra generated from 2 ps Gaussian pulses after the application of the adaptive system. More details can be found in [43]

pulses, where down-chirped pulses in the normal GVD region can experience spectral narrowing [49]. Indeed, the SPM-induced nonlinear phase shift results in a frequency downshift of the leading edge and an upshift of the trailing edge of the pulse, i.e., a positive chirp. Consequently, for a negatively chirped input pulse the linear and nonlinear phases cancel each other, leading to a redistribution of long and short wavelengths toward the center wavelength and therefore to spectral compression instead of spectral broadening. Spectral compression has been reported for various parameters [50–52]. However, in general an input pulse with a negative linear chirp, $|\psi(0,t)|\exp(ibt^2)$ (with $b > 0$), cannot be spectrally compressed to the Fourier transform limit because the instantaneous frequency (in the absence of dispersion) $\phi_t = 2bt + \gamma(|\psi(0,t)|^2)_{,t}z$ cannot in general be made equal to zero for all times, except for special cases [53] that require the balancing of the (higher-order) dispersion and chirp. This results in residual side-lobes in the compressed spectrum, as illustrated in Fig. 4.6b. On the other hand, if an input negatively linearly chirped parabolic pulse is used, the instantaneous frequency $\phi_t = 2(b - \gamma a^2 z/\tau_0^2)t$ can be chosen to yield zero for all t, for a suitable combination of γ, a, z, and b [54]. Because, at the same time, the parabolic temporal envelope with width τ_0 remains unchanged (as long as dispersion in the fiber can be ignored), this expresses the fact that spectral compression to the transform limit takes place. Experimental demonstrations at telecommunication wavelengths using a HNL fiber [55] or at the wavelengths around 1 μm using a PC fiber [56] have confirmed the potential of this method for achieving high-quality compressed spectra with very low substructures and enhanced brightness (Fig. 4.6c).

Furthermore, in-cavity nonlinear spectral compression in a mode-locked fiber laser has been recently demonstrated numerically [57]. A new concept of a fiber laser architecture has been presented, which supports self-similar pulse evolution in the amplifier fiber segment of the laser cavity and nonlinear spectral compression in the passive fiber, thereby enabling the generation of highly chirped parabolic pulses and nearly transform-limited spectrally compressed picosecond pulses from a single device.

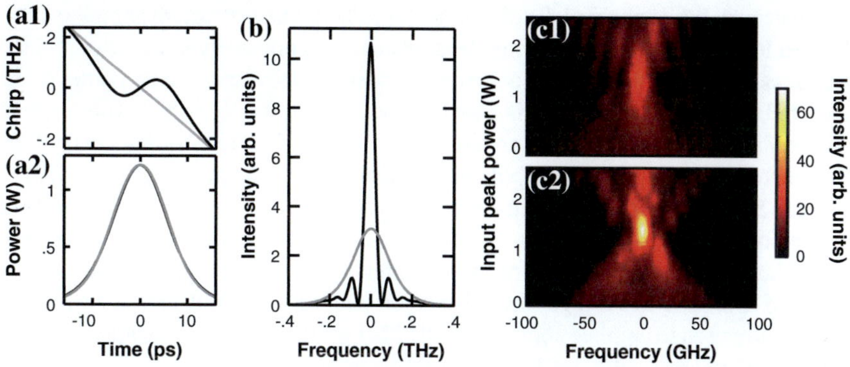

Fig. 4.6 **a** Temporal intensity and chirp profiles, and **b** spectral intensity profiles of an hyperbolic secant pulse before (*gray*) and after (*black*) spectral compression in a ND-HNL fiber (same parameters as in Fig. 4.1), as obtained by numerical integration of the NLS equation. **c** Experimentally measured dependence of the compressed spectral intensity profile on the input peak power for initially negatively chirped hyperbolic secant (*subplot c1*) and parabolic (*subplot c2*) pulses in an HNL fiber. More details can be found in [55]

4.4 Pulse Shaping in the Anomalous Dispersion Regime: From Ultrashort Temporal Structures to Ultra-Broad Spectra

4.4.1 Generation of High-Repetition-Rate Ultrashort Pulses

In this section, we review the generation of high-repetition-rate ultrashort pulses that can result from the nonlinear evolution of optical pulses in anomalously dispersive fibers.

4.4.1.1 Solitonic Temporal Compression

As mentioned in Sect. 4.2, when $\beta_2 < 0$ the standard NLS equation ((4.1) with $g = 0$) has specific pulse-like solutions, known as optical solitons, that either do not change along the fiber length (fundamental solitons, resulting from an exact balance between the GVD and SPM effects: $N = 1$) or follow a periodic evolution pattern (higher-order solitons, characterized by integers $N > 1$). A technique for generating high-quality ultrashort pulses relies on the adiabatic evolution of a fundamental soliton that takes place in an anomalously dispersive fiber with longitudinally distributed gain (e.g., a Raman amplifier) or decreasing dispersion (DD) [58–60]. Indeed, in the presence of gain, the temporal width of a fundamental soliton must decrease along the fiber length to compensate for the power increase due to gain if the pulse has to preserve its soliton character, i.e., if the requirement $N = 1$ has to be maintained. While physical distributed-amplification schemes require pumping along the fiber length, which entails an extra level of cost and complexity in the setup, soliton compression in a fiber can also be achieved by changing the dispersive properties of the fiber [61]. This passive approach is based on the fact that the NLS (4.1) with constant gain coefficient g reduces to the standard NLS equation if the anomalous GVD of the fiber increases in magnitude exponentially along the fiber length: $\beta_2(z)/\beta_2(0) = \exp(gz)$. Step-like dispersion-profiled structures [62] using a set of short segments of conventional fibers with different dispersion values, or comb-like structures [63] where the use of a large number of fiber segments with alternate signs of dispersion allows to approach an average local dispersion with the desired decrease, have been successfully used to emulate continuous DD fiber profiles, whose realization requires special, not commercially available fiber-drawing technology.

A more straightforward approach to the generation of pulses with ultrashort durations uses the stage of temporal compression that occurs in the periodic evolution of higher-order solitons. Indeed, in the case of higher-order solitons, over a period of the evolution pattern SPM dominates initially but GVD soon catches up and leads to pulse contraction, as seen in Fig. 4.1. The higher the soliton number N of the input pulse, the larger the compression achievable using this approach. However, the compressed pulse exhibits a low-amplitude structure outside the central lobe (Fig. 4.1). Moreover, given the scaling laws of the NLS equation, increasing the input pulse peak power leads to a decrease of the distance in the fiber at which compression takes place, while higher-order nonlinear and dispersive effects become important for pulse evolution. This may eventually make this method very sensitive to power fluctuations of the input pulse. Feasibility of the higher-order soliton compression scheme was first shown in 1980 [64]. Various experimental demonstrations in either HNL fibers at telecommunication wavelengths [65] or in PC fibers seeded by titanium-sapphire laser pulses [66] have been achieved since then, enabling the generation of pulses as short as a few cycles.

4.4.1.2 High-Repetition-Rate Sources

High-repetition-rate ultrashort optical pulses are widely used in various areas such as ultrafast physical processes, ultra-high-data-rate optical communications, optical frequency metrology and high-accuracy measurements. While current electronics-based and direct modulation technologies do not permit the generation of pulses at a repetition rate higher than 40 GHz, the nonlinear transformation of a dual-frequency signal in an optical fiber can provide an attractive solution in this respect. A nonlinear technique based on MI of an intense pump wave induced by a weak signal was first suggested in [67] and realized experimentally in [68]. MI in the anomalous dispersion regime of optical fibers manifests itself as breakup of the initial pump beam into a periodic pulse train whose period is inversely related to the frequency difference between the pump and signal waves, $\Omega = \omega_0 - \omega_1$, with the generation of two new spectral sidebands, and can be interpreted in terms of a FWM process that is phase-matched by SPM. Physically, the energy of two photons from the

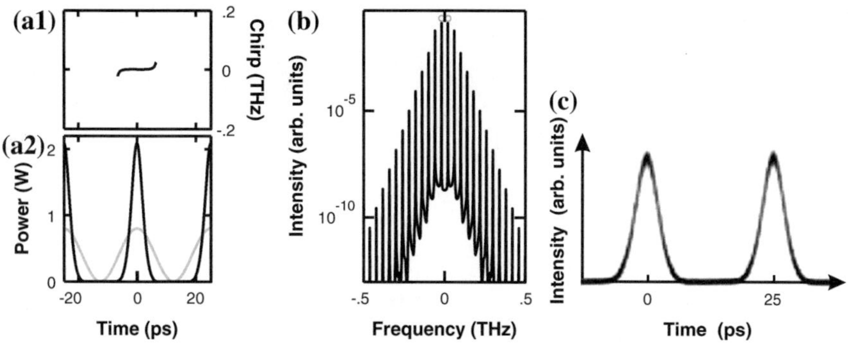

Fig. 4.7 Generation of high-repetition-rate pulse train using multiwave mixing compression of a dual-frequency beat signal. **a** Temporal intensity and chirp profiles, and **b** spectrum of the compressed pulse train as obtained by numerical simulation of the NLS equation. The initial beat signal is also shown (*gray*). **c** Intensity of the generated pulse train at 40 GHz recorded on an optical sampling oscilloscope. More details on the experiment can be found in [72]

intense pump beam is used to create two different photons, one at the probe frequency ω_1 and the other at the idler frequency $2\omega_0 - \omega_1$. Since its first demonstration, this technique has been used to create optical sources capable of producing periodic trains of ultrashort pulses at high but controllable repetition rates.

When the first-order sidebands become strong, higher-order sidebands located at $\omega_0 \pm m\Omega$, $m = 2, 3, \ldots$ are created in a FWM process. An approach based on multiple FWM in a single fiber with anomalous constant dispersion has been demonstrated experimentally for the generation of compressed pulse trains from the propagation of a dual-frequency pump field, and repetition rates ranging from a few tens of gigahertz to a few terahertz have been achieved [69–73]. The initial beat signal in these experiments was obtained via temporal superposition of two continuous waves (CWs) with slightly different frequencies delivered by external cavity lasers [70, 71], which permitted to achieve higher repetition rates than using a single CW directly modulated by an intensity modulator. Exemplary experimental results recorded on a high-speed oscilloscope show that high-quality compressed pulses with a Gaussian intensity profile and nearly constant phase can be achieved at a typical duty cycle of 1:5 [72] (Fig. 4.7c). Lower duty cycles can still be obtained with a single fiber, but at the expenses of degraded quality of the pulse train with the appearance of low-intensity pedestal components and a nonuniform phase across the pulses. Such issues can be partly avoided by using specially designed arrangements of segments of fibers [72, 73] or other comb-like structures.

4.4.2 Generation of Frequency-Tunable Pulses

One of the possibilities offered by the third-order nonlinearity of optical fibers for pulse shaping is to frequency shift the essential spectral content of ultrashort optical

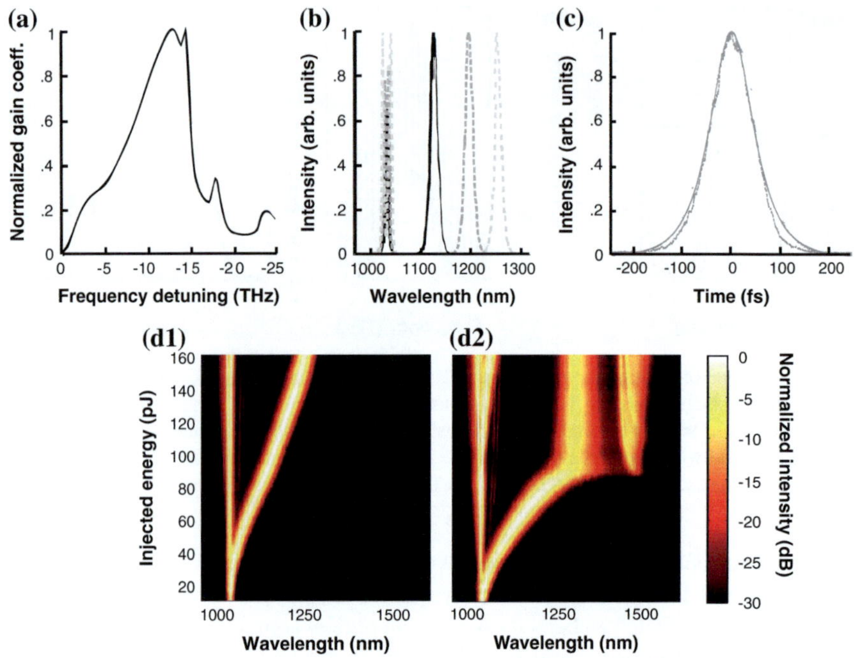

Fig. 4.8 Raman-induced self-frequency shift of solitons. **a** Raman response of silica fibers. **b** Experimental spectra obtained at the output of a 2-m long PC fiber for different input pulse energies: 83 pJ (*solid black*), 125 pJ (*dashed gray*), and 173 pJ (*dash-dot light gray*). **c** Autocorrelation traces of the frequency-shifted pulses compared with an hyperbolic secant fit. **d** Experimentally measured dependence of the output spectral intensity profile on the input energy for 2-m long PC fibers with a single (*subplot d1*) or two zero-dispersion wavelengths (*subplot d2*)

pulses whose width is close to or less than 1 ps. For pulses with a wide spectrum (>1 THz), the Raman gain of the fiber (Fig. 4.8a) can amplify the low-frequency components of a pulse by transferring energy from the high-frequency components of the same pulse. As a result of this phenomenon, called intrapulse Raman scattering, the pulse spectrum shifts toward the low-frequency (red) side as the pulse propagates inside the fiber, a phenomenon referred to as the self-frequency shift [74]. The physical origin of this effect is related to the delayed nature of the Raman (vibrational) response [75]. The effects of intrapulse Raman scattering can be included through the NLS equation by including the term $-i\gamma\tau_R\psi(|\psi|^2)_t$ on the left-hand-side of (4.1), where the parameter τ_R is related to the slope of the Raman gain spectrum [75]. For a soliton propagating inside a fiber, the amplitude is not affected by the Raman effect, but its frequency is shifted by $\delta\omega_R(z) = -8|\beta_2|\tau_R z/(15\tau_0^4)$, as illustrated experimentally in Fig. 4.8b. The temporal intensity profile remains close to the hyperbolic secant soliton shape (Fig. 4.8c).

The soliton self-frequency shift (SSFS) has been exploited in the realization of fiber-based sources of spectrally tunable ultrashort optical pulses. Frequency shifts of several hundreds of nanometers have been achieved in HNL fibers. The advent of PC fibers has made it possible to observe SSFS at the operational wavelengths of titanium sapphire lasers, and impressive shifts have been demonstrated, thereby paving the way for efficient spectrally tunable laser sources for applications in, e.g., biophotonics or spectroscopy [76, 77]. In this context, the main factors that may limit SSFS generation are the absorption due to hydroxil radicals contained in the core glass of silica fibers, the fission of higher-order solitons into multiple sub-pulses, and the existence of a second zero-dispersion wavelength for some fibers, which induces a stabilization of the soliton frequency accompanied by the generation of dispersive waves [78] (Fig. 4.8d2). Furthermore, the combined effects of delayed nonlinear response and higher-order dispersion can result in a gradual increase of the second-order dispersion during the propagation. This leads to an adiabatic temporal broadening of the soliton, thereby progressively limiting its Raman-induced frequency shift [79]. To address this issue, optimization of the dispersion landscape "seen" by the frequency-shifted pulse has been proposed as an efficient method to enhance the performance of fiber-based tunable frequency shifters [80].

4.4.3 Supercontinuum Generation and Optical Rogue Waves

Several books and review articles are now available to understand the physical mechanisms driving the continuum formation and its modeling using the NLS equation in its extended form (see, e.g., [3, 37, 38, 81]), so in this section we describe only the basic temporal and spectral features of the continuum generated in the anomalous dispersion regime of a nonlinear fiber, which stem from the involved pulse dynamics. We also discuss a possible solution to efficiently harness the large temporal and spectral fluctuations that are typically observed in this dispersion regime.

4.4.3.1 Supercontinuum Stability and Extreme Events

The nonlinear dynamics in the anomalous dispersion regime are dominated by an initial stage of MI of the input picosecond pump pulse induced by noise, followed by a stage of higher-order soliton temporal compression, and subsequent splitting of the ultrashort coherent structure at the point of maximum compression due to third-order dispersion or intra-pulse Raman scattering. Pulse splitting results in the emergence of several ultrashort solitons with significant shot-to-shot amplitude and timing jitters, which prevents this type of SC to be sliced in multiple wavelength channels for telecommunication applications. The spectral intensity profile is significantly broader than those typically obtained in ND fibers but also suffers from large fluctuations. Because the SC properties change significantly from shot to shot both in the temporal and spectral domains, the stability of the continuum obtained in the normal regime of dispersion is generally not preserved in the anomalous dispersion region [39]. The spectral phase fluctuations have been the subject of intense investigations as early as

2000 [82]. The first experimental observation of significant fluctuations in the temporal domain has been reported by Solli et al. in 2007 [83]. Spectral filtering of the most red-shifted part of the SC spectrum generated in a PC fiber by anomalous pumping resulted into a highly-skewed distribution of the statistical fluctuations of the corresponding temporal intensity. By analogy with the typical L-shaped distribution observed for extreme events, the intense and extremely rare outliers of the distribution were named "optical rogue waves". Since then, the subject has become the focus of intense research in the international optical community. This enthusiasm has indeed largely surpassed the confines of SC generation in PC fibers, and optical rogue waves have been reported in various nonlinear optical systems. A careful analysis of various configurations has disclosed that the most red-shifted events detected in the experiment of Solli et al. are fundamental solitons shifted in frequency by the delayed Raman response of silica [84]. Because such giant solitons are fully coherent and persist once they form, they are now generally referred to as "rogue solitons" [85]. The emergence of such extreme solitons results from the process of temporal collision between the substructures that are generated from the splitting of the initial pump pulse. When two ultrashort substructures collide, a temporal peak appears intermittently, thus reproducing one of the peculiar features of the infamous oceanic freak waves [86]: their unexpected appearance and decay. Such collisions are favored by convective mechanisms such as the third-order dispersion or the Raman response of silica [87]. In the process of collision, an energy exchange also occurs, leading to the progressive growth of a giant soliton, where most of the energy of the system will eventually concentrate [88, 89].

The temporal and spectral analysis of the fluctuations associated with anomalous pumping has led to reconsider the onset of SC generation where higher-order linear and nonlinear effects do not play a significant role, and pulse propagation can be described in the framework of the standard NLS equation. In fact, the stage of MI that governs the initial spectral broadening can be described analytically within the framework of the Akhmediev-Kuznetsov breather formalism. Kuznetsov derived non-stationary solutions of the NLS equation which are periodic in time in plasma in 1977 [90]. The Akhmediev breathers are space-periodic solutions of the NLS equation derived by Akhmediev and Korneev in 1986, which describe the single growth-return cycle of an initially weak sinusoidal modulation [91]. Numerical simulations and experiments have indeed confirmed that the properties of SC at the early stages of propagation can be accurately described in terms of multiple breathers corresponding to the excitation of the entire MI gain bandwidth [92].

4.4.3.2 Highly Coherent Supercontinuum Generation and Peregrine Solitons

One way to limit the degradation of the SC coherence in the anomalous dispersion regime is to use very short segments of fiber or tapered fibers [39]. An alternative solution is to reduce the spontaneous instabilities that may emerge in the early stages of propagation. By imposing an input deterministic seed located in the

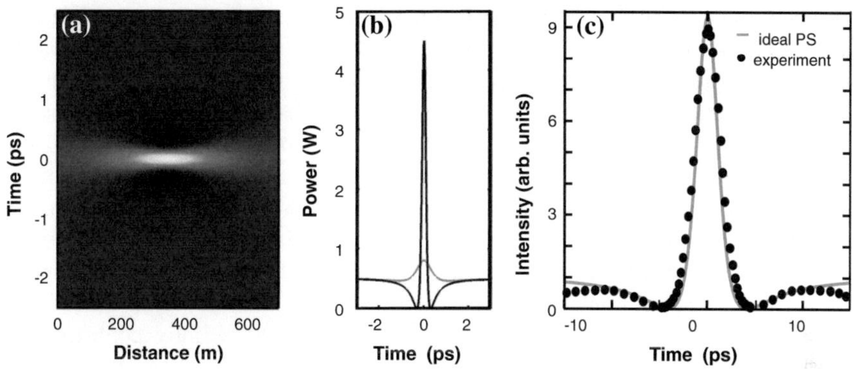

Fig. 4.9 Peregrine soliton in fiber optics. **a** Longitudinal evolution of the Peregrine soliton in an optical fiber as predicted by (4.3). **b** Temporal intensity profile of the Peregrine soliton at the point of maximum compression (*black*) compared with the input condition (*gray*). The fiber parameters are the same as those used in Fig. 4.1, and $P_0 = 0.5$ W. **c** Experimental demonstration of the Peregrine soliton. The temporal intensity profile recorded on an optical sampling oscilloscope (*circles*) is compared with the analytical waveform. More details on the experiment can be found in [95]

frequency band of the instabilities on the (coherent) pump signal, one can expect that the seed will prevail over the spontaneous effects so that enhanced stability and coherence can be ensured. Such an approach has been used in various numerical and experimental works to control the stability and spectral extent of the generated continuum [93, 94]. An accurate spectral study has confirmed that a seeded generation can be well described by the mathematical formalism introduced by Akhmediev et al. [95]. This kind of control can be used to generate Peregrine solitons, which represent a limiting case of Akhmediev breathers for maximum temporal localization or, equivalently, of Kuznetsov breathers for maximum spatial localization, and have the following analytical expression:

$$\psi(z,t) = \sqrt{P_0}\left[\frac{1-4(1+2iz/L_{\mathrm{NL}})}{1+4(t/\tau_0)^2+4(z/L_{\mathrm{NL}})^2}\right]e^{iz/L_{\mathrm{NL}}}. \tag{4.3}$$

Consequently, these pulses over a continuous background of power P_0 have the particularity to appear from nowhere and to disappear without leaving a trace, thus constituting an excellent prototype of oceanic rogue waves (Fig. 4.9). Peregrine solitons have been first predicted theoretically in 1983 in hydrodynamics [96], but have never been demonstrated experimentally before 2010 [97], and using experiments in optical fibers. From an experimental point of view, a sinusoidal beating created by the temporal overlap of two frequency-shifted CWs [97] or the direct intensity modulation of a CW [95] can reshape into a Peregrine soliton [98]. Experimental results obtained by the frequency resolved optical gating technique [97] or by an optical sampling oscilloscope [95] are well reproduced both numerically and analytically (Fig. 4.9c). Longitudinal studies performed by a cut-back technique confirm the strong spatial localization, though some replicas are manifested for large propagation distances. These replicas can be theoretically interpreted in terms of higher-order breather solutions through the Darboux transformations, and linked to the process of higher-order MIs [99]. The Peregrine soliton is only the first-order solution of a wider class of rational solitons that have been also found numerically in turbulent optical environments [84, 100] or in hydrodynamics [101]. We would also like to note that among the few other elementary analytic solutions of the NLS equation, the first soliton on a finite background solution to the NLS equation that was obtained [90] and is now known as the Kuznetsov-Ma soliton [102, 103], is of intense current interest. Indeed, the dynamics of the Kuznetsov-Ma soliton have been very recently confirmed by quantitative experiments in optical fibers [104].

4.5 Conclusions and Perspectives

In this chapter, we have provided an overview of several recent examples of the use of nonlinear phenomena in optical fibers for the shaping of optical pulses in the temporal and spectral domains. The propagation of short pulses in optical fibers is connected with a large variety of interesting and practically important phenomena. The unique dispersive and nonlinear properties of optical fibers lead to various scenarios of the pulse evolution which result in particular changes of the pulse temporal shape, spectrum and phase profile. Because the nonlinear dynamics of pulses propagating in fibers with normal GVD are generally sensitive to the initial pulse condition, it is possible to nonlinearly shape the propagating pulses through control of the initial pulse temporal intensity and/or phase profile. In fact, this approach can be flexibly applied to the generation of specialized temporal waveforms for applications in optical signal processing and manipulation, or the generation of highly coherent continua for optical telecommunications, or the generation of narrow-spectrum pulses for applications requiring high spectral resolution such as nonlinear vibrational microscopy. Application of nonlinear pulse shaping at normal dispersion has also provided significant progress in the field of high-power fiber amplifiers and lasers. The temporal and spectral features of pulses propagating in fibers with anomalous GVD are typically governed by soliton dynamics. The unique properties of optical solitons can be effectively used in the generation of ultrashort pulses with very high repetition rates, which are widely used in various scientific and technological areas, or the generation of frequency-tunable pulses, or the generation of ultra-wide spectra with many disruptive applications and breakthroughs in fields such as optical coherence tomography, metrology and spectroscopy, biophotonics, and others. The quest for a better understanding and harnessing of the temporal and spectral SC fluctuations stimulates the development of new experimental techniques for a quantitative analysis [105]. Seeding the SC generation at anomalous dispersion with a suitably chosen signal can significantly improve the SC stability. Furthermore, the same concept can be applied to stimulate the generation of Peregrine solitons, which constitute a prototype of optical rogue waves. Additional combination with linear preshaping can further widen the field of exploration of these coherent structures [106] as, for example, the recently demonstrated controlled collision of optical breathers highlights [107].

From a purely scientific point of view, the results presented in this chapter are a further illustration of how optical fiber systems can provide useful experimental testbeds for the study of a variety of nonlinear dynamical processes. Therefore, this research area is interesting in its own right. Despite substantial research in this field, qualitatively new phenomena are still being discovered. With respect to application requirements, integrated pulse shaping devices are desirable for applications such as in optical communications where size and robustness are important. The developments of microstructured fibers with extremely small effective core areas and exhibiting enhanced nonlinear characteristics, and of fibres using materials with refractive indexes higher than that of the silica glass have enabled dramatic reduction of the required fiber lengths for nonlinear interactions compared to conventional fibers, thereby paving the way for integrated fiber-based pulse shaping. With the inclusion of higher-order propagation effects, nonlinear shaping can now be observed at the integrated waveguide scale. The recent demonstration of soliton compression in a silicon photonic crystal is an excellent example of this [108].

Acknowledgments We acknowledge important contributions of our colleagues into the original papers discussed in this chapter: K. Hammani, B. Kibler, C.H. Hage, H. Rigneault, E. Andresen, P. Harper, A.I. Latkin, B.G. Bale, F. Parmigiani, P. Petroupoulos, D.J. Richardson, J.M. Dudley, G. Genty and N.N. Akhmediev. We would also like to acknowledge the financial support of the Leverhulme Trust (grant RPG-278), the Conseil Regional de Bourgogne (PARI Photcom) and the Labex ACTION program (ANR-11-LABX-01-01). The experimental work has benefited from the PICASSO Platform of the University of Burgundy.

References

1. Y.R. Shen, *Principles of Nonlinear Optics* (Wiley, New York, 1984)
2. A.M. Weiner, Femtosecond pulse shaping using spatial light modulators. Rev. Sci. Instrum. **71**, 1929–1960 (2000)
3. G.P. Agrawal, *Nonlinear Fiber Optics*, 4th edn. (Academic Press, San Diego, 2006)
4. M.E. Fermann, V.I. Kruglov, B.C. Thomsen, J.M. Dudley, J.D. Harvey, Self-similar propagation and amplification of parabolic pulses in optical fibers. Phys. Rev. Lett. **84**, 6010–6013 (2000)
5. H. Hasegawa, F. Tappert, Transmission of stationary nonlinear optical pulses in dispersive dielectric fibers I. Anomalous dispersion. Appl. Phys. Lett. **23**, 142–144 (1973)
6. L.F. Mollenauer, G.P. Gordon, *Solitons in Optical Fibers: Fundamentals and Applications* (Academic Press, San Diego, CA, 2006)
7. D. Grischowsky, A.C. Balant, Optical pulse compression based on enhanced frequency chirping. Appl. Phys. Lett. **41**, 1–2 (1982)
8. W.J. Tomlinson, R.H. Stolen, A.M. Johnson, Optical wave-breaking of pulses in nonlinear optical fibers. Opt. Lett. **10**, 457–459 (1985)
9. D. Anderson, M. Desaix, M. Lisak, M.L. Quiroga-Teixeiro, Wave-breaking in nonlinear optical fibers. J. Opt. Soc. Am. B **9**, 1358–1361 (1992)

10. C. Finot, B. Kibler, L. Provost, S. Wabnitz, Beneficial impact of wave-breaking for coherent continuum formation in normally dispersive nonlinear fibers. J. Opt. Soc. Am. B **25**, 1938–1948 (2008)
11. S. Boscolo, C. Finot, Nonlinear pulse shaping in fibres for pulse generation and optical processing. Int. J. Opt. **2012**, 159057 (2012). doi:10.1155/2012/159057
12. D. Anderson, M. Desaix, M. Karlsson, M. Lisak, M.L. Quiroga-Teixeiro, Wave-breaking-free pulses in nonlinear optical fibers. J. Opt. Soc. Am. B **10**, 1185–1190 (1993)
13. J.M. Dudley, C. Finot, D.J. Richardson, G. Millot, Self similarity in ultrafast nonlinear optics. Nature Phys. **3**, 597–603 (2007)
14. S. Boscolo, S.K. Turitsyn, V.Y. Novokshenov, J.H.B. Nijhof, Self-similar parabolic optical solitary waves. Theor. Math. Phys. **133**, 1647–1656 (2002)
15. V.I. Kruglov, J.D. Harvey, Asymptotically exact parabolic solutions of the generalized nonlinear Schrödinger equation with varying parameters. J. Opt. Soc. Am. B **23**, 2541–2550 (2006)
16. J.P. Limpert, T. Schreiber, T. Clausnitzer, K. Zöllner, H.-J. Fuchs, E.-B. Kley, H. Zellmer, A. Tünnermann, High-power femtosecond Yb-doped fiber amplifier. Opt. Express **10**, 628–638 (2002)
17. C. Billet, J.M. Dudley, N. Joly, J.C. Knight, Intermediate asymptotic evolution and photonic bandgap fiber compression of optical similaritons around 1550 nm. Opt. Express **13**, 3236–3241 (2005)
18. D.N. Papadopoulos, Y. Zaouter, M. Hanna, F. Druon, E. Mottay, E. Cormier, P. Georges, Generation of 63 fs 4.1 MW peak power pulses from a parabolic fiber amplifier operated beyond the gain bandwidth limit. Opt. Lett. **32**, 2520–2522 (2007)
19. C. Finot, G. Millot, S. Pitois, C. Billet, J.M. Dudley, Numerical and experimental study of parabolic pulses generated via Raman amplification in standard optical fibers. IEEE J. Sel. Top. Quantum Electron. **10**, 1211–1218 (2004)
20. K. Hammani, C. Finot, S. Pitois, J. Fatome, G. Millot, Real-time measurement of long parabolic optical similaritons. Electron. Lett. **44**, 1239–1240 (2008)
21. F.Ö. Ilday, J.R. Buckley, W.G. Clark, F.W Wise, Self-similar evolution of parabolic pulses in a laser. Phys. Rev. Lett. **92**, 213902(4) (2004)
22. W.H. Renninger, A. Chong, F.W. Wise, Self-similar pulse evolution in an all-normal-dispersion laser. Phys. Rev. A **82**, 021805(R) (2010)
23. B.G. Bale, S. Wabnitz, Strong spectral filtering for a mode-locked similariton fiber laser. Opt. Lett. **35**, 2466–2468 (2010)
24. B. Oktem, C. Ülgüdür, F.O. Ilday, Soliton-similariton fibre laser. Nature Photon. **4**, 307–311 (2010)
25. C. Aguergaray, D. Méchin, V.I. Kruglov, J.D. Harvey, Experimental realization of a mode-locked parabolic Raman fiber oscillator. Opt. Express **18**, 8680–8687 (2010)
26. T. Hirooka, M. Nakazawa, Parabolic pulse generation by use of a dispersion-decreasing fiber with normal group velocity dispersion. Opt. Lett. **29**, 498–500 (2004)
27. A. Plocky, A.A. Sysoliatin, A.I. Latkin, V.F. Khopin, P. Harper, J. Harrison, S.K. Turitsyn, Experiments on the generation of parabolic pulses in waveguides with length varying normal chromatic dispersion. JETP Lett. **85**, 319–322 (2007)
28. C. Finot, B. Barviau, G. Millot, A. Guryanov, A.A. Sysoliatin, S. Wabnitz, Parabolic pulse generation with active or passive dispersion decreasing optical fibers. Opt. Express **15**, 15824–15835 (2007)
29. C. Finot, L. Provost, P. Petropoulos, D.J. Richardson, Parabolic pulse generation through passive nonlinear pulse reshaping in a normally dispersive two segment fiber device. Opt. Express **15**, 852–864 (2007)
30. S. Boscolo, A.I. Latkin, S.K. Turitsyn, Passive nonlinear pulse shaping in normally dispersive fiber systems. IEEE J. Quantum Electron. **44**, 1196–1203 (2008)
31. G.I. Barenblatt, *Scaling, Self-Similarity, and Intermediate Asymptotics* (Cambridge University Press, Cambridge, 1996)

32. Y. Kodama, S. Wabnitz, Analytical theory of guiding-center nonreturn-to-zero and return-to-zero signal transmission in normally dispersive nonlinear optical fibers. Opt. Lett. **20**, 2291–2293 (1995)
33. B.G. Bale, S. Boscolo, K. Hammani, C. Finot, Effects of fourth-order fiber dispersion on ultrashort parabolic optical pulses in the normal dispersion regime. J. Opt. Soc. Am. B **28**, 2059–2065 (2011)
34. H. Wang, A.I. Latkin, S. Boscolo, P. Harper, S.K. Turitsyn, Generation of triangular-shaped optical pulses in normally dispersive fibre. J. Opt. (2010). doi:10.1088/2040-8978/12/3/035205
35. N. Verscheure, C. Finot, Pulse doubling and wavelength conversion through triangular nonlinear pulse reshaping. Electron. Lett. **47**, 1194–1196 (2011)
36. S. Boscolo, S.K. Turitsyn, Intermediate asymptotics in nonlinear optical systems. Phys. Rev. A **85**, 043811(5) (2012)
37. R.R. Alfano, *The Supercontinuum Laser Source* (Springer, New York, 2006)
38. J.M. Dudley, J.R. Taylor, *Supercontinuum Generation in Optical Fibers* (Cambridge University Press, New York, 2010)
39. M. Nakazawa, K. Tamura, H. Kubota, E. Yoshida, Coherence degradation in the process of supercontinuum generation in an optical fiber. Opt. Fiber Technol. **4**, 215–223 (1998)
40. Y. Ozeki, Y. Takushima, K. Aiso, K. Taira, K. Kikuchi, Generation of 10 GHz similariton pulse trains from 1,2 km-long erbium-doped fibre amplifier for application to multi-wavelength pulse sources. Electron. Lett. **40**, 1103–1104 (2004)
41. F. Parmigiani, C. Finot, K. Mukasa, M. Ibsen, M.A.F. Roelens, P. Petropoulos, D. J. Richardson, Ultra-flat SPM-broadened spectra in a highly nonlinear fiber using parabolic pulses formed in a fiber Bragg grating. Opt. Express **14**, 7617–7622 (2006)
42. A.M. Clarke, D.G. Williams, M.A.F. Roelens, B.J. Eggleton, Reconfigurable optical pulse generator employing a Fourier-domain programmable optical processor. J. Lightwave Technol. **28**, 97–103 (2010)
43. Y. Xin, D.J. Richardson, P. Petropoulos, Broadband, flat frequency comb generated using pulse shaping-assisted nonlinear spectral broadening. IEEE Photon. Technol. Lett. **25**, 543–545 (2013)
44. L. Boivin, B.C. Collings, Spectrum slicing of coherent sources in optical communications. Opt. Fiber Technol. **7**, 1–20 (2001)
45. Z. Yusoff, P. Petropoulos, K. Furusawa, T.M. Monro, D.J. Richardson, A 36-channel×10-GHz spectrally sliced pulse source based on supercontinuum generation in normally dispersive highly nonlinear holey fiber. IEEE Photon. Technol. Lett. **15**, 1689–1691 (2003)
46. W.J. Tomlinson, R.H. Stolen, C.V. Shank, Compression of optical pulses chirped by self-phase modulation in fibers. J. Opt. Soc. Am. B **1**, 139–149 (1984)
47. S. Demmler, J. Rothhardt, A.M. Heidt, A. Hartung, E.G. Rohwer, H. Bartelt, J. Limpert, A. Tünnermann, Generation of high quality, 1.3 cycle pulses by active phase control of an octave spanning supercontinuum. Opt. Express **19**, 20151–20158 (2011)
48. L.I. Hooper, P.J. Mosley, A.C. Muir, W.J. Wadsworth, J.C. Knight, Coherent supercontinuum generation in photonic crystal fiber with all-normal group velocity dispersion. Opt. Express **19**, 4902–4907 (2011)
49. S.A. Planas, N.L. Pires Mansur, C.H. Brito Cruz, H.L. Fragnito, Spectral narrowing in the propagation of chirped pulses in single-mode fibers. Opt. Lett. **18**, 699–701 (1993)
50. E.R. Andresen, J. Thogersen, S.R. Keiding, Spectral compression of femtosecond pulses in photonic crystal fibers. Opt. Lett. **30**, 2025–2027 (2005)
51. J.P. Limpert, T. Gabler, A. Liem, H. Zellmer, A. Tünnermann, SPM-induced spectral compression of picosecond pulses in a single-mode Yb-doped fiber amplifier. Appl. Phys. B: Lasers Opt. **74**, 191–195 (2002)
52. J.P. Limpert, N. Deguil-Robin, I. Manek-Hönninger, F. Salin, T. Schreiber, A. Liem, F. Röser, H. Zellmer, A. Tünnermann, A. Courjaud, C. Hönninger, E. Mottay, High-power picosecond fiber amplifier based on nonlinear spectral compression. Opt. Lett. **30**, 714–716 (2005)

53. B. Washburn, J. Buck, S. Ralph, Transform-limited spectral compression due to self-phase modulation in fibers. Opt. Lett. **25**, 445–447 (2000)
54. C. Finot, A. Guenot, P. Dupriez, Spectral compression of optical parabolic similaritons. Ann. Phys. France **32**, 71–74 (2007)
55. J. Fatome, B. Kibler, E.R. Andresen, H. Rigneault, C. Finot, All-fiber spectral compression of picosecond pulses at telecommunication wavelength enhanced by amplitude shaping. Appl. Opt. **51**, 4547–4553 (2012)
56. E.R. Andresen, J.M. Dudley, C. Finot, D. Oron, H. Rigneault, Transform-limited spectral compression by self-phase modulation of amplitude shaped pulses with negative chirp. Opt. Lett. **36**, 707–709 (2011)
57. S. Boscolo, S.K. Turitsyn, C. Finot, Amplifier similariton fiber laser with nonlinear spectral compression. Opt. Lett. **37**, 4531–4533 (2012)
58. S.V. Chernikov, P.V. Mamyshev, Femtosecond soliton propagation in fibers with slowly decreasing dispersion. J. Opt. Soc. Am. B **8**, 1633–1641 (1991)
59. S.V. Chernikov, D.J. Richardson, E.M. Dianov, D.N. Payne, Picosecond soliton pulse compressor based on dispersion decreasing fibre. Electron. Lett. **28**, 1842–1844 (1992)
60. M.L. Tse, P. Horak, J.H.V. Price, F. Poletti, F. He, D.J. Richardson, Pulse compression at 1.06 μm in dispersion-decreasing holey fibers. Opt. Lett. **31**, 3504–3506 (2006)
61. V.A. Bogatyrev, M.M. Bubnov, E.M. Dianov, A.S. Kurkov, P.V. Mamyshev, A.M. Prokhorov, S.D. Rumyantsev, V.A. Semenov, S.L. Semenov, A.A. Sysoliatin, S.V. Chernikov, A.N. Guryanov, G.G. Devyatykh, S.I. Miroshnichenko, A single-mode fiber with chromatic dispersion varying along the length. J. Lightwave Technol. **9**, 561–565 (1991)
62. S.V. Chernikov, J.R. Taylor, R. Kashyap, Experimental demonstration of step-like dispersion profiling in optical fibre for soliton pulse generation and compression. Electron. Lett. **30**, 433–435 (1994)
63. S.V. Chernikov, J.R. Taylor, R. Kashyap, Comblike dispersion-profiled fiber for soliton pulse train generation. Opt. Lett. **19**, 539–541 (1994)
64. L.F. Mollenauer, R.H. Stolen, J.P. Gordon, Experimental observation of picosecond pulse narrowing and solitons in optical fibers. Phys. Rev. Lett. **45**, 1095–1098 (1980)
65. B. Kibler, R. Fischer, R.A. Lacourt, E. Courvoisier, R. Ferriere, L. Larger, D.N. Neshev, J.M. Dudley, Optimised one-step compression of femtosecond fibre laser soliton pulses around 1550 nm to below 30 fs in highly nonlinear fibre. Electron. Lett. **43**, 915–916 (2007)
66. A.A. Amorim, M.V. Tognetti, P. Oliveira, J.L. Silva, L.M. Bernardo, F.X. Kärtner, H.M. Crespo, Sub-two cycle pulses by soliton self-compression in highly nonlinear photonic crystal fibers. Opt. Lett. **34**, 3851–3853 (2009)
67. A. Hasegawa, Generation of a train of soliton pulses by induced modulational instability in optical fibers. Opt. Lett. **9**, 288–290 (1984)
68. K. Tai, A. Hasegawa, A. Tomita, Observation of modulation instability in optical fibers. Phys. Rev. Lett. **56**, 135–138 (1986)
69. S. Pitois, J. Fatome, G. Millot, Generation of a 160 GHz transform-limited pedestal-free pulse train through multiwave mixing compression of a dual-frequency beat signal. Opt. Lett. **27**, 1729–1731 (2002)
70. J. Fatome, S. Pitois, G. Millot, 20-GHz-to-1-THz repetition rate pulse sources based on multiple four-wave mixing in optical fibers. IEEE J. Quantum Electron. **42**, 1038–1046 (2006)
71. J. Fatome, S. Pitois, C. Fortier, B. Kibler, C. Finot, G. Millot, C. Courde, M. Lintz, E. Samain, Multiple four-wave mixing in optical fibers: 1.5–3.4-THz femtosecond pulse sources and real-time monitoring of a 20-GHz picosecond source. Opt. Commun. **283**, 2425–2429 (2010)
72. I. El Mansouri, J. Fatome, C. Finot, M. Lintz, S. Pitois, All-fibered high-quality stable 20- and 40-GHz picosecond pulse generators for 160-Gb/s OTDM applications. IEEE Photon. Technol. Lett. **23**, 1487–1489 (2011)
73. C. Finot, J. Fatome, S. Pitois, G. Millot, All-fibered high-quality low duty-cycle 20-GHz and 40-GHz picosecond pulse sources. IEEE Photon. Technol. Lett. **19**, 1711–1713 (2007)

74. F.M. Mitschke, L.F. Mollenauer, Discovery of the soliton self-frequency shift. Opt. Lett. **11**, 659–661 (1986)
75. J.P. Gordon, Theory of the soliton self-frequency shift. Opt. Lett. **11**, 662–664 (1986)
76. E.R. Andresen, V. Birkedal, J. Thogersen, S.R. Keiding, Tunable light source for coherent anti-Stockes Raman scattering microspectroscopy based on the soliton self-frequency shift. Opt. Lett. **31**, 1328–1330 (2006)
77. T. Johnson, S. Diddams, Mid-Infrared Upconversion Spectroscopy Based on a Yb: Fiber Femtosecond Laser. Appl. Phys. B **107**, 31–39 (2012)
78. D.V. Skryabin, F. Luan, J.C. Knight, J.S. Russell, Soliton self-frequency shift cancellation in photonic crystal fibers. Science **201**, 1705–1708 (2003)
79. P.V. Mamyshev, P.G.J. Wigley, J. Wilson, G.I. Stegeman, V.A. Semenov, E.M. Dianov, S.I. Miroshnichenko, Adiabatic compression of Schrödinger solitons due to the combined perturbations of higher-order dispersion and delayed nonlinear response. Phys. Rev. Lett. **71**, 73–76 (1993)
80. R. Pant, A.C. Judge, E.C. Magi, B.T. Kuhlmey, M. de Sterke, B.J. Eggleton, Characterization and optimization of photonic crystal fibers for enhanced soliton self-frequency shift. J. Opt. Soc. Am. B **27**, 1894–1901 (2010)
81. J.M. Dudley, G. Genty, S. Coen, Supercontinuum generation in photonic crystal fiber. Rev. Modern Phys. **78**, 1135–1184 (2006)
82. K.L. Corwin, N.R. Newbury, J.M. Dudley, S. Coen, S.A. Diddams, K. Weber, R.S. Windeler, Fundamental noise limitations to supercontinuum generation in microstructure fibers. Phys. Rev.Lett. **90**, 113904(4) (2003)
83. D.R. Solli, C. Ropers, P. Koonath, B. Jalali, Optical rogue waves. Nature **450**, 1054–1057 (2007)
84. B. Kibler, K. Hammani, C. Michel, C. Finot, A. Picozzi, Rogue waves, rational solitons and wave turbulence theory. Phys. Lett. A **375**, 3149–3155 (2011)
85. J.M. Dudley, C. Finot, G. Millot, J. Garnier, G. Genty, D. Agafontsev, F. Dias, Extreme events in optics: challenges of the MANUREVA project. Eur. Phys. J. Special Topics **185**, 125–133 (2010)
86. C. Kharif, E. Pelinovsky, A. Slunyaev, *Rogue Waves in the Ocean* (Springer, Berlin, 2009)
87. M. Taki, A. Mussot, A. Kudlinski, E. Louvergneaux, M.I. Kolobov, M. Douay, Third-order dispersion for generating optical rogue solitons. Phys. Lett. A **374**, 691–695 (2010)
88. G. Genty, C.M. de Sterke, O. Bang, F. Dias, N.N. Akhmediev, J.M. Dudley, Collisions and turbulence in optical rogue wave formation. Phys. Lett. A **374**, 989–996 (2010)
89. K. Hammani, B. Kibler, C. Finot, A. Picozzi, Emergence of rogue waves from optical turbulence. Phys. Lett. A **374**, 3585–3589 (2010)
90. E.A. Kuznetsov, Solitons in a parametrically unstable plasma. Doklady Akamemii Nauk SSSR **22**, 507–508 (1977)
91. N.N. Akhmediev, V.I. Korneev, Modulation instability and periodic-solutions of the nonlinear Schrödinger equation. Theor. Math. Phys. **69**, 1089–1093 (1986)
92. J.M. Dudley, G. Genty, F. Dias, B. Kibler, N.N. Akhmediev, Modulation instability, Akhmediev breathers and continuous wave supercontinuum generation. Opt. Express **17**, 21497–21508 (2009)
93. G. Genty, J.M. Dudley, Route to coherent supercontinuum generation in the long pulse regime. IEEE J. Quantum Electron. **45**, 1331–1335 (2009)
94. D.R. Solli, C. Ropers, B. Jalali, Active control of rogue waves for stimulated supercontinuum generation. Phys. Rev. Lett. **101**, 233902(4) (2008)
95. K. Hammani, B. Kibler, C. Finot, P. Morin, J. Fatome, J.M. Dudley, G. Millot, Peregrine soliton generation and breakup in standard telecommunications fiber. Opt. Lett. **36**, 112–114 (2011)
96. D.H. Peregrine, Water waves, nonlinear Schrödinger equations and their solutions. J. Austral. Math. Soc. Ser. B **25**, 16–43 (1983)
97. B. Kibler, J. Fatome, C. Finot, G. Millot, F. Dias, G. Genty, N. Akhmediev, J.M. Dudley, The Peregrine soliton in nonlinear fibre optics. Nature Phys. **6**, 790–795 (2010)

98. M. Erkintalo, G. Genty, B. Wetzel, J.M. Dudley, Akhmediev breather evolution in optical fiber for realistic initial conditions. Phys. Lett. A **375**, 2029–2034 (2011)
99. M. Erkintalo, K. Hammani, B. Kibler, C. Finot, N. Akhmediev, J.M. Dudley, G. Genty, Higher-order modulation instability in nonlinear fiber optics. Phys. Rev. Lett. **107**, 253901(4) (2011)
100. N. Akhmediev, J.M. Soto-Crespo, A. Ankiewicz, Extreme waves that appear from nowhere: on the nature of rogue waves. Phys. Lett. A **373**, 2137–2145 (2009)
101. A. Chabchoub, N. Hoffmann, M. Onorato, N. Akhmediev, Super rogue waves: Observation of a higher-order breather in water waves. Phys. Rev. X **2**, 011015(6) (2012)
102. T. Kawata, H. Inoue, Inverse scattering method for the nonlinear evolution equations under nonvanishing conditions. J. Phys. Soc. Japan **44**, 1722–1729 (1978)
103. Y.C. Ma, The perturbed plane-wave solutions of the cubic Schrödinger equation. Stud. Appl. Math. **60**, 43–58 (1979)
104. B. Kibler, J. Fatome, C. Finot, G. Millot, G. Genty, B. Wetzel, N. Akhmediev, F. Dias, J.M. Dudley, Observation of Kuznetsov-Ma soliton dynamics in optical fiber. Sci. Rep. **2**, 463 (2012). doi:10.1038/srep00463
105. B. Wetzel, A. Stefani, L. Larger, P. Lacourt, J. Merolla, T. Sylvestre, A. Kudlinski, A. Mussot, G. Genty, F. Dias, Real-time full bandwidth measurement of spectral noise in supercontinuum generation. Sci. Rep. **2**, 883 (2012). doi:10.1038/srep00882
106. B. Frisquet, A. Chabchoub, J. Fatome, C. Finot, B. Kibler, G. Millot, Two-stage linear-nonlinear shaping of an optical frequency comb as rogue nonlinear-Schrödinger-equation-solution generator. Phys. Rev. A **89**, 023821 (2014)
107. B. Frisquet, B. Kibler, G. Millot, Collision of Akhmediev breathers in nonlinear fiber optics. Phys. Rev. X **3**, 041032(9) (2013)
108. A. Blanco-Redondo, C. Husko, D. Eades, Y. Zhang, J. Li, T.F. Krauss, B.J. Eggleton, Observation of soliton compression in silicon photonic crystals. Nature Commun. **5**, 3160 (2014). doi:10.1038/ncomms4160

Chapter 5
Optical Regeneration

Francesca Parmigiani, Radan Slavík, Joseph Kakande, Periklis Petropoulos and David Richardson

Abstract In this chapter we review the need, general principles and approaches used to regenerate mainly phase encoded signals of differing levels of coding complexity. We will describe the key underpinning technology and present the current state-of-the-art, incorporating an appropriate historic perspective throughout.

5.1 Introduction to Optical Regeneration

Data traffic on the world's core optical networks is increasing at approximately 30–40 % per annum driven largely by the emergence of new mobile and video-centric applications, cloud computing and the ever increasing number of internet/mobile users [1]. This rate of traffic growth is presenting huge challenges to network hardware vendors and operators alike, particularly since in the laboratory the ultimate capacity of conventional single mode fiber technology is already being approached as a result of the latest advances in the use of digital coherent transmission and the use of advanced modulation format signals [2]. Use of the phase (and often additionally the intensity) of an optical carrier rather than simple On-Off Keying (OOK) is already a central feature of core network deployments today, since as well as allowing for more spectrally efficient communications, it offers advantages in terms of improved resilience to certain linear and nonlinear transmission impairments [3]. However, the introduction of phase encoding means that attention needs to be paid to nonlinear phase noise, which manifests itself as a new dominant limitation to ultimate system performance [4–6]. In amplified

F. Parmigiani (✉) · R. Slavík · P. Petropoulos · D. Richardson
Optoelectronics Research Centre, University of Southampton, Southampton, Hampshire SO17 1BJ, UK
e-mail: frp@orc.soton.ac.uk

J. Kakande
Bell Laboratories, Alcatel-Lucent, Holmdel, NJ 07733, USA

S. Wabnitz and B.J. Eggleton (eds.), *All-Optical Signal Processing*,
Springer Series in Optical Sciences 194, DOI 10.1007/978-3-319-14992-9_5

transmission systems, nonlinear phase noise originates from the combined effects of noise introduced by the in-line optical amplifiers and the nonlinear interaction between the various co-propagating wavelength division multiplexing (WDM) channels. For this reason it is beneficial to periodically regenerate signals during transmission over long distances in order to increase capacity and/or reach. In currently installed transmission systems this is achieved by employing optical-to-electronic-to-optical (O/E/O) conversion and fast electronic processors. However, all-optical techniques that can perform regeneration by avoiding the need for O/E/O conversion merit attention, since they are often associated with additional attractive features, such as transparency to the symbol rate and a more favorable scaling of the energy consumption. The ability to regenerate signals also ensures maximum flexibility in network design and architecture, for example: allowing for a significant reduction in penalties due to accumulated cross-talk and filter concatenation in all optical networks based on components such as Reconfigurable Optical Add Drop Multiplexers (ROADMs), more flexible network design since performance is less route dependent, and potentially providing greater tolerance to changes in network load (adding and dropping channels).

As a consequence much research effort has been devoted over the years to the implementation of all-optical solutions for signal regeneration using either speciality optical fibers or semiconductor optical amplifier (SOA)-based devices as the nonlinear optical processing medium. In the past the primary focus was on regenerators for the then prevalent OOK modulation format signals, however in more recent times attention has turned onto the regeneration of more advanced modulation formats that incorporate phase encoding. In this chapter we first briefly review the early development of regenerators for OOK signals before turning our attention to the even thornier problem of regenerating mainly phase-based modulation format (M-ary phase shift keying, MPSK) signals with some initial efforts towards amplitude and quadrature modulation (QAM) signals.

5.2 Optical Regenerators for Simple Amplitude Encoded Signals

In accordance with the traditional optical transmission approach and technology, all early types of optical regenerator were designed to operate on OOK signals. Such regenerators were first proposed in the ‘90s and were intensively studied throughout the next decade or so [7–14].

In the most general form of regenerator, three basic signal-processing functions are required: (i) Reamplification, (ii) Reshaping and (iii) Retiming. A regenerator simultaneously providing each of these functions is referred to as a 3R-regenerator. Applying this nomenclature, optical amplifiers solely provide 1R-regeneration whereas the extended class of devices that simultaneously provide reamplification and reshaping are referred to as 2R-regenerators. A sketch of the generic layout of a typical “3R” regenerator is shown in Fig. 5.1, and ordinarily comprises:

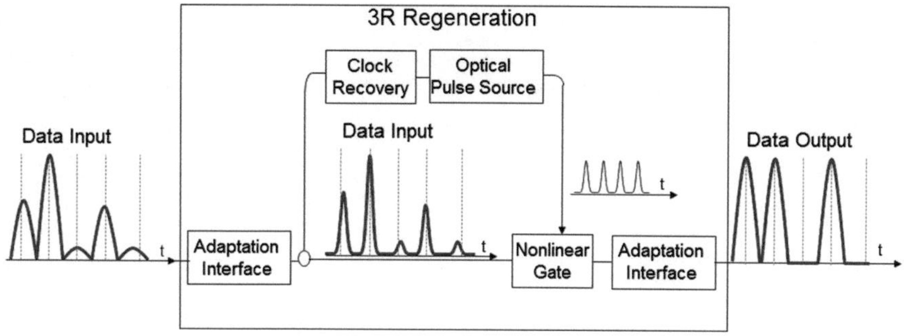

Fig. 5.1 A schematic of the generic layout of a typical "3R" regenerator with a degraded OOK signal present at the input. The input/output adaptation interfaces address issues such as input/output power adjustment, initial (linear) pulse shaping, out-of-band noise filtering, format conversion, control of the signal polarisation states, signal de-multiplexing/multiplexing (be it in the time and/or frequency domain). The clock recovery unit retrieves an accurate clock from the incoming data stream and provides appropriate synchronisation between the recovered clock and the incident data within the nonlinear (optical) gate which provides for the 3R regeneration itself. Adapted from [14, 15]

(i) input/output adaptation interfaces in order to convert the input data signal into a form suitable for use within the regenerator, and/or to convert the internally regenerated signal into the correct format for onward transmission, (ii) some form of nonlinear gate, and (iii) a clock recovery/generation device [14, 15]. The nonlinear optical gate, which ultimately provides the reshaping and retiming function, is generally the most critical element of any regenerator. Over the years many nonlinear gates have been investigated. The greatest majority of work has used an optical fiber as the nonlinear medium [7–12] and has relied upon exploitation of ultrafast nonlinear optical effects such as Self Phase Modulation (SPM), Cross Phase Modulation (XPM), Four Wave Mixing (FWM) and Stimulated Raman Scattering (SRS) to provide the required optical switching mechanism. However, a wide variety of other nonlinear materials/devices have also been explored including SOA's [15–17], various saturable absorbers (SA's) [18], periodically poled lithium niobate (PPLN) waveguides [19], silicon chips [20] and chalcogenide waveguides [21], amongst others.

The majority of fiber based gates are adaptations of devices based on either: (i) spectral broadening and offset-filtering, the so-called Mamyshev regenerator [10, 12, 16, 21], whose schematic and working principle are reported in Fig. 5.2a; or (ii) optical interferometers with a nonlinear medium in one of the arms, where one or two signals (the noisy data and the clean clock signal) might be present at its input, as shown in Fig. 5.2c [7, 8, 14, 15, 17]. While the nonlinear media and the nonlinear processes exploited might differ, all amplitude regeneration schemes aim to achieve a step-like power transfer function, defined in terms of the instantaneous output versus input signal power. The slope of the power transfer function at the logical "0" and "1" bit power levels defines the regenerative properties of the device (see Fig. 5.2b, d).

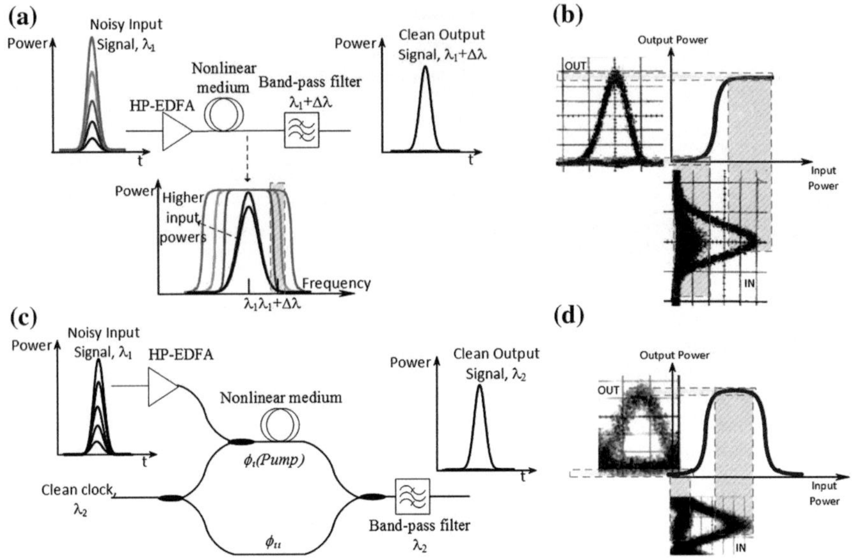

Fig. 5.2 Examples of two forms of amplitude regenerators: ((**a**) the Mamyshev 2R-regenerator and (**c**) a 3R-interferometric regenerator) and corresponding power transfer functions (**b** and **d**) together with experimental eye diagrams before and after the regenerators (Adapted from [12, 14])

Some OOK regenerators can still be used in the regeneration of modern MPSK-encoded signals. To enable this, the adaptation stage converts an incoming optical MPSK signal to one or more OOK signals, amplitude regeneration is then performed on this signal before converting the signal back to an MPSK format at the output adaptation stage [22]. Such regenerators are referred to as format conversion based regenerators. Alternatively, those classes of amplitude regenerator that preserve the phase of the optical signal, can be deployed on the basis that they suppress any amplitude fluctuations on a MPSK signal without distorting the phase information [23–30]. Such regenerators are referred to as phase preserving amplitude regenerators.

The extension to simultaneous multi-channel regeneration is a key issue in today's high capacity systems. Such an extension is very challenging in practice as this is generally limited by the occurrence of unwanted nonlinear interactions amongst WDM channels, the so-called inter-channel crosstalk, which directly competes with the nonlinear effects used for the regeneration process itself. In optical fibers, such cross-talk typically results from XPM and FWM interactions, which manifest themselves as strong spectral distortions and/or spectrally dependent power transfer functions due to the incomplete interaction of co-propagating pulses during nonlinear propagation (i.e. the pulses involved have not fully walked through each other during propagation).

Several techniques have been applied to simultaneously regenerate more than one WDM channel. The main approaches are shown in Fig. 5.3.

The first is based on wavelength-to-space demultiplexing where each wavelength channel is split out into a separate path using a wavelength demultiplexer and identical single channel regenerators are employed on a per-channel basis, see Fig. 5.3a [14, 18, 30]. After regeneration, the channels are then recombined together using a second wavelength multiplexer for further transmission. Such a per-channel scheme clearly avoids nonlinear cross-talk-induced distortion; however, the complexity (and cost) of the scheme increases linearly with the number of channels to be simultaneously regenerated, which is far from desirable.

The second approach is based on time-interleaving of the input signals [31–33]. Here the various WDM channels are aligned in time to avoid bit overlap before entering the regenerator via some sort of WDM to optical time division multiplex (OTDM) conversion process, see Fig. 5.3b. The individual WDM channel data symbols are then sequentially processed in a single nonlinear device, guaranteeing a reduction in component cost as the number of channels increases. However, this approach requires both channel synchronisation and a low duty cycle (low spectral efficiency) signal format since the maximum time-slot that a channel can occupy is now determined by the number of WDM channels simultaneously processed by the regenerator.

The third approach is based on carefully engineered dispersive 'channel walk-through maps' in the regenerator to avoid inter-channel cross-talk [32, 34–40]. In the co-propagating geometry, a high local dispersion value ensures that nonlinear inter-channel effects are reduced due to the fast walk-off time amongst the various WDM channels (thereby reducing any bit-pattern dependence of the nonlinear cross-talk), while a low overall average dispersion value (obtained via dispersion

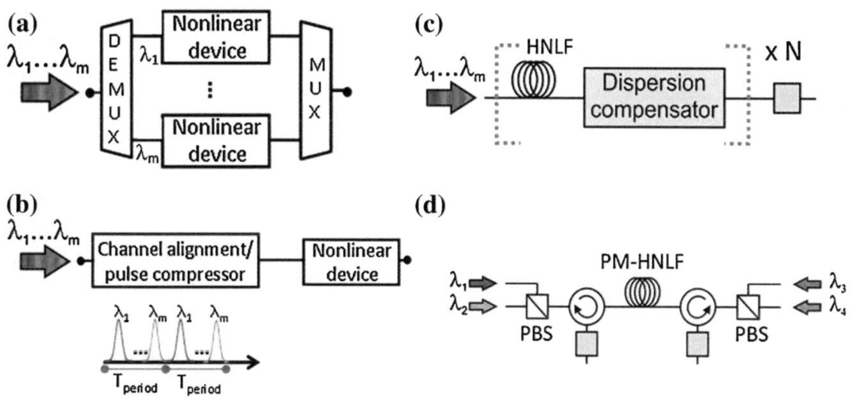

Fig. 5.3 Schematic of multi-channel 2R regenerators based on **a** wavelength-to-space demultiplexing, **b** time interleaving and dispersive walk-through in **c** co- or **d** counter-propagating geometries, exploiting SPM. *(DE)MUX* (de)multiplexer; *PBS* polarization beam splitter; *(PM-) HNLF* (polarization maintaining-) highly nonlinear fiber. *Gray squares* represent carrier-offset filters

compensation) ensures that the nonlinear effect used for the regeneration process itself (SPM for example) remains sufficiently strong, Fig. 5.3c. The regenerator, thus, consists of multiple modules incorporating appropriate combinations of highly nonlinear fibers (HNLFs) with high normal dispersion and associated periodic group delay devices (PGDDs) or dispersion compensating fibers to generate/emulate saw-tooth-like channelized dispersion profiles. This approach enables in-line processing of a (potentially quite large) number of WDM channels [32, 34–37] simultaneously in a single nonlinear device, guaranteeing a reduction in cost-per-channel as the number of channels increases. However, while the previous two approaches can potentially use any kind of nonlinear device and nonlinearity, this third approach can only be implemented using SPM in fibers and is thus only applicable to amplitude-only signals, as SPM heavily distorts the signal phase.

In the counter-propagating geometry, Fig. 5.3d, an extremely high walk-off is automatically guaranteed for counter propagating channels in the nonlinear gate. A further two-fold channel scaling can be achieved using a polarization multiplexing configuration which exploits the rapid interchannel pulse walk-off due to the large birefringence-induced differential group delay achievable within the polarisation maintaining (PM)-HNLF. A further three-fold reduction in XPM is also obtained due to the tensorial nature of the Kerr effect between orthogonally-polarized signals [38–40]. So far it has been shown that up to four channels can be processed simultaneously in a single nonlinear device [38, 39].

Having given a quick review of the status of the work on the regeneration of OOK signals, largely for contextual setting, we turn our attention to more recent work concerned with the regeneration of more modern and spectrally-efficient MPSK and QAM signals.

5.3 Regeneration of Phase-Only Encoded Signals

As we briefly mentioned in Sect. 5.2, phase regeneration can be divided into two broad categories: direct regeneration, which operates on the premise that a signal has already been significantly degraded in phase, and therefore signal quality can only be improved by suppressing these phase perturbations; and indirect regeneration, which acts to suppress amplitude fluctuations which could otherwise seed phase noise after subsequent transmission (phase preserving amplitude regeneration) [23–30]. However, to make it worthwhile, the latter process must not create significant added phase noise. Direct regeneration can be classified either as format conversion based regeneration from MPSK to OOK and vice versa (without the requirement to be phase-preserving) [22], or phase sensitive amplification (PSA) based regeneration, which has been widely recognized as an effective way to regenerate phase-encoded signals thanks to its phase squeezing capabilities [41, 42]. In the rest of the chapter PSA-based phase regenerators only will be considered.

Phase sensitive amplifiers exhibit gain characteristics that depend on the phase of the input optical signal relative to a local optical reference; they operate by amplifying one of the input signals quadratures whilst simultaneously de-amplifying the other [43]. While the theory of PSAs is not recent, historically attention focused mainly on exploiting second order nonlinearity in bulk crystals [44, 45], it is only relatively recently that much experimental effort has been placed on this topic in the context of optical fiber communications. This growth in interest has largely been driven by technological advances that have allowed a number of previously considered practical limitations to be overcome, as will be discussed below.

The implementation of PSAs using third order nonlinearity in optical fibers rather than second order nonlinearity in crystals was a first step towards practical network applications due to the increased robustness, improved power efficiency and ease of system integration that optical fibers offer. Marhic et al. successfully demonstrated the first degenerate PSA by 1991 [43], but progress in the field was quite limited up until the demonstration of amplification with a 1.8 dB noise figure by Imajuku et al. in 1999 [46]. Since then, PSAs have been demonstrated operating over wide bandwidths in excess of 20 nm [47–49] as well as with a record low noise figure of 0.8 dB [50].

Phase sensitive amplification in fibers can be accomplished either using interferometric schemes [41, 42], or schemes based on FWM [51]. The former approach is advantageous in terms of the feasibility of phase locking (signal and pump coincide) and broad bandwidth operation, however it is accompanied by severe disadvantages such as strict single-wavelength operation, guided acoustic wave Brillouin scattering (GAWBS) and low energy efficiency compared to FWM based counterparts, since the gain scales quadratically with pump power as compared to exponentially in the latter case. For these reasons, non-interferometric PSA implementations have been the preferred choice in most recent demonstrations of phase regeneration.

A FWM-based PSA can be implemented using single- or dual- pump configurations with signal, idler and pump(s) being along the same linear polarization ('scalar schemes') or different polarization axes ('vector schemes')—all possibilities cover both non-degenerate and degenerate (signal and idler at the same wavelength) configurations. In the literature, FWM-based PSAs that have been successfully used to suppress noise on a data signal can be divided into two main categories. In the first one, the same data is carried on two signals at different wavelengths—the first signal is the original data signal and the second its complex-conjugated copy (idler). The PSA amplifies equally both signal quadratures of the perfect phase-matched, power-equalized signal-idler pair, which consequently experiences 6 dB higher gain than the corresponding noise which is uncorrelated (and which consequently experiences phase insensitive amplification, PIA, only) in the high gain regime—a feature that is inherent to a PSA [52–54], see Fig. 5.4. This approach does not require any knowledge of the modulation format used since the absolute input signal phase is tracked by that of the conjugated idler, and, thus, we call it a 'Modulation Format Transparent PSA'. We intentionally do not use the term 'regenerator', in line with the existing literature. The modulation format transparent PSA inherently requires a

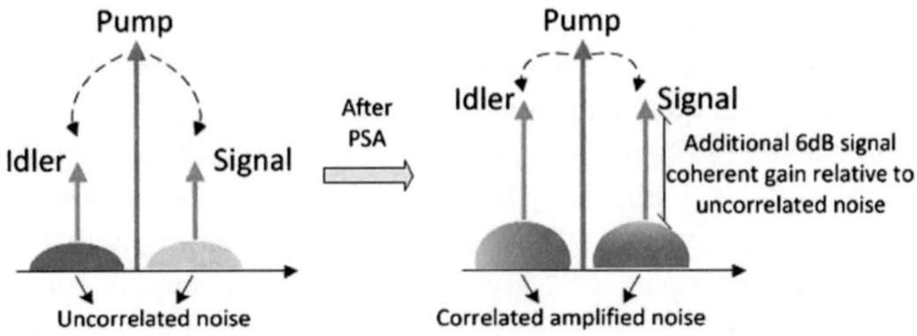

Fig. 5.4 Basic principle of the modulation format transparent PSA. Adapted from [54]

non-degenerate scheme, as the signal and idler need to propagate through the network at different wavelengths, accumulating uncorrelated noise, implying occupation of twice the bandwidth as compared to a degenerate configuration, together with the requirement of perfect phase synchronization amongst all the waves prior to the PSA. Practically, the signal-idler pairs are usually generated in a parametric PIA before the transmission link, which is generally referred to as a 'copier' given its function [53, 54]. Note that both copier and PSA should operate in a linear gain regime to avoid gain-saturation-induced distortion. Finally it should be added that modulation format transparent PSA schemes can also easily be extended to multi-channel amplification for the same pump configuration.

The second approach requires only the original incident data signal to propagate through the transmission fiber link, guaranteeing optimum bandwidth occupation, which is the standard approach used in today's telecommunications. To regenerate such a signal, the modulation format must be known and, thus, we call such schemes 'Modulation Format Specific PSA Regenerators'. In such a configuration, the idler involved in the PSA process is then generated locally (following a possible transmission link) and it carries identical or an exact multiple of the original signal phase information. Some examples of possible FWM configurations are shown in Fig. 5.5a, d, respectively. As such, one signal quadrature will be amplified, whereas the orthogonal one is de-amplified, differing in the operation principle from the modulation format transparent PSA. In the simplest case of binary phase shift keying (BPSK) modulation format regeneration, the dual-pump degenerate scalar configuration is the most commonly used, see Fig. 5.5a. Here, the in-phase component of the electric field experiences gain, g, while the quadrature component experiences de-amplification by $1/g$ in the linear regime [53]. Subsequently, the output phase of the amplified signal is more closely aligned to the amplifier's in-phase axis (see Fig. 5.5b, c). This phase quantization effect, sometimes referred to as 'phase squeezing', is inherently suitable for the regeneration of binary phase-encoded signals, as the 'phase squeezed' data bits are forced to adopt a phase of either 0 or π, thus restoring the fidelity of the signal prior to transmission [55–59]. Note that squeezing in this context refers to a classical effect, rather than one of a quantum nature. Figure 5.5b, c also show that a side-effect of the PSA action is the introduction of unwanted amplitude variations added to the signal; the larger the phase error the PSA corrects for, the larger the amplitude variation. However, this

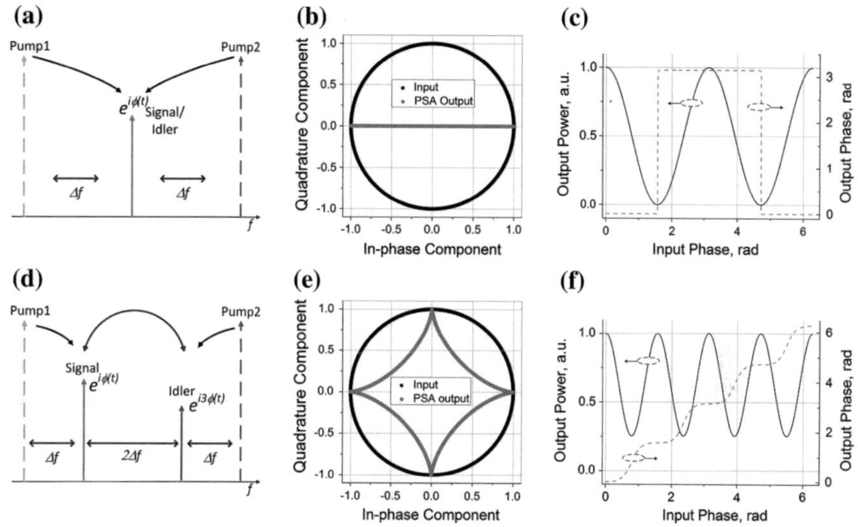

Fig. 5.5 Sketches of two modulation format specific PSA configurations (**a** and **d**), corresponding constellation diagrams (**b** and **e**) and transfer functions of the output amplitude/phase versus input phase (**c** and **f**) derived from (5.1), where $m_1 = 1/(M-1)$ and $M = 2$ (*top line*) and $M = 4$ (*bottom line*), respectively

can be reduced by operating the PSA in saturation to further suppress any amplitude noise present on the original phase-encoded signal, enabling simultaneous phase and amplitude regeneration [57–59].

As such schemes provide inherently single channel operation for the same pump configuration there have been attempts to adapt this scheme to perform multi-wavelength regeneration by employing more pumps [60, 61] (their number scales linearly with the number of channels); however, a penalty is likely to be paid due to strong and undesired pump-to-pump interactions [60, 61].

The Modulation Format Specific PSA regenerator scheme can be extended to allow the phase regeneration of an MPSK signal with an arbitrary number M of phase coding levels. This requires the realization of a staircase phase response that includes the corresponding number of steps. This can be implemented through a coherent addition of the (M − 1)-th phase harmonic of the signal with the signal itself, which mathematically can be expressed as [62–64]:

$$Ae^{i\theta(t)} \propto e^{i\phi(t)} + m_1 e^{-i(M-1)\phi(t)}, \tag{5.1}$$

where A and $\theta(t)$ are the output signal electric field amplitude and phase, respectively, $\phi(t)$ is the input phase and m_1 is the phase harmonic weight. It is worth emphasizing that also in this configuration, the (M – 1)-th phase harmonic (idler) is generated inside the regenerator, i.e. after transmission, so its phase noise is strongly correlated to that of the original signal. Figure 5.5e, f show the corresponding constellation diagrams and transfer functions derived from (5.1) for $M = 4$

levels of phase modulation and $m_1 = 1/(M - 1)$. The periodicity of the phase transfer function matches M, but the abruptness of the steps becomes less well defined, and the depth of the phase-to-amplitude conversion is also seen to reduce. A sharper step-like phase function could be achieved if more phase harmonics of the signal were coherently added to the original signal.

Umeki et al. have compared optical-signal-to-noise-ratio (OSNR) and electrical signal to noise ratio (SNR) at the input/output of the modulation format transparent PSA and modulation specific PSA regenerator [65]. They showed experimentally that, while it is possible to obtain 3 dB OSNR improvement with both schemes, only the modulation transparent PSA allows a 3 dB SNR improvement. However, it is worth emphasizing that in their study both PSAs were operating in the linear (non-saturated) regime.

In general, the implementation of any PSA raises its own practical challenges and in order to achieve phase-sensitive amplification, the phase relationship between the PSA pump(s), signal and any idlers present needs to be maintained [47]. The challenges are mainly as follows: (i) phase-encoded signals usually have a suppressed carrier field (i.e. they are 'carrier-less') and (ii) even if the carrier were extractable, it would generally still contain a component of the phase and amplitude noise of the transmitted data. Proof-of-principle PSA demonstrations have overcome this problem by ensuring that both the signal and the pump(s) originate from a common laser source, e.g. through the generation of an optical frequency comb [56, 57]. However, in a real transmission link, any pumps would have to be generated locally in the regenerating node to be truly independent of the data signal. An additional challenge, common in many parametric amplification configurations, is that strong narrow-linewidth continuous wave (CW) laser signals need to be used as PSA pumps. When the nonlinear element used for the implementation of the PSA is an optical fiber, a strong limitation on the amount of pump power that can be launched into the fiber is imposed by the onset of stimulated Brillouin scattering (SBS).

In the remaining part of this section we review work to date on PSA regeneration, mainly in chronological order, and through the description of the experimental demonstrations used, we will show how the above challenges can be overcome to realize a phase regenerator for use in real-world transmission links.

5.3.1 Modulation Format Specific PSA Regenerators

The first successful experimental demonstration of a degenerate dual pump scalar PSA regenerator operating both on phase and amplitude of a BPSK signal was reported by Croussore and Li [57]. In this instance the phase-locking between pumps and signal waves was guaranteed by generating the pumps from the same CW-laser used to produce the data signal (through comb lines generated by 40 GHz amplitude modulation). SBS was avoided by using a bismuth oxide HNLF offering an enhanced Kerr nonlinearity to SBS figure of merit. Noise was added by additional amplitude and phase modulation at a chosen single frequency to the data signal. For amplitude regeneration, saturation of the FWM process was exploited.

A major step toward the realization of a practical phase sensitive regenerator was achieved in [58, 59, 66] where the issues of phase locking the various waves, coping with real-world impairments (wideband noise) and the use of narrow-line-width CW pumps were addressed.

The schematic of the corresponding experimental configuration of the degenerate PSA-based regenerator of BPSK signals is shown in Fig. 5.6 (details of the experiment can be found in [66]). The incoming data signal is combined with locally generated sources in order to synthesize a set of suitably phase-locked waves that can be used to drive the PSA, allowing "black-box" operation. In order to achieve this, the data signal was first combined with a free-running narrow linewidth CW source (noted as Pump 1 in the figure) in a conventional parametric amplifier, producing the spectral trace shown in the inset of Fig. 5.6. This spectral trace shows that the generated idler was narrowband, i.e. any modulation existing in the data signal was stripped from it. This is a consequence of the phase relation between the idler ϕ_{idler} and the BPSK data signal ϕ, i.e. $\phi_{idler} = 2\phi - \phi_{pump1}$, through which any π phase jumps existing in the BPSK data are erased in the idler. An additional consequence of this relation is that the phases of the group of three waves (data, pump1 and idler) were now locked relative to each other, thus allowing for phase-sensitive interaction in a subsequent parametric amplifier. Furthermore any noise present in the data that was transferred through to the idler as part of this parametric process is largely removed via optical injection locking of a semiconductor laser to the noisy idler [67]. Indeed, injection locking inherently allows regeneration of signal amplitude [26], provides narrowband (few GHz) rejection of

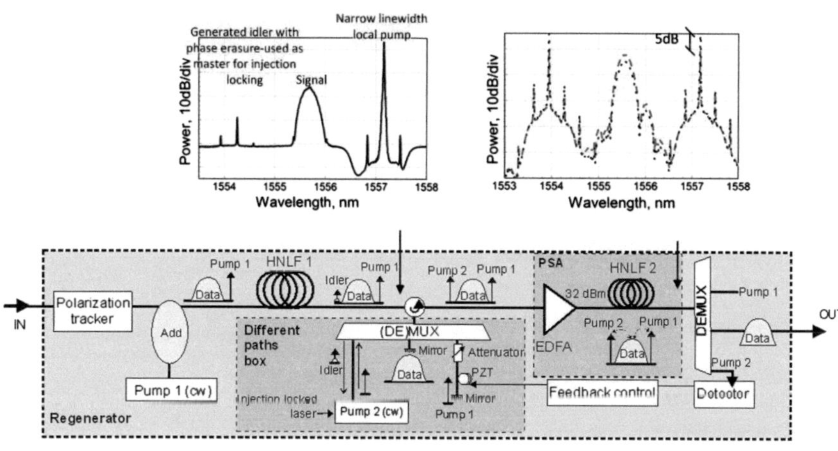

Fig. 5.6 Experimental set-up of a degenerate PSA-based regenerator for BPSK signals (*DE*)*MUX* De/multiplexer; *EDFA* Erbium-doped fiber amplifier; *PZT* piezo-electric fiber stretcher. *Insets* Spectral traces obtained at the output of HNLF1 showing the generation of the narrowband (BPSK data-free) idler and at the output of HNLF2 for maximum gain (*red, dashed*) and de-amplification (*black, dotted*). Adapted from [59, 66]

any high-frequency noise present on the idler, whilst effectively amplifying the power of the injected light. The resulting phase-locked pumps and signal then provide the inputs for the degenerate dual-pump scalar PSA.

The nonlinear medium was a 180-m sample of a fully-spliced, high SBS-threshold HNLF (HNLF 2 in Fig. 5.6) with an Al_2O_3 doped core (instead of germanium, which is the normal dopant of choice in silica-based HNLFs) and a linear strain gradient applied along its length (ranging from 400 to 20 g) to broaden the SBS gain bandwidth and thereby to reduce the peak gain (corresponding SBS threshold was about 29 dBm) [68]. Furthermore, to ensure regeneration of the signal amplitude as well as the phase, the PSA was operated in deep saturation [58, 59, 66].

The robustness of the system was tested firstly by emulating broadband phase noise [59]. Some results are presented in Fig. 5.7a, where constellation diagrams at 40 and 56 Gbit/s as well as eye diagrams (at 40 Gbit/s only) of the original signal (with and without phase noise added) and the signal at the output of the regenerator are shown, illustrating the bit rate transparency of the scheme. Secondly, the regenerator was installed in the middle of an 800-km section of the UK's Aurora dark fiber link [66], and the transmission performance of a 40 Gbit/s DPSK signal (coupled together with another 36 WDM channels operating at the same data rate) was tested in both the presence and absence of the PSA-based regenerator (Fig. 5.7b).

Following this work, a polarization diversity scheme based on such a PSA for polarization insensitive phase regeneration of single and dual polarization based phase modulation formats was proposed in [69], where the HNLF was used in a bi-directional configuration to process the two signal polarizations independently.

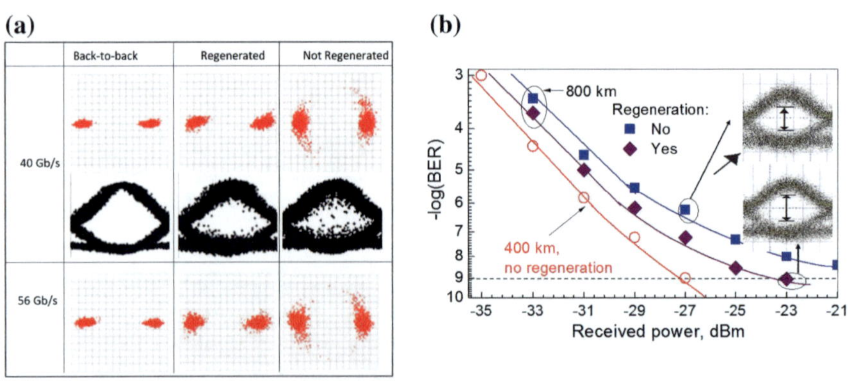

Fig. 5.7 **a** Constellation diagrams of the back-to-back, of the initial signals with added phase noise (labelled as not regenerated) and of regenerated BPSK signals for two data rates (40 and 56 Gbit/s) and the corresponding demodulated eyes at 40 Gbit/s ([59]). **b** BER curves (and eye diagrams) of the DPSK signal after transmission over 400 and 800 km of fiber, with and without PSA-based regeneration. Adapted from [59, 66]

A reduction in polarization sensitivity for BPSK and quadrature phase shift keyed (QPSK) signal regeneration was shown through simulations, however to date only preliminary experimental results have been presented [69].

The first characterization of the degenerate vector PSA in terms of gain profile as a function of input signal phase was presented in [70], demonstrating lower gain as compared to the scalar case for the same total launched pump power, but a much more compact bandwidth occupation since the vector scheme is much less affected by higher-order FWM of the pump and signals.

From (5.1), it can be noted that to achieve a binary step-like phase response (i.e. for BPSK signals), the strength of the idler must be comparable to that of the signal allowing for a large phase sensitive extinction ratio, PSER, which is defined as the difference between the maximum phase sensitive (PS) gain and the maximum PS de-amplification. Large symmetric (i.e. maximum PS gain equal to maximum PS de-amplification) PSERs are conventionally achieved only at high pump powers resulting in the generation of other FWM components across a wide spectral bandwidth. However, such sideband components, undesired in many instances, can play an important role in inducing larger PS de-amplification (asymmetric PSER) at limited pump powers. In fact, a large asymmetric PSER of 25 dB at a nonlinear phase shift (NPS) of 0.8 radians in the scalar degenerate dual-pump PSA provided efficient phase squeezing of DPSK signals [71], and QPSK phase squeezing (de-multiplexing) along the in-phase and quadrature components [72].

The first extension of the PSA regenerator scheme to allow for regeneration of, in principle, any MPSK signal was proposed and experimentally demonstrated by Kakande et al. in [62–64], based on a modification of the PSA set-up presented previously for BPSK signals. In the first part of the black-box MPSK regenerator phase multiplication and pump recovery (by injection locking the M-th harmonic, which is modulation stripped for an MPSK input) occurs and in the second part of the system the conjugated idler (bearing the temporal phase modulation $(M-1)\phi$) is coherently added to the signal, thereby regenerating the phase, as discussed previously. Figure 5.8 reports some of the corresponding results for QPSK signals at 20 and 112 Gbit/s [64]. The phase quantization was accompanied by a sinusoidal amplitude response, which is undesirable for regeneration. However, this phase-to-amplitude conversion can be eliminated: (i) by operating the PSA in saturation as was described above in the case of BPSK signals (even though it is to be appreciated the higher the number of phase harmonics involved the more challenging the task), (ii) subsequently regenerating the signal amplitude in a high dynamic range limiting optical amplifier, such as an injection-locked semiconductor laser [26], or (iii) involving a larger number of phase harmonics in the PSA process so that the phase-to-amplitude conversion can be avoided in the first place as proposed in [73] and subsequently experimentally demonstrated in [74]. This can be achieved by mixing the (M − 1)-th and the –(M + 1)-th phase harmonics together with the original signal, altering (5.1) as follows [73]:

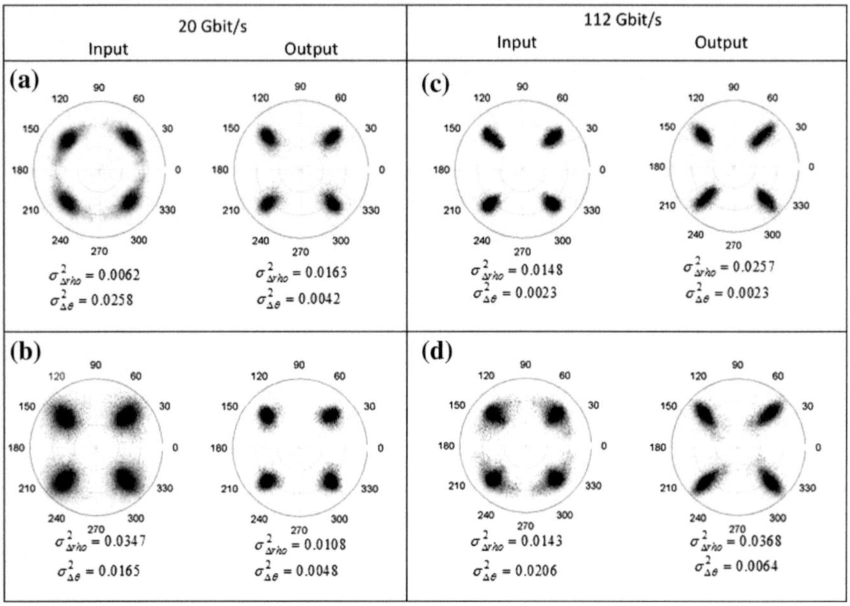

Fig. 5.8 Constellation diagrams at the input/output of the regenerator for a QPSK signal **a** at 10 Gbaud with only phase noise added, **b** at 10 Gbaud with phase and amplitude noise added, **c** at 56 Gbaud without noise, **d** at 56 Gbaud with phase noise only. $\sigma^2_{\Delta\theta}$ is the phase error variance, $\sigma^2_{\Delta\text{rho}}$ is the normalised amplitude noise variance. Adapted from [64]

$$Ae^{i\theta(t)} \propto e^{i\phi(t)} + m(e^{-i(M-1)\phi(t)} - e^{i(M+1)\phi(t)}), \tag{5.2}$$

where m is the phase harmonic weight.

Following this initial fiber-based work different nonlinear materials/processes have been investigated using the above configurations. For example, cascaded quadratic nonlinear processes in periodically poled lithium niobate (PPLN) waveguides (which offer a number of potential advantages, such as broadband operation, low-latency, no intrinsic frequency chirp and SBS resilience) were successfully exploited to achieve phase regenerative amplification with a wide phase sensitive dynamic range of 20 dB for 40 Gb/s DPSK signals, as well as to implement the in-line PSA as a repeater amplifier in a 160 km fiber link [75]. Albuquerque et al. also investigated the feasibility of using a single PPLN waveguide, used in a bidirectional configuration, for both the generation of phase correlated signals and phase-sensitive regeneration and injection-locking for carrier phase recovery for BPSK signals [76]. Cascaded PPLN waveguides were also used to achieve in line PSA for 40 Gbit/s QPSK signals [77] and some of the corresponding results are reported in Fig. 5.9, where the constellation diagrams at the input (a) and output (b) of the PSA when only phase noise was artificially added to the signal, top row, and at the output of the PSA when amplitude noise was added and different values of signal power at the input of the PPLN were chosen (3.7 and 11.8 dBm, respectively,

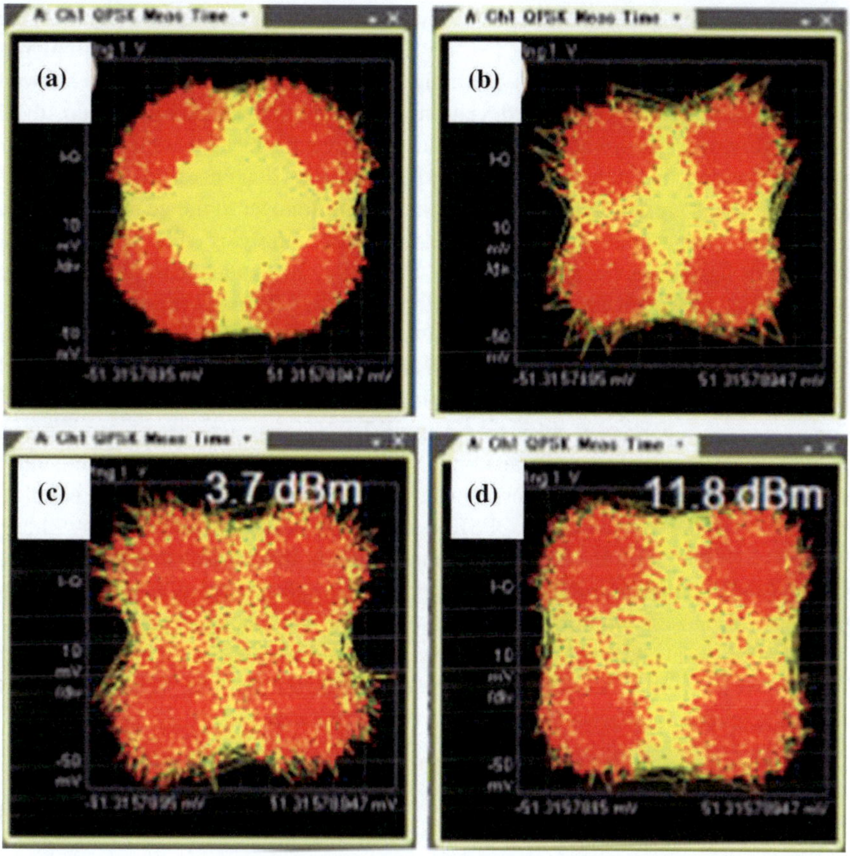

Fig. 5.9 *Top row* Constellation diagrams at the input (**a**) and output (**b**) of the PSA when phase noise was added to the signal. *Bottom row* Constellation diagrams at the output of the PSA when amplitude noise was added to the signal and different signal powers were chosen, 3.7 dBm (**a**) and 11.8 dBm (**b**), respectively at the input of the PPLN. Reproduced with permission from [77] (Copyright © 2013 IEICE)

bottom row). It can be seen that as the input power into the nonlinear media increases the amplitude noise becomes smaller highlighting the saturation regime.

Semiconductor optical amplifiers were also investigated numerically as possible nonlinear devices, confirming a suppression of phase variation in the dual-pump degenerate four wave mixing configuration [78]. An experimental implementation for two-channel regeneration using an SOA was reported in [61].

An alternative approach towards MPSK regeneration was proposed in [79]. It is based on a dual-pump degenerate PSA structure which employs bi-directional signal/conjugate generation in the first instance to generate two pumps, whose phases are conjugated as compared to the transmitted signal ($-(M/2-1)\phi$), followed by a dual-pump FWM process for phase regeneration of potentially any MPSK signal [79]. As compared to the solution proposed by Kakande et al. [62], this PSA requires lower-order signal conjugates (conjugates $[-(M/2 - 1)\phi]$ of the input MPSK signal $[\phi]$ instead of $[-(M - 1)\phi]$ and $[-M\phi]$, respectively), implying that it may have high applicability to higher level PSK. Simulation results showed the effective reduction of phase deviation for QPSK and 8PSK formats up to 50 Gbaud. However, it is worth pointing out that several practical issues need to be resolved before this scheme can be experimentally demonstrated. Firstly, any amplitude noise present in the original signal would transfer to the generated pumps and would compromise the system performance, implying that an amplitude regenerator has to precede such a scheme. Secondly, the challenge of phase-locking of the pumps to the carrier-less signal will need to be addressed prior regeneration.

Finally, complex MPSK signals can always be thought of as the combination of simpler modulation formats and as such their regeneration simplified by the parallel regeneration of several simpler modulation format signals. For example, QPSK signals can be regenerated by arranging two BPSK phase-sensitive amplifiers in parallel to separately regenerate the in-phase and quadrature components of the input signal. These regenerated streams can then be combined in parallel in a coupler, thereby re-synthesizing a cleaner QPSK signal, as proposed and simulated by Zheng et al. [80]. Note that for the correct implementation of this scheme it is essential that each individual regenerator also suppresses signal amplitude noise. Similarly, Webb et al. have proposed and numerically demonstrated the simultaneous conversion of the two orthogonal phase components of a signal to separate output frequencies, using SOAs, thus converting a QPSK signal to two BPSK outputs [81, 82], which could then be regenerated afterwards in simpler BPSK regenerators.

5.3.2 Modulation Format Transparent PSA

As discussed previously, the Modulation Format Transparent PSA scheme requires the signal and the idler (generated in the 'copier') at different wavelengths to be propagated along the transmission link together with the pump. After passing the waves through a lossy medium (such as a long length of optical fiber) to guarantee uncorrelated noise at the signal and idler wavelengths, the "cascaded "copier-PSA" pair" using HNLFs can provide low-noise phase-sensitive amplification [83]. The first experimental demonstration of amplification of three channels (100 GHz spacing), 10 Gbd DQPSK signals with improved sensitivity compared to an EDFA-based system was reported in [54]. Similar results were also demonstrated when the emulated link loss was replaced by a > 80 km fiber span carrying up to 10 Gbd DQPSK data in the linear propagation regime [84] and such an improvement may be maintained even when operating in the nonlinear transmission regime [85]. Note that in real systems, low-noise pump recovery (for example based on injection locking) is necessary to avoid noise-performance degradation, as well as polarisation trimming, and phase and dispersion/delay control to carefully align the relative temporal positions of the pump, signal and idler beams.

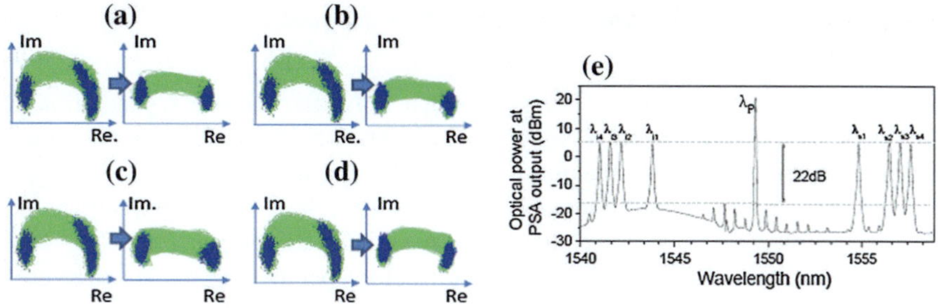

Fig. 5.10 **a–d** Constellation diagrams at the input/output of the PSA for the four channels. **e** Corresponding spectral trace at the output of the PSA. Reproduced with permission from [88] (© 2012, Optical Society of America)

The potential of such a configuration in PPLN devices was highlighted in [86] and subsequently, this concept was also employed in [87] for the first time to demonstrate the squeezing of the phase noise of a 10 GBit/s DPSK signal. Interestingly, this demonstration also considered the generation of the phase-locked signal-idler pair in a second PPLN waveguide, further miniaturizing the system. The same authors then also demonstrated multi-channel operation of the scheme with up to four signals, demonstrating negligible impact on the phase squeezing characteristics of the adjacent channels when the PPLN was operated in the linear regime [88]. Some of the corresponding experimental constellation diagrams and spectral traces at the output of the PSA are shown in Fig. 5.10. Similar schemes were also implemented by Umeki et al. to experimentally demonstrate for the very first time PSA regeneration of a 10 Gbaud multi-carrier polarisation multiplex of QAM signal as will be discussed in the following section [89].

The same configuration was also used to demonstrate PSA in a 196 μm slow-light dispersion-engineered, silicon photonic crystal providing a PSER of 11 dB [90], and in a 65 mm-long dispersion engineered chalcogenide chip obtaining a PSER of 11.6 dB [91] using pulsed pumps.

5.4 Regeneration of Amplitude and Phase Encoded Signals

Recently, there have been a few reports in the literature on how to regenerate amplitude and phase encoded signals, generally using regenerators of phase-only encoded signals as the basic building blocks.

Following the same approach as in Sect. 5.3, we divide the existing PSA-based schemes between those that are modulation format specific (requiring only the signal to be propagated through a communication channel), and those that are modulation format transparent (requiring signal and its complex conjugated copy to be simultaneously propagated).

5.4.1 Modulation Format Specific PSA Regenerators

Three main approaches have been investigated so far for the regeneration of amplitude and phase encoded signals. The first one is based on using phase regenerators to remove only the phase noise on the amplitude/phase encoded signal [73, 92, 93]. In the second approach, the amplitude/phase modulated signals are first demultiplexed into simpler phase-only modulated tributaries, e.g., 16QAM is demultiplexed into two QPSK signals [94], before parallel regeneration of each individual tributary and subsequent coherent recombination, similarly to [80–82] for QPSK input signals. The last approach takes advantages of the periodic-like transfer function of the amplitude in a nonlinear amplifying loop mirror (NALM) in conjunction with a modulation format specific PSA to simultaneously allow amplitude and phase regeneration [95].

In [73] the regenerative performances of the modified MPSK phase regenerator scheme proposed in [62] to avoid phase-to-amplitude noise conversion were numerically studied for three different modulation formats when the regenerator was placed in the middle of a transmission span (overall length of about 1200 km). The modified regenerator was found to offer a reduction of 50 % in bit error rate (BER) for a 8-PSK signal as compared to the regenerator proposed in [62] and a 20 % reduction for a star 8-QAM signal (as for this modulation format the main source of bit errors comes from the amplitude noise that remains uncorrected), see Fig. 5.11 [73].

In [92], a two level star 8QAM signal with only phase noise added was processed by the QPSK regenerator described in Sect. 5.3.1 [62]. Due to the presence of two amplitude levels, the regenerator had to operate close-to-the-linear regime, limiting the performance of the regenerator. Some experimental results are shown in Fig. 5.12, where different signal-to-idler ratios at the input of the PSA were considered.

In [93], various modulation formats were investigated theoretically in a recirculating loop considering an MPSK phase regenerator working in the linear regime [62]. The results for various 4-bit (16-level) modulation formats are shown in Fig. 5.13. As expected, the main conclusion is that single-amplitude-level phase encoded signals outperform other modulation formats of identical bits/symbol density, when a MPSK phase regenerator is used, due to the phase-to-amplitude noise conversion associated with this scheme and the uncorrected accumulated amplitude noise. This emphasizes the fact that phase regenerators work the best with phase-only encoded signals.

A more complex approach for 16QAM regeneration has been suggested recently in [94], where the 16QAM signal is demultiplexed into two QPSK data streams as shown in Fig. 5.14a, which can then be individually regenerated in simpler regenerator schemes and recombined in order to form a high fidelity 16QAM signal. The scheme is based on the 4-level quantizer proposed in [64]. The required enhanced quality of the quantisation process was guaranteed by cascading two PSAs and the choice of the relative phase between the signal and the pumps in the

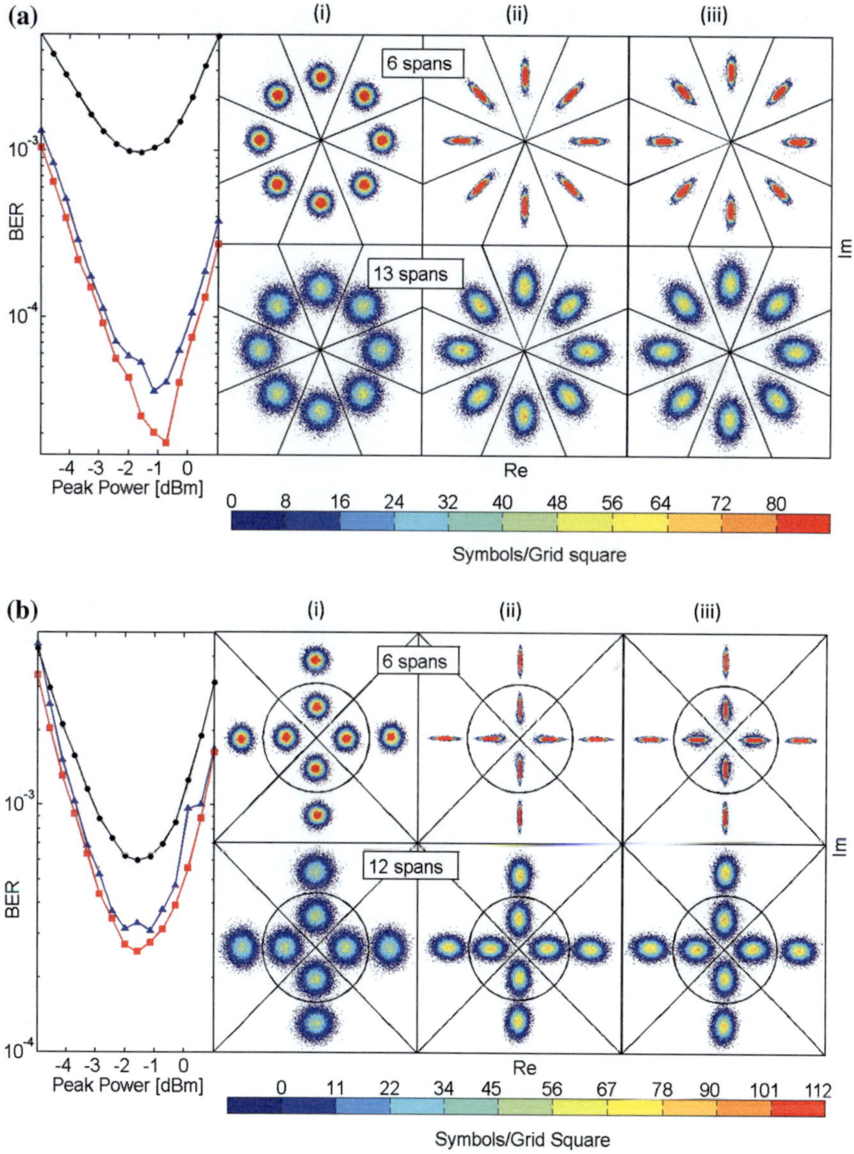

Fig. 5.11 BER curves and constellation diagrams at the end of a transmission link when no regenerator was present (*black circles* and (i)), when the regenerator proposed in [62] (*blue circles* and (ii)) and the modified regenerator proposed in [73] (*red circles* and (iii)) were placed in the middle of the transmission link, respectively, for 8-PSK (**a**) and star 8QAM (**b**) signals [73]. Figure courtesy of G. Hesketh

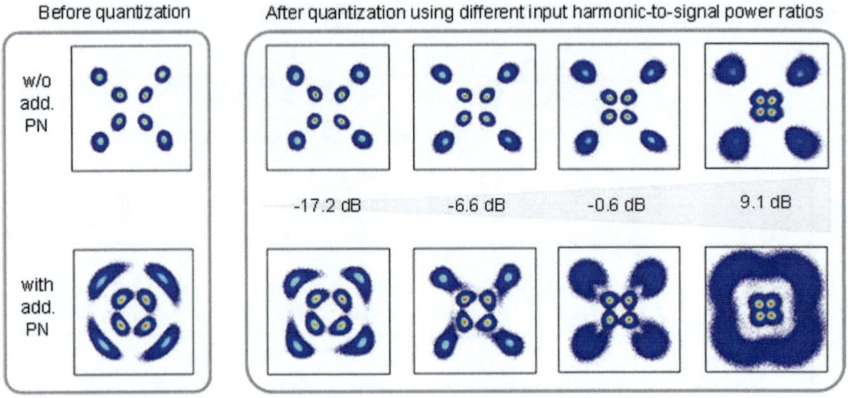

Fig. 5.12 Constellations before and after phase quantization without (*top row*) and with additional phase noise (PN) (*bottom row*). The results after the quantization are achieved at different power ratios which relate the average power of the harmonic to the average power of the signal and directly influence the optimization parameter m. The ratios are stated in logarithmic values [92]. Figure courtesy of T. Richter

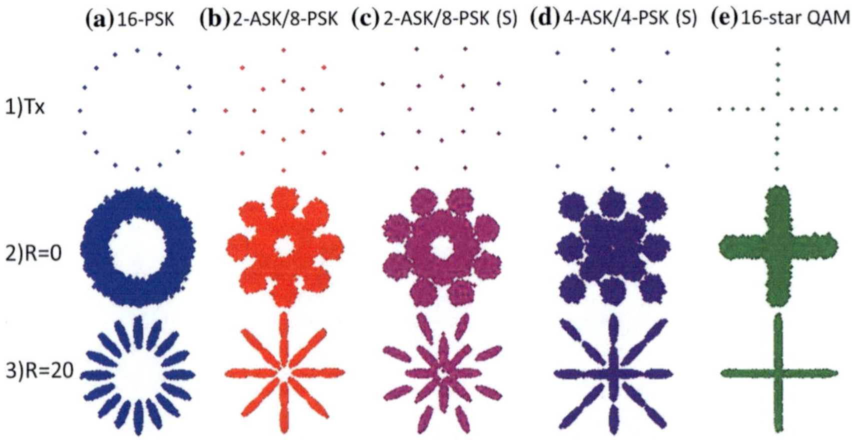

Fig. 5.13 Phase symmetric constellations diagrams for 16 symbols before (leftmost) and after transmission in linear channels (index 0) and nonlinear regenerative channels cascading 20 phase regenerators (index 0 and 20 respectively) for the fixed value of signal-to-noise ratio of 18 dB [93]. Figure courtesy of M. Sorokina

second PSA stage. As can be seen in Fig. 5.14a, in the first step, one QPSK tributary is extracted by squeezing the four points in each of the four quadrants of the constellation into single points. In the second step, the other tributary is obtained by coherent subtraction of the original 16QAM signal and the first tributary. The corresponding BER curves are shown in Fig. 5.14b, where it can be seen that the second QPSK tributary suffers from significant noise. It is worth mentioning that

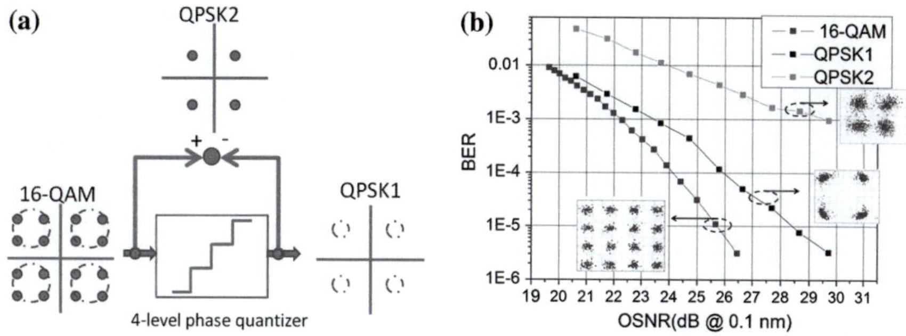

Fig. 5.14 **a** Conceptual scheme of 16QAM optical demultiplexing into two QPSK tributaries (**b**) and simulated BER curves of the 16QAM signal and its demultiplexed QPSK tributaries [94]. Figure courtesy of A. Bogris

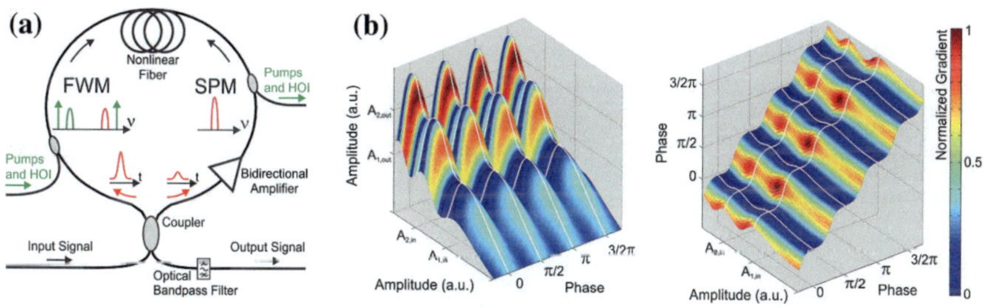

Fig. 5.15 A schematic diagram of the NALM with PSA (**a**). OBPF- optical bandpass filter, (DE) MUX—(de)multiplexer, HOI- high-order idler. **b** Amplitude (*left*) and phase (*right*) transfer characteristics of the NALM with PSA for a typical set of parameters. The signal states of a star-8QAM signal are marked [95]. Figure courtesy of T. Röthlingshöfer/ MPL

the scheme can, in principle, be extended to any format that can be divided up into QPSK tributaries (e.g., 64QAM and 256QAM) [94]. Practical considerations for this scheme include preservation of the coherence among the different paths, and Bogris et al. believe that only photonic integrated solutions could offer the required stability [96].

The last approach proposed and only simulated so far combines the characteristics of a nonlinear amplifying loop mirror (NALM), already demonstrated as a phase preserving multi-level amplitude regenerator in [24], and a directional PSA [62], Fig. 5.15a, to obtain a step-like transfer function in both the amplitude and phase domains, Fig. 5.15b, allowing for simultaneous amplitude and phase regeneration of two amplitude levels and four phase states (8-star QAM) [95]. The scheme was also compared to the cascading of a PSA, acting as a phase regenerator only, and a subsequent NALM, acting as a phase-preserving amplitude regenerator, showing a slight difference in noise conversion for similar transfer characteristics.

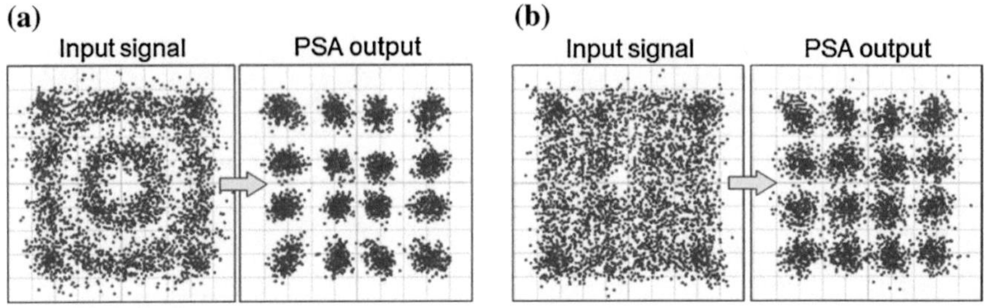

Fig. 5.16 QAM regeneration in the presence of single-tone phase noise (**a**) and ASE noise (**b**). Reproduced with permission from [89] (© 2014, Optical Society of America)

5.4.2 Modulation Format Transparent PSA Schemes

As discussed in Sect. 5.3.2, the operational principle of this approach is based on the fact that noise experienced by the signal and its copy during the propagation is uncorrelated, while the signal and copy are fully correlated. When dealing with multiple-amplitude-level signals, amplitude linearity with a high dynamic range is very important as is the generation of a high quality phase-conjugated idler. Noise reduction for a single 40 Gbit/s 16QAM signal has been experimentally demonstrated for the very first time in [89], where noise was artificially induced using either a phase modulator adding a single tone or using amplified spontaneous emission (ASE) noise to both the signal and idler, see Fig. 5.16. Also, simultaneous amplification of up to three carries and of two orthogonally polarized signals (via a polarisation diversity scheme, using four highly efficient PPLN waveguides) was demonstrated in [89]. This work addresses most of the challenges of optical regenerators; it operates multi-channel, it is in principle capable of processing any arbitrary modulation format, it is demonstrated with polarization-multiplexed signals and it operates as a black-box. However, it is worth noting that it requires the signal to propagate over two wavelength channels (signal and complex conjugated copy) and these need to be perfectly phase synchronized (i.e., with zero delay in propagation) before the PSA. It is still a hotly debated question as to this is an acceptable scenario in modern telecommunication networks.

5.5 Conclusions

In this chapter we have briefly reviewed the need, general principles and technological approaches to the regeneration of various forms of optical signal including both OOK signals and phase encoded signals of differing levels of complexity. We have discussed the evolution of the field and have presented the current state-of-the-art both in terms of the demonstrations of novel device concepts that extend the range of functionality that can be achieved (e.g. range of advanced modulation format signals that can be addressed) and enhance practicality (e.g. in providing black-box operation, WDM performance and possible routes to integration). From what we have written it should be apparent that optical regeneration remains a vibrant and potentially technologically important field of research, with significant advances made in recent years and enabled in many instances by progress in the underpinning technologies e.g. highly nonlinear fibers and high performance narrow-linewidth lasers. It should however be equally clear that many challenges remain which will need to be overcome if all-optical regeneration is ever to make it from the laboratory to commercial deployment—not least in terms of enabling significant levels of WDM capability and the development of compact, power efficient variants. Much work remains to be done and one can anticipate considerable innovation and interesting results ahead, with many other application opportunities (e.g. within test and measurement systems) opening up along the way.

References

1. S. Namiki, T. Kurosu, K. Tanizawa, J. Kurumida, T. Hasama, H. Ishikawa, T. Nakatogawa, M. Nakamura, K. Oyamada, Ultrahigh-definition video transmission and extremely green optical networks for future. IEEE Sel. Top. Quantum Electron. **17**(2), 446–457 (2011)
2. R.S. Tucker, Green optical communications—part II: energy limitations in networks. IEEE Sel. Top. Quantum Electron. **17**(2), 261–274 (2011)
3. C. Xu, X. Liu, L.F. Mollenauer, X. Wei, Comparison of return-to-zero differential phase-shift keying and on-off keying in long-haul dispersion managed transmission. IEEE Photonics Technol. Lett. **15**, 617–619 (2003)
4. A. Demir, Nonlinear phase noise in optical-fiber-communication systems. J. Lightwave Technol. **25**, 2002–2032 (2007)
5. J.P. Gordon, L.F. Mollenauer, Phase noise in photonic communication system using linear amplifiers. Opt. Lett. **15**, 1351–1353 (1990)
6. H. Kim, A.H. Gnauck, Experimental investigation of the performance limitation of DPSK systems due to nonlinear phase noise. IEEE Photonics Technol. Lett. **15**, 320–322 (2003)
7. M. Jinno, M. Abe, All optical regenerator based on nonlinear fibre Sagnac interferometer. Electron. Lett. **28**(14), 1350 (1992)
8. J.K. Lucek, K. Smith, All optical signal regenerator. Opt. Lett. **18**(15), 1226 (1993)
9. W.A. Pender, P.J. Watkinson, E.J. Greer, A.D. Ellis, 10 Gbit/s all optical regenerator. Electron. Lett. **31**(18), 1587 (1995)
10. G. Raybon, Y. Su, J. Leuthold, R.-J. Essiambre, T. Her, C. Joergensen, P. Steinvurzel, K.D.K. Feder, 40 Gbit/s pseudo-linear transmission over one million kilometers, in *OFC*, 17–22 March 2002, pp. FD10-1–FD10-3
11. S. Watanabe, F. Futami, R. Okabe, Y. Takita, S. Feber, R. Ludwig, C. Schubert, C. Schmidt, H.G. Weber, 160 Gbit/s optical 3R regenerator in a fiber transmission experiment, in *OFC* (2003), p. PD16
12. P.V. Mamyshev, All-optical data regeneration based on self-phase modulation effect, in *ECOC* (1998), p. 475
13. M. Nakazawa, E. Yamade, H. Kubota, K. Suzuki, 10 Gbit/s soliton data transmission over one million kilometers. Electron. Lett. **27**, 1270–1272 (1991)
14. O. Leclerc, B. Lavigne, E. Balmefrezol, P. Brindel, L. Pierre, D. Rouvillain, F. Seguineau, Optical regeneration at 40 Gb/s and beyond. J. Lightwave Technol. **21**, 2779–2790 (2003)

15. S. Fischer, M. Dülk, E. Gamper, W. Vogt, E. Gini, H. Melchior, W. Hunziker, D. Nesset, A.D. Ellis, Optical 3R regenerator for 40 Gbit/s networks. Electron. Lett. **35**(23), 2047–2049 (1999)
16. J. Leuthold, G. Raybon, Y. Su, R. Essiambre, S. Cabot, J. Jaques, M. Kauer, 40 Gbit/s transmission and cascaded all-optical wavelength conversion over 1000000 km. Electron. Lett. **38**(16), 890–892 (2002)
17. B. Lavigne, P. Guerber, P. Brindel, E. Balmefrezol, B. Dagens, Cascade of 100 optical 3R regenerators at 40 Gbit/s Based on all-active Mach Zehnder interferometers, in *ECOC*, vol. 3, paper We.F.2.6 (2001), p. 290
18. Q. T. Le, L. Bramerie, S. Lobo, M. Gay, M. Joindot, J.-C. Simon, A. Poudoulec, M. Van der Keur, C. Devemy, D. Massoubre, J.-L. Oudar, G. Aubin, A. Shen, J. Decobert, WDM compatible 2R regeneration device based on eight-channel saturable absorber module. Electron. Lett. **43**(23), 1305–1306 (2007)
19. A. Bogoni, X. Wu, S.R. Nuccio, A.E. Willner, 640 Gb/s all-optical regenerator based on a periodically poled lithium niobate waveguide. J. Lightwave Technol. **30**(12), 1829–1834 (2012)
20. R. Salem, M.A. Foster, A.C. Turner, D.F. Geraghty, M. Lipson, A.L. Gaeta, Signal regeneration using low-power four-wave mixing on silicon chip. Nat. Photonics **2**, 3–38 (2008)
21. V.G. Ta'eed, M. Shokooh-Saremi, L. Fu, I.C.M. Littler, D.J. Moss, M. Rochette, B.J. Eggleton, R.Y. Ruan, B. Luther-Davies, Self-phase modulation-based integrated optical regeneration in chalcogenide waveguides. IEEE Sel. Top. Quantum Electron. **12**(3), 360–370 (2006)
22. M. Matsumoto, Fiber-based all-optical signal regeneration. IEEE Sel. Top. Quantum Electron. **18**(2), 738–752 (2012)
23. K. Cvecek, G. Onishchukov, K. Sponsel, A.G. Striegler, B. Schmauss, G. Leuchs, Experimental investigation of a modified NOLM for phase-encoded signal regeneration. IEEE Photonics Technol. Lett. **18**, 1801–1803 (2006)
24. M. Hierold, T. Roethlingshoefer, K. Sponsel, G. Onishchukov, Bernhard Schmauss, G. Leuchs, Multilevel phase-preserving amplitude regeneration using a single nonlinear amplifying loop mirror. IEEE Photonics Technol. Lett. **23**(14), 1007–1009 (2011)
25. M. Matsumoto, A fiber-based all-optical 3R regenerator for DPSK signals. IEEE Photonics Technol. Lett. **19**, 273–275 (2007)
26. A. Fragkos, A. Bogris, D. Syvridis, R. Phelan, Amplitude noise limiting amplifier for phase encoded signals using injection locking in semiconductor lasers. J. Lightwave Technol. **30**(5), 764–771 (2012)
27. A. Fragkos, A. Bogris, D. Syvridis, R. Phelan, Colorless regenerative amplification of constant envelope phase-modulated optical signals based on injection-locked fabry perot lasers. IEEE Photonics Technol. Lett. **24**(1), 28–30 (2011)
28. L. Quang Trung, L. Bramerie, N. Hoang Trung, M. Gay, S. Lobo, M. Joindot, J.L. Oudar, J.C. Simon, Saturable-absorber-based phase-preserving amplitude regeneration of RZ DPSK signals. IEEE Photonics Technol. Lett. **22**, 887–889 (2010)
29. L. Quang Trung, L. Bramerie, M. Gay, M. Joindot, J. Simon, A. O'Hare, N. Hoang Trung, J. Oudar, All-optical phase-preserving amplitude regeneration of 28-Gbaud RZ-DQPSK signals with a microcavity saturable absorber in a recirculating loop experiment, in *OFC/NFOEC* (2011), pp. 1–3
30. K. Cvecek, K. Sponsel, C. Stephan, G. Onishchukov, R. Ludwig, C. Schubert, B. Schmauss, G. Leuchs, Phase-preserving amplitude regeneration for a WDM RZ-DPSK signal using a nonlinear amplifying loop mirror. Opt. Express **16**, 1923–1928 (2008)
31. N. Chi, L. Xu, K.S. Berg, T. Tokle, P. Jeppesen, All-optical wavelength conversion and multichannel 2R regeneration based on highly nonlinear dispersion-imbalanced loop mirror. IEEE Photonics Technol. Lett. **14**(11), 1581–1583 (2002)
32. C. Kouloumentas, L. Provost, F. Parmigiani, S. Tsolakidis, P. Petropoulos, I. Tomkos, D. J. Richardson, Four-channel all-fibre dispersion managed 2R regenerator. IEEE Photonics Technol. Lett. **20**(13), 1169–1171 (2008)
33. N.S.M. Shah, M. Matsumoto, 2R Regeneration of time-interleaved multiwavelength signals based on higher order four-wave mixing in a fiber. IEEE Photonics Technol. Lett. **22**(1), 27–29 (2010)

34. P.G. Patki, M. Vasilyev, T.I. Lakoba, Multichannel all-optical regeneration, in *Photonics Society Summer* Topical Meeting *Series* (2010), pp. 172–173
35. M. Vasilyev, T.I. Lakoba, All-optical multichannel 2R regeneration in a fibre-based device. Opt. Lett. **30**(12), 1458–1460 (2005)
36. C. Kouloumentas, P. Vorreau, L. Provost, P. Petropoulos, W. Freude, J. Leuthold, I. Tomkos, All-fiberized dispersion-managed multichannel regeneration at 43 Gb/s. IEEE Photonics Technol. Lett. **20**(20), 1854–1856 (2008)
37. T. Ohara, H. Takara, A. Hirano, K. Mori, S. Kawanishi, 40-Gb/s x 4- channel all-optical multichannel limiter utilizing spectrally filtered optical solitons. IEEE Photonics Technol. Lett. **15**(5), 763–765 (2003)
38. F. Parmigiani, L. Provost, P. Petropoulos, D.J. Richardson, W. Freude, J. Leuthold, A.D. Ellis, I. Tomkos, Progress in multichannel all-optical regeneration based on fiber technology. IEEE Sel. Top. Quantum Electron. **18**(2), 689–699 (2012)
39. L. Provost, F. Parmigiani, P. Petropoulos, D.J. Richardson, K. Mukasa, J. Takahashi, J. Hiroishi, M. Tadakuma, Investigation of four-wavelength regenerator using polarization- and direction- multiplexing. IEEE Photonics Technol. Lett. **20**, 1676–1678 (2008)
40. G. Zarris, E. Hugues-Salas, N. Amaya-Gonzalez, R. Weerasuriya, F. Parmigiani, D. Hillerkuss, P. Vorreau, M. Spyropoulou, S.K. Ibrahim, A.D. Ellis, R. Morais, P. Monteiro, P. Petropoulos, D.J. Richardson, I. Tomkos, J. Leuthold, D. Simeonidou, Field experiments with a grooming switch for OTDM meshed networking. J. Lightwave Technol. **28**(4), 316–327 (2010)
41. D. Levandovsky, M. Vasilyev, P. Kumar, Amplitude squeezing of light by means of a phase-sensitive fiber parametric amplifier. Opt. Lett. **24**, 984–986 (1999)
42. K. Croussore, G.F. Li, Phase and amplitude regeneration of differential phase-shift keyed signals using phase-sensitive amplification. IEEE Sel. Top. Quantum Electron. **14**(3), 648–658 (2008)
43. M.E. Marhic, C.H. Hsia, J.M. Jeong, Optical amplification in a nonlinear fibre interferometer. Electron. Lett. **27**, 210–211 (1991)
44. R.E. Slusher, B. Yurke, Squeezed light for coherent communications. J. Lightwave Technol. **8**, 466–477 (1990)
45. J.A. Levenson, I. Abram, T. Rivera, P. Grangier, Reduction of quantum noise in optical parametric amplification. J. Opt. Soc. Am. B **10**, 2233–2238 (1993)
46. W. Imajuku, A. Takada, Y. Yamabayashi, Low-noise amplification under the 3 dB noise figure in high-gain phase-sensitive fibre amplifier. Electron. Lett. **35**, 1954–1955 (1999)
47. R. Tang, P. Devgan, V.S. Grigoryan, P. Kumar, Inline frequency non- degenerate phase-sensitive fibre parametric amplifier for fibre optic communication. Electron. Lett. **41**, 1072–1074 (2005)
48. R. Tang, J. Lasri, P.S. Devgan, V. Grigoryan, P. Kumar, M. Vasilyev, Gain characteristics of a frequency nondegenerate phase sensitive fiber-optic parametric amplifier with phase self-stabilized input. Opt. Express **13**, 10483–10493 (2005)
49. J. Kakande, C. Lundström, P.A. Andrekson, Z. Tong, M. Karlsson, P. Petropoulos, F. Parmigiani, D.J. Richardson, Detailed characterisation of a fiber-optic parametric amplifier in phase-sensitive and phase-insensitive operation. Opt. Express **18**(5), 4130–4137 (2010)
50. Z. Tong, C. Lundstrom, A. Bogris, M. Karlsson, P. Andrekson, D. Syvridis, Measurement of sub-1 dB noise figure in a non-degenerate cascaded phase-sensitive fibre parametric amplifier, in *ECOC* (2009), pp. 1–2
51. C. McKinstrie, S. Radic, Phase-sensitive amplification in a fiber. Opt. Express **12**(20), 4973–4979 (2004)
52. R. Tang, P. Devgan, P.L. Voss, V.S. Grigoryan, P. Kumar, In-line frequency nondegenerate phase-sensitive fiber-optical parametric amplifier. IEEE Photonics Technol. Lett. **17**, 1845–1847 (2005)
53. M. Vasilyev, Distributed phase-sensitive amplification. Opt. Express **13**(19), 7563–7571 (2005)
54. Z. Tong, C. Lundström, P.A. Andrekson, C.J. McKinstri, M. Karlsson, D.J. Blessing, E. Tipsuwannakul, B.J. Puttnam, H. Toda, L. Grüner-Nielsen, Towards ultrasensitive optical links enabled by low-noise phase-sensitive amplifiers. Nat. Photonics **5**, 430–436 (2011)

55. A. Bogris, D. Syvridis, RZ-DPSK signal regeneration based on dual-pump phase-sensitive amplification in fibers. IEEE Photonics Technol. Lett. **18**(20), 2144–2146, (2006) Oct. 15
56. K. Croussore, G. Li, Phase Regeneration of NRZ-DPSK Signals Based on Symmetric-Pump Phase-Sensitive Amplification. IEEE Photonics Technol. Lett. **19**(11), 864–866 (2007)
57. K. Croussore, G. Li, BPSK phase and amplitude regeneration using a traveling-wave phase-sensitive amplifier, in *IEEE/LEOS Winter Topical Meeting Series* (2008), pp. 45–46
58. R. Slavik, F. Parmigiani, J. Kakande, C. Lundstrom, M. Sjodin, P.A. Andrekson, R. Weerasuriya, S. Sygletos, A.D. Ellis, L. Gruner- Nielsen, D. Jakobsen, S. Herstrom, R. Phelan, J. O'Gorman, A. Bogris, D. Syvridis, S. Dasgupta, P. Petropoulos, D.J. Richardson, All-optical phase and amplitude regenerator for next-generation telecommunications systems. Nat. Photonics **4**, 690–695 (2010)
59. R. Slavik, A. Bogris, F. Parmigiani, J. Kakande, M. Westlund, M. Skold, L. Grüner-Nielsen, R. Phelan, D. Syvridis, P. Petropoulos, D.J. Richardson, Coherent all-optical phase and amplitude regenerator of binary phase-encoded signals (Invited Paper). IEEE Sel. Top. Quantum Electron. **18**, 859–869 (2012)
60. S. Sygletos, P. Frascella, S.K. Ibrahim, L. Grüner-Nielsen, R. Phelan, J. O'Gorman, A.D. Ellis, A practical phase sensitive amplification scheme for two channel phase regeneration. Opt. Express **19**, B938–B945 (2011)
61. S. Sygletos, M.J. Power, F.C. Garcia Gunning, R.P. Webb, R.J. Manning, A.D. Ellis, Simultaneous dual channel phase regeneration in SOAs in *ECOC*, Tu.1.A.2 (2012)
62. J. Kakande, R. Slavik, F. Parmigiani, A. Bogris, D. Syvridis, L. Gruner-Nielsen et al., Multilevel quantization of optical phase in a novel coherent parametric mixer architecture. Nat. Photonics **5**, 748–752 (2011)
63. J. Kakande, A. Bogris, R. Slavik, F. Parmigiani, D. Syvridis, P. Petropoulos, D.J. Richardson, First demonstration of all-optical QPSK signal regeneration in a novel multi-format phase sensitive amplifier, in *ECOC* (2010), pp. 1–3
64. J. Kakande, A. Bogris, R. Slavik, F. Parmigiani, D. Syvridis, P. Petropoulos, D. Richardson, M. Westlund, M. Skold, QPSK phase and amplitude regeneration at 56 Gbaud in a novel idler-free nondegenerate phase sensitive amplifier, in *OFC/NFOEC*, paper OMT4 (2011)
65. T. Umeki, H. Takara, Y. Miyamoto, M. Asobe, 3-dB signal-ASE beat noise reduction of coherent multicarrier signal utilizing phase sensitive amplification. Opt. Express **20**(22), 24727–24734 (2012)
66. R. Slavik, A. Bogris, J. Kakande, F. Parmigiani, L. Gruner-Nielsen, R. Phelan, J. Vojtěch, P. Petropoulos, D. Syvridis, D.J. Richardson, Field-Trial of an all-optical PSK regenerator/multicaster in a 40 Gbit/s, 38 channel DWDM transmission experiment. J. Lightwave Technol. **30**, 512–520 (2012)
67. R. Weerasuriya, S. Sygletos, S.K. Ibrahim, R. Phelan, J. O'Carroll, B. Kelly, J. O'Gorman, A. D. Ellis. Generation of frequency symmetric signals from a BPSK input for phase-sensitive amplification, in *OFC*, Paper OWT6 (2010)
68. L. Gruner-Nielsen, S. Dasgupta, M.D. Mermelstein, D. Jakobsen, S. Herstrom, M.E.V. Pedersen, E.L. Lim, S. Alam, F. Parmigiani, D. Richardson, B. Palsdottir, A silica based highly nonlinear fibre with improved threshold for stimulated brillouin scattering, in *ECOC* (2010), pp. 1–3
69. J.-Y. Yang, M. Ziyadi, Y. Akasaka, S. Khaleghi, M.R. Chitgarha, J. Touch, M. Sekiya, Investigation of polarization-insensitive phase regeneration using polarization-diversity phase-sensitive amplifier, in *ECOC* (2013)
70. A. L. Riesgo, C. Lundström, M. Karlsson, P. A. Andrekson, Demonstration of degenerate vector phase-sensitive amplification. in *ECOC*, We.3.A.3 (2013)
71. M. Gao, T. Kurosu, T. Inoue, S. Namiki, Efficient phase regeneration of DPSK signal by sideband-assisted dual-pump phase-sensitive amplifier. Electron. Lett. **49**(2), 140–141 (2013)
72. M. Gao, T. Kurosu, T. Inoue, S. Namiki, Efficient quadrature squeezing of QPSK signals by sideband-assisted dual-pump phase sensitive amplifier, in *OFC*, OW3C.1 (2013)
73. G. Hesketh, P. Horak, Reducing bit-error rate with optical phase regeneration in multilevel modulation formats. Opt. Lett. **38**, 5357–5360 (2013)

74. K.R.H. Bottrill, G. Hesketh, F. Parmigiani, P. Horak, D.J. Richardson, P. Petropoulos, An optical phase quantiser exhibiting suppressed phase dependent gain variation, in *OFC* (2014)
75. T. Umeki, M. Asobe, H. Takenouchi, In-line phase sensitive amplifier based on PPLN waveguides. Opt. Express **21**(10), 12077–12084 (2013)
76. A.A.C. Albuquerque, B.J. Puttnam, M.V. Drummond, Á. Szabó, D. Mazroa, S. Shinada, N. Wada, R.N. Nogueira, Investigation of black-box phase regeneration using single bi-directional PPLN waveguide, in *OFC*, OW4C.2 (2013)
77. M. Asobe, T. Umeki, H. Takenouchi, Y. Miyamoto, In-line phase-sensitive amplifier for QPSK signal using multiple QPM LiNbO3 waveguide, PD2–3, OECC/PS (2013)
78. K. Saito, H. Uenohara, Analytical investigation of possibility of a phase sensitive amplifier based on four wave mixing in a semiconductor optical amplifier, in *OECC* (2012)
79. J.-Y. Yang, Y. Akasaka, M. Sekiya, Optical phase regeneration of multi-level PSK using dual-conjugate-pump degenerate phase-sensitive amplification, in *ECOC*, P3.07, 2012
80. Z. Zheng, L. An, Z. Li, X. Zhao, X. Liu, All-optical regeneration of DQPSK/QPSK signals based on phase-sensitive amplification. Opt. Commun. **281**, 2755–2759 (2008)
81. R.P. Webb, J.M. Dailey, R.J. Manning, A.D. Ellis, Phase discrimination and simultaneous frequency conversion of the orthogonal components of an optical signal by four-wave mixing in an SOA. Opt. Express **19**(21), 20015–20022 (2011)
82. R.P. Webb, M. Power, R.J. Manning, Phase-sensitive frequency conversion of quadrature modulated signals. Opt. Express **21**(10), 12713–12727 (2013)
83. Z. Tong, C. Lundström, P.A. Andrekson, M. Karlsson, A. Bogris, Ultralow noise, broadband phase-sensitive optical amplifiers, and their applications, IEEE J. Sel. Top. Quantum Electron. **18**(2), 1016–1032 (2012)
84. B. Corcoran, S.L.I. Olsson, C. Lundstrom, M. Karlsson, P.A. Andrekson, Phase-sensitive optical pre-amplifier implemented in an 80 km DQPSK link, in *OFC*, PDP5A.4 (2012)
85. S.L.I. Olsson, B. Corcoran, C. Lundström, M. Sjödin, M. Karlsson, P.A. Andrekson, Phase-sensitive amplified optical link operating in the nonlinear transmission regime, in *ECOC*, Th.2. F.1 (2012)
86. K.J. Lee, F. Parmigiani, S. Liu, J. Kakande, P. Petropoulos, K. Gallo, D. Richardson, Phase sensitive amplification based on quadratic cascading in a periodically poled lithium niobate waveguide. Opt. Express **17**, 20393–20400 (2009)
87. B.J. Puttnam, D. Mazroa, S. Shinada, N. Wada, Phase-squeezing properties of non degenerate PSAs using PPLN waveguides. Opt. Express **19**, B131–B139 (2011)
88. B.J. Puttnam, Á. Szabó, D. Mazroa, S. Shinada, N. Wada, Multi-channel phase squeezing in a PPLN-PPLN PSA, in *OFC*, OW3C.6 (2012)
89. T. Umeki, O. Tadanaga, M. Asobe, Y. Miyamoto, H. Takenouchi, First demonstration of high-order QAM signal amplification in PPLN-based phase sensitive amplifier. Opt. Express **22**(3), 2473–2482 (2014)
90. Y. Zhang, C. Husko, J. Schröder, S. Lefrancois, I. Rey, T. Krauss, B.J. Eggleton, Record 11 dB phase sensitive amplification in sub-millimeter silicon waveguides, in *CLEO-PR* (2013)
91. Y. Zhang, R. Neo, J. Schröder, C. Husko, S. Lefrancois, D.-Y. Choi, S. Madden, B. Luther-Davies, B.J. Eggleton, Pump-degenerate phase sensitive amplification in chalcogenide waveguides. CLEO-PR, ThB2-2 (2013)
92. T. Richter, R. Elschner, C. Schubert, QAM Phase-regeneration in a phase-sensitive fiber-amplifier, in *ECOC*, We.3.A.2 (2013)
93. M. Sorokina, S. Sygletos, A.D. Ellis, S. Turitsyn, Optimal packing for cascaded regenerative transmission based on phase sensitive amplifiers. Opt. Express **21**(25), 31201–31211 (2013)
94. A. Bogris, D. Syvridis, All-optical signal processing for 16-QAM using four-level optical phase quantizers based on phase sensitive amplifiers, in *ECOC*, We.3.A.6 (2013)
95. T. Roethlingshoefer, G. Onishchukov, B. Schmauss, G. Leuchs, All-optical simultaneous multilevel amplitude and phase regeneration. IEEE Photonics Technol. Lett. (2013)
96. J. Matres, G.C. Ballesteros, P. Gautier, J-M Fédéli, J. Martí, C.J. Oton, High nonlinear figure-of-merit amorphous silicon waveguides. Opt. Express **21**(4), 3932–3940 (2013)

Chapter 6
Photonic Signal Processing for Logic and Computation

Antonella Bogoni and Alan Willner

Abstract Photonic signal processing and computing take advantage of the sub-picosecond response time of nonlinear optical phenomena to perform high speed operations either on analog and digital optical signals directly in the optical domain, thus avoiding any electro-optic conversion. This chapter first provides a theoretical explanation of the main nonlinear processes and the key enabling technologies exploited in all-optical signal processing applications. In the last two section of the chapter a description of several schemes, used for signal processing, all-optical logic operation and computation, with the respective experimental characterization, are reported.

6.1 Introduction

In this chapter, after a brief description of the main nonlinear processes exploited for all-optical signal processing, a description of the main technologies used for implementing logic and computation functionalities is provided. In last two chapters the state of the art for logic functions and computing are reported.

A. Bogoni (✉)
CNIT, National Laboratory of Photonic Networks,
Via Moruzzi 1, 56124 Pisa, Italy
e-mail: antonella.bogoni@cnit.it

A. Willner
Department of Electrical and Computer Engineering,
University of Southern California, Los Angeles, CA 90089, USA

S. Wabnitz and B.J. Eggleton (eds.), *All-Optical Signal Processing*,
Springer Series in Optical Sciences 194, DOI 10.1007/978-3-319-14992-9_6

6.2 Photonic Logic and Computation Functions for Multi-Format Data Communication and Storage

Telecommunication transport networks are facing with an impressive increase in the internet traffic. Statistics show 566 % [1] worldwide increase of internet usage in the period 2001–2012. The same statistics became much more impressive if considering single world regions. In the same period Africa and Middle East experienced a growth, respectively, of about 3600 and 2600 %. This increase in the internet connection demand, pushed also by the spread of always connected smart portable devices, and together with the increased demand for high bandwidth services, forces researchers to constantly push the transmission capacity limit. In this scenario where optical links, with the use of "processing hungry" complex multi-level modulation formats (e.g., orthogonal frequency division multiplexing (OFDM), quadrature phase shift keying (QPSK), M-symbol quadrature amplitude modulation (M-QAM)), are able to provide Tbit/s channel capacity [2], opto-electronic conversion and electronic based signal processing became the system bottleneck. Femtosecond response time of optical nonlinearities makes possible process signals over bandwidths beyond THz, thus enabling photonic devices as a promising alternative [3], or even an assisting technology [4, 5] to overcome the speed limitation of electronic devices.

With respect to electronic solutions, optical processing does not requires optical-electro-optical conversions, thus performing the required processing directly in the optical domain and potentially allowing to reduce the system power consumption [6, 7]. A single photonic element can transparently process a data channel regardless of its data rate or modulation format, allowing for efficient scaling of network resources [5, 8–11]. Moreover, a single photonic element can simultaneously operate on multiple data channels, thus strongly reducing the need of redundant parallel processing structures.

Many are the materials and devices suitable for optical signal processing and photonic logic implementation, as highly nonlinear fibers, lithium niobate waveguides, silicon waveguides, and many others. All these have been largely exploited, allowing to implement all-optical logic gates [3, 5, 12], flip-flops [13, 14], buffers [15]. More complex all-optical logic functions and computation systems are also currently under investigation, as all-optical field-programmable gate arrays (FPGAs) [16] and routers [17, 18].

6.3 Overview of Nonlinear Processes

The optical nonlinear effects commonly exploited in all-optical signal processing can be divided in two categories:

- wave mixing,
- phase modulation.

Wave mixing and phase modulation can be studied by considering the material polarization induced by an externally applied electric field. When an electric field E(t) is applied on a material, a polarization field P(t) is induced. The total electric field in the medium $E_m(t)$ can be described as the superimposition of the external excitation and the induced polarization [19]:

$$E_m(t) = E(t) + \frac{P(t)}{\varepsilon_0} \tag{6.1}$$

The polarization P(t) depends on the strength of the external excitation E(t), it can be modelled by Taylor series expansion as:

$$P(t) = \varepsilon_0 \left(\chi^{(1)} E(t) + \chi^{(2)} E^2(t) + \chi^{(3)} E^3(t) + \cdots \right), \tag{6.2}$$

where ε_0 is the electrical permittivity of vacuum, $\chi^{(1)}$ is the linear electric susceptibility of the material. $\chi^{(2)}$ and $\chi^{(3)}$, that are zero in linear materials, are the second and third-order nonlinear susceptibility. If a nonlinear medium has centrosymmetric crystal structure, the second-order nonlinear coefficient $\chi^{(2)}$ vanishes, while the third-order nonlinear coefficient can be non-zero either in centrosymmetric and noncentrosymmetric materials.

6.3.1 Wave Mixing

In wave mixing multiple optical waves, at different frequencies, interact with each other and generate an idler signal at a new frequency. Wave mixing can be caused by either second ($\chi^{(2)}$) or third-order ($\chi^{(3)}$) nonlinear interactions. As it will be seen, in second-order nonlinear materials two waves can interact, thus generating a third wave (the idler), for this they are also called three wave mixing processes [20]. In third-order nonlinear materials three waves can mix, thus generating a fourth wave, this process is called four wave mixing (FWM) [21].

Let us consider a second-order nonlinear medium ($\chi^{(2)} \neq 0$), the polarization vector field is proportional to the square of the applied field:

$$P(t) \propto \chi^{(2)} E^2(t) \tag{6.3}$$

If the external excitation is given by the sum of two waves $E_1(t)$ and $E_2(t)$, with respective angular frequency ω_1 and ω_2, and phase φ_1 and φ_2, the polarization vector field can be expressed as:

$$P(t) \propto \chi^{(2)} [E_1(t) + E_2(t)]^2 \tag{6.4}$$

The mixing of the two input signals generates new waves at frequency:

$$\omega_{DFG1} = \omega_1 - \omega_2, \tag{6.5}$$

$$\omega_{DFG2} = -\omega_1 + \omega_2, \tag{6.6}$$

$$\omega_{SHG1} = 2\omega_1, \tag{6.7}$$

$$\omega_{SHG2} = 2\omega_2. \tag{6.8}$$

Amplitude and phase of the generated waves are, respectively:

$$\begin{aligned} E_{SFG}(t) &\propto E_1(t)E_2(t) \\ \varphi_{SFG} &= \varphi_1 + \varphi_2 \end{aligned} \tag{6.9}$$

$$\begin{aligned} E_{DFG1}(t) &\propto E_1(t)E_2^*(t) \\ \varphi_{DFG1} &= \varphi_1 - \varphi_2 \end{aligned} \tag{6.10}$$

$$\begin{aligned} E_{DFG2}(t) &\propto E_1^*(t)E_2(t) \\ \varphi_{DFG2} &= -\varphi_1 + \varphi_2 \end{aligned} \tag{6.11}$$

$$\begin{aligned} E_{SHG1}(t) &\propto 2E_1^2(t) \\ \varphi_{SHG1} &= 2\varphi_1 \end{aligned} \tag{6.12}$$

$$\begin{aligned} E_{SHG2}(t) &\propto 2E_2^2(t) \\ \varphi_{SHG1} &= 2\varphi_2 \end{aligned} \tag{6.13}$$

where "*" denote phase conjugation. The processes responsible of the generation of the sum, difference and double frequency/phase components are called, respectively, Sum Frequency Generation (SFG), Difference Frequency Generation (DFG) and Second Harmonic Generation (SHG).

In many applications the two input signals are in the same frequency band (e.g., C-band ~1550 nm ~193 THz), SHG/SFG term are thus generated at ~775 nm (~386 THz). Often there is an interest in keeping the generated signals in the same frequency band as the input ones, this is usually obtained by cascading two of nonlinear effects, for example SFG and DFG (cSFG/DFG) or SHG and DFG (cSHG/DFG), as shown in Fig. 6.1, respectively.

If we consider a third-order nonlinear medium, the polarization vector can be now expressed as:

$$P(t) \propto \chi^{(3)} E^3(t). \tag{6.14}$$

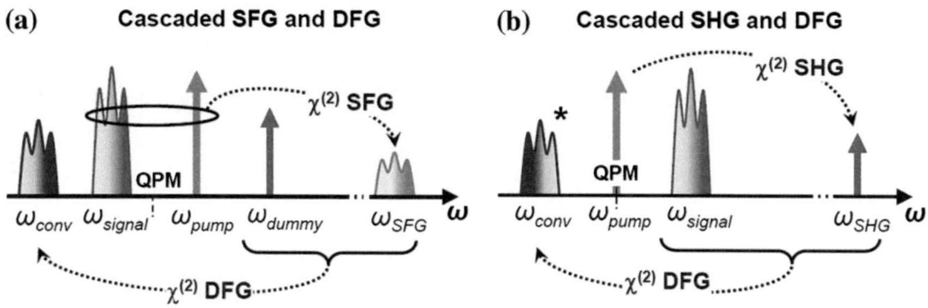

Fig. 6.1 cSFG/DFG (**a**), cSHG/DFG (**b**)

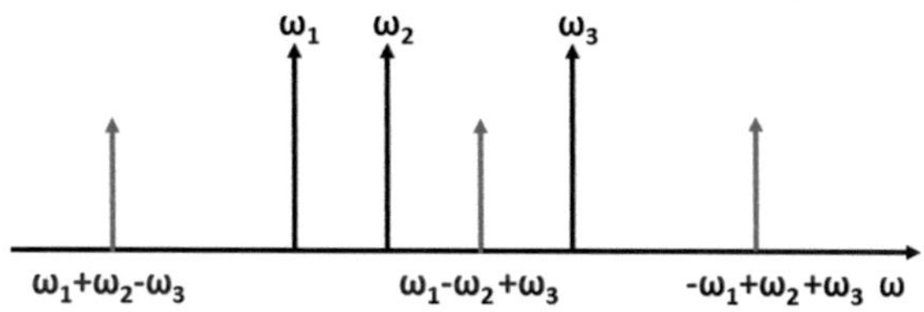

Fig. 6.2 Non degenerate FWM

Under the phase-matching condition if three optical waves $E_1(t)$, $E_2(t)$ and $E_3(t)$, with respective angular frequency and phase ω_1, ω_2, ω_3, φ_1, φ_2 and φ_3, are injected into the material, the polarization vector field can be expressed as:

$$P(t) \propto \chi^{(3)}[E_1(t) + E_2(t) + E_3(t)]^3 \tag{6.15}$$

The signals, interacting three at a time generate many mixing components, among the generated signal shown in Fig. 6.2, idlers (the fourth wave) with angular frequencies:

$$\omega_{FWM1} = \omega_1 + \omega_2 - \omega_3, \tag{6.16}$$

$$\omega_{FWM2} = \omega_1 - \omega_2 + \omega_3, \tag{6.17}$$

$$\omega_{FWM3} = -\omega_1 + \omega_2 + \omega_3, \tag{6.18}$$

Their amplitudes and phases are given by:

$$\begin{aligned} E_{FWM1}(t) &\propto E_1(t)E_2(t)E_3^*(t) \\ \varphi_{FWM1} &= \varphi_1 + \varphi_2 - \varphi_3 \end{aligned} \tag{6.19}$$

$$\begin{aligned} E_{FWM2}(t) &\propto E_1(t)E_2^*(t)E_2(t) \\ \varphi_{FWM1} &= \varphi_1 - \varphi_2 + \varphi_3 \end{aligned} \tag{6.20}$$

$$\begin{aligned} E_{FWM3}(t) &\propto E_1^*(t)E_2(t)E_3(t) \\ \varphi_{FWM1} &= -\varphi_1 + \varphi_2 + \varphi_3 \end{aligned} \tag{6.21}$$

are generated. This process, in which a fourth wave is generated by the nonlinear interaction of three input waves is called non-degenerate four wave mixing.

FWM can occur also if only two input signals are applied, thus having the so called degenerate FWM. The polarization field is described by:

$$P(t) \propto \chi^{(3)}[E_1(t) + E_2(t)]^3 \tag{6.22}$$

Thus generating idler signals at frequency (Fig. 6.3):

$$\omega_{FWM1} = 2\omega_1 - \omega_2, \tag{6.23}$$

$$\omega_{FWM2} = -\omega_1 + 2\omega_2. \tag{6.24}$$

Electric field and phase of the converted signals are thus given by:

$$\begin{aligned} E_{FWM1}(t) &\propto E_1^2(t)E_2^*(t), \\ \varphi_{DFG1} &= 2\varphi_1 - \varphi_2, \end{aligned} \tag{6.25}$$

$$\begin{aligned} E_{FWM2}(t) &\propto E_1^*(t)E_2^2(t), \\ \varphi_{DFG1} &= -\varphi_1 + 2\varphi_2. \end{aligned} \tag{6.26}$$

In the degenerate FWM, as the converted signals is proportional to the square of the input signal field, it neither conserves the phase or intensity information of the original signals.

The described wave mixing processes are often exploited in optical signal processing and logic operation, as in [3, 22], in which all-optical XOR of high speed amplitude and phase modulated signals, have been implemented by means of cascaded three wave mixing processes in periodically poled lithium niobate (PPLN) waveguides.

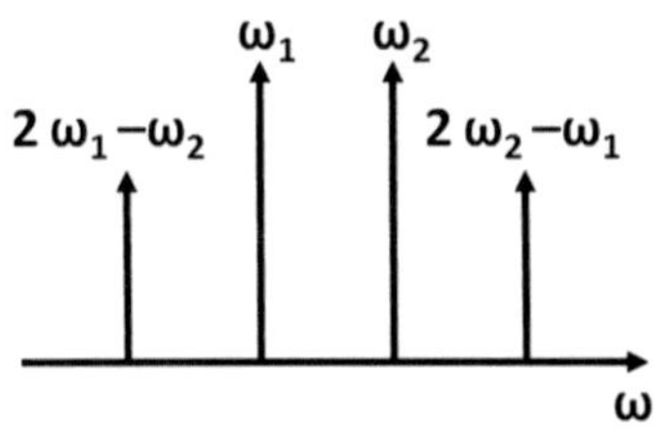

Fig. 6.3 Degenerate FWM

6.3.2 Phase Modulation

If an high intensity signal is injected into a nonlinear medium, the refractive index, experienced by the optical signal, can change. This modulation of the refractive index instantaneously changes the effective optical path length thus causing a "nonlinear phase shift" on the optical signal. The phase shift is given by [21]:

$$\varphi_{NL} = \gamma L_{eff}\left(P_{signal}(t) + 2\sum_{i \neq signal} P_i(t)\right), \tag{6.27}$$

in which γ is the nonlinear parameter, L_{eff} is the effective length of the medium, and P_i denotes the power of the i-th co-propagating wavelength. The phase modulation induced by nonlinear phase shift can be split in two terms:

$$\varphi_{SPM} = \gamma L_{eff} P_{signal}(t), \tag{6.28}$$

$$\varphi_{XPM} = 2\gamma L_{eff} \sum_{i \neq signal} P_i(t). \tag{6.29}$$

The first term, (6.28), corresponds to the nonlinear phase shift related to the signal itself, for this reason it is called self phase modulation (SPM). The second term, (6.29), takes into account the nonlinear phase shift induced by the other co-propagating signals, for this reason it is called cross phase modulation (XPM). Figure 6.4 depicts an example where an amplitude modulated data signal and a CW pump are injected into a third-order nonlinear medium. SPM causes spectral broadening on the input signal, while the CW pump experiences XPM. Thus, the information of the data signal is transferred to the CW pump through XPM and can be recovered by appropriate side-band filtering.

Because nonlinear phase shift is induced as a result of the power variations of the input signals, XPM and SPM are used in the processing of amplitude-modulated signals (e.g. OOK).

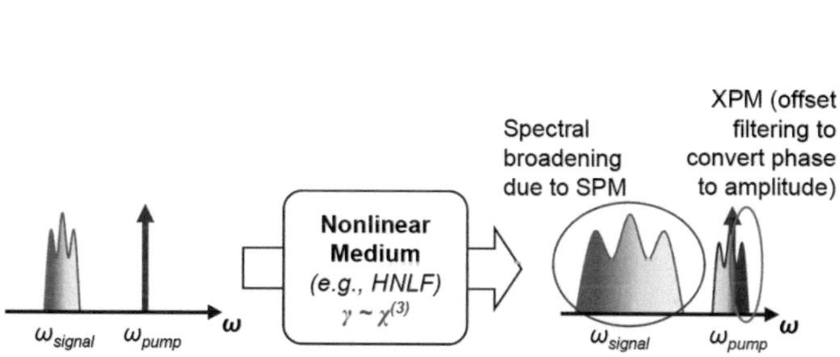

Fig. 6.4 Cross-phase modulation process

6.4 Enabling Technologies

Many are the nonlinear materials and devices available for implementing logic functions. $\chi^{(3)}$ nonlinear phenomena are shown by highly nonlinear fibers (HNLF), semiconductor optical amplifiers (SOAs) or silicon devices, while lithium niobate are used for $\chi^{(2)}$ based logic schemes. Depending on the application each material is able to meet specific system requirements, as large and flat bandwidth or integrability, just to mention some of them. In this section a brief overview on the main technologies used in all-optical signal processing is reported.

6.4.1 Optical Fiber

Optical fibers are transparent fibers made of silica. This material is a third-order nonlinear medium, thus showing FWM, XPM and SPM effects. However standard fibers have a low nonlinear parameter, which leads to low efficiency, thus making standard fibers inconvenient to be used as an optical nonlinear medium. By properly doping the silica, i.e. with germanium dioxide, or by making chalcogenide (e.g. Ag_2Se_3) or bismuth-oxide-based (Bi_2O_3) fiber, it is possible to strongly increase the $\chi^{(3)}$ parameter of the fiber, thus making it highly nonlinear. This kind of fibers are called highly nonlinear fibers (HNLF). HNLF-based logic schemes take advantage of the large phase matching bandwidth of the fiber, allowing the nonlinear interaction of multiple signals in a single device. On the other hand, HNLFs based processing usually require meter or even hundreds of meters of fiber, thus preventing any possibility of integration.

HNLF have been widely exploited for demonstrating all-optical logic functions, as NOLM-based (nonlinear optical loop mirror) 40 Gbit/s NAND, OR, NOR, XOR, and XNOR logic gates [23].

6.4.2 Semiconductor Devices

Semiconductor optical amplifiers (SOAs) are active optoelectronic devices manufactured, when designed for the 1550 nm window (C-band), with III–V compounds, as indium phosphyde (InP) or indium gallium arsenide based, as (InGaAsP). If an high intensity signal is injected into an SOA it undergoes $\chi^{(3)}$ nonlinearities, as for the HNLF [24, 25]. With respect the this last one, however, SOAs are compact in size, integrable and provide optical gain. As a drawback, the gain recovery dynamics of the SOA, typically in the order of tens of ps [26], limits the maximum working data-rate. Another common semiconductor structure is the semiconductor ring laser (SRL), which is a ring shaped laser cavity [14]. Devices based on this structure exploit the Cross Gain modulation (XGM) effect, which is a nonlinear effect where the optical gain experienced by a signal propagating into an active medium depends also on the intensity of the all co-propagating signals.

6.4.3 Photonic Crystals

Photonic crystals (PCs) are nanostructures in which the refractive index of the medium is characterized by periodic variations, in the order of the light wavelength, thus affecting the propagation properties of the medium. Proper engineering of the photonic crystal structure can enable to finely control the light propagation, e.g. negative refractive index, self-collimation etc. When applied to a nonlinear medium, PCs can strongly enhance the nonlinear behavior of such material. Among the different materials, exploiting photonic crystal structures, the most used are photonic crystal fibers (PCF) [27] and photonic crystal waveguides [28].

6.4.4 Periodically Poled Lithium Niobate Waveguides

Lithium niobate is a non-centrosymmetric crystal exhibiting second-order optical nonlinearity, this would make lithium niobate waveguides suitable for nonlinear signal processing. However, the frequency dependency of the phase velocity makes the phase relation between the interacting signals varying during the propagation through the crystal, thus causing a periodic succession of constructive and destructive interaction that makes the maximum efficiency of the nonlinear process extremely low. To solve this problem the quasi-phase matching (QPM) technique [19] is used in the waveguide designs. QPM consist of a periodic sign change in the nonlinear susceptibility of the medium, this operation is called poling, thus compensating for the velocity mismatch of the interacting waves [20] and effectively increasing the nonlinear efficiency. This device is called Periodically poled lithium niobate (PPLN) waveguide. The relatively low propagation loss, and ease of fabrication, makes this device suitable for the implementation of advanced all-optical signal-processing functions at bandwidths beyond THz [20]. As in [3] and [22], in which PPLN waveguides have been exploited to implement XOR operation among, respectively, OOK and QPSK modulated signals up to 640 Gbit/s.

6.4.5 Silicon Devices

Silicon devices for optical signal processing applications are attracting lots of research efforts. Silicon photonic technology, based on CMOS compatible fabrication processes [29], open the way for a mass production and cheap monolithic integration between optical and electronic components. Common silicon devices are based on the silicon-over-insulator (SOI) technology, which thanks to the high refractive index allows the design of ultra-small silicon wire waveguides. Silicon material shows third-order nonlinear effects, as for optical fibers, thus allowing the implementation of devices based on FWM, SPM or XPM effects [30].

6.5 State of the Art for Logic

6.5.1 Overview of Logic Functions and Achieved Results

Exactly as in the electrical domain, logic operations are basic elements for enabling digital-signal-processing. However, performing such kind of operations in the optical domain can dramatically improve system performances, both for single network operation, e.g. for OTDM demultiplexing [5, 11], or to implement more complex system, as optical routers or complex processing devices [16–18].

In Table 6.1 a brief list of optical logic functions are reported, together with the maximum achieved data rate and modulation format (in brackets). The last column reports the exploited nonlinear effect.

6.5.2 OOK 640 Gbit/s Logic Operations

PPLN waveguides can be effectively used for high speed logic gates, as in [3], in which 640 Gbit/s A and $\overline{B}$ and $\overline{A}$ and B logic gates have been implemented by exploiting pump depletion in a PPLN waveguide.

The proposed scheme of principle is shown in Fig. 6.5a.

Two signals A and B with respective wavelength (λ_A and λ_B) placed symmetrically to the Quasi-Phase Matching (QPM) wavelength Fig. 6.5b of the PPLN waveguide can nonlinearly interact through SFG nonlinear effect. As a consequence, if both input signals are high the two wavelengths undergo signal depletion, while no depletion occurs if one of the two signals is low. In this way, at the PPLN

Table 6.1 Examples of state of the art all optical Logic functions and their data rate

Logic operation	Data rate	Nonlinear effect
	40 Gbit/s (OOK)	FWM in HNLF
	1.28 Tbit/s (OOK)	FWM in ChG waveguide [5, 11]
	160 Gbit/s (DPSK)	FWM in ChG waveguide [12]
$A \cdot \overline{B}$	640 Gbit/s (OOK)	Pump depletion in PPLN waveguides [3]
Half-adder	40 Gbit/s	Cascaded SFG/DFG in PPLN waveguide [31]
Optical buffer	130 μs	

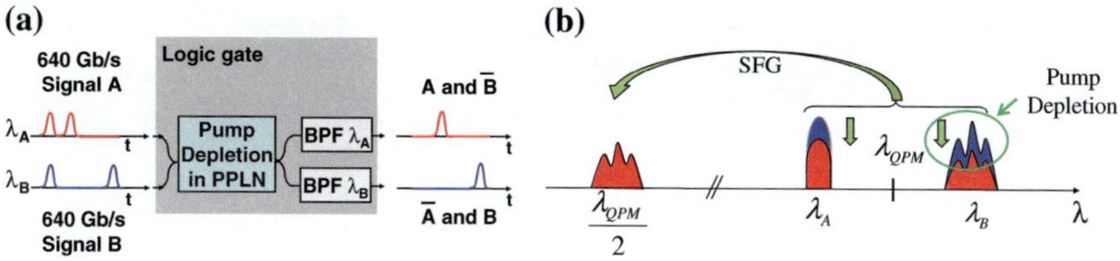

Fig. 6.5 Scheme of principle (**a**), pump-depletion (**b**)

Input signals A	Input signals B	Output signals A ($A\overline{B}$)	Output signals B ($\overline{A}B$)
0	0	0	0
0	1	0	1
1	0	1	0
1	1	0	0

OR/XOR functions

A OR B = B + $A\overline{B}$ = A + $\overline{A}B$
A XOR B = $A\overline{B}$ + $\overline{A}B$

Half-subtractor

DIFF = A XOR B = $A\overline{B}$ + $\overline{A}B$
Borrow (A-B) = $\overline{A}B$
Borrow (B-A) = $A\overline{B}$

Fig. 6.6 Truth table

output, A and $\overline{B}$ or $\overline{A}$ and B logic operation can be selected by filtering, respectively, λ_a or λ_B, as also reported by Fig. 6.6.

Moreover, as summarized in Fig. 6.6, coupling together the two logic function outputs, the XOR logic operation can be accomplished. XOR output represents the difference between A and B. Thus the output signals at λ_A and λ_B individually give the Borrow (B − A) and Borrow (A − B) functions, respectively, thus allowing to obtain an half-subtractor. Finally, coupling together the input signal B and the output signal at λ_A or the input signal A and the output signal at λ_B, the A OR B function can be obtained.

Experimentally, by exploiting this principle, two 640 Gbit/s OTDM signals have been optically processed. As shown in [32], the pump depletion effects can not be simultaneously maximized for both pumps, since simultaneous optimization will decrease the system performance. Therefore, when the pump depletion is optimized for signal A, A and $\overline{B}$ logic function is performed, otherwise $\overline{A}$ and B logic operation is obtained. The use of two PPLN waveguides will eventually allow to simultaneously optimize both logic gates. Note that pump depletion introduces little distortion on the pump signals. Figure 6.7 shows the experimental setup.

In order to generate the two input signals a 40 GHz mode locked laser (MLL), producing ~1 ps pulses at λ_C = 1542 nm, is modulated by a 40 Gbit/s PRBS sequence, and then split into two branches. The first branch is directly multiplexed, 40-to-640 Gbit/s, by using an optical fiber-based multiplexer. The second branch is used to generate a 40 Gbit/s supercontinuum spectrum through propagation in an HNLF. The supercontinuum spectrum is then filtered to obtain the 40 Gbit/s data signal at λ_s = 1560 nm. Another 40-to-640 Gbit/s optical fiber-based multiplexer

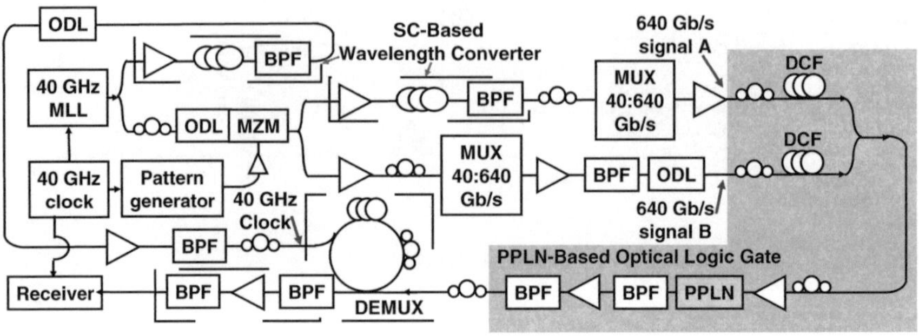

Fig. 6.7 Experimental setup

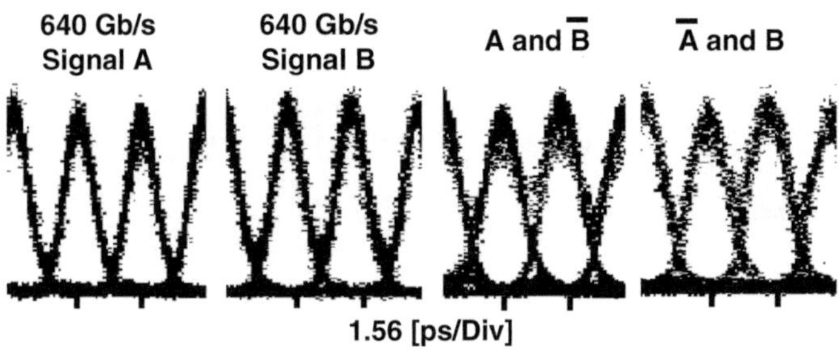

Fig. 6.8 Input and output eye diagrams

produces the second 640 Gbit/s signal. Polarization controllers are needed in the two signal paths in order to align the two signal polarizations. At the output of the PPLN waveguide, two cascaded tunable band-pass filters (BPFs) and a low noise EDFA are used to perform the desired logic function. Finally, a 640-to-40 Gbit/s nonlinear optical loop mirror (NOLM)-based optical demultiplexer is used to test the performance of the involved signals. Depending on the wavelength of the signal under test, a supercontinuum-based wavelength converter is used to tune the clock wavelength.

In Fig. 6.8, the eye diagrams of the involved signals at the PPLN input and output shows as the pump depletion effect introduces little amplitude distortions on both the OTDM frames. The slightly broadening of the involved signals pulsewidth is ascribed to chromatic dispersion.

6.5.3 PSK 160 Gbit/s Logic Functions

Phase modulated signals are an attracting solution for future optical networks, for this reason many efforts are spent for the study of all-optical logic gates able to process high speed PSK signals. In [12] an all-optical XOR an among 160 Gbit/s return-to-zero (RZ) DPSK signals is performed by exploiting four-wave mixing (FWM) in a dispersion-engineered, highly nonlinear chalcogenide (ChG) planar waveguide.

ChG planar waveguides are optical devices showing third-order nonlinear behavior. If two Differentially encoded Phase Shift Keying (DPSK) data streams with respective wavelength λ_1 and λ_2 (phases φ_1 and φ_2) and a continuous wave (CW) at λ_p (with a constant phase φ_p) are injected, in a co-propagating way, injected signals undergo non-degenerate FWM. As a result of such interaction, an optical idler at wavelength λ_i and phase $\varphi_i = \varphi_1 + \varphi_2 - \varphi_p$ is generated. As illustrated in Fig. 6.9b, showing the phase of the FWM idler at the output of the ChG chip, if the two input DPSK channels are out of phase the phase of the FWM idler is "π" while it is "0" if they are in phase. At the chip output the idler signal can be separated by means of an optical band pass filter (BPF). The XOR logic operation is verified by demodulating both input data and idler signal.

ChG waveguides exhibits a large nonlinear coefficient $\gamma \sim 104\ \mathrm{W}^{-1}\ \mathrm{km}^{-1}$, thus allowing high nonlinear interaction efficiencies. The large ChG dispersion is offset by tailored waveguide dispersion, thus yielding an anomalous dispersion of ~29 ps/nm/km. The low dispersion and short propagation length (6.8 cm) offers ultra-low walk-off and efficient phase matching between the input signals and the CW probe.

The 160 Gbit/s XOR gate has been experimentally implemented by the setup shown in Fig. 6.10. A 40 GHz pulse train (λs = 1555 nm) was spectrally broadened to ~20 nm by 400 m and a 100 mm HNLFs. The pulse train was then modulated by a Mach Zehnder modulator and interleaved to produce a 160 Gbit/s DPSK signal. A Finisar Waveshaper was used to create two replicas of the 160 Gbit/s DPSK signal at different wavelengths.

The average power of each channel inside the waveguide was ~10 mW, yielding a relatively low energy per bit (~62.5 fJ/bit). The XOR product has

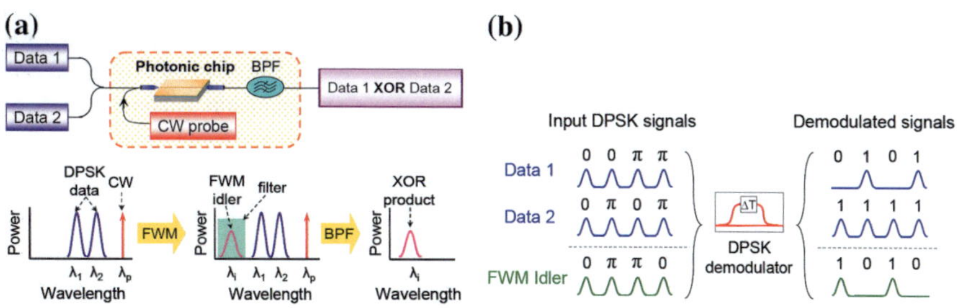

Fig. 6.9 **a** Working principle of the ChG chip based all-optical XOR gate and **b** a truth table of the phase of a FWM idler (XOR gate)

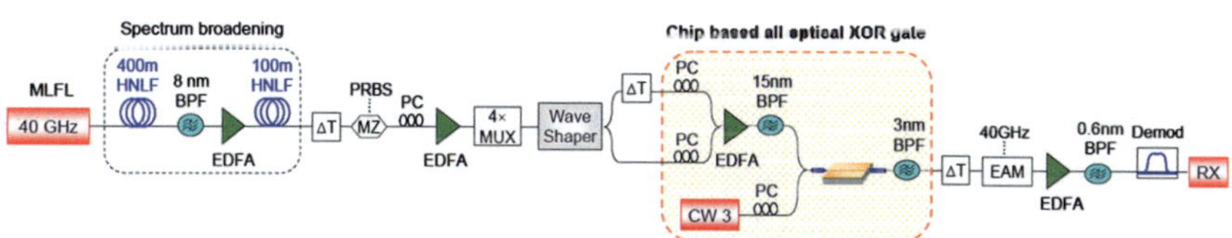

Fig. 6.10 Experimental setup for a chip based XOR gate for DPSK signals

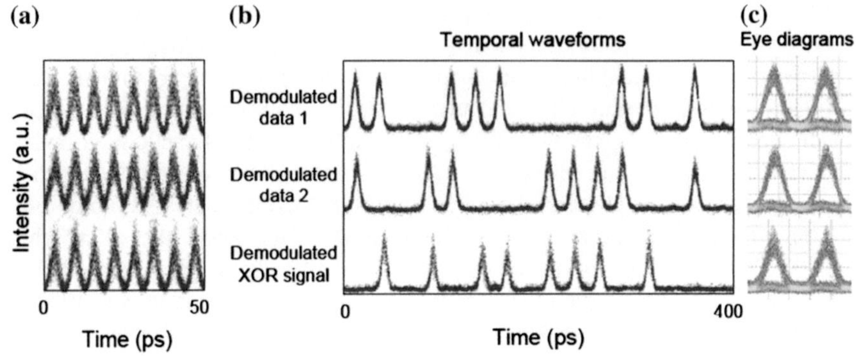

Fig. 6.11 **a** Eye diagrams of two 160 Gbit/s input channels and XOR idler. **b** waveforms and **c** eye diagrams of two demultiplexed and demodulated 40 Gbit/s input data and a XOR product

been then extracted by means of BPFs, demultiplexed to 40 Gbit/s and finally demodulated.

As illustrated by Fig. 6.11a, being the signal information encoded in the signal phase, the two 160 Gbit/s input signals and XOR product show constant pulse amplitude. XOR operation is confirmed by In Fig. 6.11b, c, where the demultiplexed and demodulated waveforms and eye diagrams of the two 40 Gbit/s DPSK data and the XOR signal are reported.

6.5.4 Hexadecimal 16PSK Addition

In binary logic functions, the information bit-rate is the same as the signal switching rate (baud rate). Optical signals can use multi-level modulation formats, such as 16-PSK, to encode multiple bits onto a symbol. Therefore, logic functions that operate on multi-level modulated signals exhibit higher bit rates than switching rates. Here, we present the work in [33] that demonstrates simultaneous addition and subtraction on 10-GBd 16-PSK signals in an HNLF through non-degenerate FWM. Figure 6.12a shows the schematic output spectrum of the HNLF. Signals A and B and a CW pump P are sent to an HNLF with low dispersion slope close to the fiber zero dispersion wavelength (ZDW). Non-degenerate FWMs occur simultaneously to produce signals proportional to $E_P^* E_A(t) E_B(t)$, $E_A^*(t) E_B(t) E_P$, and $E_B^*(t) E_A(t) E_P$. Because signals A and B are 16-PSK, the multiplication of their fields results in the addition of their phases ($\angle A + \angle B$), and the multiplication of a signal with the phase-conjugate of the other signal generates subtraction of their phases (i.e., $\angle A - \angle B$ and $\angle B - \angle A$).

The experimental constellation diagrams of the input and output signals are illustrated in Fig. 6.12b, in which the error vector magnitude (EVM) of the addition/subtraction output signals is on average ~2.5 % higher than that of the input.

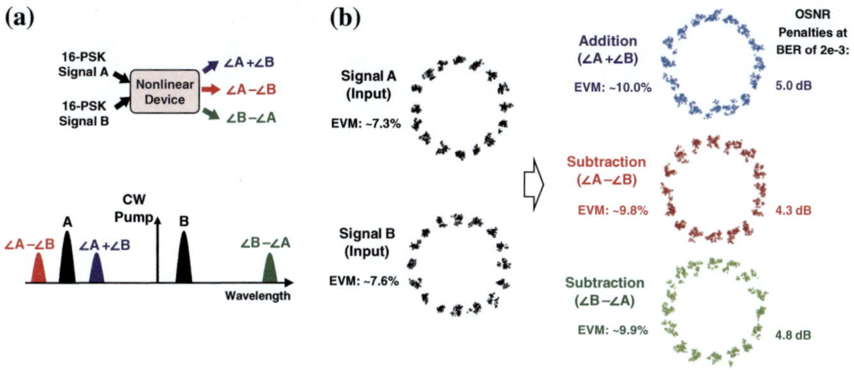

Fig. 6.12 **a** Concept of using nonlinearities (e.g., FWM) for high-base number addition and subtraction on PSK signals. **b** Experimental results for 16-PSK addition and subtraction at 10 GBd speed [33]

6.6 State of the Art for Computation

6.6.1 Overview of Tapped Delay Lines

A key building block for many signal processing applications is a tapped-delay-line [34–37]. As shown in Fig. 6.13, the incoming data signal is tapped at different time intervals; each tap is multiplied by a complex coefficient; the taps (copies) are then added together to form the output. The output, y(t), of the tapped-delay-line is a weighted sum of the delayed copies of the input signal, $x(t)$:

$$y(t) = \sum_{i=0}^{N-1} h_i x(t - T_i) \tag{6.30}$$

in which, N is the number of taps, h_i and T_i are the complex coefficient and time delay of the i-th tap, respectively. This equation represents an N-tap FIR filter that can be used for various filtering applications. The function of the tapped-delay-line (TDL) can be changed by varying its parameters.

6.6.2 Fundamental Tools to Enable Photonic TDL

An optical TDL can be realized by implementing an all-optical fan-out stage to create multiple replicas of the original signal using $\chi^{(2)}$ nonlinear processes in the periodically-poled-lithium-niobate (PPLN) waveguide, conversion/dispersion delay to induce relative delays on signal replicas [38] and coherent multiplexing in

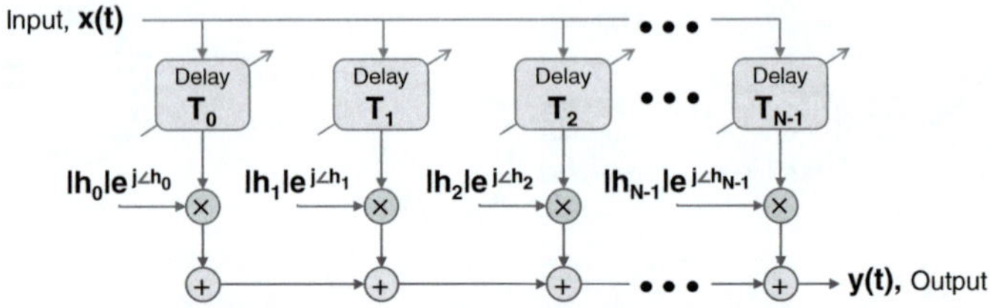

Fig. 6.13 A tapped-delay-line generic block diagram: Input signal is tapped at different time intervals, copies are weighted by a coefficient and summed to form the output

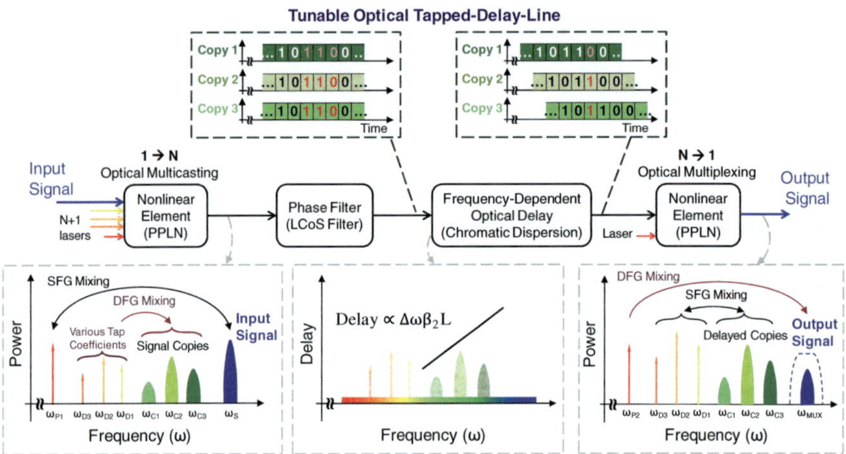

Fig. 6.14 Optical implementation of a TDL based on nonlinearities and conversion-dispersion-based delays: N copies of the input optical signal are generated at different frequencies using cascaded nonlinear wave mixings of SFG followed by DFG. Copies are sent into a chromatic dispersive medium to introduce the tap delays. Delayed and weighted signal copies are sent to a second nonlinear medium to be multiplexed and to create the output signal [36]

another PPLN waveguide [39]. The principle of operation of this tunable TDL is depicted in Fig. 6.14. The incoming data signal is first fanned-out to N signal copies using cSFG-DFG in a PPLN waveguide. The fan-out scheme uses N independent pumps to generate N copies. All signals and pumps are sent into a phase and amplitude programmable filter based on liquid crystal on silicon (LCoS) technology [40], in which the independent signal copies and their pumps are let through while the original signal and pump are blocked. The LCoS filter can also be used to apply the tap phases. Next, the signals propagate through a chromatic dispersive element such as DCF in which the N signal copies are delayed with respect to each other. The delayed signals are then sent into another PPLN waveguide to be multiplexed to one wavelength channel coherently. Each copy represents a tap in the TDL; therefore, the number of taps is the number of pump lasers, and tap amplitudes can be independently varied by changing the power of the pumps that generate the signal copies. The tap delays are based on conversion-dispersion delays; therefore, can be widely tuned by changing the wavelengths of the pumps. Finally, the tap phases can either be imposed by the LCoS filter or fine tuning of the pumps [41]. Therefore, the TDL can be completely tuned by varying the properties of the input pump lasers.

According to the wave mixing rules presented in Section II, the equivalent mathematical representation of this optical TDL, with input electric field amplitude $E_s(t)$ and output field $E_{MUX}(t)$ is [36]

$$\mathrm{E_{MUX}(t)} \propto \sum_{i=1}^{N} |\mathrm{h_i}| e^{j\angle \mathrm{h_i}} \mathrm{E_S(t - T_i)} \tag{6.31}$$

in which,

$$|\mathrm{h_i}| = |\mathrm{E_{D_i}}|^2 \tag{6.32}$$

$$\angle \mathrm{h_i} = -\mathrm{L}\ \beta_2(\omega_{\mathrm{D_i}} - \omega_{\mathrm{QPM}})^2 + \phi_{\mathrm{i}}^{\mathrm{LCoS}} + \gamma_i \tag{6.33}$$

$$\mathrm{T_i} = \mathrm{L}\ \beta_2(\omega_{\mathrm{D_i}} - \omega_{\mathrm{D_1}}) \tag{6.34}$$

In (6.22)–(6.24), E_{D_i} is the field amplitude of the i-th CW pump, β_2 is the group velocity dispersion parameter of the chromatic dispersive medium of length L, ϕ_i^{LCoS} is the phase that can be applied on i-th pump after the first PPLN stage by the LCoS filter, and γ_i represents a phase shift due to higher-order dispersion terms. Alternatively, the phase of the i-th tap can be varied by fine frequency detuning of the ω_{D_i} dummy pump by $\delta\omega_{D_i}$. This detuning can change $\angle h_i$ approximately by $-2L\ \beta_2(\omega_{D_i} - \omega_{QPM})\delta\omega_{D_i}$. Because the tap delays also depend on the frequency of the pumps, the fine-frequency detuning technique to apply tap phases can cause deviation in the tap delays $\delta T_i/T_i \approx \delta\omega_{D_i}/(\omega_{D_i} - \omega_{QPM})$. Therefore, care must be taken in choosing pump wavelength separations and dispersion values to minimize time error degrading effects.

The tunable optical TDL can be used to perform many different signal processing functions, including FIR filtering [42], equalization, correlation [36], discrete Fourier transform [43], D/A conversion, and synthesis of arbitrary optical waveforms [44], all at the line rate of optical communications.

6.6.3 Optical 1D Correlation Results Using Nonlinearities and On-Chip MZIs

A correlator can search for a pre-determined target pattern on an incoming data stream at the line rate [39, 45]. Figure 6.15a shows the concept of pattern matching to an incoming signal. A correlator can be implemented using the TDL structure.

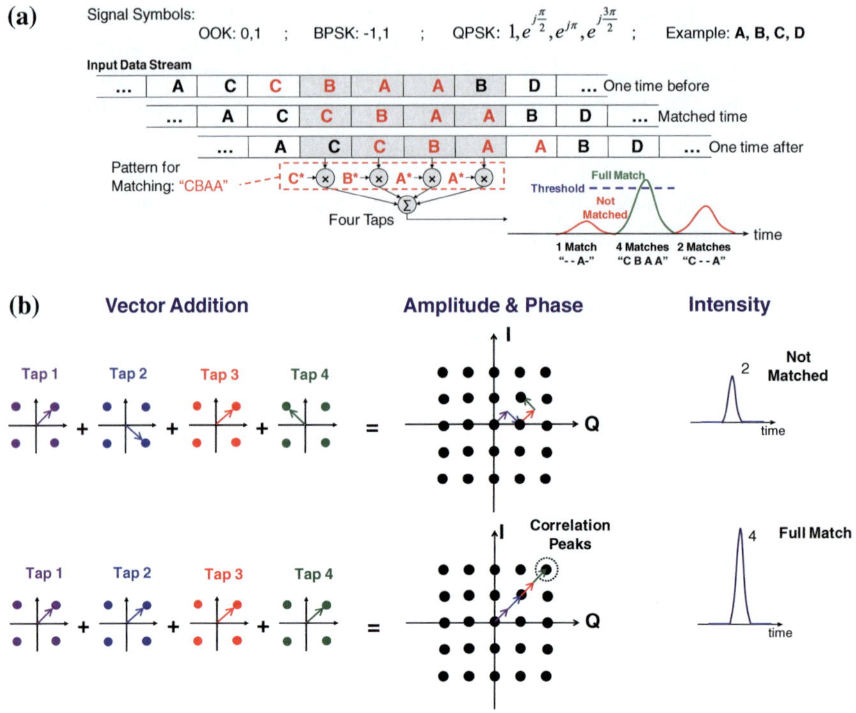

Fig. 6.15 Concept of a correlator for phase-modulated signals: **a** tap coefficients are set equal to the complex conjugate of the search pattern; so when the input data stream slides through the tapped-delay-line, a high correlation peak occurs if the patterns match. **b** Complex-coefficient taps enable correlation on QPSK signals by allowing vector addition of adjacent symbols by generating higher-order QAM with a variable intensity [36]

Here, the tap coefficients are set as the phase-conjugate of the target pattern that we want to search for. As the data stream slides through a set of taps, adjacent symbols are multiplied by the tap coefficient and added to form the output. To identify the appearance of particular consecutive values, "CBAA," in a long pattern of QPSK symbols, the correlator weights need to be set to the phase conjugate of the target pattern (i.e., "C*B*A*A*"). An N-tap correlator on QPSK signal has an output of $(n + 1)^2$-QAM constellation. Figure 6.15b shows how the 25QAM constellation can be generated by 4QPSK signals. The upper right corner of the output constellation corresponds to the target pattern. One example of matched pattern (upper right corner) and one example of a mismatched pattern are shown in Fig. 6.15b. Direct detection of the correlator output can be used to search the pattern based on output intensity. When the patterns fully match, a high intensity is created, while partial matches result in lower amplitudes. Additionally, a thresholder can be used after the correlator to identify full matches when the intensity exceeds a certain threshold. Figure 6.15b also illustrates the corresponding intensity of two sample cases of a full match (large amplitude output) and a mismatch [11]

(a) *Fixed Bit Rate 8-bit Optical Correlator* In [46], a reconfigurable silica planar-lightwave-circuit optical delay line filter is used for pattern recognition on a 40-Gbit/s BPSK signal. Figure 6.16a shows the schematic of an 8-bit correlator with tunable coefficients. The variable optical attenuators (VOA) set the amplitude of the taps to be equal, and phase shifters program each tap phase to be 0 or π to choose the target pattern. The experimental results of the 8-bit correlator are shown in Fig. 6.16b. Here, the target pattern is set to "0 π 0 π 0 π 0 0." Assuming that, of 8 bits, only m bits match, the output field of the correlator is $(+1) \times m + (-1) \times (8 - m) = 2\,m - 8$ ($m \in \{0 \ldots 8\}$). The correlator adds the optical fields; thus, when all 8 bits match, the field has a level of 8. However, the photodiode is a square law device; therefore, the amplitude of a full match after the photodiode is 64, which is higher than the amplitude of 7 matches (i.e., 36). Figure 6.16b depicts the output eye diagram after the photodiode. The optical waveform after the correlator is also shown with the electronically thresholded pattern that features a pulse where a pattern match occurs.

(b) *Tunable Correlator* The optical TDL can be used to perform optical correlation on phase- and amplitude-modulated signals (BPSK, QPSK, and OOK) [36]. Figure 6.16c–f summarize examples using this approach on 20- and 40-GBd signals. Output eye diagrams, optical correlation output waveforms, and electronically thresholded waveforms are shown for three different modulation formats and patterns. In Fig. 6.16c, an output with a "1 1 1 1" pattern search in a 40-Gbit/s RZ-OOK signal is presented. Figure 6.16d, e illustrate the experimental results for 40-Gbit/s RZ-BPSK and 80-Gbit/s RZ-QPSK signals, in which the patterns "π 0 π 0" and "$-3\pi/4$ $3\pi/4$ $\pi/4$" are searched, respectively. In Fig. 6.16f, the coherent detection of the correlator output is depicted

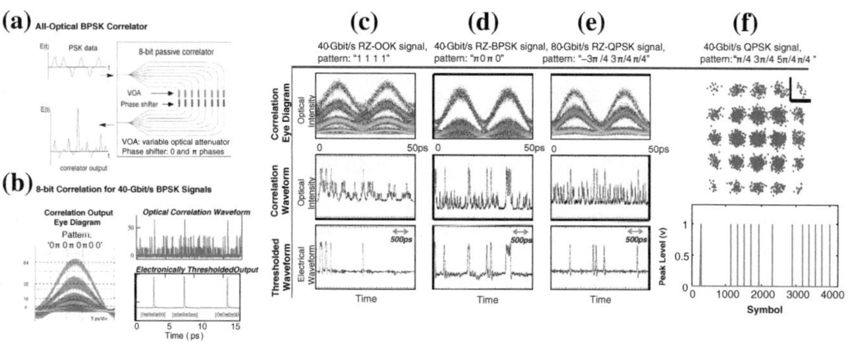

Fig. 6.16 **a** Structure of a silica planar lightwave circuit optical delay line filter for 8-bit correlation of BPSK signals. **b** Experimental eye diagram, optical correlation waveform and electronically thresholded output for searching pattern "0 π 0 π 0 π 0 0" [46]. All-optical tunable correlation based on a tunable tapped-delay-line to search for (**c**) pattern "1 1 1 1" in a 40-Gbit/s RZ-OOK signal, **d** pattern "π 0 π 0" in a 40-Gbit/s RZ-BPSK signal, and **e** pattern "−3π/4 3π/4 π/4" in a 80-Gbit/s RZ-QPSK signal [36]. **f** Coherent detection of the correlation results on 40-Gbit/s QPSK signal when "π/4 3π/4 5π/4 π/4" is the target pattern

while the incoming 20-Gbaud QPSK data is searched for the pattern "$\pi/4$ $3\pi/4$ $5\pi/4$ $\pi/4$". The correlation peaks that correspond to the upper right corner of the output constellation are also shown. These peaks indicate the position of the target pattern appearance in the first 4096 incoming symbols.

6.6.4 Optical WDM Correlator and 2D Correlation

The optical tunable correlator performs pattern searching on a single incoming data signal. In many applications, it is desired to have a single correlator that can process multiple WDM data channels simultaneously and independently. This scheme can increase the processing capacity dramatically by performing parallel computations on independent incoming data channels.

Figure 6.17a shows a multi-channel correlator. The WDM input channels are simultaneously wavelength converted in a nonlinear medium, which generates a replica for each WDM channel at a new center frequency. Because each signal and its replica are at different frequencies, a relative delay is induced between them after passing through a chromatic dispersive medium. The replicas and delayed original signals are then sent into a phase and amplitude programmable filter based on LCoS technology that applies the complex tap coefficients to the replicas. Finally, the delayed original signals and their weighted replicas are sent into another wavelength converting stage that coherently copies the original signals onto the replicas for a second time, which creates the desired two-tap processing. Thus, a same-delay through independent two-tap OTDL can be implemented on each WDM channel. A higher number of taps can potentially be implemented by cascading the two-tap tunable building blocks (see Fig. 6.17a).

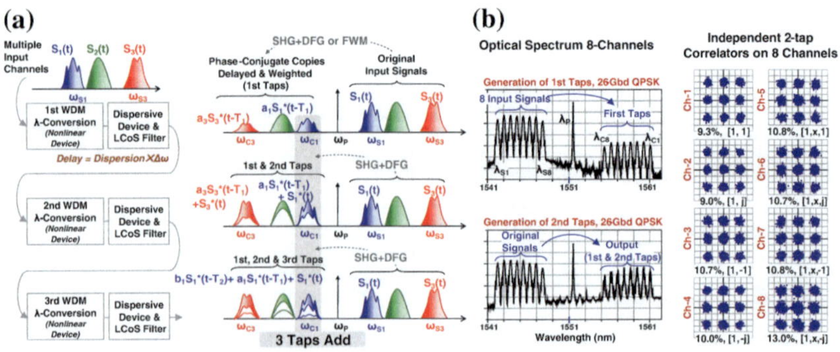

Fig. 6.17 **a** Principle of operation of the WDM-OTDL, shown for a three-tap implementation: First, taps are created by simultaneous wavelength conversion of all WDM channels in nonlinear elements, with one such element for each tap. Second, a dispersive element induces a relative delay between them. Then, tap coefficients are applied by an inline LCoS filter. Because the pump and signals are reused, the weighted taps add coherently. **b** Spectra after the first and the second nonlinear stages and simultaneous two symbol pattern search on eight 40-GBaud QPSK WDM channels, which results in an aggregate throughput of 416 Gbit/s

Figure 6.17b depicts the spectra of generating the first tap and second tap for the eight-channel 20-Gbaud QPSK data. The two-tap OTDL is in principle, an interferometer, thus destructive interference can be observed as power dips in the converted signals of the second spectrum. A correlation of a two-symbol pattern search is performed by applying various tap coefficients (see Fig. 6.17b). The processed channels are detected using a coherent receiver [47].

Using the WDM correlator scheme, a 2-D correlator can be implemented to perform pattern searching in 2-D images. The conceptual block diagram of the 2-D OTDL is shown in Fig. 6.18. First, each image row is mapped to a WDM channel by modulating the optical frequency comb fingers (phase locked sources) with QPSK symbols that correspond to the color of the image pixels. Second, a WDM-TDL is performed by generating the inner sums on each WDM channel with the given coefficients (b_i) independently. Subsequently, these processed WDM channels—which are phase coherent—are multiplexed together with column coefficients (a_i) to generate the outer sums.

The experimental setup for the 2-D correlator is depicted in Fig. 6.19a. A mode-locked laser with a 10 GHz repetition rate and a DLI with FSR 20 GHz is used to generate coherent comb fingers with 20 GHz frequency spacing. The 20 GHz comb is then passed through an HNLF fiber to generate a flat and broad spectrum. A Liquid Crystal on Silicon (LCoS-1) filter is used to select and write complex weights on the comb fingers and separate them into the signal path and pump path. A nested Mach-Zehnder modulator is used to generate the 20-Gbaud QPSK data on the two comb fingers and write the image pixels information on the signal path separated by $\sim$1.6 nm. To make the two modulated signals as two consecutive rows in a 31 × 31 image, they need to be sent to DCF-1 with 12 km length to induce $\sim$969 ps/nm dispersion (31-symbols of delay). The resulting signals are coupled with an amplified CW pump and sent to an HNLF to produce the first tap of the inner sum. All signals then travel through DCF-2 and LCoS-2 to apply tap delays, phases and amplitudes.

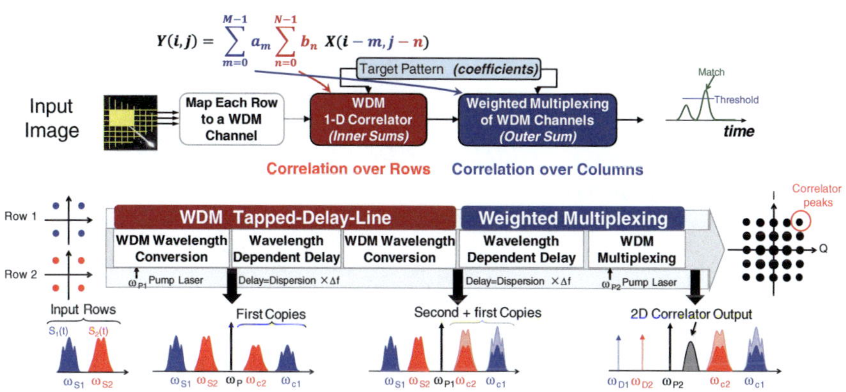

Fig. 6.18 Concept of 2-D optical tapped-delay-lines: A 2-D correlation can be achieved by cascading a WDM correlator with a multiplexing stage to realize the nested sum required for 2D operation

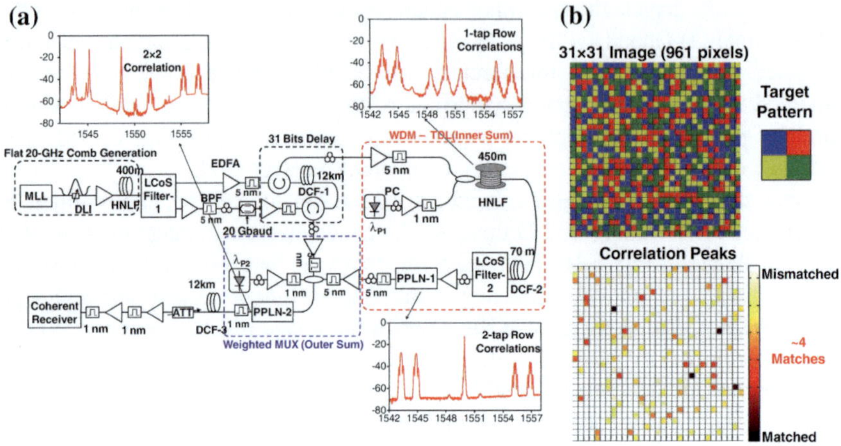

Fig. 6.19 **a** Experimental setup of 2-D optical tapped-delay-lines. First appropriate comb fingers are selected for both the signal and pump paths and the image pixel colors are mapped to the QPSK symbols. Then, using first and second wavelength conversions and a dispersive element, a WDM correlator is realized on both rows. Finally, another nonlinear stage is used to coherently combine the two processed rows with the outer sum coefficients. **b** Searching a 2 × 2 target pattern in an image with 961 pixels. The dark squares correspond to the target patterns

The signals, their copies, and the CW pump laser are then sent to the first PPLN waveguide to create the second tap. The quasi-phase matching (QPM) wavelength of the PPLN waveguide is temperature-tuned to the CW pump wavelength. These signals, along with the comb fingers selected for pump path and another CW laser pump, are sent to the second PPLN to perform coherent multiplexing of the two processed rows. The multiplexed signal is filtered and sent to the coherent receiver, after passing through another DCF of the same length to compensate for the induced dispersion. In Fig. 6.19b, a 31 × 31 image (961 pixels) is searched for a 2 × 2 target pattern. The experimental results exhibit 4 peaks (dark points) in output table at the 2 × 2 sections that matched the target patterns [48].

6.6.5 *Discrete Fourier Transforms Using Nonlinearities and On-Chip MZIs*

Optical OFDM signals involve creating a wideband data channel by modulating multiple orthogonal frequency subcarriers at a lower rate. The modulated subcarriers overlap in the frequency domain [49]. Traditionally, optical OFDM signals have been generated in electronics using electronic inverse fast Fourier transform (IFFT) techniques to calculate time-domain symbols from frequency-domain symbols, which, after parallel-to-serial and digital-to-analog conversions are conducted, create the I and Q electrical signals that modulate a laser [49]. The subcarrier modulation speed is limited to the speed at which discrete Fourier transform (DFT) or fast Fourier transform (FFT) can be implemented in electronics. Alternatively, OFDM signals can be generated optically using an optical frequency comb source as the optical subcarriers and independently modulate each comb line and combine them to create an optical OFDM channel [43, 50]. An optical OFDM channel can be demodulated using parallel electronic DFT or FFT. Alternative optical DFT/FFT approaches can allow for the use of higher baud rates for subcarrier modulation by performing the DFT/FFT at the line rate of optics [43, 50, 51].

An OFDM signal with N subcarriers separated by Δf can be represented as

$$S(t) = \sum_{n=0}^{N-1} x_n(t) e^{j2\pi(f_0+n\Delta f)t} \tag{6.35}$$

in which f_0 and $x_n(t)$ are optical carrier frequency and nth subcarrier signal, respectively [4]. Optical N-point DFT can extract the nth subcarrier signal from $S(t)$, according to

$$x_n(t) = \sum_{k=0}^{N-1} S\left(t + \frac{k}{\Delta f \times N}\right) e^{j2\pi nk/N} \tag{6.36}$$

which needs to be sampled at $1/\Delta f$ time intervals to recover the transmitted symbols. This demodulation equation follows the form of a tapped delay line in (X). In this section, we review two examples of optical OFDM demodulation involving a fixed FFT approach and a baud-rate-tunable DFT approach based on the tunable TDL presented in Sect. 6.1 [43, 50].

(a) *Optical FFT Based on Delay Interferometers* The concept of an optical FFT to separate the subcarriers of an optical OFDM signal is shown in Fig. 6.20a [50]. The structure is based on cascaded delay interferometers and is capable of performing serial-to-parallel conversion with optical FFT to decompose the optical OFDM signal into its constituent subcarriers. The data of each subcarrier is then recovered by proper time-gating using electro-absorption modulators. This configuration requires only N − 1 passive delay interferometers to realize an N-point FFT. An 8-point FFT with fixed delays is demonstrated in [50]. Furthermore, in [50], 325 comb lines of a 12.5-GHz frequency comb source are modulated by 10-GBd 16-QAM signals in two polarizations to generate an OFDM channel at 26-Tbit/s. A simplified two point FFT is performed to extract the subcarriers. Figure 6.20b shows two constellation diagrams of the extracted 16-QAM subcarriers with the spectrum and error vector magnitude (EVM) of all output subcarriers.

(b) *Adjustable Bit-Rate Optical DFT* The tunable optical TDL depicted in Fig. 6.14 has also been used to perform optical DFT with various numbers of taps and bit rates [43]. Figure 6.21a illustrates the concept of using the tunable TDL to perform DFT for extraction of OFDM subcarriers. According to OFDM demodulation (6.27), which follows the form of the tunable TDL (6.21), to

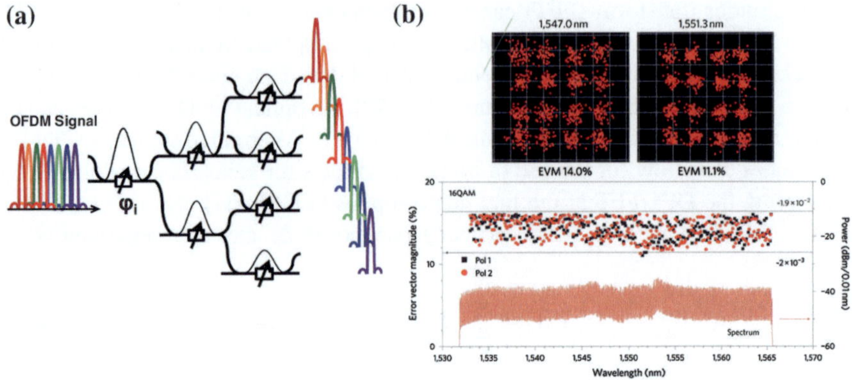

Fig. 6.20 Optical OFDM demodulation using delay-interferometer-based fast Fourier transform: **a** structure, and **b** constellations for two 10-GBd demultiplexed subcarriers of a 26-Tbit/s OFDM signal (PM-16-QAM subchannels) [50]

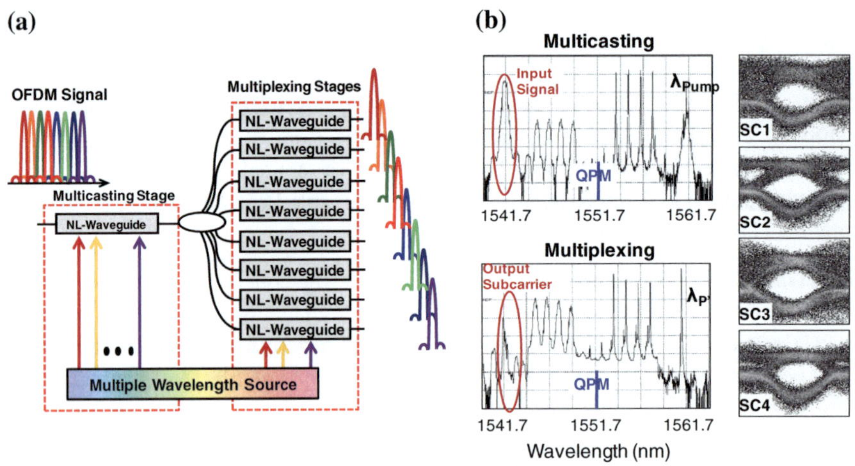

Fig. 6.21 Tunable optical OFDM demodulation using optical DFT enabled by the optical tapped-delay-line: **a** concept, and **b** multicasting and multiplexing spectra and output eye diagrams of all four subcarriers in a four 20-GBd BPSK subcarrier OFDM signal

extract the *n*th subcarrier from an *N*-subcarrier OFDM signal with Δf separation, the tunable TDL needs to be configured to have *N* taps, each separated in time by $1/(N \times \Delta f)$, with tap coefficients $e^{-j2\pi nk/N}(k \in \{0, \ldots, N-1\})$. In Fig. 6.21b, the input OFDM channel is a 80-Gbit/s four-subcarrier BPSK signal generated optically by modulating four optical comb lines separated by 20 GHz. The multicasting and multiplexing spectra for the implementation of a 4-tap TDL are shown in Fig. 6.21b, where the tap coefficients and tap delays are configured to represent the coefficients of a 4-point DFT. The TDL coefficients were reconfigured four times to demodulate each of the four 20-Gbit/s BPSK subcarriers. The eye diagrams of the four extracted subcarriers are also shown in Fig. 6.21b.

References

1. http://www.internetworldstats.com/stats.htm
2. L. Potì, G. Meloni, G. Berrettini, F. Fresi, M. Secondini, T. Foggi, G. Colavolpe, E. Forestieri, A. D'Errico, F. Cavaliere, R. Sabella, G. Prati, Casting 1 Tb/s DP-QPSK communication into 200 GHz bandwidth, in *Proceedings of the International Conference ECOC*, Amsterdam, 2012
3. A. Bogoni, X. Wu, Z. Bakhtiari, S. Nuccio, A.E. Willner, 640 Gbit/s photonic logic gates. Opt. Lett. **35**(23), 3955–3957 (2010)
4. F. Scotti, F. Laghezza, S. Pinna, P. Ghelfi, and A. Bogoni, High precision photonic ADC with four time-domain-demultiplexed interleaved channels, in *Proceedings of the International Conference Photonic in Switching*, paper TuO1-3, Kyoto, 2012
5. T.D. Vo, H. Hu, M. Galili, E. Palushani, J. Xu, L.K. Oxenløwe, S.J. Madden, D.-Y. Choi, D. A.P. Bulla, M.D. Pelusi, J. Schröder, B. Luther-Davies, B.J. Eggleton, Photonic chip based transmitter optimization and receiver demultiplexing of a 1.28 Tbit/s OTDM signal. Opt. Express **18**(16), 17252–17262 (2010)
6. J. Marti, K. Williams, Is all-optical processing green? in *presented at Workshop OFC 2010*, San Diego, CA
7. R.S. Tucker, A green internet, in *Proceedings of the IEEE Lasers Electro-Optics Society Annual Meeting* (2008), pp. 4–5
8. P. Martelli, P. Boffi, M. Ferrario, L. Marazzi, P. Parolari, R. Siano, V. Pusino, P. Minzioni, I. Cristiani, C. Langrock, M.M. Fejer, M. Martinelli, V. Degiorgio, All-optical wavelength conversion of a 100-Gbit/s polarization-multiplexed signal. Opt. Exp. **17**(20), 17758–17763 (2009)
9. M. Galili, J. Xu, H.C. Mulvad, L.K. Oxenløwe, A.T. Clausen, P. Jeppesen, B. Luther-Davies, S. Madden, A. Rode, D. Y. Choi, M. Pelusi, F. Luan, B.J. Eggleton, Breakthrough switching speed with an all-optical chalcogenide glass chip: 640 Gbit/s demultiplexing, Opt. Exp. **17**(4), 2182–2187 (2009)
10. T.D. Vo, M.D. Pelusi, J. Schroder, F. Luan, S.J. Madden, D.-Y. Choi, D.A.P. Bulla, B. Luther-Davies, B.J. Eggleton, Simultaneous multi-impairment monitoring of 640 Gbit/s signals using photonic chip based RF spectrum analyzer. Opt. Exp. **18**(4), 3938–3945 (2010)
11. H. Ji, H. Hu, M. Galili, L.K. Oxenløwe, M. Pu, K. Yvind, J.M. Hvam, P. Jeppesen, Optical waveform sampling and error-free demultiplexing of 1.28 Tbit/s serial data in a silicon nanowire, in *Presented at the OFC 2010*, San Diego, CA, paper PDPC7
12. T.D. Vo, R. Pant, M.D. Pelusi, J. Schröder, D.-Y. Choi, S.K. Debbarma, S.J. Madden, B. Luther-Davies, B.J. Eggleton, Photonic chip-based all-optical XOR gate for 40 and 160 Gbit/s DPSK signals. Opt. Lett. **36**, 710–712 (2011)
13. L. Liu, R. Kumar, K. Huybrechts, T. Spuesens, G. Roelkens, E.-J. Geluk, T. de Vries, P. Regreny, D. Van Thourhout, R. Baets, G. Morthier, An ultra-small, low-power, all-optical flip-flop memory on a silicon chip. Nat. Photonics **4**, 182–187 (2010)
14. A. Trita, G. Mezősi, M. Sorel, G. Giuliani, All-optical toggle flip-flop based on monolithic semiconductor ring laser. Phot. Tech. Lett. **26**(1) (2014)
15. F. Scotti, G. Berrettini, G. Contestabile, A. Bogoni, A regenerative Variable optical buffer for NRZ and RZ Packets. J. Lightw Technol. **30**(9), 1366–1372 (2012)
16. S. Le Beux, Z. Li, C. Monat, X. Letartre, I. O'Connor, Reconfigurable photonic switching: towards all-optical FPGAs, in *Proceedings of the 21st International Conference on VLSI-SoC* (2013)

17. H. Brahmi, G. Giannoulis, M. Menif, V. Katopodis, D. Kalavrouziotis, C. Kouloumentas, P. Groumas, G. Kanakis, C. Stamatiadis, H. Avramopoulos, D. Erasme, On the fly all-optical packet switching based on hybrid WDM/OCDMA labeling scheme. Opt. Commun. **312**, 175–184 (2014)
18. J. Bannister, J. Touch, P. Kamath, A. Patel, An optical booster for internet routers, in *Proceedings of the 8th International Conference High Performance Computing* (2001), pp. 339–413
19. R.W. Boyd, *Nonlinear Optics*, 2nd edn. (Academic, New York, 2003)
20. C. Langrock, S. Kumar, J.E. McGeehan, A.E. Willner, M.M. Fejer, All-optical signal processing using _2 nonlinearities in guided-wave devices. J. Lightw. Technol. **24**(7), 2579–2592 (2006)
21. G. Agrawal, *Nonlinear Fiber Optics* (Academic Press, New York, 2001)
22. E. Lazzeri, A. Malacarne, G. Serafino, and A. Bogoni, Optical XOR for error detection and coding of QPSK I and Q components in PPLN waveguide. 2258 PTL **24**(24), 2258 (2012)
23. Y. Miyoshi, K. Ikeda, H. Tobioka, T. Inoue, S. Namiki, K. Kitayama, Ultrafast all-optical logic gate using a nonlinear optical loop mirror based multi-periodic transfer function. Opt. Exp. **16**(4), 2570–2577 (2008)
24. G. Berrettini, A. Simi, A. Malacarne, A. Bogoni, L. Potí, Ultrafast integrable and reconfigurable XNOR, AND, NOR, and NOT photonic logic gate. IEEE PTL **18**(8) (2006)
25. Z. Li, G. Li, Ultrahigh-speed reconfigurable logic gates based on four-wave mixing in a semiconductor optical amplifier. IEEE Photonics Technol. Lett. **18**(12) (2006)
26. P.P. Bavejaa, A.M. Kaplana, D.N. Maywarb, G.P. Agrawala, Pulse amplification in semiconductor optical amplifiers with ultrafast gain-recovery times, in *Optical Components and Materials VII Proceedings of SPIE*, vol. 7598, 759817–1 (2010)
27. P.A. Andersen et al., Wavelength conversion of a 40-Gb/s RZ-DPSK signal using four-wave mixing in a dispersion-flattened highly nonlinear photonic crystal fiber. IEEE Photonics Technol. Lett. **17**(9), 1908–1910 (2005)
28. A. Yariv, P. Yeh, *Optical Waves in Crystals,* vol. 5 (Wiley, New York, 1984)
29. A.C. Turner-Foster, M.A. Foster, J.S. Levy, C.B. Poitras, R. Salem, A.L. Gaeta, M. Lipson, Ultrashort free-carrier lifetime in low-loss silicon nanowaveguides. Opt. Express **18**, 3582–3591 (2010)
30. F. Li, T.D. Vo, C. Husko, M. Pelusi, D.-X. Xu, A. Densmore, R. Ma, S. Janz, B.J. Eggleton, D.J. Moss, All-optical XOR logic gate for 40 Gb/s DPSK signals via FWM in a silicon nanowire, in *Proceedings of the IEEE Photonics Conference* (2011), pp. 593–594
31. J. Wang, J. Sun, Q. Sun, Single-PPLN-based simultaneous half-adder, half-subtracter, and OR logic gate: proposal and simulation. Opt. Express **15**(4), 1690–1699 (2007)
32. A. Bogoni, X. Wu, I. Fazal, A.E. Willner, 160 Gb/s time-domain channel extraction/ insertion and all-optical logic operations exploiting a single PPLN waveguide. J. Lightw. Technol. **27**, 4221–4227 (2009)
33. J. Wang, J.-Y. Yang, X. Wu, O.F. Yilmaz, S. R. Nuccio, A.E. Willner, 40-Gbaud/s (120-Gbit/s) octal and 10-Gbaud/s (40-Gbit/s) hexadecimal simultaneous addition and subtraction using 8PSK/16PSK and highly nonlinear fiber, in *Proceedings of the Optical Fiber Communications Conference* (2011), pp. 1–3
34. J.G. Proakis, *Digital Communications* (McGraw-Hill, New York, 2000)
35. B. Moslehi, J.W. Goodman, M. Tur, H.J. Shaw, Fiber-optic lattice signal processing. Proc. IEEE **72**(7), 909–930 (1984)
36. S. Khaleghi, O.F. Yilmaz, M.R. Chitgarha, M. Tur, N. Ahmed, S.R. Nuccio, I.M. Fazal, X. Wu, M.W. Haney, C. Langrock, M.M. Fejer, A.E. Willner, High-speed correlation and equalization using a continuously tunable all-optical tapped delay line. IEEE Photon. J. **4**(4), 1220–1235 (2012)
37. M.R. Chitgarha, S. Khaleghi, O.F. Yilmaz, M. Tur, M.W. Haney, C. Langrock, M.M. Fejer, A.E. Willner, Coherent correlator and equalizer using a reconfigurable all-optical tapped delay line. Opt. Lett. **38**, 2271–2273 (2013)

38. D.B. Hunter, R.A. Minasian, Programmable high-speed optical code recognition using fibre Bragg grating arrays. IEEE Electron. Lett. **35**(5), 412–414 (1999)
39. M.C. Hauer, J.E. McGeehan, S. Kumar, J.D. Touch, J. Bannister, E.R. Lyons, C.H. Lin, A.A. Au, H.P. Lee, D.S. Starodubov, A.E. Willner, Optically assisted Internet routing using arrays of novel dynamically reconfigurable FBG-based correlators. J. Lightw. Technol. **21**(11), 2765–2778 (2003)
40. S. Frisken, G. Baxter, D. Abakoumov, H. Zhou, I. Clarke, S. Poole, Flexible and grid-less wavelength selective switch using LCOS technology, in *Proceedings of the Optical Fiber Communications Conference* (2011), Paper OTuM3
41. S. Khaleghi, M.R. Chitgarha, O.F. Yilmaz, M. Tur, M.W. Haney, C. Langrock, M.M. Fejer, A.E. Willner, Experimental characterization of phase tuning using fine wavelength offsets in a tunable complex-coefficient optical tapped-delay-line. Opt. Lett. **39**, 735–738 (2013)
42. S. Khaleghi, M.R. Chitgarha, O.F. Yilmaz, M. Tur, M.W. Haney, C. Langrock, M.M. Fejer, A.E. Willner, Experimental performance of a fully tunable complex-coefficient optical FIR filter using wavelength conversion and chromatic dispersion. Opt. Lett. **37**, 3420–3422 (2012)
43. M.R. Chitgarha, S. Khaleghi, O.F. Yilmaz, M. Tur, M.W. Haney, C. Langrock, M.M. Fejer, A.E. Willner, Demonstration of channel-spacing-tunable demultiplexing of optical orthogonal frequency-division-multiplexed subcarriers utilizing reconfigurable all optical discrete Fourier transform. Opt. Lett. **37**, 3975–3977 (2012)
44. M.R. Chitgarha, S. Khaleghi, O.F. Yilmaz, J.-Y. Yang, A.E. Willner, Bit depth and sample rate tunable digital to analog converter using conversion/dispersion based delays, in *Presented at the European Optical Communications*, Geneva, Switzerland, 2011
45. M.S. Rasras, I. Kang, M. Dinu, J. Jaques, N. Dutta, A. Piccirilli, M.A. Cappuzzo, E.Y. Chen, L.T. Gomez, A. Wong-Foy, S. Cabot, G.S. Johnson, L. Buhl, S.S. Patel, A programmable 8-bit optical correlator filter for optical bit pattern recognition. IEEE Photon. Technol. Lett. **20** (9), 694–696 (2008)
46. I. Kang, M. Rasras, M. Dinu, M. Cappuzzo, L.T. Gomez, Y.F. Chen, L. Buhl, S. Cabot, A. Wong-Foy, S.S. Patel, C.R. Giles, N. Dutta, J. Jaques, A. Piccirilli, All-optical byte recognition for 40-Gb/s phase-shift-keyed transmission using a planar-lightwave-circuit passive correlator. IEEE Photon. Technol. Lett. **20**(12), 1024–1026 (2008)
47. S. Khaleghi, M.R. Chitgarha, M. Ziyadi, W. Daab, A. Mohajerin-Ariaei, D. Rogawski, J.D. Touch, M. Tur, C. Langrock, M.M. Fejer, A.E. Willner, Simultaneous and independent processing of multiple input WDM data signals using a tunable optical tapped delay line. Opt. Lett. **38**, 4273–4276 (2013)
48. M.R. Chitgarha, M. Ziyadi, S. Khaleghi, A. Mohajerin-Ariaei, A. Almaiman, J.D. Touch, M. Tur, C. Langrock, M.M. Fejer, A.E. Willner, Reconfigurable 2-D WDM optical tapped-delay-line to correlate 20Gbaud QPSK data, in *Presented at the European Conference Optical Communication*, Geneva, Switzerland, 2011
49. J. Armstrong, OFDM for optical communications. J. Lightw. Technol. **27**(3), 189–204 (2009)
50. D. Hillerkuss, R. Schmogrow, T. Schellinger, M. Jordan, M. Winter, G. Huber, T. Vallaitis, R. Bonk, P. Kleinow, F. Frey, M. Roeger, S. Koenig, A. Ludwig, A. Marculescu, J. Li, M. Hoh, M. Dreschmann, J. Meyer, S. Ben Ezra, N. Narkiss, B. Nebendahl, F. Parmigiani, P. Petropoulos, B. Resan, A. Oehler, K. Weingarten, T. Ellermeyer, J. Lutz, M. Moeller, M. Huebner, J. Becker, C. Koos, W. Freude, J. Leuthold, 26 Tbit s^{-1} line-rate super-channel transmission utilizing all-optical fast Fourier transform processing, Nat. Photon. **5**, 364–371 (2011)
51. J. Schröder, L.B. Du, M.A. Roelens, B. Eggleton, A.J. Lowery, Reconfigurable all-optical discrete Fourier transform in a wavelength selective switch for optical OFDM demultiplexing, in *Proceedings of the Optical Fiber Communications Conference* (2012), Paper OTh1G.6

Chapter 7
Wide-Band and Noise-Inhibited Signal Manipulation in Dispersion-Engineered Parametric Mixers

Bill P.-P. Kuo and Stojan Radic

Abstract Practical frequency generation and conversion requires highly efficient parametric interaction across wide spectral band. To achieve high efficiency, conventional devices rely on optical enhancement in highly resonant cavity structures. While optical cavity also inhibits noise generation and out-of-band amplification during the parametric process, it inherently limits frequency tuning and any band reconfiguration. Recognizing this fundamental impairment, we describe the new class of cavity-free frequency generation in heterogeneous, traveling-wave parametric mixers. Driven by a continuous-wave seeds, these devices combine inherently more stable lasers with distributed noise inhibition in dispersion-managed parametric process. The operating principle, design methodology and performance limits are outlined and discussed. The use of new technology is illustrated on examples of signal multicasting and ultrafast channel processing.

7.1 Introduction

Recent developments of new optical devices capable of wide-band optical field generation and manipulation are motivated by three general trends. Firstly, the adoption of coherent modulation schemes in terabit-per-second (Tbps) communication channels will soon require signal processing to be performed at rates exceeding that of any feasible electronic backplane [1]. Secondly, communication and computation systems relying on direct manipulation of photon states preclude, even in principle, photon-electron conversion to perform subsequent electronic processing. Finally, the need for general sensing has expanded outside the conventional bands served by sensitive imagers (visible) and detectors (near-infrared—NIR). Indeed, in the immediate vicinity of NIR band, a short-wave IR (SWIR) offers a transparent

B.P.-P. Kuo · S. Radic (✉)
Department of Computer and Electrical Engineering, University of California San Diego, 9500 Gilman Drive, La Jolla, CA 92093-0407, USA
e-mail: sradic@ucsd.edu

S. Wabnitz and B.J. Eggleton (eds.), *All-Optical Signal Processing*, Springer Series in Optical Sciences 194, DOI 10.1007/978-3-319-14992-9_7 atmospheric window, spurring the development of new LIDAR, communication and imaging systems [2]. Unfortunately, the technological complement outside of standard (NIR) window does not include either source or detector with performance comparable to the telecommunication devices. While SWIR laser and detector (imager) can arguably be improved, basic physical mechanisms responsible for 1550 nm device performance is simply inferior in this range. Recognizing this limitation, the alternative strategy can still use NIR sources and detectors by leveraging parametric mixers capable of distant spectral translation for either emission or reception of a coherent optical field. In broader terms, the combined need for high signal rates and direct, coherent field control call for specialized set of functions such as frequency referencing, beam switching and band mapping that do not rival electronics in its complexity. However, these are invariably fast and, in context of this chapter, can be realized using specialty parametric mixing device.

In addition to this application-driven impetus, a set of important technological advances has provided an additional motivation: it is now possible to design and construct parametric devices using telecommunication derived components. In contrast to conventional, crystalline parametric devices, the introduction and recent maturing of high-confinement fiber has drastically reduced pump power required for efficient frequency generation and light amplification [3]. It is no longer necessary to consider high peak intensity sources such as mode-locked lasers (MLL) in order to generate and convert signals over tens of THz: it is now possible to accomplish the same using continuous-wave (CW), Sub-Watt scale telecommunication diodes, greatly reducing the complexity and reliability of general parametric device. In addition to a great progress in reducing the loss of high confinement fiber [4], advances in precise dispersion control and measurement [5–8] are of equal importance. Indeed, combined advances in design, fabrication and characterization of highly-nonlinear fiber (HNLF) have led to the first multistage, CW-pumped parametric mixers possessing bandwidths in excess of 20 THz. New HNLF types now offer the promise of highly stabilized local dispersion parameters [4, 9–11], guaranteeing, for the first time, spectral access to combined NIR/SWIR bands approaching an octave span. While important for bandwidth engineering, localized dispersion control also assumes critical roles in noise inhibition in distributed parametric devices [12]. When combined with heterogeneous parametric design [12, 13], new generation of mixers can provide spectral purity and bandwidth that is simply not feasible by any alternative technology.

While it is not possible to accurately predict all mixer-enabled applications in the future, one can state, with high confidence, that coherent signal preprocessors, advanced emitters and sensitive receivers have widespread near-future adoption. In the first of these categories (preprocessing), coherent sampling [14], digitization [14] and spectral analysis [15] has led to performance level exceeding that of comparable electronic architectures. In the second category (emitters) parametric mixers have enabled frequency combs spanning combined C- and L-bands [16] while possessing optical signal to noise ratio (OSNR) well beyond MLL technology [17, 18]. Similarly, the new mixers were used to demonstrate fast tunable lasers operating over combined NIR/SWIR bands [19]. In the third category (receivers), it is now possible to register few-photon packets at rates exceeding 100 GHz [20].

Consequently, this chapter aims to describe basic principles, impairment mechanisms and the design principles used to generate and manipulate light over wide bandwidth and in noise-inhibited manner.

7.2 Fundamentals of Parametric Mixers

Optical parametric mixing encompasses a collection of nonlinear processes in which, the energies (as well as momenta) of the participating photons are re-distributed but not transferred to the medium. Due to the passive role of the medium, parametric mixing is the only known mechanism that provides direct access to participating photon characteristics (in terms of number, energy and polarization) since the overall states of the photons are conserved in the process. This also means parametric mixing can possess an arbitrarily-wide spectral coverage (bandwidth), which is solely limited by the transparent window of and potential multi-photon resonant processes in the medium. Furthermore, unlike stimulated-emission processes in which the finite upper-state lifetime delays the mixing response, the absence of (meta-)stable state in parametric mixing processes gives rise to its ultrafast response.

Parametric mixers of interest in this chapter are based on four-photon mixing (FPM) processes mediated by the third-order nonlinear response. A simple four-photon mixer annihilates two photons, typically from a single or two frequency-distinct pump waves, to create two photons that coincides the energy of the remaining participating waves. This photon-exchange mechanism leads to a number of important phenomena, including optical amplification, generation, and frequency conversion [21–23]. Specific wave dynamics of a parametric mixer are primarily governed by the pump configurations and the corresponding phase-matching conditions, as revealed in the following sections.

7.2.1 One-Pump Parametric Mixing

In the presence of a single pump wave possessing high intensity, the two photons donated in the FPM process are both derived from this single source. The parametric mixing process driven by this pump configuration, as shown in the energy diagram of Fig. 7.1, is known as one-pump or degenerate FPM.

In a one-pump FPM process, two photons from a single pump wave (where its relevant variables are denoted by a subscript P) are annihilated to create a photon pair known as signal and idler. Energy conservation demands that the sum of signal and idler photon energies be equal to two pump energy quanta ($\hbar\omega_S + \hbar\omega_I = 2\hbar\omega_P$), which also means the frequencies of signal and idler must be symmetric with respect

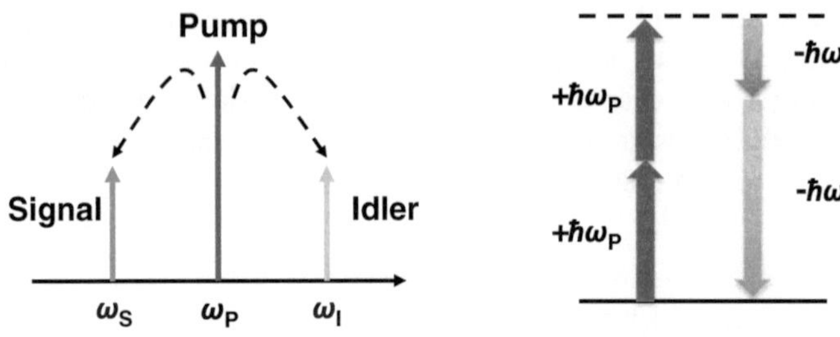

Fig. 7.1 Frequency configuration (*left*) and energy diagram (*right*) of one-pump parametric mixing

to the pump's ($\omega_S - \omega_P = \omega_P - \omega_I = \Omega$). In addition to energy conservation, the momenta of the participating photons must also be matched in order to attain the maximum mixing efficiency [24]. The extent of momentum conservation, also known as phase matching condition in parametric processes, is dictated by the propagation constants of the pump, signal and idler waves [25]:

$$\text{MI: } \Delta\beta = \beta_S + \beta_I - 2\beta_P \tag{7.1}$$

While the energy conservation is self-satisfied in absence of the idler at the input, the distributed nature of parametric mixing implies that the momentum conservation is being determined by the accumulated propagation conditions in the mixing medium. Various levels of phase matching introduce different regimes of operation distinguished by the power growth rate of signal and idler. Monotonic power growth is feasible only when the linear phase mismatch $\Delta\beta$ is bound between $-4\gamma P_P$ and zero, where γ and P_P denote the nonlinear coefficient and pump power of the mixer [25]. The signal experiences a gain scale to the square of the factor $\gamma P_P L$, referred to as nonlinear figure of merits (NFOM) hereinafter, when the gain coefficient g is considerably less than one. Meanwhile, zero phase mismatch, defined by the condition $\Delta\beta = -2\gamma P_P$ forces exponential signal power growth at a rate of $\exp(2\gamma P_P L)$ [25]. The availability of highly-nonlinear fibers with low attenuations and large nonlinear coefficients enabled one-pump parametric amplifiers providing up to 70-dB gain [26]. This level of gain is considerably higher than those demonstrated by erbium-doped fiber amplifiers (EDFA) [27], primarily due to the directivity of parametric gain which circumvented lasing due to Rayleigh scattering [28].

Accompanying the signal amplification is the creation of an idler at an efficiency $\eta = G - 1$. Unlike amplification which mandates adequately small phase mismatching ($-4\gamma P_P < \Delta\beta < 0$), idler creation occurs even with considerable phase mismatch, albeit at reduced efficiency. The idler created at a new wavelength carries the signal field envelope, but with complex conjugation [25]. This capability, known as wavelength conversion, lays the foundation of many optical signal processing functions that are unique to parametric mixers [29].

7.2.2 Two-Pump Parametric Mixing

In contrast to process driven by two photons with identical frequencies, a two-pump interaction derives the source photons from two frequency-distinct waves. The multiple FPM processes seeded by the extra pump classifies two-pump parametric mixers a distinct type of devices, rather than an extension of the one-pump architecture.

A two-pump parametric mixer supports at least three basic FPM processes [30]. Originating from one-pump mixing, the modulation instability (MI) process annihilates two photons from one of the two pumps, and creates a signal and idler photon pair symmetrical to the source pump i.e. $\omega_S + \omega_{I1} = 2\omega_{P1}$ (Fig. 7.2a). Conversely, phase conjugation (PC) mixing results from the annihilation of pump photons at distinct frequencies ω_{P1} and ω_{P2}, which generate signal and idler photons at ω_S and ω_{I2} (Fig. 7.2b). Energy conservation mandates the frequencies of signal and idler be tied to those of the two pumps as $\omega_S + \omega_{I2} = \omega_{P1} + \omega_{P2}$. The resulting frequency configuration of PC process is symmetric to the mean frequency $\omega_m = (\omega_{P1} + \omega_{P2})/2$. While both MI and PC processes provide signal gain as a result of photon addition at the signal frequency, the third process, namely Bragg scattering (BS), attenuates signal by transferring its energy to the associated idler (Fig. 7.2c). Energetically, BS process annihilates one pump photon (for instance, ω_{P1}) and one signal photon to create a photon at another pump frequency ω_{P2} and an idler photon at $\omega_{I3} = \omega_{P1} + \omega_S - \omega_{P2}$. Note, however, that the amplification and attenuation of a specific mode in the above description is fictitious for a frequency configuration that allows all three processes to occur, since then the "idler" of one process will be involved as "signal" in the other processes.

The coupling of three FPM processes in a two-pump parametric mixer defies a simple model [31]. Nevertheless, the evolution of the four signal-idler sidebands depends on the phase-matching conditions governing the individual FPM processes. The phase-matching conditions for PC, MI and BS processes are as follows:

$$\text{MI: } \Delta\beta = \beta_S + \beta_{I1} - 2\beta_{P1} \tag{7.2a}$$

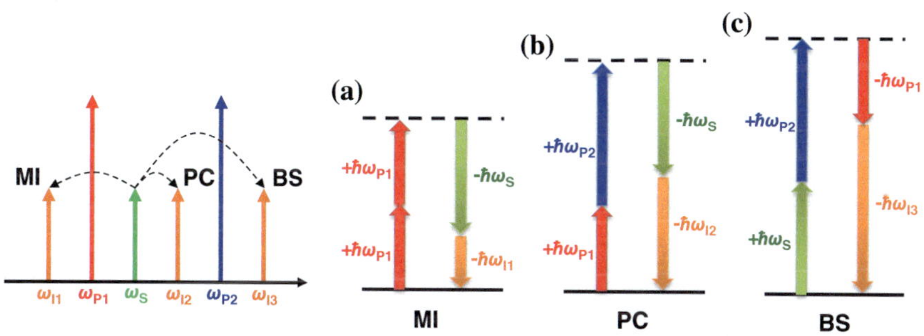

Fig. 7.2 Frequency configuration (*leftmost*) and energy diagrams (*right*) showing **a** modulation instability (MI), **b** phase conjugation (PC) and **c** Bragg scattering (BS) processes

$$\text{PC: } \Delta\beta = \beta_S + \beta_{I2} - \beta_{P1} - \beta_{P2} \tag{7.2b}$$

$$\text{BS: } \Delta\beta = \beta_S + \beta_{P2} - \beta_{I3} - \beta_{P1} \tag{7.2c}$$

The four-sideband dynamics can be solved in several notable phase-matching regimes [31]. In absence of linear phase mismatches $\Delta\beta$ for all FPM processes, all the sidebands exhibit power growth with quadratic scaling with pump powers and propagation length. Meanwhile, perfect phase-matching in which the linear phase mismatch fully compensates the nonlinear mismatch component leads to power evolutions that approaches exponential growth with respect to distance and pump power. These two regimes resemble the operation of a one-pump mixer with gain coefficients g approaching zero and maximum respectively, albeit the signal power growth is accompanied by creation of three idler sidebands in a two-pump mixer, instead of only one in its one-pump counterpart. Inherent to the coherent nature of FPM processes, the three idlers created in a two-pump mixer carry either identical modulation of the input signal through BS process, or phase-conjugated copies through MI and PC processes. The combination of three FPM processes extends the simple wavelength converting function to wavelength multicasting, in which multiple signal replicas of different wavelengths are generated [32].

7.2.3 Signal Processing Functions of Parametric Mixers

Apart from signal amplification, a unique capability of parametric mixers is the transfer of modulation (information) to a new wavelength through idler generation, namely, wavelength conversion. This attribute of parametric mixer operation is sought for applications that include wavelength switching and routing [32–34], novel light generation [19, 35, 36] and pattern recognition [37]. A particularly noteworthy phenomenon entailed by the conjugated wavelength conversion is propagation effect reversal by phase conjugation [38, 39]. In principle, propagation effects including dispersion and nonlinear phase rotation are fully reversible in presence of a spectral inverter in the middle of the otherwise homogeneous link [40]. While the temporal phase-conjugation offered by a parametric mixer does not fully resemble the spectral inversion operation required for full reversal, the phase conjugators were found effective for mitigating impairments for conventional channels [39].

Departing from continuous-wave operation, parametric mixers driven by modulated (pulsed) pumps offer an agile platform for time-domain signal manipulation—the quasi-instantaneous response inherent to parametric processes renders the performance of signal processing in parametric mixers be time-invariant. In general, resolving the FPM dynamics involving a pulsed pump require full consideration of dispersion and frequency-dependent nonlinear terms in the couple-mode equations, which will yield no closed-form solution [41]. In practice, however, parametric mixers operating with pulsed pumps are typically constructed on nearly dispersion-less medium in order to preserve the pump waveform and prevent excessive pulse walk-off [42, 43]. Consequently, the gain model developed for continuous-wave pumped parametric mixers are typically applicable to pulsed pumped counterparts, except that the gain and conversion efficiency follow the change in instantaneous pump power. This class of parametric mixers is adopted for a number of applications ranging from optical stroboscope [42], optical time-division multiplexed (OTDM) signal switching [43], analog optical signal sampling [14], to more sophisticated usage in OTDM-WDM conversion using time-frequency modulated pumps [44].

The presence of idler at the mixer input invokes a unique operating regime of parametric mixer—the signal and idler couple and modify the dynamics in the parametric mixer. Known as phase-sensitive (PS) operation, the gain (conversion efficiency) in the parametric mixer is now dependent on the initial phases of the signal, idler, as well as the pump [45], where a mutual phase difference of 0 (or π) will amplify (or attenuate) the signal and idler. In PS amplification mode, the signal and idler are theoretically devoid of excess noise due to the phase selectivity of the process. This capability of amplification and manipulation of optical signals in noise-less manner (i.e. with 0-dB noise figure) is unique to parametric mixers [45], and has recently attracted great interest in various venues, such as optical link-reach extension [46], signal regeneration [47], and wavelength multicasting [48]. Detailed discussion of PS parametric mixing can be found in [49].

7.2.4 *Effect of Chromatic Dispersion*

Chromatic dispersion of a medium is characterized by the nonlinear relationship between the phase constant and frequency. To parametric mixers, the presence of chromatic dispersion changes the phase-matching conditions according to the frequencies of the pumps, signal and resulting idlers, thereby affects the mixing efficiency, bandwidth and noise performance of the mixer. In absence of chromatic dispersion, phases of the participating waves are always balanced since the group velocities remain constant. However, typical optical media possess finite chromatic dispersion as a result of resonant (lossy) response of the material or the waveguide [50]. The frequency-dependency of group delay $d\beta/d\omega$ as a manifestation of chromatic dispersion thus restricts the frequency range in which linear phase-matching can occur. Such dependency, however, also allows perfect *total* phase matching to happen by providing a finite linear phase mismatch to balance the nonlinear phase shift terms for certain combinations of pump, signal (and idler) frequencies. Consequently, chromatic dispersion significantly shapes the spectral response of parametric mixer, which would otherwise be virtually achromatic [25].

Chromatic dispersion engineering represents one of the most important aspects in parametric mixer construction, since the intrinsic (material) dispersion characteristics of many nonlinear optical materials are far from ideal for parametric mixing. Standard single-mode fiber (SSMF), for example, possesses a chromatic dispersion in the lowest loss window at 1550 nm which allows phase-matched mixing across a band narrower than 100 GHz [52]. This contrasts the 1000 THz mixing window that the transparency window of fused silica would have provided [53]. Alternation of the material dispersion often involves the use of waveguide structures, in which waveguide dispersion is introduced to compensate for material counterpart. Optical fibers denote a particularly important class of waveguides in which chromatic dispersion engineering is practiced. While dispersion-engineered optical fibers were traditionally sought after for mitigating signal propagation impairments due to dispersion [54], the resulting dispersion-shifted waveguide designs have also provided the low-dispersion environment suitable for wide-band parametric mixing [55]. Indeed, refined waveguide structures have enhanced the phase-matched mixing bandwidth due to precise control of higher-order dispersion terms. In particular, fourth-order dispersion control in specially-tailored fiber waveguides, also known as highly-nonlinear fibers, have enabled parametric mixing over a near-octave bandwidth [10, 56]. Microstructure fiber waveguides that comprise photonic crystal or hollow guiding structures have further extended the operating spectral region and range of parametric mixers to beyond the capability of solid fiber [57], by allowing guiding in novel optical materials [58, 59] and dispersion control in regions not feasible in solid fiber waveguides [60].

Although waveguide cross-section design provides the vital dispersion control for parametric mixing, the influence on dispersion by waveguide structure means any deviation in the wavelength core size will exert significant dispersion deviation [6]. This effect signifies challenges in producing repeatable mixing performance from device to device, as well as in attaining the target performance due to the deviation of local dispersion from the global dispersion [61]. In the next section, the challenges and solutions for synthesizing dispersion-stable waveguides discussed in depth.

7.3 Dispersion-Stable Waveguide Engineering for Wide-Band Parametric Mixer Synthesis

Chromatic dispersion of the optical medium has profound effects on the efficiency as well as the attainable spectral bandwidth of a parametric mixer due to the phase matching mandates [61]. While chromatic dispersion of an optical waveguide can be arbitrarily tailored through waveguide geometry engineering, the local dispersive behavior of the waveguide can deviates from its global characteristics by considerable amount. The dependence of local dispersion for phase-matching, an unfortunate attribute of distributed nature of typical parametric mixer, only permits the dispersion to fluctuate within a small margin.

7.3.1 Wide-Band Parametric Mixing—The Atomic-Scale Challenge

Local dispersion fluctuation can be revealed from the mechanism of dispersion tailoring by waveguide geometry design [54]. Waveguiding effect shifts the overall dispersion from the material dispersion by modal area control [62–64]. In general, the modal spread of a guided optical field increases with the wavelength, consequently forcing a higher portion of the field to the cladding region and decreasing the effective index. This phenomenon gives rise to the wavelength dependency of the effective index, and thereby to the shift in dispersion. Typical dispersion-shifted fibers utilize this wavelength-dependent guiding effect to create strong waveguide dispersion for compensating the material dispersion [54]. Using a constricted core with high index contrast, a relatively abrupt change in effective index will occur in the proximity of a wavelength point which is determined by the core size. This abrupt feature in the effective index curve induces sizable second-order dispersion effect, which cancels out the large material dispersion in the region.

While the tight confinement in the core can result in ideal dispersion-free characteristics for wide-band parametric mixing [4], this waveguide geometry also sensitizes the dispersion to geometry fluctuations since the dispersion shift is controlled by the size of waveguide [6]. Microscopic geometry fluctuations are inevitably introduced during waveguide fabrication, typically amount to 0.1 % of the waveguide cross-sectional size [4]. These fluctuations stochastically deviate the dispersion characteristics of the waveguide along its propagation axis. With a typical highly-nonlinear fiber (HNLF) that is commonly deployed as a parametric mixing platform, the fiber's dispersion curve shown in Fig. 7.3 shifts by as much as 0.5 ps/nm/km for a 10-nm change to its 2 μm core radius [6].

Even though the dispersion shift might be deemed negligible when compared to the material dispersion (16 ps/nm/km), the extent is sufficiently large to provoke significant penalty to parametric mixing. The quantitative impact of dispersion fluctuations is revealed by a couple-mode model incorporating stochastic dispersion variation [65]. As an illustration of dispersion tolerance for wide-band mixing, a parametric amplifier pumped at 1.55 μm demands a dispersion accuracy better than 0.005 ps/nm/km in order to reach signal/idler pairs at 1.3 and 2.0 μm [9]. This requirement necessitates a radial control that is better than the length of a single silicon-oxygen bond [9].

7.3.2 Post-Fabrication Dispersion Fluctuations Rectification

While enforcing atomic-scale waveguide fabrication accuracy is deemed impractical, the dispersion accuracy demand can possibly be addressed through resolving the local dispersion characteristics and subsequently reversing the irregularities

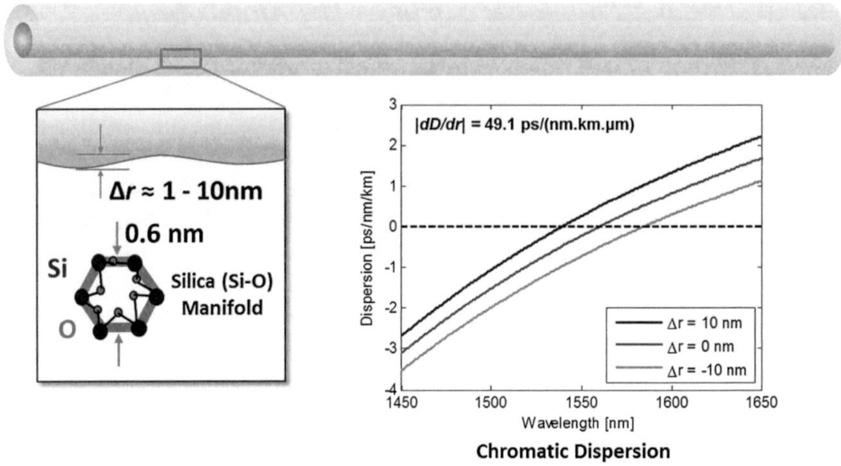

Fig. 7.3 Effect of waveguide geometry variation on chromatic dispersion in HNLF. Core-size fluctuations comparable to the elementary silica manifold are adequate to induce sizable dispersion fluctuations

after fabrication. This method will require local dispersion measurement techniques with fs/ps/km dispersion and meter-scale longitudinal resolutions [9, 61].

The evolution of FPM products depends on the aggregated local phase walk-off between participating waves. While this presents a challenge in parametric device construction, this mechanism also provides means to resolve the local dispersion characteristics of the mixing medium.

7.3.2.1 Modified Optical Time-Domain Reflectometry

The first class of methods relies on a modified optical time-domain reflectometry (OTDR) [66–68]. Owing to dispersion fluctuations along the waveguide's propagation axis, a pump-signal wavelength pair may experience phase-matching at a certain region which consequently leads to amplification of the signal and creation of an idler [67]. By detecting the "photon echo" of the pulsed signal produced by Rayleigh scattering, the captured power evolution of the signal and/or created idler along the waveguide can then infer the position of phase-matched region in the waveguide. Subsequently, the complete dispersion map of the waveguide is generated by sweeping the pump and signal wavelengths in order to capture all possible phase-matching wavelength combinations and their respective position in the waveguide. The major drawback of this approach rests on its sensitivity—due to the low Rayleigh scattering coefficients in typical waveguides (~ -40 dB for optical fibers), the detection of signal gain and idler generation typically requires long signal pulse width ($\sim \mu$s), as well as long averaging period [67]. The first operating requirement necessarily compromises the spatial resolution to 100-m level, while the second imposes long measurement time for obtaining accurate results.

7.3.2.2 Space-Frequency Domain Retrieval

Alternatively, one may obtain the localized FPM behavior by resolving the field evolution from the spectral variation of wavelength conversion efficiency [5]. With two waves of similar power propagating in a nonlinear medium of interest, the power of the FPM product is given by:

$$P_{FPM} = \left| \int_{0}^{L} e^{j(\Delta\beta z + \phi(z))} dz \right|^2 \tag{7.3}$$

The mean phase mismatch and its local fluctuations are denoted by the terms $\Delta\beta z$ and $\phi(z)$ respectively. Particularly, if the mean phase mismatch is controlled such that it spans a linear space by, for instance, changing the wavelength pair of the two waves, the corresponding variation of FPM product will then resemble the squared amplitude of the Fourier transform of the local phase mismatch evolution $\exp[j\phi(z)]$. In the spectral region close to the zero-dispersion wavelength (ZDW), the linear phase-mismatch scaling can be performed by fixing the wavelength separation between the two input waves while the mean wavelength is being tuned in linear manner, since the dispersion curve in this region is nearly proportional to wavelength by its dispersion slope $dD/d\lambda$.

By generating the power spectrum from the power of FPM products, the local phase mismatch term $\phi(z)$ is subsequently resolved by applying phase-retrieval algorithms to the power spectrum of $\exp[j\phi(z)]$. While a number of phase-retrieval algorithm exists for problems of this kind [5, 69], the spatial resolution of this approach can be estimated by regarding the mixing process as sampling of the phase mismatch evolution $\exp[j\phi(z)]$ [5]. The spatial resolution, or equivalently, the bandwidth of the "sampler", is given by the maximum range of mean phase mismatch $\Delta\beta$ one can introduce. Taking the tunable range of typical lasers into consideration, the 100-nm tunable range will limit the spatial resolution to 40 m for a HNLF with 0.02 ps/nm^2/km dispersion slope. This resolution level remains deficient for capturing meter-scale fluctuations that are pertinent to wide-band parametric mixing [61].

7.3.2.3 FPM Localization by Dispersive Pulse Walk-off

The third class of methods measures the local FPM evolution by utilizing spatial walk-off between modulations on two distinct wavelengths [70]. The architecture involves two pulsed sources at two tunable wavelengths, where one serves as the pump and the other as the signal. The pump and signal pulses travel at different group velocities in the dispersive medium and subsequently overlap (collide) at a certain region of the medium, whereas the location of collision is controlled by the initial delay between the pump and signal pulses. The collided pump and signal then create an idler whose generation efficiency is determined by the local dispersion (phase-matching) properties of the medium. Consequently, the dispersion map of the medium is aggregated by capturing the power of idler generated at various pump-signal wavelength pairs and differential delays, which define the dispersion value and position of the dispersion map respectively. However, the reliance of the medium's dispersion to localize pump-signal collision sets a compromise between the smallest dispersion and the finest spatial resolution one can attain with this method. In essence, the spatial resolution is proportional to the inverse of the dispersion, meaning that the lower the dispersion, the coarser the spatial resolution will be. Therefore the adoption of this approach is limited to highly dispersive waveguides such as standard single-mode fibers (SSMF), but is not applicable to low-dispersion media that are necessary for wide-band parametric mixing.

7.3.2.4 Counter-Collision Method

Recognizing the compromise between dispersion and spatial resolution that plagues conventional mapping approaches, a new approach was introduced in which the spatial resolution is controlled by active means [6].

The key to achieving fine spatial resolution unrestricted by the dispersion of the medium rests in local optical amplification. The architecture of the new approach, as shown in Fig. 7.4, involves two wavelength-tunable pulse streams, referred to as probe (*P*) and signal (*S*) hereinafter. The probe and signal pulses are both launched into the waveguide-under-test at a low power, at which nonlinear effects are

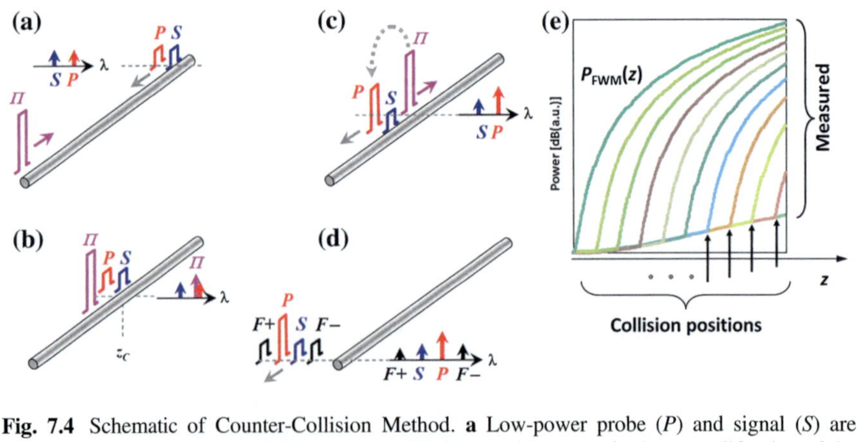

Fig. 7.4 Schematic of Counter-Collision Method. **a** Low-power probe (*P*) and signal (*S*) are launched on contra-directionally to the pump (*Π*); **b**, **c** collision at z_C leads to amplification of the probe; **d** idlers *F* are generated due to mixing between signal and amplified probe. **e** The aggregated idler power evolution depicts the local phase matching conditions

negligible. Meanwhile, high-power pulses, denoted as pump pulses (*Π*), are launched into the waveguide at the other end. The wavelength of the pump pulses is controlled such that the pump pulse will create a Brillouin gain peak at the probe wavelength, whereas the timing of the pump pulses controls the location of collision. When the forward propagating pulses collide with a pump pulse at location z_C, the probe pulse is selectively amplified through stimulated Brillouin scattering. The probe pulse, now possessing significantly higher power, interacts with the co-propagating probe pulse and creates idler ($F+$ and $F-$). The output power of the idler is governed by the phase-matching condition from location z_C to the output of the waveguide L. Consequently, aggregation of idler power for various pump-gate collision locations yield the phase mismatch with resolution only constrained by the timing resolution of gate generation.

The combination of spatial and dispersion resolution gives rise to the new level of mapping accuracy—the scheme provided meter-scale length resolution and fs/nm/km accuracy [6, 7]. This level of performance enabled measurement and subsequent rectification of dispersion fluctuations in low-dispersion HNLF for the first time [6]. Using the dispersion map captured by a counter-collision setup, the dispersion fluctuations in the HNLFs were mitigated by splicing sections of similar dispersion profiles [74], or by applying strain to shift the local dispersion characteristics [7, 8]. The reduction in dispersion fluctuations in the treated HNLFs was manifested by the enhancement in both the gain and bandwidth of the parametric amplifier constructed upon [7, 71].

7.3.3 *Waveguide Design Methods for Achieving Intrinsic Dispersion Stability*

Rather than post-fabrication correction, the required dispersion stability can achieved by an alternative waveguide design that provides reduced dispersion sensitivity to geometric fluctuations. This design-based approach offers an unprecedented level of dispersion stability out of reach by post-fabrication rectification, and eliminates the production that is limited by device-basis trimming.

7.3.3.1 Numerical Optimization-Based Design Approach

Early attempts to address dispersion stability issues in fiber waveguide relied on numerical optimization to search for an optimal refractive index profile. Particularly noteworthy achievements in optimization-based fiber design was reported by Gabrielli et al., in which a genetic algorithm was used to search for refractive index profiles [72]. Using a “fitness” function that comprised zero-dispersion wavelength and higher-order dispersion terms (β_4 and β_6) for the genetic evolution, the resulting multi-layer index profiles provided control over higher-order dispersion at up to the sixth order, and improvement in zero-dispersion wavelength stability. However, the extensive computation associated with repeated mode solving for evaluating possible index profiles forbade consideration on certain important parameters (e.g. nonlinear coefficient and β_2 fluctuations). Consequently the optimization converged to solutions that possessed low nonlinearity (3–4 W^{-1} km^{-1}), and counterintuitively, high dispersion fluctuations (2 ps/nm/km/%) due to elevated dispersion slope.

7.3.3.2 Semi-theoretic Design Approach

Apart from optimization-based design method, a semi-theoretical approach has recently been developed for dispersion-stable waveguide synthesis [9]. This method stemmed from the observation that the single-layered core structure deployed in typical high-confinement waveguides produces two distinct responses in waveguide dispersion when the core is scaled radially. In conventional high-confinement waveguides such as HNLF, the field confinement and dispersion control are provided both by a single delta-like core structure. As shown in Fig. 7.5, scaling the radius of the single-layer core causes the V-shaped normalized waveguide dispersion curve to shift laterally, thereby changing the dispersion in the low and high frequency regions in opposite directions [9].

Recognizing this opposing behavior, a dual-layer core structure was developed to leverage the dispersion shifts in two distinct core layers for overall dispersion change cancellation. Instead of relying on one core layer to control field confinement and dispersion, the additional layer in the new structure shown in Fig. 7.6 provides an additional degree of freedom to manipulate the wavelength-dependent behavior of the

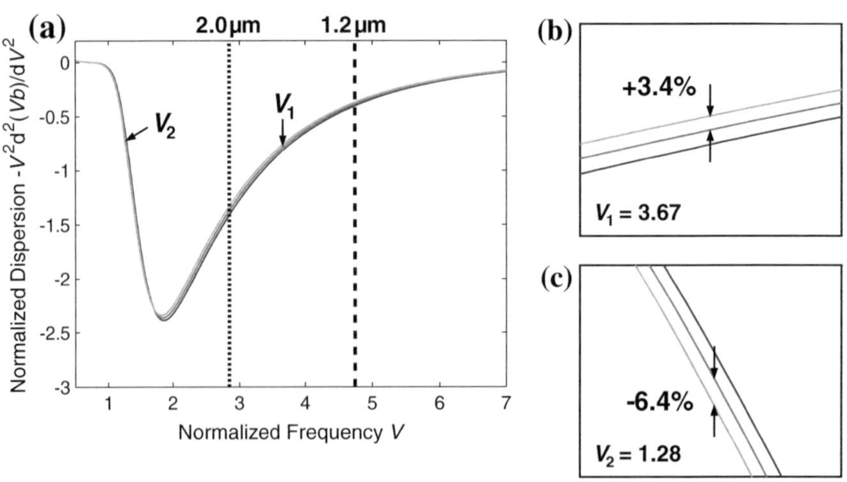

Fig. 7.5 Normalized waveguide dispersion of a conventional HNLF. Sub-graphs (**b**) and (**c**) show the dispersion shift in the high frequency and low frequency regions respectively

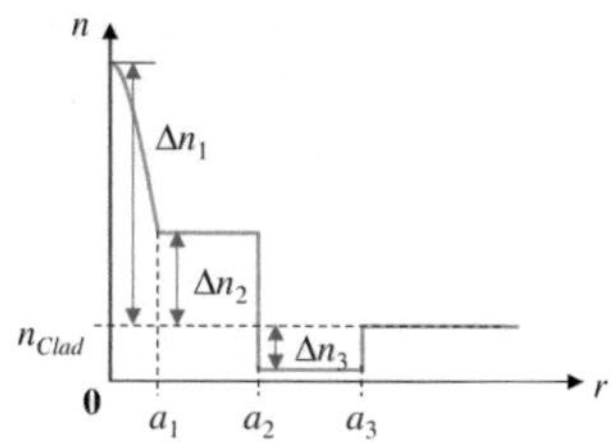

Fig. 7.6 Refractive index profile of the dispersion-stabilized waveguide design. Radii a_1 and a_2 denote the size of the inner and outer cores. Trench between $\{a_2, a_3\}$ provides further control of guiding condition

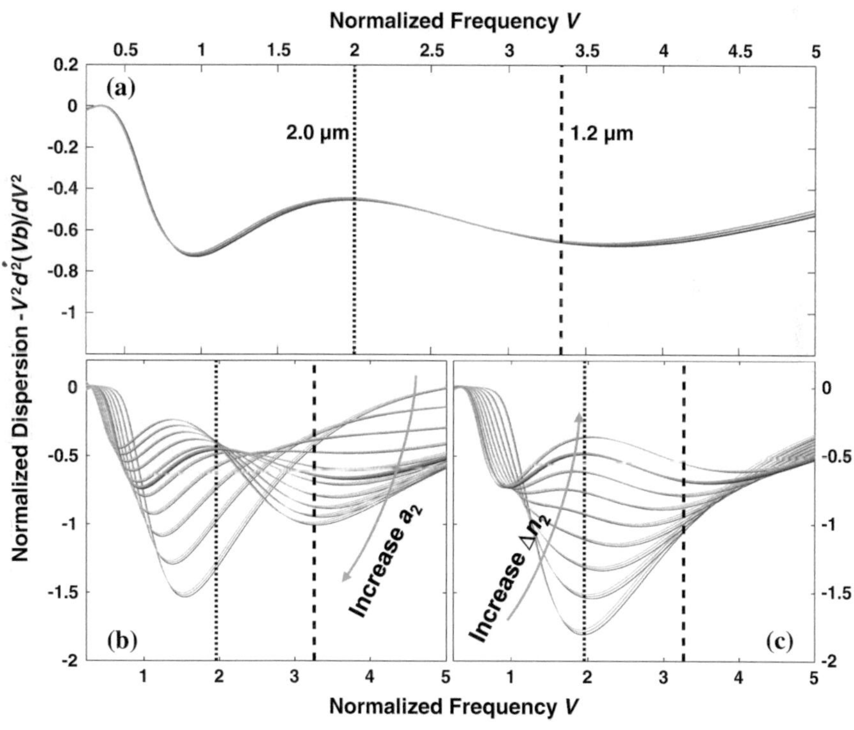

Fig. 7.7 Normalized waveguide dispersion of the new waveguide. **a** Optimal outer core parameters results in complete cancellation of dispersion fluctuations; **b**, **c** outer core size and refractive index affects overall waveguide dispersion in different manner

central core, thus enabling control of dispersion stability. The origin of dispersion stability in the new core design is revealed in Fig. 7.7. In the waveguide dispersion curve shown in Fig. 7.7a, the dual minima characteristic reflects two distinct guiding regimes enacted by the dual-layer core. At low frequencies, the spread of the optical field over both cores implies the field is guided by both cores and cladded by the cladding layers. At high frequencies, on the other hand, the optical field is primarily confined to the central core, meaning that the outer core layer now partially behaves as a cladding layer. The low-contrast guiding provided by the inner-outer core structure for high-frequency fields subsequently introduces the additional dip in the high-frequency regime. With proper selection of the outer core index and radius, the opposite dispersion shift exerted by the two guiding regimes can eliminate the dispersion change in the wide-range saddle region, in which a parametric mixing can operate. As a result, the overall dispersion of the waveguide remains stable in the presence of geometrical fluctuations. HNLF as an example, the fiber dispersion shift due to core size scaling is reduced by 100-fold when compared to a conventional single-core HNLF design.

HNLFs using the new waveguide design had demonstrated improved dispersion stability while maintaining high field confinement. Experimental characterization shows that the dispersion fluctuations were reduced by 92 % [10]. The improved dispersion stability consequently led to enhancement in parametric mixing efficiency, manifested by an increased parametric gain and spectral reach of the amplifier. Specifically, the new HNLFs reduced the threshold pump power by 60 % for fiber parametric oscillators operating in the short-wave infrared (SWIR) band [10], and allowed, for the first time, tunable lasing from 1.2 to 2.15 μm in a fiber parametric oscillator (Fig. 7.8a) [73]. In addition to parametric mixing efficiency and bandwidth enhancement, dispersion resilient to core deformation in the new HNLFs also allowed stimulated Brillouin scattering (SBS) management by fiber stretching without compromising phase-matching (Fig. 7.8b) [11]. This attribute enabled the first continuous-wave (CW) pumped coherent and modulation-agnostic wavelength conversion across 700 nm [11, 74].

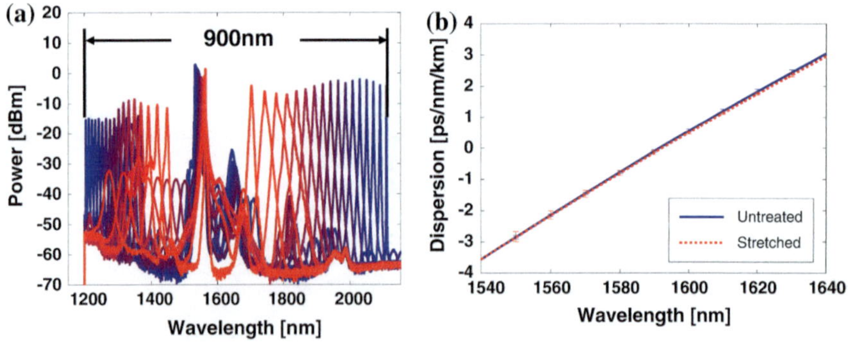

Fig. 7.8 **a** Tuning spectrum of a parametric oscillator constructed with dispersion-stabilized HNLFs; **b** dispersion curves of dispersion-stabilized HNLFs under various level of longitudinal stretch

7.4 Inhomogeneous Dispersion Engineering for Noise-Inhibited Parametric Mixing

In a two-pump parametric mixer, combination of mixing processes replicate a single input to multiple frequency-distinct locations. In addition to the pump-signal interactions, the presence of two strong pump waves intuitively leads to efficient pump-pump mixing which produces additional intense waves. These newly created secondary pumps further duplicate the mixing process driven by the primary pumps and consequently create more replica of the input. Known as self-seeded or cascaded parametric mixing, this class of cascaded interaction is actively sought in recent years for low-noise signal processing as well as other advanced time-frequency domain processing [13, 14, 43, 48, 49].

The conversion efficiency, spectral reach and noise performances of a self-seeded mixer are primarily determined by the secondary pump generation processes. The following section reviews the principle of self-seeded mixing and derives its limits.

7.4.1 Self-seeded Two-Pump Parametric Mixing—Homogeneous Limit

In general, the wide-band and complicated generation dynamics of secondary pumps require full account of the nonlinear interplay between FPM, dispersion and loss using nonlinear Schrodinger models. However, a simple model can be derived by knowing the optimal environment for homogeneous self-seeded mixing [13]—nonlinear media characterized by negligible chromatic dispersion are chosen to facilitate bandwidth-uninhibited interaction between pumps [13, 43]. Under this condition, the pump-pump mixing process is reduced to mere nonlinear phase modulation due to the sinusoidal power evolution of the pump field in time [13]. The number of secondary pumps created is consequently predicted by a modified Carson's rule for phase modulation, in which the empirical law $N = 2\lceil m + 1\rceil$ states the tone count scales proportionally to the nonlinear figure of merit (NFOM) $m = \gamma PL$ [13]. This rule implies an important consequence—the number of tones obtainable in a homogeneous mixer is pathetically bounded to a small count by nonlinear power limits, defined by various mechanisms including Brillouin scattering and material absorption [75, 76]. As an illustration, the state-of-the-art silica HNLFs possess a Brillouin-scattering bounded NFOM of 0.5 [75], meaning the maximum number of tone creatable is lower than 4. While the Brillouin-scattering strength can be suppressed by various means [77–79], the level of suppression is inadequate to enable generation beyond four tones [13, 43]. In addition to NFOM bound, high pump power demand for secondary pump generation will lead to initiation of multiple noise generation processes, thereby severely degrades the fidelity of the secondary pumps and the ensuing photon mixing processes [80].

7.4.2 Inhomogeneous Dispersion Engineering

Recognizing the limits of homogeneous mixer for cascaded parametric mixing, a new mixer design based on stage-wise dispersion engineering was devised to reduce the necessary pump power [11, 12]. The new mixer design, as shown in Fig. 7.9, consists of at least two elementary blocks—a seeding and a mixing stage [11]. Two input pump waves are launched to the first seeding stage, which comprises a section of nonlinear waveguide and followed by a dispersive element. In this first seeding stage, the pump waves create new tones through FPM in the nonlinear waveguide section. In the time domain, the nonlinear interaction of pumps is equivalent to phase modulation due to SPM, thus resulting in frequency chirping of the pump field. Consequently when the pump field propagates in the dispersion section with anomalous dispersion, the frequency-chirped pump field will experience temporal compression of its originally sinusoidal envelope to pulses with high peak power [81]. Subsequent to the seeding stage, efficient cascaded mixing occurs in the mixing stage composed of a nonlinear waveguide. Since nonlinear interaction strength increases at higher peak power and spectral span [82], the compressed pump field from the first stage experiences extensive SPM and results in massive spectral broadening. The spectral broadening of the pump pulses is manifested as generation of optical tones, with their frequency spacing strictly following the input pump frequency separation.

The efficiency of cascaded mixing and fidelity of the generated tones are solely determined by the dispersion management in each of the mixer's constituent stages. The following sections will define the dispersion criteria for optical mixing.

7.4.2.1 Inhomogeneous Mixer Design Considerations—Seeding Stage

The seeding stage converts relatively low intensity pump field into high peak power pulses. Proper dispersion and nonlinearity managements are required in this stage to preserve tone fidelity, as well as to attain the ideal temporal and spectral pulse characteristics for mixing in subsequent stages.

In the first stage, the ideal mixer should comprise an achromatic nonlinear section characterized by the absence of chromatic dispersion, which is followed by a dispersive section with appropriate chromatic dispersion to convert the chirped

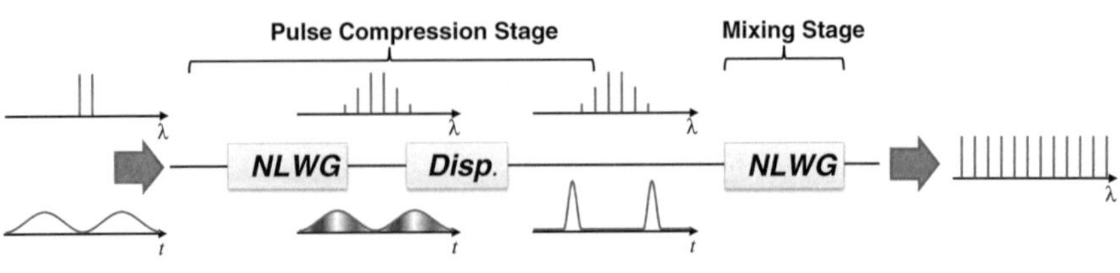

Fig. 7.9 Multi-stage mixer architecture for achieving massive cascaded mixing. NLWG and Disp denote nonlinear waveguides and dispersive element respectively

field into transform-limited pulses. Following the derivation in [13], it is found that the phase profile of the chirped field obtained at the output of the first nonlinear section is largely quadratic. Consequently, a dispersive element providing appropriate negative second-order (anomalous) dispersion will be adequate to remove the quadratic nature of the spectral phase profile, thereby converting the field into transform-limited pulses. In practice, the need for anomalous dispersion can be fulfilled by standard SMFs. However, the spectral span at high phase-modulation depth m becomes excessively wide, so that the higher-order dispersion terms of standard SMFs will cause considerable pulse shape deviation from its transform-limited form [83]. Nevertheless, the dispersion provided by SMFs can adequately satisfy the need in the seeding (pulse compression) stage, over the practical operating condition constrained by SBS [75].

7.4.2.2 Inhomogeneous Mixer Design Considerations—Mixing Stage

Subsequent wide-band generation in the second stage mandates a considerably more stringent dispersion control in the nonlinear medium [12]. Practical mixing stage construction deploys dispersion-flattened fibers to provide low-dispersion environment that enables wide-band mixing. Indeed, the combination of intense peak power in the mixing stage and complex dispersion characteristics of a dispersion-flattened waveguide guarantees rich nonlinear dynamics and tight dispersion requirements [12]. Depicted by the nonlinear Schrodinger model shown in Fig. 7.10, a miniscule dispersion deviation of 0.1 ps/nm/km is adequate to overturn the tone fidelity possessed by a properly dispersion managed mixer. The spectral collapse observed in an anomalous dispersion as low as 0.1 ps/nm/km is attributed to the perturbation of

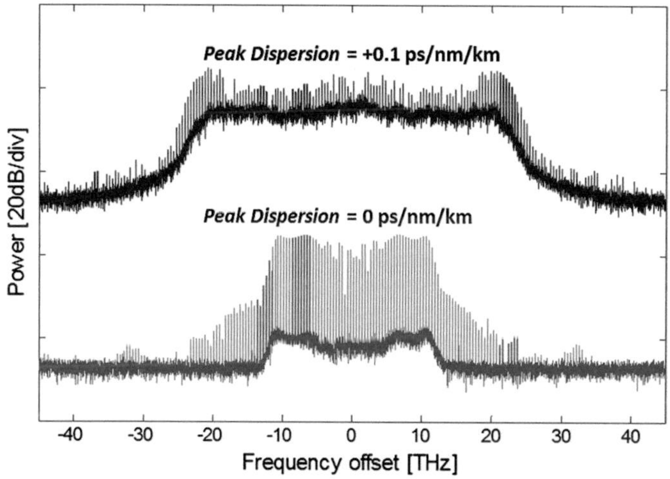

Fig. 7.10 Output spectra of the mixing stage with anomalous (*top*) and normal (*bottom*) dispersion

solitons by modulation instability [84, 85]. With the extreme power built-up by soliton formation, vacuum fluctuations are vigorously amplified which consequently leads to destruction of the soliton train [85]. In sharp contrast, a mixing stage with strictly normal dispersion ($D_{peak} \leq 0$) produces pristine tones characterized by high carrier-to-noise ratio. Perhaps more importantly, equalized tone power distribution is possible only in the normal dispersion regime.

The low-noise, power-equalized tone generation is attributed to the existence of similariton-like asymptotic state for pulse propagation in medium possessing normal parabolic dispersion profile [86]. Figure 7.11 shows the evolution of the Gaussian pulse in the presence of only 4th-order dispersion, which emulates the nonlinear dynamics in an optimal mixing stage shown in Fig. 7.10. Following the initial dispersion-free propagation, the Gaussian pulse gradually evolves into an arch-shape, featured with steep edges and a near-triangular top portion. Known as optical shock-wave formation, the steepening of edges is responsible for the rapid spectral broadening before the normalized position 40 due to Kerr-induced frequency chirping [87]. The onset of wave breaking at 50 marks the transition of pulse propagation regime to asymptotic (steady) state, as the pulse edges start to collapse as a result of the continued action of the shear force by the chirp and dispersion [88]. Further propagation beyond this point merely leads to pulse converging into a rectangular shape. The rectangular portion of the pulse contains a linear frequency chirp, of which the spectral width of the pulse equate the chirp

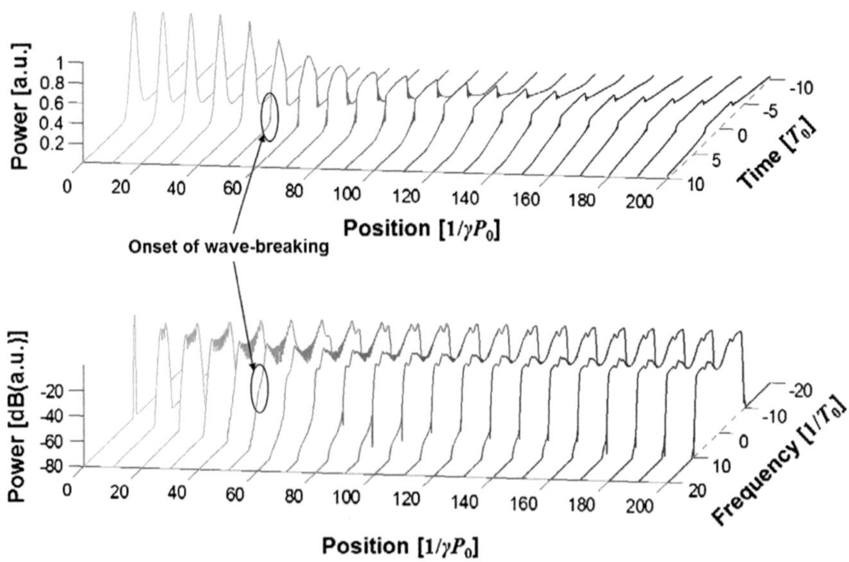

Fig. 7.11 Nonlinear evolution of a Gaussian pulse in the time and spectral domains. The position, time and frequency are normalized to the nonlinear coefficient γ, pulse peak power P_0 and Gaussian pulse width T_0. Simulation parameters: $\gamma P_0 = 1\,\mathrm{m}^{-1}$; zero second- and third-order dispersion, and a fourth-order dispersion $b_4 = \beta^{(4)}/(\gamma P_0 T_0^4) = 2.56 \times 10^{-6}$

extent [87]. The mechanisms for low-noise and spectrally flattened generation become apparent from this study. First, convergence to an asymptotic state implies the propagation dynamics is immune to noise perturbations, contrary to the unstable solitonic regime mediated by anomalous dispersion [85]. Second, the fourth-order dispersion that produces parabolic dispersion profile forces the pulse to converge into rectangular shape. The flat-top shape of the chirped pulse means the frequency components of the pulse are equally distributed across its span, thus giving rise to the equalized spectral shape.

The combination of multi-stage design and precise dispersion engineering for low-noise mixing allowed generation of a wide-band, low-noise frequency comb using two low-power continuous-wave pumps [11, 12]. Creation of high-OSNR optical frequency tones further enables a plethora of applications in coherent communications and generalized signal processing, as discussed in the next section.

7.4.3 Applications

7.4.3.1 Coherent Carrier Generation

The adoption of coherent modulation has radically changed the landscape of optical transport. Full-field recovery of signals by interdyne/intradyne detection using optical local oscillators alleviates the need in physical mitigation of channel impairments comprising chromatic dispersion, polarization mode dispersion (PMD) and nonlinear penalties [89–93]. By carrying information on both amplitude and phase quadratures of a light-wave carrier using multi-level modulations, coherent optical transmission allows expansion of conventional, direct-detection based system capacity as a result of improved spectral efficiency [94]. On the other hand, however, coherent transmission challenges conventional optical carrier generation approaches due to strict carrier-phase and frequency stability requirements [95]. As an illustration of the challenge, transporting a 100 Gbit/s, 16QAM channel requires an optical carrier with a linewidth below 30 kHz to guarantee optimal link budget [95]. Distributed feedback (DFB) diode laser that is a mainstay of conventional direct-detection systems, however, typically possess a linewidth exceeding 200 kHz [96]. Worse, even if one is allowed to construct a transmission system solely based on narrow linewidth sources, the emission frequencies of these individual sources wander in random manner and, thus, requires power-consuming frequency control on per-channel basis.

The low-noise optical frequency comb generated by multi-stage parametric mixer provides a practical solution in phase noise and frequency stability management for high capacity optical links [16, 17, 94]. Using a three-stage mixer, an optical frequency comb comprising 120 tones on 100-GHz grid or 1500-tones ultra-dense WDM comb at 12.5 spacing were generated using a total optical pump of less than 1 W. The mixer was seeded by a single narrow-linewidth source, where its coherence was transferred to all generated tones characterized by a maximum

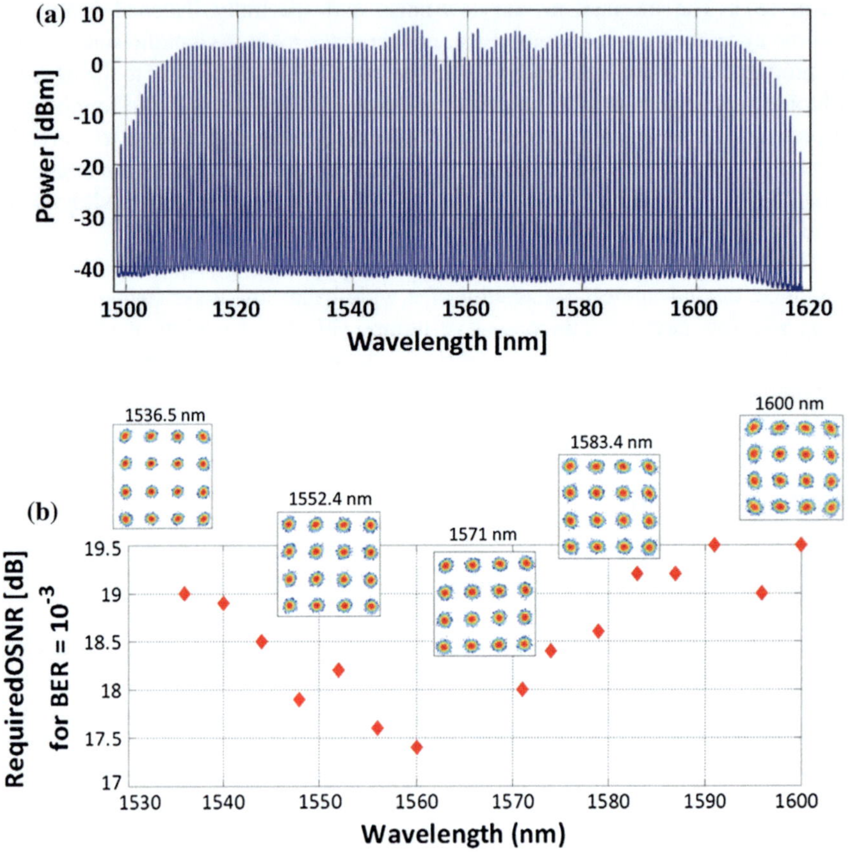

Fig. 7.12 **a** Spectra of comb spaced by 100 GHz; **b** sensitivity of coherent channels carried by the comb at up to 64QAM

linewidth of 25 kHz. The 100-GHz comb in Fig. 7.12a shows the tones possessed power of 5 dBm average and optical signal-to-noise ratio (OSNR) beyond 50 dB. The low-noise comb consequently permitted transport of complex modulation at performance near the theoretical bound (Fig. 7.12b). Low-noise parametric comb generation further allowed coherent transport at an aggregated rate of 15.6 Tbps, using complex modulations of up to 64QAM [94]. The absence of frequency control for each carrier in parametric frequency comb rendered considerable advantage in power consumption over individual emitter architecture—while the parametric comb generator consumed a total of 40 W, an equivalent emitter bank easily consumes ten times higher power due to active temperature control.

7.4.3.2 High-Count Wavelength Multicasting

Due to the presence of three parametric mixing modes, a two-pump parametric mixer is capable of delivering at least replicas of the input signal at three distinct wavelength locations, resulting in a pair of side-bands adjacent to each pump carrying modulations identical to that of the signal [98]. With the creation of secondary pumps in a multi-stage mixer, the new pumps will duplicate the side-bands of the primary pumps and thereby, creating more signal copies as the number of pump tones grows [98]. This mixer architecture thus allows generation of massive number of spectrally-distinct replicas which preserve both amplitude and phase of the signal, and align to a frequency grid that is strictly defined by two pump seeds. Such operation enables novel spectral analysis and time-frequency domain signal processing [14, 15], alongside with the classical usages in data stream multicasting [99]. Further description on spectral analysis and signal processing applications will be given in the next section.

The concept of signal replication in multi-stage mixer has been demonstrated with both amplitude and phase modulated signals [11]. Shown in Fig. 7.13, a 10-dBm input signal to a mixer comprising two nonlinear stages produced nearly 70 copies, with positive conversion gain to a majority of the duplicates. The copies were characterized by a nearly identical fidelity to the original signal, as depicted by a 0.2 and 0 dB average sensitivity penalties for amplitude- and phase-modulated inputs, respectively.

In additional to the classical, phase-insensitive mixing modality, the multi-stage parametric mixer architecture further allows non-classical, noise-less interaction by phase-sensitive mixing. Typical phase-insensitive optical amplification processes cause a noise figure of at least 3 dB as a consequence of quantum noise amplification in both in-phase and out-of-phase signal quadratures [100]. Parametric mixers are inherently phase-sensitive devices as a consequence of phase

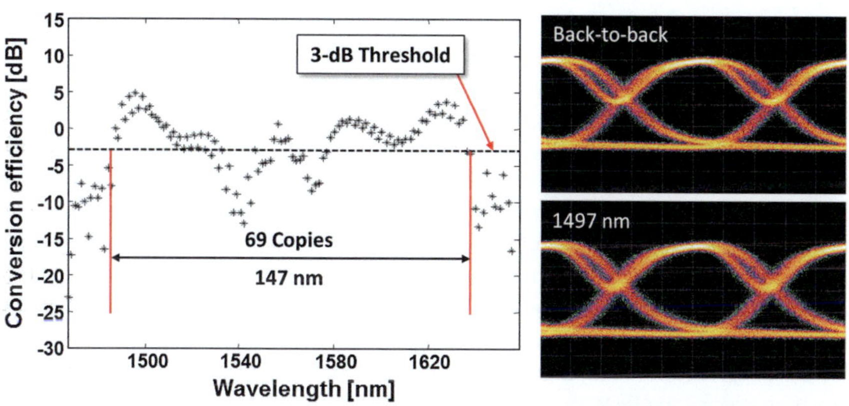

Fig. 7.13 Multicast conversion efficiencies in a two-stage dual-pump parametric mixer. The *right column* shows the eye diagrams of the back-to-back signal and its replica on the short-wavelength edge of the conversion spectrum

preservation in parametric processes. When operating in phase-sensitive regime by, for instance, injection of signal and idler at the input or allowing signal and idler frequencies to overlap, an ideal parametric mixer produces no excess noise since only the in-phase component of the signal receives amplification, thereby resulting in a zero noise figure.

Multi-stage parametric mixer extends the conventional phase-sensitive amplification concepts by generation of spectrally-distinct copies with no additional noise, in addition to optical amplification localized to the injected input modes [48]. Using a two-stage design with strictly normal dispersion in the mixing stage, the wavelength multicasting mixer with near 0-dB noise figure managed to distribute signals to more than 12 separate wavelength channels from its four-wavelength inputs [48]. Indeed, the noise-less multicasting operation provided performance elevation over conventional multicasting device comprising of power-splitting architecture. In a direct comparison between the receiver sensitivity metrics of a phase-sensitive multicasting mixer and power-splitting (attenuating) device, the phase-sensitive mixer improved the overall receiver sensitivity by 11 dB, equivalent to at least three-fold increase in copy counts in power-splitting architecture. The phase-sensitive multicasting architecture consequently provides a compelling means to extend conventional passive optical network (PON) and generic signal processor performance, in which signal replication plays critical roles [49].

7.4.3.3 Hybrid Time-Frequency Domain Signal Processing and Recognition

The ability to produce massive count of signal replicas in parametric mixers permits extensive parallelism for signal processing. This allows, for the first time, optical signal processors matching the of high-speed (THz-wide) channel.

Two general classes of signal processors relying on parametric signal replication have been demonstrated to date. The first kind enables real-time acquisition and data conversion of fast analog or digital signals by parallelized sampling operations, known as copy-and-sample-all (CASA) architecture [14]. This photonic-aided analog-to-digital conversion (ADC) processing scheme, as shown in Fig. 7.14, comprises two parametric mixers to perform parallel (flash) sampling. The first mixer is driven by two pumps to multicast the input signal into a number of color-distinct copies. The copies are subsequently sampled by the second parametric mixer driven by a pulsed pump, in which all copies at distinct frequencies mix with the pump pulses to create corresponding idlers. An array of electronic digitizers receives the idlers carrying temporal snapshots of the copies at a rate equal to the pump repetition frequency, thus producing the digital discrete-time equivalence of the input signal. The maximum signal bandwidth is ultimately tied to the number of copies at which the mixer chain can generate and sample. While single-stage mixers generating ten copies have satisfactorily fulfilled the demand for 40 GHz capture [14], reception of THz-wide signals will require a copy-count in the order of

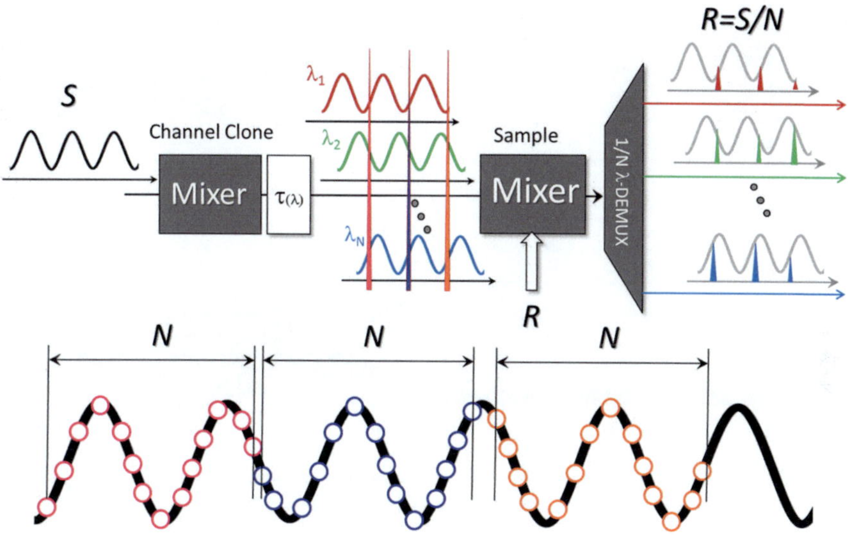

Fig. 7.14 Architectural schematic of CASA photonic processor

hundreds since commercial converter chips typically possess bandwidths of 10 GHz, thus rendering the use of multi-stage mixers mandatory.

Frequency traceability of signal replicas in parametric multicasters creates a new signal-processing regime, in which signal manipulation occurs in the frequency domain. This second class of signal processors, known as channelizer, granulates signal contents into sufficiently narrow frequency bands that permit low-latency, low-rate signal recognition in the (electronic) reception back plane [15]. The general concept illustrated in Fig. 7.15a utilizes spectral replication function of a two-pump mixer to create a set of signal copies falling on a known periodic frequency grid. Frequency-domain slices are subsequently extracted out of each copy by resonant filtering [101] or digital filter bands synthesized by coherent field capture and digital signal processing [15]. The spectral positions of the slices relative to their originating copies are determined jointly by the mixer pump separation and free-spectral range of the filter. Consequently, reception of these spectrally-constricted slices allows full reconstruction of the original signal using band-restricted receivers. Beyond signal reception, channelizers naturally support arbitrary channel response synthesis and, perhaps more importantly, instantaneous spectral analysis. Conventional spectral analysis relies on frequency-scanned heterodyning or discrete Fourier transform (DFT), but both of these mechanisms constrain the spectral acquisition rate due to fundamental limits in frequency source tuning and digital computations [102]. On the other hand, single-step frequency decomposition in a photonic channelizer is not limited by the aforementioned barriers, thereby allowing acquisition rate reaching the fundamental uncertainty bound [103]. Indeed, the advantages of photonic signal channelization would have remained elusive without a scalable generation capacity for spectrally separated yet

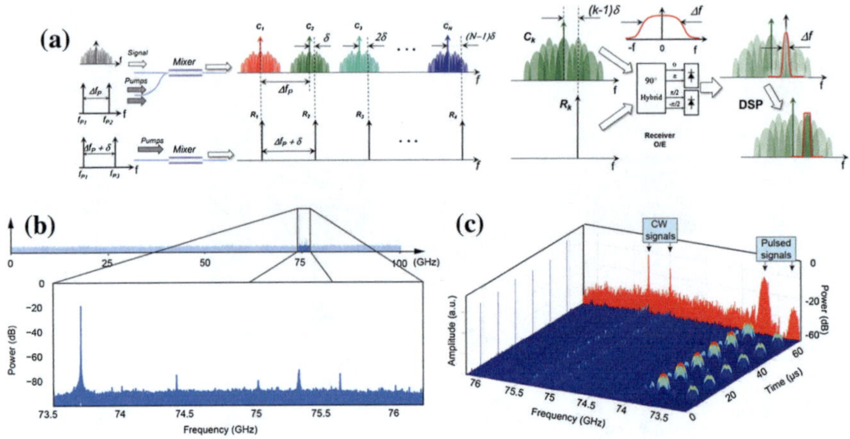

Fig. 7.15 **a** Photonic signal channelizer principle; spectrum and spectrogram of a 100-GHz-wide signal captured by a photonic channelizer are shown in (**b**) and (**c**) respectively

disciplined signal replicas, the ability uniquely provided by multi-stage mixers. Validated in [15], a multi-stage mixer capable of creating 18 copies enabled spectral analysis across 100 GHz in microsecond refresh rates. The high-speed, high spectral resolution acquisition had made capture of fast but rare events possible (Fig. 7.15b, c).

7.5 Conclusions

This chapter reviewed the physics, design practice and applications of wide-band signal processors realized dispersion-engineered parametric mixers. While the ubiquity of third-order nonlinear response implies that parametric mixing is feasible in a plethora of nonlinear optical platforms, tight dispersion specifications driven by phase-matching conditions renders many of those unsuitable for signal processing purposes when efficiency and noise performance metrics are considered. Recognizing the fundamental challenges in attaining the dispersion requirements for synthesizing low-noise and efficient photon mixers, two classes of dispersion engineering schemes were reviewed.

Based on waveguide cross-section geometry control, the first class of dispersion engineering techniques mitigates the dispersion fluctuations that impacts phase-matching stability over an extended nonlinear interaction length. Chromatic dispersions of waveguides are inevitably subject to perturbation due to geometry variation in atomic scales introduced during the fabrication processes. Instead of demanding unphysical fabrication accuracy, the miniscule dispersion fluctuations can be measured with meter-level position resolution, thereby enabling post-fabrication reversal. Alongside with post-fabrication corrections, dispersion stability can be guaranteed by design. New waveguide design methodologies have now reduced dispersion fluctuations level due to size variation by hundred fold, thus allowing nearly perfect phase-matched parametric amplification and wavelength conversion over 900 nm for the first time.

Besides homogeneous dispersion control, multi-section dispersion and nonlinearity engineering provides a new two-pump parametric mixing regime in which massive frequency tone generation is allowed. While homogeneous mixer prohibits extensive pump-pump mixing due to the power demanded for this interaction type, a tailored cascade of nonlinear and dispersive lines can scale the mixing efficiency by orders of magnitude due to temporal compression of pump fields. The consequent massive generation of secondary pumps enabled a plethora of signal processing applications, comprising of coherent transport, high-count spectral replication, ultra-fast signal reception and instantaneous spectral analysis.

In summary, dispersion engineering techniques have redefined the conventional notions of inefficiency and functional deficiency commonly associated with parametric mixer technology, providing the viable path for bridging high-speed photonic signals with complex electronic processing. Further advances in dispersion engineering, perhaps extension to material platforms other than silica, are expected and will ultimately allow direct photon access in arbitrary spectral regions and seamless interface with large-scale electronic backplane.

References

1. G. Raybon, A. Adamiecki, P.J. Winzer, S. Randel, L. Salamanca, A. Konczykowska, F. Jorge, J.-Y. Dupuy, L.L. Buhl, S. Chandrashekhar, C. Xie, S. Draving, M. Grove, K. Rush, R. Urbanke, High symbol rate coherent optical transmission systems: 80 and 107 GBaud. J. Lightwave Technol. **32**, 824–831 (2014)
2. M. Ebrahim-Zadeh, I.T. Sorokina (eds.), *Mid-Infrared Coherent Sources and Applications* (Springer, Berlin, 2007)
3. S. Radic, C.J. McKinstrie, Optical amplification and signal processing in highly nonlinear optical fiber. IEICE Trans. Electron **E88-C**, 859–869 (2005)
4. M. Hirano, T. Nakanishi, T. Okuno, M. Onishi, Silica-based highly nonlinear fiber and their application. IEEE J. Select. Topics Quant. Electron. **15**, 103–113 (2009)
5. I. Brener, P.P. Mitra, D.D. Lee, D.J. Thomson, High-resolution zero-dispersion wavelength mapping in single-mode fiber. Opt. Lett. **23**, 1520–1522 (1998)
6. E. Myslivets, N. Alic, J.R. Windmiller, S. Radic, A new class of high-resolution measurements of arbitrary-dispersion fibers: localization of four-photon mixing process. J. Lightwave Technol. **27**, 364–375 (2009)
7. E. Myslivets, C. Lundström, J.M. Aparicio, S. Moro, A.O.J. Wiberg, C.-S. Brès, N. Alic, P.A. Andrekson, S. Radic, Spatial equalization of zero-dispersion wavelength profiles in nonlinear fibers. IEEE Photon. Technol. Lett. **21**, 1807–1809 (2009)
8. M. Takahashi, M. Tadakuma, T. Yagi, Dispersion and Brillouin managed HNLFs by strain control techniques. J. Lightwave Technol. **28**, 59–64 (2010)
9. B.P.-P. Kuo, S. Radic, Highly nonlinear fiber with dispersive characteristics invariant to fabrication fluctuations. Opt. Express **20**, 7716–7725 (2012)
10. B.P.-P. Kuo, J.M. Fini, L. Grüner-Nielsen, S. Radic, Dispersion-stabilized highly-nonlinear fiber for wideband parametric mixer synthesis. Opt. Express **20**, 18611–18619 (2012)

11. B.P.-P. Kuo, M. Hirano, S. Radic, Continuous-wave, short-wavelength infrared mixer using dispersion stabilized highly-nonlinear fiber. Opt. Express **20**, 18422–18431 (2012)
12. E. Myslivets, B.P.-P. Kuo, N. Alic, S. Radic, Generation of wideband frequency combs by continuous-wave seeding of multistage mixers with synthesized dispersion. Opt. Express **20**, 3331–3344 (2012)
13. B.P.-P. Kuo, E. Myslivets, N. Alic, S. Radic, Wavelength multicasting via frequency comb generation in a bandwidth-enhanced fiber optical parametric mixer. J. Lightwave Technol. **29**, 3515–3522 (2011)
14. A.O.J. Wiberg, L. Liu, E. Myslivets, B.P.-P. Kuo, N. Alic, S. Radic, Demonstration of 40 GHz analog-to-digital conversion using copy-and-sample-all parametric processing, in *Proc. OFC/NFOEC 2012*, paper OW3C.2
15. A.O.J. Wiberg, D.J. Esman, L. Liu, V. Ataie, E. Myslivets, B.P.-P. Kuo, N. Alic, S. Radic, S. Zlatanovic, J. R. Adleman, E.W. Jacobs, Agile wideband microwave/millimeter-wave analysis by coherent filterless channelizer, in *Proc. 2013 IEEE International Topical Meeting on Microwave Photonics (MWP)*, Post-deadline paper
16. B.P.-P. Kuo, E. Myslivets, V. Ataie, E.G. Temprana, N. Alic, S. Radic, Wideband parametric frequency comb as coherent optical carrier. J. Lightwave Technol. **31**, 3414–3419 (2013)
17. E.G. Temprana, V. Ataie, B.P.-P. Kuo, E. Myslivets, N. Alic, S. Radic, Low-noise parametric frequency comb for continuous C-plus-L-band 16-QAM channels generation. Opt. Express **22**, 6822–6828 (2014)
18. A.H. Gnauck, B.P.-P. Kuo, E. Myslivets, R.M. Jopson, M. Dinu, J.E. Simsarian, P.J. Winzer, S. Radic, Comb-based 16-QAM transmitter spanning C and L bands. IEEE Photon. Technol. Lett. **26**, 1041–1135 (2014)
19. B.P.-P. Kuo, N. Alic, P.F. Wysocki, S. Radic, Simultaneous NIR and SWIR wavelength-swept generation over record 329-nm range using swept-pump fiber optical parametric oscillator. J. Lightwave Technol. **29**, 410–416 (2011)
20. R. Nissim, A. Pejkic, E. Mylivets, B.P.-P. Kuo, N. Alic, S. Radic, Ultrafast optical control by few photons in nanometer scale tailored fiber, in review
21. J. Hansryd, P.A. Andrekson, M. Westlund, J. Li, P. Hedekvist, Fiber-based optical parametric amplifiers and their applications. IEEE J. Select. Topics Quant. Electron. **8**, 506–520 (2002)
22. S. Radic, Parametric amplification and processing in optical fibers. Laser Photonics Rev. **2**, 498–513 (2008)
23. M.E. Marhic, *Fiber Optical Parametric Amplifiers, Oscillators and Related Devices* (Cambridge University, Cambridge, 2007)
24. J.A. Armstrong, N. Bloembergen, J. Ducuing, P.S. Pershan, Interaction between light waves in a nonlinear dielectric. Phys. Rev. **127**, 1918–1939 (1962)
25. R.H. Stolen, J.E. Bjorkholm, Parametric amplification and frequency conversion in optical fibers. IEEE J. Quant. Electron. **18**, 1062–1072 (1982)
26. T. Torounidis, P.A. Andrekson, B.-E. Olsson, Fiber-optical parametric amplifier with 70-dB gain. IEEE Photon. Technol. Lett. **18**, 1194–1196 (2006)
27. R. Laming, M.N. Zervas, D.N. Payne, Erbium-doped fiber amplifier with 54 dB gain and 3.1 dB noise figure. IEEE Photon. Technol. Lett. **4**, 1345–1347 (1992)
28. M.N. Zervas, R.I. Laming, Rayleigh scattering effect on the gain efficiency and noise of erbium-doped fiber amplifiers. IEEE J. Quant. Electron. **31**, 468–471 (1995)
29. S. Radic, Parametric signal processing. J. Selected Topics Quant. Electron. **18**, 670–680 (2012)
30. C.J. McKinstrie, S. Radic, A.R. Chraplyvy, Parametric amplifiers driven by two pump waves. IEEE J. Select. Topics Quant. Electron. **8**, 538–547 (2002)
31. C.J. McKinstrie, S. Radic, M.G. Raymer, Quantum noise properties of parametric amplifiers driven by two pump waves. Opt. Express **12**, 5037–5066 (2004)
32. Q. Lin, R. Jiang, C.F. Marki, C.J. McKinstrie, R. Jopson, J. Ford, G.P. Agrawal, S. Radic, 40-Gb/s optical switching and wavelength multicasting in a two-pump parametric device. IEEE Photon. Technol. Lett. **17**, 2376–2378 (2005)

33. S. Radic, C.J. McKinstrie, R.M. Jopson, A.H. Gnauck, J.C. Centanni, A.R. Chraplyvy, Multiple-band bit-level switching in two-pump fiber parametric devices. IEEE Photon. Technol. Lett. **16**, 852–854 (2004)
34. S. Watanabe, F. Futami, R. Okabe, R. Ludwig, C. Schmidt-Langhorst, B. Huettl, C. Schubert, H.-G. Weber, An optical parametric amplified fiber switch for optical signal processing and regeneration. IEEE J. Select. Topics Quant. Electron. **14**, 674–680 (2008)
35. M.E. Marhic, K.K.-Y. Wong, L.G. Kazovsky, T.-E. Tsai, Continuous-wave fiber optical parametric oscillator. Opt. Lett. **27**, 1439–1441 (2002)
36. M. Fiorentino, P.L. Voss, J.E. Sharping, P. Kumar, All-fiber photon-pair source for quantum communications. IEEE Photon. Technol. Lett. **14**, 983–985 (2002)
37. D.M.F. Lai, C.H. Kwok, K.K.-Y. Wong, All-optical picoseconds logic gates based on a fiber optical parametric amplifier. Opt. Express **16**, 18362–18370 (2008)
38. S. Watanabe, G. Ishikawa, T. Naito, T. Chikama, Generation of optical phase-conjugate waves and compensation for pulse shape distortion in a single-mode fiber. J. Lightwave Technol. **12**, 2139–2146 (1994)
39. B.P.-P. Kuo, E. Myslivets, A.O.J. Wiberg, S. Zlatanovic, C.-S. Brès, S. Moro, F. Gholami, A. Peric, N. Alic, S. Radic, Transmission of 640-Gb/s RZ-OOK channel over 100-km SSMF by wavelength-transparent conjugation. J. Lightwave Technol. **29**, 516–523 (2011)
40. R.A. Fisher (ed.), *Optical Phase Conjugation* (Academic, New York, 1983)
41. A.V. Yulin, D.V. Skryabin, P.St.J. Russell, Four-wave mixing of linear waves and solitons in fiber with higher-order dispersion. Opt. Lett. **29**, 2411–2413 (2004)
42. J. Li, J. Hansryd, P.O. Hedekvist, P.A. Andrekson, S.N. Knuden, 300-Gb/s eye-diagram measurement by optical sampling using fiber-based parametric amplification. IEEE Photon. Technol. Lett. **13**, 987–989 (2001)
43. C.-S. Brès, A.O.J. Wiberg, B.P.-P. Kuo, J.M. Chavez-Boggio, C.F. Marki, N. Alic, S. Radic, Optical demultiplexing of 320 Gb/s to 8×40 Gb/s in single parametric gate. J. Lightwave Technol. **28**, 434–442 (2010)
44. E. Palushani, H.C.H. Mulvad, M. Galili, H. Hu, L.K. Oxenlowe, A.T. Clausen, P. Jeppesen, OTDM-to-WDM conversion based on time-to-frequency mapping by time-domain optical Fourier transformation. IEEE. J. Select. Topics Quant. Electron. **18**, 681–688 (2012)
45. C.J. McKinstrie, S. Radic, Phase-sensitive amplification in a fiber. Opt. Express **12**, 4973–4979 (2004)
46. Z. Tong, C. Lundstrom, P.A. Andrekson, C.J. McKinstrie, M. Karlsson, D.J. Blessing, E. Tipsuwannakul, B.J. Puttnam, H. Toda, L. Gruner-Nielsen, Towards ultrasensitive optical links enabled by low-noise phase-sensitive amplifiers. Nat. Photon. **5**, 430–436 (2011)
47. R. Slavik, F. Parmigiani, J. Kakande, C. Lundstrom, M. Sjodin, P.A. Andrekson, R. Weerasuriya, S. Sygletos, A.D. Ellis, L. Gruner-Nielson, D. Jakobsen, S. Herstrom, R. Phelan, J. O'Gorman, A. Bogris, D. Syvridis, S. Dasgupta, P. Petropoulos, D.J. Richardson, All-optical phase and amplitude regenerator for next-generation telecommunications systems. Nat. Photon. **4**, 690–695 (2010)
48. Z. Tong, A.O.J. Wiberg, E. Myslivets, B.P.-P. Kuo, N. Alic, S. Radic, Broadband parametric multicasting via four-mode phase-sensitive interaction. Opt. Express **20**, 19363–19373 (2012)
49. Z. Tong, S. Radic, Low-noise optical amplification and signal processing in parametric devices. Adv. Opt. Photon. **5**, 318–384 (2013)
50. M. DiDomenico Jr., Material dispersion in optical fiber waveguides. Appl. Opt. **11**, 652–654 (1972)
51. D. Gloge, Dispersion in weakly guiding fibers. Appl. Opt. **10**, 2442–2445 (1971)
52. N. Shibata, R.P. Braun, R.G. Waarts, Phase-mismatch dependency of efficiency of wave generation through four-wave mixing in a single-mode optical fiber. IEEE J. Quant. Electron **QE-23**, 1205–1210 (1987)
53. H.R. Philipp, Silicon Dioxide (SiO_2) (Glass), in *Handbook of Optical Constants of Solids*, ed. by E.D. Palik (Academic, New York, 1985)

54. B.J. Ainslie, C.R. Day, A review of single-mode fibers with modified dispersion characteristics. J. Lightwave Technol. **LT-4**, 967–979 (1986)
55. K. Inoue, Four-wave mixing in an optical fiber in the zero-dispersion wavelength region. J. Lightwave Technol. **10**, 1553–1561 (1992)
56. J.M. Chavez-Boggio, S. Moro, B.P.-P. Kuo, N. Alic, S. Radic, Tunable parametric all-fiber short-wavelength IR transmitter. J. Lightwave Technol. **28**, 443–447 (2010)
57. J.E. Sharping, C. Pailo, C. Gu, L. Kiani, J.R. Sanborn, Microstructure fiber optical parametric oscillator with femtosecond output in the 1200 to 1350 nm wavelength range. Opt. Express **18**, 3911–3916 (2010)
58. V.V. Ravi Kanth Kumar, A.K. George, J.C. Knight, PStJ Russell, Tellurite photonic crystal fiber. Opt. Express **11**, 2641–2645 (2003)
59. L. Brilland, F. Smektala, G. Renversez, T. Chartier, J. Troles, T.N. Nguyen, N. Traynor, A. Monteville, Fabrication of complex structures of holey fibers in chalcogenide glass. Opt. Express **14**, 1280–1285 (2006)
60. P. Russell, Photonic crystal fibers. Science **17**, 358–362 (2003)
61. F. Yaman, Q. Lin, S. Radic, G.P. Agrawal, Impact of dispersion fluctuations on dual-pump fiber-optical parametric amplifiers. IEEE Photon. Technol. Lett. **16**, 1292–1294 (2004)
62. P. Sansonetti, Modal dispersion in single-mode fibres: simple approximation issued from mode spot size spectral behaviour. Electron. Lett. **18**, 647–648 (1982)
63. K. Petermann, Constraints for fundamental-mode spot size for broadband dispersion-compensated single-mode fibres. Electron. Lett. **19**, 712–714 (1983)
64. C. Pask, Physical interpretation of Petermann's strange spot size for single-mode fibres. Electron. Lett. **20**, 144–145 (1984)
65. M. Farahmand, M. de Sterke, Parametric amplification in presence of dispersion fluctuations. Opt. Express **12**, 136–142 (2004)
66. S. Nishi, M. Saruwatari, Technique for measuring the distributed zero dispersion wavelength of optical fibres using pulse amplification caused by modulation instability. Electron. Lett. **31**, 225–226 (1995)
67. L.F. Mollenauer, P.V. Mamyshev, M.J. Neubelt, Method for facile and accurate measurement of optical fiber dispersion maps. Opt. Lett. **21**, 1724–1726 (1996)
68. J. Gripp, L.F. Mollenauer, Enhanced range for OTDR-like dispersion map measurements. Opt. Lett. **23**, 1603–1605 (1998)
69. J.B. Schlager, Zero-dispersion wavelength distribution in optical fibers from CW four-wave mixing. Lasers and Electro-Optics Society Annual Meeting, 1998. LEOS'98. IEEE **2**, 309–310 (1998)
70. M. Eiself, R.M. Jopson, R.H. Stolen, Nondestructive position-resolved measurement of the zero-dispersion wavelength in an optical fiber. J. Lightwave Technol. **15**, 135–143 (1997)
71. S. Moro, E. Myslivets, J.R. Windmiller, N. Alic, J.M. Chavez-Boggio, S. Radic, Synthesis of equalized broadband parametric gain by localized dispersion mapping. IEEE Photon. Technol. Lett. **20**, 1971–1973 (2008)
72. L.H. Gabrielli, H.E. Hernández-Figueroa, H.L. Fragnito, Robustness optimization of fiber index profiles for optical parametric amplifiers. J. Lightwave Technol. **27**, 5571–5579 (2009)
73. B.P.-P. Kuo, Recent advances in wide-band optical parametric mixer synthesis, in *Proc. OECC/PS 2013*, paper TuS2-1
74. F. Gholami, B.P.-P. Kuo, S. Zlatanovic, N. Alic, S. Radic, Phase-preserving parametric wavelength conversion to SWIR band in highly nonlinear dispersion stabilized fiber. Opt. Express **21**(9), 11415–11424 (2013)
75. J.H. Lee, T. Tanemura, K. Kikuchi, T. Nagashima, T. Hasegawa, S. Ohara, N. Sugimoto, Experimental comparison of a Kerr nonlinearity figure of merit including the stimulated Brillouin scattering threshold for state-of-the-art nonlinear optical fibers. Opt. Lett. **30**, 1698–1700 (2005)
76. H.K. Tsang, C.S. Wong, T.K. Liang, I.E. Day, S.W. Roberts, A. Harpin, J. Drake, M. Asghari, Optical dispersion, two-photon absorption and self-phase modulation in silicon waveguides at 1.5μm. Appl. Phys. Lett. **80**, 416–418 (2002)

77. N. Yoshizawa, T. Imai, Stimulated Brillouin scattering suppression by means of applying strain distribution to fiber with cabling. J. Lightwave Technol. **11**, 1518–1522 (1993)
78. K. Shiraki, M. Ohashi, M. Tateda, Suppression of stimulated Brillouin scattering in a fibre by changing the core radius. Electron. Lett. **31**, 668–669 (1995)
79. E. Lichtman, R.G. Waarts, A.A. Friesem, Stimulated brillouin scattering excited by a modulated pump wave in single-mode fibers. J. Lightwave Technol. **7**, 171–174 (1989)
80. A. Cerqueira-Sodre Jr., J.M. Chavez Boggio, A.A. Rieznik, H.E. Hernandez-Figueroa, H.K. Fragnito, J.C. Knight, Highly-efficient generation of broadband cascaded four-wave mixing products. Opt. Express **16**, 2816–2828 (2008)
81. D. Grischkowsky, A.C. Balant, Optical pulse compression based on enhanced frequency chirping. Appl. Phys. Lett. **41**, 1–3 (1982)
82. G. Yang, Y.R. Shen, Spectral broadening of ultrashort pulses in a nonlinear medium. Opt. Lett. **9**, 510–512 (1984)
83. P.K.A. Wai, C.R. Menyuk, Y.C. Lee, H.H. Chen, Nonlienar pulse propagation in the neighborhood of the zero-dispersion wavelength of monomode optical fibers. Opt. Lett. **11**, 464–466 (1986)
84. K. Tai, A. Hasegawa, A. Tomita, Observation of modulational instability in optical fibers. Phys. Rev. Lett. **56**, 135–138 (1986)
85. M. Nakazawa, H. Kubota, K. Tamura, Random evolution and coherence degradation of a higher-order optical soliton train in the presence of noise. Opt. Lett. **24**, 318–320 (1999)
86. B.P.-P. Kuo, *Agile Optical Frequency Synthesis via Parametric Processes*, Ph.D. Dissertation (University of California, San Diego, 2011)
87. J.E. Rothenberg, D. Grischkowsky, Observation of the formation of an optical intensity shock and wave breaking in the nonlinear propagation of pulses in optical fibers. Phys. Rev. Lett. **62**, 531–534 (1989)
88. W.J. Tomlinson, R.H. Stolen, A.M. Johnson, Optical wave breaking of pulses in nonlinear optical fibers. Opt. Lett. **10**, 457–459 (1985)
89. M.G. Taylor, Coherent detection method using DSP for demodulation of signal and subsequent equalization of propagation impairments. IEEE Photonics Technol. Lett. **16**, 674–676 (2004)
90. S.J. Savory, G. Gavioli, R.I. Killey, P. Bayvel, Electronic compensation of chromatic dispersion using a digital coherent receiver. Opt. Express **15**, 2120–2126 (2007)
91. E. Ip, J. Kahn, Digital equalization of chromatic dispersion and polarization mode dispersion. J. Lightwave Technol. **25**, 2033–2043 (2007)
92. D.S. Millar, S. Makovejs, C. Behrens, S. Hellerbrand, R.I. Killey, P. Bayvel, S.J. Savory, Mitigation of fiber nonlinearity using a digital coherent receiver. IEEE J. Select. Topics Quant. Electron. **16**, 1217–1226 (2010)
93. S. Savory, Digital coherent optical receivers: algorithms and subsystems. IEEE J. Select. Topics Quant. Electron. **16**, 1164–1179 (2010)
94. R. Essiambre, G. Kramer, P.J. Winzer, G.J. Foschini, B. Goebel, Capacity limits of optical fiber networks. J. Lightwave Technol. **28**, 662–701 (2010)
95. E. Ip, J.M. Kahn, Feedforward carrier recovery for coherent optical communications. J. Lightwave Technol. **25**, 2675–2692 (2007)
96. I. Fatadin, D. Ives, S.J. Savory, Differential carrier phase recovery for QPSK optical coherent systems with integrated tunable lasers. Opt. Express **21**, 10166–10171 (2013)
97. V. Ataie, E. Myslivets, B.P.-P. Kuo, N. Alic, S. Radic, Spectrally equalized frequency comb generation in multistage parametric mixer with nonlinear pulse shaping. J. Lightwave Technol. **32**, 840–846 (2014)
98. C.-S. Brès, A.O.J. Wiberg, B.P.-P. Kuo, N. Alic, S. Radic, Wavelength-multicasting of 320 Gb/s channel in self-seeded parametric amplifier. IEEE Photon. Technol. Lett. **21**, 1002–1004 (2009)
99. L.H. Sahasrabuddhe, B. Mukherjee, Light-trees: optical multicasting for improved performance in wavelength–routed networks. IEEE Commun. Mag. **37**, 67–73 (1999)
100. H.P. Yuen, Reduction of quantum fluctuation and suppression of the Gordon-Haus effect with phase-sensitive linear amplifiers. Opt. Lett. **17**, 73–75 (1992)

101. C.-S. Brès, S. Zlatanovic, A.O.J. Wiberg, J.R. Adleman, C.K. Huyuh, E.W. Jacobs, J.M. Kvavle, S. Radic, Parametric photonic channelized RF receiver. IEEE Photon. Technol. Lett. **23**, 344–346 (2011)
102. H. Sorensen, M. Heideman, C. Burrus, On computing the split-radix FFT. IEEE Trans. Acoust. Speech Signal Process. **34**, 152–156 (1986)
103. W.J. Williams, M.L. Brown, A.O. Hero III, Uncertainty, information, and time-frequency distributions, *Proc. SPIE 1566, Advanced Signal Processing Algorithms, Architectures and Implementations II*, 144–156 (1991)

Chapter 8
All-Optical Pulse Shaping for Highest Spectral Efficiency

Juerg Leuthold and Camille-Sophie Brès

Abstract Pulse shaping gives communications engineers another degree of freedom in designing a link. It holds promise to allow extending transmission reach, achieve optical multiplexing at highest spectral efficiency or to limit nonlinear distortions. A variety of pulse shapes—rectangular, sinc, raised cosine to cite just a few—have been investigated but the important question is how optical transmitters can generate such pulses at the necessary speed. Should the transmitter be realized in the digital domain, the all-optical domain or can it be implemented as a hybrid? In this chapter, the fundamentals for pulse shaping in transmitters and receivers are reviewed. A particular emphasis is on orthogonal frequency division multiplexing (OFDM) where the system's data are encoded onto subcarriers with a rectangularly shaped impulse response, and Nyquist pulse shaping where the symbols are carried by Nyquist pulses. Electronic, digital and optical processors are described and recent experimental demonstrations are reported.

8.1 Introduction

Pulse shaping has been used for many years in optical communications. However, pulse shaping is an effort and therefore it was only used for medium and long-haul transmission systems. In practice, conventional pulse shaping would be performed this way: One would encode information in the form of an on-off keying signal onto a cw laser. This way a non-return-to-zero (NRZ) encoded signal would be formed. NRZ signals were good enough to transmit 10 Gbit/s signals over hundreds of

J. Leuthold (✉)
Institute of Electromagnetic Fields, ETH-Zurich, ETZ K 81, Gloriastrasse 35, CH-8092 Zurich, Switzerland
e-mail: JuergLeuthold@ethz.ch

C.-S. Brès
Institute of Electrical Engineering, Ecole Polytechnique Fédérale de Lausanne (EPFL), Station 11, CH-1015 Lausanne, Switzerland

S. Wabnitz and B.J. Eggleton (eds.), *All-Optical Signal Processing*, Springer Series in Optical Sciences 194, DOI 10.1007/978-3-319-14992-9_8 kilometers. If longer reach was needed, one would carve the pulse by means of a Mach-Zehnder modulator onto the NRZ signal. For long reach communications one would typically use a 33 % return-to-zero pulse shape. In some instance, when nonlinearities would be an issue one could as well opt for a 66 % carrier-suppressed return to zero pulse shape.

More recently pulse shaping is not only used to extend the reach but rather as a means to achieve optical multiplexing at highest spectral efficiency. Two most prominent examples are OFDM [1–10] and Nyquist [11–13] multiplexed signals. And indeed, a recent transmission experiment shows that signal encoding with 18 bit/s/Hz and beyond is feasible [14]. Prerequisites for such multiplexing schemes are advanced optical transmitters that encode the information as complex-valued, specially shaped pulse envelopes onto optical carriers. The shapes are selected such that signals are orthogonal in time and frequency. This allows for demultiplexing of appropriately generated signals even if they overlap in time and/or in the frequency domain.

A variety of pulse shapes have been investigated: Rectangularly-shaped, sinc-shaped and raised-cosine shaped pulses are just a few. The important question now is how optical transmitters can generate such pulses at the necessary speed.

Pulse shaping is most easily performed by digital signal processing. Shaping signals with digital signal processing (DSP) has become possible due to the availability of massive parallel high-speed electronics. We now cannot only encode any complex value onto an optical carrier, but we even have sufficient processing bandwidth to shape it real-time within the time slot of a symbol with sufficient precision. And indeed, the trend is towards software defined optical transmitters that can adapt modulation formats, pulse shapes and symbol rates at the push of a button [15]. However, electronic pulse shaping costs DSP signal processing resources and comes at the price of a higher power consumption. Moreover, electronic pulse shaping will always be limited to electronic signal processing speed. Conversely, all-optical techniques can be extremely energy efficient and there is hardly any speed limitation. Also, with all-optical techniques quantization and clipping of pulses with high peak-power can be avoided. Generally, it is believed that "all-DSP" pulse shaping approaches are most reasonable when lower-speed tributaries are combined into a super-channel at a moderate aggregate data rate. In turn, all-optical approaches are attractive when combining higher-bit-rate channels into multiple, spectrally efficient Tbit/s superchannels. In practice, hybrid implementations, where transceivers are neither "all-DSP" nor "all-optical" might make most sense for encoding superchannels.

8.2 Fundamentals

Transmitters (Tx) and receivers (Rx) employing pulse shaping techniques are somewhat different from more conventional transceivers. In the following section we follow and to some extend summarize a discussion recently published in [16], where one may also find a more extensive review.

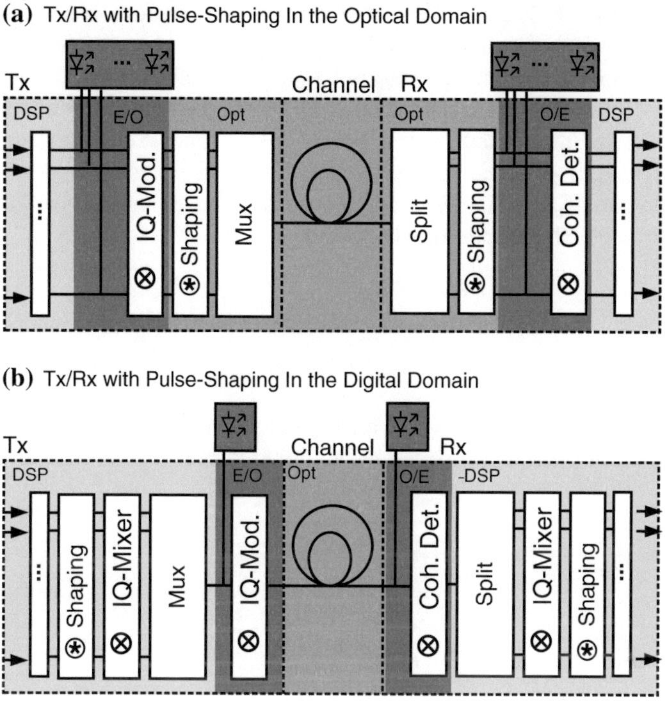

Fig. 8.1 Optical and electrical transmitter-receiver concepts for multiplexing at highest spectral efficiencies. In **a** the symbols are first converted into the optical domain by IQ-modulators and then optically multiplexed. In **b** the signals are modulated on different radio frequency carriers and multiplexed electronically before being encoded onto an optical carrier (figure modified from [16])

Two implementations of a Tx/Rx pair using pulse shaping are depicted in Fig. 8.1. In Fig. 8.1a pulse shaping is performed in the optical domain. Here the complex information in the form of complex symbols are first converted from the electronic to the optical domain (E/O-box) by IQ-mixers ($\otimes$, in optical communications also known as "optical IQ-modulators"). A particular pulse shape $h_s(t)$ is then carved onto the optical signal. Finally, several tributaries may be wavelength multiplexed. The optical transmitter scheme of Fig. 8.1a is best suited to generate Tbit/s superchannels [8]. In Fig. 8.1b pulse shaping is implemented in the digital domain. In the digital domain, pulse shaping, signal encoding and multiplexing are performed by a processor. The whole electrical signal with all the subcarriers is then encoded by means of an IQ-Modulator (E/O) onto a single laser. Such digital signal processing power has only recently become available and in fact it is only recently, that 100 Gbit/s OFDM [17] and Nyquist multiplexing [13] has been demonstrated. A key argument for using pulse shaping is the possibility to increase the spectral efficiency dramatically. If the pulse shape in the transmitter and receiver are properly chosen record spectral efficiencies of as much 18 bit/s/Hz with a net spectral efficiency of 15 bit/s/

Hz [14] have already been demonstrated. Such world records are only possible by clever pulse shaping that allow multiplexing of subcarrier data without guard bands.

We now more closely review the pulse-shaping functions implemented in the transmitter (Tx) and receiver (Rx). Pulse-shaping in the transmitter and filtering in the receiver are interrelated. When the impulse response of the signal in the transmitter is $h_s(t)$ then the received signal needs to be convolved with a filter response function $h_r(t)$, where $h_s(t)$ and $h_r(t)$ need to be orthogonal, i.e. they need to obey the orthogonality condition [16]

$$T_s \int_{-\infty}^{+\infty} h_s(t - t_m) h_r(t_{m'} - t)\,\mathrm{d}t = \delta_{mm}, \tag{8.1}$$

where $t_m = mT_s$ with integers m, m' and T_s the symbol duration.

Important functions that meet an orthonormality relations are e.g.

$$\begin{aligned}
&\frac{1}{T}\int_{-\infty}^{+\infty} \mathrm{rect}\left(\frac{t}{T} - m\right)\mathrm{rect}\left(\frac{t}{T} - m'\right)\mathrm{d}t = \delta_{mm'} \\
&\frac{1}{T}\int_{-\infty}^{+\infty} \mathrm{sinc}\left(\frac{t}{T} - m\right)\mathrm{sinc}\left(\frac{t}{T} - m'\right)\mathrm{d}t = \delta_{mm'} \\
&\frac{1}{T}\int_{T_o-T/2}^{T_o+T/2} \exp\left(+\mathrm{j}2\pi m\frac{t}{T}\right)\exp\left(-\mathrm{j}2\pi m'\frac{t}{T}\right)\mathrm{d}t = \delta_{mm'} \\
&\frac{1}{F}\int_{F_o-F}^{F_o+F} \cos\left(\frac{\pi}{2}\frac{f}{F}\right)\exp\left(+\mathrm{j}2\pi m\frac{f}{F}\right)\cos\left(\frac{\pi}{2}\frac{f}{F}\right)\exp\left(-\mathrm{j}2\pi m'\frac{f}{F}\right)\mathrm{d}f = \delta_{mm'}
\end{aligned} \tag{8.2}$$

In (8.2) the functions $\mathrm{rect}(t/T)$ and $\mathrm{sinc}(t/T)$ are defined as follows

$$\mathrm{rect}\left(\frac{t}{T}\right) = \begin{cases} 1 & \text{for } |t| < T/2 \\ 0 & \text{else} \end{cases} \tag{8.3}$$

$$\mathrm{sinc}\left(\frac{t}{T}\right) = \begin{cases} 1 & \text{for } t = 0 \\ \frac{\sin(\pi t/T)}{\pi t/T} & \text{else} \end{cases}. \tag{8.4}$$

Similar to sinc-pulses, raised cosine pulses share the property that impulse maxima and zeros of neighbouring pulses can be made to coincide, but such pulses are not necessarily orthogonal in the sense of (8.2) and as such do not allow ISI-free detection with spectral efficiencies at the Nyquist limit.

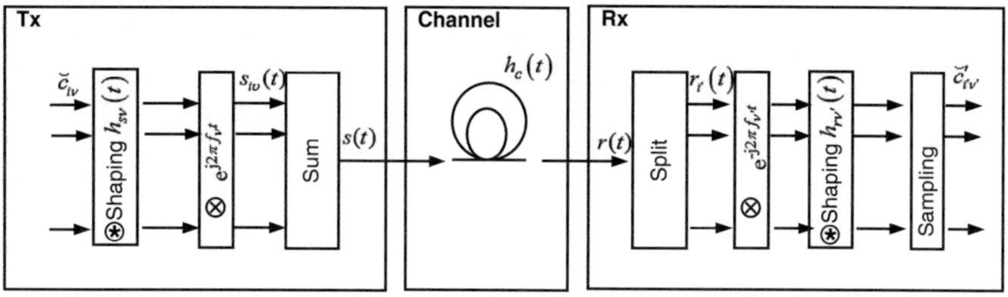

Fig. 8.2 Generic transmission system with transmitter (Tx), channel, and receiver (Rx). A Tx with N branches v is shown. Each branch is fed by a time series of complex data coefficient $\breve{c}_{iv}$ at time intervals iT. They are convoluted by impulse-forming filter with impulse response $h_{sv}(t)$ and up converted to the sub-channel frequency f_v by a complex (in-phase/quadrature, IQ) mixer with a local oscillator (LO). The summing node $\sum_v$ adds the up-converted signals $s_{iv}(t)$ in time frame i for all branches v. The resulting total signal $s(t)$ consists of concatenated time frames and enters the linear channel. On the *right side* we have a Rx with N branches v'. The total received signal $r(t)$ that leaves the channel enters a splitting node s_v and is distributed with equal amplitudes to the receiver branches v'. The received signal $r_{i'v'}(t)$ in time frame i' and branch v' is frequency down-converted with an IQ mixer and a LO $\exp(-j2\pi f_{v'}(t))$. After a receiver filter with impulse response $h_{rv'}(t)$, a sampler extracts at time $t = i'T + t_{gv'}$ the complex data coefficient $\breve{c}_{i'v'}$ (figure modified from [16])

We now take a more mathematical approach to the implementation of a pulse-shaping transmitter and receiver pair, see Fig. 8.2. In Fig. 8.2 the coefficients $\breve{c}_{iv}$ represent the complex baseband data signal in branch v at time $t = iT$. A complex mixer (in-phase/quadrature mixer, or a so-called IQ mixer) converts the baseband signal to the frequency f_v of a complex subcarrier represented by the analytic signal $\exp(j2\pi f_v t)$,

$$f_v = vF_s + f_c, \quad v = 0, 1, \ldots, N-1, \quad F_s = 1/T_s, \quad \text{channel frequency } f_c \cdot \tag{8.5}$$

As a result, a signal at a subcarrier frequency f_v is generated. It enters the node "Sum" in Fig. 8.2 where the subcarrier signals from all branches are summed (or multiplexed) to form the signal $s_i(t)$.

$$s_i(t) = \sum_{v=0}^{N-1} \breve{c}_{iv} T_s h_{sv}(t - iT) \exp(j2\pi f_v t) \cdot \tag{8.6}$$

Finally, the infinitely many consecutive time frames i constitute the signal $s(t)$ that is sent across the channel

$$s(t) = \sum_{i=-\infty}^{+\infty} s_i(t) \cdot \tag{8.7}$$

The effect of the channel can be taken into account by delaying the signal's impulse response $h_{sv}(t)$ by the group delay t_{gv}, and by adapting a retardation onto the carrier phase. Of course, the signal will also be distorted by dispersion, which can be undone in the receiver. The channel will further introduce some nonlinear distortions. Some of which can be undone by clever DSP algorithms. Other nonlinear distortions need to be minimized by e.g. reducing the launch power of the signal. To keep things simple we only introduce the group delay t_{gv} onto the received signal and indicate potential distortions due to transmission by modifying the signal impulse response $h_{sv}(t)$ into a modified response $h'_{sv}(t)$. The received signal $r(t)$ then reads

$$r(t) = \sum_{i=-\infty}^{+\infty} r_i(t), \quad r_i(t) = T_s \sum_{v=0}^{N-1} \breve{c}_{iv} h'_{sv}\left(t - \left(iT + t_{gv}\right)\right) \exp[\mathrm{j}(2\pi f_v t - \varphi_v)]. \quad (8.8)$$

The ultimate goal of a receiver is the extraction of the data coefficients $\breve{c}_{iv}$. A schematic of a generic receiver that performs this operation is shown at the right-hand side of Fig. 8.2. Again, only one out of N branches v' is depicted. First, the received signal $r(t)$ is split in a splitting node "Split" with equal magnitudes and phases to N receiver branches v'. In a second step the received signal $r_{i'}(t)$ is then down converted in branch v' for each time frame i' with a complex mixer and a LO $\exp(-\mathrm{j}2\pi f_{v'} t)$. In a third step, the down-converted signal needs to be convolved with the receiver's impulse response $h_{rv'}(t)$. If sampling is performed at times $t = i'T + t_{gv'}$ then one can extract the coefficient $\breve{c}'_{i'v'}$. The total operation can be expressed by

$$\breve{c}'_{i'v'} = \sum_{i=-\infty}^{+\infty} \int_{-\infty}^{+\infty} r_{i'}(t') \exp(-j2\pi f_{v'} t') h_{rv'}(t - t') \mathrm{d}t' \quad \text{with } t = i'T + t_{gv'}. \quad (8.9)$$

To find the proper sampling time $t = i'T + t_{gv'}$ and frequency $f_{v'}$ a timing, frequency and phase estimation unit is needed, which is not shown in the simplified receiver depicted in Fig. 8.2.

Inserting (8.8) into (8.9) gives

$$\begin{aligned} \breve{c}'_{i'v'} &= \sum_{i=-\infty}^{+\infty} \int_{-\infty}^{+\infty} \left[\sum_{v=0}^{N-1} \breve{c}_{iv} T_s h_{sv}\left(t' - \left(i'T + t_{gv}\right)\right) \exp(\mathrm{j}2\pi f_v t')\right] \exp(-2\pi f_{v'} t') h_{rv'}(t - t') \mathrm{d}t' \\ &= \sum_{i=-\infty}^{+\infty} \sum_{v=0}^{N-1} \breve{c}_{iv} T_s \int_{-\infty}^{+\infty} \left[h_{sv}\left(t' - \left(i'T + t_{gv}\right)\right) h_{rv'}(t - t')\right] \exp(\mathrm{j}2\pi [f_v - f_{v'}] t') \mathrm{d}t' \end{aligned} \quad (8.10)$$

In the ideal case where the frequency of the local oscillator in the receiver $f_{v'} = f_v$ one can see from (8.10) that the exponential term reduces to a "1". Further, if the pulse shape in the transmitter h_{sv} and the filter response $h_{rv'}$ in the receiver are orthogonal and the sampling time in the receiver is chosen as $t = i'T + t_{gv'}$, then (8.10) reduces to

$$\breve{c}'_{i'v'} = \breve{c}_{iv}\delta_{i'i}\delta_{v'v}. \tag{8.11}$$

However, if frequencies or pulse shapes are not properly chosen there might be interference. Equation (8.10) allows for interferences from sub-channels $v \neq v'$ or inter channel interference (ICI) from channels other than the present receiving branch v'. Distortions might also come in through neighboring data inside the same sub-channel v' but belonging to different time slots $i \neq i'$, i.e. inter-symbol interference (ISI).

8.3 Orthogonal Frequency Division Multiplexing (OFDM)

8.3.1 OFDM Tx and Rx Implementations

8.3.1.1 Pulse-Shaping Implementation

In orthogonal frequency division multiplexed (OFDM) systems data is encoded onto subcarriers with a rectangularly shaped impulse response. The minimum pulse duration is T_s, with T_s the symbol duration. In OFDM the pulse maintains its constant value for the whole duration of the symbol time. The symbol period T, i.e. the time it takes for the next symbol to start, however might be longer than the symbol duration T_s, i.e. $T > T_s$. The symbol period defines the clock rate $f_T = 1/T \leq F_s$ at which the symbols are sent. In practical implementations the pulse at best maintains its value not only for T_s but the whole period T. The extension of the pulse shape beyond T_s is known as the cyclic extension. However, this extension not necessarily needs to be used as a cyclic extension. As a matter of fact the pulse shape may take on any value in the interval beyond T_s. In many cases this extension is simply needed because it takes some time for a practical modulator to produce a rectangularly shaped symbol with a certain value.

In frequency domain the rectangularly shaped time domain subcarriers of the OFDM signal have sinc-shaped spectra and the sub-carriers are spaced at the inverse of the symbol duration T_s.

Mathematically, we can describe an OFDM impulse response by

$$h_{sv}(t) = \begin{cases} 1/T_s & \text{for} \quad 0 < t < T_s \\ \text{unspecified} & \text{for} \quad T_s < T \\ 0 & \text{else.} \end{cases} \tag{8.12}$$

The receiver's filter impulse response needs to meet the orthogonality condition (8.1) in order to guarantee interference-free reception with highest spectral efficiency. For this the filter needs to be chosen such that it is non-zero only in a region $0 < t < T_s$, see equation.

$$h_{rv}(t) = \frac{1}{T_s}\mathrm{rect}\left(\frac{t}{T_s} - \frac{1}{2}\right) = \begin{cases} 1/T_s & \text{for } 0 < t < T_s \\ 0 & \text{else} \end{cases}. \tag{8.13}$$

To encode an OFDM signal we now only need to follow the recipe from (8.6). We first shape the pulse with the rectangularly shaped pulse from (8.12). To keep things simple we disregard any cyclic redundancy and potential guard bands. We then mix it with the subcarrier frequencies f_v and sum all subcarriers. Finally, the signal is mapped onto an optical carrier at frequency f_c by means of a complex mixer, i.e. an IQ-modulator. This leads us to the following expression for an OFDM signal

$$s_i(t) = \exp(j2\pi f_c t)\sum_{v=0}^{N-1} \breve{c}_{iv}\,\mathrm{rect}\left(\frac{t - iT_s}{T_s}\right)\exp(j2\pi f_v t)\cdot \tag{8.14}$$

We then notice, that the N subcarriers of an OFDM signal are equipspaced by F_s and $f_v = vF_s + f_c$

$$s_i(t) = \exp(\mathrm{j}2\pi f_c t)\cdot\mathrm{rect}\left(\frac{t - iT_s}{T_s}\right)\sum_{v=0}^{N-1} \breve{c}_{iv}\exp(\mathrm{j}2\pi v F_s t). \tag{8.15}$$

In literature, the rect expression is often dropped as people usually assume that the coefficients $\breve{c}_{iv}$ are only defined for the duration of a symbol (Fig. 8.3).

The shape and spectrum of an OFDM signal with three subcarriers is visualized in Fig. 8.3.

Next we built a pulse-shaping OFDM receiver. It is a hypothetical receiver as this type of a receiver is not really implemented in practical systems. Still—it is

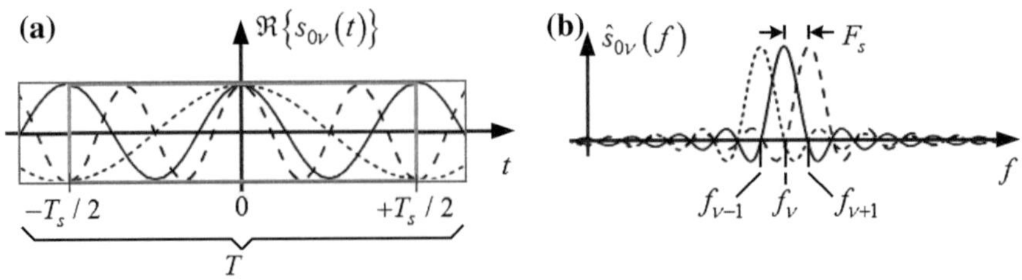

Fig. 8.3 OFDM pulses in **a** time domain, $s_{iv}(t)$ and **b** frequency domain $\hat{s}_{iv}(f)$. The *green box* marks the finite symbol time. **a** The *left graph* shows the real parts of *three rectangularly shaped* subcarriers oscillating at a different carrier frequency each. The sum over all three carriers would represent one time-domain OFDM symbol of duration T_s. **b** The *right graph* shows the corresponding spectrum $\hat{s}_{0v}(f)$ of three subcarriers with a spectral separation of $|f_{v+1} - f_v| = F_s = 1/T_s$ (figure derived from [16])

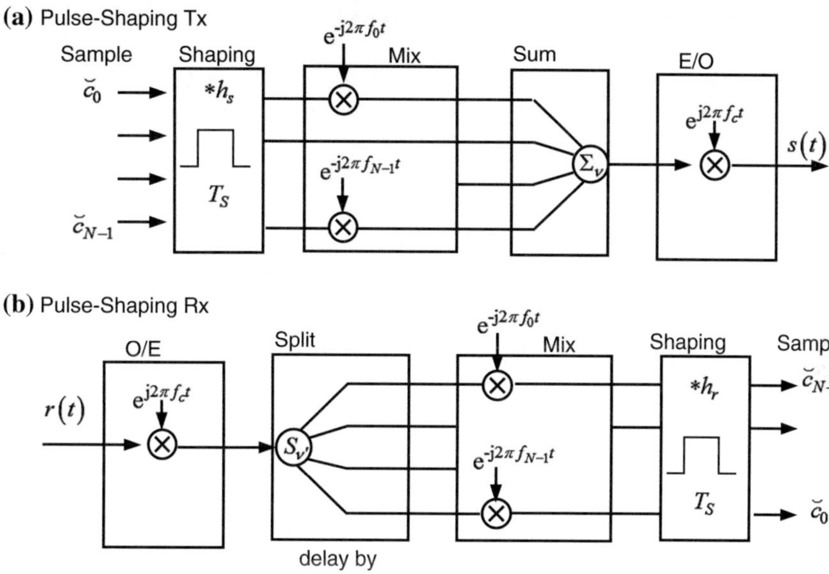

Fig. 8.4 Possible implementation **a** OFDM transmitter and **b** receiver by means of the pulse-shaping technique

quite educational to discuss it. Now, in this hypothetical receiver, the signal would first be down converted by the local oscillator frequency f_c by means of an optical hybrid from the optical to the electrical domain. Subsequently, the electrical signal would be split onto N copies. Each of which would then be mixed by the subcarrier frequency and integration over a symbol duration would be performed. The mathematical process according to (8.9) leads to

$$\breve{c}'_{i'v'} = \exp\left(-j2\pi f_c\left(i'T_s + t_{gv'}\right)\right) \sum_{i=-\infty}^{+\infty} \int_{i'T_s+t_{gv'}}^{(i'+1)T_s+t_{gv'}} r_{i'}(t') \exp(-j2\pi f_{v'} t')\mathrm{d}t'. \quad (8.16)$$

A graphical sketch of the OFDM transmitter and receiver is shown in Fig. 8.4.

The OFDM transmitter and receiver of Fig. 8.4 are electrical OFDM implementations since all the relevant processing is done in the electrical domain—except for the E/O and O/E up- and down conversion. The scheme represents a more in-detail description of the concept mentioned by Fig. 8.1b. However, it should be emphasized, that the electrical up- and down mixers could be replaced by E/O and O/E converters, i.e. the electrical mixers could be replaced by lasers and IQ-modulators. This would then lead to an optical implementation as Fig. 8.1a. We cannot discuss any possible transceiver implementation here. However, we will see such an implementation further below when we discuss our example.

8.3.1.2 Discrete Fourier Transform Implementation

We now take a closer look at the OFDM signal in (8.15). We notice that in practice one usually works with digital signals and thus generates and detects signals by time-discrete sampling at times $t_n = iT_s + nT_s/N$ with $n = 0, 1, 2, \ldots N-1$. Replacing t by t_n in (8.15) leads to

$$\begin{aligned} s_i(t_n) &= \exp(\mathrm{j}2\pi f_c t)\cdot \mathrm{rect}\left(\frac{t_n - iT_s}{T_s/N}\right)\sum_{\nu=0}^{N-1}\breve{c}_{i\nu}\exp\left(\mathrm{j}2\pi\nu F_s\frac{nT_s}{N}\right)\\ &= \exp(\mathrm{j}2\pi f_c t)\cdot \mathrm{rect}\left(\frac{t_n - iT_s}{T_s/N}\right)\sum_{\nu=0}^{N-1}\breve{c}_{i\nu}\exp\left(\mathrm{j}2\pi\frac{\nu n}{N}\right)\\ &= \exp(\mathrm{j}2\pi f_c t)\cdot \mathrm{rect}\left(\frac{t_n - iT_s}{T_s/N}\right)c_{in}. \end{aligned} \quad (8.17)$$

The last term thus happens to be the expression for the Inverse Discrete Fourier Transform (IDFT) and the coefficients c_{in} are thus the discrete time samples of the Fourier coefficients $\breve{c}_{i\nu}$. Thus, an alternative way to encode the OFDM signal is by applying the IDFT onto the coefficients $\breve{c}_{i\nu}$ and to directly encode the signal $s_i(t_n)$ in time domain onto an optical carrier with frequency f_c.

An implementation of an IDFT transmitter is depicted in Fig. 8.5a. The coefficients $\breve{c}_{i\nu}$ are first gated to a duration T_s/N before being fed into the IDFT processor. The IDFT processer then performs the inverse DFT operation on the coefficients $\breve{c}_{i\nu}$

$$c_{in} = \sum_{\nu=0}^{N-1}\breve{c}_{i\nu}\exp\left(\mathrm{j}2\pi\frac{\nu n}{N}\right). \quad (8.18)$$

Subsequently, the time samples c_{in} of duration T_s/N are concatenated (a parallel-to-serial conversion is performed) in order to produce an OFDM symbol of duration T_s. Eventually, the signal is E/O converted by means of an IQ-modulator, see Fig. 8.5a.

At the receiver, the inverse process is performed. The incoming OFDM data stream $r(t)$ is split onto N signals that are delayed such that N OFDM time samples c_{in} within a symbol T_s are simultaneously fed into the DFT processor (fed in parallel into the DFT processor). The DFT processor generates the Fourier coefficients $\breve{c}_{i\nu}$. The Fourier coefficients $\breve{c}_{i\nu}$ are yielded only for the short duration of T_s/N, i.e. the short time during which all N time samples have been fed in parallel into the processor. We thus perform a gating operation at the output in order to suppress output signals from the DFT processor that are based on wrong input signals.

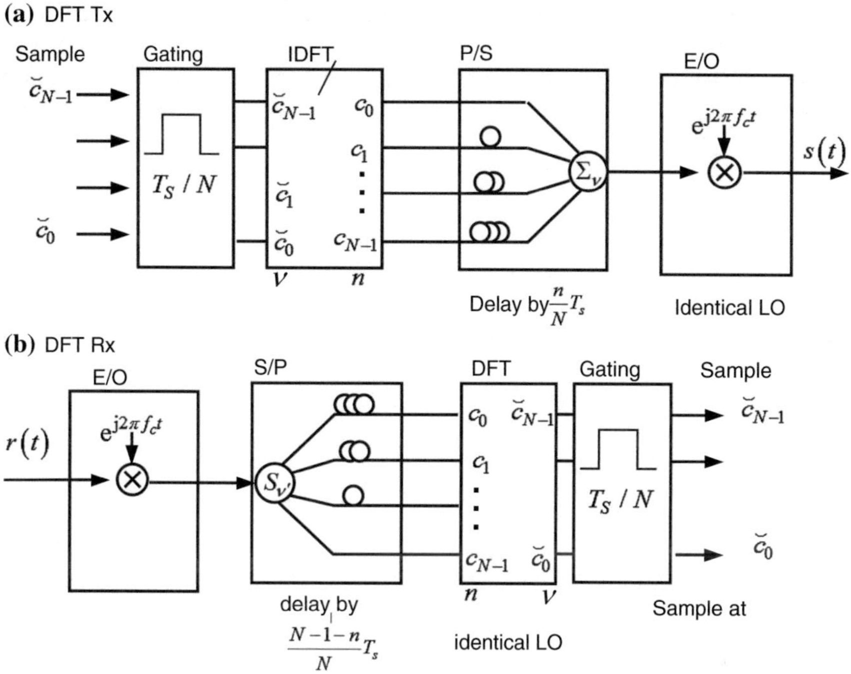

Fig. 8.5 Implementation of an **a** OFDM transmitter and **b** receiver based on the IDFT and DFT

8.3.2 Optical Fourier Transform Processors and Optical OFDM

We just have seen that Discrete Fourier Transform (DFT) processors are important building blocks. The generation and reception of OFDM is just one possible application. DFT processors also play an important role in many applications where spectral components of a signal are needed. Meanwhile there exist efficient electronic DFT processors that allow operation up to highest speed. And while there is an impressive progress to the extent that 100 Gbit/s real-time processing has become possible [18] the speed is ultimately limited by electronics. Fortunately, unprecedented DFT processing speed can be obtained by optical means. We will now discuss such optical realizations of the DFT operation.

8.3.2.1 FFT with Butterfly Topology and Optical OFDM

A fast Fourier Transform (FFT) processor yields the same input-output operation as a DFT processor. Yet, a FFT processor performs the DFT with less complexity and thus fewer elements. Since the number of elements should be minimized in an optical implementation it makes sense to implement an FFT rather than a DFT.

More precisely, if the expression for a DFT of order N is given by

$$\breve{c}_{iv} = \sum_{n=0}^{N-1} c_{in} \exp\left(-\mathrm{j}2\pi \frac{vn}{N}\right), \tag{8.19}$$

where $N = 2^p$ with an integer p then the equivalent expression can be decimated into two interleaved DFTs of order $N/2$, where

$$\breve{c}_{iv} = \begin{cases} E_v + \exp\left(-\mathrm{j}2\pi \frac{v}{N}\right) O_v & \text{if } v < \mathrm{N}/2 \\ E_{v-N/2} - \exp\left(-\mathrm{j}2\pi \frac{v-N/2}{N}\right) O_{v-N/2} & \text{if } v \geq \mathrm{N}/2, \end{cases} \tag{8.20}$$

where the quantities E_v and O_v are the even and odd DFT of size $N/2$ for even and odd inputs $c_{i,2n}$ and $c_{i,2n+1}$ ($n = (0, 1, 2, \ldots, N/2 - 1)$), respectively. The FFT of (8.20) is then performed recursively to the end.

In Fig. 8.6 we have depicted the configuration for an optical OFDM receiver with a built in optical 4 × 4 FFT processor. The overall OFDM implementation closely follows the scheme outlined in Fig. 8.5b. The OFDM symbols r_i are first fed into a serial-to-parallel converter. This can be obtained by splitting the signal into four copies and delaying each of the copies by $T_s/4$ with respect to the previous copy. This way the symbol is split in this example into four time samples. The signals c_{in} (still encoded on an optical carrier) are then launched into said optical FFT processor. The FFT depicted here is a direct implementation of the mathematical FFT expression (8.20). The optical FFT scheme shown here was first suggested by Marhic as early as 1987 [19]. A first realization of a 2 × 2 FFT was presented by Sanjoh et al. in 2002 [20] and a most impressive OFDM implementation with an 8 × 8 FFT was demonstrated in 2010 by Takiguchi et al. [21]. In the optical implementation shown here, the down conversion into the electronic domain (O/E conversion) by means of a coherent receiver is only performed at the end. This

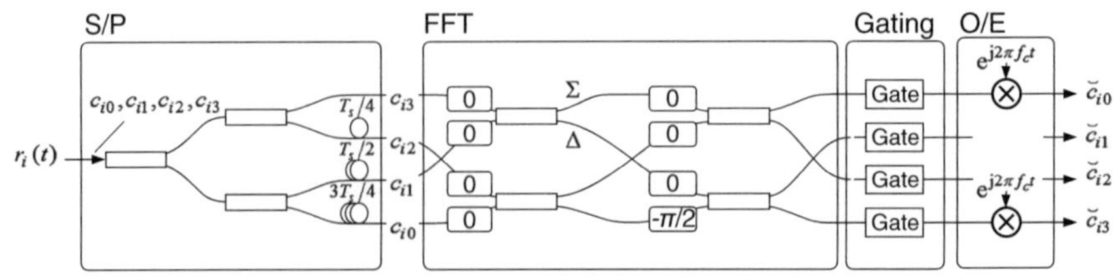

Fig. 8.6 Plot of an optical OFDM processor with an optical 4 × 4 FFT. The $N = 4$ butterfly-type FFT processor comprises of a serial-to-parallel converting delay stage, a FFT stage and a gating stage. The unit processes signals with a symbol duration T_s. *Rectangular sharp-edged boxes* stand for optical couplers with an (unphysical) amplitude split ratio of 1. *Loops* symbolize time delays by fractions of the symbol duration T_s. *Rounded boxes* with numbers mark the respective phase shifts φ according to (8.18). See [10, 16]

is in contrast to an electrical FFT where the O/E conversion would be performed in a first stage.

With an FFT algorithm the DFT complexity of order $C_{\mathrm{DFT}} = N^2$ can typically be reduced to a complexity of order $C_{\mathrm{FFT}} = N \log_2 N$. Despite of this simplification the direct FFT implementation still scales very badly with the order of the FFT. An integrated version of order 8 is already quite a task and requires multiple couplers and precisely tuned phase-shifter elements [21].

8.3.2.2 FFT with Cascaded Delay Interferometers and Optical OFDM

In 2010 Hillerkuss et al. demonstrated that the optical FFT could be performed with a much simpler delay interferometer cascade [10]. Here we follow the line of arguments in said paper where it is demonstrated how the cascaded delay interferometer approach can be obtained by applying some simple transformations on the aforementioned optical FTT. In 2011, a record 26 Tbit/s OFDM signal was encoded onto a single laser and the cascaded delay interferometer FFT approach was used to demultiplex a total of 325 subcarriers [9].

The cascaded delay interferometer FFT in [10] is based on clever rearrangements of both the S/P and FFT elements. To derive the new configuration we follow the drawings in Fig. 8.7. In a first step we rearrange the delays in the S/P conversion stage as indicated in Fig. 8.7a and re-label the outputs accordingly. Please note that we started with the very same configuration as in Fig. 8.6. After this rearrangement, we end up with an optical FFT consisting of two parallel delay interferometers (DI) with the same free spectral range (FSR) but different absolute delays, see Fig. 8.7b. By moving the common delay $T_s/4$ in both arms of the lower DI to its outputs, one obtains two identical DIs that are fed with the same input signal, Fig. 8.7c. This redundancy can be eliminated by replacing the two DI with a single DI, and by splitting its output, Fig. 8.7d. These simplification rules (a mathematical proof is available [10]) can be iterated for FFTs of any size.

The new optical FFT processor consists of $N - 1$ cascaded DI with a complexity of only $C_{\mathrm{FFT-DI}} = 2(N - 1)$ couplers and $N - 1$ phase shifters. Thus, we not only got rid of the S/P converter by integrating it into the FFT but also have now an optical FFT that scales linearly with the order. The new FFT always requires less complexity than the previous FFT—except for $N = 2$, where the complexity is identical, since the two schemes by chance lead to an identical configuration.

Once the concept has been understood, one may go back to literature and will find, that cascaded delay-interferometers have already been studied by Li in 1998 for narrow spectral slicing. No gating elements were present for these so-called Fourier filters though as these elements were mostly considered for narrow spaced filtering. The gates, however are important for digital signal processing because outside of the window one will not detect the proper coefficients. These filters were also put in a wider context in connection with the wavelet transform by Cincotti [22, 23] in 2002 and 2004 who also integrated the serial-to-parallel converter into a network.

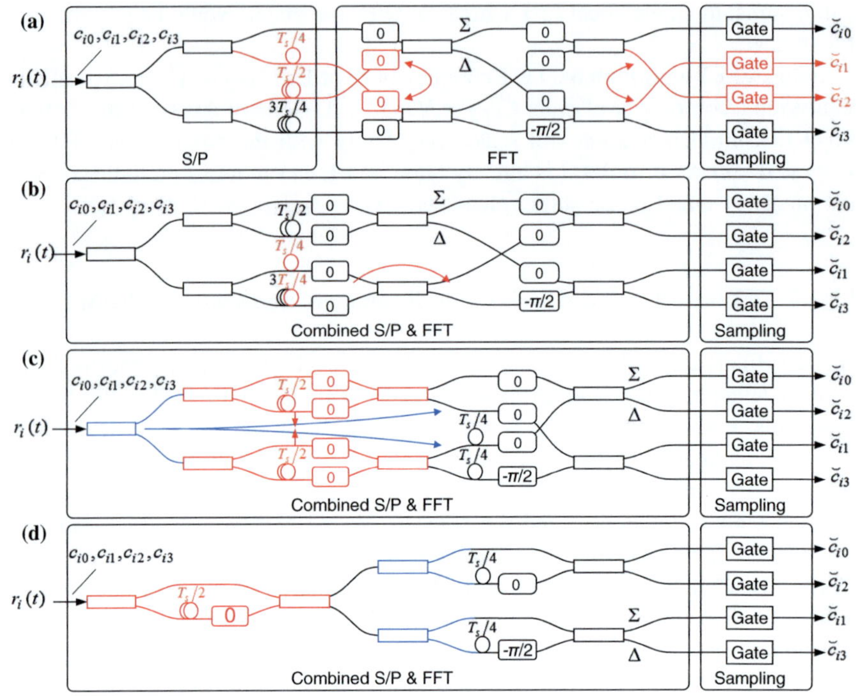

Fig. 8.7 Simplification of optical FFT by rearranging and combining the S/P and all-optical FFT elements illustrated on a four-point FFT circuit. **a** Traditional FFT implementation with separate S/P conversion stage as discussed in Fig. 8.6. The *arrows* in *red* indicate as of how the coupler elements can be rearranged without an effect on the output signals. **b** New combined S/P and FFT structure consisting of two delay interferometers (DI) with the same differential delay. The *red arrow* indicates that the additional $T_s/4$ delay can be moved out of the second DI. **c** The structure comprises of two identical DI which can be replaced by a single DI followed by signal splitters. **d** Resulting low-complexity scheme with combined S/P Conversion and FFT. In this picture, *rectangular sharp-edged boxes* stand for optical directional couplers with an (unphysical) amplitude split ratio of 1. We follow the convention that the field in the upper output arm ($\sum$) results from summing the complex amplitudes of the two coupler inputs, while the field in lower output arm (Δ) stands for the difference (scheme modified from [10, 16])

8.3.2.3 FFT Approximated by DI and Conventional Filters

The optical FFT schemes discussed above are perfect optical analog implementations of the FFT operation. However, in many instances an approximation would be sufficient. The cascaded delay interferometer FFT as derived by Hillerkuss [10] will lead us to approximations of the FFTs. We again follow [10, 16] to derive said FFT approximations.

An approximation to an FFT is a filter that as closely as possible approximates the ideal frequency response of the FFT. To find an approximation to the FFT it is useful to study effects of the delay interferometers (DI) in the FFT DI cascade. We consider

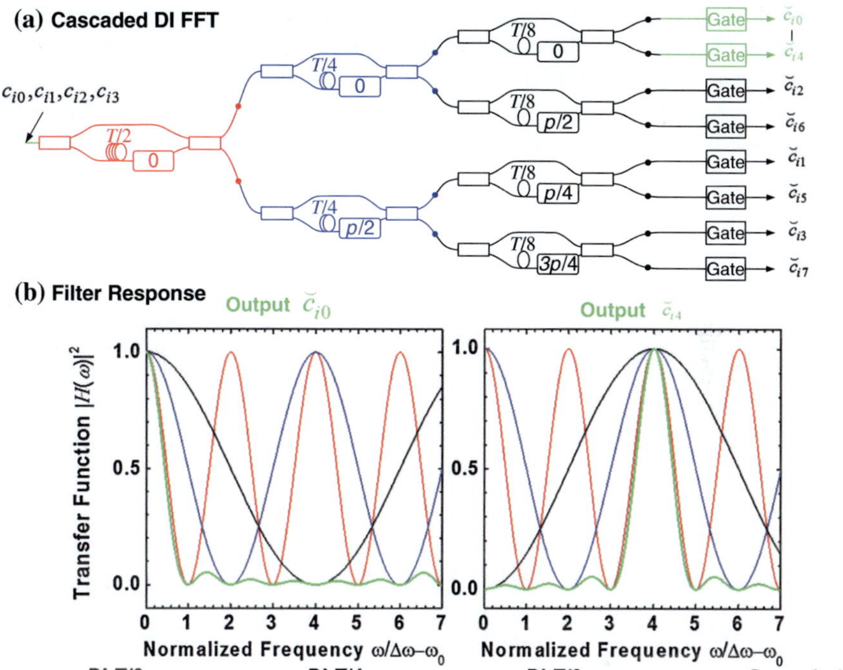

Fig. 8.8 **a** Cascaded delay interferometer (DI) FFT of order $N = 8$. **b** Frequency response of FFT. The frequency response of the first DI is plotted in *red*, the response of the second DI stage is plotted in *blue* and the response of the third stage is plotted in *black*. The overall response of the output with response $\breve{c}_{i0}$ and $\breve{c}_{i4}$ is plotted in *green*. It can be seen that the first DI stage has the largest impact on the overall response, see [10]

the FFT of order $N = 8$ in Fig. 8.8a. The frequency response of the 0th and 4th FFT coefficients are plotted in Fig. 8.8b. From the figure one can see the effect of the response of the first DI stage (plotted in red), the second DI stage (plotted in blue) and the third DI stage (plotted in black) and the overall response of all DI providing the elements $\breve{c}_{i0}$ and $\breve{c}_{i4}$. If we consider e.g. the response for the element $\breve{c}_{i4}$, it can be seen that the largest effect on the response is from the first DI stage. The subsequent 2nd stage still changes the overall response but it mainly serves as filter to suppress the elements 2 and 6. Similarly the 3rd stage is needed to suppress the 0th element. One could thus conclude that any filter combination that provides a sufficiently good approximation to a DI stage might serve as a good approximation to the FFT.

By the arguments of the discussion with Fig. 8.8 it follows that in most cases a good approximation to an FFT of any order may be found by a simple single-stage DI together with a bandpass filter and a subsequent sampling gate [10]. In the spirit of the arguments given above and derived in [10] one can search for other filter schemes with similar frequency responses. One will then find that e.g. ring filters [24] might provide a sufficient approximation as well.

The effect on the eye diagram and the signal quality as a function of the number of DI stages is shown in Fig. 8.9. Calculations have been performed for a 20 GBd OFDM signal with 8 subcarriers equispaced by 25 GHz.

In Fig. 8.9a we show the eye diagram and signal quality as a Q^2-factor in dB. One can see that the DI stage FFT provides an almost ideal eye diagram with close

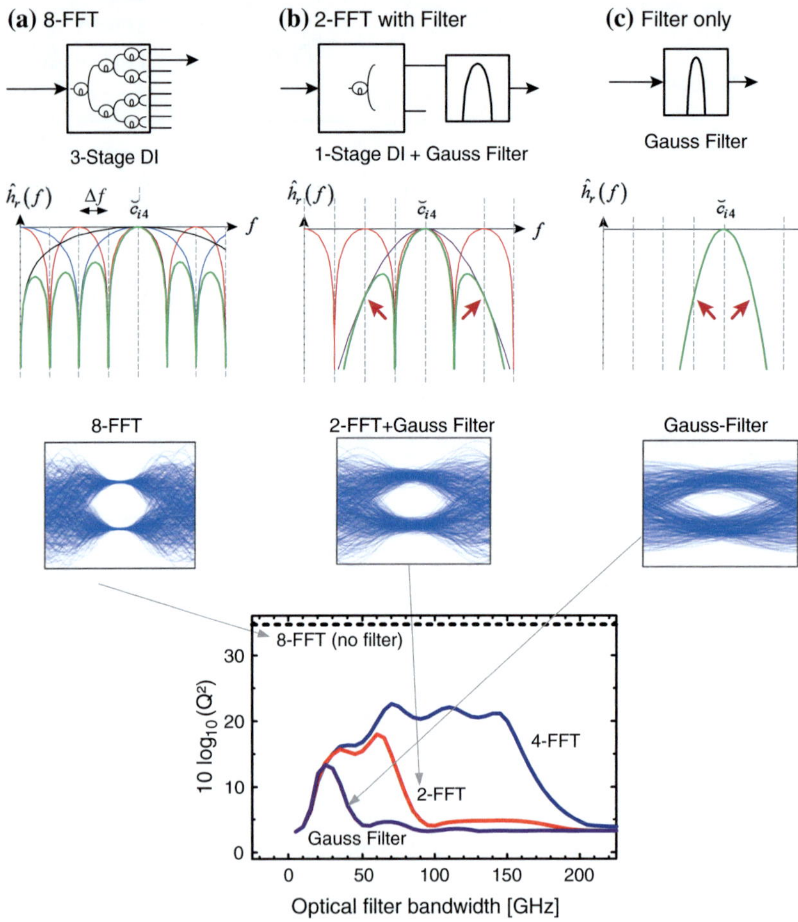

Fig. 8.9 Three different implementations to an 8 point FFT. **a** 8 point FFT with 3 delay interferometers (DI) providing an ideal FFT operation providing a wide and open eye diagram and a close to perfect signal quality. **b** Approximation to the FFT with a single stage DI and a subsequent Gaussian filter to approximate stage 2 and 3 of the DIs. The eye is still open and the signal quality is still close to 20 dB if the Gaussian bandpass filter is chosen with a proper bandwidth. **c** The DI stages are replaced by a single Gaussian bandpass filter. The frequency response is quite degraded—even if an optimum optical filter bandwidth is chosen. (Experiment performed for a 20 GBd NRZ signal with a 25 GHz subcarrier spacing). The figure has been modified from [10, 16]

to perfect signal quality. When the stages 2 and 3 of the DIs are replaced by a Gaussian filter, the eye degrades and the signal quality drops, see Fig. 8.9b. Yet the quality is still sufficient, particularly, if the optical filter bandwidth of the Gaussian filter is properly chosen (around 50 and 75 GHz). If the DI stages are replaced by a Gaussian filter the FFT still provides an eye diagram. However, the eye is degraded and the signal quality is rather low—even at the optimum filter bandwidth around 25 GHz, see Fig. 8.9c.

8.3.2.4 Free Space Fourier Transform

A lens is a Fourier transform element on its own. For a lens to work as a Fourier element the temporal samples of the input signal need to be serial-to- parallel (S/P) converted and mapped onto an object plane x. The lens then maps the object into the image plane, which represents a Fourier image of the object.

In Fig. 8.10 we have depicted the situation. The input time samples c_{in} need to be mapped (S/P converted) onto the positions x_n in the focal plane at $z = 0$ of the input object. The lens with focal length l then maps the input points onto an image plane at $z = 2l$. Each image point in the image plane is the product of a coherent superposition of all input image samples. The superposition of the discrete input images happens to be the DFT as discussed above. A thorough mathematical analysis has been given in [16]. Figure 8.10 has been repeated here for the sake of a complete discussion of the topic.

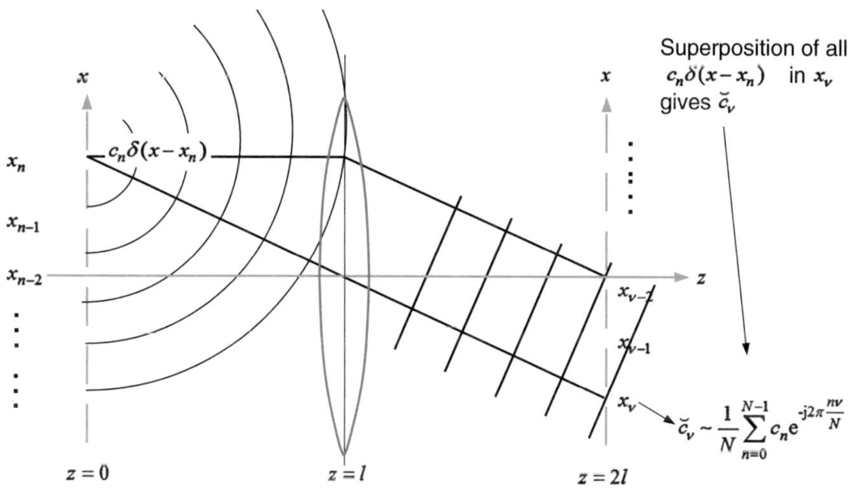

Fig. 8.10 Fourier transform lens with a focal length l mapping the input samples c_{in} at positions x_n of the input image at $z = 0$ onto output images at $z = 2l$, where each image happens to be the superposition of the input images with the phase weights of the DFT. The lens therefore acts as a Fourier lens as long as the paraxial approximation holds. The figure has been modified from [16]

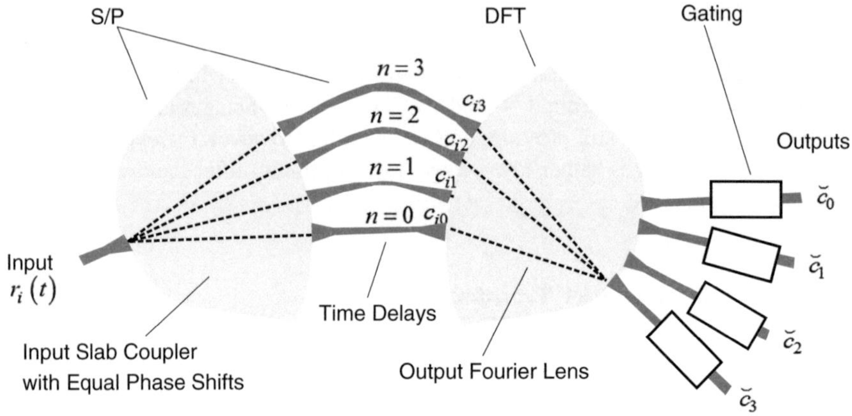

Fig. 8.11 DFT based on an arrayed waveguide grating router (AWGR). The input slab coupler with the arrayed waveguides acts as a S/P converter. The subsequent slab coupler at the output acts as a Fourier lens that superposes the input coefficients according to the DFT expression (figure from [16])

8.3.2.5 The Arrayed Waveguide Grating Router DFT

The function of a lens can be mimicked in integrated optics by means of an arrayed waveguide grating router (AWGR) [25]. A discussion of the AWGR as a Fourier lens for a discrete set of input samples has been given by Lowery [26].

In Fig. 8.11, the AWGR input slab coupler illuminates all grating waveguides with equal amplitude and phase. The increasing lengths of grating waveguides provide time delays incremented by T_s/N for providing a S/P conversion. This provides the N time samples $\breve{c}_{in}$ to specific locations of the second slab couplers. The slab coupler is then designed such that it acts as a Fourier lens mapping and superposing the N input samples $\breve{c}_{in}$ onto the output plane of the slab coupler. If the locations are properly chosen, the input samples will have equal weight with the phase coefficients of the DFT. Consequently, one can extract the DFT coefficients at the output of the AWGR.

Experimental demonstrations based on the AWGR concept have been given by various groups, so for instance by Lee et al. [27], Lowery et al. [28] and others. A challenge of the AWGR concept is that all the $N-1$ phase shifters have to be accurately adjusted relative to each other.

8.3.3 OFDM Tx and Rx—Experimental Implementations

In the previous sections we have studied the various possibilities to generate and receive OFDM signals. OFDM signals can be generated and detected by either using pulse-shaping techniques or by applying DFT techniques. Likewise we can either encode or decode OFDM in the electronic domain or the optical domain.

8.3.3.1 Electronic OFDM Implementations

Electronic OFDM transmitters are preferred when many lower bit-rate subcarrier signals are to be multiplexed to a total data-rate of a few 100 Gbit/s. A complete real-time OFDM transceiver operating at 11 Gbit/s has already been shown [29] and more recently the first 256 Gbit/s OFDM transmission has been demonstrated [30]. It is expected that electronic approaches will reach Tbit/s capacity in the future. Electronic implementations typically generate OFDM signals by carrying out a serial-to-parallel conversion of the input data stream and then performing the IDFT electronically as outlined in in the previous section. After performing the IDFT, the signal is usually clipped before being passed on to digital-to-analogue (DAC) converters. The real and the complex part of the signal (the so-called inphase and quadrature part) are then separately fed to an optical IQ-modulator in order to encode the electrical signal on an optical carrier. Electronic implementations may flexibly adapt the order of the FFT and the symbol rate to changing channel conditions and customer requirements. With such transmitters OFDM signals with very few up to a 1024 of subcarriers have been implemented in optical communication systems [31]. On the other hand, electronic OFDM transmitters have a speed limitation which is due to given electronic circuit speeds. Electronics also scales not very favorable with power consumption, particularly at higher speed.

Conversely, optical OFDM concepts have advantages when data channel speed exceed the speed offered by electronics. In this context they are most interesting in view of a next generation Tbit/s superchannel implementation. Various concept with 2, 4 or 8 OFDM subcarriers are currently discussed as possible candidates of a 1.2 Tbit/s superchannel. Likely scenario include the generation and reception of the higher speed subcarriers in the optical domain whereas the subcarriers themselves could carry lower speed data signals based on electronic OFDM or Nyquist multiplexed sub-sub carriers.

8.3.3.2 Optical OFDM Implementations

Optical OFDM subcarrier generation and reception however will only be pursued if signals can be generated and received at highest quality with the least effort. We have evaluated various all optical concepts such as discussed above and now pick the one concept that in our opinion works most reliably with the least effort.

For the all-optical generation of an OFDM signal we have selected the OFDM transmitter based on optical pulse shaping which follows the general concept outlined in Fig. 8.4a. In this concept the OFDM signal is generated by applying (8.6). In Fig. 8.12a we have depicted an example for an implementation of such an optical OFDM transmitter. The transmitter comprises of lasers emitting at frequencies f_0, f_1, ..., f_{N-1} equispaced by F_s. The carriers are encoded with information $\breve{c}_{i0}$, $\breve{c}_{i1}$ and $\breve{c}_{iN-1}$ by means of an optical IQ-modulator. In this implementation, the IQ-modulators serve two purposes. They encode information and carve a

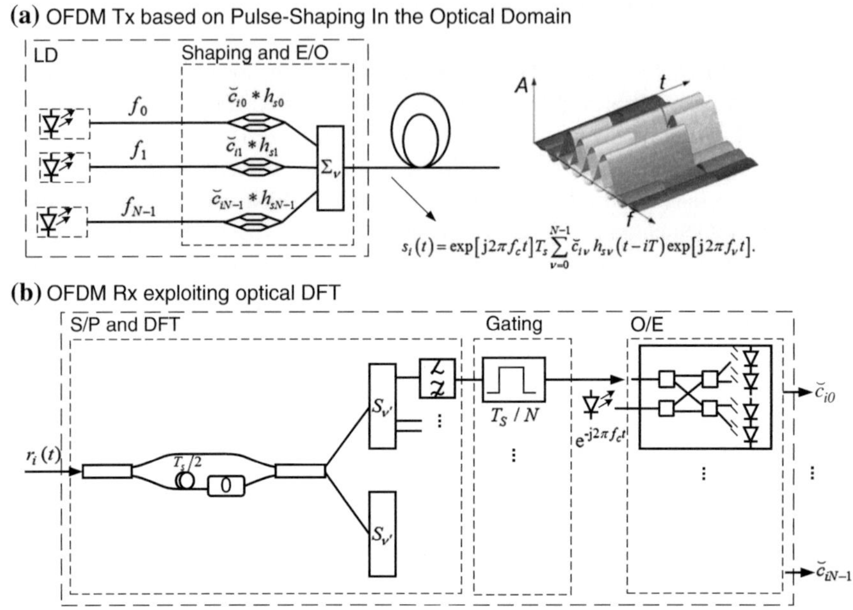

Fig. 8.12 Schematic of an optical OFDM transmitter and receiver. **a** Optical OFDM transmitter based on the optical pulse shaping concept. **b** OFDM receiver relying on the cascaded delay interferometer approach with a Gaussian filter to perform the FFT in the optical domain

rectangularly shaped pulse h_s onto the carriers. Finally, the subcarriers are multiplexed e.g. by guiding them through an optical combiner.

The OFDM receiver on the other hand is based on an implementation of the DFT. We have chosen to do so, because receiver pulse shaping in the optical domain is not straight forward and digital signal processing at highest speed is not really an option. Implementation of the all-optical DFT approach is however quite straight forward and works at almost any speed. In Fig. 8.12b we show the implementation that has first been introduced in [10] and that ultimately was used to process a 26 Tbit/s OFDM superchannel [9]. In this work a single stage DI followed by a flat-top passpand has been used, see Fig. 8.9b. The advantage of this concept is that this way a single DI is sufficient to first demultiplex the OFDM signal into even and odd channels. This DI also is the first stage in the cascaded delay interferomter FFT approach discussed in Fig. 8.9b. Subsequently, the second stage filter only needs to suppress the subcarriers in every 2nd channel. A simple Gaussian bandpass filter or flat-top fill can do it. Finally, gating with a Gating window of a length $T_s/2$ is applied and the optical signal is down converted into the electronic domain by means of a coherent receiver.

8.3.3.3 Optical OFDM of a 26 Tbit/s Signal

To demonstrate the strength of the optical OFDM techniques we would like to conclude this section with a brief discussion of the 26 Tbit/s optical OFDM generation, transmission and reception experiment published in [9].

The experimental setup is shown in Fig. 8.13. Rather than multiple lasers, a single mode-locked laser (MLL) was used to generate an optical comb with a total of 325 subcarriers and a frequency separation $\Delta f = 12.5$ GHz. The frequency comb with its large amount of subcarriers has been generated by launching the MLL in a highly nonlinear fibre for broadening [32]. Subsequently, a programmable optical filter equalizes the power of the subcarriers and separates them into even and odd subcarriers. Both sets of subcarriers are then individually modulated with independent 16QAM pseudo-random bit sequences (PRBS 2^{15}-1).

The OFDM signal with a power of 9.5 dBm is then launched into a single-mode fibre SMF-28 with a length of 50 km and a span loss of 10.2 dB. After transmission, the signal is amplified and the dispersion of the link is compensated in a dispersion compensation module. The total insertion loss of the module was 4.8 dB and the residual dispersion of the link was 17 ps/nm.

The receiver is based on the optical FFT using a single delay interferometer followed by a filter in the form of a wavelength-selective switch (Finisar Waveshaper). The time gating was performed by an electro absorption modulator (EAM—CIP 40G-PS-EAM-1550) at 12.5 GHz. The subcarriers are then converted to the electrical domain by a coherent receiver in the form of an OMA (Agilent N4391A) with an electrical bandwidth of 16 GHz (Agilent Infiniium DSOX93204A). In the OMA, the signal is equalized with a 21 tap equalizer. Typical constellation diagrams are depicted in Fig. 8.13c. To assess the quality of the received signal in each polarization, the error vector magnitude is calculated and averaged over 10 received sequences of 1024 symbols. The error vector magnitude (EVM) is an important quality metric for complex signals—similar to what the quality factor is for on-off keying signals [33]. The EVM measurement was carried out for all 325 subcarriers of the OFDM signal, see Fig. 8.13d. It can be seen that the EVM for all subcarriers after transmission are between 11 and 16 % with an average EVM of 14.2 %. The estimations indicate that all 325 subcarriers are below the BER $= 1.9 \times 10^{-2}$ third generation FEC limit with 20 % overhead [34]. The demonstrated line rate of 26 Tbit s^{-1} therefore corresponds to a net error-free data rate of 21.7 Tbit s^{-1}. The achieved spectral efficiency for the modulated subcarriers is 6.3 bit/s/Hz.

8.4 Nyquist Pulse Shaping

The combination of several lower-rate channels with high spectral efficiency into Tbit/s super-channel can be achieved in the time domain by appropriate pulse shaping. For Nyquist transmission with a maximum spectral efficiency, symbols are

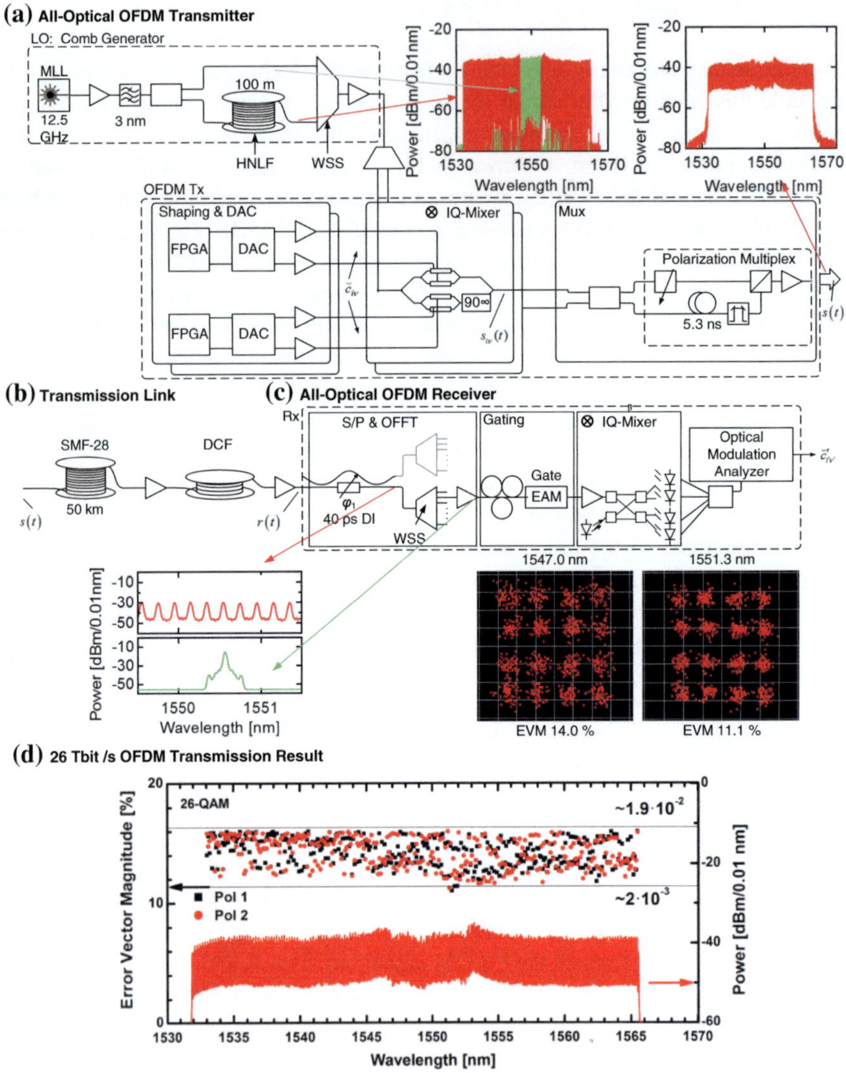

Fig. 8.13 Setup of the single laser 26 Tbit/s OFDM experiment with **a** optical OFDM transmitter, **b** transmission link, **c** all-optical FFT receiver and **d** a plot with error-vector magnitudes of each and every subcarrier. Figure modified from [9, 16]

carried by Nyquist pulses [11] that meet the condition of (8.1). If the maximum spectral efficiency is not needed, the Nyquist criterion for intersymbol interference (ISI) is good enough, according to which impulse maxima coincide with the zeroes of adjacent pulses such that the communication channel is ISI free. Nyquist pulses that are shaped to meet the Nyquist criterion but do not offer the maximum spectral efficiency have already been used to demonstrate some quite impressive data rate transmission experiments [35–38]. Compared to OFDM, Nyquist pulse shaping has several unique advantages: reduction of the transmitter and receiver complexity [39–41], lower receiver bandwidths requirement [42], lower peak-to-average power ratios [11, 40], reduced sensitivity to fibre nonlinearities [41]. Nyquist signaling is also known as an orthogonal time-division-multiplexing technique, because although each symbol pulse spreads into many adjacent time slots, no ISI occurs if the sampling is performed at the center of each symbol. The Nyquist criterion for a channel impulse response $h(t)$ satisfying the zero ISI implies that, for a particular sampling period T_S and any integer i:

$$h(iT_s) = \begin{cases} 1; & i = 0 \\ 0; & i \neq 0 \end{cases} \tag{8.21}$$

When such a signal is periodically sampled with a period T_S, a nonzero value is obtained only at the time of origin. In the frequency domain, the Nyquist criterion is equivalent to

$$\frac{1}{T_S} \sum_{k=-\infty}^{\infty} H\left(f - \frac{k}{T_s}\right) = 1 \tag{8.22}$$

with $H(f)$ the Fourier transform of $h(t)$. From (8.22) it is intuitively understood that aliased versions of the frequency domain $H(f)$ must add up to a constant. These two concepts, time and frequency Nyquist criterion, are illustrated in Fig. 8.14a, b, respectively. The pulse in Fig. 8.14a is at its maximum when $t = 0$, while null at all other sampling locations corresponding to $t = iT_S$, with i an integer. The maxima of adjacent pulses coincide with zeroes of the initial pulses while the sum of all other pulses at $t = 0$ is equal to zero, thus leading to no ISI at this specific sampling time. In the frequency domain, i/T_S frequency shifted versions of the pulse spectrum add up to a constant. From such concept, it can right away be inferred that the minimum bandwidth Nyquist pulse would have a rectangular spectrum.

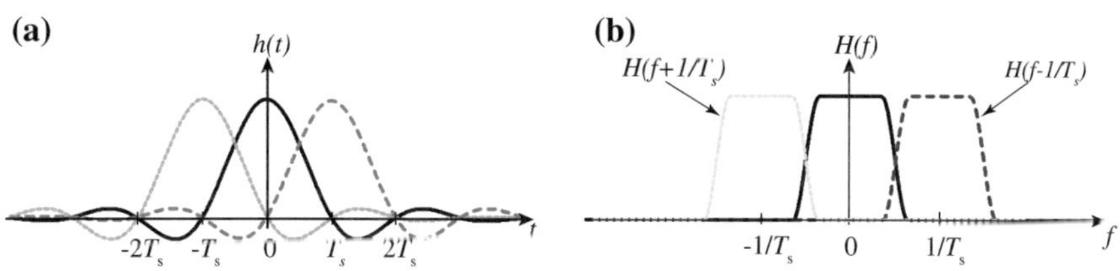

Fig. 8.14 **a** Raised cosine response (*solid line*) meets the Nyquist ISI criterion. The maximum is at $t = 0$ while $h(t)$ is zero at all other sampling locations given by $t = iT_S$, i an integer. Consecutive raised cosine impulses (*dashed lines*) demonstrate the zero ISI property: at $t = 0$, the sum of all other impulses is zero. **b** Aliased versions of the frequency domain impulse add up to a constant

The general expressions for the pulse waveform $h(t)$ and corresponding spectrum $H(f)$ are given by the impulse response and transfer function of a raised cosine Nyquist filter, respectively [43, 44]:

$$h(t) = \frac{\sin\left(\frac{\pi t}{T_s}\right)}{\frac{\pi t}{T_s}} \frac{\cos\left(\frac{\beta \pi t}{T_s}\right)}{1 - \left(\frac{2\beta t}{T_s}\right)^2} \tag{8.23}$$

$$H(f) = \begin{cases} T_s & |f| \le \frac{1-\beta}{2T_s} \\ \frac{T_s}{2}\left[1 + \cos\left(\frac{\pi T_s}{\beta}\left[|f| - \frac{1-\beta}{2T_s}\right]\right)\right] & \frac{1-\beta}{2T_s} \le |f| \le \frac{1+\beta}{2T_s} \\ 0 & \text{otherwise} \end{cases} \tag{8.24}$$

The factor β is known as the roll off factor [43], which is in the range $0 \le \beta \le 1$. Amongst the class of Nyquist pulses [11], the sinc-shaped pulse obtained for $\beta = 0$ is of particular interest owing to its rectangular spectrum. It is therefore the minimum bandwidth Nyquist pulse and such feature allows minimizing the guard band between optical channels. As the roll off increases, the required bandwidth increases but a smoother time response is obtained, see Fig. 8.15.

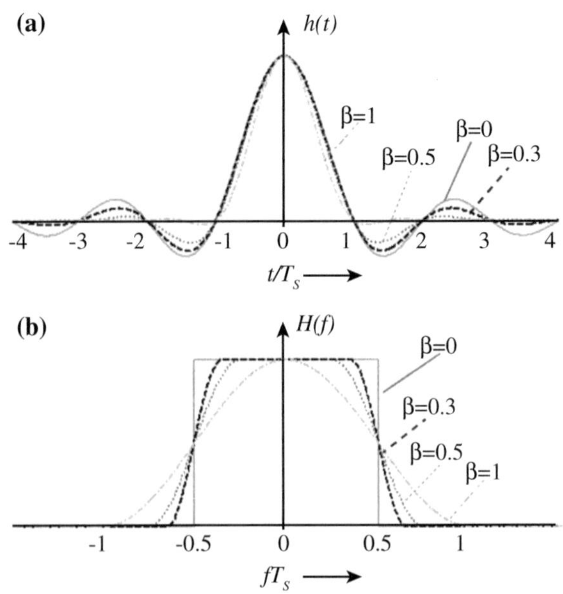

Fig. 8.15 **a** The pulse waveform h(t) is given by the impulse response of a raised cosine filter shown here for different values of the roll-off factor β. The time response becomes smoother as the roll-off increases. **b** The corresponding spectrum H(f), also shown for different values of β, shows that the required bandwidth increases with β while a rectangular spectrum is obtained for $\beta = 0$, corresponding to the sinc-shaped pulse

Theoretically for a sinc-pulse Nyquist transmission, each symbol consists of a time unlimited sinc-pulse. As a consequence of causality, the sinc function $\text{sinc}(t) = \sin(\pi t)/\pi t$ is therefore only a theoretical construct. Instead, periodic pulses are typically used in most experimental demonstrations of Nyquist pulse transmission [13, 39–42, 44–46]. Several approaches for the generation of Nyquist pulses have been suggested and can be divided into three main categories depending on whether pulse shaping is performed in the electrical [38], digital [35] or optical domain [40]. Electronic Nyquist processing consists in utilizing analog electrical filters to form an appropriate output signal with an ideal rectangular shaped spectrum. Digital signal processing is based on the ISI free shaping of a baseband signal in a digital signal processor (DSP) that is part of the transmitter. Finally, Nyquist pulses can be directly shaped in the optical domain by relying on optical filtering or frequency comb synthesis. While at low symbol rates (<25 Gbaud) all shaping methods are available, at higher symbol rates, only analog or optical methods are at hand [47]. Additionally, optical pulse shaping can lead to pulses with much shorter duration (in the picosecond or less). The three methods are described in the following sections.

8.4.1 Electronic Nyquist Processing

Pulse shaping can be done in the electrical domain by solely relying on an electrical filter to approximate sinc-shaped pulses by analog means. The electrical filter shapes the electrical drive signals that are then fed to the optical modulator as shown in Fig. 8.16 for a software defined transmitter.

This technique was applied to demonstrate the long haul transmission of 28 Gbaud 16QAM signals on a 50 GHz grid, achieving a spectral efficiency of 4 bits/s/Hz [38]. Passive low pass electrical filters with bandwidths of 14 GHz were used for the pulse shaping in order to reduce spectral overlap between adjacent wavelength division multiplexed channels. For the generation of sinc-shaped pulses, the electrical filter should have very steep slope, in order to get as close as possible to a rectangular spectrum. However such filters are not easily available while a complex transmitter would necessitate multiple of these analog filters.

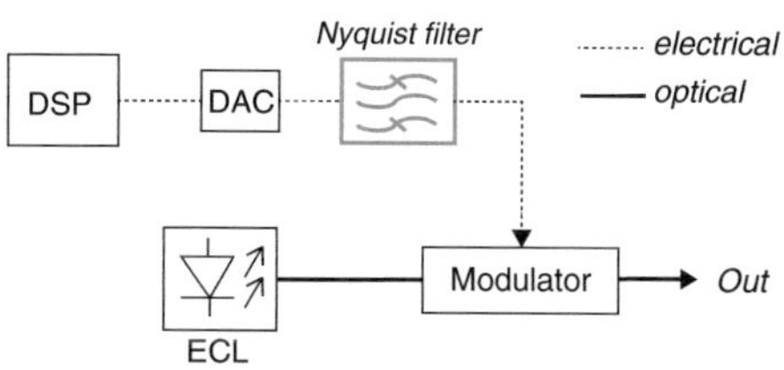

Fig. 8.16 Setup for Nyquist pulse shaping in the electrical domain using an electrical filter. The electrical filter solely performs pulse shaping of the signals. The electrical signals are then fed to an optical modulator which encodes the data onto an external cavity laser (ECL)

In [47], pulse shaping in the electrical domain was investigated in a setup generating 20 GBd QPSK. In addition to Nyquist pulse shaping at the transmitter, the receiver was optimized with a sophisticated equalization technique based on Nyquist brick-wall filtering. The filter used in the experiment had 3 and 20 dB cut-off frequencies of 12.3 and 12.5 GHz, respectively as from the simulated S-parameters in Fig. 8.17a. A typical received and assembled averaged power spectrum of a 20 GBs QPSK signal at an optical signal to noise ratio of 30 dB before and after the receiver DSP is shown in Fig. 8.17b, top and bottom respectively. The individual pulse form was derived from the un-equalized signal spectrum. The electrical pulse shaper produces sinc-typical side lobes but they decay rapidly as seen in Fig. 8.17c.

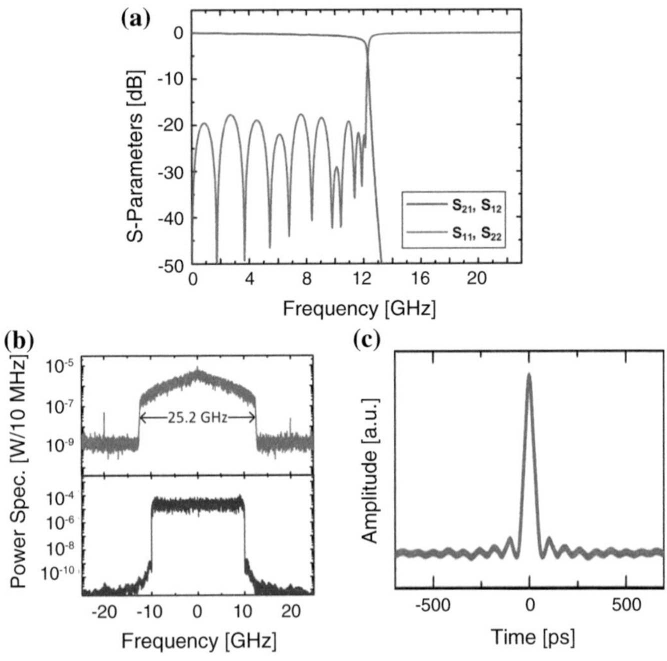

Fig. 8.17 **a** Simulated S-parameters of the electrical low pass filter employed in [47]. The measured cut-off frequency is ≈12.3 GHz and reflections are suppressed by at least 20 dB. **b** Measured and assembled averaged spectrum of a 20 GBd electrically generated QPSK signal before (*top*) and after (*bottom*) receiver DSP. As expected the overall bandwidth corresponds to two times the electrical filter bandwidth. **c** Impulse response measured from the received un-equalized spectrum. The electrical pulse shaper produces sinc-typical side lobes but with a faster decay. Figure modified from [47]

8.4.2 Digital Signal Processing Based Generation

The use of electronics within the optical transceiver, such as field programmable gate arrays (FPGA), high speed digital-to-analog and analog-to-digital converters (DAC, ADC), has enabled real time signal processing and has facilitated the use of advanced modulation formats in optical communication systems. Such approach is commonly applied to optical orthogonal frequency division multiplexing (O-OFDM) for the demonstration of real time transmission with very compact spectrum [9, 48]. Similarly, software defined optical transmitters acting as arbitrary waveform generator (AWG) programs can be used for the real-time generation and transmission of Nyquist pulses [13]. Nyquist pulses are generated by ISI free shaping of a baseband signal in a digital signal processor (DSP) that is part of the transmitter as shown in Fig. 8.18.

Electrically generated Nyquist pulses are shaped with a finite duration impulse response (FIR) filter. While an elementary Nyquist shaped impulse with minimum spectral width is a sinc-function extending infinitely in time, real Nyquist pulses need to be approximated by a finite length representation. In addition, the baseband spectrum needs to be separated from its periodic repetitions (aliases) using realizable filters. To such ends, oversampling by a factor q is needed. The standard oversampling factor of $q = 2$ saves FPGA resources since sampling points of adjacent symbols fall onto the same time slot but limits the symbol rate to half of the sampling frequency of DACs. Smaller oversampling factors, such as $q = 1.33$, have been demonstrated but at the cost of increased processing complexity [30].

An FIR filter of order R can be constructed by a sequence of R delay elements of value T_s/q and $R + 1$ taps in between as shown in Fig. 8.19. The tapped signals are weighted by the filter coefficients k_r, also called tap weights, and summed up to form the filter output. When $R = 0$, the filter simply reproduces the input. As the filter order gets larger, the closer the spectrum can approach a rectangle and thus reaching roll-off factors close to 0.

The order R of the filter has a significant effect on the spectrum. Signal generation with various FIR filter orders R and $q = 2$ are shown in Fig. 8.20, see [30]. The left column shows the simulated impulse response of each filter. The effective

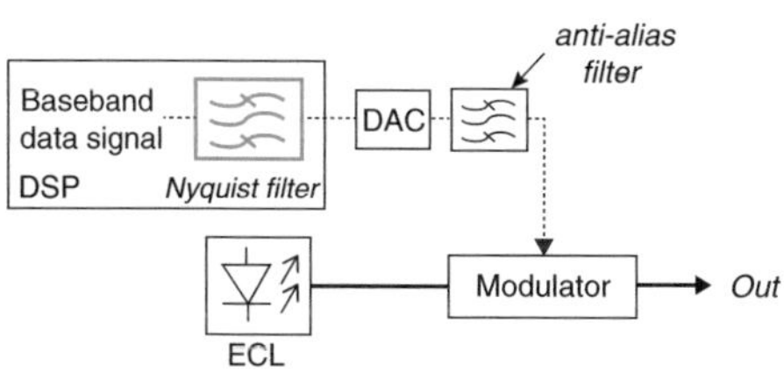

Fig. 8.18 Shaping (pre-filtering) of the baseband data signal for Nyquist pulse shaping in the digital domain. The digital filter is part of the digital signal processing (DSP) in the transmitter. Pulse shapes can be programmed offline. An off-the-shelf analog electrical filter is used after the digital-to-analog conversion to remove spurious spectra. The electrical signal is then fed to an optical modulator which encodes the data onto an external cavity laser (ECL)

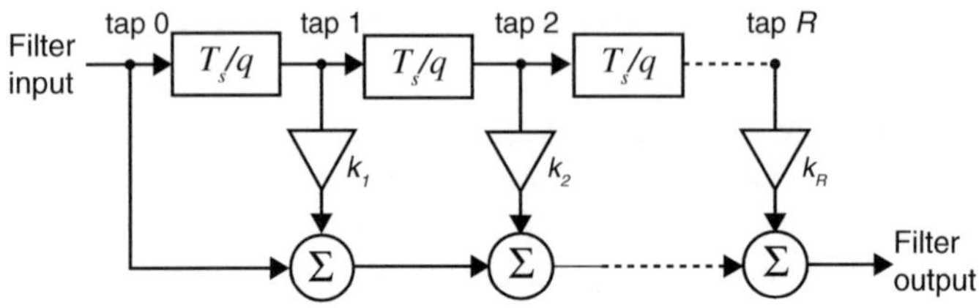

Fig. 8.19 A finite impulse response filter (FIR) of order R. The *top part* is the R stage delay line with $R + 1$ taps. A delay of T_s/q is positioned between each tap. The tapped signals are weighted by R tap weights k_r and summed to form the filter output. Figure modified from [30]

windowing is indicated by a green rectangle. The corresponding simulated transfer functions on a linear and logarithmic scale are shown in the second and third column, respectively. An FPGA based DSP was implemented to realize such real time digital Nyquist pulse transmitter. Within the FPGA, complex Nyquist pulses are calculated from a 2^{15}-1 pseudo random bit sequence in real time and signals pre-shaped by FIR filters with order 16 and 32 were implemented.

The convolution of the rectangular spectrum of an infinitely extended temporal sinc-pulse with the sinc shaped spectrum of the rectangular time window leads to the power spectra shown in the two middle columns. As the filter order increases, the spectrum evolves towards an ideal rectangle with a spectral width equal to the Nyquist bandwidth. For $R = 32$, a significant increase in spectral efficiency can be already obtained in comparison to NRZ modulation. Experimentally, this method can provide a very good roll-off factor of $\beta = 0.0024$ and has been applied for the demonstration of 512QAM Nyquist sinc-pulse transmission at 54 Gbit/s in an optical bandwidth of 3 GHz [14]. This method is however restricted by the speed of the electronics because of the limited sampling rate and limited processor capabilities, whereas the quality of the Nyquist pulse highly depends on the resolution of the digital-to-analog converters.

8.4.3 Optical Processors

Compared to pulse shaping in the electrical or digital domain, optical pulse shaping is not entirely limited by electrical components or processors capabilities. Optical techniques can also lead to pulse sequences with much shorter duration which can subsequently be multiplexed in time to reach an ultrahigh symbol rate. To optically generate Nyquist pulses, liquid crystal spatial modulators have been used to shape Gaussian pulses from a mode-locked laser into raised-cosine Nyquist pulses [44]. It is also possible to generate Nyquist pulses using fiber optical parametric amplification pumped by parabolic pulses in combination with a phase modulator to compensate the pump-induced chirp [46]. Additionally, sequences of very high

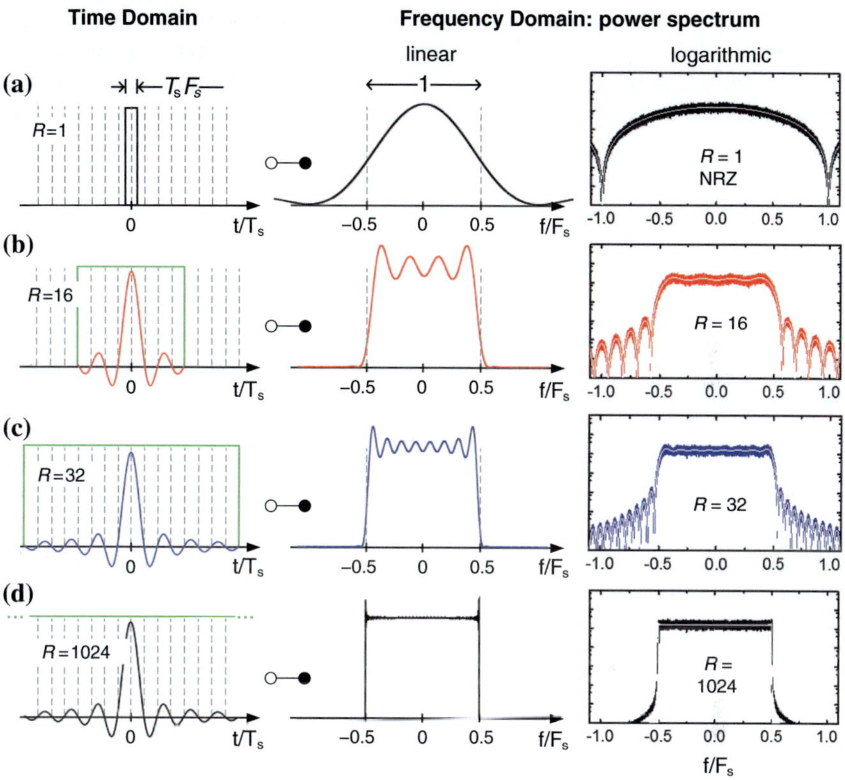

Fig. 8.20 Impulse responses and transfer functions of FIR filters with various orders R. The *first column* shows the simulated impulse responses. The *second column* shows plots of simulated power spectra on a linear scale and the *third column* shows the same spectra on a logarithmic scale. For **a** a single NRZ impulse shaped by a filter of order $R = 0$. **b** An ideal sinc pulse truncated by a filter with order $R = 16$. **c** An increased order filter to $R = 32$ resulting in a spectrum evolving towards an ideal *rectangular shape*. **d** A filter with very high order $R = 1024$ closely approximating a *rectangular shaped* spectrum. Figure has been modified from [30]

quality Nyquist pulses (β close to 0) can be obtained by the direct synthesis of a flat phase-locked frequency comb with high suppression of out-of-band components [49]. In the following sections, filtering based pulse shaping and frequency comb synthesis are discussed in more details.

8.4.3.1 Filtering Based Pulse Shaping

Filtering of the optical signal can be done in order to achieve the best possible transmission quality. As for electrical filters, the optical filter must result in a spectrum of rectangular shape by removing side lobes and flattening the region of interest. An optical band pass filter can be directly applied to a conventionally modulated signal [47]. In this configuration, the spectrum of the data signal is reshaped as a bandwidth limited data signal and the obtained pulse width remains close to identical to the allocated bit interval. Optical interleavers can also be used for fixed-frequency optical filters with a nearly rectangular frequency response. An alternative for a more flexible system is the use of a programmable optical filter, which comprises of general imaging optics, a diffraction grating and liquid crystal on silicon (LCoS) cells for shaping the phase of the optical spectrum [50].

In view of the high potential of optical filtering methods for the generation of ultra-short duration pulses, Nakasawa et al. proposed a slightly different approach consisting in the synthesis of a Nyquist pulse train at a repetition rate F_s in which the pulse width is set to be much shorter than the bit interval given by $1/F_s$ [44]. The output of an initial optical pulses source, such as a modelocked laser, with a repetition rate F_s and Gaussian pulses, is shaped by the liquid crystal spatial modulators into raised-cosine Nyquist pulses. The concept is illustrated in Fig. 8.21. For a spectrum reshaped around the optical carrier f_c with a width of NF_s, the symbol period defined as the peak-to-first-zero time interval is then given by $T_s = 1/(NF_s)$.

Optical Nyquist pulse generation based on this filtering technique produces much higher roll-off factors compared with the electrical method because optical

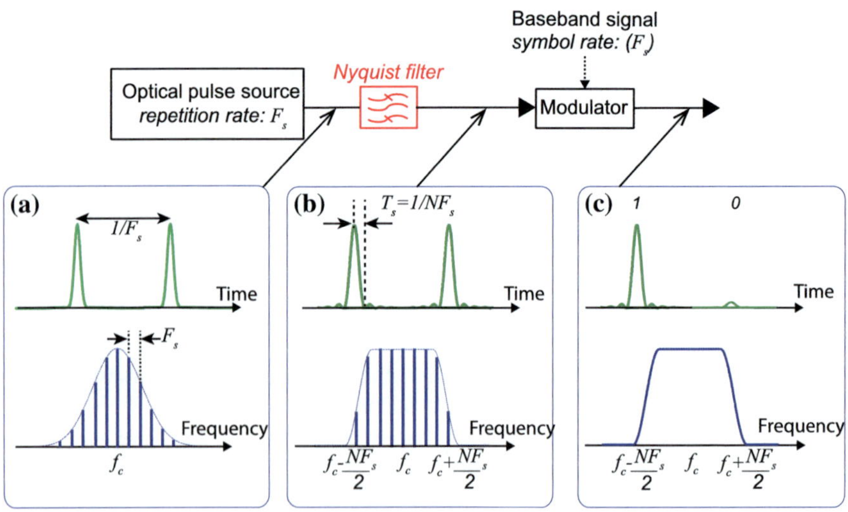

Fig. 8.21 Schematic of pulse shaping of the output of a pulses source with Gaussian shaped pulses into sinc-shaped Nyquist pulse train using an optical filter. **a** An optical pulse source with repetition rate F_s outputs a train of Gaussian shaped pulses with repetition period $1/F_s$ and corresponding optical spectrum comprising of several spectral lines spaced by F_s and centered at the carrier frequency f_c. **b** The optical filter reshapes line by line the spectrum of the pulse train such that it is confined to a bandwidth of NF_s, with N the number of spectral lines. The corresponding pulse train shows sinc-like oscillations with a peak-to-first-zero spacing of $1/NF_s$. **c** The obtained pulse train can then be modulated by going through an optical modulator driven by the electrical signal, here on-off-keying shown

filters only produce slopes with limited steepness. A programmable optical filter like a Waveshaper has a minimum 3 dB bandwidth of around 12 GHz and a 30 dB bandwidth of around 40 GHz. The roll-off factor also strongly depends on the overall bandwidth of the signal and typically ranges between 0.15 and 0.5 [51, 52]. However the obtained pulses are characterized by a peak-to-first zero-crossing separation of $1/NF_s$ for a $1/F_s$ repetition period, such that the Nyquist pulse stream can be multiplexed in time domain to increase the overall transmission rate by a factor of N. This is designated as orthogonal optical time-division multiplexing (OTDM). The generated sequence is split into N channels, which are then delayed and modulated to transport the channel corresponding data. This requires N modulators, with N being the number of branches or the number of time domain channels. However compared with direct modulation, the baud rate of each modulator is N times reduced which drastically relaxes the requirements on modulators and electronics (Fig. 8.22).

Let's define the time interleaved optical Nyquist pulses as $g_i(t) = h(t - iT_s)$, where $h(t)$ is given by (8.23), and the symbol data at $t = iT_s$ as c_i. For example, if we consider on-off-keying modulation c_i can take value 1 or 0. When $\beta = 0$, according to the sampling theorem, the time division multiplexed Nyquist data train $s(t) = \sum c_i g_i(t)$ is equivalent to an analog signal $s(t)$ given by the interpolation of c_i

$$s(t) = \sum_i s(iT_s) g_i(t) \tag{8.25}$$

The relationships between $s(t)$, $g_i(t)$ and $c_i = s(iT_s)$ are shown in Fig. 8.23, for the case $i = 4$. Time division multiplexing of Nyquist pulses can be viewed as a digital-to-analog conversion from c_i to $s(t)$ in the optical domain. Based on this property, the symbol data c_i can be extracted from the time division multiplexed stream by use of narrow optical sampling gate at $t = iT_s$, as illustrated by the grey box in Fig. 8.23. Optical sampling gates based on nonlinear effects have been successfully used for the demultiplexing of time division multiplex Nyquist pulses proving the potential of such technique, despite the fact that the sampling gate cannot be an ideal delta function thus leading to leakage from adjacent time channels. The simple example drawn in the figure, also illustrates the orthogonality of the time

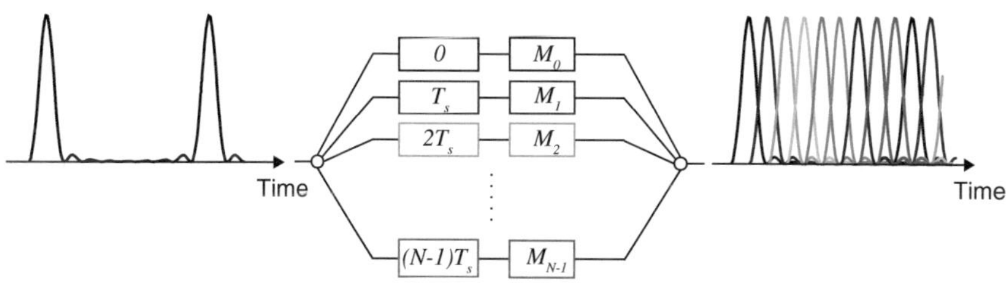

Fig. 8.22 Periodic sinc-pulse sequences can be split into N branches, each of which corresponds to an independent channel. In the nth branch, the sequence is delayed by n times the interval $T_s = 1/NF_s$, with $n = 0, 1\ldots N-1$. Each channel can be modulated independently with a modulator $M_0,\ldots M_{N-1}$ with any modulation formats. The shown multiplexing is carried out in the time domain at one carrier wavelength

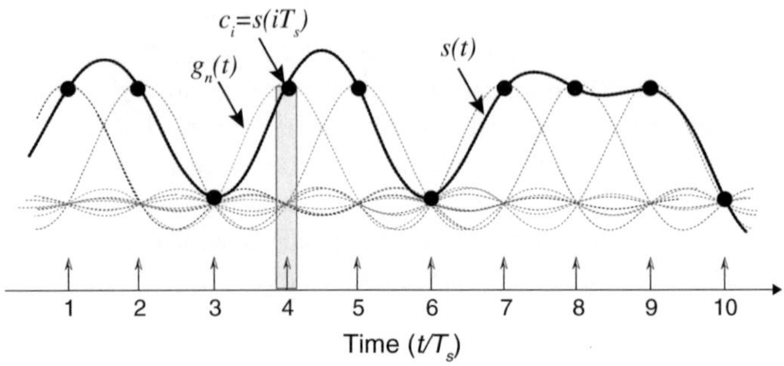

Fig. 8.23 Illustration of orthogonal time division multiplexing of Nyquist pulse train. The individual channels $g_i(t)$ (*dashed lines*) are time multiplexed with a delay multiple of $T_s = 1/NF_s$. At times $t = iT_s$, the symbol data is given by c_i (in this example 0 or 1). The time division multiplexed Nyquist data train is $s(t)$. The symbol data d_i can be extracted by optical sampling of $s(t)$ at $t = iT_s$. The narrow gate (shown here for $i = 4$) and the orthogonal properties of the Nyquist channels enable recovering of the data despite strong overlap between channels

interleaved optical Nyquist pulses in spite of the strong overlap with adjacent pulses. Indeed the time interleaved optical Nyquist pulses $g_i(t) = s(t - iT_s)$ satisfy the orthogonality condition when $\beta = 0$

$$\frac{1}{T_S}\int_{-\infty}^{+\infty} g_i(t)g_m(t)dt = \begin{cases} 0\,(i \neq m) \\ 1\,(i = m) \end{cases} \tag{8.26}$$

For a general value of β, the integral of (8.26) has generally a non-zero value for $i \neq m$ and the orthogonality is not rigorously satisfied theoretically. However, the integral converges to 0 for large $|i - m|$ such that the quasi-orthogonality property is maintained even when $\beta \neq 0$.

8.4.3.2 Frequency Comb Based Nyquist Pulse Train Generation

Considering that owing to physical limitations the ideal sinc pulse with perfect rectangular optical spectrum has not been demonstrated yet, a different optical approach was proposed by Soto et al. [49]. The technique is a straightforward way to realize sinc-shaped Nyquist pulses in the optical domain and is based on the time-frequency duality described by Fourier domain. It is well known that a sinc pulse can be represented by a rectangular spectrum in the Fourier domain, while the frequency content of a train of sinc pulses corresponds to a flat comb with equally spaced components within the bandwidth defined by the single-pulse spectrum (Fig. 8.24). Therefore instead of shaping a single sinc pulse, a sequence of sinc pulses is directly produced by the generation of an optical frequency comb having uniformly

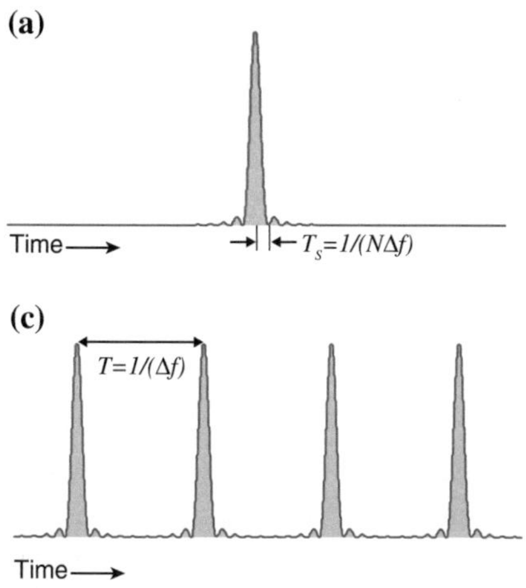

Fig. 8.24 Time (*left*) and frequency (*right*) representation of a single-sinc pulse (*top*) and a sinc-pulse sequence (*bottom*). The figure shows the intensity in the time domain since the directly observed quantity is proportional to the intensity. The Fourier domain representation of a sinc pulse (**a**) is a rectangular function (**b**), while the spectrum of an unlimited sing-pulse sequence (**c**) is a frequency comb with uniform phase under a rectangular envelope (**d**)

spaced components with narrow linewidth, equal amplitude and linear phased locked phase, together with strong out-band suppression [26]. While properties similar to that of the filtering based methods are expected, such as the possibility of utilizing such generated Nyquist pulse train for orthogonal time division multiplexing, this method leads to the generation of a sequence of very high quality Nyquist pulses with an almost rectangular spectrum (β close to 0).

The time-domain representation of the optical field of a frequency comb with N lines, having the same amplitude E_0/N and frequency spacing Δf around the central frequency f_0 can be expressed as

$$\begin{aligned} E(t) &= \frac{E_0}{N} \sum_{n=-\frac{N-1}{2}}^{\frac{N-1}{2}} e^{2i\pi(f_0+n\Delta f)t+i\phi} = \frac{E_0}{N} e^{2i\pi f_0 t+i\phi} \sum_{n=-\frac{N-1}{2}}^{\frac{N-1}{2}} e^{2i\pi n\Delta ft} \\ &= E_0 \frac{\sin(\pi N\Delta ft)}{N\sin(\pi\Delta ft)} e^{2i\pi f_0 t+i\phi} \end{aligned} \tag{8.27}$$

This mathematical demonstration if given for an odd number of frequency lines and the derivation for an even number can be straightforwardly obtained following the same procedure. For simplicity, it is assumed that all frequency components have the same phase ϕ. It is however sufficient that all frequency components be locked showing a linear dependence. From (8.27), the normalized envelope of the optical field is calculated to be $\sin(\pi N\Delta ft)/N\sin(\pi\Delta ft)$ referred thereafter as periodic sinc function. Using Fourier transform, the frequency domain representation is given by

$$\Im\left\{\frac{\sin(\pi N\Delta ft)}{\sin(\pi\Delta ft)}\right\} = \Im\left\{\sum_{n=-\frac{N-1}{2}}^{\frac{N-1}{2}} e^{2i\pi n\Delta ft}\right\} = \sum_{n=-\frac{N-1}{2}}^{\frac{N-1}{2}} \delta(f - n\Delta) \tag{8.28}$$

By introducing the rectangular function $\Pi(n/N)$ that is 1 for all integers n where $|n| \leq (N-1)/2$ and 0 elsewhere, (8.28) can be written as

$$\sum_{n=-\frac{N-1}{2}}^{\frac{N-1}{2}} \delta(f - n\Delta) = \Pi\left(\frac{f}{N\Delta f}\right) \sum_{n=-\infty}^{+\infty} \delta(f - n\Delta f) \tag{8.29}$$

where the rectangular spectrum $\Pi(f/N\Delta f)$, covering a bandwidth $N\Delta f$, is represented in the time domain by a sinc-pulse $N\Delta f\,\mathrm{sinc}(N\Delta ft)$. The temporal dependence of the above expression can then be obtained by taking its inverse Fourier transform and using the Poisson summation formula

$$\Im^{-1}\left\{\Pi\left(\frac{f}{N\Delta f}\right) \sum_{n=-\infty}^{+\infty} \delta(f - n\Delta f)\right\} = N\,\mathrm{sinc}(N\Delta ft) \otimes \sum_{n=-\infty}^{+\infty} \delta\left(t - \frac{n}{\Delta f}\right) \tag{8.30}$$

where $\otimes$ denotes the convolution operation. Considering the right hand side of (8.30)

$$N\,\mathrm{sinc}(N\Delta ft) \otimes \sum_{n=-\infty}^{+\infty} \delta\left(t - \frac{n}{\Delta f}\right) = \sum_{n=-\infty}^{+\infty} N\,\mathrm{sinc}\left(N\Delta f\left(t - \frac{n}{\Delta f}\right)\right) \tag{8.31}$$

From (8.28)–(8.31) we can conclude that

$$\frac{\sin(\pi N\Delta ft)}{N\sin(\pi\Delta ft)} = \sum_{n=-\infty}^{+\infty} \mathrm{sinc}\left(N\Delta f\left(t - \frac{n}{\Delta f}\right)\right) \tag{8.32}$$

Following similar procedure, the envelope of the optical field in the case of an even number of frequency lines can also be expressed as a train of sinc pulses. The generalized expression for the normalized envelope of the optical field resulting from a flat frequency comb, independent of the parity of N

$$\frac{\sin(\pi N\Delta ft)}{N\sin(\pi\Delta ft)} = \sum_{n=-\infty}^{+\infty} (-1)^{(N-1)n}\,\mathrm{sinc}\left(N\Delta f\left(t - \frac{n}{\Delta f}\right)\right) \tag{8.33}$$

The $(-1)^n$ factor appearing for even values of N comes from the absence of a spectral line at the central frequency of the comb, eliminating the direct current (DC) component on the optical field envelope. The difference in the periodic sinc function $x(t) = \sin(\pi N\Delta ft)/N\sin(\pi\Delta ft)$ for even and odd number of frequency components N is depicted in Fig. 8.25. For odd values of N, all sinc pulses of the

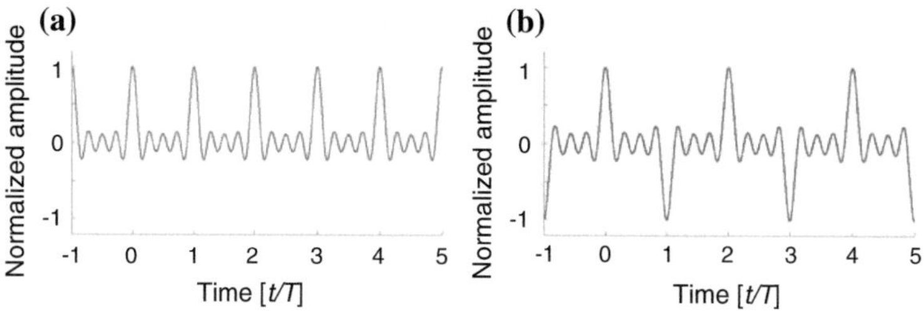

Fig. 8.25 Normalized field envelope of a frequency comb for **a** an odd (N = 9) and **b** an even (N = 8) number of spectral lines. An odd number of lines leads to a sequence of in-phase sinc shaped Nyquist pulses, while an even number results in a sequence of alternated π-phase-modulated pulses

pulse train show the same phase and $x(t_s) = 1$ at every sampling instant $t_s = n/\Delta f$ for all integer n. For even N, each pulse envelope is of opposite sign with its preceding and following pulses and $x(t_s) = (-1)^n$. However, for either odd or even N the optical intensity measured by a photodetector is the same and given by $I(t) = |E(t)|^2$.

Consequently the field envelope of the time domain representation of a frequency comb of N identical and equally spaced lines corresponds to an infinite summation of sinc-shaped Nyquist pulses with period $1/\Delta f$ and symbol period $1/N\Delta f$. Considering that the pulse repetition period $T_s = 1/\Delta f$ is a multiple of the time interval $T = 1/N\Delta f$, the resulting time-domain envelope $x(t)$ satisfies the following condition for any integer m

$$x(m\tau) = \begin{cases} (-1)^{(N-1)m/N} & m = \ldots, -2N, -N, 0, N, 2N\ldots \\ 0 & \text{otherwise} \end{cases} \tag{8.34}$$

The sequence of sinc pulses resulting from a phase-locked, rectangular frequency comb satisfies the Nyquist criterion for free ISI within every pulse repetition period T_s. The generated sinc-pulse sequence can be multiplexed in time without ISI, as described in the previous section. In addition, since the time multiplexed channel shows a sharp-edged spectrum, the next wavelength channel can be directly adjacent to the previous with almost not guard band while being multiplexed in the time domain the same way, reaching high temporal and spectral densities together.

There are several approaches for the generation of a frequency comb. They can be obtained from conventional femtosecond lasers [32, 53–55] or from continuous wave optical source exploiting Kerr nonlinearities in optical resonators [56, 57], employing a combination of intensity and phase modulation [58], intensity and phase together with dispersive medium [59], chirped Bragg gratings [60] or highly nonlinear fibers [61–63]. However not every comb does necessarily result in a sequence of Nyquist pulses. The sinc pulse sequence can only be obtained under specific conditions, namely line amplitudes as equal as possible, linear phase dependence through all lines and strong out-of-band lines rejections. Close to ideal rectangular shaped optical frequency combs can be produced from a non-optimal frequency comb in combination with a spectral line-by-line manipulation of the optical Fourier components. A simple method consisting of a cascade of intensity modulator will be presented in the following section.

8.4.4 Implementations

In this section, we describe the implementations of optical pulse shaping using a programmable filter [44, 52, 64, 65] and frequency comb synthesis [66].

The transmitter in [44] consists of a 40 GHz modelocked fiber laser. The output pulses of the source are generated at 1540 nm and are characterized by a Gaussian shape with 1.8 ps width. The spectrum was initially broadened by self-phase modulation (SPM) in a highly nonlinear fiber in order to enable the generation of short Nyquist pulse. The Nyquist pulses with a peak-to-first-zero time $T_s = 1.56$ ps and roll-off $\beta = 0.5$ are generated from this pulsed source by manipulating the spatial intensity and phase of the modelocked laser spectral components using a liquid crystal spatial modulator (Waveshaper). The output of the modelocked laser shown in Fig. 8.26a is therefore reshaped according to the optical filter transfer function shape depicted in Fig. 8.26b. The resulting Nyquist pulse spectrum and waveform, Fig. 8.26c, d, respectively, have the typical raised-cosine profile. Due to the limited steepness of the optical filter, the spectrum deviates from the ideal rectangular spectrum resulting in damped sinc-like oscillations in the time domain. These pulses where multiplexed to a 1.28 Tbit/s polarization-multiplexed DPSK of 640 GBd Nyquist OTDM and transmitted over 525 km.

In addition to Nyquist OTDM, the technique proposed in [44] was also applied for Nyquist OTDM-WDM [51, 52]. The output of a 43 GHz, 400 fs optical pulse generator is modulated by a 43 Gbit/s PRBS sequence and multiplexed in time to generate the OTDM stream. The multiplexing scheme can be varied with factor N from 1 to 8 in order to obtain signals from 43 to 344 GBd. A Waveshaper is used for Nyquist filtering with a raised cosine function of bandwidth $(N + 1) \times 43$ GHz to shape the $N \times 43$ GBd signal. The extra bandwidth is used for the purpose of clock recovery. The WDM channels are generated by applying the Nyquist filtering at different wavelengths, using the same Waveshaper. In addition the signals are polarization multiplexed to double the total capacity. The transmitter is shown in Fig. 8.27.

The spectrum of the mixed bitrate Nyquist OTDM-WDM shown in Fig. 8.28a reflects the typical raised-cosine filtering performed by the Waveshaper. As mentioned in Sect. 4.3.1, due to the limited resolution of the programmable filter, the steepness of the filtering transfer function varies with the overall bandwidth. Wider Nyquist pulses are therefore characterized by a larger roll-off than shorter ones and the performance between the different OTDM channels slightly varies. The corresponding waveforms show the typical sinc-oscillations and a good fitting with the raised-cosine approximation (Fig. 8.28b). Due to the non-ideal rectangular spectra,

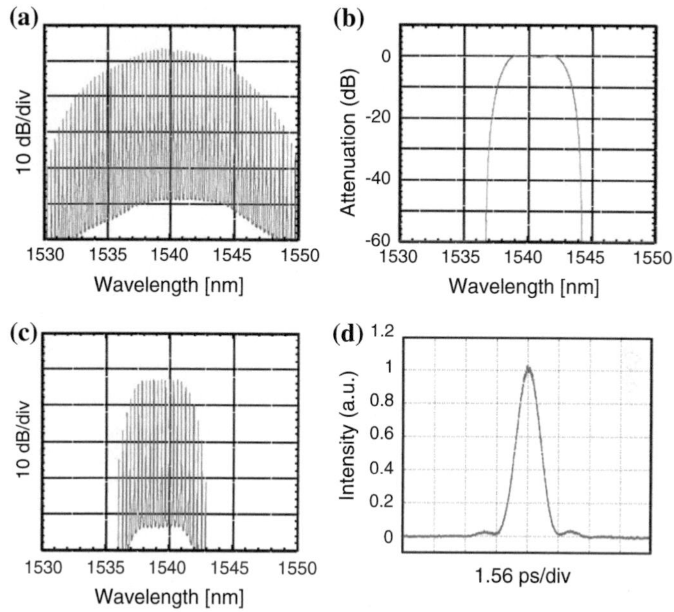

Fig. 8.26 Generation of Nyquist pulse from shaping the output of a modelocked laser. **a** Optical spectrum of 40 GHz Gaussian pulse after self-phase modulation induced spectral broadening. **b** Transfer function of the programmable optical filter based on liquid crystal spatial modulators used to shape the spectrum shown in (**a**). **c** Optical spectrum after the programmable filter and **d** corresponding waveform of the generated optical Nyquist pulses characterized by a peak-to-first-zero of 1.56 and 25 ps repetition time (adapted from [64])

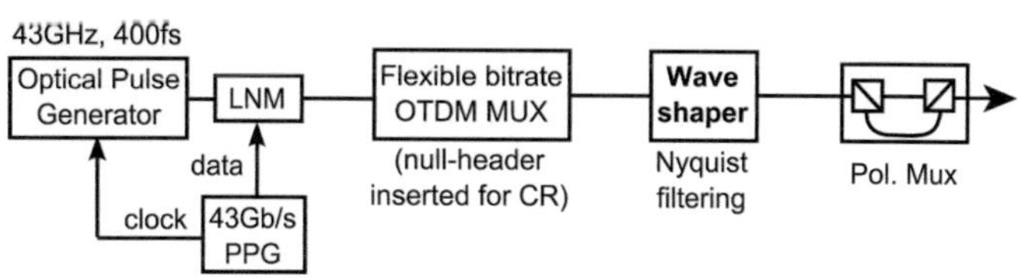

Fig. 8.27 Optical transmitter for Nyquist TDM-WDM. The output of 43 GHz optical pulse generator is modulated by a PRBS sequence and time multiplexed by a factor $N (N = 1\text{–}8)$. A Waveshaper performs the Nyquist filtering at various wavelengths simultaneously before polarization multiplexing. The output consists of a Pol-mux Nyquist OTDM-WDM (adapted from [27])

small guard bands are introduced between the WDM channels to avoid inter-channel-interference. This obtained data stream was successfully used for the transmission and pass-drop operations of Nyquist OTDM.WDM signal with baudrate of 43 GBd to dual polarization 344 GBd over 320 km of super large area—inverse dispersion fiber (SLA-IDF) in combination with wavelength selective switch nodes.

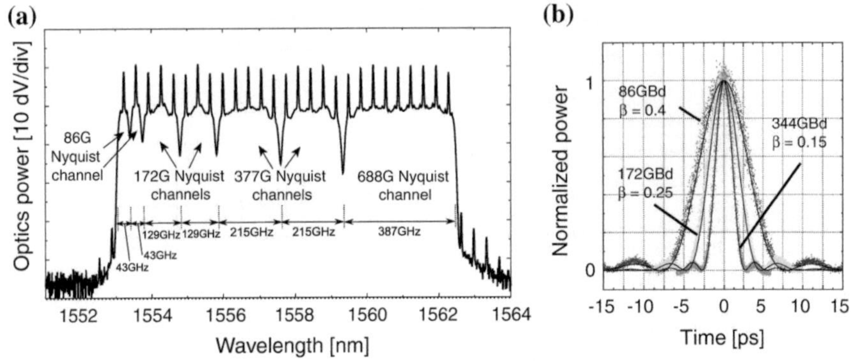

Fig. 8.28 Nyquist OTDM-WDM demonstration. **a** Optical spectrum of the dual-polarization Nyquist OTDM-WDM signal with spectral arrangement of mixed baud rate from 43 to 344 GBd/pol. **b** The waveforms captured on a high resolution (1 ps) optical sampling oscilloscope for the different baud rates. The raised cosine fitting reflects the decreasing roll-off factor β as the bandwidth of the pulse increases. Figure adapted from [52]

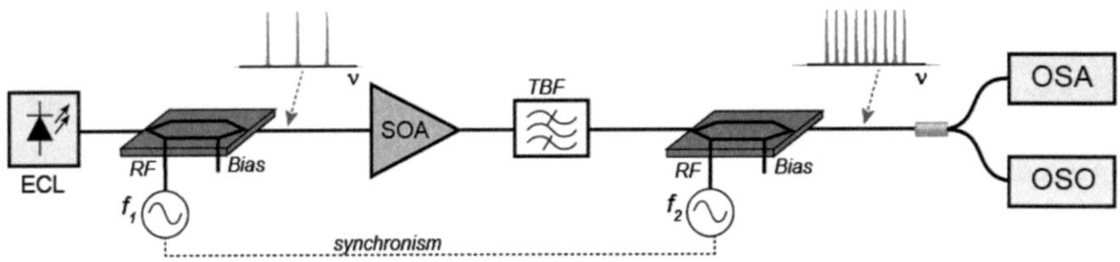

Fig. 8.29 Optical Nyquist pulse generation through frequency comb synthesis using two cascaded Mach Zehnder modulators (MZM). An external cavity laser (ECL) generates a narrow linewidth continuous wave light at 1550 nm. The first MZM generates spectral lines separated by a frequency $f_1 = 30$ GHz. The second MZM re-modulates these seeding components with a frequency $f_2 = 10$ GHz. RF power and DC bias of both modulators are adjusted so that a rectangular frequency comb is obtained. The RF generators are synchronized on a common time base. The resulting output is amplified and observed on as optical spectrum analyzer (OSA) with a 0.01 nm spectral resolution and on an optical sampling oscilloscope with a 500 GHz bandwidth

The experimental setup for the generation of sinc-shaped Nyquist pulses through frequency comb synthesis is shown in Fig. 8.29. The proof-of-concept experiment relied on two cascaded Mach-Zehnder modulators. While the first modulator, driven by a radio frequency (RF) f_1 is adjusted to generate three seeding spectral components, the second modulator re-modulates those lines using an RF signal at f_2. For instance to generate a comb with $N = 9$ spectral lines, the condition $f_1 = 3f_2$ or $f_2 = 3f_1$ has to be satisfied, resulting in a frequency spacing $\Delta f = \min(f_1, f_2)$. The number of spectral lines can be easily varied by operating the modulators in carrier-suppressed mode and/or driving the modulators with multiple RF signals such that and many experimental variants can be implemented using a similar approach.

A high quality rectangular-shaped frequency comb can be obtained by tuning the DC bias and the RF voltage amplitude of each modulator while adjusting the phase delay between the different RF signals. The results for the case $f_1 = 30$ GHz and $f_2 = 10$ GHz are shown in Fig. 8.30. A frequency comb spanning 90 GHz, comprising of 9 spectral lines, spaced by 10 GHz is obtained. An out-of-band rejection of at least 27 dB and a power difference between the spectral lines of less than 0.2 dB was measured. The corresponding time domain trace, observed on an optical sampling scope with 500 GHz bandwidth, confirms the resulting high quality sinc-shape Nyquist pulses. The repetition period of 100 GHz and the peak-to-zero crossing of 11.1 ps are in agreement with the spectral features. The power error with respect to the theoretical curve was less than 1 %, the roll off of the experimentally obtained pulse is close to zero. Close to ideal rectangular spectrum, such as seen in fig. (8.30a) can be compared to the ones obtained with the filter based approach and a no guard-band Nyquist OTDM-WDM can be now envisioned.

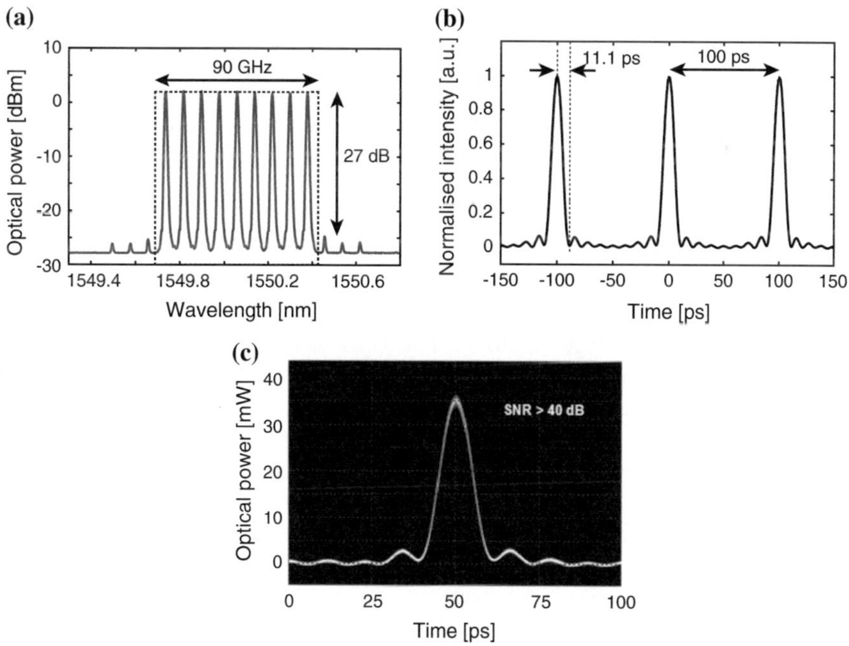

Fig. 8.30 Sinc-shaped Nyquist pulse from frequency comb synthesis. **a** Spectrum and **b** respective time-domain waveform of a comb generated with $N = 9$ spectral components separated by 10 GHz and expending over a 90 GHz bandwidth. A power difference between spectral components lower than 0.2 dB and out-of-band rejection of more than 26 dB is obtained. The *dashed box* represents the theoretical Nyquist bandwidth of the generated sinc pulses. **c** Color grade figure of the experimentally obtained sinc-pulse. Measurements indicate a jitter of 82 fs and a signal-to-noise greater than 40 dB [49]

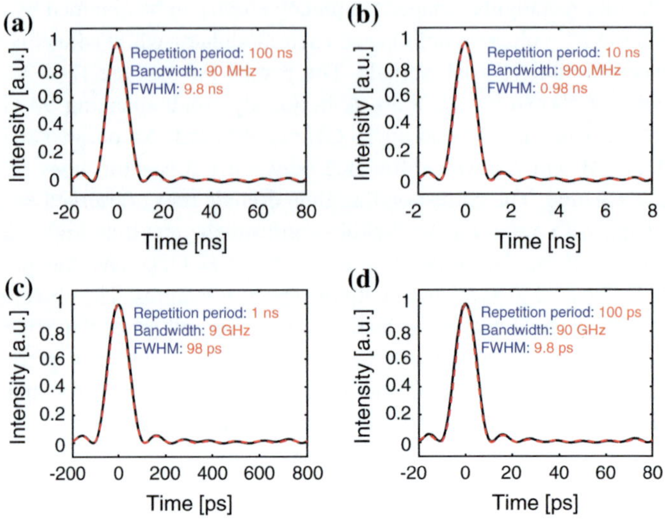

Fig. 8.31 Sinc-shaped Nyquist pulses measured using a 500 GHz optical sampling oscilloscope. The calculated waveforms (*dashed lines*) are compared with the measured pulses (*black straight lines*) for different bandwidth conditions. Nyquist pulses are obtained from the generation of a rectangular frequency comb with 9 phase-locked components spanning over a spectral width between 90 MHz and 90 GHz, using modulating frequencies **a** $f_1 = 30$ MHz and $f_2 = \Delta f f = 10$ MHz, **b** $f_1 = 300$ MHz and $f_2 = \Delta f = 100$ MHz, **c** $f_1 = 3$ GHz and $f_2 = \Delta f = 1$ GHz, and **d** $f_1 = 30$ GHz and $f_2 = \Delta f = 10$ GHz. The maximum difference between measured pulses and theoretical ones remained in all cases below 1 % showing the high flexibility of this method [49]

The last major advantage of this method is its flexibility which lies in the fact that by changing the modulating signal frequencies different frequency combs can be generated while maintaining the same quality in terms of pulse characteristics. Figure 8.31 shows measured sinc pulses when the frequency spacing between the spectral components of the comb was varied over many decades (10 MHz and 10 GHz). The generated pulses coincide very well with the ideal ones over four frequency decades, showing the root mean square error below 1 % for all cases.

Acknowledgments Juerg Leuthold acknowledges contributions from the EU-Project FOX-C and support by ETH Zurich, Switzerland. Juerg Leuthold is also very much indebted to Wolfgang Freude from Karlsruhe Institute of Technology (KIT) with whom he collaborated in the past and who contributed a lot to the OFDM section in this chapter. Camille Brès acknowledges support by the EPFL Lausanne, Switzerland.

References

1. W. Shieh, C. Athaudage, Coherent optical orthogonal frequency division multiplexing, in *Electronics Letters* (Institution of Engineering and Technology, 2006), pp. 587–589
2. R.W. Chang, Synthesis of band-limited orthogonal signals for multichannel data transmission. Bell Syst. Tech. J. **45**, 1775–1796 (1966)
3. A.J. Lowery, L. Du, J. Armstrong, Orthogonal Frequency Division Multiplexing for Adaptive Dispersion Compensation in Long Haul WDM Systems. in *Optical Fiber Communication Conference and Exposition and The National Fiber Optic Engineers Conference* (Optical Society of America, Anaheim, California, 2006), p. PDP39
4. I.B. Djordjevic, B. Vasic, Orthogonal frequency division multiplexing for high-speed optical transmission. Opt. Express **14**, 3767–3775 (2006)
5. N. Cvijetic, L. Xu, T. Wang, Adaptive PMD compensation using OFDM in long-haul 10 Gb/s DWDM systems, in *Optical Fiber Communication Conference and Exposition and The National Fiber Optic Engineers Conference* (Optical Society of America, Anaheim, California, 2007), p. OTuA5
6. S.L. Jansen, I. Morita, T.C. Schenk, H. Tanaka, Long-haul transmission of16 × 52.5 Gbits/s polarization-division- multiplexed OFDM enabled by MIMO processing (Invited). J. Opt. Netw. **7**, 173–182 (2008)
7. E. Yamada, A. Sano, H. Masuda, E. Yamazaki, T. Kobayashi, E. Yoshida, K. Yonenaga, Y. Miyamoto, K. Ishihara, and Y. Takatori, 1 Tb/s (111 Gb/s/ch × 10ch) no-guard-interval CO-OFDM transmission over 2100 km DSF, in *Opto-electronics and Communications Conference, Sydney* (2008)
8. S. Chandrasekhar, X. Liu, B. Zhu, D. Peckham, Transmission of a 1.2-Tb/s 24-carrier no-guard-interval coherent OFDM super channel over 7200-km of ultra-large-area fiber, in *ECOC 2009* (2009)
9. D. Hillerkuss, R. Schmogrow, T. Schellinger, M. Jordan, M. Winter, G. Huber, T. Vallaitis, R. Bonk, P. Kleinow, F. Frey, M. Roeger, S. Koenig, A. Ludwig, A. Marculescu, J. Li, M. Hoh, M. Dreschmann, J. Meyer, S. Ben Ezra, N. Narkiss, B. Nebendahl, F. Parmigiani, P. Petropoulos, B. Resan, A. Oehler, K. Weingarten, T. Ellermeyer, J. Lutz, M. Moeller, M. Huebner, J. Becker, C. Koos, W. Freude, J. Leuthold, 26 Tbit s(-1) line-rate super-channel transmission utilizing all-optical fast Fourier transform processing. Nat. Photonics **5**, 364–371 (2011)
10. D. Hillerkuss, M. Winter, M. Teschke, A. Marculescu, J. Li, G. Sigurdsson, K. Worms, S. Ben Ezra, N. Narkiss, W. Freude, J. Leuthold, Simple all-optical FFT scheme enabling Tbit/s real-time signal processing. Opt. Express **18**, 9324–9340 (2010)
11. H. Nyquist, Certain topics in telegraph transmission theory. Trans. Am. Inst. Electrical Eng. (A. I. E. E.) **47**, 617–644 (1928)
12. G. Bosco, V. Curri, A. Carena, P. Poggiolini, F. Forghieri, On the performance of Nyquist-WDM terabit superchannels based on PM-BPSK, PM-QPSK, PM-8QAM or PM-16QAM subcarriers. J. Lightwave Technol. **29**, 53–61 (2011)
13. R. Schmogrow, M. Winter, M. Meyer, D. Hillerkuss, S. Wolf, B. Baeuerle, A. Ludwig, B. Nebendahl, S. Ben-Ezra, J. Meyer, M. Dreschmann, M. Huebner, J. Becker, C. Koos, W. Freude, J. Leuthold, Real-time Nyquist pulse generation beyond 100 Gbit/s and its relation to OFDM. Opt. Express **20**, 317–337 (2012)
14. R. Schmogrow, D. Hillerkuss, S. Wolf, B. Bauerle, M. Winter, P. Kleinow, B. Nebendahl, T. Dippon, P.C. Schindler, C. Koos, W. Freude, J. Leuthold, 512QAM Nyquist sinc-pulse transmission at 54 Gbit/s in an optical bandwidth of 3 GHz. Opt. Express **20**, 6439–6447 (2012)
15. R. Schmogrow, D. Hillerkuss, M. Dreschmann, M. Huebner, M. Winter, J. Meyer, B. Nebendahl, C. Koos, J. Becker, W. Freude, J. Leuthold, Real-time software-defined multiformat transmitter generating 64QAM at 28 GBd. IEEE Photonic Tech. L **22**, 1601–1603 (2010)

16. J. Leuthold, W. Freude, Optical OFDM and Nyquist multiplexing, in *Optical Fiber Telecommunications VIB*, ed. by I.P. Kaminow, T. Li, A.E. Willner (Academic Press, 2013)
17. R. Schmogrow, M. Winter, D. Hillerkuss, B. Nebendahl, S. Ben-Ezra, J. Meyer, M. Dreschmann, M. Huebner, J. Becker, C. Koos, W. Freude, J. Leuthold, Real-time OFDM transmitter beyond 100 Gbit/s. Opt. Express **19**, 12740–12749 (2011)
18. R. Freund, No, x, M. lle, C. Schmidt-Langhorst, R. Ludwig, C. Schubert, G. Bosco, A. Carena, P. Poggiolini, Oxenl, x00F, L. We, M. Galili, H.C.H. Mulvad, M. Winter, D. Hillerkuss, R. Schmogrow, W. Freude, J. Leuthold, A.D. Ellis, F.C.G. Gunning, J. Zhao, P. Frascella, S.K. Ibrahim, N.M. Suibhne, Single- and multi-carrier techniques to build up Tb/s per channel transmission systems, in *Transparent Optical Networks (ICTON), 2010 12th International Conference on* (2010), pp. 1–7
19. M.E. Marhic, Discrete Fourier transforms by single-mode star networks. Opt. Lett. **12**, 63–65 (1987)
20. H. Sanjoh, E. Yamada, Y. Yoshikuni, Optical orthogonal frequency division multiplexing using frequency/time domain filtering for high spectral efficiency up to 1 bit/s/Hz, in *Optical Fiber Communication Conference and Exhibit, 2002*. OFC **2002**, 401–402 (2002)
21. K. Takiguchi, M. Oguma, H. Takahashi, A. Mori, Integrated-optic eight-channel OFDM demultiplexer and its demonstration with 160 Gbit/s signal reception, in *Electronics Letters* (Institution of Engineering and Technology, 2010), pp. 575–576
22. G. Cincotti, Fiber wavelet filters [and planar waveguide couplers for full-wavelength demultiplexers]. IEEE J. Quantum Electron. **38**, 1420–1427 (2002)
23. G. Cincotti, Full optical encoders/decoders for photonic IP routers. J. Lightwave Technol. **22**, 337–342 (2004)
24. M. Nazarathy, D.M. Marom, W. Shieh, Optical comb and filter bank (De)Mux enabling 1 Tb/s orthogonal sub-band multiplexed CO-OFDM free of ADC/DAC limits, in *Optical Communication, 2009. ECOC '09. 35th European Conference on* (2009), pp. 1–2
25. M.K. Smit, C. Van Dam, PHASAR-based WDM-devices: principles, design and applications. IEEE J. Sel. Top. Quantum Electron. **2**, 236–250 (1996)
26. A.J. Lowery, Design of arrayed-waveguide grating routers for use as optical OFDM demultiplexers. Opt. Express **18**, 14129–14143 (2010)
27. K. Lee, C.T.D. Thai, J.-K.K. Rhee, All optical discrete fourier transform processor for 100 Gbps OFDM transmission. Opt. Express **16**, 4023–4028 (2008)
28. A.J. Lowery, L. Du, All-optical OFDM transmitter design using AWGRs and low-bandwidth modulators. Opt. Express **19**, 15696–15704 (2011)
29. R.P. Giddings, X.Q. Jin, E. Hugues-Salas, E. Giacoumidis, J.L. Wei, J.M. Tang, Experimental demonstration of a record high 11.25 Gb/s real-time optical OFDM transceiver supporting 25 km SMF end-to-end transmission in simple IMDD systems. Opt. Express **18**, 5541–5555 (2010)
30. R. Schmogrow, M. Meyer, P.C. Schindler, B. Nebendahl, M. Dreschmann, J. Meyer, A. Josten, D. Hillerkuss, S. Ben-Ezra, J. Becker, C. Koos, W. Freude, J. Leuthold, Real-time Nyquist signaling with dynamic precision and flexible non-integer oversampling. Opt. Express **22**, 193–209 (2014)
31. B. Inan, S. Adhikari, O. Karakaya, P. Kainzmaier, M. Mocker, H. von Kirchbauer, N. Hanik, S.L. Jansen, Real-time 93.8-Gb/s polarization-multiplexed OFDM transmitter with 1024-point IFFT. Opt. Express **19**, B64–B68 (2011)
32. D. Hillerkuss, T. Schellinger, M. Jordan, C. Weimann, F. Parmigiani, B. Resan, K. Weingarten, S. Ben-Ezra, B. Nebendahl, C. Koos, W. Freude, J. Leuthold, High-quality optical frequency comb by spectral slicing of spectra broadened by SPM. IEEE Photonics J. **5**, 7201011 (7201011 pp) (2013)
33. R. Schmogrow, B. Nebendahl, M. Winter, A. Josten, D. Hillerkuss, S. Koenig, J. Meyer, M. Dreschmann, M. Huebner, C. Koos, J. Becker, W. Freude, J. Leuthold, Error vector magnitude as a performance measure for advanced modulation formats. IEEE Photonic Technol. Lett. **24**, 61–63 (2012)
34. Fujitsu, *Soft-Decision FEC Benefits for 100G* (Fujitsu Network Communications Inc., 2012)

35. Z. Xiang, L.E. Nelson, P. Magill, R. Isaac, Z. Benyuan, D.W. Peckham, P.I. Borel, K. Carlson, PDM-Nyquist-32QAM for 450-Gb/s Per-Channel WDM Transmission on the 50 GHz ITU-T Grid. J. Lightwave Technol. **30**, 553–559 (2012)
36. T. Kobayashi, A. Sano, A. Matsuura, Y. Miyamoto, K. Ishihara, Nonlinear tolerant spectrally-efficient transmission using PDM 64-QAM single carrier FDM with digital pilot-tone. J. Lightwave Technol. **30**, 3805–3815 (2012)
37. C. Jin-Xing, C.R. Davidson, A. Lucero, Z. Hongbin, D.G. Foursa, O.V. Sinkin, W.W. Patterson, A.N. Pilipetskii, G. Mohs, N.S. Bergano, 20 Tbit/s transmission over 6860 km with sub-nyquist channel spacing. J. Lightwave Technol. **30**, 651–657 (2012)
38. A.H. Gnauck, P.J. Winzer, S. Chandrasekhar, X. Liu, B. Zhu, D.W. Peckham, Spectrally efficient long-haul WDM transmission using 224-Gb/s polarization-multiplexed 16-QAM. J. Lightwave Technol. **29**, 373–377 (2011)
39. R. Schmogrow, R. Bouziane, M. Meyer, P.A. Milder, P.C. Schindler, R.I. Killey, P. Bayvel, C. Koos, W. Freude, J. Leuthold, Real-time OFDM or Nyquist pulse generation–which performs better with limited resources? Opt. Express **20**, B543–B551 (2012)
40. T. Hirooka, P. Ruan, P. Guan, M. Nakazawa, Highly dispersion-tolerant 160 Gbaud optical Nyquist pulse TDM transmission over 525 km. Opt. Express **20**, 15001–15007 (2012)
41. T. Hirooka, M. Nakazawa, Linear and nonlinear propagation of optical Nyquist pulses in fibers. Opt. Express **20**, 19836–19849 (2012)
42. G. Bosco, A. Carena, V. Curri, P. Poggiolini, F. Forghieri, Performance limits of Nyquist-WDM and CO-OFDM in high-speed PM-QPSK systems. IEEE Photonics Technol. Lett. **22**, 1129–1131 (2010)
43. J.G. Proakis, M. Salehi, *Digital Communications* (McGraw Hill, 2008)
44. M. Nakazawa, T. Hirooka, P. Ruan, P. Guan, Ultrahigh-speed "orthogonal" TDM transmission with an optical Nyquist pulse train. Opt. Express **20**, 1129–1140 (2012)
45. D. Hillerkuss, R. Schmogrow, M. Meyer, S. Wolf, M. Jordan, P. Kleinow, N. Lindenmann, P. C. Schindler, A. Melikyan, X. Yang, S. Ben-Ezra, B. Nebendahl, M. Dreschmann, J. Meyer, F. Parmigiani, P. Petropoulos, B. Resan, A. Oehler, K. Weingarten, L. Altenhain, T. Ellermeyer, M. Moeller, M. Huebner, J. Becker, C. Koos, W. Freude, J. Leuthold, Single-Laser 32.5 Tbit/s Nyquist WDM transmission. J. Opt. Commun. Netw. **4**, 715–723 (2012)
46. A. Vedadi, M.A. Shoaie, C.-S. Brès, Near-Nyquist optical pulse generation with fiber optical parametric amplification. Opt. Express **20**, B558–B565 (2012)
47. R. Schmogrow, S. Ben-Ezra, P.C. Schindler, B. Nebendahl, C. Koos, W. Freude, J. Leuthold, Pulse-shaping with digital, electrical, and optical filters-A comparison. J. Lightwave Technol. **31**, 2570–2577 (2013)
48. L. Xiang, S. Chandrasekhar, Z. Benyuan, P.J. Winzer, A.H. Gnauck, D.W. Peckham, 448-Gb/s reduced-guard-interval CO-OFDM transmission over 2000 km of ultra-large-area fiber and five 80-GHz-Grid ROADMs. J. Lightwave Technol. **29**, 483–490 (2011)
49. M.A. Soto, M. Alem, M. Amin Shoaie, A. Vedadi, C.-S. Brès, L. Thévenaz, T. Schneider, Optical sinc-shaped Nyquist pulses of exceptional quality. Nat. Commun. **4** Article No. 2898. (2013)
50. G. Baxter, S. Frisken, D. Abakoumov, Z. Hao, I. Clarke, A. Bartos, S. Poole, Highly programmable wavelength selective switch based on liquid crystal on silicon switching elements, in *Optical Fiber Communication Conference, 2006 and the 2006 National Fiber Optic Engineers Conference. OFC 2006* (2006), p. 3
51. H.N. Tan, K. Tanizawa, T. Inoue, T. Kurosu, S. Namiki, No guard-band wavelength translation of Nyquist OTDM WDM signal for spectral defragmentation in an elastic add–drop node. Opt. Lett. **38**, 3287–3290 (2013)
52. T. Hung Nguyen, T. Inoue, T. Kurosu, S. Namiki, Transmission and pass-drop operations in all-optical elastic network using Nyquist OTDM-WDM up to 2×344 Gbaud/channel, in *OptoElectronics and Communications Conference held jointly with 2013 International Conference on Photonics in Switching (OECC/PS), 2013 18th* (2013), pp. 1–2

53. B. Washburn, R. Fox, N. Newbury, J. Nicholson, K. Feder, P. Westbrook, C. Jørgensen, Fiber-laser-based frequency comb with a tunable repetition rate. Opt. Express **12**, 4999–5004 (2004)
54. A. Ruehl, A. Marcinkevicius, M.E. Fermann, I. Hartl, 80 W, 120 fs Yb-fiber frequency comb. Opt. Lett. **35**, 3015–3017 (2010)
55. A. Bartels, R. Gebs, M.S. Kirchner, S.A. Diddams, Spectrally resolved optical frequency comb from a self-referenced 5 GHz femtosecond laser. Opt. Lett. **32**, 2553–2555 (2007)
56. T.J. Kippenberg, R. Holzwarth, S.A. Diddams, Microresonator-based optical frequency combs. Science **332**, 555–559 (2011)
57. P. Del'Haye, A. Schliesser, O. Arcizet, T. Wilken, R. Holzwarth, T.J. Kippenberg, Optical frequency comb generation from a monolithic microresonator. Nature **450**, 1214–1217 (2007)
58. A.K. Mishra, R. Schmogrow, I. Tomkos, D. Hillerkuss, C. Koos, W. Freude, J. Leuthold, Flexible RF-Based Comb Generator. IEEE Photonic Technol. Lett. **25**, 701–704 (2013)
59. Y. Yamamoto, T. Komukai, K. Suzuki, A. Takada, Multicarrier light source with flattened spectrum using phase modulators and dispersion medium. J. Lightwave Technol. **27**, 4297–4305 (2009)
60. T. Yamamoto, T. Komukai, K. Suzuki, A. Takada, Spectrally flattened phase-locked multi-carrier light generator with phase modulators and chirped fibre Bragg grating, in *Electronics Letters* (Institution of Engineering and Technology, 2007), pp. 1040–1042
61. T. Yang, J. Dong, S. Liao, D. Huang, X. Zhang, Comparison analysis of optical frequency comb generation with nonlinear effects in highly nonlinear fibers. Opt. Express **21**, 8508–8520 (2013)
62. V. Ataie, B.P.P. Kuo, E. Myslivets, S. Radic, Generation of 1500-tone, 120 nm-wide ultraflat frequency comb by single CW source, in *Optical Fiber Communication Conference/National Fiber Optic Engineers Conference 2013* (Optical Society of America, Anaheim, California, 2013), p. PDP5C.1
63. Z. Tong, A.O.J. Wiberg, E. Myslivets, B.P.P. Kuo, N. Alic, S. Radic, Spectral linewidth preservation in parametric frequency combs seeded by dual pumps. Opt. Express **20**, 17610–17619 (2012)
64. K. Harako, D. Seya, T. Hirooka, M. Nakazawa, 640 Gbaud (1.28 Tbit/s/ch) optical Nyquist pulse transmission over 525 km with substantial PMD tolerance. Opt. Express **21**, 21062–21075 (2013)
65. H. Hu, D. Kong, E. Palushani, J.D. Andersen, A. Rasmussen, B.M. Sørensen, M. Galili, H.C. H. Mulvad, K.J. Larsen, S. Forchhammer, P. Jeppesen, L.K. Oxenløwe, 1.28 Tbaud Nyquist signal transmission using time-domain optical fourier transformation based receiver, in *CLEO: 2013 Postdeadline* (Optical Society of America, San Jose, California, 2013), p. CTh5D.5
66. M.A. Soto, M. Alem, M.A. Shoaie, A. Vedadi, C.S. Brès, L. Thévenaz, T. Schneider, Generation of Nyquist Sinc pulses using intensity modulators, in *CLEO: 2013* (Optical Society of America, San Jose, California, 2013), p. CM4G.3

Chapter 9
Energy-Efficient Optical Signal Processing Using Optical Time Lenses

Leif Katsuo Oxenløwe, Michael Galili, Hans Christian Hansen Mulvad, Hao Hu, Pengyu Guan, Evarist Palushani, Mads Lillieholm and Anders Clausen

Abstract This chapter describes advanced functionalities for optical signal processing using optical time lenses. A special focus is devoted to functionalities that allow for energy-savings. In particular, we find that optical signal processing, where the processing is broadband and capable of handling many bits in a single operation allows for sharing the processing energy by the many bits, and hence the energy per bit is reduced. Such functionalities include serial-to-parallel conversion in a single time lens, where a large number of parallel demultiplexers may be substituted by a single time lens. Combining time lenses into telescopic arrangements allows for more advanced signal processing, such as temporal or spectral compression or magnification. A spectral telescope may for instance allow for conversion of OFDM signals to DWDM-like signals, which can be separated passively, i.e. without additional energy. This is opposed to the DFT OFDM receivers otherwise suggested, where a temporal active gate is required for each tributary. With the spectral telescope, only two active time lenses are required, irrespective of how many tributaries are used. This chapter describes how optical time lenses function and by showing examples of some advanced functionalities points to future scenarios where energy consumption may be considerably reduced.

9.1 Introduction: Energy-Efficient Solutions Using Nonlinear Optical Signal Processing

Some of the biggest challenges in optical communications today are the relentless demands for more capacity and the accompanying demands for more energy-efficient solutions. Current technologies cannot keep up with the bandwidth demands without

L.K. Oxenløwe (✉) · M. Galili · H.C.H. Mulvad · H. Hu · P. Guan · E. Palushani · M. Lillieholm · A. Clausen
DTU Fotonik, Department of Photonics Engineering, Technical University of Denmark, Ørsteds Plads 343, 2800 Kongens Lyngby, Denmark
e-mail: lkox@fotonik.dtu.dk

S. Wabnitz and B.J. Eggleton (eds.), *All-Optical Signal Processing*,
Springer Series in Optical Sciences 194, DOI 10.1007/978-3-319-14992-9_9

resorting to extensive parallelism, where the number of components and their combined power consumption, simply scale with the bandwidth. The actual traffic on the internet grows approximately 50 % per year, pointing to the looming imminent "capacity crunch", where the traffic outgrows the available capacity [1, 2], and is responsible for about 2 % of the global CO_2 emissions [3] today. These are CO_2 emissions comparable to that of the aviation industry, and it thus remains one of the biggest technological challenges to find sustainable solutions for the future internet. There is therefore a strong need for groundbreaking developments in optical science enabling novel breakthrough technologies that will allow for increased capacity with a reduced energy consumption. There are indications that nonlinear optical signal processing may in some cases offer the higher bandwidth at lower energy. However one needs to be very careful with choosing the right functionalities to harvest the benefits from both electronics and optics to globally benefit from what is possible with electronics and what is possible with optics. For instance, in [4, 5], it is concluded that optical signal processing may be appropriate for certain functionalities where the number of signal processing elements is small, and where the signal processing itself is simple. This could be the case for the ultrafast nonlinear optical signal processing of ultrahigh bit rates in few but ultrafast devices, as will become apparent in the following sections. Furthermore, this chapter will describe a large number of advanced ultrafast functionalities that are achievable today using mostly commercially available components.

9.1.1 Energy-Efficient Optical Signal Processing: Many Bits per Operation

Certain functionalities are promising for low-energy signal processing, for instance functionalities where many bits are processed per operation, so that many bits share the used pump energy. But one has to carefully choose the right type of functionalities and nonlinear effects. An example demonstrating reduced energy consumption per bit is cross-phase modulation (XPM) based nonlinear signal processing relying on the Kerr effect in e.g. highly nonlinear fibre (HNLF). Here the amount of phase modulation depends on the peak power of the pump signal as well as the nonlinear coefficient and the length of the fibre: $\Delta\varphi \propto \gamma L P_{peak}$. This fact is useful for designing energy-efficient switching schemes. Consider a train of pulses with a certain width and repetition rate at a given average power resulting in a certain peak power. If the repetition rate is now doubled, and the pulse widths are correspondingly twice as narrow, then the peak power remains the same for the same average power. If this pulse train is used as the pump pulses, then it is clear that the switching energy per bit is inversely proportional to the bit rate: $\frac{E}{bit} \propto 1/N$, where N is the multiplication factor on the pulse rate. So basically, a 20 Gbit/s data signal could be switched with half the energy/bit compared to a 10 Gbit/s signal, and when going to very high bit rates like 640 Gbit/s or 1.28 Tbit/s, this advantage is multiplied manifold. One concrete example of this idea is the use of XPM-based sideband generation on a CW wave [6]. A data pulse stream exerts XPM on the CW, and every time a 1-bit, i.e. a pulse, arrives, a sideband occurs as opposed to when a 0-bit arrives, where nothing happens. Hence by filtering out the sidebands, one gets a wavelength converted, or in some cases an additionally regenerated, copy of the original data stream. To create a certain sideband requires a certain peak power, and as outlined above, the higher the bit rate, the less energy/bit is needed for this wavelength conversion scheme.

In the following, we will address another scheme, where we will show in detail that the switching energy per bit decreases with increased bit rates, ultimately resulting in total system power reduction.

Figure 9.1 shows a scheme to perform serial-to-parallel conversion based on optical time lenses. The technical principles of operation will be described in later sections, and here we will simply discuss the implications on power consumption, and for now just assume that it works. The nonlinear effect used for this scheme is four-wave mixing (FWM). In FWM, two pump photons are used to create an idler photon together with the signal photon. As long as the pump is not depleted, which is most cases of any practical use, there are enough pump photons to create the idlers, and hence for a given (undepleted) pump power, one may simply increase the bit rate and thus use more of the pump and thereby reduce the pump energy/bit. This is confirmed in [7] where detailed simulations of the serial-to-parallel conversion scheme demonstrated that increasing the data bit rate from 160, over 320–640 Gbit/s resulted in the same bit error rate for the same pump power, see Fig. 9.1c.

In [9], power consumption for various data capacities is calculated for OTDM and WDM point-to-point systems. No in-line functionalities other than amplification and passive dispersion compensation are considered. The total system power

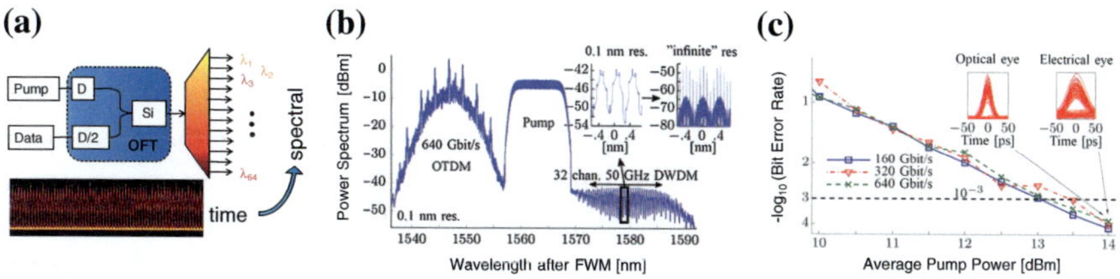

Fig. 9.1 Serial-to-parallel conversion using time lens based time-domain optical Fourier transformation (OFT). **a** Principle of serial-to-parallel conversion, in this sketch using a silicon nanowire as nonlinear medium. **b**, **c** Simulation results of converting a 640 Gbit/s serial optical data signal into 64 parallel DWDM data channels each at 10 Gbit/s, from [7]. **b** Output spectra of the OFT showing the input broad-spectrum OTDM data signal, the flat-top-spectrum pump signal and the resulting DWDM signal, in this case 50 GHz spaced. **c** Resulting BER performance, revealing that for the same average pump power as used in the experimental demonstration in [8] the bit rate is not important, i.e. for bit rates of 160, 320 and 640 Gbit/s, the same BER is obtained for the same pump power, revealing that the pump energy per bit is inversely proportional to the bit rate. The functionality may be carried out using a highly nonlinear fibre or a silicon nanowire as the nonlinear medium creating the optical phase modulation required in the time lens

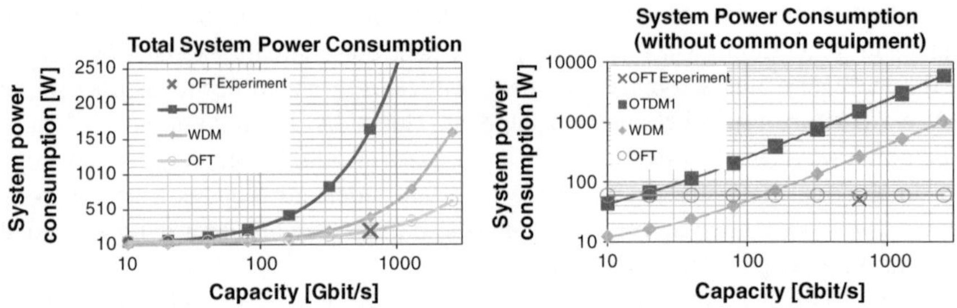

Fig. 9.2 Total system power consumption comparison between WDM and OTDM scenarios, including the case of OTDM using time lens based time-domain optical Fourier transformation (*OFT*) for demultiplexing all OTDM channels simultaneously. *Left* Total system power consumption including all components. *Right* Disregarding common equipment (data modulators, in-line EDFA, and photodiodes) to more easily spot the differences between the scenarios

Table 9.1 Assumed power values for power consumption calculation

	P (W)	Cool (W)	
CW	1	3	× N WDM
Data mod	2.2		× N WDM/OTDM
RZ mod	2.2		× 1 WDM
RZ laser	3.2	3	× 1 OTDM
Demux-laser	3.2	3	× N OTDM
Line-EDFA	2.5	6	2 pump diodes × 1
MUX-EDFA	2.5	6	2 pump diodes × 1
Demux-EDFA	5	12	4 pump diodes × N OTDM
PD	0.04		× N WDM/OTDM

Values correspond to settings required for BER 1E-9, and are either taken from actual laboratory settings or from data specifications from manufacturers, cf. [9]

consumption is shown in Fig. 9.2, based on numbers derived for the results presented in [9], with the addition of a data point corresponding to an experimental demonstration of OTDM-to-WDM conversion at 640 Gbit/s. The total system power is calculated for a system performance yielding BER = 1E-9 for each channel, including cooling of all laser devices, and electrical power to all amplifiers, lasers etc. [9]. The assumed values for power both for driving lasers a.o. and for cooling lasers are realistic in the sense that they are taken from tabletop commercial products actually used in the experimental demonstrations, see Table 9.1.

Looking at Table 9.1, one may easily derive how many of a certain component is used in the classical WDM and OTDM setups. In this simple calculation, only a single in-line EDFA is used, corresponding to only transmitting over a single span of fibre. Both the WDM and the OTDM setups require the same number of data modulators ("data mod" in Table 9.1) and photodiodes ("PD"), which do not need to be temperature controlled. In the WDM case, N parallel CW lasers are used, and a single RZ modulator is assumed ("RZ mod"), which does not change much. In the OTDM transmitter a single pulse source ("RZ laser") is used, corresponding in power to a CW laser and an RZ modulator. The OTDM setup also includes a booster amplifier to compensate for the loss in the OTDM split-and-delay-and-combine multiplexer ("MUX EDFA"), N parallel demultiplexers based on N × HNLFs and N × control pulse sources ("demux laser") and N × EDFAs to boost these control pulse sources ("demux-EDFA"), corresponding to a worst case scenario. The demux-EDFA is assumed to be a high-power EDFA requiring 4 pump laser diodes, where the other EDFAs have 2 pump diodes. For cooling, it is assumed that only the lasers require cooling and that each laser will require 3 W for cooling—numbers taken from standard laboratory equipment. In addition to this, the serial-to-parallel scheme setup is added, assuming two HNLF-based OFTs, sharing a single pump source and using the pump power used in the actual experimental demonstration described in [8]. The experimental demonstration at 640 Gbit/s is also added.

Figure 9.2 shows the outcome of these calculations. Generally WDM is less power consuming due to the passive demultiplexing. The worst case is by far the N parallel HNLFs and pump sources, being an order of magnitude worse than the WDM case. However for serial-to-parallel conversion at high bit rates (above 640 Gbit/s) the OFT-OTDM system is almost an order of magnitude less power consuming compared to the WDM system, and almost two orders of magnitude better than the worst-case parallel HNLF receiver OTDM setup. Furthermore, this benefit increases for higher bit rates as can be seen in Fig. 9.2, right, which is because the OFT-receiver scheme uses constant power irrespective of the bit rate.

This finding agrees well with the above postulation that processing more bits in fewer components is where optical signal processing has its strongest advantage [4, 5]. For serial-to-parallel conversion, the energy per bit scales inversely with the bit rate, as most power is used for the control signal. And irrespective of the total OTDM bit rate, only two OFTs will be required, and to a large extent, the same pump power will be needed, which ultimately leads to a better energy-efficiency for higher bit rates. It even outperforms the simple point-to-point WDM systems.

These calculations are realistic in the sense that they are based on extracted parameters from the experimental demonstration. One should of course be cautious with putting too much trust into calculations like these—they are based on assumptions of certain power values, and these may vary a lot, and the actual curves representative of the different system scenarios will change correspondingly. This sort of investigation is not hard science and the results should be seen as speculative, and we mostly focus on the observed trends. In that respect, we can observe that the use of time lenses for serial-to-parallel conversion has a tendency to increase in system power more slowly than the WDM and classical OTDM systems. Where exactly the curves will cross each other is obviously strongly dependent on the parameter values used.

In the following section, a detailed description of the time lens based OFT scheme is given.

9.2 Time-Domain Optical Fourier Transformation/Time Lens Principle

In [10] Kolner presents the space-time duality principle, which accounts for the equivalence between diffractive imaging and dispersive propagation. Based on this space-time duality formalism, temporal imaging becomes a simple and versatile tool with many applications, such as scaling temporal waveforms using simple systems, or indeed scaling of spectral shapes, as well as converting between temporal and spectral domains.

9.2.1 Time Lens Principle and Time-Domain OFT

In the space-time duality description, it is shown how diffraction and dispersion operate similarly on spatial objects and temporal waveforms, respectively. The diffractive far field image (Fraunhofer diffraction) is equivalent to a dispersed waveform, in that the far field image becomes the Fourier transform of the spatial object, and similarly the dispersed waveform becomes the Fourier transform of the

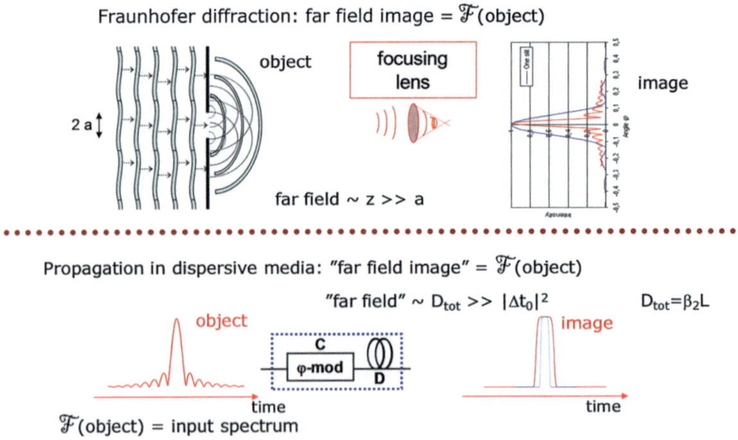

Fig. 9.3 Equivalence of diffractive imaging and dispersive propagation. *Top* A plane wave propagating through a slit in a wall, corresponding to a rectangular "object" is diffracted to generate the Fourier transform of the object as the image in the far field (propagation distance z ≫ a, object size), known as Fraunhofer diffraction. A spherical lens will add a quadratic phase modulation on the diffracted wavefront and may be used to manipulate the size and focal plane of the image, e.g. by focusing the image down. *Bottom* Dispersive propagation of an optical pulse is equivalent to diffractive imaging in the time domain. For large dispersion ($D = \beta_2 L \gg |\Delta t_0|^2$), i.e. much greater than the original pulse width (object size) squared, the transmitted pulse shape ("image") becomes the Fourier transform of the input pulse shape ("object"). A parabolic phase modulation, corresponding to a linear chirp acts as a lens in time ("time lens"), and may be used to focus the image to smaller extent, or more advanced temporal imaging [10]

input waveform. Figure 9.3 shows this basic equivalence. So diffraction and dispersion are equivalent. One may also include lenses. In spatial optics, a spherical lens will result in a quadratic phase modulation of the diffracted wavefront, which can be used to focus or enlarge an image. In the time domain, a temporal parabolic phase modulation on the waveform will impose a linear chirp, which, when balanced with proper amount of dispersion, can be used to focus or enlarge the temporal waveform, and such a system is therefore called a time lens.

In the following, a derivation of the simple time lens based OFT system shown in Fig. 9.3 (bottom) and in Fig. 9.4 is given. This derivation shows that a parabolic phase modulation followed by appropriate amount of dispersion will result in an output waveform, which is exactly equal to the Fourier transformation of the input waveform, which again corresponds to the input spectrum [11]. For the reverse system, i.e. dispersion followed by phase modulation, the solution may readily be derived the same way, and result in an output where the spectral output is equal to the input waveform, i.e. a temporal-to-spectral conversion.

In Fig. 9.4, the input and output electric field amplitudes in the spectral and temporal domains are shown. They relate to the optical power as

$$I_i(t) = |A_i(t)|^2, \quad I_i(\omega) = \left|\tilde{A}_i(\omega)\right|^2$$

Figure 9.4 shows how the electric fields evolve through the different parts of the set-up. In the phase modulator, a linear chirp, C, is added to the electric field yielding an electric field waveform of

$$A_c(0,t) = A_0(0,t).\, e^{\frac{i}{2}Ct^2}$$

Note that in order to obtain a linear chirp, it is necessary to add a parabolic phase modulation, as the chirp is the time derivative of the phase. The chirped spectrum is the Fourier transform of the chirped waveform, i.e.

$$\tilde{A}_c(0,\omega) = \mathcal{F}\{A_c(0,t)\}$$

The chirped waveform is now propagated through a length of dispersive medium, such as a fibre with accumulated dispersion $D = \beta_2 L$, where β_2 is the GVD parameter of the fibre and L is the total length of the fibre. The linear transfer function for such a fibre is $H(\omega) = e^{\frac{i}{2}\beta_2\omega^2 L}$, see e.g. [12]. Thus, the spectral output of the dispersive element becomes

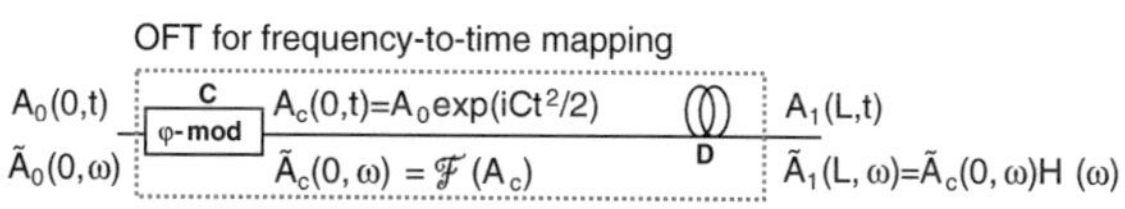

Fig. 9.4 Principle sketch for deriving the time-domain optical Fourier transformation (OFT) relations for the case of frequency-to-time conversion

$$\tilde{A}_1(L,\omega) = \tilde{A}_c(0,\omega)\cdot H(\omega) = \tilde{A}_c(0,\ \omega)\cdot e^{\frac{i}{2}\beta_2\omega^2 L}$$

and the output temporal shape is just the inverse Fourier transform of this, i.e.

$$\begin{aligned} A_1(L,t) &= \mathcal{F}^{-1}\{\tilde{A}_1(L,\omega)\} = \mathcal{F}^{-1}\left\{\tilde{A}_c(0,\omega)\cdot e^{\frac{i}{2}\beta_2\omega^2 L}\right\} \\ &= \int_{-\infty}^{\infty} A_c(0,t')\cdot h(t-t')dt \end{aligned}$$

where

$$h(t) = \mathcal{F}^{-1}\{H(\omega)\} = \frac{1}{2\pi}\int_{-\infty}^{\infty} e^{\frac{i}{2}\beta_2\omega^2 L} e^{-i\omega t}\, d\omega = \sqrt{\frac{i}{2\pi D}} e^{\frac{-i}{2D}t^2}$$

i.e. the output temporal shape is the convolution of the chirped input pulse to the dispersive element with the fibre response function h(t). Now, assuming that the accumulated dispersion is matched to the chirp added to the pulse (D = 1/C), and using the transformation t/D = ω, the output waveform becomes

$$\begin{aligned} A_1(L,t) &= \int_{-\infty}^{\infty} A_c(0,t')\cdot h(t-t')dt \\ &= \sqrt{\frac{i}{2\pi D}}\int_{-\infty}^{\infty} A_0(0,t')\cdot e^{\frac{i}{2}Ct'^2}\cdot e^{\frac{-i}{2D}(t-t')^2}dt' \\ &= \sqrt{\frac{i}{2\pi D}}\cdot e^{\frac{-i}{2D}t^2}\int_{-\infty}^{\infty} A_0(0,t')\cdot e^{\frac{it}{D}t'}dt' = \sqrt{\frac{i}{2\pi D}} e^{\frac{-i}{2D}t^2}\int_{-\infty}^{\infty} A_0(0,t')e^{i\omega t'}dt' \\ &= \sqrt{\frac{i}{2\pi D}}\cdot e^{\frac{-i}{2D}t^2}\cdot\tilde{A}_0(0,\omega) \end{aligned}$$

Finally, the intensity at the output of the OFT becomes

$$I_1(t) = |A_1(t)|^2 = \frac{1}{2\pi D}\cdot\left|\tilde{A}_0(\omega)\right|^2 = \frac{1}{2\pi D}\cdot I_0(\omega)$$

i.e. the output temporal waveform is exactly given by the input spectral shape [13] scaled with a constant depending on dispersion D.

Example: A supergaussian spectrum: It is often easier to create special shapes in the spectral domain than in the time domain, and thus this OFT scheme can be seen as a tool to transfer e.g. grating-based spectral shapes into temporal waveform shapes. If for instance a supergaussian spectrum is generated, $I_0(\omega) = K\cdot e^{-(\omega/\Delta\omega)^{2m}}$ with amplitude K, then the temporal output pulse shape becomes $I_1(t) = \frac{1}{2\pi D} \cdot e^{-(t/D\Delta\omega)^{2m}}$, for K = 1, when using the aforementioned transformation t/D = ω. That is, exactly the same shape as the spectral shape and with a tuneable 1/e-width given by $\Delta t_{1/e} = D\Delta\omega$. The width is tuneable, as D = 1/C is tuneable, or in other words, by tuning the chirp rate (i.e. the quadratic phase modulation) of the time lens, the size of the image can be controlled.

From the above equations, it is seen that it is also possible to compensate for the temporal impairments induced by dispersion in transmission fibres by filtering and applying OFT, thus transforming the Gaussian unaffected spectrum to a clean temporal waveform, as investigated extensively by Nakazawa et al. [11]. In [11], m = 1, and the pulse is thus Gaussian. As a benchmark, the dispersion is set to be equal to the original pulse width squared, i.e. $|D| = \Delta t_0^2$, which leads to an output pulse width after OFT of $\Delta t_{1/e} = \Delta t_0^2 \Delta\omega$, which is equal to Δt_0 for a Gaussian, and hence the original pulse is obtained.

The above derivation is for frequency-to-time mapping, and a similar approach will result in the formalism for time-to-frequency mapping [14], where the order of dispersion and phase modulation is simply reversed, yielding

$$I_c(\omega) = \left|\tilde{A}_c(\omega)\right|^2 = \frac{2\pi}{|C|} \cdot |A_1(0,t)|^2 = \frac{2\pi}{|C|} \cdot I_0(t)$$

Here again, t = ωD = ω/C, and C is seen to determine the scaling factor between time and frequency. Thus a temporal spacing Δt will be mapped to a spectral spacing of Δω = ΔτC.

The phase modulation may be obtained in several ways, e.g. by electro-optic phase modulation as sketched above, or by four-wave mixing (FWM) to obtain very high chirp rates. In [15], a good comparison between different types of phase modulation is given.

In the case of FWM, the dispersed waveform to be transformed in a time-to-frequency mapping acts as signal $E_s(t)$ and the phase modulation is applied using linearly chirped pump $E_p(t)$ pulses, generated by propagation of transform limited pulses in a dispersive medium. In the FWM process, the signal is converted to an idler $E_i(t)$, which combines the phases of both pump and signal $(E_i \propto E_p^2 E_s^*)$. In our case, the signal is $A_0(L, t)$, and the idler is $A_c(t)$. As a result of the FWM process, the time information contained in the signal is mapped onto the power spectrum of the generated idler, see more detailed example below for serial-to-parallel conversion.

9.2.2 Important OFT/Time Lens References

Several groups worldwide have contributed enormously to the development of time lenses and their use for different applications. It is beyond the scope of this chapter to give a full account of all the work that has been done, but some of the most important references that will be useful to learn more about time lenses are mentioned here. In [10], Kolner first described the equivalence between spatial and temporal optics as mentioned above. In [16], the time-to-frequency case is described in detail, and in [17], the use of the space-time duality is described by Azaña and Muriel in the context of the application *spectral analysis*. The ability of a time lens to focus a waveform to a certain focal point in time is suggested by Mollenauer and Xu to be used for timing jitter compensation [18]. Jalali and co-workers describe the time lens usage for time-stretching in [19]. This scheme may be used to investigate spectral profiles/fingerprints in the time domain, by simple Fourier transformation and photonic stretching. In [11], Nakazawa et al. details the time-domain optical Fourier transformation (OFT) principle used for distortion-free transmission, where a frequency-to-time conversion is utilized. Weiner's group has used the time lens concept for arbitrary waveform generation, e.g. for ultra-wide-band communication [20]. In [21], a description of various ultra-fast applications of time lenses are given, and in [22], the first demonstration of time lenses based on a silicon chip is presented. Finally, in [15], Salem, Foster and Gaeta give a good overview of recent ultrahigh-speed optical signal processing applications using time lenses.

9.3 Serial-to-Parallel Conversion

Serial-to-parallel conversion is a case of temporal to spectral conversion, i.e. where dispersion is preceding phase modulation. One implementation of this is to transform a serial OTDM data signal into separate parallel WDM channels as already introduced in Fig. 9.1

9.3.1 640 Gbit/s OOK OTDM-to-DWDM Conversion

For an efficient solution, the electro-optic phase modulation is exchanged with an optical phase modulation, which may be provided in various ways. Using four-wave mixing (FWM) allows for very strong phase modulation [15]. Figure 9.5 shows a schematic of OTDM-to-WDM conversion using FWM.

The data and pump are first dispersed in standard single mode fibre (SMF) to create a linear chirp on the pulses. The rectangular pump spectrum is sent through a fibre of dispersion D, starting at its Fourier transform position. After D, it will have acquired a flat top waveform. To obtain OTDM-to-WDM conversion with spectral compression of the individual WDM channels, allowing for DWDM creation, the data pulses must traverse a fibre with dispersion exactly equal to half of the pump's dispersion, i.e. D/2. Now, these two signals are injected into a nonlinear medium, e.g. a HNLF, where they interact via FWM. The pump and data signal create an

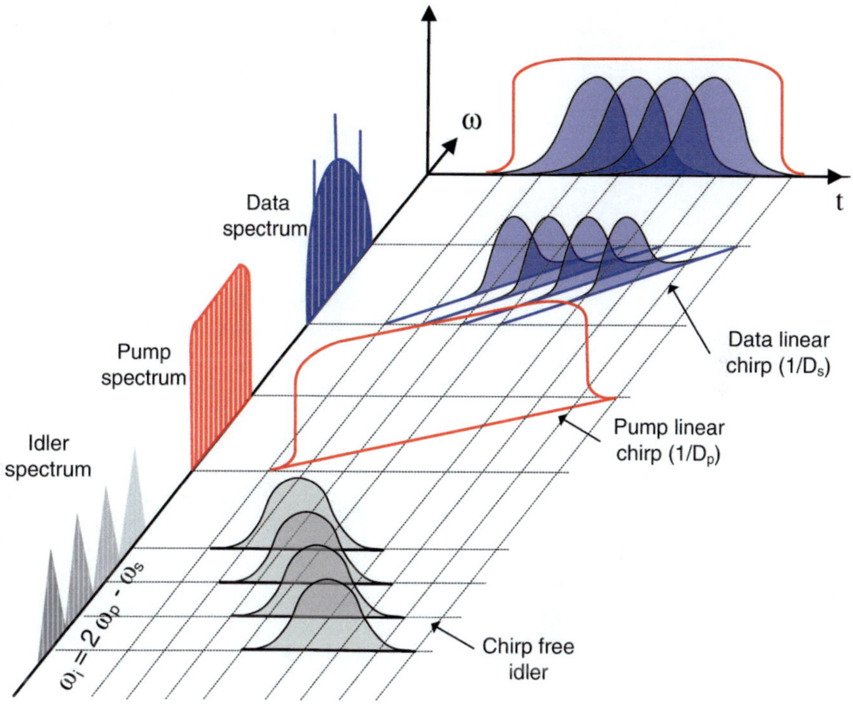

Fig. 9.5 OTDM-to-WDM conversion using FWM. Spectrographic representation of the basic principle. Dispersed waveforms interact in a four-wave mixing process yielding non-chirped idlers at separate wavelengths

idler at the optical frequency $\omega_i = 2\omega_p - \omega_s = 2\pi c/\lambda_i$. Because the pump is chirped and covers all the OTDM channels, different channels will overlap in time with different parts of the pump spectrum and thus give rise to idlers at different wavelengths. And because the data is chirped with D/2 compared to the pump being chirped by D, each OTDM pulse will be converted to an idler with a compressed spectrum. This makes it possible to place the idlers very densely, i.e. to create a DWDM signal. Each DWDM channel now originates from a specific OTDM channel, and they may simply be filtered out by passive filters. This way, all OTDM channels are demultiplexed simultaneously.

Figure 9.6 shows the basic schematic setup for OFT-based OTDM-to-WDM conversion. In principle all OTDM channels can be demultiplexed in one device [23], but in practice it will be very difficult to create the required sharp edged rectangular pump pulses, and thus some data pulses may overlap with two neighbouring pump pulses giving some impairment. It is therefore more practical to assume that two of these OFT-based demultiplexers or OTDM-DWDM converters in parallel are needed to demultiplex all channels. Still, this is potentially a substantial reduction in active components in an OTDM receiver. As described in the introduction, this gives rise to an enormous power consumption reduction,

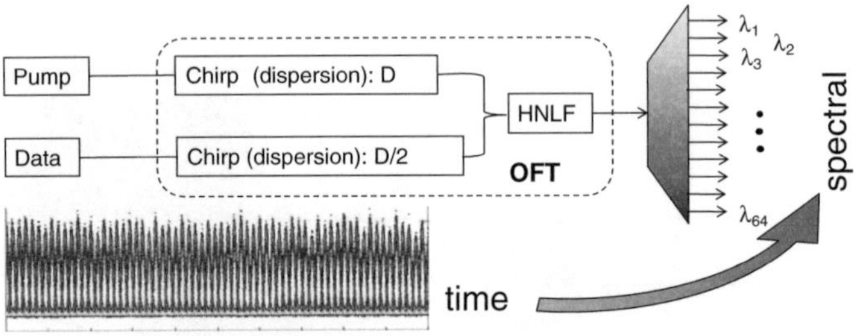

Fig. 9.6 Schematic setup for OTDM-to-DWDM conversion using the optical Fourier transformation technique (OFT) by four-wave mixing in a HNLF. In practice, two OFT units may be preferred. From [8]

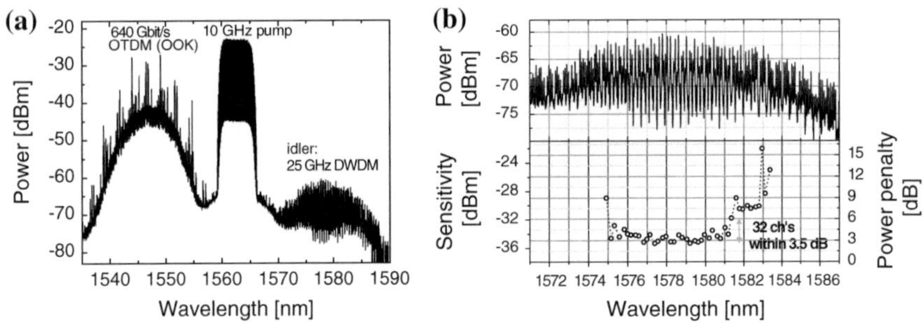

Fig. 9.7 Experimental demonstration of 640 Gbit/s OTDM-DWDM conversion by OFT using FWM in HNLF. **a** Output spectrum from the HNLF showing the OTDM data, the flat-top pulse and the DWDM idler channels. **b** Zoom in on the DWDM idler spectrum (*top*) and the receiver sensitivity and power penalty at BER 10^{-9} for the demultiplexed channels (*bottom*)

rendering an OTDM point-to-point system less power consuming than a WDM point-to-point system for high bit rates.

Figure 9.7 shows results from an experimental demonstration where a 640 Gbit/s OTDM signal is converted into a DWDM signal with only 25 GHz spacing between the DWDM channels [8, 24]. 40 of the 64 tributaries are simultaneously demultiplexed with a BER less than 1E-9. The channels to the sides of the spectrum, correspond to the OTDM channels overlapping with the edges of the pump pulse, and thus have lower conversion efficiency, and ultimately lower OSNR, and therefore higher penalty. The most extreme channels cannot get a BER less than 10^{-9}. The receiver sensitivity at BER 10^{-9} is measured for the 40 centre-most channels and 32 of these have a sensitivity spread within 3.5 dB. This clearly demonstrates the effectiveness of this scheme. With only two of such OFT-units in parallel using temporally detuned pumps, all OTDM channels should be attainable, irrespective of the OTDM bit rate. In the example shown in Fig. 9.7, the non-linear element was a HNLF but in [24], it was also shown that this scheme could be carried out using a silicon nanowire as the nonlinear medium.

9.3.2 Serial-to-Parallel Conversion of Data with Advanced Modulation Formats

Using FWM for phase modulation, one may benefit from the phase-preserving nature of FWM, i.e. that the idler retains the phase information of the signal, albeit in conjugate form. This means that the above described serial-to-parallel conversion scheme may also be used on phase-modulated data signals without losing any information. Figure 9.8 shows experimental results from [25], where a 160 or 320 Gbaud OTDM signal carrying QPSK or 16QAM modulation is converted to DWDM 10 Gbaud channels. In this experiment, it was confirmed that the 16 QAM data could be successfully received to below the forward error correction (FEC) limit of 3.8E-3, corroborating that this scheme is independent of the modulation format.

9.3.3 Wavelength-Preserving Serial-to-Parallel Conversion

In the two cases mentioned above, degenerate FWM is used, where the two pump photons are at the same wavelength, originating from the same source. In that case the idler appears at the other side of the pump from the signal. However, nondegenerate FWM may equally well be used. This adds more design freedom to where the desired idler wavelength may be positioned. In [26], the pump photons are in orthogonal polarisation and at different wavelengths.

This scheme allows one to choose the idler wavelength to be for instance exactly where the input signal was. In [26], the orthogonally polarized pumps are symmetrically positioned around the input signal wavelength, and one pump is aligned in polarisation with the input data signal. Owing to energy and momentum conservation, the idler will now be generated at the same wavelength as the data signal, but in orthogonal polarisation to the data signal. One may thus filter the idler away

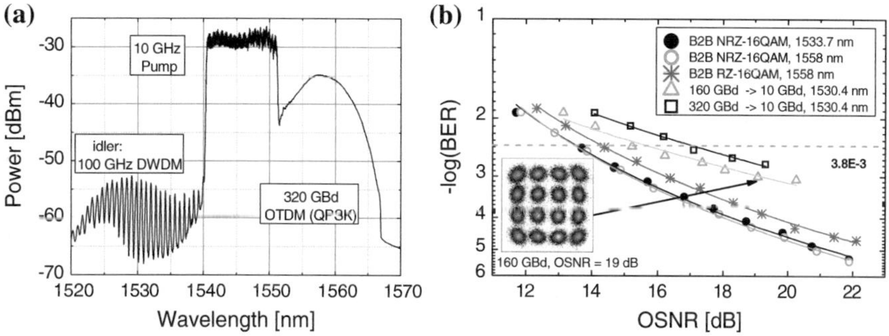

Fig. 9.8 QPSK and 16QAM data on up to 640 Gbaud converted to DWDM 10 Gbaud channels. From [25]

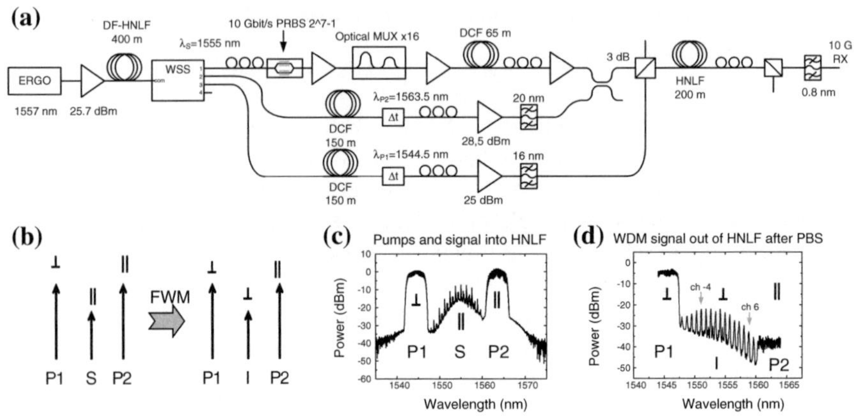

Fig. 9.9 Setup and principle of wavelength-preserving serial-to-parallel conversion using two orthogonally polarized pumps symmetrically spaced about the data channel. **a** The setup: two orthogonal pumps are merged with the data signal into a HNLF. **b** Principle of orthogonally polarised idler. **c** Spectrum of the pumps and the signal into the HNLF. **d** Spectrum of the output of the HNLF and after a polarisation beam splitter. From [26]

from the signal using e.g. a polarisation beam splitter. Figure 9.9 shows a schematic setup of this scheme, where a 160 Gbit/s OTDM signal is converted to a DWDM signal at the same wavelength as the original data. This can reduce the required spectral bandwidth otherwise needed for FWM OFT, and shows that OFT is wavelength transparent.

9.3.4 Nyquist-OTDM to "OFDM"-Like Signal Conversion

In addition to the pursuit of higher spectral efficiency by the use of higher order modulation, signal formats like orthogonal frequency division multiplexing (OFDM) and Nyquist WDM are also very interesting. They both rely on shaping of either the spectral or temporal domains to create sinc-function-like shapes and very dense multiplexing. In a Nyquist channel, the temporal waveform is a sinc-function, and each symbol goes to zero exactly at the neighbouring symbol position, resulting in orthogonality. The spectral shape is rectangular, and it is therefore simple to put WDM channels shoulder-by-shoulder and thus increase the spectral efficiency almost to 1 bit/Hz (for binary modulation). If one combines Nyquist channels with the potential of ultra-high symbol rate OTDM, one may simply create ultra-high symbol rate Nyquist channels by the use of rectangular optical filtering directly on an OTDM signal. This is simple and straightforward, and additionally helps to reduce spectral redundancy from guard bands that one would otherwise encounter in low-rate Nyquist WDM channels [27, 28]. In fact, one may say that the higher the symbol rate, the lower the waste of valuable spectrum. This is because optical filters will give relatively steeper flanks of the filter compared to the width of the filter shape for broader spectra. In the time domain, the pulses will become sinc-function shaped, and if the filter has the width of the symbol rate, then each sinc-function shaped pulse will exactly give a null in the neighbouring symbol. In [29], it was shown that a 1.28 Tbaud Nyquist channel could be generated this way using only 1.28 THz bandwidth, i.e. with no spectral redundancy within the 1.28 THz bandwidth. This type of spectrally efficient serial data signal can also be demultiplexed using a time lens serial-to-parallel converter [29].

Figure 9.10 shows the experimental results for the 1.28 Tbaud Nyquist channel generation and detection. It is observed that the spectral extent is dramatically reduced to 1.28 THz, about 10 nm. The characteristic time-domain waveform is observed in Fig. 9.10b, with the minimum interference points, corresponding to where the sinc-shaped waveforms go to null. The serial-to-parallel conversion spectra are shown in Fig. 9.10c. The N-OTDM spectrum is converted to an OFDM-like spectrum for each channel—the individual spectral channels are not discernable

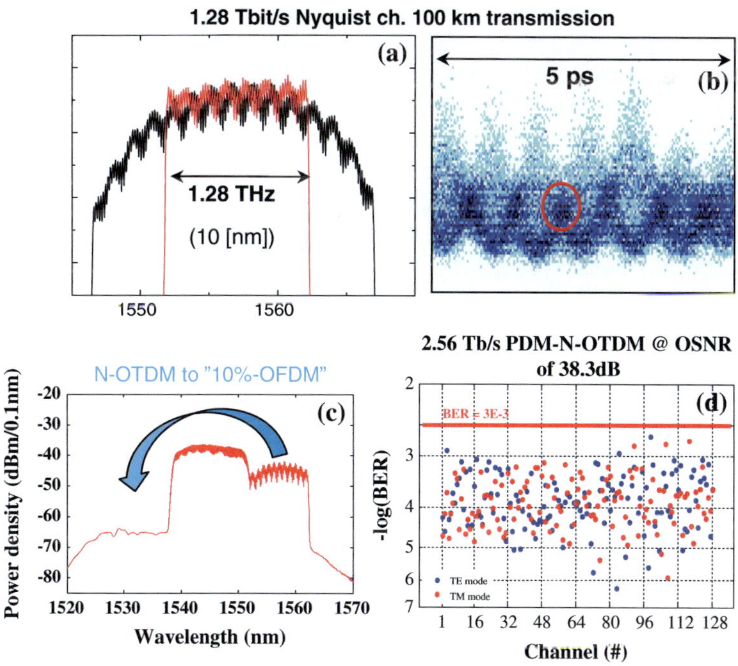

Fig. 9.10 Nyquist-OTDM serial-to-parallel conversion to "OFDM"-like signal. **a** Spectral profile of the original 1.28 Tbaud RZ signal and the filtered Nyquist signal. **b** The generated temporal profile of the 1.28 Tbaud Nyquist channel. **c** The output spectral profile of the time lens based serial-to-parallel converter. **d** Measured bit error rate points for all 128 tributaries in two polarisations, corresponding to 256 points in total yielding 2.56 Tbit/s transmission. From [29]

in this figure, due to the spectral overlap of the channels. The OFT is tailored such that the generated OFDM-like signal has a large guard interval (around 90 %). This enables extraction of the tributaries simply using passive filtering, as the associated pulse broadening is accommodated for by the guard interval. The 1.28-Tbaud N-OTDM signal is passed through a dispersion compensation fibre (DCF) having $\beta_2 L_{DCF} = 1.243\ ps^2$ and subsequently launched into a 100 m polarisation-maintaining highly nonlinear fibre (PM-HNLF, zero dispersion at 1545-nm and $S = 0.025\ ps/nm^2km$). The desired chirp rate $C = 1/(\beta_2 L) = \Delta\omega/\Delta t$ depends on the temporal spacing and desired frequency spacing. In this work, the target was to map a 1.28 Tbaud signal with $\Delta t = 0.781$ ps to a frequency spacing of 100 GHz. The generated pump pulses having a super-Gaussian spectrum (FWHM of 1.6 THz) centred at 1545 nm are linearly chirped with a chirp rate of C/2 by propagation through some DCF length and merged with the data in the FWM process in the PM-HNLF. Since the frequency spacing (100 GHz) is much larger than the baud rate (10 Gbaud for each tributary), a tuneable OBF with a bandwidth of 40 GHz is used to extract the subcarriers. The chosen bandwidth is a trade-off between minimisation of interchannel interference (ICI) and inter-symbol interference (ISI). In this particular experiment, there was no attempt to obtain many simultaneous converted channels, but nevertheless 14 tributaries at different wavelengths could still be simultaneously received with a BER < 3E-3. As explained above, at least half of the channels should be possible under the right circumstances. Finally it should be mentioned that the N-OTDM signal was transmitted over 100 km SLA-IDF transmission fibre and that forward error correction coding (FEC) was employed, resulting in a net 2.4 Tbit/s truly error free data rate after transmission, i.e. with zero errors occurring in the measurement time.

In [30], it is furthermore shown that polarisation-independent serial-to-parallel conversion is possible using a polarisation diversity scheme placing the above-mentioned PM-HNLF in a polarisation diversity loop.

In summary, serial-to-parallel conversion using time lens based OFT has been described in this section, and various demonstrations have been mentioned. In particular, it has been experimentally demonstrated that this functionality is transparent to symbol-rate, modulation format, data format, wavelength and polarisation, and indeed that the nonlinear medium can successfully be a silicon nanowire as well as a highly nonlinear fibre. In addition to this, it should be mentioned that the opposite process may equally be implemented, and a demonstration of WDM-to-OTDM conversion has also been demonstrated by converting 16 WDM channels into 16 OTDM channels [31]. This is a case of spectral-to-time conversion, and other examples of this include distortion-free pulse transmission [11] and flat-top pulse generation [12, 32]. Having control of both time-to-frequency and frequency-to-time conversion opens up for a number of interesting applications, where more advanced time lens systems may be created using more than one time lens. In the following some such "telescopic arrangements" will be described.

9.4 Spectral Telescopes and Applications

Just like in spatial optics, where the combination of several lenses can make more advanced imaging systems such as telescopes, the same is the case for time lenses. Both temporal telescopes and spectral telescopes can be constructed [15]. In the following we will focus on spectral telescopes used for optical processing of spectrally parallel data signals such as WDM and OFDM signals.

9.4.1 WDM Nonlinear Optical Signal Processing—WDM Grid Manipulation

Nonlinear optical signal processing directly on WDM channels is traditionally challenging due to the nonlinear intermixing of the WDM channels. Therefore many schemes require a separation of the WDM channels followed by an individual signal processing of each channel separately. This is not considered a power-efficient and viable solution as discussed in Sect. 9.1, because one would not be processing the entire WDM signal in one device. Using optical spectral telescopes this may change. One could imagine converting a WDM signal to a TDM signal, where the bits do not overlap in time, and then do some form of optical signal processing on that TDM signal. Adding a second time lens stage could then bring the OSP'ed signal back to the original WDM state or some manipulated form of the WDM signal.

In this section, we will describe how to use two time lenses to construct a spectral telescope to manipulate the grid spacing of a WDM signal. This may e.g. find use in flexible grid applications.

Figure 9.11 shows a principle spectrographic representation of the use of FWM for WDM grid manipulation. This representation tool is a simple and illustrative way to design such time lens systems. An incoming WDM signal, in this case represented by three wavelengths filling up most of the available time slot and shoulder-to-shoulder in wavelength, is aligned with a chirped pump pulse for phase modulation.

FWM occurs between the pump and the signal and creates chirped idlers following the relation $\omega_i = 2\omega_p - \omega_s$, with the indices i, p, s corresponding to the idler, the pump and the signal (WDM inputs), respectively. In this mapping schematic, each temporal component of the signal is "mirrored" on the spectral axis (vertical) through the pump to a position on the other side of the pump at the same spectral distance to the pump. Thus it is easy to predict the spectrographic profile of each idler. The idlers are now dispersed, e.g. by transmission through fibre with accumulated dispersion $D_1 = \beta_2 L_1$. This has the effect of temporally positioning the individual frequency components of the idlers following the linear mapping relation $\Delta\tau = (\omega_1 - \omega_2)\beta_2 L_1$. In this schematic spectrographic representation, the linear mapping due to dispersion is illustrated by a dashed straight line in which

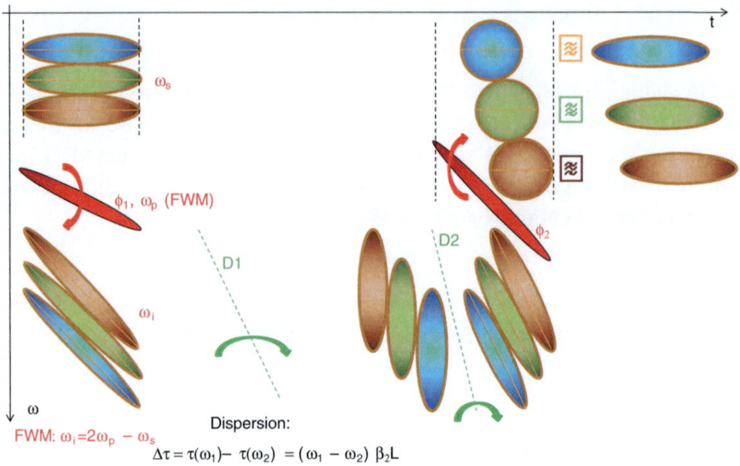

Fig. 9.11 Spectrographic schematic of a spectral telescope, consisting of two time lenses, based on FWM, and accompanying dispersion. In this schematic, 3 WDM channels are closely positioned in optical frequency, ω, vertical axis and they each fill out their time slots (space between *dotted vertical lines*). By FWM using a chirped pump, doubly chirped idler waves are generated (*lower left*). These are then dispersed to compensate for the chirp to yield 3 separate temporal channels. Thus a parallel-to-serial conversion has been obtained. Please note that they are also slightly displaced in frequency, corresponding to the original spacing between the WDM channels. ϕ_1 and D_1 constitute the first time lens stage. Now, the converted signal is sent through the second time lens stage with a different "focal length" corresponding to a different phase (chirp rate) and dispersion. In this case, ϕ_2 and D_2 are a factor of two off from ϕ_1 and D_1, yielding a telescopic magnification factor of 2. After serial-to-parallel conversion in the second time lens stage, a magnified WDM spectrum is obtained. The WDM channels are now twice as wide spectrally compared to the original, and spaced twice as far from each other, and twice as narrow in the time domain. This makes is possible to use a narrow filter, with the original WDM signal filter bandwidth on each channel, and the result is a WDM signal but with twice the grid spacing

each spectral component of the generated idler is now simply "mirrored" horizontally along the time axis to a point on the other side of the dispersion line with the same distance to this line as the original idler point. This completes the first step from parallel to serial conversion, and the WDM channels are now completely separated in time, i.e. being a serial signal. This first time lens system is now followed by a second time lens system, where the dispersion and chirp values may be chosen freely.

They could be the same, which would lead back to the original signal, or as here, differ by a factor of 2. Going through the second dispersion (note that the slope of the D_2-line is twice that of the D_1-line) and phase modulation, ϕ_2, with the FWM pump-2 being chirped twice as much as pump-1 for ϕ_1, and being spectrally twice as broad, results in a mapping of the WDM channels to different frequencies and with different pulse and spectral widths. Essentially, as $D_1/D_2 = 2$, and the spectral width of pump-1 is half that of pump-2, the output spectra are twice as broad and the pulse widths twice as narrow compared to the input. However, the magnification factor can be dimensioned with a large degree of freedom and does not need to be an integer number. Just like for a spatial telescope, where the magnification is given by the ratio of the focal lengths of the objective lens (f_1) and the eye piece lens (f_2), $M = f_1/f_2$, the magnification for a spectral telescope is given by the equivalent *focal lengths*, i.e. the accumulated dispersions: $M = D_1/D_2$. In the above example, M = 2, but can be tailored to a specific application.

For M < 1, the spectrum is compressed. In [33], an example of spectral compression is given. There, a 100 GHz spaced WDM signal consisting of 8 wavelengths is compressed to 50 GHz spacing. Each wavelength carries 10 Gbit/s DPSK, and thus already fits into a 50 GHz grid. The BER performance is very good with only 1.4 dB sensitivity penalty at a BER = 1E-9.

Figure 9.12 shows an example of M = 2. In this case, 16 WDM channels spaced at 50 GHz each carrying 10 Gbit/s DPSK data are spectrally magnified to a grid spacing of 100 GHz. To simplify the setup, a folded time lens scheme is employed —the same nonlinear device (here an HNLF) is used for both time lenses operating in opposite directions.

The input 16-channel 50 GHz spacing WDM signal enters the grid manipulator from the top left through a circulator together with a chirped pump ($Pump_1$). Together they travel from left-to-right through the HNLF into a second circulator sending the pump, signal and generated idler down into the loop. Here a filter removes all but the idler, which is now transmitted through a dispersive medium, either SMF or DCF (anyone can be chosen as long as the chirp is properly matched to it). After the dispersive medium, the chirped idler is joined with $Pump_2$ through a coupler and send back through the right-most circulator and into the HNLF from right-to-left. The now magnified WDM signal is extracted through the circulator and an optical filter suppressing pump and idler from the loop. Figure 9.12 bottom

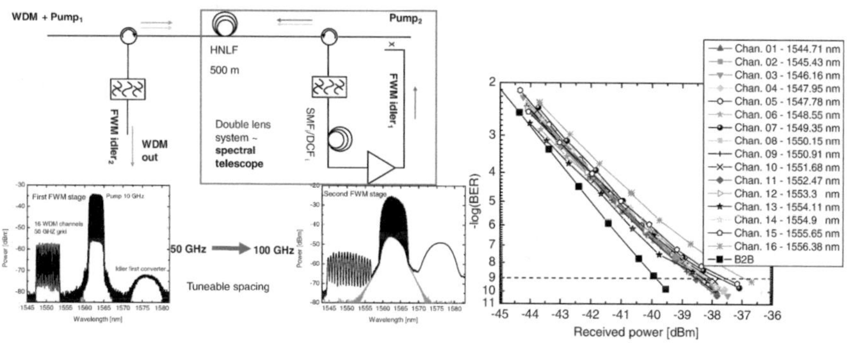

Fig. 9.12 WDM grid manipulation using spectral magnification. *Left* A folded counter propagation configuration is used for the two time lenses. *Bottom left* Output of first FWM stage (time lens), showing the 16 WDM channels 50 GHz spaced together with $Pump_1$ and the first idler. *Bottom right* Output of the second time lens showing the now magnified WDM signal with 16 channels 100 GHz spaced along with the output spectrum of the second pump and the first idler. *Right* Bit error rate performance. The back-to-back (B2B) is an average of all 16 original WDM channels

left shows the spectra involved with the 16 WDM channels being 50 GHz spaced at the input and the output being 100 GHz spaced. A detailed BER characterisation shows that all channels are error free, Fig. 9.12, right. There is a roughly 2 dB penalty at BER = 1E-9. By changing the chirp rate of the pump pulses, and matching the dispersion to it, the magnification may be changed. This may find use in e.g. flexible grid networks, where data channels may be easily fitted into available bandwidths and according to required performance criteria.

In the following, an alternative application of spectral magnification is presented. The idea here is to spectrally magnify an orthogonal frequency division multiplexing (OFDM) signal in order to be able to merely receive it by passive optical filtering.

9.4.2 OFDM-to-"DWDM" Conversion by Spectral Magnification

OFDM is a very spectrally efficient modulation format with sinc-function shaped spectra with a null in the neighbouring spectral channel. These sinc-shaped spectra are thus overlapping and it is not straightforward to separate these OFDM channels spectrally, as it is for e.g. DWDM signals. One could say that both a spectral and temporal filtering is most often required, either in the electrical or the optical domain. Most work on *all-optical* OFDM demultiplexing has focused on the Discrete Fourier transformation (DFT), where up to 26 Tbit/s OFDM demultiplexing has been achieved [34]. In optical DFT, the Fourier transformation is accomplished by matched filtering using cascaded delay-interferometers as suggested by Marhic [35] and subsequent optical gating in a narrow time window where the interference is absent. This scheme can give very good performance [36, 37]. However, optical DFT requires phase-stabilisation of the optical paths in the delay-interferometer, and an optical gate per subcarrier—in other words, the receiver complexity increases with the number of subcarriers, and as the optical gating requires active switching and time alignment, the power consumption thus also scales with the number of subcarriers.

If instead of DFT, one uses time-domain OFT in a single spectral magnifier, the OFDM signal may be spectrally magnified in a transform-limited way. As above this means that the spectral components are spread out, whereas the temporal components are squeezed tighter together within the available time slots. This now makes it possible to apply bandpass filtering with a filter bandwidth equal to the symbol rate and actually extract the data information without excessive inter-symbol and inter-carrier interference (ISI and ICI), as also sketched in Fig. 9.11. In Fig. 9.11, after spectral magnification, bandpass filters are applied, which broadens the time waveform and narrows the spectral extent of each channel. As suggested in Fig. 9.13, this corresponds to a conversion from OFDM to a DWDM-like signal, which may be received merely by passive filtering, i.e. avoiding optical gating of

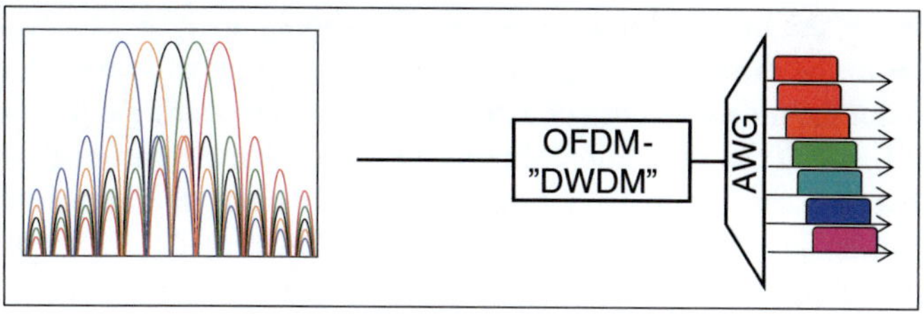

Fig. 9.13 Reception of OFDM signals using a WDM-type receiver requires an OFDM-to-"DWDM" data format conversion. This may be achieved with a spectral telescope

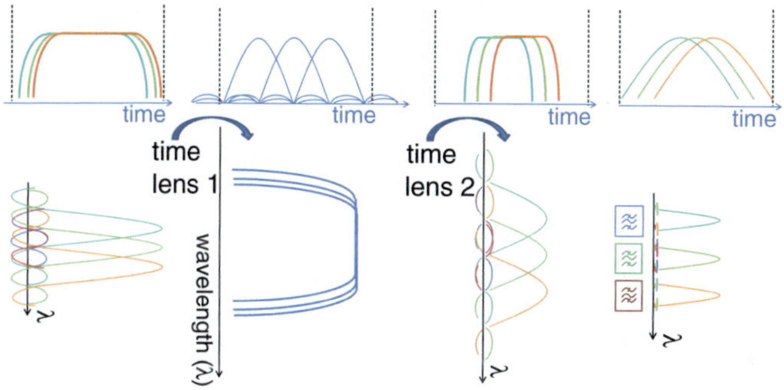

Fig. 9.14 Principle of OFDM conversion to a WDM-like data signal using a spectral telescope [38]. The two time lens stages will create a spectral magnification of the OFDM signal, allowing for tight direct filtering using e.g. a WDM filter

each channel. As discussed in the introduction for the OTDM-case, there is almost a two-order of magnitude power reduction when going from N parallel gates to a single (or two) time lens based serial-to-parallel converters. Here a similar situation is observed: many optical gates may be replaced by a single spectral magnifier and yield power savings. For AO-OFDM, the gates do not need to be as fast and power consuming as for the OTDM scheme, so the actual power saving may not be as dramatic as for OTDM, but the power and complexity scaling with the number of tributaries can be eliminated this way.

Figure 9.14 shows a principle spectrographic sketch, similar to Fig. 9.11, of the OFDM-to-WDM-like converter using spectral magnification and subsequent passive filtering.

Figure 9.14 shows an OFDM signal with its characteristic overlapping sinc-shaped spectral channels and rectangular waveforms. This OFDM signal is sent through the spectral magnifier consisting of two time lenses separated by dispersion. After the first time lens, the OFDM signal is basically converted to a Nyquist OTDM signal with sinc-shaped waveforms. The second time lens converts the signal back to an OFDM signal but with a magnification factor, in this sketch roughly set to M = 2. Since the time domain waveforms do not fill out the time slots and the spectra are broad, it is now possible to simply apply a filter with a bandwidth equal to the symbol rate, i.e. $\Delta f \sim$ timeslot^{-1}. This will broaden the waveform to fill out the time slot, but not extend into the neighbouring time slots, i.e. not generate ISI, and at the same time filter away a considerable part of the spectra of the neighbouring carriers to reduce the ICI. In principle, the larger the spectral magnification factor M, the less ICI. However, for the same Δf, the larger M is the more spectral content is filtered away, and hence the less OSNR remains for useful detection. Therefore, an optimum exists, and empirically M $\sim$ 4 seems like a good compromise.

Figure 9.15 shows a schematic setup for of 4× magnification of an OFDM signal and simulation results in terms of eye diagrams. The spectral telescope consists of two time lenses and dispersion in between. The simulation assumes 5 × 10 Gbaud subcarriers, with 12.5 GHz sinc spectra and 12.5 GHz spacing with DQPSK modulation. The lower part of Fig. 9.15 shows the case where a bandpass filter (BPF) is applied directly to the original OFDM signal. The filter bandwidth is tuned to yield the best possible eye opening under these circumstances. It is evident that the eye opening is not very good even for an optimum filter bandwidth of 8 GHz. Figure 9.15 top shows the magnification ×4 results. After the second phase modulation, the spectrum is 4 times broader, and therefore it is seen to be possible to apply a filter with a bandwidth close to the baud rate to strongly suppress the neighbouring carriers. In this case the optimum is 15 GHz, and clear and open eye diagrams are obtained. Hence only a single spectral magnifier is required in conjunction with bandpass filtering. The filter shapes are Gaussian in both cases.

Figure 9.16 shows experimental results in terms of spectra at two points within the spectral telescope, after the first time lens and after the second time lens. 10 carriers are used in this experiment, and they are 12.5 GHz spaced and carry 10 Gbit/s DPSK modulation at 10 Gbaud. The OFDM carriers are generated by spectral sinc-function shaped filtering directly on the 10 Gbit/s DPSK continuous spectrum, in a scheme adapted from the method proposed and demonstrated in [39]

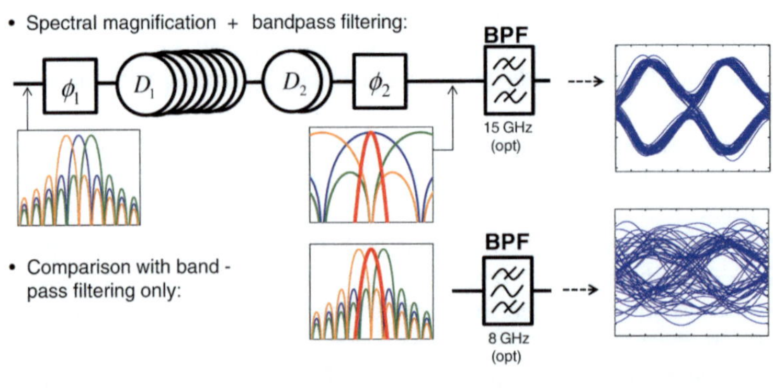

Fig. 9.15 Comparison of spectral magnification and bandpass filtering (*top*) to the case without magnification (*bottom*). Simulations show a clear improvement of the eye diagram when spectrally magnifying ×4. From [38]

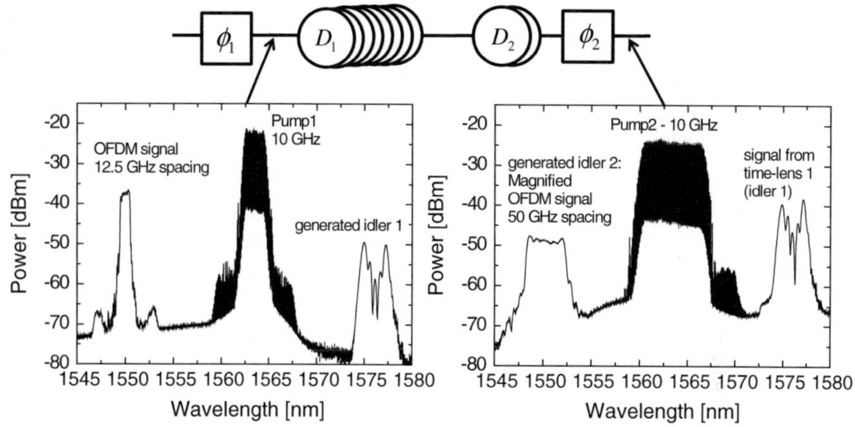

Fig. 9.16 OFDM-to-"DWDM" conversion for a 100 Gbit/s OFDM superchannel (10 × 10 Gbit/s carriers 12.5 GHz spaced). Spectra after the first time lens (*left*) and after the second time lens (*right*) revealing the 4× broader OFDM spectrum

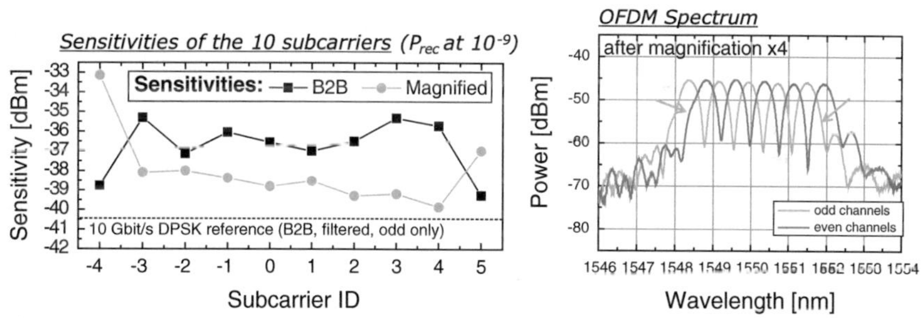

Fig. 9.17 BER evaluation of spectral magnification of an OFDM signal. The sensitivities are measured at BER 1E-9 and show a clear improvement with spectral magnification compared to simply filtering on the unmagnified OFDM signal. The outer channels suffer somewhat due to spectral broadening caused by imperfections in the time lens. From [38]

by Du et al. The OFDM signal is emulated by separately generating even and odd subcarriers, each consisting of five 12.5 GHz sinc functions with 25 GHz spacing, using a wavelength selective switch (WSS). The sign is reversed between neighbouring sinc subcarriers (for both even and odd), in order to overcome the limited WSS resolution of ~10 GHz and thus obtain the highest possible contrast ratio in the generated sinc spectra.

The time lenses are based on FWM in HNLF and the ×4 magnification is clearly observed in Fig. 9.16.

Figure 9.17 right shows a zoom in on the magnified spectra. For visualisation purposes the even and odd channels are shown individually, but they are magnified as a combined signal for BER characterisation. Figure 9.17 left, shows the BER performance for all 10 channels, before and after magnification and bandpass filtering, in terms of receiver sensitivities at a BER of 1E-9. The subcarriers are individually filtered out using an optical tuneable filter, with a Gaussian profile of 0.12 nm full-width at half maximum (FWHM) for the magnified case and 0.08 nm FWHM for the unmagnified case. In nearly all cases the bandpass filtered magnified spectra are better than the original signals after bandpass filtering. Note that as the data signal used in this case is an emulated OFDM signal, as described above with even and odd channels and every other carrier being π out of phase with its neighbour, there is a greater isolation between the carriers than in a real OFDM signal. Therefore it is even possible to get very low BER values by simply bandpass filtering directly on this emulated OFDM signal. This will not be the case for a real OFDM signal. None the less, when applying spectral magnification to this signal, the sensitivity is improved on average with about 2 dB. As seen above in Fig. 9.15, the simulated eye diagrams would suggest a much greater improvement if used on real OFDM signals, where one would not expect to be able to detect the bits, with the eye being completely closed. The outer channels in Fig. 9.17 are worse for the magnified case, which is due to small aberrations of the time lenses. These cause distortions on the magnified spectra of the outer channels (see arrows in Fig. 9.17 right), and these then cause ICI with their neighbours. With better optimisation of the time lenses, these aberrations are expected to disappear.

This application may become useful for AO-OFDM superchannels with many subcarriers, where one may wish to eliminate the need for equally many optical gates by introducing a single active element, namely the spectral telescope. The telescope is in its current implementation more complex and power consuming than some simple optical gates, like the eletroabsorption modulator used in [34] or low-energy variants as described in [40]. But there may be potential for low-energy time lenses based on e.g. amorphous silicon nanowires [41], which may indeed also hold potential for ultra-wideband operation over several hundred nm [42, 43]. This would allow for reduced power consumption and handling of ultra-broadband superchannels.

Other applications of time lens telescopic arrangements include temporal telescopes, as e.g. demonstrated in [44] for data packet compression. In that case, a short data packet is compressed a factor ×27 in time to bring the bit rate from 10 to 270 Gbit/s. In some applications, a smaller modification of the bit rate may be desired, and in that case a simplified time lens system may be used, so instead of a full telescope, a single lens may be used. In the following, such an example is given.

9.5 Time-Domain Processing

One of the early communication applications utilizing time lenses was for timing jitter compensation or retiming [18], which was basically based on an electro-optic phase modulator with sinusoidal modulation and a dispersive element, as sketched in Fig. 9.4. The same basic setup was also used for dispersion compensation [11] and later pulse shaping [13, 32] and pulse position locking [45]. With this implementation, the sinusoidal modulation implies that one can only operate at a bit rate at or below the rate of the modulation. This is in contrast to the FWM based time lens using flat-top pulses. If the e/o-phase modulation could be truly parabolic, then the situation would be similar to the FWM cases above. However, one of the attractions of the sinusoidal modulation is its simplicity. It is very convenient to obtain sinusoidal modulation.

Figure 9.18 shows an example where the time lens system is set to operate on an Ethernet data packet. Ethernet data packets/frames are very challenging to handle optically, as the packet length may vary from 64 to 1518 bytes, they arrive asynchronously and they have a certain allowed repetition rate variation, which in the 10 GE WAN PHY standard can vary with up to ±20 ppm of the nominal transmission rate, i.e. ±200 kHz frequency offset between transmitter and receiver for 10 Gbit/s Ethernet packets. The idea in Fig. 9.18 is to use a time lens to synchronise the 10 Gbit/s Ethernet data packets in such a strict manner that they can be optically time multiplexed together to form a Tbit/s multiplexed Ethernet data signal (TbE). The requirements to Tbit/s timing jitter are very strict ($\sim$50 fs) and thus the synchronisation has to be very effective. Furthermore, it needs to be very flexible to accommodate for the varying types of data packets and include NRZ-to-RZ conversion.

Figure 9.18 left shows the schematic setup. An NRZ data packet arrives to the simple time lens system, where an additional amplitude modulator based on a Mach-Zehnder modulator (MZM) is introduced. The MZM carves part of the NRZ bit out, thus effectively converting from NRZ to RZ, and the phase modulator adds the required chirp to the bit, and the following dispersive element aligns the now RZ bit to the focal point within the time slot, i.e. at the peak of the sinusoidal modulation, where it approaches a parabolic phase modulation and the chirp is linear, being the derivative of the phase modulation. The phase modulator modulation depth is 4π, and it and the MZM are both driven by a master clock at 9.9536 GHz (200 kHz offset from the input data packet clock), and then launched into a 400 m DCF. The 10 Gbit/s input asynchronous Ethernet packet with a data

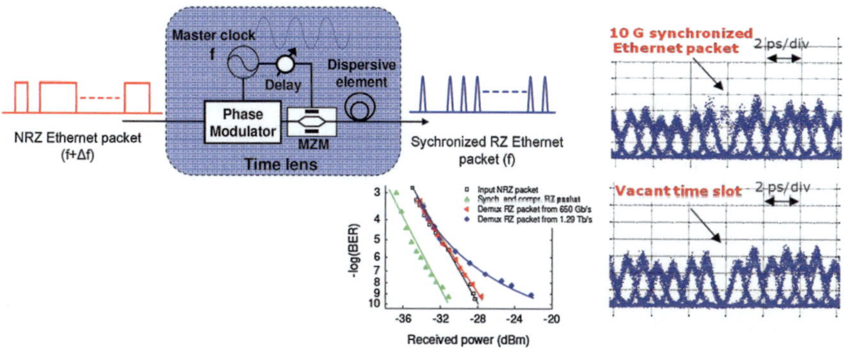

Fig. 9.18 Time lens based Ethernet packet synchronisation for 1.29 Tbit/s data streams. From [46]

rate of 9.9534 Gbit/s is thus converted into a synchronized Ethernet packet with a data rate of 9.9536 Gbit/s. At the same time, the Ethernet packet is format converted into an RZ signal with a full width at half maximum (FWHM) of 6 ps, due to the pulse carving of the MZM and the chirping. Additionally, the converted RZ signal is further pulse compressed to a FWHM of 400 fs in a 500 m dispersion-flattened highly nonlinear fibre (DF-HNLF), not shown here [46]. Alongside this synchronisation part of the demonstration setup, a standard 1.28 Tbit/s OTDM data signal is generated, but with the time channels moved together a small fraction so as to make room for an additional empty time slot. The now synchronised RZ data packet is then multiplexed together with the 1.28 Tbit/s OTDM signal to form a 1.29 Tbit/s aggregate signal, as seen to the right in Fig. 9.18. The Ethernet packet channel fits into the vacant OTDM channel, and when demultiplexing and BER characterizing, the performance is sufficiently good to reach a BER of 1E-9.

Thus this time lens scheme does fulfil the 10 GE WAN PHY requirements and is able to operate on asynchronous standard Ethernet packets with 200 kHz offset, and a maximum packet length of 1518 bytes to the extent that they may be multiplexed to Tbit-Ethernet, TbE.

9.6 Summary

This chapter has addressed the use of optical time lenses for efficient optical signal processing. We have identified and described scenarios where time lens based OSP with its ultrafast potential can offer power-efficiency. This goes for applications where many bits are processed simultaneously, such as for serial-to-parallel conversion. Other applications where all the data channels are processed simultaneously include the possibility of processing WDM signals to e.g. manipulate grid spacing using a spectral telescope. The same approach may be used for spectral magnification of OFDM signal in order to receive AO-OFDM signals directly using WDM-like receivers relying on passive bandpass filtering. This, like serial-to-parallel conversion, would break the scaling of complexity and power with number of tributaries, since one would only need a single spectral telescope. Finally, we showed how time lenses would enable the optical synchronisation and multiplexing of standard Ethernet packets to create a new multiplexed type of Tbit/s Ethernet with ultrahigh timing precision and format conversion.

All in all, we find that time lens based optical Fourier transformation is a very exciting and versatile tool for optical communications, allowing for a range of optical signal processing applications, which may offer energy efficiency as well as advanced functionality otherwise unobtainable, relying on the ultrafast nature of e.g. four-wave mixing and implementations offering transparency to data and signal formats, wavelength, polarisation and bit rate. Ultra-broadband operation may also be obtainable with nonlinear waveguides, such as nanowires, and with future developments in this field, ultra-low energy switching using time lenses may be realistic.

References

1. M. Mauldin, The state of the global internet, Webinare (2011). www.telegeography.com
2. E.B. Desurvire, Capacity demand and technology challenges for lightwave systems in the next two decades. J. Lightwave Technol. **24**(12), 4697–4710 (2006)
3. http://www.redorbit.com/news/science/1112756793/internet-produces-greenhouse-gas-010313/
4. K. Hinton, G. Raskutti, P.M. Farrell, R.S. Tuckeret, Switching energy and device size limits on digital photonic signal processing technologies. J. Sel. Top. Quantum Electron. **14**(3), 938–945 (2008)
5. R.S. Tucker, K. Hinton, Energy consumption and energy density in optical and electronic signal processing. IEEE Photonics J. **3**(5), 821–833 (2011)
6. M. Galili, L.K. Oxenlowe, H.C.H. Mulvad, A.T. Clausen, P. Jeppesen, Optical wavelength conversion by cross-phase modulation of data signals up to 640 Gb/s. J. Sel. Top. Quantum Electron. **14**(3), 573–579 (2008)
7. M. Lillieholm, H.C.H. Mulvad, E. Palushani, C. Peucheret, P. Jeppesen, L.K. Oxenløwe, in *IEEE Photonics Conference (IPC)*. Numerical Investigation of Power Requirements for Ultra-high-speed Serial-to-parallel Conversion, Burlingame, California, paper MM5, 23–27 Sept 2012
8. H.C.H. Mulvad, E. Palushani, H. Hu, H. Ji, M. Lillieholm, M. Galili, A.T. Clausen, M. Pu, K. Yvind, J.M. Hvam, P. Jeppesen, L.K. Oxenløwe, Ultra-high-speed optical serial-to-parallel data conversion by time-domain optical Fourier transformation in a silicon nanowire. Opt. Express **19**(26), B825–B835 (2011)
9. J. Xu, C. Peucheret, P. Jeppesen, in *Proceedings of the ICTON 2010*. Power Consumption Comparison Between Point-to-point WDM and OTDM Systems, Munich, Germany, paper Th. A1.1 (2010)
10. B.H. Kolner, Space-time duality and the theory of temporal imaging. J. Quantum Electron. **30** (8), 1951–1963 (1994)
11. M. Nakazawa, T. Hirooka, F. Futami, S. Watanabe, Ideal distortion-free transmission using optical Fourier transformation and Fourier transform-limited optical pulses. Photonics Technol. Lett. **16**(4), 1059–1061 (2004)
12. G.P. Agrawal, *Fiber-Optic Communication Systems* (Wiley, New York, 1997). ISBN: 0471175404/0-471-17540-4
13. E. Palushani, L.K. Oxenløwe, M. Galili, H.C. Hansen Mulvad, A.T. Clausen, P. Jeppesen, Flat-top pulse generation by the optical Fourier transform technique for ultrahigh speed signal processing. J Quantum Electron. **45**(11), 1317–1324 (2009)
14. E. Palushani, H.C. Hansen Mulvad, M. Galili, H. Hu, L.K. Oxenløwe, A.T. Clausen, P. Jeppesen, OTDM-to-WDM conversion based on time-to-frequency mapping by time-domain optical Fourier transformation. IEEE J. Sel. Top. Quantum Electron. **18**(2), 681–688 (2012)
15. R. Salem, M.A. Foster, A.L. Gaeta, Application of space–time duality to ultrahigh-speed optical signal processing. Adv. Opt. Photonics **5**(3), 274–317 (2013)
16. J. Azaña, Time-to-frequency conversion using a single time lens. Opt. Commun. **217**(1), 205–209 (2003)
17. J. Azaña, M.A. Muriel, Real-time optical spectrum analysis based on the time-space duality in chirped fiber gratings. IEEE J. Quantum Electron. **36**(5), 517–526 (2000)
18. L.F. Mollenauer, C. Xu, Time-lens timing-jitter compensator in ultra-long haul DWDM dispersion managed soliton transmissions. CLEO, paper CPDB1 (2002)
19. Y. Han, B. Jalali, Photonic time-stretched analog-to-digital converter: fundamental concepts and practical considerations. J. Lightwave Technol. **21**, 3085–3103 (2003)
20. I.S. Lin, J.D. McKinney, A.M. Weiner, Photonic synthesis of broadband microwave arbitrary waveforms applicable to ultra-wideband communication. IEEE Microwave Wireless Compon. Lett. **15**, 226–228 (2005)

21. J. van Howe, Chris Xu, Ultrafast optical signal processing based upon space-time dualities. J. Lightwave Technol. **24**(7), 2649 (2006)
22. R. Salem, M.A. Foster, A.C. Turner, D.F. Geraghty, M. Lipson, A.L. Gaeta, Optical time lens based on four-wave mixing on a silicon chip. Opt. Lett. **33**(10), 1047–1049 (2008)
23. K.G. Petrillo, M.A. Foster, Scalable ultrahigh-speed optical transmultiplexer using a time lens. Opt. Express **19**(15), 14051–14059 (2011)
24. H.C.H. Mulvad, E. Palushani, H. Hu, H. Ji, M. Galili, A.T. Clausen, M. Pu, K. Yvind, J.M. Hvam, P. Jeppesen, L.K. Oxenløwe, in *Proceedings of the ECOC 2011*. Ultra-high-speed Optical Serial-to-parallel Data Conversion in a Silicon Nanowire, Geneva, Switzerland, paper Th.13.A.2 (2011)
25. E. Palushani, T. Richter, R. Ludwig, C. Schubert, H.C. Hansen Mulvad, A.T. Clausen, L.K. Oxenløwe, in *Proceedings of the OFC 2012*. OTDM-to-WDM Conversion of Complex Modulation Formats by Time-domain Optical Fourier Transformation, Los Angeles, California, USA, paper OTh3H.2 (2012)
26. M. Galili, E. Palushani, H.C.H. Mulvad, H. Hu, L.K. Oxenløwe, Wavelength preserving optical serial-to-parallel conversion, OFC/NFOEC technical digest. Opt. Soc. Am. OM2G.4 (2013)
27. M. Nakazawa, T. Hirooka, P. Ruan, P. Guan, Ultrahigh-speed 'orthogonal' TDM transmission with an optical Nyquist pulse train. Opt. Express **20**, 1129–1140 (2012)
28. H.N. Tan, T. Inoue, T. Kurosu, S. Namiki, Transmission and pass-drop operations of mixed baudrate Nyquist OTDM-WDM signals for all-optical elastic network. Opt. Express **21**, 20313–20321 (2013)
29. H. Hu, D. Kong, E. Palushani, J.D. Andersen, A. Rasmussen, B.M. Sørensen, M. Galili, H.C.H. Mulvad, K.J. Larsen, S. Forchhammer, P. Jeppesen, L.K. Oxenløwe, in *CLEO Technical Digest*. 1.28 Tbaud Nyquist Signal Transmission Using Time-domain Optical Fourier Transformation Based Receiver, Optical Society of America, Postdeadline paper CTh5D.5 (2013)
30. H. Hu, D. Kong, E. Palushani, M. Galili, H.C.H. Mulvad, L.K. Oxenløwe, in *ECOC 2013*, Detection of 320 Gb/s Nyquist OTDM Received by Polarisation-insensitive Time-domain Optical Fourier Transformation, paper We.1.C.4 (2013)
31. H.C.H. Mulvad, H. Hu, M. Galili, H. Ji, E. Palushani, A.T. Clausen, L.K. Oxenløwe, P. Jeppesen, in *Proceedings of the ECOC 2011*. DWDM-to-OTDM Conversion by Time-domain Optical Fourier Transformation, Geneva, Switzerland, paper Mo.1.A.5 (2011)
32. L.K. Oxenløwe, M. Galili, A.T. Clausen and P. Jeppesen, in *European Conference on Optical Communication, ECOC' 06*. Generating a Square Switching Window for Timing Jitter Tolerant 160 Gb/s Demultiplexing by the Optical Fourier Transform Technique, Cannes, France, Paper We2.3.4, Sept 2006
33. E. Palushani, H.C. Hansen Mulvad, M. Galili, F.D. Ros, H. Hu, P. Jeppesen, L.K. Oxenløwe, in *OptoElectronics and Communications Conference, OECC 2013*. Spectral Compression of a DWDM Grid Using Optical Time-lenses, Kyoto, Japan, Paper ThO2-1, July 2013
34. D. Hillerkuss, R. Schmogrow, T. Schellinger, M. Jordan, M. Winter, G. Huber, T. Vallaitis, R. Bonk, P. Kleinow, F. Frey, M. Roeger, S. Koenig, A. Ludwig, A. Marculescu, J. Li, M. Hoh, M. Dreschmann, J. Meyer, S. Ben Ezra, N. Narkiss, B. Nebendahl, F. Parmigiani, P. Petropoulos, B. Resan, A. Oehler, K. Weingarten, T. Ellermeyer, J. Lutz, M. Moeller, M. Huebner, J. Becker, C. Koos, W. Freude, J. Leuthold, 26 Tbit/s line-rate super-channel transmission utilizing all-optical fast Fourier transform processing. Nat. Photonics **5**, 364–371 (2011)
35. M.E. Marhic, Discrete Fourier transforms by single-mode star networks. Opt. Lett. **12**(1), 63–65 (1987)
36. H. Sanjoh et al., in *Proceedings of the OFC 2009*.Optical Orthogonal Frequency Division Multiplexing Using Frequency/time Domain Filtering for High Spectral Efficiency up to 1 bit/s/Hz, ThD1 (2009)
37. A.J. Lowery, Design of arrayed-waveguide grating routers for use as optical OFDM demultiplexers. Opt. Express **18**, 14129 (2010)

38. E. Palushani, H.C.H. Mulvad, D. Kong, P. Guan, M. Galili, L.K. Oxenløwe, All-optical OFDM demultiplexing by spectral magnification and band-pass filtering. Opt. Express, **22**(1), 136–144 (2014). (ISSN: 1094-4087) (doi:http://dx.doi.org/10.1364/OE.22.000136)
39. L.B. Du, J. Schroeder, J. Carpenter, B. Eggleton, A.J. Lowery, in *Proceedings of the OFC'13*. Flexible All-Optical OFDM using WSSs, PDP5B.9 (2013)
40. L.K. Oxenløwe, Optical communications: single-laser superchannel, news and views. Nat. Photonics **5**, 329–331 (2011)
41. K.-Y. Wang, K.G. Petrillo, M.A. Foster, A.C. Foster, Ultralow-power all-optical processing of highspeed data signals in deposited silicon waveguides. Opt. Express **20**(22), 24601–24606 (2012)
42. M. Pu, H. Hu, M. Galili, H. Ji, L.K. Oxenløwe, K. Yvind, P. Jeppesen, J.M. Hvam, in *Conference on Lasers and Electro-Optics, CLEO' 11*.Ultra-Broadband Tunable Wavelength Conversion of Sub-Picosecond Pulses in a Silicon Nanowire, Baltimore, Maryland, U.S.A., Paper CMAA1, May 2011
43. N. Ophir, R.K.W. Lau, M. Menard, X. Zhu, K. Padmaraju, Y. Okawachi, R. Salem, M. Lipson, A.L. Gaeta, K. Bergman, Wavelength conversion and unicast of 10-Gb/s data spanning up to 700 nm using a silicon nanowaveguide. Opt. Express **20**(6), 6488–6495 (2012)
44. M.A. Foster, R. Salem, Y. Okawachi, A.C. Turner-Foster, M. Lipson, A.L. Gaeta, Ultrafast waveform compression using a time-domain telescope. Nat. Photonics **3**, 581–585 (2009)
45. C.W. Chow, A.D. Ellis, F. Parmigiani, Time-division-multiplexing using pulse position locking for 100 Gb/s applications. Opt. Express **17**(8), 6562–6567 (2009)
46. H. Hu, J. Laguardia Areal, H.C. Hansen Mulvad, M. Galili, K. Dalgaard, E. Palushani, A.T. Clausen, M.S. Berger, P. Jeppesen, L.K. Oxenløwe, in *European Conference on Optical Communication, ECOC'11*. Synchronisation, Retiming and OTDM of an Asynchronous 10 Gigabit Ethernet NRZ Packet Using a Time Lens for Terabit Ethernet, Geneva, Switzerland, Paper Tu.3.K.4, Sept 2011

Chapter 10
Signal Processing Using Opto-Electronic Devices

Mary McCarthy, Simon Fabbri and Andrew Ellis

Abstract All-optical signal processing is a powerful tool for the processing of communication signals and optical network applications have been routinely considered since the inception of optical communication. There are many successful optical devices deployed in today's communication networks, including optical amplification, dispersion compensation, optical cross connects and reconfigurable add drop multiplexers. However, despite record breaking performance, all-optical signal processing devices have struggled to find a viable market niche. This has been mainly due to competition from electro-optic alternatives, either from detailed performance analysis or more usually due to the limited market opportunity for a mid-link device. For example a wavelength converter would compete with a reconfigured transponder which has an additional market as an actual transponder enabling significantly more economical development. Never-the-less, the potential performance of all-optical devices is enticing. Motivated by their prospects of eventual deployment, in this chapter we analyse the performance and energy consumption of digital coherent transponders, linear coherent repeaters and modulator based pulse shaping/frequency conversion, setting a benchmark for the proposed all-optical implementations.

10.1 Introduction

The demand for communication capacity has continued to increase exponentially over the last 35 years [1] regardless of repeated projections in the intervening period that this growth rate will slow down or stop. Since the introduction of optical fibre [2] the optical communications industry initially met this growth with symbol rate increases, followed by with wavelength division multiplexing (WDM) [3]. As we

---

M. McCarthy (✉) · S. Fabbri · A. Ellis (✉)
Aston Institute of Photonic Technologies, School of Engineering and Applied Science,
Aston University, Aston Triangle, Birmingham B4 7ET, UK
e-mail: andrew.ellis@aston.ac.uk

S. Wabnitz and B.J. Eggleton (eds.), *All-Optical Signal Processing*,
Springer Series in Optical Sciences 194, DOI 10.1007/978-3-319-14992-9_10

approach the ($\sim$5 THz) bandwidth limit of conventional amplification technology, research has focused on enabling higher complexity modulation formats to access fibre capacity [4]. As total volume of data transported across the network has grown, energy consumption, as well as increased capacity, have become a critical drivers in deploying new technologies [5]. In addition to transponders at the end terminals of an optical network, certain signal processing functions, such as regeneration, format conversion, wavelength conversion and arbitrary waveform generation are often proposed. As described extensively elsewhere in this book, many of these intermediate functions may be performed all optically, with the primary advantage of increased bandwidth and consequent resource sharing. However these functions may be performed using the very optoelectronic devices used in the transponders themselves, either including decision circuitry and/or FEC or as linear media converters. In this chapter, we analyse the performance and energy consumption of digital coherent transponders and modulator based pulse shaping/frequency conversion, setting a benchmark for the proposed all-optical implementations.

10.2 Coherent Transponders

After low loss single mode fibres, the most transformative invention in optical communications was the Erbium Doped Amplifier (EDFA), enabling wavelength division multiplexed (WDM) transmission over long distances [6]. This has fuelled much of the increase in capacity up to the mid-2000s. In the first coarse WDM systems, where only a few wavelengths were amplified by each EDFA, the power consumption of the system was dominated by the amplifiers but as the information spectral density of the systems have increased the link power budget has become dominated by the transponders.

The power consumption of EDFAs should ideally be dominated by the current required for the semiconductor un-cooled pumps (650 mW) with a simple electrical to optical power conversion efficiency of about 25 % achievable in practice. However, in terrestrial systems, cooling, control loops and supervisory overhead using robust technologies have come to dominate [7]. Terrestrial WDM amplifiers consume between 10 and 25 W of electrical power in order to amplify optical signals up to 26 dBm. It is likely that, once the power consumption of transponders has been considered and as the restrictions on power consumption become ever more stringent, a trade-off between remote supervision of links and power consumption of amplifier nodes will be considered. Similar control and supervisory overheads should be considered for any all-optical processing nodes, all of which are likely to require at least one additional optical amplifier to overcome device insertion loss.

Once the available spectrum is allocated to its maximum permitted extent by a combination of WDM and increased symbol rate, it is necessary to increase number of bits per symbols, many different constellations are possible, but here we focus on

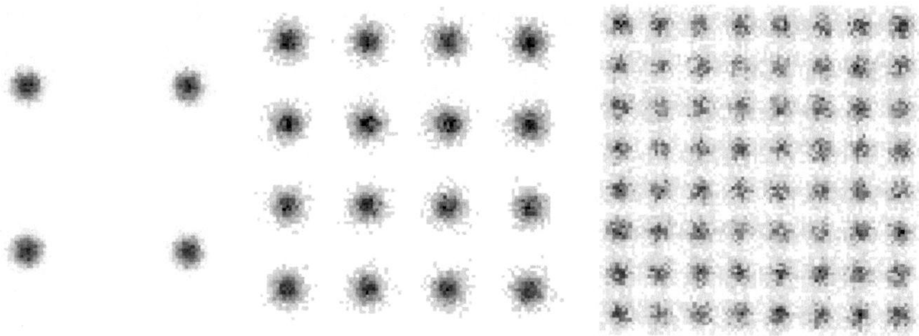

Fig. 10.1 Constellation diagrams for QPSK (*left*), 16QAM (*centre*) and 64QAM (*right*) with a signal to noise ratio of 20

the well-known rectangular QAM family as shown in the constellation (amplitude and phase) diagrams of Fig. 10.1 for an increasing number of constellation points M, where each symbol coveys $\log_2 M$ bit per symbol giving a total bit rate per wavelength R_b of.

$$R_b = R_s \log_2 M \tag{10.1}$$

where R_s is the symbol rate. Increasing the number of constellation points that can be encoded increases the required digital to analogue converter resolution and the complexity of the receiver digital signal processing (DSP). This is mitigated by the increased number of bits that are processed simultaneously. In what follows, we optimise the constellation for minimum energy consumption.

Whilst dual quadrature formats were originally detected and simultaneously demodulated using optical interferometers [8], coherent detection is currently used to enable enhanced receiver sensitivity and digital impairment compensation. For a coherent receiver, the requirement for polarisation diversity [9] implies that both polarisation multiplexing and compensation of polarisation mode dispersion (PMD) in the DSP are straightforward, doubling the bit rate shown in (10.1).

10.2.1 Power Consumption of Opto-Electronic Devices

Narrow linewidth lasers are required for coherent detection to both minimise phase noise induced error floors and reduce the complexity of the phase recovery of the coherent receiver digital signal processing. Whilst there is much development in narrow linewidth widely tunable lasers, especially for use in integrated devices, current state of the art lasers [10] have a power consumption of 3.5–5 W representing an approximately 10 % electrical to optical power conversion efficiency. Integration will reduce insertion loss by several dBs and consequently the required output powers [11] and in certain circumstances one laser can be shared between transmitter and receiver part of a coherent transponder module.

Optical modulators with bandwidths well in excess of 28 GHz have been commercially available for many years. For binary systems they may be driven by limiting amplifiers which may operate with high power efficiency. For coherent systems, dual polarisation IQ modulators with low drive voltages are required and are currently based on several key technologies including $LiNbO_3$ (3.5 V) [12], Silicon (3.25 V) [13], InP ($\sim$2.5 V) [14] and polymer (2.5 V) based photonics [15]. In production systems automatic bias control circuitry is required, but power consumption is dominated by drive amplifier power consumption (typically determined by the 50 Ω impedance of the connector standard selected) and any DAC required for pulse shaping or pre compensation of channel impairments. Ideally modulator drive voltages for these modulators should be reduced to match the DAC output.

In order to implement modulation formats with high cardinality (16QAM and above), linear amplifiers are required. Whilst switched mode amplifiers are useful for narrowband wireless systems [20], for broadband amplifiers the linearity requirement results in a substantial increase in power consumption to minimise the signal distortion. Figure 10.3 shows the power consumption of typical commercially available drivers. For the limiting amplifiers, the power consumption is dominated by the amplifier output power (3 V in 50 Ω is $\sim$0.18 W) and consequently is independent of the required bandwidth. This power consumption should be at least doubled for linear drivers. Consequently we allocate 3.3 W per quad drive amplifier (one driver for each modulator in a of a PM-IQ modulator).

The Optical Internetworking Forum (OIF-DPC-RX-01.0) have agreed a standardised design for the coherent receiver [21]. It consists of a polarisation beam splitter, two 90° Hybrids, four balanced photodiodes and a transimpedance amplifier (TIA) array as shown in Fig. 10.2. The main source of power dissipation is the TIA. Figure 10.3 shows the power dissipation of currently available TIAs and TIAs with appropriate automatic gain control circuitry, typically required to match the output voltage the optimum range of the analogue to digital converters. Here the output power is a negligible contribution to the power consumption, which illustrates the anticipated frequency dependence for devices without AGC. However, for ease of implementation, we assume the use of AGC circuits, and consequently a fixed power consumption of around 0.8 W per TIA.

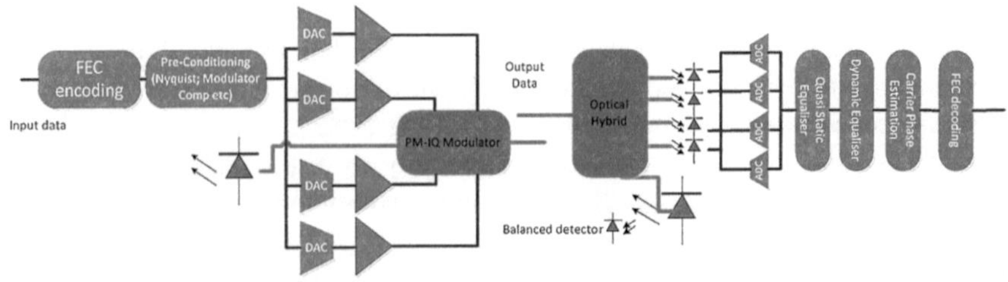

Fig. 10.2 Schematic diagram of a typical coherent transponder showing FEC, DSP, drivers, passive and active optics. Framers, gearboxes and buffers are omitted for simplicity

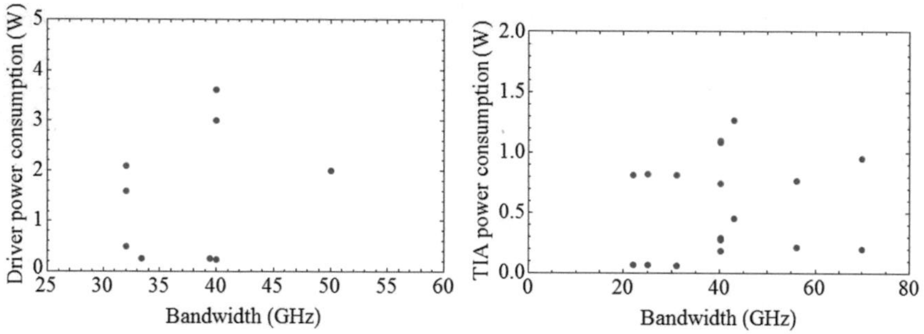

Fig. 10.3 *Left* Power consumption of commercially available modulator drivers showing limiting amplifiers (*blue*) and analogue amplifiers (*red*) [16–19]. *Right* Power consumption of trans impedance amplifiers with (*red*) and without (*blue*) automatic gain control [19]

10.2.2 Energy Consumption of DSP

Consideration of the opto-electronic subsystems alone reveals that the power consumption is largely independent of the symbol rate, suggesting that the highest possible rate should be deployed as the appropriate technologies become available, and indeed this was the case for much of the history of optical communications up to systems operating at 40 Gbit/s. However, the generation and, in particular, the detection of optical signals using coherent receivers requires the use of digital signal processing which adds significantly to the overall energy consumption. As shown when coherent detection was first investigated, DSP can clearly be avoided [22], however this requires widely tuneable lasers with greater stability and limits the opportunities to compensate for linear impairments. To account for the signal processing, the signal processing and DAC/ADC requirements should be considered.

Accurate design of CMOS DSP chips is a lengthy process requiring several steps after optimised algorithms have been identified, and a full analysis is beyond the scope of this book. Full details are design processes and the potential efficiency gains are now appearing following rigorous analysis, as shown in, for example [23]. Many detailed design choices are required, including choice of algorithms, parallelism [24], layout, resolution, target latency and performance [25]. Whilst such exact calculations are complex, estimation of power consumption through simple counting of the required basic building blocks of digital signal processing (adders and multipliers) is readily performed. The total resources need to take into account both the number of operations presented in the algorithm, and also the precision to which this operation is carried out. An adder with N bit precision requires N − 1 stages and an N bit multiplier requires N − 1 adders hence complexity of a multiplier scales as $(N - 1)^2$. This holds for both carry-save and Wallace Tree multiplier architectures. As the precision N increases, multipliers will dominate in terms of power consumption so we will further simplify our cost analysis to only include complex multiplication which will require 3 real multipliers, two additions and Boolean overhead per operation [26]. It has been demonstrated that the required resolution for rectangular QAM constellations with M constellation points varies slightly faster than the number of bits carried by the format [27, 28], is shown in Fig. 10.4 and may be approximated as

$$N_{ADC} \approx 2.8 + \log_2 \sqrt{M} \tag{10.2}$$

We assume that a radix-2 algorithm FFT requires $N/2 \log_2 N$ complex multiplications and that trigonometric functions are estimated using CORDIC algorithm with 3N adders [29]. Some commercially available digital coherent receivers contain transmitter DSP and digital to analogue converters. These allow for the generation of high cardinality modulation formats, pulse shaping (including OFDM and Nyquist WDM), compensation of transmitter and/or a line impairments such as dispersion and to limited extent nonlinearity [30]. DAC performance is limited by the linearity of the circuit and quantization noise which is evaluated in terms of effective number per bits (ENOB). ENOB is limited by integral nonlinearity, differential nonlinearity, total harmonic distortion and noise [31] which limits both the accuracy and speed of the DAC. An ENOB of at least 3.8 is required to generate QPSK and at least 4.9 to generate 16QAM.

As can be seen from Fig. 10.5, DAC power consumption has remained relatively constant with respect to sample rate. State of the art converters [33] consume 0.75 W and offer 8-bit physical resolution and sample rates up 65 GS/s which should allow 2.3× oversampling for 28 Gbaud signals.

With regards to the transmitter DSP, some of the functions will be specific to the transmitter hardware and can readily be implemented by a factory set look up table (LUT) which will have relatively low power consumption but others will be link dependent and may need to be tailored dynamically. These will take the form of finite input response (FIR) filter whose length will be determined by the complexity of the signal pre-conditioning required. The complexity of the filter for chromatic

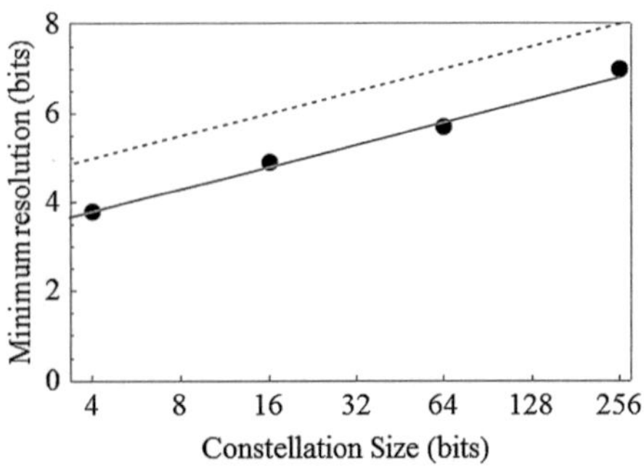

Fig. 10.4 Minimum required ADC resolution for digital coherent receivers (points) [27], approximation (10.2, *solid line*) and 1 bit margin for ENOB degradation (*dashed line*)

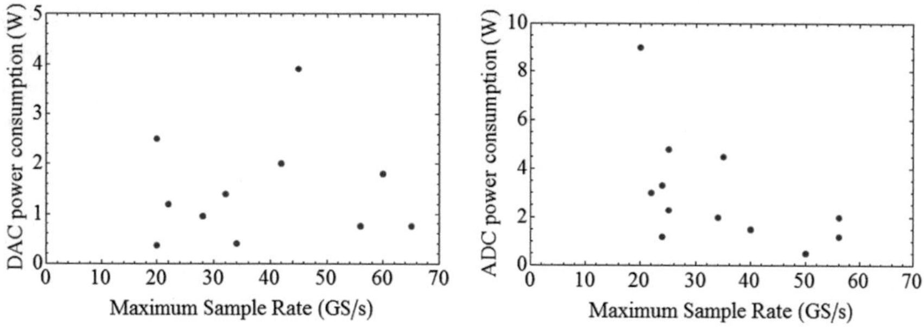

Fig. 10.5 *Left* Commercially available DAC power consumption as a function of sample rate [19, 32, 33]. *Right* Power consumption of ADCs [19, 32, 34–38]

dispersion compensation will be discussed in the receiver digital signal processing section and may be arbitrarily shared between transmitter and receiver with negligible implications for overall energy consumption. The most common signal conditioning is currently Nyquist filtering and its variants, offering the highest spectral efficiency. Ideally the filter impulse response would be a Sinc function, however this requires infinite taps. In practice, a RRC filter gives a very good approximation to a truncated Sinc function. Use of a RRC filters results in a filter at the transmitter and a "matched" filter at the receiver.

$$\mathrm{RRC(t)} = \mathrm{sinc}\left(\frac{\mathrm{t}}{T}\right)\frac{\cos\left(\frac{\pi\alpha t}{T}\right)}{1-\frac{4\alpha^2t^2}{T^2}}, \quad (10.3)$$

where T is the symbol period and α is the roll-off factor. Typically the limiting factor in the implementation of Nyquist filter will be the ENOB and sample rate of the DAC rather than the number of taps in the FIR filter in the receiver so we will limit the length of the filter considered in this work to four filters of sixty four taps each for 28 GBaud signal [39]. All current systems employ some form of forward error correction. The complexity of forward error correction encoding depends on the latency constraints and the complexity of the code. However for RS(255,239) encoding, this can be implemented using a shift register so the increase in transmitter side power consumption is small compared to other functions.

Digital Signal Processing in the receiver is the major consideration for any digital coherent receiver, allowing for polarisation mode dispersion compensation and carrier frequency and phase error correction. Receiver DSP discussed in detail in many, many publications and details of the algorithms may be found in [40–42]. Its implementation can be divided in five main areas in terms of power consumption.

Optimisation of receiver DSP is currently an active area of research so we will draw on recent published papers and bench mark examples with acknowledgement that further, potentially substantial, improvement is expected. We consider each of the five major blocks individually (Fig. 10.6).

10.2.2.1 Analogue to Digital Conversion

High speed analogue to digital converters (ADCs) have been the single most important key technology in the introduction of DSP to optics. There are many architectures [27] which trade off speed, latency, NOB and implementation size. They are generally specified in ENOB similar to DACs as reported in the previous section. The main source of reduction of ENOB at high frequencies is clock jitter which results in ADC being unable to sample at precise intervals. Figure 10.5 shows details of power consumption for ADCs versus Sampling Rate over the last decade or so, again the best available devices show negligible variation in power consumption with sampling rate. The latest result demonstrates the availability of 56 Gbit/s ADC with 8 bit resolution at 1.2 W power consumption. This allows for 2× sampling at 28 GHz.

Progress in ADC performance usually tracks the development of successive CMOS nodes, however the input referred noise has been increasing as the feature sizes reduce below 90 nm and so it is possible that we are very close to the optimum design for electronic ADCs and further improvement is ENOB and power consumption will be incredibly challenging [31, 43]. This may result in a stagnation of coherent DSP based systems at $\sim$28 Gbaud.

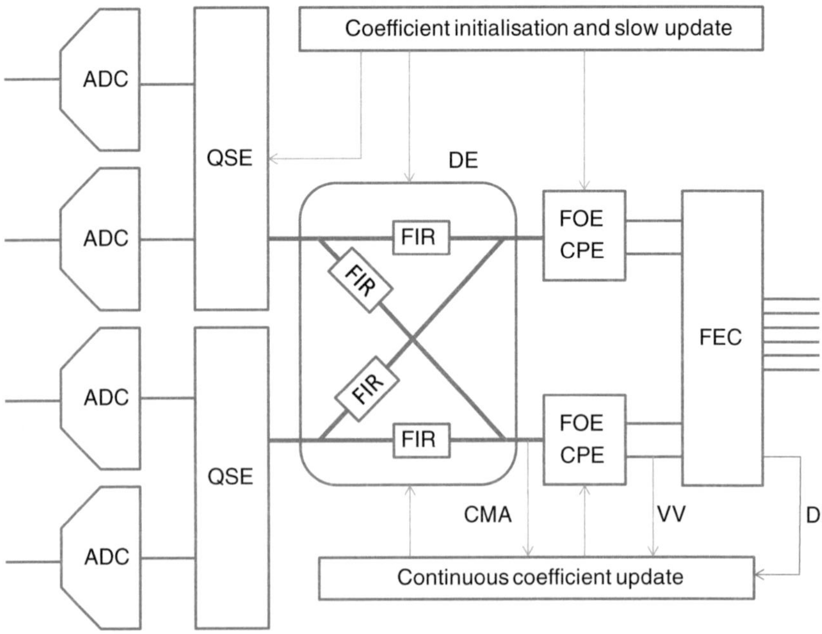

Fig. 10.6 Schematic diagram of common DSP configuration including (1) analogue to digital converters (ADC), (2) quasi static equaliser (QSE) for matched filtering, chromatic dispersion compensation and PMD, (3) dynamic equaliser comprising time domain filters (FIR) in a butterfly configuration, (4) frequency offset estimation (FOE) and carrier phase estimation (CPE) and (5) forward error correction decoder (FEC)/symbol decisions. Also shows options for updating filters and estimators based on constellation analysis (e.g. constant modulus algorithm (CMA)), nonlinearly processes signal analysis (e.g. Viterbi and Viterbi phase estimation (VV)) and decision directed (DD) feedback

10.2.2.2 Quasi-static Equaliser [44]

Chromatic dispersion and quasi-static PMD can be compensated for using a combination of a slowly updating (quasi static) filter and a rapidly updating dynamic filter. In the quasi static filter, the expected chromatic dispersion is equalised (with a memory length of τ_{CD}) and the bulk of the polarisation mode dispersion (typically up to three times the mean value of the differential group delay $3\langle\tau_{DGD}\rangle$) are equalised with a long averaging time [45]. Fine tuning, particularly of the polarisation mode dispersion, is performed in the dynamic equaliser (see below). Hardware efficient clock recovery can also be implemented as part of this structure [46].

The PMD tolerance that may be compensated in the quasi-static equalizer is a variable design parameter depending on what is acceptable level of outage (typically 10^{-15}) when the dynamic equaliser is included. Thus it is possible to estimate a conservative bound for the length of the filter, N_{CIR}, to compensate the channel impulse response (CIR) as

$$N_{CIR} = \frac{n}{T}(\tau_{CD} + 3\langle\tau_{DGD}\rangle) \tag{10.4}$$

where n is the over sampling ratio and T is the symbol period. A time domain equalizer to include some PMD compensation would require 4 such butterfly FIR filters giving a total number of multiplications per bit

$$\cos t_{TDE} = 4\left(\frac{N_{CIR}}{2\log_2(M)}\right) \tag{10.5}$$

However it has been shown extensively that for channels with realistic memory lengths (10.4), the complexity of the time domain filter far out strips that of a Fourier or frequency domain implementation [47, 48]. The complexity of frequency domain algorithm is dominated by the FFT/iFFT operations. For a frequency domain compensator for two polarisation, including the need for overlap and save algorithm and assuming that $\tau_{CD} \gg \tau_{DGD}$, the number of multiplications per bit is;

$$\cos t_{FDE} = \frac{N_{FDE}(\log_2(N_{FDE}) + 1)n}{(N_{FDE} - N_{CD} + 1)\log_2(M)} \tag{10.6}$$

where $N_{CD} = \tau_{CD}n/T$ is the number of overlap bits required to ensure correct frequency domain filtering. The choice of FFT size is an optimisation of $N_{FDE} = N_{CD} + 1 + m$ where m is the number of new data per FFT. If data throughput is at least 85 % N_{CD},

$$N_{FDE} > \frac{n}{T}\frac{1}{2\log_2(M)}\frac{1}{0.85}N_{CD} \tag{10.7}$$

for 2 polarisations. FFT implementations are usually more efficient in multiples of radix 2–4 in an ASIC implementation and implementation cost must be balanced with the acceptable latency. Tap coefficients are at setup when a channel is first established, taking approximately $10N_{CIR}$ multiplications for a zero-forcing algorithm [49] and are updated infrequently (so that overhead is negligible) over the lifetime of the system. Dynamic changes in the network will be compensated by the subsequent dynamic equalizer.

10.2.2.3 Dynamic Equaliser

To compensate for channel parameters that change dynamically, it is necessary to implement a short dynamic equaliser to compensate for residual CD and fast changing PMD. It can be assumed that generally a channel is time invariant with a period of 5 μs [50]. As the length of the equaliser is short it is more power efficient to implement it as a time domain butterfly FIR filter consisting of four individual filters [41]. The taps must be updated continuously or quasi-continuously to track the signal polarisation and other fluctuations in the link transfer function. Many algorithms can be used to update the taps [40–42]. The Constant Modulus algorithm [51] has been widely applied and to illustrate the required power consumption, we assume CMA, and limit the length of the individual filters to $N_{CMA} = 20$ taps [40]. For the time domain equalizer, we require 4 FIR filters in a butterfly configuration, with the number of multiplications per bit given by the FIR filter length, plus an additional multiplication per symbol and per tap in order to update the tap weight, giving a total multiplication per bit of

$$\cos t_{TDE} = 4\overbrace{\left(\frac{N_{CMA}}{2\log_2(M)}\right)}^{filter} + 2\overbrace{\frac{N_{CMA}}{\log_2(M)}}^{update} \tag{10.8}$$

10.2.2.4 Phase Estimation

After the mixing of the signal and the local oscillator (LO), it is almost certain that there will be a frequency and phase offset which results in rotation of the constellation of the incoming signal. There are several possible methods of phase estimation including maximum a posteriori (MAP), decision directed, power law-Wiener filter and Viterbi-Viterbi estimation [52, 53]. The VV algorithm are considered to have the lowest computation complexity. Here we consider the Viterbi-Viterbi algorithm where an estimate, ϕ_k, averaged over k symbols of the actual relative phase between signal and local oscillator ϕ is given by

$$\phi_k = \frac{1}{4}\arg\left(\sum_{i=-k}^{k}(x_i w(i)(x_i)^4)\right) \tag{10.9}$$

where w is an optimised weighting function dependent on the relative impact of additive noise and laser phase noise and x_i the received signal sample. As the modulation format complexity increases, it is possible to only consider "Class 1" symbols those that lie at modulation angles of $(\frac{\pi}{4} + \{0, 1, 2..\} * \frac{\pi}{2})$. The different symbol levels are detected using amplitude discrimination. However this would result in only utilising a reducing percentage of the received signals being used as even some Class 1 signals must be neglected as their modulus is very close to other symbols. Several methods have been suggested for partitioning higher modulation formats into sets of QPSK [54] which can then be processed in stages. Another method [55] partitions by considering that sets of Class 2 (symbols with rotated by $\{0, 1, 2..15\} * \frac{\pi}{8} \pm 4°$) which is only suitable to 16QAM. Another method is to raise 16QAM Class 2 symbol by eight [56] and including outer most 64QAM Class 1 symbols to generate the phase estimate. CPE is an active area of research as performance is traded off against complexity. In Table 10.1, there is an estimate of the minimum number of multiplications to implement CPE consisting of dual polarisation, number of QPSK decomposition and updates.

Using these approximations, it is possible to generate an estimate of operations to

$$\cos t_{VV} = 2 * k * (7.2 \log_e(M) - 3.35) \tag{10.10}$$

The complexity scales logarithmically with M up to 64. It is assumed that the same number of symbols will be used for each modulation format. This may result in different maximum combined line-width symbol duration product for different modulation format as a smaller percentage of the symbols are suitable for processing with VV at 64QAM. It is likely that, other more computationally intensive methods will be necessary to track for higher modulation formats in practice. Also each of these methods will have to implement a method to prevent cyclic slips either using differential encoding or an intelligent phase unwrap. Also performance can be improved by use of a maximum likelihood estimation stage.

It is assuming that the arg(*) operation is carried out efficiently but with latency cost using a CORDIC algorithm, lookup table or equivalent. A correctly configured phase estimator may also have the beneficial effect of cancelling some SPM and XPM effects in WDM transmission environments. Such beneficial effects are neglected here.

Table 10.1 Estimation of CPE

$\cos t_{VV}$	QPSK [54]	16QAM [55]	64QAM [56]
Complexity	$2((4+1)+1)*k$	$2*3(4+1+1)*\text{k}$	$2*\frac{20}{64}\text{k}\{(4+1+1)+(8+1+1)\}$

10.2.2.5 Forward Error Correction (FEC)

Whilst research results demonstrate the clear value of soft decision codes such as low density parity check codes (LDPC), Reed-Solomon (RS) codes are typically deployed in optical communications either as standalone codes or as concatenated or turbo codes. The RS decoder consists of:

- Syndrome computation block (SC)
- Key equation Solver (KES)
- Chien search error evaluation block (CSEE)

The implementation of these algorithms has received much effort over the years but the challenge of implementing them for optical communication speeds is an on-going concern. It is often a trade-off between area and latency and iterations to allow optimum performance. For minimum latency the complexity is proportional to ($t = (N - K)/2 = 8$ for RS(N,K) [57])

$$C_{FEC\_ME} = 4 * \log(M) * (50 + 2t) \tag{10.11}$$

Using a systolic modified Euclidean algorithm which has a latency of 80 clock cycles or for minimum complexity but with a latency of 288 clock cycles

$$C_{FEC\_BM} = 4 * \log(M) * (6 + 2t) \tag{10.12}$$

using the inversion-less Berlekamp-Massey algorithm. FEC decoding is implemented at one sample per bit.

10.3 Link Power Consumption

Having established the power consumption of the essential transponders and optical amplifiers, estimation of the exact power consumption of an optical network requires detailed knowledge of the node locations and link length and properties. Using a closed form expression for the nonlinear Shannon limit [58–61] with a 3 dB additional operating margin as a network planning tool, we estimate the physical layer power consumption of an arbitrary network with a total capacity of 1 Pbit/s, requiring both multiple fibres and, for modulation formats with high cardinality, multiple regenerators. For fully coherent systems, optimum link configurations are using non dispersion shifted single mode fibre (ideally ultra-low loss) without in-line dispersion compensation, since dispersion may be compensated for without additional noise penalty in the digital domain. We use the formalism of [58], but the results are equally valid for all forms.

Figure 10.7 illustrates the energy consumption of the 1 Tbit/s point to point link, illustrating that the expected monotonic rise with transmission distance is dominated by the addition of regenerator sites. We observe that the energy consumption

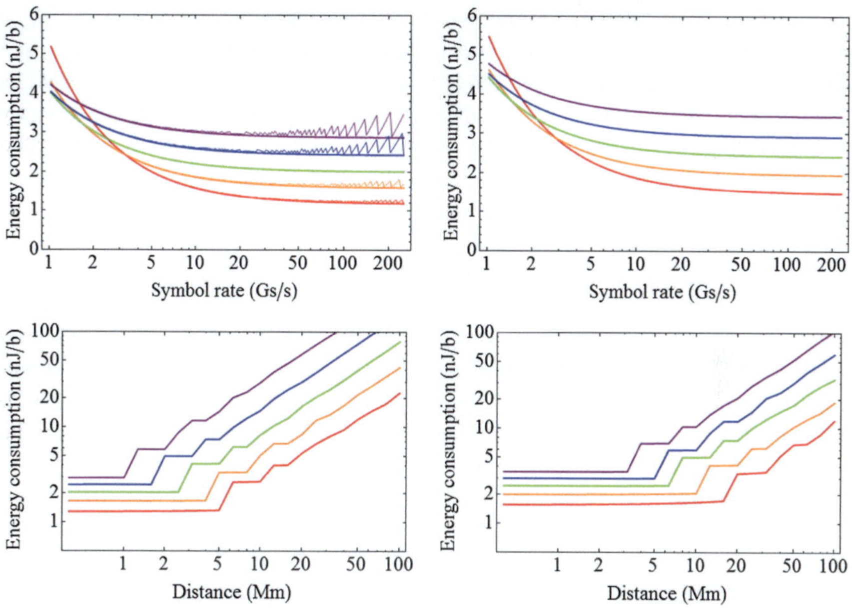

Fig. 10.7 Energy consumption of an optical system transporting 1 Pb/s data between two points using coherently detected PM-QAM formatted signals spaced at the Nyquist rate, 100 km amplifier spacing (fibre with 0.2 dB/km loss, 16 ps/nm/km dispersion and 1.4/W/km nonlinearity) and 5 dB noise figure amplifiers with 5 THz bandwidth. Showing QPSK (*red*), 8QAM (*orange*), 16QAM (*green*), 32QAM (*blue*) and 64QAM (*purple*) homogenous traffic for hard decision FEC with an FEC overhead of 7 % (*left*) and an FEC overhead of 25 % (*right*), both assuming a 3 dB signal to noise ratio margin, as a function of link length at a spacing of 33 GHz (*bottom*) and symbol rate at a link length of 1000 km (*top*). *Thin lines* in *top right hand figure* show similar calculations for a 10 Tbit/s transported capacity

also increases dramatically with the cardinality of the modulation format; this is effect is dominated by the choice of phase estimator. We also observe a variation in energy consumption as a function of symbol rate, associated with the inefficient utilisation of inherent energy consumption individual channels such as modulator drivers and individual lasers. Low symbol rates and low cardinality formats significantly increase the number of transponders and hence total energy consumption. For higher symbol rates, the reduced granularity in networks of low overall capacity (thin solid lines in Fig. 10.7) may result in inefficient occupancy and slightly enhanced energy consumption implying an optimum symbol rate in the region of 30 Gbaud. However, the optimum value appears to favour the highest implementable symbol rate. Overall, we anticipate that the link energy consumption of a digital coherent link will be less than ~0.2 pJ/bit/km for ultra-long links and a few nJ/b for shorter links. Links employing coherently detected single quadrature modulation formats (e.g. m-ASK, or m-PSK) will have similar transmission characteristics, but would typically require twice the number of transponders as their QAM counterparts (m^2QAM) significantly increasing the overall energy consumption. Two alterative modulation schemes also bear attention from an overall power consumption point of view. Firstly, transmission with a pilot tone and associated all optical pilot tone recovery system [62–64], where we neglect the energy consumption associated with the phase recovery block, replacing this with a fixed 5 W energy consumption for the all optical phase recovery system and adding a 10 % guard band to accommodate the pilot tone.

For terrestrial distances, the optimum energy consumption switches from a low order format to the highest available modulation format, whilst for transoceanic distances the difference in overall energy consumption becomes negligible with a net consumption of ~0.1 pJ/b/km for ultra-long haul links. Secondly we consider a direct detection ASK system spaced at twice the Nyquist rate, significantly higher signal to noise ratios, in-line dispersion compensation (requiring dual stage amplifiers with 1 dB higher noise figures) and complex optical receiver filters. It might be considered that these significant disadvantages would be offset by the reduced transponder power consumption since the DSP required for coherent detection is omitted. The resulting energy consumption is shown in Fig. 10.8. Total energy consumptions are significantly greater than 0.5 pJ/b/km (at 1000 km) for all

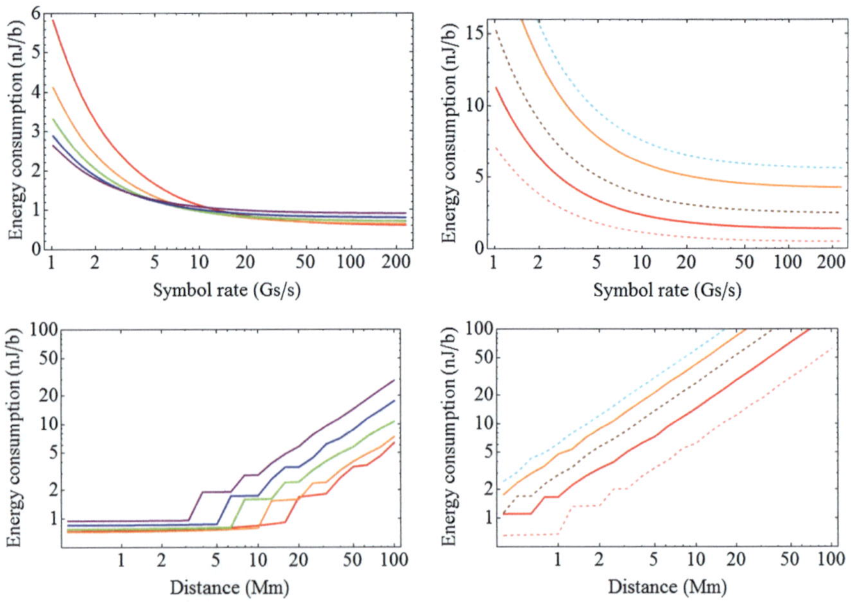

Fig. 10.8 Energy consumption of an optical system transporting 10 Pbit/s gross data rate over 1000 km as a function of symbol rate (at 1000 km) and distance (at 33 Gbaud) using (*left*) pilot assisted PM-QAM with a 10 % guard band, all other parameters as above and (*right*) directly detected binary PM-ASK formatted signals spaced at the Nyquist rate with 80 km amplifier spacing (fibre with 0.2 dB/km loss, 1.6 ps/nm/km dispersion and 1.4/W/km nonlinearity) and 6 dB noise figure amplifiers with 5 THz bandwidth for homogenous traffic for hard decision FEC with a 25 % overhead. Colours for *solid lines* represent the same number of constellation points as shown in Fig. 10.7, *dashed lines* represent; *pink* (2-ASK), *brown* (6 ASK), *cyan* (10 ASK)

formats, except simple on-off keyed signals at symbol rates above 20 Gbaud, where polarisation mode dispersion would become a key limitation. Furthermore, the above calculations have been performed assuming a 65 nm CMOS processing node, with an energy consumption of 1.5 pJ per multiplication [65]. With 14 nm CMOS processes currently entering production, enhanced CMOS energy consumption is perfectly possible; although not without difficulties as smaller gates appear to have increased noise properties.

The dominance of digital signal processing functions suggests that energy reduction through ever more efficient CMOS processing will be of benefit. The level of improvement which may be expected is shown in Fig. 10.9, which quantifies the level of performance gain which may be achieved for a given improvement in CMOS gate efficiency. We have also estimated the required CMOS node required to achieve three specific efficiency enhancements: consistent with the issues identified in the International Technology Roadmap for Semiconductors [66], the reduction in total power consumption per gate is beginning to saturate [67]. Note that Fig. 10.9 neglects the increase in ADC noise (decrease in effective number of bits) associated with small CMOS feature sizes [68]. Nevertheless, the potential energy saving of the order of 50 % will be welcome, and will result in similar savings in optical networks transporting high cardinality modulation formats. The benefit will be lower for PM-QPSK based networks.

Many signal processing functions, such as wavelength conversion, format conversion (aggregation) etc. may be readily achieved using a transponder with the receiver electrical output connected to the electrical input to the transmitter resulting again in energy consumptions of the order of 1 nJ/b for a full functionality transponder. This reduces to around 200 pJ/b for a pure transparent wavelength converter or phase conjugator [69–71] without digital signal processing. The performance, versatility and low energy consumption of the transparent opto-electronic wavelength converter sets a benchmark for alternative schemes. To match the available performance of conventional systems, it is essential that any alternative system is

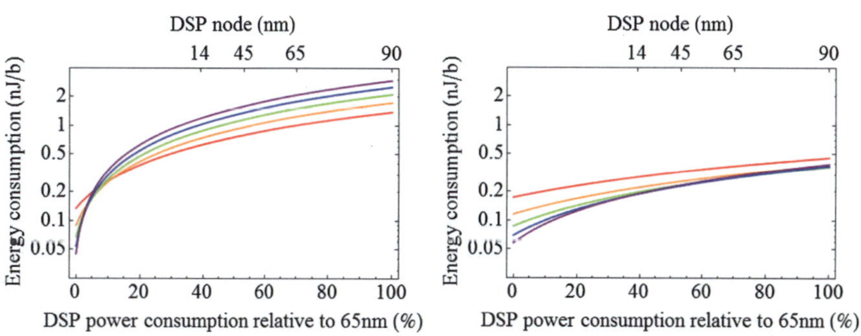

Fig. 10.9 Impact of DSP processing power on overall link energy consumption for full digital coherent system (*left*) and pilot aided coherent system (*right*) with 7 % FEC overhead. All other parameters as specified in Fig. 10.7

compatible with high cardinality coherently detected polarisation multiplexed systems and should offer lower energy consumption than the conventional scheme (0.2–1 nJ/b per operation).

10.4 Optical Signal Modulation Using Electro-Optic Modulators

This section will cover the waveform processing capabilities of electro-optic devices in the context of optical frequency translation and optical pulse shaping. The section will identify the performance limitations, optical and electrical power efficiencies as a function of frequency shift or pulse width. We will consider two types of electro-optic modulator for these applications; the phase modulator and its direct derivatives of the Mach Zehnder modulator and IQ modulators, and the electro-absorption modulator. The simplest mean to generate an optical waveform using an electro-optic modulator is to simply prepare a drive waveform of the required temporal shape, amplify to a suitable level and apply to the modulator. This works well up to a large faction of the modulator bandwidth, and is utilised extensively for basic signal generation such as data encoding. We consider here the capabilities of such modulators to generate signals with optical bandwidths which exceed the modulator bandwidth, focussing on pulse generation and optical comb generation.

10.4.1 Pulse Generation Using Electro-Absorption Modulators

Electro-absorption modulators may be readily used for pulse generation by applying a simple sine wave drive [72–75]. They generate pulses whose temporal intensity profile closely approximates a sech^2 shape, and the same configuration may be readily employed for optical gating or demultiplexing. A pulse width as low as 7 % of the drive frequency may be readily achieved using appropriately designed devices.

A typical absorption characteristic is shown in Fig. 10.10, and for this example, the packaged device performance included, typically, an additional 8.4 dB insertion loss [76]. To first order, the modulation characteristic may be expressed by considering an idealised relationship between the absorption (in dB) and the applied reverse bias (V);

$$\alpha_{dB} = \begin{cases} 0 & V \le V_0 \\ k(V - V_0) & ,\cdots V_0 < V < V_1 \\ XR = k(V_1 - V_0) & V_1 \le V \end{cases} \tag{10.13}$$

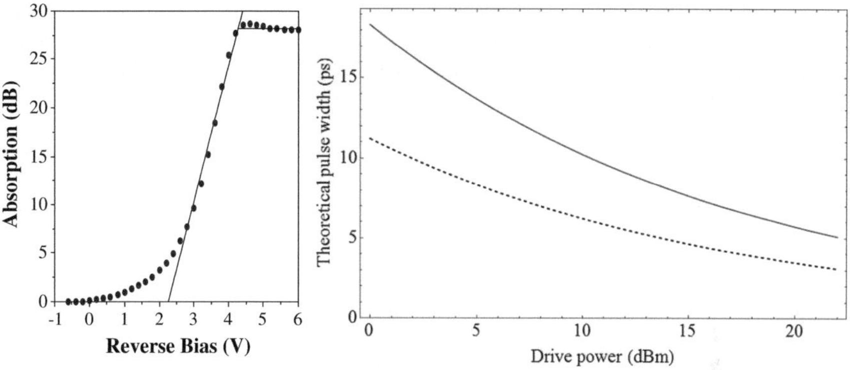

Fig. 10.10 *left* Typical absorption characteristic of an electro-absorption modulator (after [76]) showing exponential loss characteristic for reverse bias between 2 and 4 V. *Right* Predicted pulse widths for a 10 GHz drive frequency as a function of available drive power for a modulator driven at 10 GHz with an absorption slope of 11.8 dB/V (*solid*) and at 20 GHz with an absorption slope of 8 dB/V (*grey*)

where k represents the gradient of the absorption characteristic in dB/V, V_0 is the maximum reverse bias for zero absorption and V_1 is the reverse bias at which maximum absorption is achieved. XR represents the maximum extinction ratio (dB), although for high values this may be neglected ($V_1 \rightarrow \infty$). For a sinusoidal drive signal, with a peak voltage of V_1 below V_0, the output of the modulator is given by;

$$E(t) = E_0 e^{-\frac{k}{2\log_{10}(e)}\left(V_b + V_{pp}\cos(2\pi.t)/2 - V_0\right)} \tag{10.14}$$

From which is it is easy to show [77] that for $V_b > V_0 + V_{pp}/2$ the resultant pulse width is given by;

$$\tau_{FWHM} = \frac{2}{\omega}\cos^{-1}\left(1 - \frac{2\log_{10}(2)}{kV_{pp}}\right) \approx \frac{2}{\omega}\left(\frac{1.2}{kV_{pp}}\right)^{\frac{1}{2}} \tag{10.15}$$

where V_{pp} represents the peak to peak amplitude of the drive signal of angular frequency ω. From this equation, we may deduce that the shortest pulses are produced for the highest drive frequencies, and for the steepest absorption characteristics. The suitability of sine wave driven EAMs as a compact, stable, low jitter source of short pulses and polarisation insensitive time gates suitable for use in high bit-rate OTDM systems has been confirmed in an 80 Gbaud all electro-absorption modulator system [78] and in system experiments at 160 Gbaud.

For $V_b > V_0 + V_{pp}/2$, part of the sinusoidal drive falls within the low loss region of the absorption characteristic ($V < V_0$). This has the effect of flattening the peak of the generated pulse (or switching window) as shown in Fig. 10.11a. This mode of operation has several advantages, including improved de-multiplexer jitter tolerance [79], and for devices able to withstand high forward bias voltages, channel drop [80].

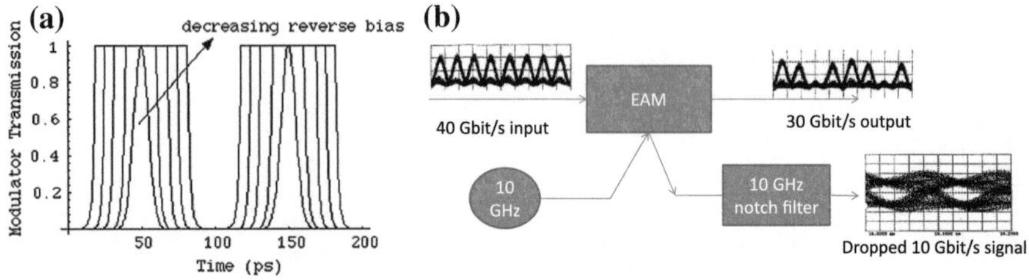

Fig. 10.11 **a** Theoretical switching window of electroabsorption modulator as a function of reverse bias. **b** Concept of add drop multiplexer with simultaneous channel detection, taken from two separate experiments [81]

A further advantage of this operating scheme is that the absorbed signal generates a photocurrent in the device, which may be used as a channel selective receiver. Thus a device driven with a high RF amplitude and low DC bias may simultaneously clear an OTDM channel from a data stream and allow the detection of the dropped channel (see Fig. 10.11b).

10.4.2 Pulse Generation and Optical Sampling Using Mach Zehnder Modulators

Pulse generation using optical devices with sinusoidal transfer functions [82], and in particular Mach Zehnder modulators has been considered by several authors [83–87]. The technique relies on the sinusoidal transfer function of the modulator, when driven by sine wave signals. It is well known that the output of a Mach Zehnder modulator (E_{out}) is governed by the following equation [88];

$$E_{out} = E_{in} \cos\left(\frac{\pi}{2} \frac{V - V_0}{V_{AM}}\right) \tag{10.16}$$

and illustrated in Fig. 10.10 in terms of the power transmission as a function of normalised applied voltage. This results in a temporal output of the modulator of

$$E_{out} = E_{in} \cos\left(\frac{\pi}{2} \frac{V_b - V_0 + V_{pp} \sin(\omega.t)/2}{V_{AM}}\right) \tag{10.17}$$

where we have expressed the minimum on-off switching voltage of the modulator (often referred to as V_{π}) as V_{AM}. In general the modulator bias and (V_b) and drive amplitude (V_{pp}) may be set independently, however there are a number of common alignment strategies as shown in Fig. 10.12. In these strategies the modulator is biased to such that, one extreme of the sine wave corresponds to maximum or minimum transmission, quadrature, or to give frequency doubling. In three of these

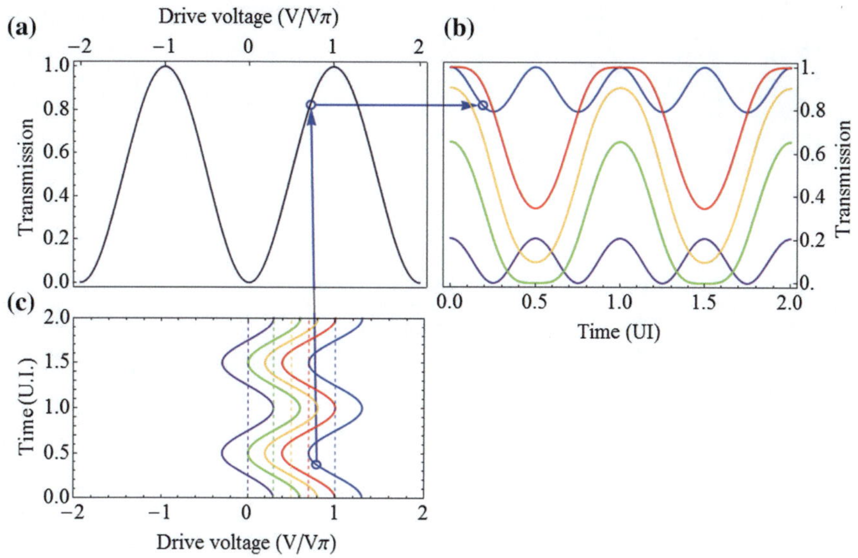

Fig. 10.12 **a** Normalised power transfer characteristic of Mach Zehnder modulator along with **b** drive voltages and **c** output waveforms for different bias conditions and a drive voltage of 0.6 V_{AM} corresponding to the following bias conditions: minimum bias (*green* $V_b - V_0 = V_{pp}/2$), maximum bias (*red* $V_b - V_0 = V_{AM} - Vpp/2$), quadrature bias (*orange* $V_b - V_0 = V_{AM}/2$), carrier suppressed (*purple* $V_b - V_0 = 0$), and frequency doubled (*blue* $V_b - V_0 = V_{AM}$). Carrier suppressed and frequency doubled modes result in pulse trains a twice the drive frequency

conditions if the bias is set such that the sine wave extreme or crossing corresponds with minimum transmission, then the drive amplitude determines the pulse width and amplitude (left hand panels of Fig. 10.13). For the other conditions, pulses of acceptable extinction ratio are only generated for specific drive amplitudes (right hand panels of Fig. 10.13). Achievable pulse widths, as a function of drive power assuming V_{AM} = 4 V are shown in Fig. 10.14.

The shortest pulse width is obtained for the frequency doubled mode (maximum transmission corresponding to the crossing point of the sine wave drive), but which obviously results in a frequency doubling operation. The fundamental frequency may be achieved either by cascading with a second modulator, driven with $V_{pp} = V_{AM}$ (at the expense of increased excess loss, typically around 4 dB) [89], or by halving the drive frequency (at the expense of doubling the pulse width from 16 to 32 % of the target frequency). The most rectangular pulse is achieved using V_{AM} drive, and whilst increasing the drive amplitude slightly reduced the extinction ratio, it results in a more rectangular pulse shape, which may be advantageous for demultiplexing operations [90]. Note that in addition to sampling and pulse carving operations described above, Mach Zehnder modulators may also be used to carve out dark pulses for various applications such as serial TDM multiplexing and add-drop multiplexing [91, 92].

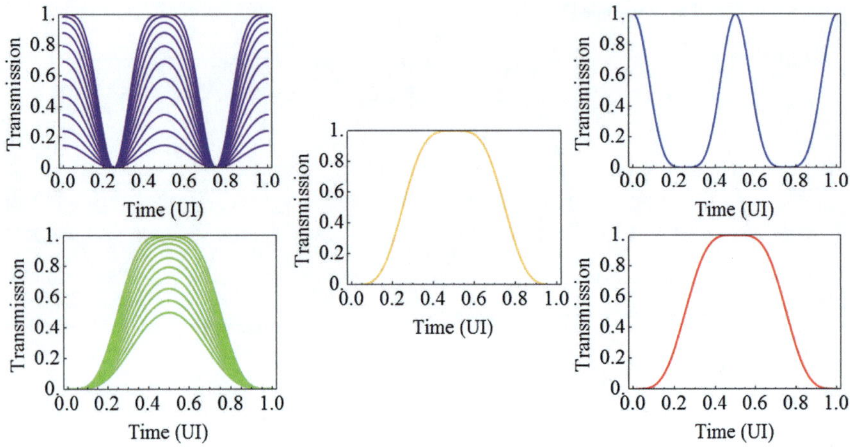

Fig. 10.13 Variation of Mach Zehnder modulator switching window with RF amplitude for minimum bias (*green* $V_{AM}/2 < V_{pp} < V_{AM}$), maximum bias (*red* $V_{pp} = V_{AM}$), quadrature bias (*orange* $V_{pp} = V_{AM}$), carrier suppressed (*purple* $V_{AM}/2 < V_{pp} < 2\ V_{AM}$), and frequency doubled (*blue* $V_{pp} = 2\ V_{AM}$)

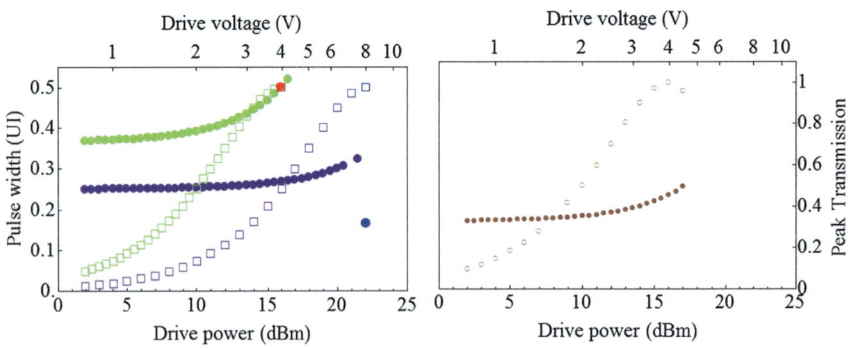

Fig. 10.14 Variation of pulse width (*filled circles*), expressed as a fraction of the cycle period of the sine wave drive and peak transmission loss (*open squares*) as a function of applied drive power (for a modulator $V_{AM} = 4$ V) for minimum bias (*green*) maximum bias (*red*), carrier suppressed (*purple*) and frequency doubled (*blue*) modes, and for a modulator driven with a pulsed drive signal obtained from a frequency multiplier (*brown*, *right*). Power and drive voltages assume a V_{AM} of 4 V

10.4.3 Optical Comb Generation

Optical combs have a wide variety of uses, ranging from various forms of metrology, frequency translation and carrier generation for multi-carrier transmission schemes including orthogonal and conventional frequency division multiplexing (OFDM and FDM respectively) [93–95]. A wide variety of comb generation schemes are possible, including mode-locked solid state [96], fibre [97, 98] and semiconductor ring [99] lasers, semiconductor lasers [100–102], non-linear beat frequency generation [103–106], or gain switching [107] some of which are discussed in Chaps. 2 and 6 of this book. In this section, we will analyze the frequency manipulation capabilities of sine wave driven modulators, showing that direct analysis in the frequency domain allows optimized comb generation performance.

We consider first a simple sine wave driven phase modulator, where we observe that the output spectrum may be obtained by considering [108];

$$E_{out} = |E_{in}| \cos\left(\omega.t + \vartheta_{in} + \frac{\pi}{V_\pi}\{x_i + v_i \sin(\Delta\omega.t + \varphi_i)\}\right) \tag{10.18}$$

$$E_{out} = |E_{in}| \sum_{n=-\infty}^{\infty} J_n\left(\frac{\pi}{V_\pi} v_i\right) \cos\left(\omega t + \vartheta_{in} + \frac{\pi}{V_\pi} x_i + n\Delta\omega t + n\varphi_i\right) \tag{10.19}$$

where ω represents the input optical carrier frequency, ϑ_{in} its phase, V_π the voltage required to modulate the phase by π [109], x_i the dc bias applied to phase modulator section i, v_i the drive amplitude of an applied sinusoidal signal of frequency Δω and phase ϕ_i. For a continuous light source input the amplitudes of each frequency component are given by a series of Bessel functions (10.19, J_n represents the nth order Bessel function). The amplitudes of each component are dependent only on the relative drive amplitude whilst the phases are dependent on the input optical and electrical phases, the dc bias voltage and the line number. Two typical intensity spectra from such a comb is shown in Fig. 10.16a with drive amplitudes of 0.6 and 1.9 V_π. It is impossible to achieve an identically flat comb spectrum for more than three comb lines, the characteristic spectrum being two wings with a central section with strongly varying amplitude in between. A common enhancement to a phase modulator based comb intended to improve the width and/or flatness of the comb is to cascade multiple phase modulators in series [110] or parallel respectively [111–115] or combinations of both [116, 117]. For brevity, we consider here an arbitrary 2 dimensional array of phase modulators, as shown in Fig. 10.15.

A cascade of phase modulators would be analysed by considering a single row of phase modulators, whilst a single Mach Zehnder or IQ modulator would be analysed by considering 2 or 4 parallel paths respectively, each comprising a single stage of phase modulation. The performance of more complex combinations such as cascaded MZMs may be calculated by an appropriate combination of modulators, with defined relationships between the drive signals. To calculate the response of an arbitrary configuration of modulators, a number of rows equal to the number of possible independent paths through the structure are selected, and a number of columns equal to the maximum number of independent drive voltages which may be experienced on these paths. The corresponding drive voltages are assigned to each of the modulator sections. For the pth phase modulator cascade, the overall response may be emulated by a single phase modulator with a drive signal of

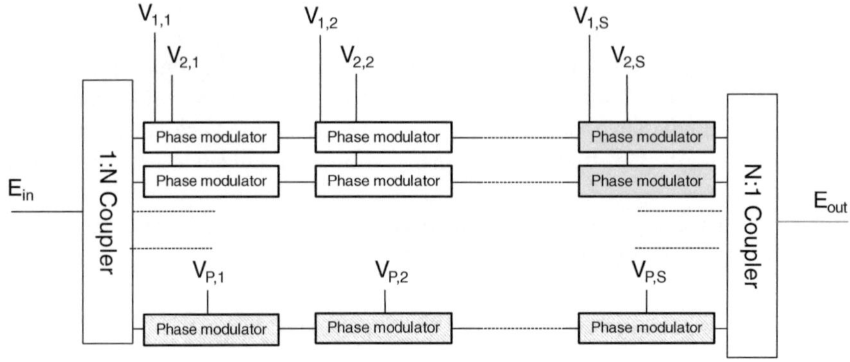

Fig. 10.15 A two dimensional array of phase modulator segments in a simple interferometric structure, illustrating the approach taken for the calculation of various combinations of phase and amplitude modulators. Cascaded phase modulators would use, for example, the diagonally hashed modulator segments, dual drive Mach Zehnder two parallel segments (*filled grey*). For cascaded amplitude modulators, four rows of modulator sections are used to represent the four possible paths through the structure, with two columns allowing the different drive voltages to be applied with appropriate signs

amplitude Vp, dc bias X_p and phase Φ_p, where the parameters are calculated by summing phasors for each modulator section and are given by;

$$
\begin{aligned}
X_p &= \sum_{s=1}^{N_S} \lambda_{s,p} x_{s,p} V_p^2 = \frac{\pi^2}{V_\pi^2} \sum_{s=1}^{N_S} \left\{ v_{s,p}^2 + 2 \sum_{t=s+1}^{N_S} v_{s,p} v_{t,p} \cos\left(\varphi_{s,p} - \varphi_{t,p}\right) \right\} \\
\Phi_p &= \mathrm{ArcTan2}\left(\sum_{s=1}^{N_S} v_{s,p} \cos\left(\varphi_{s,p}\right), \sum_{s=1}^{N_S} v_{s,p} \sin\left(\varphi_{s,p}\right) \right) \mu_p = \max_s\left(\lambda_{s,p}\right)
\end{aligned}
\tag{10.20}
$$

Having reduced the 2 dimensional arrays of phase modulators to a column of individual phase modulators, the outputs are summed to a single virtual modulator, again using phasor summation, with amplitude V' and phase Φ' given by

$$
\begin{aligned}
V' &= \sqrt{\sum_{p=1}^{N_P} \left\{ \mu_p J_n^2\left(V_p\right) + 2 \sum_{q=p+1}^{N_P} \mu_p \mu_q J_n\left(V_p\right) J_n\left(V_q\right) \cos\left(X_p + n\Phi_p - X_q - n\Phi_q\right) \right\}} \\
\Phi' &= \mathrm{ArcTan2}\left(\sum_{k=1}^{N_P} \mu_p J_n\left(V_p\right) \cos\left(X_p + n\Phi_p\right), \sum_{p=1}^{N_P} \mu_p J_n\left(V_p\right) \sin\left(X_p + n\Phi_p\right) \right)
\end{aligned}
\tag{10.21}
$$

such that the overall output of the combination of modulators is given by

$$E_{out} = \frac{1}{2 \cdot N_P} |E_{in}| \sum_{n=-\infty}^{\infty} V' \cdot \cos(\omega t + \vartheta_{in} + n\Delta\omega t + \Phi') \tag{10.22}$$

where N_P represents the number of modulators in parallel, N_S the number in series, $x_{s,p}$, $v_{s,p}$, and $\phi_{s,p}$ the dc voltage, amplitude and phase of the RF signal applied to the phase modulator in the pth row and sth column of the modulator array, $\lambda_{s,p}$ a binary flag representing the presence (1) or absence (0) of the phase modulator section, and ArcTan2 the two argument inverse tangent function. ArcTan2 is used to return the appropriate quadrant for the computed angle. The outputs of five alternative comb generators are also shown in Fig. 10.16. The spectra are the optimised outputs, subject to either a maximum drive amplitude to each electrode of 0.6 V_π (red), or to a maximum total drive power of 1.6 W, assuming a V_π of 4 V for each individual phase modulator (blue). Analysing these results, we can clearly see that the number of comb lines with significant amplitudes increases both with the number of stages, and the available drive power. On the other hand the uniformity (flatness) of the comb is enhanced by additional parallel paths, with identically flat spectra available with 5 lines from a MZM, and 7 for a cascade of 2 MZM's.

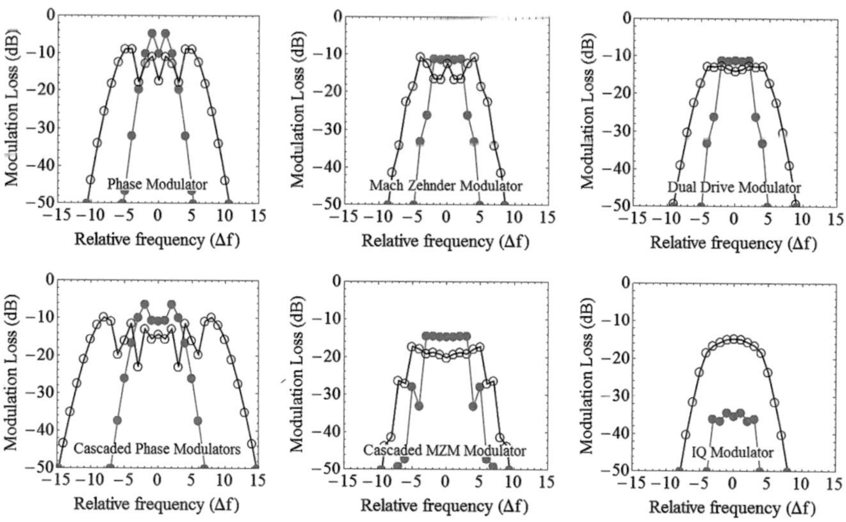

Fig. 10.16 Amplitude of optical comb line versus frequency shift (in multiples of applied drive frequency) for a maximum drive amplitudes per modulator section of 0.6 V_π (filled discs) or a maximum total RF power of 1.6 W (open circles) for *top left* phase modulators, *top centre* single balanced Mach Zehnder modulator, *top right* dual drive Mach Zehnder modulator, *bottom left* cascaded phase modulators, *bottom centre* two cascaded Mach Zehnder modulators, *bottom right* an IQ modulator (four phase modulators in parallel). Indicative drive powers are shown assuming a modulator V_π of 4V

It is straightforward to show that provided all modulators are driven at the same frequency and phase, a cascade of phase modulators results in the same optical comb as a single phase modulator, but with the argument of the Bessel function in (10.19) replaced with the sum of the drive amplitudes of the modulator cascade (see 10.20). This results in an enhancement to the electrical power efficiency in direct proportion to the number of phase modulators cascaded. However, in practice, cascading phase modulators results in a trade-off between optical output power and conversion efficiency due to the excess loss of each modulator, as illustrated in Fig. 10.17. The optimum number of stages may be readily determined to be 10 $\log_{10}(e)/IL_{dB}$ where IL_{dB} is the static insertion loss of the modulator in decibels, implying that a series of modulators represents the most energy efficient solution provided the modulator insertion loss is less than 3 dB. For higher modulator insertion losses, the most efficient strategy is to use the smallest number of stages allowed by the phase modulators power handling limits, usually determined by bond wires and/or on chip electrical terminations. Use of optical amplifiers and arbitrary cascades of phase modulators is clearly useful, and is optimally implemented in a ring configuration [118].

Different definitions of comb flatness have been proposed by various authors. However, since it is likely that line amplitudes will be adjusted by programmable optical filters, the most valid measure is the maximum optical insertion loss encountered by any given comb line. This particular measure is shown in Fig. 10.17b, where we may observe that the additional flatness offered by the dual drive MZM excess loss (over the theoretical optimum of sharing the input power between comb lines) of less than 5 dB, with drive powers of up to 1.5 W (as noted previously, the direct power consumption associated with the drive signal should be

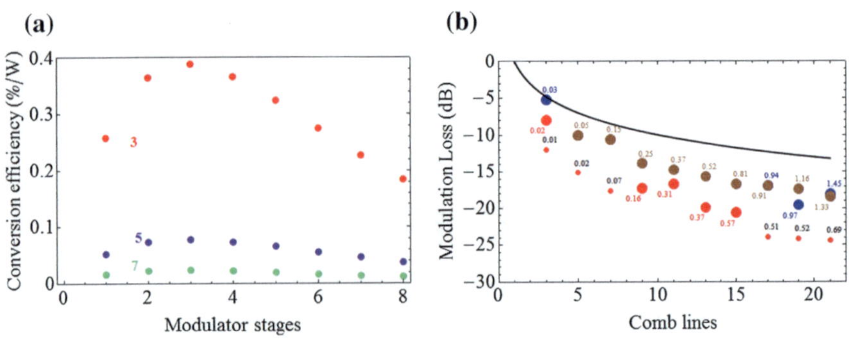

Fig. 10.17 **a** Comb line conversion efficiency (minimum power per line divided by applied input optical and electrical powers) for 3 (*red*), 5 (*purple*), and 7 (*green*) line combs generated by a cascade of phase modulators, each of 1.5 dB insertion loss. **b** Modulation loss per line for a number of different modulator based comb generators showing the configurations offering either the lowest loss or the lowest total RF drive power as a function of the number of comb lines generated. *Solid line* represents an ideal comb generator which shares the available optical power equally between lines. Large coloured dots represent modulator configuration: (*blue* MZM followed by a phase modulator, *red* phase modulator, *brown* dual drive modulator). Small *red dots* represent cascade of 2 phase modulators with 4 dB additional insertion loss

doubled or tripled to take into account the efficiency of electrical power amplifiers). Significantly lower power requirements are found for single and cascaded phase modulators with a power consumption of ~2/3 W for 21 comb lines (Fig. 10.17).

State of the art devices offer bandwidths of up to 30 GHz, suggesting that a total frequency shift of 300 GHz (450 GHz/W) could be achieved using such modulator based comb sources and optical filters. Operation beyond the modulator bandwidth is also possible, albeit at the expense of increasing the power consumption. As travelling wave modulators typically have a linearly decreasing frequency response, doubling the operating frequency would result in double the required drive power. To avoid using optical filters, an IQ modulator may be biased to give single side band frequency shifts of Δf with a drive power of ~4 mW and an excess insertion loss of 6 dB (for a 24 dB extinction ratio of the next highest harmonic). As seen in Sect. 3.2.1, MZM may also be used for frequency doubling, and such properties translate directly to comb generation, with balanced and dual drive MZM allowing frequency doubled operation (by setting DC bias to the carrier suppressed point), and IQ modulators offer frequency multiplication of 2.n.Δf where n is an integer. It was recently shown that 4 [119] and 6 [120] fold frequency multiplication (2 sub carriers spaced at 6.Δf) could be readily achieved using commercial IQ modulators with less than 1 W drive per modulator (6 V V_π) and that 12 fold frequency multiplication was theoretically possible with drive voltages of 6.5 V_π.

Comb generation is also possible using multiple drive frequencies, either slightly offset to avoid field addition in modulator cascades [121] or with larger offsets [114, 122–127]. Following the same analysis as described above, the output field of a Mach Zehnder modulator driven by a number of different RF drive signals is given by;

$$E_{out} = E_{in}(-1)^{N_d} \sum_{\rho_1=-\infty}^{\infty} \cdots \sum_{\rho_N=-\infty}^{\infty} \cos\left(a + \sum_{i=1}^{N_d} \frac{\pi}{2}\rho_i\right) \cos\left(\sum_{i=1}^{N_d} \rho_i \Delta\omega_i + \phi_i\right) \prod_{i=1}^{N_d} J_{\rho_i}\left(\frac{\pi b_i}{V_\pi}\right)$$
$$\phi_i = \left(1 + i + (-1)^i\right)\frac{\pi}{2} \tag{10.23}$$

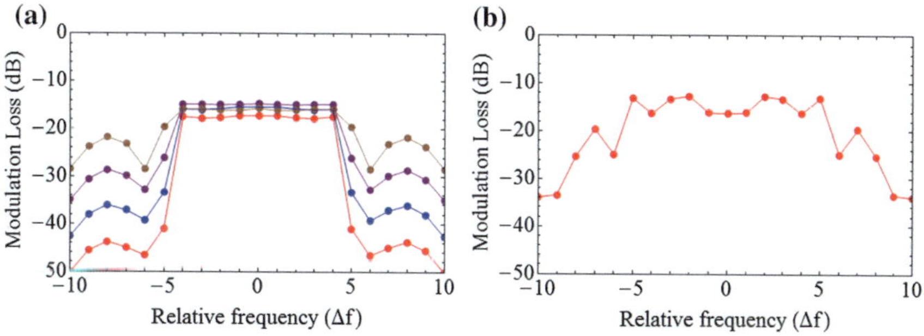

Fig. 10.18 Comb line spectra using high and low power electrical drive signals. **a** Four equally spaced equal amplitude drive signals of low power, *red* 34 mW, *blue* 0.1 W, *purple* 0.4 W, *brown* 1.5 W. **b** Large signal drive combined with its third harmonic (1.6 W), (powers calculated assuming phase aligned drive signals)

where $N_d i$ represents the number of drive signals, a the dc bias of the modulator and $\Delta\omega_I$, ϕ_I, and b_i are the frequency phase and amplitude of the ith drive signal respectively. To calculate the amplitude of a particular frequency harmonic Δω, it is necessary to sum all terms where $\sum_{i=1}^{N_d} \rho_i \Delta\omega_i = \Delta\omega$ which is equivalent to asserting the condition that $\rho_1 \Delta\omega_1 = \Delta\omega - \sum_{i=2}^{N_d} \rho_i \Delta\omega_i$. Typically the infinite summations are restricted to a few times the expected number of optical comb lines without significant loss of accuracy. Two strategies are typically used to generate an optical frequency comb from a multi-tone drive signal; low signal amplitudes to avoid unwanted intermodulation products (Fig. 10.18) [124] or with optimised large signal amplitudes to exploit the intermodulation products to generate new frequencies.

In a similar way to the linear power reduction associated with modulator cascades, additional line generation using a multi-frequency drive typically reduces the required electric drive power without sacrificing optical insertion loss. However unlike cascaded phase modulators, it does not extend the frequency range of the generated comb.

10.5 Conclusions

In this chapter, we have considered the implementation of various signal processing functions in an optical communications network using either electro-optic modulators or a combination of coherent detection and digital signal processing. Digital coherent transmission systems, using a 90 nm processing node, offer fully regenerated signal processing including FEC for an energy consumption of less than 4 nJ/b for modulation formats up to 64QAM. This is dominated by carrier phase recovery, and the sacrifice of a small level of spectral efficiency to enable the use of pilot tones reduces this to less than 1 nJ/band unsurprisingly increasing FEC overheads is most effective for longer reach systems. Optimisation of the nonlinear performance is beneficial primarily by reducing the number of transponders/regenerators in a link, and the contribution of optical amplifiers to the overall energy consumption of these systems is negligible. Due to the dominance of DSP energy consumption, the benefits associated with improved energy consumption of particular CMOS technology nodes translate almost directly to benefits in physical layer energy consumption. However, this also opens the way for optical techniques, such as phase recovery based on pilot tones to enable significant reductions in energy consumption.

We find that pulse generation with duty cycles down to ~16 (7) % of the maximum drive frequency are readily achievable using Mach Zehnder (electro-absorption) modulators, whilst optical frequency comb generation with an excess modulation loss of less than 6 dB are possible, with total drive powers less than 1 W. Excess insertion loss is typically in the region of 4–8 dB.

References

1. A.D. Ellis, N.J. Doran, Are few mode fibres a practical solution to the capacity crunch? in *ICTON 2013*, paper Tu.C2.1 (2013)
2. Optical fiber link for police computer. Electronics Weekly, Sept. 24, 3 (1975)
3. I.P. Kaminov, T. Li, A.E. Wilner, *Optical Fiber Telecommunications V* (Academic Press, Burlington, 2008)
4. R. Essiambre, G. Kramer, P.J. Winzer, G.J. Foschini, B. Goebel, Capacity limits of optical fiber networks. J. Lightwave Technol. **28**(4), 662 (2010)
5. R.S. Tucker, Green optical communications-Part I: energy limitations in transport. J. Sel. Top. Quantum Electron. **17**(2), 245 (2011)
6. Information Gatekeepers Inc., Lucent technologies helps BT increase network capacity and offer enhanced services. Fiber Opt. Wkly Update **17**(39), 1 (1997)
7. P.J. Winzer, Power consumption of SDM systems, in *ECOC Workshop on Integration of Optical Devices for SDM Transmission, ECOC 2013*, http://www.ecoc2013.org/docs/peter-winzer.pdf (2013)
8. A.H. Gnauck, P.J. Winzer, Optical phase-shift-keyed transmission. J. Lightwave Technol. **23**, 115 (2005)
9. T. Okoshi, K. Kikuchi, Heterodyne type optical communications. J. Opt. Commun. **2**(3), 82 (1981)
10. http://www.emcore.com/fiber-optics/telecom-tunable-dwdm/itla-integrable-tunable-laser-assembly/ (July 2014)
11. B. Collings, Photonics for metro and client 100G and beyond—wants in 100G stream…!, in *Next Generation Optical Networking 2014*, paper day 3, stream 2, 9:40 (2014)
12. http://jp.fujitsu.com/group/foc/en/services/optical-devices/100gln/
13. P. Dong, L. Chen, Y.K. Chen, High-speed low-voltage single-drive push-pull silicon Mach-Zehnder modulators. Opt. Express **20**(6), 6163–6169 (2012)
14. N. Kikuchi, R. Hirai, Y. Wakayama, High-speed optical 64QAM signal generation using InP-based semiconductor IQ modulator, in *Optical Fiber Communication Conference*, M2A-2 (2014, March)
15. D. Korn, R. Palmer, H. Yu, P.C. Schindler, et al., Silicon-organic hybrid (SOH) IQ modulator for 16QAM at 112 Gbit/s, in *The European Conference on Lasers and Electro-Optics* (p. CK_9_2) (Optical Society of America, 2013, May)
16. Quad-Channel 32 Gbps Linear MZ Modulator Driver, http://www.macom.com/products/product-detail/MAOM-003408 (July 2014)
17. Inphi IN3212SZ, 32 Gbps Single-Ended Mach-Zehnder Driver. http://www.inphi.com/products/in3212sz.php (July 2014)
18. Gigoptix 4x32Gb/s MZ Linear Modulator Driver, http://www.gigoptix.com/products/gx62472/ (July 2014)
19. M. Möller, High-speed electronic circuits for 100 Gb/s transport networks, in *OFC*, (2010)
20. B. Berglund, J. Johansson, T. Lejon, High efficiency power amplifiers. Ericsson Rev. **6**, 92 (2006)
21. W. Forysiak, D.S. Goven, Progress towards 100 G digital coherent pluggables in InP-based photonics. J. Lightwave Technol. doi:10.1109/JLT.2014.2329754 (2014). Alternative would be W. Forysiak, Progress in InP-based photonic components and sub-systems for digital coherent systems at 100Gbit/s and beyond, in *ECOC 2013*, Mo.3.C.2, Vol. 32, 16, pp. 2925−2934 (2014)
22. D.J. Malyon, Digital fibre transmission using optical homodyne detection. Electron. Lett. **20** (7), 281 (1984)
23. Y. Audzevich, P.M. Watts, A. West, A. Mujumdar, S.W. Moore, A.W. Moore, Power optimised transceivers for future switched networks. IEEE Trans. Very Large Scale Integr. VLSI Syst. 22(10), 2081−2092 (2014).doi:10.1109/TVLSI.2013.2283300

24. M. Horowitz, E. Alon, D. Patil, S. Naffziger, K. Rajesh, K. Bernstein, Scaling power and the future of CMOS, in *Electron Devices Meeting, 2005*. IEDM Technical Digest. IEEE International (2005), pp. 7, 15
25. M. Kuschnerov, T. Bex, P. Kainzmaier, Energy efficient digital signal processing, in *OFC 2014*, paper Th3E.7 (2014)
26. D. Kudithipudi, E. John, Implementation of low power digital multipliers using 10 transistor adder blocks. J. Low Power Electron. **1**, 1 (2005)
27. C. Laperle, M. O'Sullivan, Advances in high speeds DACs, ADCs and DSP for optical coherent transceivers. J. Lightwave Technol. **32**(4), 629 (2014)
28. T. Pfau, S. Hoffmann, R. Noe, Hardware-efficient coherent digital receiver concept with feedforward carrier recovery for M-QAM Constellations. J. Lightwave Technol. **27**(8), 989 (2009)
29. H. Dowid, H. Meyr, CORDIC algorithms and architectures, in *Digital Signal Processing for Multimedia Systems* (1999), pp. 623–655
30. M. O'Sullivan, C. Laperle, A. Borowiec, K. Farley, A 400G/1T high spectral efficiency technology and some enabling subsystems, in *OFC 2012*, paper OM2H1 (2012)
31. B.E. Jonsson, A/D-converter performance evolution, in *Converter Passion, ADM Design* (2012)
32. C. Laperle, M. O'Sullivan, Advances in high-speed DACs, ADCs, and DSP for optical coherent transceivers. J. Lightwave Technol. **32**(4), 629–643 (2014)
33. M. Nagatani, H. Nosaka, K. Sano, K. Murata, K. Kurishima, M. Ida, A 60-GS/s 6-Bit DAC in 0.5-im InP HBT technology for optical communications systems, in *Compound Semiconductor Integrated Circuit Symposium* (CSICS) (2011)
34. Y. Greshishchev, et al., A 40GS/s 6b ADC in 65 nm CMOS, in *ISSCC 2010* (2010), pp. 390–392
35. D. Crivelli, M. Hueda, H. Carrer, J. Zachan, V. Gutnik, M. Del Barco, R. Lopez, G. Hatcher, J. Finochietto, M. Yeo, A. Chartrand, N. Swenson, P. Voois, O. Agazzi, A 40 nm CMOS single-chip 50 Gb/s DP-QPSK/BPSK transceiver with electronic dispersion compensation for coherent optical channels, in *ISSCC 2012* (2012), pp. 328–330
36. F. Lang, T. Alpert, D. Ferenci, M. Grozing, M. Berroth, A 6 bit 25 GS/s flash interpolating ADC in 90 nm CMOS technology. in *Ph.D. Research in Microelectronics and Electronics (PRIME), 2011 7th Conference on. IEEE, 2011* (2011), p. 117
37. Fujitsu 56GSa/s 8-bit analogue-to-digital converter Factsheet, http://www.fujitsu.com/downloads/MICRO/fme/documentation/c45.pdf (July 2014)
38. T. Pfau, R. Peveling, V. Herath, S. Hoffmann, C. Wördehoff, O. Adamczyk, M. Porrmann, R. Noé, Towards real-time implementation of coherent optical communication, in *OFC 2009*, paper OThJ4 (2009)
39. D. Hillerkuss, R. Schmogrow, M. Meyer, S. Wolf, M. Jordan, P. Kleinow, N. Lindenmann, P.C. Schindler, A. Melikyan, Y. Xin S. Ben-Ezra, B. Nebendahl, M. Dreschmann, J. Meyer, F. Parmigiani, P. Petropoulos, B. Resan, A. Oehler, K. Weingarten, L. Altenhain, T. Ellermeyer, M. Moeller, M. Huebner, J. Becker, C. Koos, W. Freude, J. Leuthold, Single-laser 32.5 Tbit/s Nyquist WDM transmission. Opt. Commun. Netw. **4**(10), 715–723 (2012)
40. E. Ip, J.M. Khan, Digital equalisation of chromatic dispersion and polarisation mode dispersion. J. Lightwave Technol. **25**(8), 2033 (2007)
41. S.J. Savory, Digital coherent receivers: algorithms and subsystems. IEEE J. Sel. Top. Quantum Electron. **16**(5), 1164 (2010)
42. K. Kikuchi, Digital coherent optical communication systems: fundamentals and future prospects. IEICE Electron. Express **8**(20), 1642 (2011)
43. B.E. Jonsson, On CMOS scaling and A/D-converter performance, in *28th IEEE CAS Norchip Conference 2010, 15–16 Nov 2010, Tampere, Finland, Tu8.1.1* (2010), p. 284
44. B. Spinnler, Equalizer design and complexity for digital coherent receivers. IEEE J. Sel. Top. Qunatum. Electron. **16**(5), 1180 (2010)

45. K. Roberts, M. O'Sullivan, K.T. Wu, H. Sun, A. Awadalla, D.J. Krause, C. Laperle, Performance of dual-polarization QPSK for optical transport systems. J. Lightwave Technol. **27**(16), 3546–3559 (2009)
46. H. Sun and K.-T. Wu, A novel dispersion and PMD tolerant clock phase detector for coherent transmission systems, in *OFC*, Paper OMJ4 (2011)
47. J.C. Geyer, C.R. Fludger, T. Duthel, C. Schulien, B. Schmauss, Efficient frequency domain chromatic dispersion compensation in a coherent Polmux QPSK-receiver, in *Optical Fiber Communication Conference* (Optical Society of America, 2010), p. OWV5
48. M.S. Faruk, K. Kikuchi, Adaptive frequency-domain equalization in digital coherent optical receivers. Opt. Express **19**(13), 12789–12798 (2011)
49. R.A. Soriano, F.N. Hauske, N.G. Gonzalez, Z. Zhang, Y. Ye, I.T. Monroy, Chromatic dispersion estimation in digital coherent receivers. J. Lightwave Technol. **29**(11), 1621 (2011)
50. T. Pfau, C. Wördehoff, R. Peveling, S.K. Ibrahim, S. Hoffmann, O. Adamczyk, S. Bhandare, M. Porrmann, R. Noé, A. Koslovsky, Y. Achiam, D. Schlieder, N. Grossard, J. Hauden, H. Porte, Ultra-fast adaptive digital polarization control in a realtime coherent polarization-multiplexed QPSK receiver, in *OFC2008*, paper OTuM3 (2008)
51. D.N. Godard, Self-recovering equalization and carrier tracking in two-dimensional data communication systems. IEEE Trans. Commun. **28**, 11 (1980)
52. M.G. Taylor, Phase estimation methods for optical coherent detection using digital signal processing. J. Lightwave Technol. **27**(7), 901 (2009)
53. Y. Gao, A.P.T. Lau, S. Yan, C. Lu, Low-complexity and phase noise tolerant carrier phase estimation for dual-polarization 16-QAM systems. Opt. Express **19**(22), 21717–21729 (2011)
54. S.M. Bilal, G. Bosco, P. Poggiolini, A. Carena, Low-complexity linewidth-tolerant carrier phase estimation for 64-QAM systems based on constellation transformation, in *ECOC* (2013), p. 966
55. I. Fatadin, I. David, S.J. Savory, Laser linewidth tolerance for 16-QAM coherent optical systems using QPSK partitioning. Photonics Technol. Lett. **22**(9), 631–633 (2010)
56. S.M. Bilal, A. Carena, C. Fludger, G. Bosco, Dual stage CPE for 64-QAM optical systems based on a modified QPSK-partitioning algorithm. IEEE Photonics Technol. Lett. **26**(3), 267 (2014)
57. H.-C. Chang, C.B. Shung, C.-Y. Lee, A reed-solomon product-code (RS-PC) decoder chip for DVD applications. IEEE J. Solid-State Circuits **36**(2), 229–238 (2001)
58. P.P. Mitra, J.B. Stark, Nonlinear limits to the information capacity of optical fibre communications. Nature **411**, 1027–1030 (2001)
59. X. Chen, W. Shieh, Closed-form expressions for nonlinear transmission performance of densely spaced coherent optical OFDM systems. Opt. Express **18**, 19039 (2010)
60. P. Poggiolini, Modeling of non-linear propagation in uncompensated coherent systems, in *OFC 2013*, paper OTh3G1 (2013)
61. A.D. Ellis, M.A. Sorokina, S. Sygletos, S.K. Turitsyn, Capacity limits in nonlinear fiber transmission, in *Proceedings of Asia Communications and Photonics*, Paper AW4F.1 (2013)
62. Y. Koizumi, K. Toyoda, M. Yoshida, M. Nakazawa, 1024 QAM (60 Gbit/s) singe-carrier coherent optical transmission over 150 km. Opt. Express **20**(11), 12508 (2012)
63. S. Adhikari, S. Sygletos, A.D. Ellis, B. Inan, S.L. Jansen, W. Rosenkranz, Enhanced self-coherent OFDM by the use of injection locked laser, in *OFC 2012*, paper JW2A.64 (2012)
64. B.J. Puttnam, R.S. Luís, J.M.D. Mendinueta, J. Sakaguchi, W. Klaus, Y. Kamio, M. Nakamura, N. Wada, Y. Awaji, A. Kanno, T. Kawanishi, T. Miyazaki, Self-homodyne detection in optical communication systems. Photonics **1**(2), 110 (2014)
65. S.J. Savory, Digital signal processing for coherent systems, in *OFC 2012*, paper OTh3C7 (2012)
66. International Roadmap for Semiconductors, http://www.itrs.net/Links/2013ITRS/Home2013.htm (June 2014)

67. D.C. Kilper, G. Atkinson, S.K. Korotky, S. Goyal, P. Vetter, D. Suvakovic, O. Blume, Power trends in communication networks, IEEE J. Sel. Top. Quantum Electron. **17**(2), 275 (2011)
68. B.E. Jonsson, A/D converter performance evolution, ADMS Design AB, Doc. No.: T-13:001 (2013), http://converterpassion.wordpress.com/articles/ad-converter-performance-evolution/
69. E.F. Mateo, X. Zhou, G. Li, Electronic phase conjugation for nonlinearity compensation in fiber communication systems, in *OFC*, Los Angeles, CA, 2011, Paper JWA025 (2011)
70. B.-E. Olsson, C. Larsson, J. Martensson, A. Alping, Experimental demonstration of electro-optic mid-span spectrum inversion for mitigation of nonlinear fiber effects, in *ECOC 2012*, paper Th.1.D (2012)
71. R. Asif, G. Shabbir, A. Akram, Hybrid mid link spectral inverter to compensate for linear and nonlinear distortions for efficient transmission performance in 224 Gbit/s DP-16QAM system. Opt. Eng. **52**(9), 096103 (2013)
72. M. Suzuki, H. Tanaka, K. Utaka, N. Edagawa, Y. Matsushima, Transform limited 14 ps optical pulse generation with 15 GHz repetition rate by InGaAsP electroabsorption modulator. Electron. Lett. **28**(11), 1007 (1992)
73. D.G. Moodie, A.D. Ellis, A.R. Thurlow, M.J. Harlow, I.F. Lealman, S.D. Perrin, L.J. Rivers, M.J. Robertson, Multiquantum well electroabsorption modulators for 80 Gbit/s OTDM systems. Electron. Lett. **31**(16), 1370 (1995)
74. K. Wakita, K. Sato, I. Kotaka, M. Yamamoto, M. Asobe, Transform limited 7 ps optical pulse generation using a sinusoidally driven InGaAsP/InGaAsP stained multiple quantum well DFB laser/modulator monolithically integrated. IEEE Photonics Technol. Lett. **5**(8), 899 (1993)
75. G. Raybon, M.G. Young, U. Koren, B.I. Miller, M. Chien, M. Zirngibl, C. Dragone, N.M. Froberg, C.A. Burrus, Five channel WDM soliton pulse generation using sinusoidally driven electroabsorption modulators in 16 × 1 laser/modulator array. Electron. Lett. **31**(14), 1147 (1995)
76. D.G. Moodie, A.D. Ellis, C.W. Ford, Generation of 6.3 ps optical pulses at a 10 GHz repetition rate using a packaged electroabsorption modulator and dispersion decreasing fibre. Electron. Lett. **30**(20), 1700 (1994)
77. D.G. Moodie, Electroabsorption modulators and their impact on future telecommunication networks, in *MSc Thesis, Electronic and Electrical Engineering Department* (University College London, London, UK, 1994)
78. D.D. Marcenac, A.D. Ellis, D.G. Moodie, 80 Gbit/s OTDM using electroabsorption modulators. Electron. Lett. **34**(1), 101–103 (1998)
79. L.F. Mollenauer, E. Lichtman, G.T. Harvey, M.J. Neubelt, B.M. Nyman, Demonstration of error-free soliton transmission over more than 15000 km at 5 Gbit/s, single-channel, and over more than 11000 km at 10 Gbit/s in two-channel WDM. Electron. Lett. **28**(8), 792 (1992)
80. I.D. Phillips, A. Gloag, D.G. Moodie, N.J. Doran, I. Bennion, A.D. Ellis, Drop and insert multiplexing with simultaneous clock recovery using an electroabsorption modulator. Phonics Technol. Lett. **10**(2), 291–293 (1998)
81. A.D. Ellis, All optical networking beyond 10 Gbit/s, in *PhD Thesis, School of Engineering and Applied Science* (Aston University, Birmingham, UK, 1997)
82. T.S. Kinsel, R.T. Denton, Terminals for a high speed optical pulse code modulation communication system: II optical multiplexing and demultiplexing. Proc. IEEE **56**(2), 146–154 (1968)
83. H.A. Haus, S.T. Kirsch, K. Mathysek, F.J. Leonberger, Picosecond optical sampling. IEEE J. Quantum Electron. **16**(8), 870 (1980)
84. J.J. Veselka, S.K. Korotky, P.V. Mamyshev, A.H. Gnauck, G. Raybon, N.M. Froberg, A soliton transmitter using a cw laser and an NRZ driven Mach-Zehnder modulator. IEEE Photonics Technol. Lett. **8**(7), 950 (1996)
85. P.A. Krug, D.M. Spirit, L.C. Blank, Transmission study of optical pulses at 6.2 Gbit/s generated using LiNbO3 modulators to externally modulate cw lasers, in *Proceedings of 17th ACOFT* (1992)

86. H. Haga, M. Izutsu, T. Sueta, An integrated 1x4 high speed optical switch and its applications to a time demultiplexer. J. Lightwave Technol. **3**(1), 116 (1985)
87. G. Ishikawa, H. Ooi, Y. Akiyama, S. Taniguchi, H. Nishimoto, 80 Gbit/s (2x40 Gbit/s) transmission experiments over 667 km dispersion shifted fibre using Ti:$LiNbO_3$ OTDM modulator and demultiplexer, in *Proceedings of ECOC'96*, Paper ThC.3.3 (1996)
88. F. Koyama, K. Iga, Frequency chirping in external modulators. J. Lightwave Technol. **6**(1), 87 (1988)
89. A.D. Ellis, T. Widdowson, X. Shan, G.E. Wickens, D.M. Spirit, Transmission of a true single polarisation 40 Gbit/s soliton data signal over 205 km using a stabilised erbium fibre ring laser and 40 GHz electronic timing recovery. Electron. Lett. **29**(11), 990 (1993)
90. A.D. Ellis, C.W. Chow, Serial OTDM for 100 Gbit-Ethernet applications. Electron. Lett. **42** (8), 485–486 (2006)
91. T. Matsumoto, M. Jinno, K. Kimura, K. Noguchi, Studies on optical digital cross-connect systems for very-high-speed optical communications networks, in *SUPERCOMM/ICC '94, IEEE International Conference on 'Serving Humanity Through Communications*, vol. 2 (1994), p. 1060
92. A.D. Ellis, T. Widdowson, X. Shan, D.G. Moodie, A 3 Node 40 Gbit/s OTDM network experiment using electro optic switches. Electron. Lett. **30**(16), 1333 (1994)
93. V. Torres-Company, A.M. Weiner. Optical frequency comb technology for ultra-broadband radio-frequency photonics. Laser Photonics Rev. **8**(3), 368 (2014)
94. J. Pfeifle, V. Brasch, M. Lauermann, Y. Yu, D. Wegner, T. Herr, K. Hartinger, P. Schindler, J. Li, D. Hillerkuss, R. Schmogrow, C. Weimann, R. Holzwarth, W. Freude, J. Leuthold, T. J. Kippenberg, C. Koos, Coherent terabit communications with microresonator Kerr frequency combs. Nat. Photonics **8**, 375 (2014)
95. N.R. Newbury, Searching for applications with a fine-tooth comb. Nat. Photonics **5**(4), 186 (2011)
96. H.R. Telle, G. Steinmeyer, A.E. Dunlop, J. Stenger, D.H. Sutter, U. Keller, Carrier-envelope offset phase control: A novel concept for absolute optical frequency measurement and ultrashort pulse generation. Appl. Phys. B **69**(4), 327–332 (1999)
97. Y. Fukuchi, K. Hirata, H. Ikeoka, Wavelength-tunable and bandwidth-variable ultra-flat optical frequency comb block generation from a bismuth-based actively mode-locked fiber laser. IEEE Photonics J. **6**(1), 1–9 (2014)
98. A.D. Ellis, R.J. Manning, I.D. Phillips, D. Nesset, 1.6 ps pulse generation at 40 GHz in a phase locked ring laser incorporating highly nonlinear fibre for application to 160 Gbit/s OTDM networks. Electron. Lett. **35**(8), 645 (1999)
99. M. Foster, J. Levy, O. Kuzucu, K. Saha, M. Lipson, A. Gaeta, Silicon-based monolithic optical frequency comb source. Opt. Express **19**, 14233–14239 (2011)
100. C.A. Williamson, M.J. Adams, A.D. Ellis, A. Borghesani, Mode locking of semiconductor laser with curved waveguide and passive mode expander. Appl. Phys. Lett. **82**(3), 322 (2003)
101. A. Akrout, A. Shen, R. Brenot, F. Van-Dijk, O. Legouezigou, F. Pommereau, F. Lelarge, A. Ramdane, D. Guang-Hua, Separate error-free transmission of eight channels at 10 Gb/s using comb generation in a quantum-dash-based mode-locked laser. Photonics Technol. Lett. **21** (23), 1746–1748 (2009)
102. S. Gee, F. Quinlan, S. Ozharar, P.J. Delfyett, J.J. Plant, P.W. Judoawlkis, Optical frequency comb generation from modelocked diode lasers—techniques and applications, in *Digest of the LEOS Summer Topical Meetings* (2005), pp. 71–72
103. D.J. Richardson, R.P. Chamberlin, L. Dong, D.N. Payne, A.D. Ellis, T. Widdowson, D.M. Spirit, Demonstration of 205 km transmission of 35 GHz, 5 ps pulses generated from a diode driven, low jitter, beat signal to soliton conversion source. Electron. Lett. **31**(6), 470 (1995)
104. J.J. Veselka, S.K. Korotky, A multiwavelength source having precise channel spacing for WDM systems. Photonics Technol. Lett. **10**(7), 958–960 (1998)
105. Z. Tong, A. Wiberg, E. Myslivets, B. Kuo, N. Alic, S. Radic, Spectral linewidth preservation in parametric frequency combs seeded by dual pumps. Opt. Express **20**, 17610–17619 (2012)

106. R.P. Scott, N.K. Fontaine, J.P. Heritage, B.H. Kolner, S.J.B. Yoo, 3.5-THz wide, 175 mode optical comb source, in *Optical Fiber Communication and the National Fiber Optic Engineers Conference, 2007*. OFC/NFOEC 2007. PaperOWJ3 (2007)
107. R. Zhou, S. Latkowski, J. O'Carroll, R. Phelan, L. Barry, P. Anandarajah, 40 nm wavelength tunable gain-switched optical comb source. Opt. Express **19**, B415–B420 (2011)
108. M. Kourogi, K.I. Nakagawa, M. Ohtsu, Wide-span optical frequency comb generator for accurate optical frequency difference measurement. J. Quantum Electron. **29**(10), 2693–2701 (1993)
109. C. Cox, E. Ackerman, H. Roussell, P. Staecker, R. Osgood, T. Izuhara, D. Djukic, N. Jain, N. Design of a broadband electro-optic modulator with very low Vπ, in *Technical Paper, Photonic Systems* (2003)
110. M. Fujiwara, M. Teshima, J. Kani, H. Suzuki, N. Takachio, K. Iwatsuki, Optical carrier supply module using flattened multicarrier generation based on sinusoidal amplitude and phase hybrid modulation. IEEE J. Lightwave Technol. **21**, 2705–2714 (2003)
111. J.J. O'Reilly, P.M. Lane, R. Heidemann, R. Hofstetter, Optical generation of very narrow linewidth millimetre wave signals. Electron. Lett. **28**, 2309 (1992)
112. T. Sakamoto, T. Kawanishi, M. Izutsu, Widely wavelength-tunable ultra-flat frequency comb generation using conventional dual-drive Mach-Zehnder modulator. Electron. Lett. **43**(19), 1039–1040 (2007)
113. I.L. Gheorma, G.K. Gopalakrishnan, Flat frequency comb generation with an integrated dual-parallel modulator. Photonics Technol. Lett. **19**(13), 1011 (2007)
114. A.K. Mishra, R. Schmogrow, I. Tomkos, D. Hillerkuss, C. Koos, W. Freude, J. Leuthold, Flexible RF-based comb generator. Photonics Technol. Lett. **25**(7), 701–704 (2013)
115. W. Li, J. Yao, Investigation of photonically assisted microwave frequency multiplication based on external modulation. Trans. Microwave Theory Tech. **58**(11), 3259–3268 (2010)
116. T. Healy, F.C. Garcia.Gunning, A.D. Ellis, J. Bull, Multi-wavelength source using low drive-voltage amplitude modulators for optical communications. Opt. Express **15**(6), 2981–2986 (2007)
117. Q. Chang, J. Gao, Y. Su, Generation of optical comb frequency signal with high spectral flatness using two cascaded optical modulators, in *Optical Fiber Communication & Optoelectronic Exposition & Conference 2008*, paper SaB.4 (2008)
118. K.P. Ho, J.M. Kahn, Optical frequency comb generator using phase modulation in amplified circulating loop. Photonics Technol. Lett. **5**(6), 721–725 (1993)
119. C.T. Lin, P.T. Shih, J. Chen, W. Xue, P.C. Peng, S. Chi, Optical millimeter-wave signal generation using frequency quadrupling technique and no optical filtering. Photonics Technol. Lett. **20**(12), 1027–1029 (2008)
120. C. O'Riordan, S.J. Fabbri, A.D. Ellis, Variable optical frequency comb source using a dual parallel Mach-Zehnder modulator, in *ICTON 2011*, Paper Tu.B1.5 (2011)
121. J. Zhang, J. Yu, N. Chi, Z. Dong, X. Li, Y. Shao, J. Yu, L. Tao, Flattened comb generation using only phase modulators driven by fundamental frequency sinusoidal sources with small frequency offset. Opt. Lett. **38**, 552–554 (2013)
122. T. Saitoh, M. Kourogi, M. Ohtsu, Expansion of span-width of an optical frequency comb using a higher harmonic wave modulation. Photonics Technol. Lett. **8**(10), 1379–1381 (1996)
123. J. Zhang, J. Yu, L. Tao, Y. Fang, Y. Wang, Y. Shao, N. Chi, Generation of coherent and frequency-lock optical subcarriers by cascading phase modulators driven by sinusoidal sources. J. Lightwave Technol. **30**(24), 3911–3917 (2012)
124. S. Fabbri, S. Sygletos, A. Ellis, Multi-harmonic optical comb generation, in *ECOC 2012*, paper Mo2A2 (2012)
125. Y. Lu, Y. Xing, Y. Dong, Equal-amplitude optical comb generation using multi-frequency phase modulation in optical fibers. Chin. Opt. Lett. **8**, 316–319 (2010)

126. S. Ozharar, F. Quinlan, I. Ozdur, S. Gee, P.J. Delfyett, Ultraflat optical comb generation by phase-only modulation of continuous-wave light. Photonics Technol. Lett. **20**(1), 36–38 (2008)
127. M. Yamamoto, Y. Tanaka, T. Shioda, T. Kurokawa, K. Higuma, Optical frequency comb generation using dual frequency optical phase modulation, in *Integrated Photonics Research and Applications/Nanophotonics for Information Systems, Technical Digest* (Optical Society of America, 2005), paper ITuF5 (2005)

Chapter 11
Optical Information Capacity Processing

Mariia Sorokina, Andrew Ellis and Sergei K. Turitsyn

Abstract The never-stopping increase in demand for information transmission capacity has been met with technological advances in telecommunication systems, such as the implementation of coherent optical systems, advanced multilevel multidimensional modulation formats, fast signal processing, and research into new physical media for signal transmission (e.g. a variety of new types of optical fibers). Since the increase in the signal-to-noise ratio makes fiber communication channels essentially nonlinear (due to the Kerr effect for example), the problem of estimating the Shannon capacity for nonlinear communication channels is not only conceptually interesting, but also practically important. Here we discuss various nonlinear communication channels and review the potential of different optical signal coding, transmission and processing techniques to improve fiber-optic Shannon capacity and to increase the system reach.

11.1 Introduction: Information Capacity in Optical Communications

The capacity of global and local communication systems has increased dramatically over past decades, with recent laboratory experiments approaching the order of Pbits/s data rates. However, the data demands rise with time at an even higher rate at the moment calling for new technology advances. The efficiency of current transmission systems are limited by a number of physical effects, with optical noise and nonlinearity being the major challenges [1–15]. It is well recognized nowadays that current fiber-optic communication systems and technologies are facing very specific hurdles due to the nonlinear properties of optical fiber channels, problems not existing in wireless and other linear channels. In optical fiber, the dominant

M. Sorokina (✉) · A. Ellis · S.K. Turitsyn
Aston Institute of Photonic Technologies, School of Engineering and Applied Science, Aston University, Aston Triangle, Birmingham B4 7ET, UK
e-mail: sorokinm@aston.ac.uk

S. Wabnitz and B.J. Eggleton (eds.), *All-Optical Signal Processing*,
Springer Series in Optical Sciences 194, DOI 10.1007/978-3-319-14992-9_11

nonlinearity is induced by the Kerr effect, manifesting itself as an intensity dependent refractive index. Intensities in an optical fiber are enhanced by the confinement of the optical field in a small core area and by long interaction distances during propagation. As a result, the output of fiber communication channel with increasing signal power is distorted not only by noise, as in linear channels, but also by nonlinearity originating from the Kerr effect, specific manifestations include: self-phase modulation (SPM)—nonlinear effect at the same frequency; cross-phase modulation (XPM)—nonlinear interactions between two waves having different frequencies; and four-wave mixing (FWM)—nonlinear interaction between three or four waves [16–20]. This significantly reduces the efficiency of fiber transmission channel, and a variety of mitigation techniques have been proposed and demonstrated [21–39]. Nevertheless there remains strong interest in the development of signal processing techniques that would take into account specific nonlinear properties of fiber-optic channels.

Considering the impact of nonlinearity on a transmitted signal, one should distinguish between deterministic nonlinear effects, which, in principle, can be fully compensated, e.g. by digital backward propagation (even if at the moment some of those techniques might appear impractical) and noise-mixed nonlinear distortion (including nonlinear signal-to-noise beating) that are non-recoverable. Thus, the transmission efficiency is limited by noise and nonlinearity, but also by practical conditions and constraints, impacting the ability to compensate deterministic nonlinearity [32], such as availability, complexity, cost and power consumption of signal processing elements and stochastic effects, such as polarization mode dispersion [40]. In this chapter we will discuss and overview both the practical and fundamental limits imposed by nonlinearity on fiber-channel capacity [1–15]. The definition of nonlinear limits is not yet well established, and we hope that this chapter will contribute to the definition of nonlinear capacity limits and a better understanding of the required signal processing techniques to provide for highest system capacity.

First of all, it is important to distinguish between use of a daily engineering term "system capacity" and the strictly defined mathematical term "Shannon channel capacity" [41]. The former used to describe the total available throughput (typically at some very low bit error rate level) of the communication systems, *measured* either in bits/second for the whole bandwidth used, or in bits/Hz/second for spectral efficiency. The Shannon channel capacity defines the maximum error-free transmission rate *achievable* in communication channels obtained by the maximization of the mutual information over all possible input signal distributions (modulation formats and error-correcting codes), and is also measured in bits/second (total capacity over certain bandwidth) or in bits/Hz/second—Shannon capacity per unit of bandwidth.

In fiber-optic channels it was observed and reported in many publications that the fiber nonlinearity limits the spectral efficiency at high enough signal power. Though the exact Shannon capacity of nonlinear fiber channels has not yet been calculated, due to the lack of appropriate analytical channel models, it was shown in [42] that in case of full nonlinear compensation and signal optimization over the whole used bandwidth (single-user channel model analysis) and neglecting parametric noise amplification, the Shannon capacity should grow monotonically with signal power [4, 42]. However, the calculation of the Shannon capacity in a multi-user system, for instance, with independent spectral channels in wavelength-division multiplexed (WDM) transmission is more difficult. The currently favoured approach in the optical communication community is to consider such multi-userchannel system from the perspectives of a selected channel, assuming that the other customer only impose a noise-like interference on the channel under consideration. This approach originates from the fact that the full compensation of the "deterministic" impact of nonlinearity is not always technically feasible, and in multi-channel networks that use of the full bandwidth for post-processing is not possible in practice. In this case, one can define practical "nonlinear system capacity limits" that occur due to decay of spectral efficiency with signal power ([2, 9] and references therein). In particular, the conventional multiplicative noise approximation [2, 9, 10, 43] shows that for uncompensated un-optimized link the spectral efficiency is a convex function of signal power.

Studies based on multiplicative noise typically calculate a so-called "nonlinear signal-to-noise ratio" and determine predictions of Shannon capacity or bit error rate (BER) of a given format, by substitution of the SNR with the nonlinear SNR in the appropriate linear equation. However, a demonstrable difference between the achievable rate (no optimization), the constrained capacity (limited optimization), and the Shannon capacity (optimized over possible signal distributions) exists, highlighting the importance of analytical calculation of the optimized mutual information functional—Shannon capacity. Note that the channel capacity of most nonlinear channels is still unknown, but upper and lower bounds can be estimated. We would also like to note that there are a vast number of various nonlinear fiber channels, for example any specific signal power distribution along the fiber link or any additional in-line elements may change the performance and constitute a new nonlinear channel. Therefore, the nonlinear Shannon limit remains an open and acute problem.

Approaching the Shannon channel capacity requires a range of efforts in multiple research directions: advances in amplification schemes and forward error-correction codes, development of new signal transmission and processing techniques, including new methods of nonlinearity mitigation or, indeed, positive use of nonlinearity, progress in optical signal regeneration, as well as, investigation of additional degrees of freedom for more advanced data multiplexing. The latter includes combining spectral multiplexing with polarization and spatial multiplexing. Modern fiber-optic communication systems already employ spectral-, time-, and polarization-division multiplexing to increase data rates through conventional single-mode fiber. As these degrees of freedom become saturated, new directions, such as space-division multiplexing (SDM) with multicore [44] and multimode [45] fibers come into play. Indeed, it was shown that SDM technology enables several fold capacity increase [46]. Note, however, that installation of novel fiber base assumes an enormous cost of new infrastructure, and such spatially-demultiplexed systems may be also affected by the nonlinear crosstalk between the cores/modes and may require digital signal processing (DSP) with similar complexity as digital nonlinear compensation.

Overall, signal processing plays a critical role for further progress in this field. This includes both digital signal processing and linear/nonlinear analog all-optical processing [21–32]. Using digital signal processing one can pre- or post-compensate deterministic transmission distortions. The immense challenge of a system capacity crunch, coupled with the availability of new materials and advances in integrated optics has re-ignited interest in previously studied methods in new context of coherent optical communications. These include: optical phase conjugation [47–49], phase-conjugated twin waves [30], optical regeneration [50–52]. Recent experiments have shown that employing phase-conjugation one can efficiently mitigate nonlinearity and dispersion [49, 53–55], though it reduces the flexibility of the system due to required symmetry conditions. This was addressed by proposing different system configurations [56, 57]. In [30] it was proposed to transmit a pair of symmetric phase-conjugated twin waves with the additional requirement of dispersion-symmetry along the transmission path, resulting the canceling of nonlinearity impairments by adding the recovered twin waves at the receiver. This enabled demonstration of 400 Gbit/s super-channel with a record distance of 12,800 km. This, however, reduces system capacity by at least a factor of two, since two polarizations were exploited carrying the same data [22].

Currently, the most widely studied approach to compensate for nonlinearities is digital back propagation [58], which enables full compensation of the "deterministic" nonlinear and dispersive effects within the compensation bandwidth. Digital backward propagation requires extensive signal processing and optimization of computational efforts is under development right now. One of the most important impacts of nonlinearity is that it mixes otherwise orthogonal modes affecting multiplexing. In some special designs fiber channels (e.g. [59, 60]) that can be approximated by the lossless nonlinear Schrödinger equation (NLSE) and the so-called Manakov equations with additive noise, there is an interesting possibility of signal multiplexing on an alternative basis that is not affected by the Kerr nonlinearity. Recently, digital signal processing based on the inverse scattering transform [61–63] demonstrated that it is possible to expand signal on a special basis, within which the propagation of individual "nonlinear modes" is effectively linear. The proof of this is dependent on the integrability of NLSE along similar bases [64]. The change of basis—the inverse scattering transform technique is a nonlinear analog of the Fourier transform. All these techniques offer efficient nonlinearity compensation, though differing in computational expends and complexity.

The nonlinearity is an essential factor in the design of advanced fiber communication systems, but it is often shunned by engineers in view of its complexity. Here we will review different nonlinear optical techniques and their impact on Shannon capacity.

11.2 Fundamentals

We will start discussion of the fiber channel capacity with a brief reminder of the well known classical results of the information theory [41]. Note that we do not aim to provide in this section a comprehensive overview of all relevant information theory, but rather refresh well known facts, referring readers for more details to the original seminal work [41] or to an excellent recent review paper [8].

11.2.1 Shannon Channel Capacity

Shannon showed that any noisy channel can still be used for reliable communication at non-zero rate (when the input and output are correlated) and introduced the expression for the maximum possible error-free transmission rate. Let us consider a communication channel as a system linking two random variables x and y representing the channel input and output. A communication channel is characterized by a probability distribution function $P(y|x)$ that defines the probability of receiving symbol y of the output alphabet $\mathscr{Y}$ given the transmitted symbol x of the input alphabet $\mathscr{X}$. Further random variables and their deterministic outcomes are denoted by upper Y and lower y letters correspondingly. The channel can be considered as memoryless, when the output probability is defined by the input at that particular time and is independent on previous channel inputs and outputs. Shannon formulated the following theorem for the Shannon channel capacity C of the memoryless discrete noisy channel.

Shannon stated [41] that "It is possible to send information at the rate C through the channel with as small a frequency of errors or equivocation as desired by proper encoding. This statement is not true for any rate greater than C."

Here information channel capacity C (in bits per degree of freedom) for a memoryless channel involves maximizing the mutual information I [41]:

$$C = \sup_{p(x)} I(X, Y) = \max_{p(x)} \int dxdyp(x)p(y|x) \log_2 \frac{p(y|x)}{\int dxp(x)p(y|x)}, \qquad (11.1)$$

over all valid input probability distributions $p(x)$ subject to the power constraint $\int dx\, p(x)\, |x|^2 \leq S$. The capacity calculation requires knowledge of the statistical properties of communication channel that are given by the conditional input-output probability density function (PDF) $p(y|x)$. Shannon capacity analysis is not possible without a full statistical description of the transmission for *all possible signal inputs*.

Achievability of the capacity through appropriate coding and signal modulation is covered by Shannon's theorem 12 [41].

Let us denote by $n(T, q)$ the maximum codebook size of duration T and error probability q (i.e. the maximum number of signals selected from the alphabet with probability of error less or equal to q), then the channel capacity is defined as follows:

$$C = \lim_{n\to\infty} \frac{\log_2 n(T,q)}{T}. \tag{11.2}$$

It means that one can reliably decode about CT bits in time interval T, when T is sufficiently large.

This theorem states that even in the presence of noise one can recover the original signal from the corrupted channel output. The construction of capacity achieving codebooks over time interval T is a challenging problem involving various approaches to modulation, coding and digitizing of an analog signal.

In case of channels with memory, the capacity is defined as [65]

$$C = \lim_{n\to\infty} \sup \frac{1}{n} I(X_1^n, Y_1^n), \tag{11.3}$$

with optimization over all joint distributions of $X_1, .., X_n$ subject to the power constraint $\int dx\ p(x)\ |x_1^n|^2 \le nS$.

11.2.2 Numerical Computation of Shannon Capacity

The analytical derivation of the Shannon capacity is possible only for very limited number of communication channels. However, one can compute Shannon capacity by numerically maximizing the mutual information over all valid input signal distributions for a discrete memoryless channel with fixed input and output alphabets. The Shannon capacity has a unique local maximum which is finite for a finite alphabet. For complex channels, numerical computations is the only practical way to obtain the Shannon capacity. The following key numerical methods are typically used for capacity calculations:

- Constrained maximization using the Kuhn-Tucker conditions [66, 67]. The method generalizes the method of Lagrange multipliers for inequality constraints
- The gradient search algorithm of Frank-Wolfe [68], an iterative first-order optimization algorithm for constrained convex optimization, also known as the conditional gradient method, reduced gradient algorithm and the convex combination algorithm. It is based on a linear approximation of the objective function in each iteration and moves slightly towards a minimum of this linear function (taken over the same domain).
- The Arimoto-Blahut algorithm [69, 70], an iterative method, which maximizes MI of arbitrary finite input/output alphabet sources. The algorithm was modified for faster convergence using natural-gradient-based or accelerated Blahut-Arimoto

algorithm [71]. A number of algorithms were proposed for discrete alphabet [72] or continuous: using computation of a sequence of finite sums [73] or particle based Blahut-Arimoto algorithm [74]. In the latter, the a particles x_k are moved to increase the relative entropy while keeping the output probability fixed.

- For discrete-input and discrete-output channels with memory an efficient technique for calculation of transmission rate was proposed [75–77], later generalized for continuous channels in [78].

11.3 Linear Additive White Gaussian Noise Channel

11.3.1 Capacity of Linear Additive White Gaussian Noise Channel

The linear Gaussian noise channel is a basic model of a communication channel that has a number of applications. In particular, it is used in radio and satellite, wired and wireless communications. A fiber-optic channel can be approximated as a linear Gaussian channel in the low signal power regime.

In the linear additive white Gaussian noise (AWGN) channel the output y and input x are random variables (here both are real functions, but generalizations to complex variables and the multi-dimensional case are straightforward):

$$Y = X + \eta \tag{11.4}$$

Often, it is convenient to consider a time discrete linear Gaussian channel: $y_l = x_i + \eta_i$ where the channel output and input at the time i are denoted by y_i and x_i, whereas the channel is disturbed by a zero-mean additive white Gaussian noise with a variance N. The additive noise is independent of the signal. Though, the origin of the noise may be different, using the central limit theorem it can be shown that the cumulative effect of a large number of small random effects, each with a well-defined expected value and well-defined variance, very often can be well approximated by a normal distribution [79].

If the noise variance is zero or there are no constraints on the input signal, the capacity is infinite. However, in practice one has input constraints. In particular, constraint on the average input power S results in the following:

$$\langle |x|^2 \rangle \le S \tag{11.5}$$

The information capacity of power constrained system is given by:

$$C = \max_{p_x, \langle |x|^2 \rangle \le S} I(X, Y) \tag{11.6}$$

For the linear AWGN power-constrained channel the Shannon capacity can be found analytically:

$$C = \log_2\left(1 + \frac{S}{N}\right) \tag{11.7}$$

This is one of the most celebrated formula in the information and communication theory. It can be generalized further for more complex channel models with various noise distributions.

11.3.1.1 Constant-Intensity Modulation with Coherent Detection [80–82]

A channel using constant-intensity modulation formats with coherent detection is a linear AWGN channel:

$$Y = X + \eta, \tag{11.8}$$

with complex x and y and the additional constraint $|x|^2 = S$

$$p(y|x = \sqrt{S}e^{i\phi_0}) = \frac{1}{\pi\langle|\eta|^2\rangle}\exp\left[-\frac{|y - \sqrt{S}e^{i\phi_0}|^2}{\langle|\eta|^2\rangle}\right] \tag{11.9}$$

where the input phase ϕ_0 is used for coding. In [79] Shannon capacity was calculated numerically and the asymptotic expression was found for high SNR:

$$\lim_{SNR\to\infty} C = \frac{1}{2}\log_2\left(\frac{4\pi}{e}SNR\right) \tag{11.10}$$

As pointed out in [80] the intuitive explanation lies in the fact that the phase difference between input and output at high SNR can be approximated as a Gaussian random variable with variance $(2SNR)^{-1}$, then its entropy is given by (11.10).

11.3.1.2 Capacity of Bandlimited Channels

The possibility of using discrete alphabets (digitized analog signals) to approach the capacity of noisy communication channels is based on the observation that in practice any communication channel has a limited bandwidth. In bandwidth limited communication a continuous-time and amplitude/phase signal $x(t)$ can be fully characterized by a discrete-time and discrete amplitude/phase signal x_n. This is possible due to fundamental mathematical results that have been developed in a communication context [83–87].

Shannon-Nyquist Sampling Theorem

Any function $f(t)$ bandlimited in frequency to W Hz can be completely determined by samples $f(\frac{n}{2W})$ taken at the Nyquist rate of $2W$ samples per second. Taking the sum of the samples, one reconstruct the original time-continuous signal using:

$$f(t) = \sum_{n=-\infty}^{\infty} f\left(\frac{n}{2W}\right)\frac{\sin[\pi(2Wt-n)]}{\pi(2Wt-n)} \tag{11.11}$$

It is easy to check that the spectrum of the reconstructed function lies in the band W and is zero outside. The Shannon-Nyquist theorem [41, 83, 84] defines that a bandlimited function has $2W$ degrees of freedom per second.

In the bandlimited Gaussian noise channel each sample can be considered as an independent, identically distributed Gaussian random variable. If the noise has power spectral density $N_0/2$ and occupies the same bandwidth as the signal, the noise power is $N_0 W$. The expression for Shannon capacity of discrete-time linear AWGN channel:

$$[C]_{\text{bits per modulation symbol}} = \log_2\left(1+\frac{S}{N}\right) \tag{11.12}$$

Given that there are $2W$ samples per second and $N = N_0 W$, one obtains:

$$[C]_{\text{bits per second}} = 2W\log_2\left(1+\frac{S}{N_0 W}\right) \tag{11.13}$$

There is a number of well developed codes which allow the Shannon capacity of the linear AGWN channel to be approached (see e.g. [88] and references therein).

11.4 Nonlinear Fiber Channel

11.4.1 Basic Models

The communication channel using optical fiber cable as a transmission medium is significantly different from linear AWGN due to nonlinear dependence of the refractive index on the signal power [89]. In general, the dynamics of an optical field $E(z,t)$ in silica fiber is governed by the nonlinear Schrödinger equation (NLSE):

$$\frac{\partial E}{\partial z} + i\frac{\beta_2}{2}\frac{\partial^2 E}{\partial t^2} + \frac{\alpha}{2}E = i\gamma E|E|^2, \tag{11.14}$$

Here β_2, α, and γ denote dispersion, attenuation, and nonlinear coefficients correspondingly.

This equation governs only key deterministic linear (dispersion and attenuation) and nonlinear effects. In standard single mode fiber the attenuation coefficient is 0.2 dB/km in the 1550-nm wavelength region [8], consequently, optical amplification is required (in the absence of opto-electronic regeneration) for long transmission distances. This induces optical noise, that depends on the amplification scheme and includes amplitude spontaneous emission (ASE). The noise can be incorporated into the model as an additive Langevin-noise source term η that accounts for the stochastic perturbations of the signal with the ASE being the main contributor. It has been shown that ASE can be represented as a circularly symmetric additive Gaussian noise [90, 91], which is fully defined by an autocorrelation function [8, 89]. Without going into details (such as redefinition of the nonlinear coefficient) that have been discussed already in [8] we define the principal fiber channel model as:

$$\frac{\partial E}{\partial z} + i\frac{\beta_2}{2}\frac{\partial^2 E}{\partial t^2} - i\gamma E|E|^2 = 0. \tag{11.15}$$

The (11.15) apart from being the principal master channel model, can also be directly derived in two practically important periodic amplification schemes: (i) periodically spaced discrete point amplifiers (e.g. Erbium-doped fiber amplifiers). In this case, a path-average model (11.15) occurs as a result of the averaging over periodic gain and loss variation (valid when the amplifier spacing is much smaller compared with the length scale of nonlinear effects) leading to effectively conservative signal dynamics governed by the re-normalized NLSE (11.15) (the nonlinear coefficient is modified due to the averaging), see e.g. [92, 93]; and (ii) quasi-lossless distributed Raman amplification where the Raman gain continuously compensates for the fiber loss and the signal average power remains a constant along the entire transmission span [7, 59, 60]. The papers [7, 8] provide a detailed account of how noise intensity is related to the parameters of the line.

The similar vector NLSE model that takes into account propagation polarization effects is also applied, especially in polarization-based modulation formats. However, in general, fiber polarization properties vary at the scale of several hundred meters, therefore, full modeling, accounting for these fast rotations of polarizations with propagation over long distances is time consuming. The model that is used instead, is an average system of equations (here averaging is over fast changes of polarization) was derived by Manakov [94] and subsequently verified in [95–97]:

$$\frac{\partial E_1}{\partial z} = -\frac{\alpha}{2}E_1 - i\frac{\beta_2}{2}\frac{\partial^2 E_1}{\partial t^2} + i\frac{8\gamma}{9}\left(|E_1|^2 + |E_2|^2\right)E_1, \tag{11.16}$$

$$\frac{\partial E_2}{\partial z} = -\frac{\alpha}{2}E_2 - i\frac{\beta_2}{2}\frac{\partial^2 E_2}{\partial t^2} + i\frac{8\gamma}{9}\left(|E_1|^2 + |E_2|^2\right)E_2 \tag{11.17}$$

Note that NLSE and Manakov equations are integrable by inverse scattering method [92, 98].

11.4.2 Simplified Nonlinear Channel Models

The nonlinear propagation equations (NLSE and Manakov system) presented above describe complex nonlinear continuous-time communication channels. Corresponding conditional discrete-time probabilities are not known in the general case for these channels. Instead we consider simplified models with particular applications in which capacity can be estimated or computed. The continuous-time model can be transformed into discrete-time model by expanding the continuous-time signal (complex input $X(t)$ and output $Y(t)$) over a complete set of orthogonal functions [99]. This results in discrete-time complex variables $X_{[k]}$ and $Y_{[k]}$. Further we omit indexes for memoryless channels, except Sects. 11.4.2.4 and 11.4.2.5. Input X and output Y can take deterministic values $x = r_{in}e^{i\varphi_{in}}$ and $y = r_{out}e^{i\varphi_{out}}$ chosen from the alphabets X and Y correspondingly.

11.4.2.1 Intensity-Modulated Direct-Detection (IMDD) Channel [100, 101]

The simplest communication systems are intensity-modulated direct-detection (IMDD), where signal power is modulated and detected. A photo-detector responds to the square of the incident electric field, and so the output and input are connected as follows:

$$Y = |X + \eta|, \tag{11.18}$$

The conditional pdf for this model was found to be the Rice distribution [100]:

$$p(y|x) = \frac{y}{\pi\langle|\eta|^2\rangle}\exp\left(-\frac{y^2 + |x|^2}{\langle|\eta|^2\rangle}\right) \tag{11.19}$$

where I_0 is zero-order modified Bessel function of the first kind. It was shown in [100, 101], that the capacity-achieving distribution has a number of infinite mass points, also lower and upper bounds on capacity were presented, including the asymptotically tight lower bound [102]:

$$C_{ub}^{IMDD} = \frac{1}{2}\log_2\left(1 + SNR\right) - 0.5 \tag{11.20}$$

where the square root operation required to recover the output from the detected photo-current eventually translates to an offset of approximately $1/2\log(1/2)$. Thus, the Shannon capacity of the IMDD channel is a monotonically increasing function of SNR and the capacity-achieving distribution has an infinite number of mass points and does not contain an infinite set of mass points on any bounded interval [101].

11.4.2.2 Partially Coherent AWGN [101–104]

The discrete-time partially coherent (phase detection is applied) AWGN channel can be introduced as:

$$Y = Xe^{i\Theta} + \eta, \tag{11.21}$$

here the additive noise η is as above a complex Gaussian random variable with zero mean. Variable Θ mimics the nonlinear phase noise and can be modelled as real-valued wrapped AWGN with variance σ_Φ^2 as in [103, 104].

The conditional pdf for such channel can be also found:

$$p(y|x) = \int\limits_{-\pi}^{\pi} \frac{d\theta}{\pi\langle|\eta|^2\rangle} \exp\left[-\frac{|y - xe^{i\theta}|^2}{\langle|\eta|^2\rangle}\right] \frac{1}{\sqrt{2\pi\sigma_\Phi^2}} \sum_{k=-\infty}^{\infty} \exp\left[\frac{(\theta - 2\pi k)^2}{2\sigma_\Phi^2}\right] \tag{11.22}$$

If a phase tracking device (such as phase-locked loop) is employed, than Θ might be modelled by Thikhonov distribution [101] with positive parameter ρ:

$$p(\theta) = \frac{e^{\rho\cos(\theta)}}{2\pi \mathrm{I}_0(\rho)} 1 \tag{11.23}$$

The conditional pdf for such channel was derived as:

$$p(y|x) = \frac{r_{out}}{\pi N} \exp\left[-\frac{r_{out}^2 + r_{in}^2}{N}\right] \frac{\mathrm{I}_0(v)}{\mathrm{I}_0(\rho)} \tag{11.24}$$

$$v = \sqrt{4\frac{r_{out}^2 r_{in}^2}{N^2} + 4\rho\frac{r_{out} r_{in}}{N}\cos(\varphi_{in} - \varphi_{out}) + \rho^2}$$

The capacity-achieving distribution was shown to be circularly symmetric [105] and (similarly to the result in the previous subsection) it does not contain an infinite set of mass points on any bounded interval [101].

In the case when $\Theta(x)$ is not a random variable [106] (describing deterministic nonlinear phase noise, e.g. due to SPM effect) the conditional pdf is simplified to:

$$p(y|x) = \frac{1}{\pi\langle|\eta|^2\rangle} \exp\left[-\frac{|y - xe^{i\theta(x)}|^2}{\langle|\eta|^2\rangle}\right] \tag{11.25}$$

In [106] it was demonstrated that the Shannon capacity (11.1) with the conditional pdf given by (11.26) is identical to linear Shannon limit—Shannon capacity of a linear AWGN channel $\log_2(1 + SNR)$. This is because Θ is deterministic, so there is no loss of information. Following optimization procedure on can find the optimum input pdf as:

$$p(x) = \frac{1}{\pi S}\left[1 + i\gamma\left(x^* \frac{\partial\theta(x)}{\partial x^*} - x\frac{\partial\theta(x)}{\partial x}\right)\right]\exp\left[-\frac{|x|^2}{S}\right] \tag{11.26}$$

If, however, (11.26) is approximated by Gaussian distributing (drop first bracketed term), the obtained result (mutual information with non-optimal $p(x)$) is lower than the channel capacity.

11.4.2.3 Nonlinear Phase Noise Channel [4, 107–111]

In systems predominantly limited by a nonlinear phase noise channel, such as systems operating with low path average dispersion, the output is given by:

$$Y = \left(X + \eta\right)e^{-i\Phi_{NL}} \tag{11.27}$$

Here η is the total AWGN accumulated on the transmission path. The equation models NLSE (11.15) with zero dispersion $\beta_2 = 0$ and periodically inserted ASE and describes the phase noise caused by the SPM effect. The nonlinear phase noise is given as [107]:

$$\Phi_{NL} = \frac{\gamma L}{K}\sum_{i=1}^{K}|x + \eta_i|^2 \tag{11.28}$$

here summation is performed over the ASE noise accumulated at each of $i = 1, \ldots, K$ fiber segments: $\eta_i = N_1 + \cdots + N_i$ (N_i denotes ASE noise aroused at i-th span) and L is the propagation distance. The conditional pdf for such channel (after proper normalization described in [4, 107, 111]) reads:

$$p(y = r_{out}e^{i\varphi_{out}}|x = r_{in}e^{i\varphi_{in}}) = \frac{p(r_{out}|r_{in})}{2\pi} + \frac{1}{\pi}\sum_{m=1}^{\infty}\mathrm{Re}\left(e^{im(\varphi_{out}-\varphi_{out})}K_m(r_{out}, r_{in})\right) \tag{11.29}$$

where $p(r_{out}|r_{in})$ is the Ricean pdf of the received power:

$$p(r_{out}|r_{in}) = \frac{2r_{out}}{N}\exp\left(-\frac{r_{out}^2 + r_{in}^2}{N}\right)\mathrm{I}_0\left(\frac{2r_{out}r_{in}}{N}\right) \tag{11.30}$$

and the analytical expressions for the coefficients $C_m(r_{out}, r_{in})$:

$$K_m(r) = r_{out}b_m\exp\left[-a_m(r_{out}^2 + r_{in}^2)\right]I_m(2b_m r_{out}r_{in}) \tag{11.31}$$

$$a_m = \frac{\sqrt{im\gamma}}{\sigma}\coth\sqrt{im\gamma\sigma^2 z}$$

$$b_m = \frac{\sqrt{im\gamma}}{\sigma} \frac{1}{\sinh \sqrt{im\gamma\sigma^2} z}$$

here I_m is the modified Bessel function.

The impact of optimization of symbol error probability over amplitude phase-shift keying constellations was studied in [107–110]. The lower bound on channel capacity was derived in [4]:

$$C \geq \frac{1}{2} \log_2(SNR) \tag{11.32}$$

11.4.2.4 Infinite-Memory Gaussian Noise Channel Model [1, 3, 8–10, 32, 43, 112, 113]

The general channel model NLSE can be rewritten in the case of wavelength-division multiplexing (WDM) in a form that shows the evolution of the signal in each WDM channel (the optical field of k-th channel) with account of SPM and XPM effects:

$$\frac{\partial E_k}{\partial z} = -i\frac{\beta_2}{2}\frac{\partial^2 E_k}{\partial t^2} + i\gamma E_k \left(|E_k|^2 + \sum_{j \neq k} 2|E_j|^2 \right) + \eta(t, z), \tag{11.33}$$

here β_2 and γ denote dispersion and nonlinearity coefficients and ideal compensation of fiber losses is assumed, with $\eta(t, z)$ being an amplifier spontaneous emission noise, which results from the signal amplification required to compensate fiber losses during propagation.

This presentation allows one to introduce another simplified channel model by omitting intra-channel effects. This may be justified in the regime when the inter-channel effects are dominant nonlinear effects or when self-phase modulation is compensated [114–116]. In this case the resulted model equation has the form:

$$\frac{\partial E_k}{\partial z} = -i\frac{\beta_2}{2}\frac{\partial^2 E_k}{\partial^2 t} + i\gamma E_k V_k + \eta(t, z), \tag{11.34}$$

where the effective potential V_k is

$$V_k = \sum_{j \neq k} 2|E_j|^2 \tag{11.35}$$

In dispersion uncompensated transmission the potential V_k can be modelled (approximately) as a Gaussian stochastic process short-range correlated in space and time (assuming high dispersion and moderate to no spread in symbol rates).

This leads to the following simplified channel model with multiplicative Gaussian noise:

$$Y_i = AY_i + BH_{ij}X_j + \eta \tag{11.36}$$

where the output spectrum is denoted as $Y_i = \tilde{E}_s(\omega_i, z)$, whereas the input signal is $X_i = \tilde{E}_s(\omega_i, 0)$. The inter-channel interference is given by the matrix H_{ij}, which is assumed to be a zero-mean, Gaussian independent and identically distributed complex random matrix with variance $\langle H_{ik}H_{jl}\rangle = m^{-1}\delta_{ij}\delta_{kl}$, here m is the dimensionality of the signal. Factor B defines the strength of signal-signal interactions and A is a normalization parameter for conservation of signal power. The conditional pdf of such model is given in [2]:

$$p(y|x) = \frac{1}{(2\pi\sigma^2)^m} e^{-|y-Ax|^2/\sigma^2} \tag{11.37}$$

where variance is given by:

$$\sigma = m^{-1}(\langle|\eta|^2\rangle + B^2|x|^2) \tag{11.38}$$

Various authors have followed a similar generic approach which shares the same assumptions but include a variety of different physical effects. These produce slightly different lower bounds on capacity (including impairments from SPM, XPM, and FWM effects), which can be generalized by the following formula:

$$C_{LB} = \log_2\left(1 + \frac{S}{N + \mu S^3 + \nu S^2 N}\right) \tag{11.39}$$

which summarizes the a number of results for a variety of communication channels, such as OFDM or WDM, with arbitrary dispersion, distributed or lumped amplification, for each of which the coefficients are found taking into account of different nonlinearities: μ reflects signal-signal interactions, such as FWM and XPM ([1, 3, 8–10, 43]) and ν arises from signal-noise interactions ([32, 113]).

Equation (11.39) is important for systems with low signal to noise ratio or for systems employing nonlinearity compensation (such that μ is small) [32, 117], and, as originally suggested in [2], signal depletion should be taken into account by subtracting μS^3 from the numerator of (11.39). It has been shown to have limitations for coded transmission [118] and for QPSK with low accumulated dispersion [108, 119], which may correspond to a dominance of nonlinear phase noise.

11.4.2.5 Finite-Memory Gaussian Noise Channel Model [118]

In the infinite-memory Gaussian noise channel model it was implied that the input is independent and identically distributed (i.i.d.). Under this assumption one can derive the channel model where nonlinear signal-signal interactions are modelled as Gaussian noise with variance dependent on average input signal power, so that the noise affecting the k-th symbol is given by:

$$\eta_{IMGN} = \bar{\eta}_{GN}\sqrt{N + \Gamma\left(\lim_{M\to\infty}\frac{1}{2M+1}\sum_{i=k-M}^{k+M}|X_i|^2\right)^3} = \bar{\eta}_{GN}\sqrt{N + \Gamma S^3} \quad (11.40)$$

where $\bar{\eta}_{GN}$ is i.i.d. zero-mean unit-variance circularly symmetric complex Gaussian random variable, Γ is nonlinear parameter, and M is memory. This simplified channel model yields simple and experimentally verified expression for the lower bound on channel capacity (see 11.39).

However, in practice channels have a finite memory defined by physical properties of the channel, such as dispersion, nonlinearity, etc. The finite memory Gaussian noise channel can be modelled with its nonlinear noise given by [118]:

$$\eta_{FMGN} = \bar{\eta}_{GN}\sqrt{N + \Gamma\left(\frac{1}{2M+1}\sum_{i=k-M}^{k+M}|X_i|^2\right)^3} \quad (11.41)$$

The model enables the study of non i.i.d. non-stationary inputs, where the channel capacity definition given by (11.3) can be applied. For the case of uncoded signals the proposed channel model yields similar results as the infinite memory GN channel, whereas, the case of coded signals new improved lower bounds on Shannon capacity saturate at high signal powers to a non-zero constant independent on channel memory M [118]. Though, the model is simplified and does not take into account a number of effects, such as signal-noise interactions or effects in multichannel systems, it highlights the importance of channel coding and analytical modelling for capacity estimations, it clearly demonstrates that it is possible to transmit information at high signal power via coded transmission.

11.4.2.6 Regenerative Channel [120, 121]

Consider a model transmission system of fixed total length that consists of R spans with linear AWGN channels interleaved with nonlinear regenerators. The ideal regenerators assign each transmitted symbol to the closest element of the given alphabet without signal decoding. We assume that the regenerator performs this transformation independently for each signal quadrature, therefore the problem for multidimensional signal can be reduced to one-dimensional independent lattices. We write down formulae for the one-dimensional signal, but they can be generalized in a straightforward manner to the multi-dimensional case.

The ideal regenerator assigns diffused point x' (originated from the input point x_l) to the closest neighbor x_k in the decision area S_k. Thus, the conditional probability for the output of regenerator to be a symbol x_k given the input signal x_l is the probability that the distorted point x' falls within the decision boundary S_k. Therefore, the conditional pdf of such a system is defined through the matrix elements [120]:

$$P(y = x_k|x_l) = \int_{S_k} dx' p_G(x'|x_l) = W_{kl}, \quad p_G(x'|x_l) = \frac{1}{\sqrt{2\pi N/R}} \exp\left(-\frac{(x' - x_l)^2}{2N/R}\right) \tag{11.42}$$

It may readily be shown by considering the probability of insertion and omission errors for each constellation point that the transmission matrix is given by

$$W_{kl} = \frac{1}{2}\left(\mathrm{erf}[\Delta_{kl}^{+}] - \mathrm{erf}[\Delta_{kl}^{-}]\right), \Delta_{kl}^{\pm} = (x_k + x_{k\pm1} - 2x_l)\sqrt{\frac{R}{8N}}, \tag{11.43}$$

where delta is a statistical parameter describing the spread of the constellation points and is normalized the closest neighbor distance as follows $\Delta = d\sqrt{R/8N}$. Due to the Markovian nature of the stochastic system (the regenerative transformations are statistically independent), the overall transition matrix after R regenerative segments reads as $\mathbf{M} = \mathbf{W}^R$.

If, instead of an ideal transfer function, the regenerator has a smooth transfer function T, the output of the n-th regenerator can be expressed through the recurrence relation:

$$Y_{n+1} = T(Y_n) + \eta_n; \; Y_1 = x. \tag{11.44}$$

Here $T(y)$ is a one dimensional projection of the nonlinear transfer function (see Fig. 11.1a), which is required to satisfy the following [121, 122]:

$$T[x] = x, \quad T''[x] = 0, \quad |T'[x]| < 1 \tag{11.45}$$

The first expression implies that an undistorted input point (alphabet), x, is unaffected by transformation, the second ensures that the alphabet point is in the center of the regenerative region. The solution of the first two expressions identifies the alphabet points—the set of points, which are further used for the construction of modulation format. Finally, the third expression means that the distortion is effectively suppressed and it determines the flexibility in the transfer function parameters. When the first derivative is equal to zero, the alphabet is superstable and one can observe plateau around the alphabet points (the solution of $T'[x] = 0$ is the best parameter choice), however plateau is not necessary as long as $|T'[x]| < 1$ (suboptimal parameters choice).

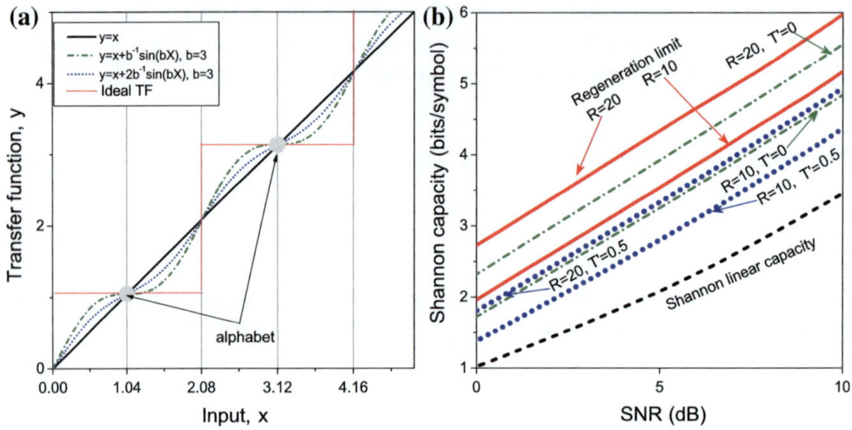

Fig. 11.1 **a** Example of a regenerative TF: $y = x + a\sin(bx)$ and TF for the ideal regenerator; **b** the regenerative limit—Shannon capacity of the channel with ideal regenerators plotted alongside Shananon capacity of regenerative channel with smooth TF: $y = x + a\sin(bx)$ plotted for different number of regenerators $R = 10$ and 20 and different choice of parameters: superstable $T'(x) = 0$ (corresponds to $ab = 1$) and stable $T'(x) = 1/2$ ($ab = 1/2$), the value of T' defines parameter a, whilst b is subject of optimization of MI in (11.1)

The conditional pdf for the output at the n-th regenerator for each quadrature y_{n+1} given the input y_n is found to be

$$p(y_{n+1}|y_n) = \frac{1}{\sqrt{2\pi N_n}}\exp\left[-\frac{(y_{n+1} - T(y_n))^2}{2N_n}\right]N_n = <|\eta_n|^2> \quad (11.46)$$

Because of the Markovian property of the process, the conditional pdf of the received signal after propagation through R links, y_R, given the input, x, is expressed by a product of single-step conditional probabilities

$$p(y_{R+1}|y_1 = x) = \int dy_R..dy_2 p(y_{R+1}|y_R), \ldots, p(y_2|y_1 = x) \quad (11.47)$$

Using the conditional pdf given in (11.47), one can calculate Shannon capacity using (11.1). The Shannon capacity analysis of the system with the ideal regenerators defines the upper bound of regeneration efficiency—maximum Shannon capacity that can be achieved by implementation of regeneration, which was calculated in [121] (Fig. 11.1b). Shannon capacity calculations for a regenerative channel with smooth transfer functions is plotted in Fig. 11.1b. The benefit of using regeneration becomes more pronounced when the number of regenerators is increased (results are plotted for $R = 10$ and $R = 20$). The figure demonstrates that the regenerative elements need to be properly optimized using aforementioned optimization procedure, which defines the optimum parameter values and suboptimal ones (within the operating margins of the element)—compare the Shannon capacity for two cases $T'(x) = 0$ (TF with plateaus) and $T'(x) = 0.5$. The results illustrate that by suppression of signal distortions one can achieve Shannon capacity higher than a system without regenerators operating with the same signal launch power and with the same power of noise added to the signal during transmission [121].

11.5 Optical Signal Processing to Improve Signal Transmission

11.5.1 Introduction

Knowledge of the optical phase (in addition to field intensity) after coherent detection has dramatically changed the whole concept of compensation and management of impairments in fiber-optic communication. In particular, linear effects such as dispersion and polarization-mode dispersion are now routinely compensated entirely in the coherent receiver. Electronic signal processing using back propagation has also been applied to the compensation of nonlinear fiber impairments. However, the drawback of the methods of reconstruction of the transmitted data from the received signal based on nonlinear back propagation is that they rely on computationally-intensive techniques. Real time nonlinear digital backward propagation is still a challenging problem especially for SNR gains of more than a DB.

In this section we will review methods, which have potential to either increase the system capacity or to simplify DSP.

11.5.2 Quasilossless Transmission

As channel capacity is limited by noise, for long haul communication the amplification scheme is an important part of the system design. Since 1987 the Erbium doped fiber amplifier [123] enabled high capacity transmission in a broad region of wavelengths, which led to the development of WDM systems [123].

Distributed Raman amplification via stimulated Raman scattering (which enables energy transfer from the pump to the signal) was proposed [124] to create distributed gain. Moreover, it was experimentally demonstrated [59, 60] that ultralong fiber Raman lasers enable quasilossless long-distance transmission. As was indicated in [59, 60] the distributed gain is achieved for a broad spectral bandwidth.

The resulted OSNR improvement (due to low noise) and, corresponding, capacity improvement was estimated in [7, 8]. Furthermore, the lossless NLSE is a nonlinear integrable equation, which simplifies DSP and offers potential for a number of techniques, including soliton transmission and employment of inverse scattering method. Also, quasilossless transmission offers symmetric power variation which is of particular importance for the technique based on optical phase conjugation.

11.5.3 Optical Phase Conjugation

To compensate nonlinear impairments optically without extensive DBP simulations a number of methods were proposed. In particular, in 1979 [47] it was shown that performing optical phase conjugation (OPC) in the middle of the link one can achieve effective mitigation of deterministic nonlinear and dispersive impairments with improved system performance over longer transmission distances. Since then a number of experimental and theoretical works were performed [48, 49]. The OPC technique sets stringent requirements for the symmetry of the system and near perfect phase conjugation of the wave near the middle of the link. As a result, a formal time reversal is achieved in the NLSE and deterministic nonlinear signal-signal distortions are cancelled during propagation the second half-link. The generalized condition for effective compensation was given in [125]:

$$\frac{\beta_2(-z_1)}{\gamma(-z_1)S(-z_1)} = \frac{\beta_2(z_2)}{\gamma(z_2)S(z_2)} \tag{11.48}$$

which means that the ratio of the dispersion and nonlinearity has to be equal at the corresponding positions $-z_1$ and z_2. However, the stringent symmetry requirements result in complexity of system design and high sensitivity to symmetry violations (in particular, third order dispersion in highly nonlinear regime). To achieve symmetric power map Raman amplification scheme was used [126]. Recently, it was experimentally demonstrated that OPC accompanied with second-order ultra-long Raman fiber laser enables to increase system reach by 30 % by compensating 70 % of the nonlinear distortions. To avoid mid-link OPC, an optical tunable compensation method was proposed [56], where the nonlinear phase shift (sign-reversed to that acquired by propagation through the link) is added to the signal with consequent OPC at before transmission.

Removal of deterministic distortions increases the achievable rate [32, 127, 128]:

$$C_{LB}^{OPC} = \log_2\left(1 + \frac{S}{N + \frac{n_s}{2(n_{OPC}+1)}\nu S^2 N}\right) \tag{11.49}$$

where n_s and n_{OPC} (assumed to be odd) are the numbers of amplifiers and OPCs correspondingly, assuming OPC compensates entirely deterministic distortions.

11.5.4 Phase-Conjugated Twin Waves

To compensate nonlinear distortions without adding an OPC element, but exploiting the symmetry available through the use of polarization, the twin wave model was proposed, where the signal and its conjugated copy are transmitted simultaneously on either different polarization [30] or different wavelength [129]. Using the symmetric properties of Manakov equation and a symmetric dispersion map: $D(z) = -D(L - z)$ (this can be achieved by applying pre-dispersion at the transmitter), the nonlinear perturbations are correlated and, therefore, the original wave can be restored at the receiver by adding the wave and its conjugated twin. Compared to DBP, which requires a large number of additional multiplications per bit [130], PCTW only requires one multiplication and addition per bit. Alternatively, it may be implemented optically using a PSA [131]. Thus, the system requirements are relaxed compared to other schemes (e.g. OPC or DBP), which increases the flexibility and reduces cost, also no additional DSP is required.

As a result, it was demonstrated that the nonlinear distortions were reduced by more than 8.5 dB with the high speed 400 Gbit/s over 12,800 km. The technique offers greater communication reach when compared to existing schemes. However, loading two polarizations with the same information affects the system capacity. As was indicated in [22], as polarization multiplexing is impossible, the information capacity of the proposed model will be reduced (unless, as can readily be shown using (11.49) with $n_{OPC} = 1$ and accounting for the doubling of the SNR arising from the transmission of duplicate information, the capacity of the equivalent dual polarization system is less than 1.7 dB).

The effective improvement was observed for a wide range of parameters, however, is limited by crosstalk induced by polarization mode dispersion polarization-dependent loss. As indicated by the authors, the effect is analogous to nonlinear noise squeezing [30], which demonstrates the beneficial interplay of nonlinearity and dispersion.

The technique offers a simple and energy efficient scheme (without requirement of additional mid-link devices) for fast ultra long-haul transmission, however it reduces spectral efficiency of the channel by the factor of 2 due to the exploitation of the spectrum or polarization to carry a conjugated copy of the same signal. Recently, dual PCTW technique was proposed [132], where two sets of PCTW were combined a single carrier. Compared to single PCTW with transmitted signal $E_x = E$ and $E_y = E^*$ and recovered $E_x + E_y^*$, dual PCTW two signals E_1 and E_2 are combined at the transmitter as $E_x = E_1 + E_2$ and $E_y = E_1^* - E_2^*$ and the recovered signals are $E_1 = E_x + E_y^*$ and $E_2 = E_x - E_y^*$. This enables nonlinearity mitigation (demonstrated Q-factor improvement of 1.2 dB compared to conventional DP-QPSK performance.

The idea to use phase conjugation to reduce nonlinear impairments was also implemented in coding [133] and Q-factor improvement up to 1.5 dB was achieved for QPSK CO-OFDM signals with phase-conjugated subcarrier coding (even number of OFDM subcarriers is required), when pairs of neighboring subcarriers were encoded as $S_k = E_k + E_{k+1}$ and $S_{k+1} = E_k^* - E_{k+1}^*$ with decoding at the receiver: $R_k = (E_k + E_{k+1}^*)/2$ and $R_{k+1} = (E_k - E_{k+1}^*)/2$.

11.5.5 Optical Regeneration

As described above only regenerative channels can reduce the accumulation of amplified spontaneous emission noise and introduce contractive effects that can combat fiber nonlinear impairments. Regeneration may take many forms, ranging from Decode-and-Forward model (allowing the use of FEC in each stage), hard decision (without FEC) [134], transformation with smooth nonlinear transfer functions (TF), which results in noise squeezing [51] and distributed pulse and noise shaping [135]. System capacities for Decode-and-Forward model can be readily calculated from the capacity of a single link channel [136], whilst another approach is required for hard decision and smooth TF based regenerators, these were discussed in Sect. 11.4.2.6.

Optical regeneration enables partial removal of the optical noise and, consequently, increases channel capacity. We would like to stress that impact of regeneration on capacity does not require using the Decode-and-Forward method that effectively means that the whole link presents a chain of optical transmitter and receivers. Instead, regenerative effect can be achieved by passive distributed nonlinear filtering with signal regeneration without requiring a hard decision. The regeneration is based on the constructive use of nonlinear signal transformations which result in noise squeezing. A number of promising all-optical regenerators have been reported recently, falling into the class of nonlinear transformations based regenerators: black-box WDM [137], 4-phase-shift keying (PSK) [117, 138–142], multilevel modulation formats 6-PSK [50], 8-PSK [143], and 16-quadrature amplitude modulation (QAM) [144], which prove the possibility to remove noise from both quadratures of the signal. Another important feature of regeneration is cascadability, as being placed in cascades along the line regeneration enables accumulation of noise squeezing effect and, as a result, improves a received signal. Thus, optical regeneration in coherent communication systems has emerged as fast developing and promising technique (e.g. recently demonstrated cascaded PPLN based phase sensitive amplifiers [145]).

Recently, the impact of regeneration on Shannon capacity has been studied [120, 121] proving that regenerative channels have Shannon capacity higher than a system without regenerators operating with the same signal launch power and with the same noise per amplifier. Recent advances opened regeneration for coherent communication, e.g. multilevel phase regeneration [142, 146]. Nevertheless, implementation of these techniques faces a number of challenges as regenerators are designed to operate with isolated pulses, so dispersion compensation and demultiplexing of WDM channels is required. Although a number of works were proposed for multichannel regeneration (for the case coherent communication see [147]), it still remains an open problem.

11.6 Conclusion

This chapter gives an overview of optical signal processing in the context of the Shannon channel capacity. The most important and challenging limit on fiber channel spectral efficiency is imposed by fiber nonlinearity. Therefore, exploitation of inherently "linear" techniques in a nonlinear communication fiber channel is not optimal. The increase in the signal power to improve the signal-to-noise ratio that is the standard procedure in a linear channel, leads to power-dependent nonlinear transmission distortions. Though, the Shannon capacity for the nonlinear fiber-optic channel was not yet determined, numerous techniques have been proposed for increasing achievable rates.

Here we reviewed basic analytical models of communication channels, which allow to estimate bounds on Shannon capacity. We discussed various techniques to increase the achievable transmission rate. It was shown that quasilossless transmission enables communication with lower amplification noise, OPC and PCTW efficiently suppress nonlinearity mitigation and might be considered as all-optical competitor to DBP, whilst all-optical regeneration offers possibility to reduce both destructive nonlinearity and stochastic noise, however it requires additional efforts towards practical implementability. There is a clear challenge to develop radically new approaches to coding, transmission and processing of information in fiber communication channels that will take into account the nonlinear properties of the transmission medium.

Acknowledgments We are grateful to S. Sygletos and K.S. Turitsyn for useful discussions. This work has been supported by the EPSRC project UNLOC (Unlocking the capacity of optical communications) EP/J017582/1.

References

1. A. Splett, C. Kurtzke, K. Petermann, Ultimate transmission capacity of amplified optical fiber communication systems taking into account fiber nonlinearities, in *Technical Digest of European Conference on Optical Communication* (1993), paper MoC2.4
2. P.P. Mitra, J.B. Stark, Nonlinear limits to the information capacity of optical fibre communications. Nature **411**, 1027–1030 (2001)
3. J. Tang, The Shannon channel capacity of dispersion-free nonlinear optical fiber transmission. J. Lightw. Technol. **19**(8), 1104–1109 (2001)
4. K.S. Turitsyn, S.A. Derevyanko, I.V. Yurkevich, S.K. Turitsyn, Information capacity of optical fiber channels with zero average dispersion. Phys. Rev. Lett. **91**, 203901 (2003)

5. B. Wu, E. E. Narimanov, Information Capacity of Nonlinear Fiber-Optical Systems, in *Optical Fiber Communication Conference and Exposition and The National Fiber Optic Engineers Conference, Technical Digest (CD)* (Optical Society of America, 2005), paper OME58
6. E.B. Desurvire, Capacity demand and technology challenges for lightwave systems in the next two decades. J. Lightwave Technol. **24**, 4697–4710 (2006)
7. R.-J. Essiambre, G. Foschini, G. Kramer, P. Winzer, Capacity limits of information transport in fiber-optic networks. Phys. Rev. Lett. **101**, 163901 (2008)
8. R.J. Essiambre, G. Kramer, P.J. Winzer, G.J. Foschini, B. Goebel, Capacity limits of optical fiber networks. J. Lightwave Technol. **28**, 662–701 (2010)
9. A.D. Ellis, J. Zhao, D. Cotter, Approaching the non-linear Shannon limit. J. Lightwave Technol. **28**, 423–433 (2010)
10. X. Chen, W. Shieh, Closed-form expressions for nonlinear transmission performance of densely spaced coherent optical OFDM systems. Opt. Express **18**, 19039–19054 (2010)
11. D.J. Richardson, Filling the pipe. Science **330**, 327 (2010)
12. R.I. Killey, C. Behrens, Shannon's theory in nonlinear systems. J. Mod. Optics **58**, 1–10 (2011)
13. A. Mecozzi, Limits to the long haul coherent transmission set by the Kerr nonlinearity and noise of in-line amplifiers. J. Lightwave Technol. **12**, 1993–2000 (1994)
14. A. Mecozzi, R.-J. Essiambre, Nonlinear Shannon limit in pseudo-linear coherent systems. J. Lightwave Technol. **30**, 2011–2024 (2012)
15. K.S. Turitsyn, S.K. Turitsyn, Nonlinear communication channels with capacity above the linear Shannon limit. Opt. Lett. **37**, 3600–3602 (2012)
16. D.J. Malyon, T. Widdowson, E.G. Bryant, S.F. Carter, J.V. Wright, W.A. Stallard, Electron. Lett. **27**(2), 120–121 (1991)
17. H.J. Thiele, R.I. Killey, P. Bayvel, Influence of fiber dispersion and bit rate on cross-phase-modulation-induced distortion in amplified optical fiber links. Electron. Lett. **34**(21), 2050–2051 (1998)
18. A.D. Ellis, W.A. Stallard, Four wave mixing in ultra long transmission systems incorporating linear amplifiers. IEE Colloquium 159 (1990), http://ieeexplore.ieee.org/xpl/freeabsall.jsp?arnumber=190875
19. D.A. Cleland, A.D. Ellis, C.H.F. Sturrock, Precise modelling of four wave mixing products over 400 km of step-index fibre. Electronics Letrers **28**, 1171–1172 (1992)
20. R.-J. Essiambre, B. Mikkelsen, G. Raybon, Intra-channel cross-phase modulation and four-wave mixing in high-speed TDM systems. Electron. Lett. **35**(18), 1576–1578 (1999)
21. X. Liu, D.A. Fishman, A fast and reliable algorithm for electronic pre-equalization of SPM and chromatic dispersion, in *OFC* (Optical Society of America, 2006), paper OThD4
22. E. Ip, J. Kahn, Compensation of dispersion and nonlinear impairments using digital backpropagation. J. Lightwave Technol. **26**, 3416–3425 (2008)
23. M. Nazarathy, J. Khurgin, R. Weidenfeld, Y. Meiman, P. Cho, R. Noe, I. Shpantzer, V. Karagodsky, Phased-array cancellation of nonlinear FWM in coherent OFDM dispersive multi-span links. Opt. Express **15**, 15777–15810 (2008)
24. X. Li, X. Chen, G. Goldfarb, E. Mateo, I. Kim, F. Yaman, G. Li, Electronic post-compensation of WDM transmission impairments using coherent detection and digital signal processing. Opt. Express **16**(2), 880–888 (2008)
25. D.S. Millar, S. Makovejs, C. Behrens, S. Hellerbrand, R.I. Killey, P. Bayvel, S.J. Savory, Mitigation of fiber nonlinearity using a digital coherent receiver. IEEE J. Sel. Topics Quantum Electron. **16**, 1217–1226 (2010)
26. E. Ip, Nonlinear compensation using backpropagation for polarization-multiplexed transmission. J. Lightwave Technol. **28**(6), 939–951 (2010)
27. L.B.Y. Du, A.J. Lowery, Pilot-based XPM nonlinearity compensator for CO-OFDM systems. Opt. Express **19**, B862–B869 (2011)

28. E.F. Mateo, X. Zhou, G. Li, Improved digital backward propagation for the compensation of inter-channel nonlinear effects in polarization-multiplexed WDM systems. Opt. Express **19**, 570–583 (2011)
29. L.B. Du, M.M. Morshed, A.J. Lowery, Fiber nonlinearity compensation for OFDM super-channels using optical phase conjugation. Opt. Express **20**, 19921–19927 (2012)
30. X. Liu, A.R. Chraplyvy, P.J. Winzer, R.W. Tkach, S. Chandrasekhar, Phase-conjugated twin waves for communication beyond the Kerr nonlinearity limit. Nat. Photonics **7**, 560–568 (2013)
31. H. Cheng, W. Li, Y. Fan, Z. Zhang, S. Yu, Z. Yang, A novel fiber nonlinearity suppression method in DWDM optical fiber transmission systems with an all-optical pre-distortion module. Opt. Comm. **290**, 152–157 (2013)
32. D. Rafique, A.D. Ellis, Impact of signal-ASE four-wave mixing on the effectiveness of digital back-propagation in 112 Gb/s PM-QPSK systems. Opt. Express **19**, 3449–3454 (2011)
33. S. Shieh, I. Djordjevic, *OFDM for Optical Communications* (Academic Press, Amsterdam, 2010)
34. W. Shieh, H. Bao, Y. Tang, Coherent optical OFDM: theory and design. Opt. Express **16**, 841–859 (2008)
35. A.D. Ellis, J.D. Cox, D. Bird, J. Regnault, J.V. Wright, W.A. Stallard, 5 Gbit/s soliton propagation over 350 km with large periodic dispersion coefficient perturbations using erbium doped fiber amplifier repeaters. Electron. Lett. **27**(10), 878 (1991)
36. I. Morita, K. Tanaka, N. Edagawa, M. Suzuki, Impact of the dispersion map on long-haul 40 Gbith single-channel soliton transmission with periodic dispersion compensation, in *Proceedings of OFC99*, San Diego, Paper FD1, 1999
37. P.V. Mamyshev, L.F. Mollenauer, Pseudo-phase-matched four-wave mixing in soliton wavelength-division multiplexing transmission. Opt. Lett. **21**(6), 396–398 (1996)
38. N.J. Smith, N.J. Doran, Modulational instabilities in fibers with periodic dispersion management. Opt. Lett. **21**(8), 570–572 (1996)
39. E. Pincemin, A. Tan, A. Bezard, A. Tonello, S. Wabnitz, J.-D. Ania-Castanon, S. Turitsyn, Robustness of 40 Gb/s ASK modulation formats in the practical system infrastructure. Opt. Express **14**(25), 12049–12062 (2006)
40. T. Tanimura, T. Hoshida, S. Oda, T. Tanaka, C. Ohsima, J.C. Rasmussen, Systematic analysis on multi-segment dual-polarisation nonlinear compensation in 112 Gb/s DP-QPSK coherent receiver, in *Technical Digest of European Conference on Optical Communication* (2009), paper 9.4.5
41. C.E. Shannon, A mathematical theory of communication. Bell Syst. Tech. J. **27**(379–423), 623–656 (1948)
42. E. Agrell, M. Karlsson, WDM channel capacity and its dependence on multichannel adaptation models, in *Optical Fiber Communication Conference and Exposition and the National Fiber Optic Engineers Conference, OFC/NFOEC* (2013)
43. P. Poggiolini, A. Carena, V. Curri, G. Bosco, F. Forghieri, Analytical modeling of non-linear propagation in uncompensated optical transmission links. IEEE Photon. Technol. Lett. **23**, 742–744 (2011)
44. J. Sakaguchi, Y. Awaji, N. Wada, A. Kanno, T. Kawanishi, T. Hayashi, T. Taru, T. Kobayashi, M. Watanabe, Space division multiplexed transmission of 109-Tb/s data signals using homogeneous seven-core fiber. J. Lightwave Technol. **30**, 658–665 (2012)
45. R. Ryf, S. Randel, A.H. Gnauck, C. Bolle, A. Sierra, S. Mumtaz, M. Esmaeelpour, E.C. Burrows, R.-J. Essiambre, P.J. Winzer, D.W. Peckham, A.H. McCurdy, R. Lingle, Mode-division multiplexing Over 96 km of few-mode fiber using coherent 6 6 MIMO processing. J. Lightwave Technol. **30**, 521–531 (2012)
46. A. Mecozzi, C. Antonelli, M. Shtaif, Nonlinear propagation in multi-mode fibers in the strong coupling regime. Opt. Express **20**, 11673–11678 (2012) (York, 2002)
47. A. Yariv, D. Fekete, D.M. Pepper, Compensation for channel dispersion by nonlinear optical phase conjugation. Opt. Lett. **4**(2), 52–54 (1979)

48. R.A. Fisher, B.R. Suydam, D. Yevick, Optical phase conjugation for time-domain undoing of dispersive self-phase-modulation effects. Opt. Lett. **8**, 611–613 (1983)
49. I. Phillips, M. Tan, M.F. Stephens, M. McCarthy, E. Giacoumidis, S. Sygletos, Exceeding the nonlinear-shannon limit using raman laser based amplification and optical phase conjugation, in *Optical Fiber Communication Conference*, M3C. 1
50. J. Kakande, R. Slavik, F. Parmigiani, A. Bogris, D. Syvridis, L. Gruner-Nielsen, R. Phelan, P. Petropoulos, D.J. Richardson, Multilevel quantization of optical phase in a novel coherent parametric mixer architecture. Nat. Photonics **4**(12), 748–752 (2011)
51. M. Matsumoto, Fiber-based all-optical signal regeneration. IEEE J. Sel. Top. Quantum Electron. **18**(2), 738–752 (2012)
52. M. Vasilyev, T.I. Lakoba, Opt. Lett. **30**, 1458 (2005)
53. R.A. Fisher, B.R. Suydam, D. Yevick, Optical phase conjugation for time-domain undoing of dispersive self-phase-modulation effects. Opt. Lett. **8**, 611–613 (1983)
54. S. Watanabe, T. Chikama, G. Ishikawa, T. Terahara, H. Kuwahara, Compensation of pulse shape distortion due to chromatic dispersion and Kerr effect by optical phase conjugation. IEEE Photon. Technol. Lett. **5**, 1241–1243 (1993)
55. X. Chen, X. Liu, S. Chandrasekhar, B. Zhu, and R. W. Tkach, Experimental demonstration of fiber nonlinearity mitigation using digital phase conjugation, in *Optical Fiber Communication Conference (OFC)*, paper OTh3C.1 (2012)
56. M.D. Pelusi, B.J. Eggleton, Optically tunable compensation of nonlinear signal distortion in optical fiber by end-span optical phase conjugation. Opt. Express **20**, 8015–8023 (2012)
57. D. Rafique, A.D. Ellis, Nonlinearity compensation in multi-rate 28 Gbaud WDM systems employing optical and digital techniques under diverse link configurations. Opt. Express **19**, 16919–16926 (2011)
58. X. Li, X. Chen, G. Goldfarb, E. Mateo, I. Kim, F. Yaman, G. Li, Electronic post-compensation of WDM transmission impairments using coherent detection and digital signal processing. Opt. Express **16**(2), 880–888 (2008)
59. J.D. Ania-Castañón, T.J. Ellingham, R. Ibbotson, X. Chen, L. Zhang, S.K. Turitsyn, Ultralong Raman fiber lasers as virtually lossless optical media. Phys. Rev. Lett. **96**, 023902 (2006)
60. S.K. Turitsyn, J.D. Ania-Castañón, S.A. Babin, V. Karalekas, P. Harper, D. Churkin, S. Kablukov, A. El-Taher, E. Podivilov, V.K. Mezentsev, 270-km ultralong Raman fiber laser. Phys. Rev. Lett. **103**, 133901 (2009)
61. E.G. Turitsyna, S.K. Turitsyn, Digital signal processing based on inverse scattering transform. Opt. Lett. **38**, 4186–4188 (2013)
62. J.E. Prilepsky, S.A. Derevyanko, S.K. Turitsyn, Opt. Express **21**, 24344 (2013)
63. S.K. Turitsyn, Capacity-achieving techniques in nonlinear channels, in *Optical Communication (ECOC 2014), 39th European Conference and Exhibition on Optical Communications*, paper Mo4.3.5
64. V.E. Zakharov, A.B. Shabat, Sov. Phys. JETP **34**, 62 (1972)
65. S. Verdu, T.S. Han, A general formula for channel capacity. IEEE Trans. Inf. Theory **4**, 1147–1157 (1994)
66. W. Karush, *Minima of Functions of Several Variables with Inequalities as Side Constraints*. M.Sc. Dissertation. Department of Mathematics, University of Chicago, Chicago, Illinois (1939)
67. H. W. Kuhn, A.W. Tucker, Nonlinear programming, in *Proceedings of 2nd Berkeley Symposium* (University of California Press. Berkeley, 1951), pp. 481–492
68. M. Frank, P. Wolfe, An algorithm for quadratic programming. Naval Res. Logistics Q. **3**, 95 (1956). doi:10.1002/nav.3800030109
69. S. Arimoto, An algorithm for computing the capacity of arbitrary discrete memoryless channels. IEEE T. Inform Theory **18**, 14–20 (1972)
70. R. Blahut, Computation of channel capacity and rate-distortion functions. IEEE T. Inform Theory **18**, 460–473 (1972)

71. G. Matz, P. Duhamel, "Information geometric formulation and interpretation of accelerated Blahut-Arimoto-type algorithms, in *Proceedings of the 2004 IEEE Information Theory Workshop*, San Antonio, TX, USA, 24–29 Oct 2004
72. N. Varnica, X. Ma, A. Kavcic, Capacity of power constrained memoryless AWGN channels with fixed input constellations, in *Proceedings of the IEEE Global Telecommunications Conference (GLOBECOM)*, Taipei, Taiwan, China, 2002, pp. 1339–1343
73. C.-I. Chang, L.D. Davisson, On calculating the capacity of an infinite-input finite (infinite) - output channel. IEEE Trans. Inf. Theory **34**, 1004–1010 (1988)
74. J. Dauwels, Numerical computation of the capacity of continuous memoryless channels, in *Proceedings of the Symposium on Information Theory in the Benelux*, Brussels, Belgium, 2005, pp. 221–228
75. D. Arnold, H.-A. Loeliger, On the information rate of binary-input channels with memory, in *Proceedings of the 2001 IEEE International Conference on Communications*, Helsinki, Finland, 11–14 June 2001, pp. 2692–2695
76. V. Sharma, S.K. Singh, Entropy and channel capacity in the regenerative setup with applications to Markov channels, in *Proceedings of the 2001 IEEE International Symposium. Information Theory*, Washington, DC, USA, 24–29 June 2001, p. 283
77. H.D. Pfister, J.B. Soriaga, and P.H. Siegel, "On the achievable information rates of finite-state ISI channels, in *Proceedings of the 2001 IEEE Globecom*, San Antonio, TX, 25–29 Nov 2001, pp. 2992–2996
78. J. Dauwels, H.-A. Loeliger, Computation of information rates by particle methods. IEEE Trans. Inf. Theory **54**, 406–409 (2008)
79. J. Rice, *Mathematical Statistics and Data Analysis*, 2nd edn. (Duxbury Press, ISBN 0-534-20934-3, 1995)
80. Certain topics in telegraph transmission theory, AIEE Trans. **47**, 617–644 (1928)
81. J.M. Geist, Capacity and cutoff rate for dense M-ary PSK constellations, in *Military Communications Conference*, vol. 2 (1990) pp. 768–770
82. J.P. Aldis, A.G. Burr, The channel capacity of discrete time phase modulation in AWGN. IEEE Trans. Inf. Theor. **39**(1) (1993)
83. K. Ho, J.M. Kahn, Channel capacity of WDM systems using constant-intensity modulation formats, in *Optical Fiber Communications Conference*, ed by A. Sawchuk, vol. 70 of OSA Trends in Optics and Photonics (Optical Society of America, 2002), paper ThGG85
84. E.T. Whittaker, On the functions which are represented by the expansions of the interpolation theory. Proc. R. Soc. Edinburgh **35**, 181–194 (1915)
85. V.A. Kotelnikov, On the carrying capacity of the ether and wire in telecommunications, in *Material for the First All-Union Conference on Questions of Communication*, Izd. Red. Upr. Svyazi RKKA, Moscow, 1933 (Russian)
86. H. Nyquist, Certain factors affecting telegraph speed. Bell Syst. Tech. J. **3**, 324–352 (1924). http://ieeexplore.ieee.org/stamp/stamp.jsp?arnumber=989875
87. D. Gabor, Theory of communication. J. Inst. Elec. Eng. **93**, 429–457 (1946)
88. T.M. Cover, J.A. Thomas, *Elements of Information Theory* (Wiley, New York, 1991)
89. S. Verdu, Fifty years of shannon theory. IEEE Trans. Inf. Theor. (Special Commemorative Issue) **44**(6), 2057–2078 (1998)
90. G.P. Agrawal, *Fiber-Optic Communication Systems*, 3rd edn. (Wiley, New York, 2002)
91. J.P. Gordon, W.H. Louisell, L.R. Walker, Quantum fluctuations and noise in parametric processes II. Phys. Rev. **129**, 481–485 (1963)
92. J.G. Proakis, *Digital Communications* (McGraw-Hill, New York, 2001)
93. A. Hasegawa, Y. Kodama, *Solitons in Optical Communications* (Oxford University Press, Oxford, 1995)
94. E. Iannone, F. Matera, A. Mecozzi, M. Settembre, *Nonlinear Optical Communication Networks* (Wiley, New York, 1998)
95. S.V. Manakov, On the theory of two-dimensional stationary self-focusing of electromagnetic waves. Sov. Phys. JETP **38**, 248–253 (1974)

96. D. Marcuse, C.R. Menyuk, P.K.A. Wai, Application of the Manakov-PMD equation to studies of signal propagation in optical fibers with randomly varying birefringence. J. Lightwave Technol. **15**, 1735–1746 (1997)
97. C. Menyuk, Application of multiple-length-scale methods to the study of optical fiber transmission. J. Eng. Math. **36**, 113–136 (1999)
98. I.V. Kolokolov, K.S. Turitsyn, Effective equation of nonlinear pulse evolution in a randomly anisotropic medium. J. Exp. Theor. Phys. **98**(2), 348351 (2004)
99. J.P. Gordon, L.R. Walker, W.H. Louisell, Quantum statistics of masers and attenuators. Phys. Rev. **130**, 806–812 (1963)
100. V.E. Zakharov, A.B. Shabat, Sov. Phys.-JETP **34**, 62 (1972)
101. Keang-Po Ho, Exact evaluation of the capacity for intensity-modulated direct-detection channels with optical amplifier noises. IEEE Photonics Technol. Lett. **174**, 858–860 (2005)
102. M. Katz, S. Shamai (Shitz), On the capacity-achieving distribution of the discrete-time noncoherent and partially coherent AWGN channels. IEEE Trans. Inf. Theor. **50**(10), 2257–2270 (2004)
103. A. Lapidoth, Capacity bounds via duality: a phase noise example, in *Proceedings of the 2nd Asian-European Workshop on Information Theory*, Breisach, Germany, 26–29 June 2002, pp. 58–61
104. B. Goebel, R.-J. Essiambre, G. Kramer, P.J. Winzer, N. Hanik, Calculation of mutual information for partially coherent Gaussian channels with applications to ?ber optics. IEEE Trans. Inf. Theor. **57**(9), 5720–5736 (2011)
105. K. Ho, *Phase-modulated Optical Communication Systems* (Springer, New York, 2005)
106. P. Leoni, S. Calabro, B. Lankl, Constellation expansion for 100G transmission. IEEE Photon. Technol. Lett. **25**, 1904–1907 (2013)
107. E. E. Narimanov, P. Mitra, The channel capacity of a fiber optics communication system: perturbation theory. J. Lightw. Technol. **20**, 3 (2002)
108. P. Hou, B. Belzer, T.R. Fischer, Shaping gain of partially coherent additive white Gaussian noise channel, *in Proceedings of the Conference Information Sciences and Systems, Princeton*, NJ, 15–17 Mar 2000, pp. wp6–17–wp6–22
109. P. Hou, B. Belzer, T.R. Fischer, Shaping gain of the partially coherent additive white Gaussian noise channel. IEEE Commun. Lett. **6**, 175–177 (2002)
110. A.P.T. Lau, J.M. Kahn, Signal design and detection in presence of nonlinear phase noise. J. Lightw. Technol. **25**(10), 3008–3016 (2007)
111. C. Hager, A. Graell i Amat, A. Alvarado, E. Agrell, "Constellation optimization for coherent optical channels distorted by nonlinear phase noise, in *Global Communications Conference (GLOBECOM)* (2012), pp. 2870–2875
112. A. Mecozzi, C.B. Clausen, M. Shtaif, S.-G. Park, A.H. Gnauck, Cancellation of timing and amplitude jitter in symmetric links using highly dispersed pulses. IEEE Photon. Technol. Lett. **13**, 445–447 (2001)
113. K. Kikuchi, Electronic post-compensation for nonlinear phase fluctuations in a 1000-km 20-gbit/s optical quadrature phase-shift keying transmission system using the digital coherent receiver. Opt. Express **16**(2), 889–896 (2008)
114. M.I. Yousefi, F.R. Kschischang, On the per-sample capacity of nondispersive optical fibers. IEEE Trans. Inf. Theor. **57**(11), 7522–7541
115. J.B. Stark, P. Mitra, A. Sengupta, Information capacity of nonlinear wavelength division multiplexing fiber optic transmission line. Opt. Fiber Technol. **7**, 275–288 (2001)
116. X. Liu, X. Wei, R.E. Slusher, C.J. McKinstrie, Improving transmission performance in differential phase-shift-keyed systems by use of lumped nonlinear phase-shift compensation. Opt. Lett. **27**, 1616–1618 (2002)
117. M. Nakazawa, E. Yamada, H. Kubota, K. Suzuki, 10 Gbit/s soliton data transmission over one million kilometres. Electron. Lett. **27**(14), 1270–1272 (1991)

118. L. F. Mollenauer, A. Grant, X. Liu, X. Wei, C. Xie, I. Kang, C. Doerr, Demonstration of 109 X 10G dense WDM over more than 18,000 km using novel, periodic-group-delay complemented dispersion compensation and dispersion managed solitons, in *Proceedings of ECOC 03*, Rimini, Post-deadline Paper Th4.3.4, 2003
119. E. Agrell, A. Alvarado, G. Durisi, M. Karlsson, Capacity of a nonlinear optical channel with finite memory. J. Lightwave Technol. **16**, 2862–2876 (2014)
120. R. Dar, M. Shtaif, M. Feder, New bounds on the capacity of the nonlinear fiber-optic channel. Opt. Lett. **39**, 398–401 (2014)
121. M. Shtaif, R. Dar, A. Mecozzi, M. Feder, Nonlinear interference noise in WDM systems and approaches for its cancelation, in *Optical Communication (ECOC 2014), 39th European Conference and Exhibition on Optical Communications*, paper We1.3.1
122. K.S. Turitsyn, S.K. Turitsyn, Nonlinear communication channels with capacity above the linear Shannon limit. Opt. Lett. **37**(17), 3600–3602 (2012)
123. R.J. Mears, L. Reekie, I.M. Jauncey, D.N. Payne, Low-noise Erbium-doped fibre amplifier at 1.54 μm. Electron. Lett. **23**, 1026–1102 (1987)
124. R.H. Stolen, E.P. Ippen, Raman gain in glass optical waveguides. Appl. Phys. Lett. **22**, 276 (1973)
125. M.A. Sorokina, S.K. Turitsyn, Regeneration limit of classical Shannon capacity. Nat. Commun. **5**, Article number: 3861 (2014)
126. M.A. Sorokina, S. Sygletos, S.K. Turitsyn, Optimization of cascaded regenerative links based on phase sensitive amplifiers. Opt. Lett. **38**, 4378–4381 (2013)
127. S. Watanabe, M. Shirasaki, Exact compensation for both chromatic dispersion and Kerr effect in a transmission fiber using optical phase conjugation. J. Lightwave Technol. **14**(3), 243–248 (1996)
128. I. Brener, B. Mikkelsen, K. Rottwitt, W. Burkett, G. Raybon, J. B. Stark, K. Parameswaran, M.H. Chou, M.M. Fejer, E.E. Chaban, R. Harel, D.L. Philen, S. Kosinski, Cancellation of all Kerr nonlinearities in long fiber spans using a LiNbo3 phase conjugator and Raman amplification, in *Optical Fiber Communication Conference, OSA Technical Digest Series* (Optical Society of America, 2000), paper PD33
129. D. Rafique, A.D. Ellis, Nonlinearity compensation via spectral inversion and digital back-propagation: a practical approach, in *Optical Fiber Communication Conference, OSA Technical Digest* (Optical Society of America, 2012), paper OM3A.1
130. S.L. Jansen, D. van den Borne, B. Spinnler, S. Calabro, H. Suche, P.M. Krummrich, W. Sohler, G.-D. Khoe, H. de Waardt, Optical phase conjugation for ultra long-haul phase-shift-keyed transmission. J. Lightwave Technol. **24**, 54 (2006)
131. Yue Tian, Yue-Kai Huang, Shaoliang Zhang, Paul R. Prucnal, Ting Wang, Demonstration of digital phase-sensitive boosting to extend signal reach for long-haul WDM systems using optical phase-conjugated copy. Opt. Express **21**, 5099–5106 (2013)
132. B. Corcoran, S. Olsson, C. Lundström, Mitigation of Nonlinear Impairments on QPSK Data in Phase-Sensitive Amplified Links, in *39th European Conference on Optical Communication and Exhibition, ECOC 2013*, London, 22–26 Sept 2013
133. X. Liu, S. Chandrasekhar, P.J. Winzer, A.R. Chraplyvy, R.W. Tkach, B. Zhu, T.F. Taunay, M. Fishteyn, D.J. DiGiovanni, Scrambled coherent superposition for enhanced optical fiber communication in the nonlinear transmission regime. Opt. Express **20**, 19088–19095 (2012)
134. T. Yoshida, T. Sugihara, K. Ishida, T. Mizuochi, Spectrally-efficient dual phase-conjugate twin waves with orthogonally multiplexed quadrature pulse-shaped signals, in *Optical Fiber Communication Conference, OSA Technical Digest (online)* (Optical Society of America, 2014), paper M3C.6
135. S.T. Le, E. Giacoumidis, N. Doran, A.D. Ellis, S.K. Turitsyn, Phase-conjugated subcarrier coding for fibre nonlinearity mitigation in CO-OFDM transmission, in *Optical Communication (ECOC 2014), 39th European Conference and Exhibition on Optical Communications*, paper M3C.6

136. T. Widdowson, J.P. Hueting, A.D. Ellis, D.J. Malyon, P.J. Watkinson, Global fibre transmission using optically amplified regenerators for maximised repeater spacing. Electron. Lett. **30**, 2056–2057 (1994)
137. O. Leclerc, U.E. Desurvire, O. Audouin, Synchronous WDM soliton regeneration: toward 80–160 Gbit/s transoceanic systems. Opt. Fiber Technol. **3**(2), 97–116 (1997)
138. A.D. Ellis, J. Zhao, Channel capacity of non-linear transmission systems, in *Impact of Nonlinearities on Fiber Optic Communications*, ed. by S. Kumar. (New York, Springer, 2011), ch. 13, pp. 507–538
139. S. Sygletos, P. Frascella, S.K. Ibrahim, L. Gruner-Nielsen, R. Phelan, J. O"Gorman, A.D. Ellis, A practical phase sensitive amplification scheme for two channel phase regeneration. Opt. Express **19**(26), B938–B945 (2011)
140. R.P. Webb, J.M. Dailey, R.J. Manning, A.D. Ellis, Phase discrimination and simultaneous frequency conversion of the orthogonal components of an optical signal by four-wave mixing in an SOA. Opt. Express **19**, 20015–20022 (2011)
141. S. Watanabe, S. Kaneko, T. Chikama, Long-haul fiber transmission using optical phase conjugation. Opt. Fiber Technol. **2**(2), 169178 (1996)
142. G. Goldfarb, M.G. Taylor, G. Li, Experimental demonstration of fiber impairment compensation using the split-step infinite impulse response method. Digest of the IEEE/LEOS Summer Topical Meetings **2008**, 145–146 (2008)
143. D. Rafique, J. Zhao, A.D. Ellis, Digital back-propagation for spectrally efficient WDM 112 Gbit/s PM m-ary QAM transmission. Opt. Express **19**, 5219–5224 (2011)
144. R. Slavik, F. Parmigiani, J. Kakande, C. Lundstrom, M. Sjodin, P. Andrekson, R. Weerasuriya, S. Sygletos, A.D. Ellis, L. Gruner-Nielsen, D. Jakobsen, S. Herstrom, R. Phelan, J. O'Gorman, A. Bogris, D. Syvridis, S. Dasgupta, P. Petropoulos, D. Richardson, All-optical phase and amplitude regenerator for next generation telecommunications systems. Nat. Photonics **4**(10), 690–695 (2010)
145. G. Hesketh, P. Horak, Reducing bit-error rate with optical phase regeneration in multilevel modulation formats. Opt. Lett. **38**, 5357–5360 (2013)
146. A. Bogris, D. Syvridis, All-optical signal processing for 16-QAM using four-level optical phase quantizers based on phase sensitive amplifiers, in *Proceedings of European Conference in Optical Communications (ECOC 2013)*, London, paper We.3A.6, (2013)
147. T. Umeki, M. Asobe, H. Takara, Y. Miyamoto, H. Takenouchi, Multi-span transmission using phase and amplitude regeneration in PPLN-based PSA. Opt. Express **21**, 18170–18177 (2013)

Chapter 12
Nonlinear Optics for Photonic Quantum Networks

Alex S. Clark, Lukas G. Helt, Matthew J. Collins, Chunle Xiong, Kartik Srinivasan, Benjamin J. Eggleton and Michael J. Steel

Abstract By harnessing quantum mechanical effects we can create technology with greatly improved functionality including enhanced sensing, exponentially faster computing, the simulation of previously inaccessible quantum systems, and the secure transfer of information. In our ever more connected world, it is important that we can communicate securely, and as technology develops into the quantum regime it will become important for distant quantum processors to exchange information over a quantum network. Here we discuss how quantum information can be encoded and transferred around a such a network. We concentrate on the use of nonlinear optics to generate and manipulate photons, and present schemes for fully secure quantum communication.

A.S. Clark (✉) · M.J. Collins · C. Xiong · B.J. Eggleton
Centre for Ultrahigh bandwidth Devices for Optical Systems (CUDOS),
Institute of Photonics and Optical Science (IPOS), School of Physics,
University of Sydney, Camperdown, NSW 2006, Australia
e-mail: a.clark@physics.usyd.edu.au

M.J. Collins
e-mail: mcoll@physics.usyd.edu.au

C. Xiong
e-mail: chunle@physics.usyd.edu.au

B.J. Eggleton
e-mail: egg@physics.usyd.edu.au

L.G. Helt · M.J. Steel
MQ Photonics Research Centre and CUDOS, Department of Physics and Astronomy,
Macquarie University, North Ryde, NSW 2109, Australia
e-mail: luke.helt@mq.edu.au

M.J. Steel
e-mail: michael.steel@mq.edu.au

K. Srinivasan
Center for Nanoscale Science and Technology, National Institute of Standards
and Technology (NIST), Gaithersburg, MD 20899, USA
e-mail: kartik.srinivasan@nist.gov

S. Wabnitz and B.J. Eggleton (eds.), *All-Optical Signal Processing*,
Springer Series in Optical Sciences 194, DOI 10.1007/978-3-319-14992-9_12

12.1 Introduction: Photonic Quantum Networks

Photonics is an enabling technology for global communication systems and is of profound importance in the evolution of society as we know it. In classical communication networks many optical pulses of light, containing billions of photons, are sent through optical fibres and routed to different locations using a variety of technologies. In classical networks these optical pulses can be amplified and frequency converted with relative ease, allowing extremely high transmission rates for information. A quantum network is intrinsically different, in that data is encoded in an individual quantum system such as a *single photon*, which obeys inherently different physics. It is in fact quantum physics that guarantees fully secure communication schemes when using these quantum states. A quantum network is therefore useful for secure communication and to move quantum information between quantum information processing nodes. Unfortunately these quantum states cannot be easily shifted in frequency due to the addition of noise to the original and converted channels, and unknown arbitrary quantum states can definitely not be amplified due to the famous *no-cloning theorem* developed by Wootters and Zurek [1], and separately Dieks [2], in 1982. This has limited current demonstrations of quantum communication to around 200 km. There are a number of methods currently suggested to extend this distance, most of which require small scale quantum information processing (QIP). These include quantum repeaters, which store quantum information until it is ready to be used, and quantum relays, which allow for the teleportation of quantum information.

Nonlinear optical processes and devices are used in many different forms in quantum information experiments and technology. One of the most intuitive is the use of a nonlinear material to allow a single photon to impart a π phase shift on another single photon. This, however, has not been realised to date due to the limited strength of nonlinear media meaning that the device would have to extend for thousands of kilometres. There have, however, been a number of proposals that show that quantum information processing schemes could be possible with weak nonlinearities [3]. Such schemes rely on a single photon imparting a measurable cross-phase shift on a high intensity beam of light. Experiments with optical fibres [4] and gas-filled hollow core photonic bandgap fibres [5] have shown promising routes to achieving such phase shifts. In parallel to these developments there have been arguments against such nonlinear quantum computing schemes which state that if a photon imparts an appreciable phase shift then it must also undergo self-phase modulation which can destroy the encoded quantum state and cause unwanted noise in the system [6]. The debate over whether such schemes can work continues. Another more recently developed QIP scheme employing nonlinear processes is that of coherent photon conversion (CPC) [7]. In this scheme a third-order nonlinear medium is pumped with a high intensity laser in a particular mode which creates an effective second-order nonlinearity which acts on a further three modes. This can be thought of as a four-wave mixing interaction that creates a tunable second-order nonlinearity. Theoretically, this process enables the creation of numerous QIP components including a controlled phase gate, photon-doubling for efficient detection of photons, and clearing higher-order noise terms from photon sources; however current demonstrations have been limited to very weak effective nonlinearities. These nonlinear QIP schemes are in their infancy and will not be discussed in this chapter. Instead we focus on how nonlinear processes are already in use in many QIP experiments across the world and will be integrated into future quantum networks to provide enhanced functionality and agility. In the following sections we will present advances in the use of nonlinear optics for the generation of photons, first looking at the theory behind such sources before turning to experimental implementations with both second- and third-order nonlinear media. We will discuss how to characterise these photon sources using simple measurements which provide us with measures of spectral purity, noise metrics, and photon statistics. We then describe how many of these spontaneous, probabilistic sources can be actively multiplexed together to create a deterministic source of photons. An integral part of a classical communications network is wavelength division multiplexing and the ability to convert between different channels. In a quantum network this wavelength conversion cannot be performed using stimulated four-wave mixing. We describe methods for performing quantum frequency conversion and review the state-of-the-art in this field. Finally, we describe how a quantum communications network can be built and how nonlinear optical processes are integrated into such networks to enhance functionality and improve performance. We use the remainder of this section as a short introduction to what a quantum bit, or *qubit*, is, how such qubits can be manipulated and how they can be made to interact to form two qubit operations.

12.1.1 Photonic Qubits, Gates and Algorithms

12.1.1.1 Qubits

Classical computation uses bits with two possible values, 0 and 1, to encode and process information. This is commonly the off or on state of a switch or transistor. As the physical size of this switch decreases quantum mechanical effects will come into play. At this point we are left with one quantum mechanical object, in this case a single electron, causing the switch to close, however it becomes difficult to distinguish between the on and off states or make measurements on such objects. This need not necessarily be a hindrance to computation. Such quantum mechanical systems may in fact allow computational speed-up for certain operations.

In quantum computation the bit is replaced with what is termed a qubit. This is a two level quantum mechanical system which may take the value $|0\rangle$ or $|1\rangle$, where we use the Dirac notation to show that these are states of a quantum system. This qubit may also be placed in any superposition of $|0\rangle$ and $|1\rangle$ such that the most general state $|\Psi\rangle$ is

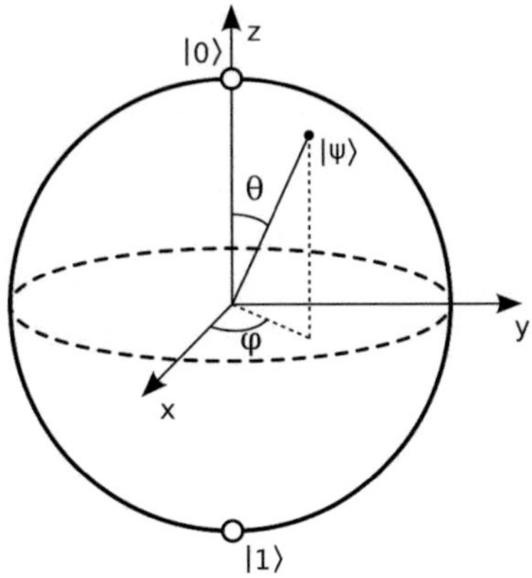

Fig. 12.1 A schematic of the Bloch sphere mapping all possible qubit states

$$|\Psi\rangle = \cos\frac{\theta}{2}|0\rangle + e^{i\phi}\sin\frac{\theta}{2}|1\rangle, \tag{12.1}$$

where $\theta \in [0, \pi]$ and $\phi \in [0, 2\pi]$. We can also write this arbitrary state in the following vector format

$$|\Psi\rangle \rightarrow \begin{pmatrix} \cos\frac{\theta}{2} \\ e^{i\phi}\sin\frac{\theta}{2} \end{pmatrix}, \tag{12.2}$$

which we may find useful when performing transformations of qubits from one state to another.

Here θ and ϕ define a point on a sphere, known as the Bloch sphere, as seen in Fig. 12.1. One can think of the surface of the Bloch sphere as a map of states that a qubit can take. $|0\rangle$ and $|1\rangle$ are at the poles of the sphere, and therefore an equal superposition of $|0\rangle$ and $|1\rangle$ can be found on the equator, with the phase ϕ varying around the equator. Other positions on the sphere can be reached by changing the values of θ and ϕ through qubit transformations. Superposition is one of the properties of qubits that allow the speed up of some algorithms compared to classical methods.

12.1.1.2 Entanglement

Another property of qubits that is often hailed as the most counter-intuitive is that of entanglement. Let us take two qubits which can each take the state $|0\rangle$ or $|1\rangle$. These two qubits can therefore be in the states $|00\rangle$, $|01\rangle$, $|10\rangle$ or $|11\rangle$ (where $|ab\rangle$ is equivalent to $|a\rangle|b\rangle$) but they could also be in an arbitrary superposition state with variable amplitudes α_{ij} where i and j can take the value 0 or 1. The state is then

$$|\Psi\rangle = \alpha_{00}|00\rangle + \alpha_{01}|01\rangle + \alpha_{10}|10\rangle + \alpha_{11}|11\rangle, \tag{12.3}$$

for which there are special cases that we will now discuss. For particular amplitudes we can create states which demonstrate entanglement, where one photon of the pair will be correlated in some way to the other photon of a pair. The states we will consider are known as the Bell states after John S. Bell presented them in 1964 to use as a test of non-locality [8].

$$|\Psi^+\rangle = \frac{1}{\sqrt{2}}(|01\rangle + |10\rangle), \tag{12.4}$$

$$|\Psi^-\rangle = \frac{1}{\sqrt{2}}(|01\rangle - |10\rangle), \tag{12.5}$$

$$|\Phi^+\rangle = \frac{1}{\sqrt{2}}(|00\rangle + |11\rangle), \tag{12.6}$$

$$|\Phi^-\rangle = \frac{1}{\sqrt{2}}(|00\rangle - |11\rangle). \tag{12.7}$$

The Bell states are maximally entangled states in that they cannot be factorised into a state of the form $(\alpha_1|0\rangle + \beta_1|1\rangle)_1(\alpha_2|0\rangle + \beta_2|1\rangle)_2$ where the subscripts represent two separate quantum particles 1 and 2. It is this property of entanglement that underlies much research into quantum communication and cryptography where the correlations observed when measuring an entangled qubit can be used to build up a key to securely transfer information [9, 10].

There are many avenues currently being explored to find which two level system, or qubit, is most suited to the building of a quantum computer. Some suggested routes are the use of the electron levels of cooled trapped atoms [11] or ions [12] or the electron spins in colour centres in diamond [13]. We will concentrate on the use of photons as carriers of quantum information. This provides many advantages such as the longevity of the storage of information due to the low interaction of photons with their environment, characterised by their long coherence times, and the ease of application of single qubit operations, performed with the use of birefringent crystals or wave plates.

12.1.1.3 Encoding and Measurement

There are many different two level systems available to a photon which can be used to encode a qubit. One of the most common is the use of path encoding shown in Fig. 12.2a, where one path, say the upper path, can be labeled $|0\rangle$ whilst an alternate path, the lower in this case, is labeled $|1\rangle$. This encoding scheme is especially useful

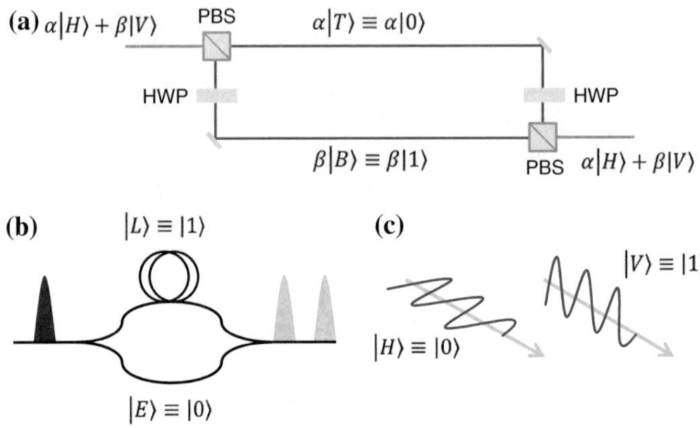

Fig. 12.2 Possible qubit encoding schemes using photons including **a** path encoding, **b** time-bin encoding, and **c** polarisation encoding

in media where polarisation is either not discriminated or not preserved, such as monolithic waveguide structures [14]. A scheme that is often used for long distance quantum communication, where only one channel exists between the two users is that of time-bin encoding [15], shown in Fig. 12.2b. In this scheme a photon that travels through the short arm of the Mach-Zehnder interferometer is in the state $|0\rangle$ whilst a photon that has travelled through the long arm is in the state $|1\rangle$. These photons are then sent and, as there is no dependence upon polarisation, are relatively free from decoherence over long distances in optical fibre. The reverse process can be performed at the receiving end and analysis can be performed by measuring the arrival times of the photons. Whilst this encoding is very robust against decoherence it suffers from the fact that is hard to construct gates between multiple qubits, making it more suited to communication schemes rather than computation, although some recent schemes do exist [16].

For this discussion we will use one of the most common and well researched encoding schemes, namely polarisation encoding. In this scheme, shown in Fig. 12.2c, $|0\rangle$ and $|1\rangle$ are represented by horizontally and vertically polarised photons, $|H\rangle$ and $|V\rangle$ respectively. Other polarisation states represent superpositions of these states and can be easily generated through the use of birefringent plates or wave plates, which will be discussed in the next section. It can be seen from Fig. 12.2a that it is relatively easy to convert between polarisation encoding and path encoding, and vice versa, with the use of polarising beamsplitters and wave plates. Wave plates allow us to perform single qubit operations on polarisation encoded photons. This birefringent media formed into plates causes light travelling through in one polarisation to travel faster than the orthogonal polarisation so a phase is gained between them. It is this extra phase that allows the wave plate to change the polarisation of the light as it propagates. There are commonly two types of wave plates used, the half-wave plate and the quarter-wave plate, so named for the proportion of a wavelength of the light that they retard one polarisation compared to the other.

Using the vector notation for polarisation states as in (12.2) we can write the matrix representations for quarter- and half-wave plate operators as

$$\widehat{HWP}(\varphi) = \begin{pmatrix} \cos(2\varphi) & \sin(2\varphi) \\ \sin(2\varphi) & -\cos(2\varphi) \end{pmatrix}, \quad (12.8)$$

$$\widehat{QWP}(\varphi) = \frac{1}{\sqrt{2}} \begin{pmatrix} 1 + i\cos(2\varphi) & i\sin(2\varphi) \\ i\sin(2\varphi) & i - \cos(2\varphi) \end{pmatrix}, \quad (12.9)$$

where φ is the angle the fast axis of the waveplate has been rotated from horizontal in both cases, which is the same as the Jones matrices used in classical polarisation optics. The NOT operation is performed by setting the angle of a half-wave plate to $\varphi = 45°$ while the Hadamard operation corresponds to a setting of $\varphi = 22.5°$. With a combination of three wave plates, namely a quarter-, half-, quarter-wave plate, any arbitrary rotation of the state of the qubit can be performed (Fig. 12.3).

Quantum information processing (QIP) requires more than just single qubit operations. We also have the need for an interaction between qubits, known as two qubit operations. We often refer to these operations as controlled gates as the state of one qubit often decrees what will happen to the second qubit in the operation. One of the simplest of such gates is the controlled-Phase gate, often shortened to CZ gate. This operation applies a phase shift of π to one qubit, the target, conditional on the other qubit, the control, being in the $|1\rangle$ state. To perform such operations using photons as qubits we require two photons to interact with one another; an inherently difficult task. The interaction associated with the CZ gate is equivalent to a cross-phase modulation of π on one photon, induced by another. There has been much research into nonlinear materials that may allow the presence of one photon in a medium to have an effect on another photon present, but currently no material has been found that could provide the full π phase shift required to perform a CZ or CNOT gate in a sensible amount of media. In the absence of such a strong nonlinearity one may think that photonic quantum gates are not realizable, but a publication in 2001 by Knill et al. [17] suggested a feasible route. This paper shows that by using "ancilla" (meaning simply extra) qubits,

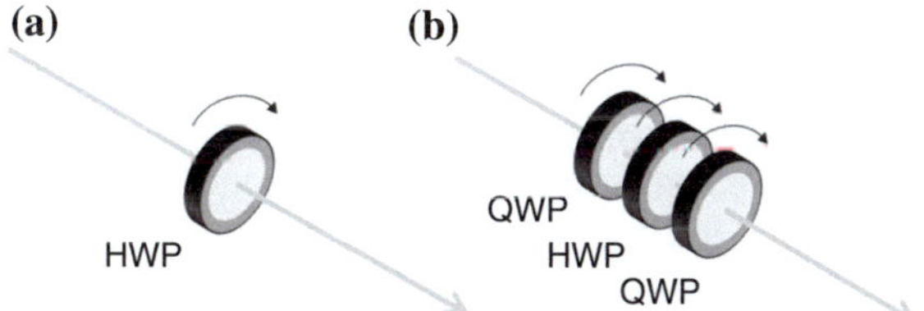

Fig. 12.3 The use of wave plates to perform **a** a Hadamard operation and **b** the generation of an arbitrary qubit state

detection, and linear optical networks a measurement-induced nonlinearity is achieved, allowing the CZ and CNOT to be realised in a probabilistic manner.

At the heart of this scheme is the injection of two photons into a beam splitter at which point they interfere with one another in a non-classical way. If we take a beam splitter with a reflectivity of 0.5 and input one photon into each of its input ports then, if the photons are completely indistinguishable in all degrees of freedom including spectral, temporal, spatial and polarisation modes, they will both exit through the same output port. This is what is known as quantum interference and it arises from the fact that when we are using quantum particles we must calculate their probability amplitudes for undergoing particular operations and not just the probabilities.

Experiments to verify this effect involve taking a particular mode of one of the photons and adjusting it in order to make it distinguishable from the other photon involved and then scanning this through the point where the modes of the two photons match. This effect can be observed by monitoring output ports and counting the number of coincident events, confirming that there are none at the point of complete indistinguishability. The first such experiments were performed by Hong et al. in 1987 [18] and demonstrated independently by Rarity and Tapster [19] in which they varied the temporal mode by delaying the arrival at the beam splitter of one photon compared to the other. As the delay is reduced a drop in coincidences is observed which should reach zero at the point of zero delay if all other photon modes are indistinguishable. As the delay is then increased again the photon coincidences will rise. This characteristic dip in counts is often referred to as the Hong-Ou-Mandel (HOM) dip and is at the heart of many linear optical QIP schemes.

Whether we wish to perform some small scale quantum information processing using linear optics, or investigate other nonlinear optical schemes, we will always require sources of single photons which often rely on nonlinear interactions. In the next section we will discuss how such photon sources can be constructed and how the use of different nonlinear media can lead to differences in the generated photons.

12.2 Nonlinear Optics for Single Photon Generation

12.2.1 Single Photon Sources and the Heralding of Single Photons

In the previous section, we have established the importance of developing high quality sources of single photons. We now turn to the methods by which this might be accomplished and the realistic description of both the ideal states and what is possible today.

12.2.1.1 Ideal Single Photons

The ideal single photon should typically have a specified spatial and polarisation state, and a temporal/spectral profile that is near Fourier-transform limited. Classically we might represent such a state by an electric field wave packet written in the frequency domain as

$$\mathbf{E}(\mathbf{r},t) = A\int_0^\infty \mathrm{d}\omega f(\omega)\mathscr{E}_\omega(\mathbf{r})\mathrm{e}^{-i\omega t} + \mathrm{c.c}, \tag{12.10}$$

where $f(\omega)$ is the spectral envelope, $\mathscr{E}_\omega$ is the spatial mode profile and c.c. denotes the complex conjugate. The normalisation A would be chosen so that the energy of the wave packet corresponds to $\hbar\bar{\omega}$ where $\bar{\omega}$ is the centre frequency of the pulse.

However, many of the applications [17, 20, 21] discussed within Sect. 12.1 can only be described in terms of the fully quantised electromagnetic field, and as we see below, the generation of few photon states by nonlinear optical processes must also be described in this language. Therefore we introduce our ideal single photon state using the formalism of field quantisation and boson creation and annihilation operators. A suitable single photon state can thus be written

$$|\psi_\alpha\rangle = \int_0^\infty \mathrm{d}\omega f(\omega)\hat{a}^\dagger_{\alpha\omega}|\mathrm{vac}\rangle, \tag{12.11}$$

where $\hat{a}^\dagger_{\alpha\omega}$ is the creation operator for the mode α at frequency ω, and satisfies the commutation relation $[\hat{a}_{\alpha\omega}, \hat{a}^\dagger_{\beta\omega'}] = \delta_{\alpha\beta}\delta(\omega - \omega')$, generalising the states $|H\rangle$ and $|V\rangle$ of Sect. 12.1.1.3. Further $|\mathrm{vac}\rangle$ is the vacuum state, and the photon wave function $f(\omega)$ is normalised according to

$$\int_0^\infty \mathrm{d}\omega |f(\omega)|^2 = 1. \tag{12.12}$$

For the remainder of Sect. 12.2, we are concerned with understanding and developing physical techniques for creating light fields that approach the ideal of (12.11).

12.2.1.2 Approximating Single Photons

There are three methods commonly used to produce approximations to this idealisation.

Fig. 12.4 Schematic illustration for the generation of photons using atom—like sources

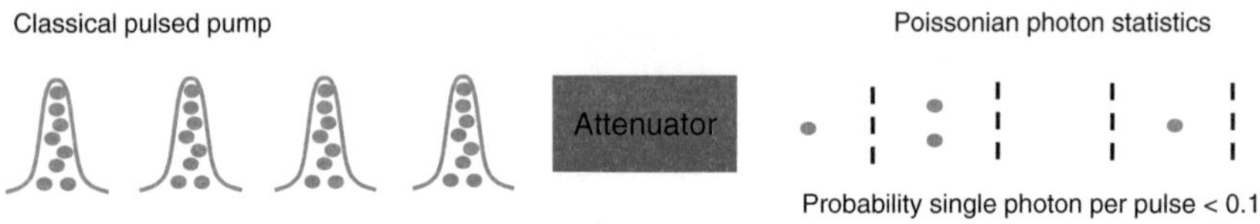

Fig. 12.5 Schematic illustration for the generation of photons using attenuation

The first approach is triggered or "on-demand" sources (see Fig. 12.4), which rely on the spontaneous decay of "atom-like systems"—usually atoms, ions, colour centres or quantum dots in a cavity—to truly emit one photon at a time. Examples of such systems include quantum dots [22, 23] and nitrogen vacancy colour centres [24, 25]. The challenge here is that the systems available often require cryogenic conditions and have poor emission directivity. System stability and reproducibility of the emission wavelength can also be a problem, though they are rapidly improving. Nevertheless while triggered sources represent an ultimate goal, it remains highly challenging to produce indistinguishable photons from distinct atom-like systems [26].

Alternatively, an approximation to a single photon source can be created by strongly attenuating a laser, leaving only a small probability that a single photon exits in a given time window (see Fig. 12.5). The likelihood of two or more photons exiting the laser is even lower. However, as a laser field corresponds to a coherent state in quantum optics [27], there is a nonzero probability of emitting n photons, determined by the Poisson distribution

$$P_{\text{Poisson}}(n,\mu) = \mathrm{e}^{-\mu}\frac{\mu^n}{n!}, \tag{12.13}$$

where μ is the average number of emitted photons per pulse. The ratio of multi to single photon emission probabilities $P_{\text{Poisson}}(>1,\mu)/P_{\text{Poisson}}(1,\mu) \approx \mu/2$ can only be reduced by lowering the net flux, limiting the usefulness of these sources.

As neither of these two classes depend on nonlinear optical effects we will not discuss them further here. An excellent review may be found in [28].

The final method is to use probabilistic or "heralded" sources, in which a pair of photons is generated together and one member of the pair detected to herald the presence of the other (see Fig. 12.6). This approach usually relies on the production of photon pairs via spontaneous parametric downconversion (SPDC) [29–36], or spontaneous four-wave mixing (SFWM) [37–42]. The concept of a single photon state described by (12.11) should now be replaced by the "bi-photon" state

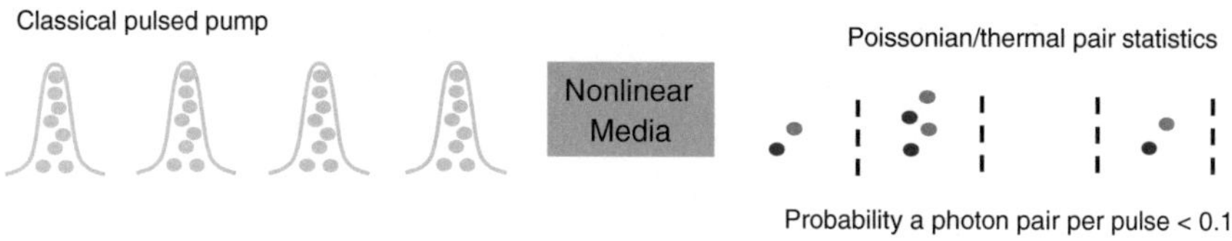

Fig. 12.6 Schematic illustration for the generation of photons using photon pair generation in a nonlinear medium

$$|\psi_{\mathrm{II}}\rangle = \int_0^\infty \mathrm{d}\omega_{\mathrm{s}} \int_0^\infty \mathrm{d}\omega_{\mathrm{i}}\, \phi_{\alpha\beta}(\omega_s, \omega_i)\, \hat{a}^\dagger_{\alpha\omega_{\mathrm{s}}} \hat{a}^\dagger_{\beta\omega_{\mathrm{i}}} |\mathrm{vac}\rangle, \tag{12.14}$$

characterised by the bi-photon wavefunction $\phi_{\alpha\beta}(\omega_{\mathrm{s}}, \omega_{\mathrm{i}})$. Apart from a normalisation condition, the function $\phi_{\alpha\beta}(\omega_{\mathrm{s}}, \omega_{\mathrm{i}})$ is in principle unconstrained, and indicates that the generated bi-photon state may have strong and non-trivial frequency and time correlations between the two photons. The correct quantum description of the heralded single photon is found by "tracing" over the detected photon by a procedure described in Sect. 12.2.5.3.

While the probabilistic nature of these sources is perhaps their greatest impediment, we see in Sect. 12.2.6 that schemes for enabling near-deterministic behaviour are emerging rapidly [43, 44]. We therefore focus on probabilistic sources throughout the rest of this chapter, focusing on theory in the next Sect. 12.2.2, before turning to practice in Sects. 12.2.3 and 12.2.4. We then delve more deeply into the quantum description of the generated states in Sect. 12.2.5.

12.2.2 Photon Pair Sources and Photon Statistics

The two nonlinear processes of SPDC and SFWM are actually quantum analogs of familiar classical nonlinear wave-mixing effects [45, 46]: difference frequency generation and four wave mixing. In the classical processes, photons are transferred in wavelength from one or two pump fields onto a weak input "signal" field and a second "idler" channel, the frequency of which is determined by energy conservation. The spontaneous quantum processes are essentially identical except that there is no input signal field—that role is taken by random quantum fluctuations. It is for this reason that we require a quantum formalism to describe their operation [47].

In SPDC a second-order ($\chi^{(2)}$) nonlinear optical material is used to probabilistically convert an incident pump photon at ω_{p} into a photon at each of ω_{s} and ω_{i} where $\omega_{\mathrm{p}} = \omega_{\mathrm{s}} + \omega_{\mathrm{i}}$. Similarly, in SFWM a third-order ($\chi^{(3)}$) nonlinear material is used to probabilistically convert two incident pump photons at ω_{P} into a photon at each of ω_{s} and ω_{i} where $2\omega_{\mathrm{p}} = \omega_{\mathrm{s}} + \omega_{\mathrm{i}}$ (see Fig. 12.7). However, since there is no pre-existing signal field, in both cases the generated state is actually a superposition

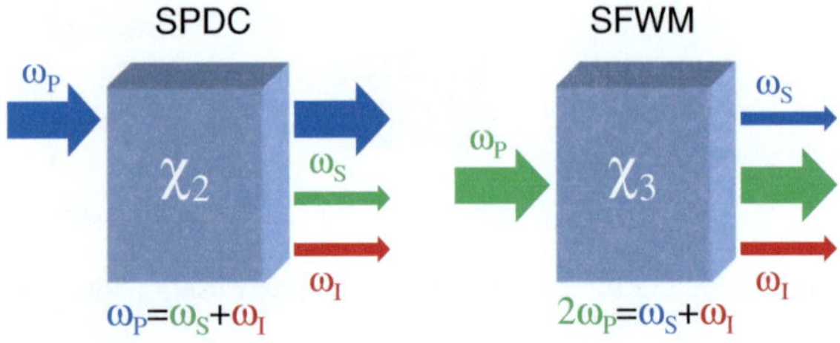

Fig. 12.7 Schematics of the processes of spontaneous parametric downconversion (*SPDC*) and spontaneous four-wave mixing (*SFWM*)

Fig. 12.8 Typical photon pair production probabilities per pump pulse as given by a thermal distribution. The vast majority of events produce no pairs, and the multiple pair events cannot be eliminated

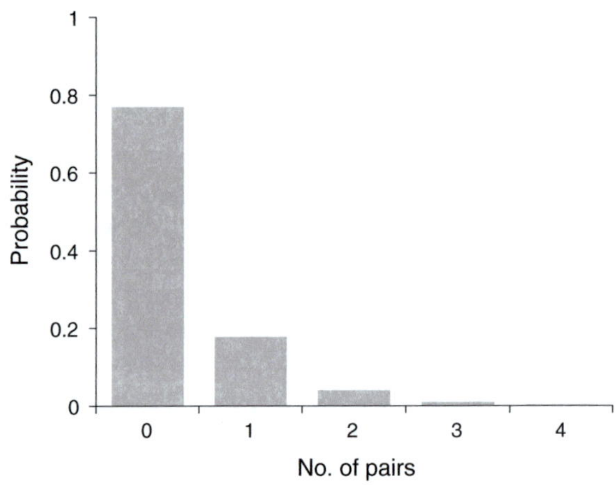

over all frequency pairs ω_s and ω_i which satisfy energy conservation. This leads to frequency correlations between the two photons as we explore in detail in Sect. 12.2.5.1. Note that as with classical four-wave mixing, it is also possible to achieve degenerate photon pair production by using two pump fields satisfying $\omega_{p1} + \omega_{p2} = 2\omega_s$.

Unfortunately a heralded photon source is still far from the ideal of being able to produce a single photon on demand, for a number of reasons we will explore in later sections. Perhaps the most fundamental is that we can't be sure of obtaining a single pair (just as the attenuated laser cannot guarantee a single photon). In fact, the simplest description of the output of such a source is as a two-mode squeezed vacuum. In that approximation, the probability of obtaining n pairs per pump pulse is given by the thermal distribution [48]

$$P_{\mathrm{Thermal}}(n, \mu) = \frac{\mu^n}{(1+\mu)^{n+1}}, \tag{12.15}$$

where μ is now the average number of *pairs* per pulse (see Fig. 12.8). This description assigns only a single frequency mode each to the signal and idler, a deficiency we amend below. However, it gives useful insight into the photon statistics.

From (12.15), it is easily found that the relative likelihood of generating two or more pairs is given by $P_{\text{Thermal}}(>1, \mu)/P_{\text{Thermal}}(1, \mu) = \mu$, and so to keep this rate small, the mean photon flux must be limited to $\mu \ll 1$. As a result, the most likely event for any given pump pulse is for the source to produce no photons at all! Moreover, it is impossible to predict which pump pulses will produce pairs. What then is the advantage over the attenuated laser? The utility of heralded sources is that when photons are produced, they always appear in pairs.

Thus following the separation of signal and idler photons along different spatial paths (or one of the different encoding schemes of Sect. 12.1.1.3), the detection of a single photon serves as a good indication that another is present (subject to absorption and scattering losses etc.). We cannot predict when a single photon from this type of source is coming, but at least we get a warning. In Sect. 12.2.6, we see that this advance notice is a crucial tool for turning an unpredictable spontaneous source into a quasi-deterministic one.

We now turn to the laboratory characterisation of many of the photon statistics concepts including the impact of losses and detector imperfections. Reference [49] also discusses a number of these ideas in detail. The characterisation of sources begins with the photon statistics: the rate of pairs, coincidence statistics and effects of noise. This is the realm of the second order correlation function and the *coincidence-to-accidental ratio* or CAR, with which we begin.

12.2.2.1 Noise in Photon Pair Sources and the Coincidence-to-Accidental Ratio (CAR)

An ideal heralded pair source would produce a steady succession of equally spaced single photon pairs, all identical. As we've seen however, both SPDC and SFWM actually give rise to squeezed vacuum states with a nonzero rate of multiple pairs and characterising this rate is important. Moreover in addition to noise from multi-pair production, we must consider contributions from spurious photons generated by spontaneous Raman scattering (especially in amorphous materials), and imperfections of the measurement system including channel losses, dark counts, after-pulsing and detector efficiency [49].

The standard tool for this kind of characterisation is a coincidence measurement setup, illustrated in Fig. 12.9. The output photons from the pair source under study are filtered to remove residual pump light, and then separated into distinct output channels, typically optical fibres, labelled as signal and idler modes (based on frequency, wavevector or polarisation as the case may be). In practice, this separation may be accomplished with a dichroic mirror, arrayed waveguide grating (AWG) or polarising beam splitter. Each output channel may contain additional filtering elements such as Bragg gratings to remove as much of the residual pump light as possible (a total of 120 dB pump suppression is common for SFWM sources). Finally, each channel is connected to a single photon detector with one

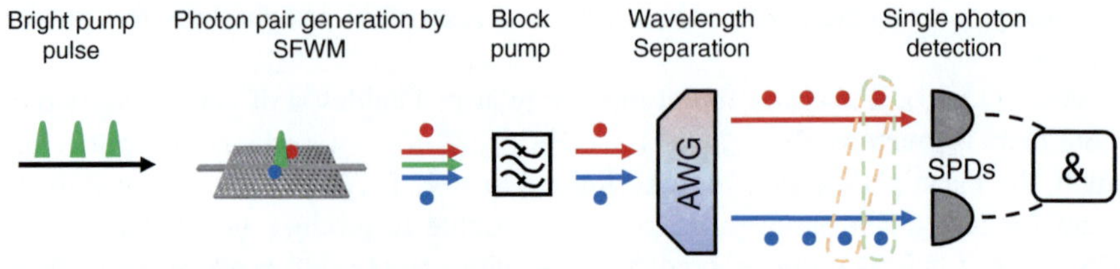

Fig. 12.9 Schematic of a coincidence setup for CAR measurement. The output of a pair source is filtered to remove excess pump light, and then passed through a device to separate signal and idler channels. The two channels are detected separately with single photon detectors connected to a time-interval analyser or other device that identifies detection coincidences within some time window

arm containing a variable time delay. The two detectors are connected to a time-interval analyser which records coincidence events when the two detectors fire within a given interval, usually on the order of a few nanoseconds. The experiment measures the number of coincidence events as the time delay is varied, either continuously or discretely depending on whether a CW or pulsed laser is used.

If there were perfect separation of signal and idler, complete extinction of the pump, negligible loss, no other sources of stray photons, and noiseless detectors of perfect efficiency, one would see both detectors click in coincidence when a pair is present, and an absence of clicks (whether in coincidence or as "singles" firing in only one channel) otherwise. In practice, most of these assumptions are invalid: the pump filtering does not remove every pump photon from the signal and idler bands, particularly for SFWM sources where the pump-signal separation may be only a few THz; scattering or absorption loss in waveguides may be considerable; spontaneous Raman scattering can introduce a significant level of uncorrelated photons, particularly in amorphous materials; out-coupling efficiencies from on-chip devices may be only a few percent; and while single photon detectors are improving rapidly, it is routine to work with efficiencies in the range of $\eta = 5$–$25\ \%$, especially in the telecom band. Finally, there is the fact that the photon statistics guarantee that some events are multiple pairs.

To quantify these imperfections, a natural signal to noise metric is the ratio of the rate of "true" coincidences, due only to the desired single pair events, to the coincidence rate from all other causes: dark counts and after-pulsing, multiple pairs and noise photons. Operationally, this motivates the introduction of the *coincidence to accidental* ratio

$$\mathrm{CAR} = \frac{C_t}{A} \approx \frac{C - A}{A}, \tag{12.16}$$

where C_t is the true rate of coincidences due to single pairs, C is the measured rate of coincidences due to all causes, and A is the expected rate of "accidental" coincidences due to all other causes, such as say the coincidence of a photon generated by spontaneous Raman scattering and a detector dark count. We have no way of identifying the "good" coincidences from the "bad", so the accidentals rate is determined by a statistical trick in which the coincidence rate measurement is repeated with a time delay inserted into one arm. For a pulsed pump, this delay corresponds to one or more multiples of the pump repetition time τ_p. We explain why this works shortly.

Expressions for the CAR can be obtained with different levels of rigor [48, 50]. Suppose for a moment that pair creation events are entirely uncorrelated (a point to which we return in a moment) and occur according to Poisson statistics with a mean rate μ per pump pulse. Ignoring dark counts and assuming $\mu \ll 1$, the coincidences are given by $C_t \approx \mu\eta_s\eta_i$, where η_s, η_i are the total efficiencies including coupling losses etc. of each detector. The rate for accidental coincidences in adjacent pump pulses is $A = N_s N_i$ where the "singles" rate of clicks at each detector individually are $N_s = \mu\eta_s$ and $N_i = \mu\eta_i$. For uncorrelated creation events, this accidentals rate is *precisely* the same as the rate for two pair creation events to occur in the same pulse window. Thus the CAR obeys a simple $1/\mu$ law indicating that multiple pair events are more likely at higher pair rates (see Fig. 12.10). Observe that while there is no particular interest in the coincidences between successive pump pulses per se, in the Poisson photon statistics regime, the accidentals rate is a convenient estimator of the contribution of multiple pairs and other imperfections to the overall coincidence rate, a contribution we cannot observe directly.

In practice, dark counts and other spurious effects such as spontaneous Raman noise must be included, and a modified expression [50] is commonly used:

$$\mathrm{CAR} = \frac{\mu\eta_s\eta_i}{(\mu\eta_s + d_s)(\mu\eta_i + d_i)} + \sigma, \qquad (12.17)$$

where d_s, d_i are the rate of dark counts per pump pulse of each detector, and σ is a correction factor we set as $\sigma = 0$ for now. Dark counts for near infra-red detectors typically range from a few hundred Hz for superconducting devices to a few kHz for InGaAs based avalanche photodiodes. As might be expected, the dark counts suppress the CAR when the effective singles rate $\mu\eta$ approaches the dark count rate from above. Indeed assuming the detectors have identical characteristics, there is a

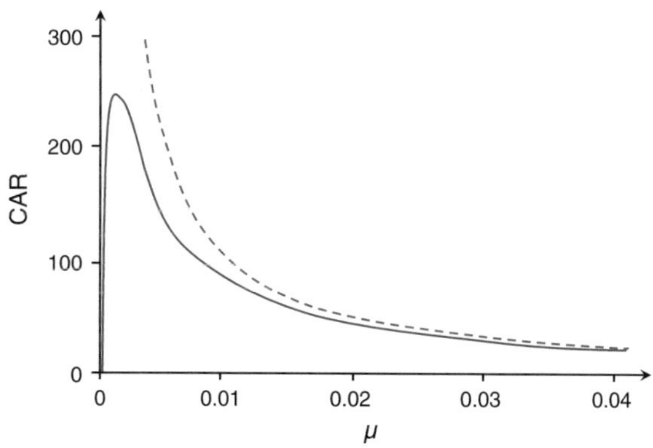

Fig. 12.10 Coincidence-to-accidental ratio CAR according to (12.17) for efficiency $\eta = 0.1$, and dark count rate $d = 10^{-4}$/pulse (*solid blue line*), and the limiting case $1/\mu$ for negligible dark counts (*dashed red*)

maximum CAR of $\eta/(4d)$ at $\mu_{\mathrm{max}} = d/\eta$ (see solid blue line in Fig. 12.10). In practice, researchers work in a range where $\mu \gtrapprox \mu_{\mathrm{max}}$ and the $1/\mu$ fall due to multipair generation (dashed red line in Fig. 12.10) is the dominant effect. It is apparent that detector inefficiency, dark counts and other imperfections play a key role in determining the CAR which is in some ways a strange metric: it is as much a measure of the detectors as the source itself.

The argument above regarding the accidentals relied on the generation of uncorrelated events with Poisson statistics. How does the identification of the accidentals with the rate of uncorrelated events in successive pulses stand up when the photon pair statistics are not Poissonian? Note that correlated pairs can be generated with different numbers of independently correlated frequency modes. With just a single pair of modes, the distribution of the number of generated pairs per pulse is thermal according to (12.15). A more careful analysis of the photon statistics leading to (12.17) has been performed by Takesue and Shimizu [48], and shows that for $\mu \gg \mu_{\mathrm{max}}$ the same equation holds with a correction $\sigma = 2$. On the other hand, with a large number of frequency modes, while the distribution of pairs *within* each mode is thermal, there is no correlation between different modes and as multiple pair events are much more likely to involve two different modes rather than the same mode, the total distribution of pairs approaches a Poisson distribution. In this case, the correction $\sigma = 1$. For useful sources, researchers seek to obtain CARs in the range 100 or larger [51], and the difference in the expected CAR between the two regimes is unimportant, if even measurable.

Note that the question of whether a source is in the regime of a large or small number of frequency modes, is very important, because of the connection to the purity of the heralded photon, but this is best characterised by other measurements: see Sect. 12.2.5.5.

12.2.3 Pair Generation Processes in Second-Order Nonlinear Media

Today, the workhorse process for pair generation is spontaneous parametric downconversion (SPDC) in $\chi^{(2)}$ materials, still usually in bulk optics configurations. In SPDC, one pump photon is annihilated to produce two daughter photons at lower frequencies, (which may or may not be degenerate). As with the corresponding classical process of difference frequency generation, for efficient SPDC energy and momentum must be conserved (or "phase-matched"), so that

$$\omega_{\mathrm{p}} = \omega_{\mathrm{s}} + \omega_{\mathrm{i}},$$

$$\mathbf{k}_{\mathrm{p}} = \mathbf{k}_{\mathrm{s}} + \mathbf{k}_{\mathrm{i}}. \tag{12.18}$$

Here $\omega_{p,s,i}$ and $\mathbf{k}_{p,s,i}$ are the frequencies and wave vectors of the pump, signal and idler photons, respectively. As usual, $|\mathbf{k}| = n(\omega)\omega/c$, where $n(\omega)$ is the refractive index of the nonlinear medium at frequency ω and c is the speed of light in vacuum. Such equations are familiar from the classical processes of difference and sum frequency generation. In fact, as the nonlinear medium has finite length L and the pump has nonzero bandwidth $\delta\omega_p$, these conditions are more accurately expressed as inequalities of the form

$$|\Delta\omega| \equiv |\omega_p - \omega_s - \omega_i| \lesssim \delta\omega_p,$$

$$|\Delta k| \equiv |\mathbf{k}_p - \mathbf{k}_s - \mathbf{k}_i| \lesssim \frac{\pi}{L}. \tag{12.19}$$

Equation (12.19) show that the bandwidth of generated photons is determined by the pump bandwidth and chromatic dispersion.

As the refractive index at ω_p is usually much higher than those at ω_s and ω_i, and different values of chromatic and group velocity dispersion can be found at these frequencies, it is non-trivial to achieve the required phase-matching and the non-linear devices and experimental apparatus must be carefully designed to do so. Depending on the design, the two generated photons can be emitted either collinearly or at separated angles to the pump beam, and be either at different frequencies or degenerate. With non-collinear generation in bulk crystals, the formal representation of the state generalising (12.14) has the rather complex form

$$|\psi_{II}\rangle = \int_0^\infty d\omega_s \int_0^\infty d\omega_i \int d^2\mathbf{k}_{t,s} \int d^2\mathbf{k}_{t,i} \Phi(\omega_s, \omega_i, \mathbf{k}_{t,s}\mathbf{k}_{t,i})\, \hat{a}^\dagger_{\omega_s \mathbf{k}_{t,s}} \hat{a}^\dagger_{\omega_i \mathbf{k}_{t,i}} |\text{vac}\rangle, \tag{12.20}$$

where the bi-photon correlations may depend on the transverse wavevectors of the two photons as well as their frequencies [52]. In our later theoretical discussion, we restrict to the notationally simpler case of collinear emission.

Photon pairs generated through SPDC have found numerous applications in quantum photonic information processing such as entanglement generation [53], quantum logic gates [14], quantum walks [54] and quantum simulation [55]. In the following, we present schemes for their production in both bulk nonlinear crystals and waveguide structures.

12.2.3.1 SPDC in Bulk Crystals

SPDC in bulk nonlinear crystals is the most often used approach to generate single photons or entangled photons for quantum information processing. In this section we briefly describe SPDC in bulk crystals such as beta-barium borate (BBO) crystals. We discuss the phase matching angle, emission in cones and the different types of SPDC. Reference [52] provides a recent and very detailed review of this topic.

BBO is a negative uniaxial crystal, i.e., it has only one optic axis and $n_o > n_e$, where n_o and n_e are the ordinary and extraordinary refractive indices respectively. Based on this, there are two possible types of phase matching: type-I (e $\rightarrow$ o + o) and type-II (e $\rightarrow$ o + e). In type-I SPDC, the two emitted photons have the same polarisation which is orthogonal to that of the pump; in type-II SPDC, one of the emitted photons has the same polarisation as the pump.

A typical type-I SPDC process in a BBO crystal is schematically shown in Fig. 12.11a. The incident pump beam is generally normal or slightly tilted to the surface of the crystal, and the generated signal and idler photons are emitted along cones centered on the pump. To meet the phase-matching condition as shown in Fig. 12.11b, the photons of each pair must lie on opposite sides of the pump beam. Through the cut of the crystal surface, the angle θ_{pm} between the pump beam and the crystal optic axis can be varied to control the phase-matching condition. By adjusting θ_{pm}, the photons may be chosen to be degenerate and thus are emitted on a single cone, as indicated by the squares in Fig. 12.11a. Alternatively they can have different frequencies in which case they are emitted along two different cones, as shown by the circles and triangles in Fig. 12.11a. The precise geometry of emission is beyond the scope of this chapter and the details can be found in [56].

When θ_{pm} is adjusted properly, one can also have type-II phase matched SPDC processes, in which the down-converted photons are emitted into two cones, one ordinarily polarised, the other extraordinarily polarised. By adjusting the phase matching angle, the relative orientation of the two cones can be adjusted. If they are

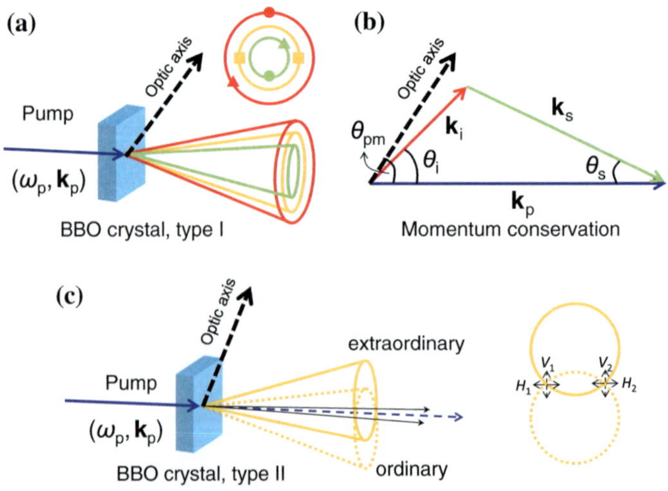

Fig. 12.11 The diagram for SPDC processes in BBO. **a** Type-I SPDC, **b** momentum conservation for phase matching, and **c** type-II SPDC for polarisation entanglement generation

made to overlap [52], (Fig. 12.11c) the photons emitted along the intersection directions can be described in the simplified picture of Sect. 12.1 by the entangled state [53]:

$$|\Psi\rangle = (|H_s V_i\rangle + e^{i\alpha}|V_s H_i\rangle)/\sqrt{2}, \tag{12.21}$$

where H and V indicate horizontal (extraordinary) and vertical (ordinary) polarisation, respectively, and the subscripts denote the signal and idler directions. The relative phase α derives from the crystal birefringence. Using an additional birefringent phase shifter, α can be set as desired, for example, to the values of 0 or π to produce standard Bell states.

In experiments, the entanglement generation is complicated by the walk-off between the ordinary and extraordinary photons since they have different velocities inside the birefringent crystal. Two crystals identical to the down-conversion crystal but only half as long, may be used to compensate the walk-off and remove the temporal distinguishability of the ordinary and extraordinary photons [53].

Other nonlinear crystals that are often used to generate heralded single photons and entangled photons are BiBO (BiB_3O_6) and potassium titanyl phosphate ($KTiOPO_4$ or KTP). Compared with BBO, BiBO exhibits two times higher nonlinearity and has two optic axes. As a biaxial crystal, phase matching in BiBO becomes more complicated because it depends on the azimuthal pump angle in addition to the polar pump angle. However such increased complexity offers more flexibility to tune the phase-matching condition and thus allows broadband angle tuning [57, 58]. The exceptionally high nonlinearity and versatile phase matching make BiBO very attractive for multi-photon entanglement generation. In contrast to BBO and BiBO, which can only be used as bulk crystals, KTP is usually periodically poled to provide quasi-phase-matching for efficient SPDC [59, 60] and waveguides can be formed in this material through titanium indiffusion or soft proton exchange. The details will be covered in next section.

12.2.3.2 SPDC in Integrated Structures

The bulk nonlinear crystals just described have played a very important role in generating heralded single photons and entangled photons for quantum experiments. However there are intrinsic limitations that constrain the performance of the sources based on bulk crystals. For example, the length of crystals is usually limited to a few mm or sub-mm to avoid severe walk-off between generated photons [53, 57, 58]. The crystal length and damage threshold set the bottleneck of photon source brightness. The other limiting factor is the need to use many free-space optical components, which are environmentally sensitive and scale poorly for more complicated quantum operations. To overcome these hurdles, integrated waveguide devices based on both second-order and third-order nonlinear materials for heralded single photon and entangled photon generation have attracted intensive research.

In waveguides, where all fields propagate collinearly, new strategies for achieving phase-matching are required. These may include designing waveguides that support highly distinct mode-profiles at different wavelengths as has been done with Bragg reflection waveguides in AlGaAs structures [61–63]. Another elegant approach is to use ring resonator structures [64] in which the variation of the orientation of the crystal axes and field polarisation guarantees phase-matching. However, the most common approach is to adopt quasi-phase matching (QPM) [65] in periodically poled second-order nonlinear waveguides such as periodically poled KTP (PPKTP) [59, 60], lithium niobate (PPLN) [31, 66, 67] and lithium tantalate (PPLT) [68].

QPM operates essentially identically for spontaneous processes as for classical frequency conversion. Figure 12.12, shows a homogeneous single crystal (a) and a periodically poled waveguide (b), and the evolution (c) of the field amplitude of the generated wave along the propagation direction z for three different phase-matching conditions. It can be seen from Fig. 12.12c that in the case of perfect phase-matching, the amplitude of the generated wave grows linearly with propagation distance; while in the situation of no phase-matching, the amplitude oscillates with distance. If the poling period Λ is set equal to twice the coherence length L_{coh} of the nonlinear interaction, each time the field amplitude of the generated wave is about to begin to decrease as a consequence of the phase-mismatch, a reversal of the sign of the nonlinear coefficient occurs. This allows the field amplitude to continue to

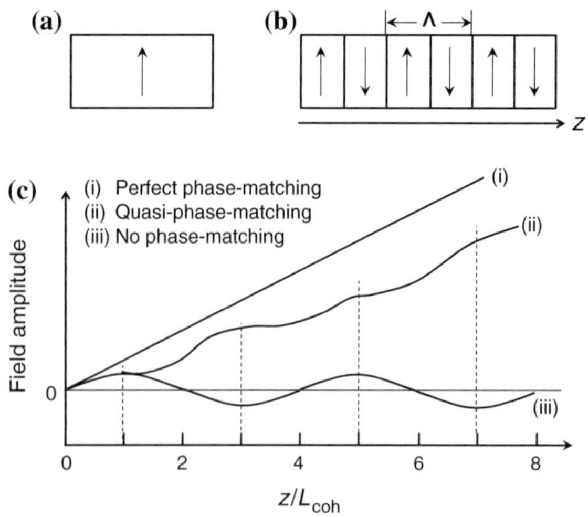

Fig. 12.12 Schematic of a second-order nonlinear optical material in the form of **a** a homogeneous single crystal and **b** a periodically poled waveguide in which the axis of the nonlinear material alternates in orientation with period Λ so that its nonlinear coefficient changes sign periodically. **c** Comparison of the evolution of the field amplitude of the generated wave along the propagation direction z for three different phase matching conditions: (*i*) perfect phase-matching, (*ii*) quasi-phase-matching and (*iii*) no phase-matching. L_{coh} indicates the coherent buildup length [65]

grow monotonically and is equivalent to adding an additional term $2\pi/\Lambda$ to (12.18) to compensate the phase-mismatch. In practice, the smallest poling period is limited to a few μm due to fabrication constraints and this determines the maximum phase-mismatch that can be compensated.

Compared with the perfect phase-matching via birefringence in bulk crystals, QPM leads to a lower SPDC efficiency per unit device length if the nonlinear coefficient is the same: the effective nonlinearity is reduced by a factor of $2/\pi$ [65]. However, QPM makes it possible to use the same polarisation direction for all interacting waves, and this often allows a larger element of the nonlinear tensor to be accessed. Furthermore, QPM allows SPDC in waveguide structures which confine all the interacting waves in the same spatial waveguide mode and thus increases their nonlinear coupling. As a result, the SPDC efficiency in poled waveguides can be significantly higher than that in bulk crystals. For example, the SPDC efficiency in a PPKTP waveguide is more than 50 times higher than that in a bulk crystal [59, 60].

Moreover, a poled waveguide can be more than ten times longer than a bulk crystal without decreasing the SPDC efficiency, further enhancing the source brightness. QPM in poled waveguides makes phase-matching so flexible that it allows not only SPDC through three-wave mixing [31, 67], but also four-wave mixing type SPDC via cascaded second-order nonlinear processes [66, 68]. This makes it possible to generate photons in the telecom-band using readily available telecom-band pump lasers. Finally, variation in the poling period can be introduced to engineer the precise bi-photon state emitted from the structure. This can be used to improve the purity of the resulting heralded single photon [69].

12.2.3.3 Cavity Enhanced SPDC

Since the bandwidth is determined by the chromatic dispersion, as was previously mentioned, photons generated via SPDC in either bulk crystals or periodically poled waveguides usually have a spectral bandwidth on the order of THz. This is significantly beyond the working bandwidth of Hong-Ou-Mandel quantum interference experiments [18] and available quantum memories (MHz to GHz) [70], both of which are at the heart of schemes for linear optical quantum computation [17]. The use of SPDC sources in these applications requires narrow-band frequency filters. In particular, to store SPDC photons directly in atomic memories, the required filters will be extremely narrow and introduce large losses [71]. To solve this problem, cavity enhanced SPDC has been proposed and demonstrated [70, 72, 73].

The idea is illustrated in Fig. 12.13. A second-order nonlinear device, either a bulk crystal or periodically poled waveguide, is placed in a Fabry-Perot cavity. The cavity is designed such that both input and output mirrors transmit the pump, while in the down-conversion band, the input mirror is highly reflective, but the output mirror is only partially reflective. The bandwidth of the down-converted photons is determined by $\Delta\nu = \frac{FSR}{\mathscr{F}}$, where FSR and $\mathscr{F}$ are the free spectral range and finesse of the cavity, respectively.

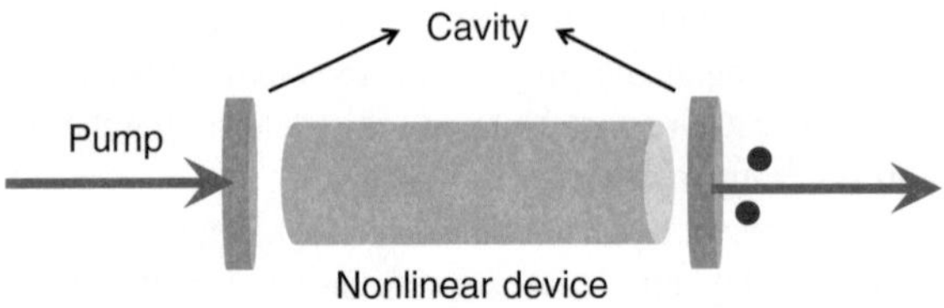

Fig. 12.13 A second-order nonlinear device is placed in a Fabry-Perot cavity for narrow-band SPDC. Both input and output mirrors of the cavity transmit the pump, while the input and output mirrors highly and partially reflect the down-converted photons, respectively

12.2.4 Pair Generation Processes in Third-Order Nonlinear Media

We now turn to the corresponding third-order ($\chi^{(3)}$) nonlinear process, spontaneous four-wave mixing (SFWM), and its applications in quantum photonics. The third order nonlinearity is far weaker than the second order one, but is attractive because it is present in all platforms including fibre, semiconductor waveguides and planar glass technologies. In contrast to SPDC in which only three photons are involved, SFWM is a four-photon interaction: two pump photons are annihilated to produce a signal photon and an idler photon at new frequencies. The signal and idler photons generated in this way are correlated and can be used for entanglement generation and heralded single-photon sources in the same fashion as pairs from SPDC. Efficient SFWM requires energy conservation and phase-matching according to [46]:

$$\omega_{\mathrm{p}_1} + \omega_{\mathrm{p}_2} = \omega_{\mathrm{s}} + \omega_{\mathrm{i}}, \tag{12.22a}$$

$$\mathbf{k}_{\mathrm{p}_1} + \mathbf{k}_{\mathrm{p}_2} = \mathbf{k}_{\mathrm{s}} + \mathbf{k}_{\mathrm{i}} + \Delta k_{NL}. \tag{12.22b}$$

Here $\omega_{\mathrm{p}_1,\mathrm{p}_2,\mathrm{s,i}}$ and $\mathbf{k}_{\mathrm{p}_1,\mathrm{p}_2,\mathrm{s,i}}$ are the frequencies and wave vectors of pump, signal and idler photons, respectively; Δk_{NL} is the nonlinear phase shift introduced by self-phase modulation and cross-phase modulation. Like (12.18), these conditions are more accurately represented by inequalities of the form of (12.19). Depending on the polarisation states of the four waves and whether the two pump waves have the same frequency, the form of Δk_{NL} varies. When $\omega_{\mathrm{p}_1} \neq \omega_{\mathrm{p}_2}$ and all four waves are linearly co-polarised, $\Delta k_{NL} = \gamma(P_1 + P_2)$, where P_1 and P_2 are the pump power, $\gamma = n_2\omega/(cA_{\mathrm{eff}})$ is the nonlinear coefficient, n_2 is the nonlinear refractive index at frequency ω and A_{eff} is the effective mode area of pump waves in a nonlinear waveguide. When $\omega_{\mathrm{p}_1} \neq \omega_{\mathrm{p}_2}$ and two pumps are linearly co-polarised but signal and idler photons are orthogonally polarised with respect to pumps, $\Delta k_{NL} = -\frac{5}{3}\gamma(P_1 + P_2)$. In the case of $\omega_{\mathrm{p}_1} = \omega_{\mathrm{p}_2}$ and pump power of P, $\Delta k_{NL} = 2\gamma P$ when all waves are linearly co-polarised, and $\Delta k_{NL} = -\frac{2}{3}\gamma P$ when the pump is linearly polarised while signal and idler photons are polarised orthogonally to the pump polarisation.

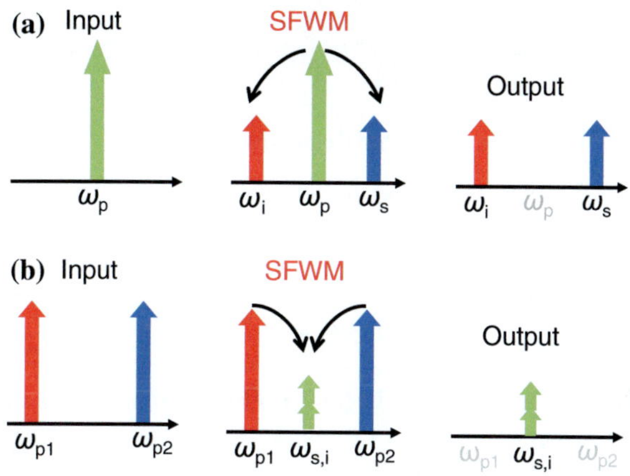

Fig. 12.14 Schematic of pump frequency **a** degenerate and **b** non-degenerate SFWM processes

The case of $\omega_{\mathrm{p}_1} = \omega_{\mathrm{p}_2}$ is usually referred to as pump frequency degenerate SFWM, and the case of $\omega_{\mathrm{p}_1} \neq \omega_{\mathrm{p}_2}$ as non-degenerate SFWM. Both are schematically shown in Fig. 12.14. In the pump degenerate SFWM process shown in Fig. 12.14a, the signal and idler photons appear at higher and lower frequency than the pump, respectively. For non-degenerate pumps, as shown in Fig. 12.14b, the signal and idler photons can be generated at the same frequency between two pump frequencies. In this case the two pump frequencies are usually very close, so the frequency dependency of γ can be ignored [46]. In the following sections, we discuss the implementation of these two types of SFWM in optical fibres, integrated waveguides and resonant structures.

12.2.4.1 SFWM in Optical Fibres

Here we describe previous single or entangled photon generation experiments using SFWM in standard optical fibres and photonic crystal fibres (PCFs). Most work has been done with pump degenerate SFWM, so we focus on this type of SFWM and then briefly introduce the demonstration of non-degenerate pump SFWM. For a recent review of pair generation in fibres, see [74].

As mentioned above, for a single pump we have $\omega_{\mathrm{p}_1} = \omega_{\mathrm{p}_2}$, and (12.22a, 12.22b) become

$$2\omega_{\mathrm{p}} = \omega_{\mathrm{s}} + \omega_{\mathrm{i}}, \quad 2k_{\mathrm{p}} = k_{\mathrm{s}} + k_{\mathrm{i}} + B\gamma P. \tag{12.23}$$

Here $B = 2$ when all waves are linearly co-polarised while $B = -\frac{2}{3}$ when the polarisation of signal and idler photons is orthogonal to that of the linearly polarised pump. For the case of all fields co-polarised, expanding $k_{\mathrm{p,s,i}}$ in a Taylor series about the pump frequency ω_{p} and keeping the first few terms, the phase-matching equation becomes

$$\beta_2\Omega^2 + \frac{2}{4!}\beta_4\Omega^4 + \frac{2}{6!}\beta_6\Omega^6 + B\gamma P = 0, \tag{12.24}$$

where $\beta_n = \left(\frac{d^n k}{d\omega^n}\right)_{\omega_p}$ is the nth-order dispersion parameter of an optical fibre and $\Omega = \omega_s - \omega_p = \omega_p - \omega_i$. When the signal and idler are polarised orthogonally to the pump $\Delta\beta$ includes two contributions: phase mismatch $\delta\beta_D$ induced by dispersion and phase mismatch $\delta\beta_B$ induced by modal birefringence such that

$$\delta\beta_D + \delta\beta_B + B\gamma P = 0 \tag{12.25}$$

It can be seen that phase-matching in optical fibres is determined by the chromatic and polarisation dispersion of the fibre mode. Using (12.23), (12.24) and (12.25), phase-matching curves calculated for a weakly birefringent PCF are shown in Fig. 12.15 [75]. There are several features in this plot: (1) phase-matching can be achieved when the generated photons are either co-polarised or cross-polarised to the pump photons; (2) for the co-polarisation phase-matching (ff→ff or ss→ss), the wavelength of generated photons is very close to that of pump photons when pumping in the anomalous dispersion regime, while the wavelength of generated photons is well separated from that of pump photons when pumping in the normal dispersion regime; (3) for the cross-polarisation phase-matching (ss→ff or ff→ss), there is a zero slope region of the curves, which means the bandwidth of generated photons is independent of the pump bandwidth; (4) the phase-matching for ss→ff only occurs in the normal dispersion regime, while for ff→ss it occurs in both normal and anomalous dispersion regimes and the zero slope region is in the anomalous dispersion regime. These features have found many applications in single and entangled photon generation.

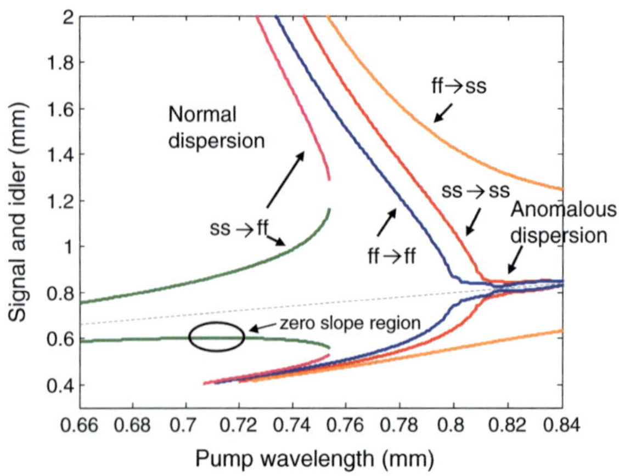

Fig. 12.15 Calculated co-polarisation and cross-polarisation phase-matching curves for a weakly birefringent PCF. ss→ss and ff→ff mean that both the signal and idler photons are in the same slow/fast polarisation mode as the pump. ss→ff and ff→ss mean that both the signal and idler photons are in the polarisation orthogonal to the pump [75]

Initial attempts at photon generation in optical fibres concentrated on phase-matching in the anomalous dispersion regime because fibre design and fabrication technology was not available to introduce enough higher-order dispersion β_4 and β_6 to compensate the phase-mismatching induced by the positive β_2 term for normal dispersion [76–79]. As can be seen from Fig. 12.15, the wavelengths of the generated photons are very close to that of the pump photons. Since silica has a broad Raman gain spectrum and many uncorrelated photons close to the pump wavelength are also generated from spontaneous Raman scattering (SpRS), this introduces a high level of noise to the photon sources. To reduce the SpRS noise, fibres have to be cooled using liquid nitrogen or helium [80, 81], which requires cryostatic conditions and is undesirable for many applications.

Breakthroughs in this regard were made as PCF fabrication technology matured. Compared with standard step-index fibres, PCFs have periodically arranged air holes in the cladding, which significantly increase the index difference between the fibre core and cladding and thus introduce relatively large higher-order dispersion over a broad wavelength range. This makes phase-matching possible when pumping in the normal dispersion regime and many demonstrations have shown low-noise single or entangled photon generation at room temperature in this regime [38, 50, 82–85]. In these demonstrations, the frequencies of generated photons are typically over 70 THz away from that of pump. This is far beyond the Raman gain band of silica and therefore the generated photons are uncontaminated by the SpRS noise. The structure of a PCF is so flexible that a weak birefringence can be introduced by making two holes symmetrically around the fibre core slightly smaller than the other holes. This is particularly interesting because cross-polarisation phase-matching can be achieved in such a fibre, as shown in Fig. 12.15, and the generated photons are naturally narrow-band without the use of narrow-band filters [84, 85]. This significantly reduces the loss introduced by narrow-band filters and thus greatly increases the source brightness. More importantly the heralded photons are in a pure single-photon quantum state (see Sect. 12.2.5.3).

Hong-Ou-Mandel quantum interference of indistinguishable photons lies at the heart of linear optical quantum computation [14, 17, 18]. In the pump frequency degenerate SFWM process, the signal and idler photons appear at different wavelengths and are therefore distinguishable. To perform Hong-Ou-Mandel quantum interference, two heralded single-photon sources based on such a SFWM process are required [83–85]. This involves four-photon coincidence events, which are significantly less likely to happen and so experiments are much longer and more challenging than those involving two-photon coincident events. There have been previous demonstrations of pump frequency non-degenerate SFWM to generate photons at the mean frequency of two pump frequencies in optical fibres [86, 87]. Since pulsed pumps are required for efficient SFWM in optical fibres, when the two pulsed pumps are used, they must be synchronised temporally so that they can interact. Fan et al. used a pump frequency degenerate four-wave mixing to generate the two pump pulses [86], while Chen et al. spectrally sliced a femtosecond laser to form the two pump pulses [87].

12.2.4.2 SFWM in Integrated Waveguides

Optical fibre based photon sources are naturally compatible with current fibre communication systems and therefore they are very useful for implementing quantum key distribution for secure communication. On the other hand, photon sources based on photonic chips are desired for quantum computation because scalability and miniaturisation are essential for this particular application. We next describe single or entangled photon generation via SFWM in direct write silica waveguides [42], chalcogenide ridge waveguides and nanowires [88–90], and silicon nanowires [91–95].

The integrated platform most similar to optical fibre is the femtosecond laser direct write waveguide. The approach to generating heralded single photons in direct write waveguides is similar to that in optical fibres described in Sect. 12.2.4.1 as the waveguiding properties are rather similar. A 4 cm long waveguide was fabricated by femtosecond laser writing in an undoped silica chip [42]. An elliptical transverse mode was produced to introduce a birefringence so that cross-polarisation phase-matching similar to that in Fig. 12.15 was achieved in the zero slope region. The single-photon purity (see Sect. 12.2.5.2) was measured to be 0.86 from the source number statistics without narrow spectral filtering [42]. Although the intrinsic nonlinearity of silica is very low, the low-loss feature of silica waveguides and the ease of coupling generated photons to silica fibres make this platform promising.

Compared with silica, chalcogenide glasses have up to 1000 times higher nonlinear refractive index and can be made into compact waveguides [96]. In particular the waveguide dispersion can be engineered so that the waveguides exhibit anomalous dispersion in the telecommunication 1550 nm band and these glasses exhibit very low values of nonlinear losses such as two-photon absorption (TPA) and free-carrier absorption (FCA). These features have found many applications in nonlinear optical signal processing [96]. In 2011, chalcogenide As_2S_3 ridge waveguides were used for quantum-correlated photon pair generation [88]. Similar to the phase-matching in the anomalous dispersion regime of silica fibres, very high SpRS noise was observed in this As_2S_3-based photon source. Fortunately As_2S_3 glass has a low-Raman gain window at 7.4 THz detuning and this was subsequently used for low-Raman noise photon generation [89]. Another chalcogenide glass, namely $Ge_{11.5}As_{24}Se_{64.5}$, exhibits more than 10 times higher nonlinearity than As_2S_3 and can be made into more compact photonic nanowires rather than ridge waveguides [90]. The $Ge_{11.5}As_{24}Se_{64.5}$ glass also has a low-Raman gain window, but at 5.1 THz, which has allowed low Raman-noise photon generation [90]. Due to the large index contrast between chalcogenide glasses (n = 2.4) and silica (n = 1.5), the coupling between chalcogenide waveguides and silica fibres remains a challenge.

Silicon nano-photonic waveguides are the most widely used platform for single and entangled photon generation via SFWM because of the mature silicon nanofabrication technology and unique optical properties of silicon [91–95]. There are two forms of silicon: amorphous and crystalline silicon, both exhibiting as high a nonlinearity as the chalcogenide glasses. Amorphous silicon has a broadband Raman gain spectrum and shows very low TPA and FCA losses. Crystalline silicon has a very narrow Raman gain band (<1 THz) with its peak at a frequency detuning of more than 15 THz. Therefore photon sources based on crystalline silicon [91–94] have a much higher quantum signal-to-noise ratio than those based on amorphous silicon [95]. In the following sections, we focus on crystalline silicon unless stated otherwise.

Silicon can be made into nanowires [91–94] and resonant structures including micro-rings [93, 97], photonic crystal waveguides [98], coupled resonator optical waveguides (CROWs) [41, 99], and micro-disks [100]. We address silicon nanowires in this section and silicon resonant structures in next section. A typical silicon nanowire's cross section is 450 nm wide and 220 nm high. Such a small area results in an extremely high nonlinearity and anomalous dispersion in the telecommunication 1550 nm band. This enables efficient SFWM for photon generation in waveguide devices of only a few mm [91–94], which is much smaller than the platforms based on other materials such as silica [42]. More importantly, silicon fabrication techniques allow the addition of inverse tapers at both input and output ends of a waveguide for fibre pigtailing, which significantly improve the coupling efficiency and stability [94].

12.2.4.3 SFWM in Resonant Structures

To further reduce the size of a nonlinear device while maintaining the SFWM efficiency without increasing the pump intensity, resonant structures are the natural solution. Here we discuss the use of resonant structures, such as micro-rings [93, 97], photonic crystal waveguides [98, 101, 102], CROWs [41, 99] and micro-disks [100] to enhance SFWM.

In a micro-ring or a micro-disk structure photons can travel along a closed loop for many orbits and thus the nonlinear interaction between photons and the device can be enhanced. This requires that all three photon frequencies satisfy the resonance condition $n_i L = m\lambda_i$, where n_i is the effective mode index of the micro-ring or micro-disk at vacuum wavelength λ_i, L is the circumference, c is the speed of light in vacuum, and m is a different positive integer for pump, signal and idler.

A typical micro-ring or micro-disk structure is made from silicon and shown in Fig. 12.16. The structure has a ring or disk and a bus waveguide. In a micro-ring structure, the ring can be either a perfect circle as shown in Fig. 12.16a [97] or a racetrack cavity [93]. Both the ring and bus waveguides are dispersion-engineered silicon nanowires. The gap between the bus and ring waveguides can be optimised so that maximum pump power is stored in the ring enhancing the nonlinear interaction. This optimised gap is called the critical-coupling distance. In summary, the nanowire dimension determines the basic SFWM properties such as phase-matching bandwidth, and the ring circumference together with the ring-bus gap determines the SFWM enhancement. Photons generated in these devices can have high signal-to-noise values, showing CAR values over 600 [51], albeit after sweeping out the free carriers generated through two-photon absorption.

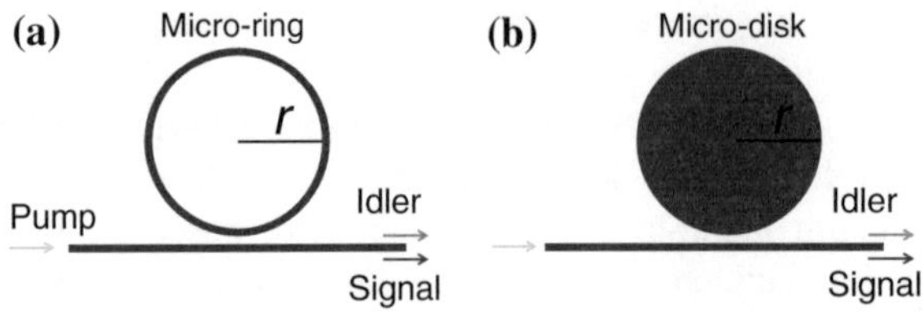

Fig. 12.16 A typical micro-ring (**a**) and micro-disk (**b**) structure for photon generation via enhanced SFWM. Three frequencies: idler, pump, signal are all resonant in the structure, and are equally spaced by the FSR

Similar to a micro-ring, a micro-disk can resonantly enhance the SFWM efficiency by the resonant effect given optimisation of the disk circumference and disk-bus waveguide gap. However a micro-disk differs from a micro-ring in several aspects. Unlike a silicon ring that sits on a silica substrate, a silicon disk sits on a silica pedestal. This means that the cladding is air, so that the index contrast is fixed and significantly larger than that between silicon and silica. Therefore the dispersion is determined largely by the disk thickness. For a disk thickness of 260 nm, the zero-dispersion wavelength is shifted to the telecommunication band around 1.55 μm [100]. To access a micro-disk, a tapered optical fibre is used as the bus waveguide to couple light in and out through evanescent coupling [100]. Since both ends of a tapered fibre are just single mode fibre, the input and output coupling efficiency can be very high, resulting in unprecedented spectral brightness [100]. Taking advantage of the resonantly enhanced SFWM, the pump power required for photon generation in micro-rings and disks can be as low as tens of μW [97, 100], which is five orders of magnitude lower than that required for silicon nanowires [91, 92]. Another feature of micro-rings and disks is that the resonance bandwidth is extremely narrow, so the resulting heralded photons can be obtained in a spectrally pure quantum state without the use of narrow-band filters. As with ring resonators, these devices can show very high CAR values, accentuated by the low-loss tapered coupling technique, reaching values over 1000 [100].

Compared with silicon micro-rings and disks, a silicon photonic crystal waveguide (PhCW) lacks a cavity, but nevertheless is still able to increase interaction times compared to straight waveguides. A PhCW has a triangular lattice of air holes etched in a suspended silicon membrane with a row of holes missing along the ΓK direction, as shown in Fig. 12.17a [98]. The row without holes acts as the waveguide core and the air-hole lattice as the cladding. The photonic crystal cladding forms a photonic bandgap for guiding light. Near the bandgap edge, the dispersion can be very large and the group index n_g can be much greater than the native material refractive index n_0.

This "slow-light effect" can be characterised by the slow-down factor $S = n_g/n_0$. For efficient SFWM, the group index n_g should be nearly constant across a broad bandwidth. This is realised by laterally shifting the first row of holes on both sides of the waveguide core away from the core by tens of nanometres [103]. By adjusting the lattice period, hole diameter and the shift of the first row of

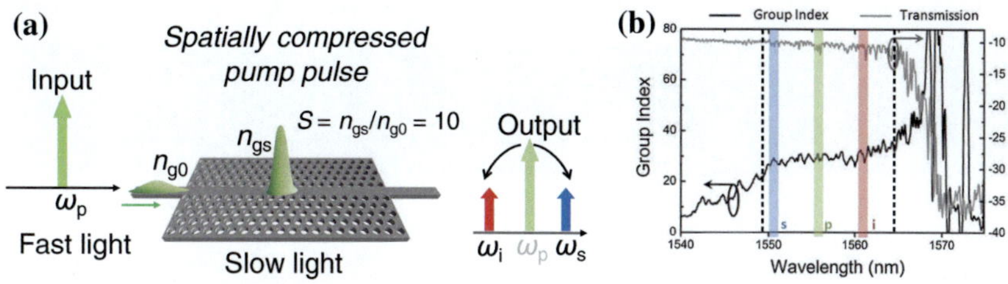

Fig. 12.17 **a** Schematic of SFWM in a silicon slow-light PhC waveguide. **b** Group index and total transmission of light in a 96 μm long silicon PhC waveguide. The window (*between dotted lines*) with a flat group index of 30 and slightly increased loss defines the slow-light regime. The pump, signal, and idler bands are represented by *green* (*middle*), *blue* (*left*), and *red* (*right*) lines, respectively

holes, both n_g and the slow-light bandwidth can be varied [103, 104]. Figure 12.17b shows the measured group index and transmission for a 96 μm long silicon PhCW with a lattice period of 404 nm, hole radius of 115 nm and first row shift of 50 nm [98]. The wavelength window (between dotted lines) with a flat group index of around 30 and slightly increased loss defines the slow-light regime. The low loss is achieved by silicon access waveguides which incorporate inverse tapers terminated by wide polymer waveguides at both the input and output of the PhCW region.

Photon generation through SFWM in such a silicon PhCW structure with $n_g = 30$ is illustrated in Fig. 12.17a. When a coherent pulse of light enters the PhCW region, the pump travels at a group velocity of $c/30$ and is strongly spatially compressed by the slow-light effect. Compared with silicon nanowires, this effectively enhances the nonlinear interaction per unit length per unit pump power, γ, approximately two orders of magnitude, and in turn the SFWM efficiency, which is proportional to γ^2, by four orders of magnitude [105]. After accounting for changes in losses and mode area, this enhancement enabled single photon generation in a 100 times shorter nonlinear device than silicon nanowires operated at similar pump power [98], which now often achieve a CAR in excess of 200. It is well known that nonlinear losses such as two-photon absorption (TPA) and TPA-induced free-carrier absorption (FCA) occurs at around 1550 nm in silicon when the optical power reaches a certain level. These effects are also enhanced by slow light in a silicon PhCW and have been observed as a limiting factor for photon generation when operating at high power [102]. However, if the PhCW is made from a larger bandgap material such as GaInP, the nonlinear losses in photon generation can be eliminated [101]. The challenge with GaInP is to further improve fabrication technology to reduce waveguide propagation losses and therein improve the CAR from the previously reported value of 30.

Finally we consider photon pair generation in coupled resonator optical waveguides (CROWs), which consist of a sequence of coupled resonators with high quality factors [106]. Each individual resonator forms a unit cell of the CROW and can be a micro-ring [41], a micro-disk or a photonic crystal cavity (PhCC) [99]. Figure 12.18 shows an image of a micro-ring CROW and a schematic of a typical PhCC CROW.

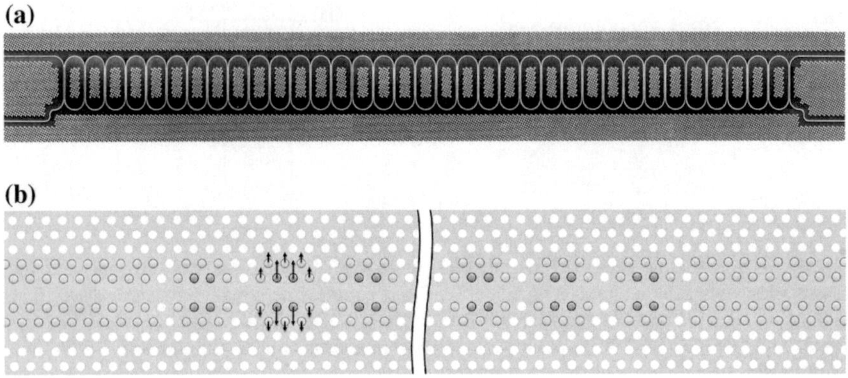

Fig. 12.18 **a** Micro-ring CROW [41], and **b** PhCC CROW [99]. The micro-ring CROW is accessed by a silicon bus waveguide, while the PhCC CROW is accessed by a standard PhCW

Light propagates through a micro-ring CROW via evanescent-field coupling between adjacent rings, and in a PhCC CROW through evanescent Bloch waves [106]. In both cases the separation between two adjacent resonators is sufficiently large that the resonators are weakly coupled. Consequently, the eigenmode of the electromagnetic field in such a CROW structure is a linear combination of the high-Q mode of the individual resonators and forms a collectively resonant supermode.

Compared with a single micro-ring, a micro-ring CROW exhibits much larger group index and broader resonant bandwidth. In a heralded single photon generation experiment, a CROW consisting of N rings can offer an N^2 enhancement in the generation rate [41, 107]. At the same time, the wider passbands have the benefits of ease of experimental alignment and increased thermal stability. Compared with a single PhCC, a PhCC CROW shows a broadband transmission rather than multiple narrow band resonant peaks [99]. The transmission band behaves like a slow-light band in a standard PhCW, but with smaller area and greater group index. The nonlinear coefficient of a PhCC CROW can be as high as 13,000/Wm. This can significantly enhance the SFWM efficiency. The limiting factor of a PhCC CROW is that the slow-light bandwidth is narrower than a standard slow-light PhCW, making the isolation of generated photons from the pump challenging, which thus far have restricted CAR values to be below 10.

12.2.5 Quantum Description of Photon Pair States

Having explored a range of experimental structures for photon generation, we now return to the development of the formal quantum description of pair states produced by the SPDC or SFWM processes.

12.2.5.1 The Bi-photon Wave Function

We begin by constructing a more complete description of the quantum state that is produced in spontaneous pair sources. Here we restrict ourselves to the discussion of integrated sources which, as they use a collinear configuration, can be represented in the frequency domain, rather than working on the full wavevector version described in (12.20). Ignoring multi-pair and parasitic effects such as spontaneous Raman generation or scattering loss, the full state generated per pump pulse in a waveguide structure in the undepleted pump approximation is [47, 108–110]

$$\left|\psi_{\text{gen}}\right\rangle = \exp\left(\frac{v}{\sqrt{2}}\sum_{\alpha,\beta}\int_0^\infty \mathrm{d}\omega_1 \int_0^\infty \mathrm{d}\omega_2 \phi_{\alpha\beta}(\omega_1,\omega_2)\hat{a}^\dagger_{\alpha\omega_1}\hat{a}^\dagger_{\beta\omega_2} - \text{H.c}\right)|\text{vac}\rangle, \tag{12.26}$$

where H.c. denotes the Hermitian conjugate. This complicated expression can be seen as a continuous frequency multi-mode squeezed vacuum. In the limit of a low probability of pair production, $|v| \ll 1$, the full expression may be approximated by

$$\left|\psi_{\text{gen}}\right\rangle \approx |\text{vac}\rangle + v|\psi_{\text{II}}\rangle, \tag{12.27}$$

where the photon pair state

$$|\psi_{\text{II}}\rangle = \frac{1}{\sqrt{2}}\sum_{\alpha,\beta}\int_0^\infty \mathrm{d}\omega_1 \int_0^\infty \mathrm{d}\omega_2 \phi_{\alpha\beta}(\omega_1,\omega_2)\hat{a}^\dagger_{\alpha\omega_1}\hat{a}^\dagger_{\beta\omega_2}|\text{vac}\rangle. \tag{12.28}$$

Here Greek subscripts indicate the mode type of each photon (e.g. lowest order TE-like mode), and $\phi_{\alpha\beta}(\omega_1,\omega_2) = \phi_{\beta\alpha}(\omega_2,\omega_1)$ is known as the "bi-photon wave function" or "joint spectral amplitude" for the photon pair in modes α and β. Provided we normalise the components of the bi-photon wavefunction according to

$$\sum_{\alpha,\beta}\int_0^\infty \mathrm{d}\omega_1 \int_0^\infty \mathrm{d}\omega_2 \left|\phi_{\alpha\beta}(\omega_1,\omega_2)\right|^2 = 1, \tag{12.29}$$

then, in this limit, $|v|^2$ is just the average number of pairs generated per pulse.

Note that (12.28) applies for both SPDC and SFWM. The differences in the outputs of these processes sits in $\phi_{\alpha\beta}$, along with issues such as phase-matching and power dependence. Following terminology of classical nonlinear optics, a "type-I" process creates photon pairs in which both photons are of the same mode type and is represented by a single $\alpha\alpha$ term in the sums above. A type-II process creates photon pairs in which each photon is of a different type and thus has an $\alpha\beta$ as well as a $\beta\alpha$ term. An additional possibility involves the use of two type-I processes back to back in a "sandwich" configuration [111] to produce an $\alpha\alpha$ as well as a $\beta\beta$ term. Such schemes are useful for making polarisation-entangled states.

Also note that in (12.26) the integrals are over the frequencies of the generated photon pair rather than the wavevectors. The latter is normally required to treat bulk systems such as those discussed in Sect. 12.2.3.1. In waveguides, the requirement of transverse wavevector conservation is eliminated by the spatial confinement, and the frequency representation is typically more convenient. Nevertheless, the general theory that follows holds in the bulk case, with suitable transformations to a wavevector expansion.

12.2.5.2 Frequency Correlations in Bi-photon States

The importance of the bi-photon wave function comes in understanding the frequency correlations between the signal and idler photons. The implication of (12.28) is that except for certain "separable" states (see Sect. 12.2.5.3), the two photons are frequency-entangled and it is not meaningful to talk of the frequency content of one or the other. Such a description is vital in quantum engineering, since different quantum information applications demand different correlations. If the $\phi_{\alpha\beta}(\omega_1, \omega_2)$ for all α and β in the sum of (12.28) are equal, then there is no possible frequency measurement that can determine the mode type of either photon. For photon pair states in modal superpositions (i.e. with more than one term in the sum of (12.28)), such equality implies that the photons are maximally entangled in mode type [112], and enables maximal violation of Bell-like inequalities [113, 114]. On the other hand, high fidelity quantum computation tasks may be spoiled by any sort of frequency correlation [115]. Thus, it is useful to be able to quantify the degree of spectral correlations.

The most common measure for quantifying the degree of frequency correlations is through a Schmidt decomposition of the bi-photon wave function [116]. As a complex function of two variables, it can always be expressed in the form

$$\sqrt{N_{\alpha\beta}}\phi_{\alpha\beta}(\omega_1, \omega_2) = \sum_j \sqrt{p_j}\Phi_{\alpha j}(\omega_1)\Phi_{\beta j}(\omega_2), \tag{12.30}$$

where $\sqrt{N_{\alpha\beta}}$ is a normalisation factor such that $N_{\alpha\beta}\int_0^\infty \mathrm{d}\omega_1\mathrm{d}\omega_2 \left|\phi_{\alpha\beta}(\omega_1, \omega_2)\right|^2 = 1$. In addition, the complex Schmidt functions $\Phi_{\alpha j}(\omega)$ and $\Phi_{\beta j}(\omega)$ form complete orthonormal sets, the weighting coefficients p_j are real, and $\sum_j p_j = 1$. If there is only one nonzero Schmidt coefficient p_0, i.e. $p_0 = 1$ with $p_j = 0$, $\forall j \neq 0$, then the bi-photon wave function contains no spectral correlations and the state is termed *separable*. If there is more than one Schmidt coefficient, the Schmidt number [116]

$$K = \sum_j \frac{1}{p_j^2}, \tag{12.31}$$

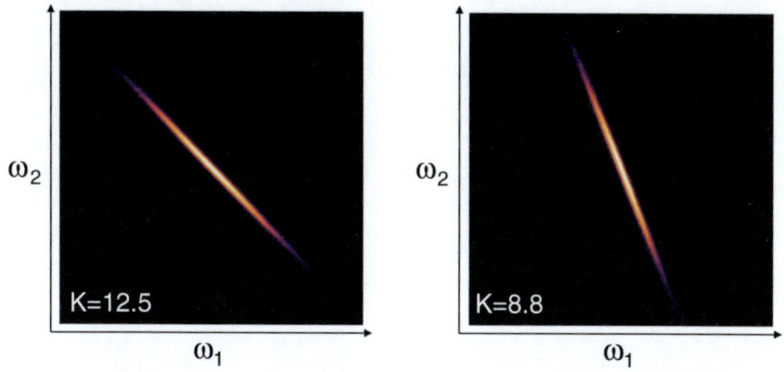

Fig. 12.19 Example bi-photon wave functions and their Schmidt numbers

quantifies the effective number of Schmidt modes involved in the decomposition. Note that $K > 1$ when the bi-photon wave function contains spectral correlations, and reaches a minimum of 1 when it contains no spectral correlations. It is also the inverse of the heralded single photon spectral purity, in the limit of negligible multipair production. By way of example, we consider two real-valued bi-photon wave functions in Fig. 12.19, and note how K is reduced the more $\phi_{\alpha\beta}(\omega_1, \omega_2)$ is aligned with the ω_1–ω_2 axes. Energy conservation and phase matching tend to produce a bi-photon wave function that is rather narrow and runs from the top left to the bottom right of ω_1–ω_2 space, but clever pump and source engineering can change this and thus reduce K, as seen in the next section.

12.2.5.3 The Factorability of Bi-photon States

Separable bi-photon states are not easily produced, but they play a special role in the design of photon sources. When a herald photon of a pair is detected, the remaining photon is represented by a single photon state. However in general, if the heralding photon is observed by a broadband detector, the single photon state that remains is not represented by the ideal *pure* state form in (12.11). Instead, it is projected onto a quantum *mixed state* that accounts for the classical ignorance of the detected photon's wavelength. In the quantum density operator formalism, we find the heralded state by tracing over the frequency expansion of the detected photon as

$$\hat{\rho}_\alpha = \int_0^\infty \mathrm{d}\omega \langle \omega_\beta | \hat{\rho} | \omega_\beta \rangle \tag{12.32}$$

where the photon pair density operator is

$$\hat{\rho} = 2 \int_0^\infty d\omega_1 d\omega_2 d\omega_1' d\omega_2' \phi_{\alpha\beta}(\omega_1, \omega_2) \phi^*_{\alpha\beta}(\omega_1', \omega_2') \left| \omega_{1\alpha}, \omega_{2\beta} \right\rangle \left\langle \omega_{1\alpha}', \omega_{2\beta}' \right|. \quad (12.33)$$

Here we have defined $\left|\omega_{1\alpha}, \omega_{2\beta}\right\rangle = |\omega_{1\alpha}\rangle \left|\omega_{2\beta}\right\rangle$ where $|\omega_\theta\rangle = \hat{a}^\dagger_{\theta\omega}|\text{vac}\rangle$, and assumed our initial bi-photon state to be $|\psi\rangle = \sqrt{2}\int_0^\infty d\omega_1 d\omega_2 \phi_{\alpha\beta}(\omega_1, \omega_2) \left|\omega_{1\alpha}, \omega_{2\beta}\right\rangle$ for convenience. Performing the trace in (12.32) gives

$$\hat{\rho}_\alpha = 2 \int_0^\infty d\omega_1 d\omega_2 d\omega \, \phi_{\alpha\beta}(\omega_1, \omega) \phi^*_{\alpha\beta}(\omega_2, \omega) |\omega_{1\alpha}\rangle \langle \omega_{2\alpha}|. \quad (12.34)$$

If the bi-photon wave function can be written in the separable form $\sqrt{2}\phi_{\alpha\beta}(\omega, \omega') = f(\omega)g(\omega')$ then (12.34) collapses to the pure state $\hat{\rho}_\alpha = |\psi_\alpha\rangle\langle\psi_\alpha|$ with

$$|\psi_\alpha\rangle = \int_0^\infty d\omega f(\omega) |\omega_\alpha\rangle, \quad (12.35)$$

which has the form of (12.11). However, in general this is not possible and the heralded photon occupies a mixed state. This is a practical problem, since most quantum protocols involving single photons assume pure state inputs.

There are two approaches to resolve the problem. First, one can simply filter the outgoing photons strongly. Filters with bandwidth comparable to the minor axis length of the bi-photon wave function will generally produce near-pure states, but at the cost of discarding most of the generated pairs. Alternatively, one can attempt to *engineer* the bi-photon state produced by the source as follows.

In waveguide sources, the shape of the bi-photon wave function in the ω_1–ω_2 plane is largely governed by just two things: the spectral shape of the incident pump pulse, and the dispersion and nonlinearity of the waveguide. In $\chi^{(2)}$ structures this is often expressed as

$$\phi^{(2)}_{\alpha\beta}(\omega_1, \omega_2) = \phi_P(\omega_1 + \omega_2) F^{(2)}_{\alpha\beta}(\omega_1, \omega_2), \quad (12.36)$$

where $\phi_P(\omega)$ is the pump spectral amplitude and $F^{(2)}_{\alpha\beta}(\omega_1, \omega_2)$ is determined by material properties and the waveguide geometry. Similarly, in $\chi^{(3)}$ structures we can write

$$\phi^{(3)}_{\alpha\beta}(\omega_1, \omega_2) = \int_0^{\omega_1+\omega_2} d\omega \phi_P(\omega) \phi_P(\omega_1 + \omega_2 - \omega) F^{(3)}_{\alpha\beta}(\omega_1, \omega_2, \omega), \quad (12.37)$$

where again $F^{(3)}_{\alpha\beta}(\omega_1, \omega_2, \omega)$ describes the waveguide materials and geometry. Thus, many attempts to produce frequency uncorrelated states (that is states with a Schmidt number $K = 1$), have focused on pump pulse shaping and waveguide property engineering.

Researchers have examined the group velocity mismatch, crystal length, and pump bandwidth conditions necessary to achieve factorable bi-photon wave functions created via type-II SPDC in bulk crystals [117], as well as via SFWM in photonic crystal fibre [118]. These studies were experimentally verified in [33] and [119] respectively. More exotically, pump pulse shaping [120], and engineering of the strength of an SPDC waveguide nonlinearity itself [69], to achieve factorability have also been investigated, in addition to resonant effects during SFWM in microring resonators [121]. There is more than one way to skin Schrödinger's cat.

12.2.5.4 Measuring the Joint Spectral Intensity

Although the bi-photon wave function is quite an important quantity, it has proven difficult to measure directly. Nonetheless, there are a number of experimental techniques available to measure a related quantity, the joint spectral intensity (JSI) of generated photon pairs, defined as $|\phi_{\alpha\beta}(\omega_1, \omega_2)|^2$. One of the simplest methods is to separate the signal and idler photons at the output of the generation device and scan a tunable filter in both the signal and idler arms. At each combined position $\{\omega_s, \omega_i\}$ one measures the number of coincident arrivals of photons, as shown schematically in Fig. 12.20a. By plotting these values as ω_s versus ω_i one can create a contour plot of the JSI. As previously discussed the shape of the JSI implies different amounts of entanglement in the system and therefore conveys the purity with which photons can be heralded. This filtering technique has been used both manually [33, 42, 122] with a low resolution and more recently using an automated liquid-crystal-on-silicon (LCOS) tunable filter [123] with a high resolution. An example of a JSI measured using this method from a chalcogenide fibre photon pair source is shown in Fig. 12.21. The LCOS device can also be used to create arbitrary filter functions which allows the use of different measurement acquisition types, including Hadamard coding [124], which can reduce the measurement time required to reconstruct a full JSI.

Another method of measuring the JSI replaces the tunable filters in the signal and idler arms with dispersive media, such as long lengths of dispersive optical fibres, and detectors with a high resolution photon arrival time measurement [125], as shown in Fig. 12.20b. As the photon propagates through the dispersive medium it becomes elongated in time, with the time of arrival directly related to the frequency of the input photon. The coincidences arrive with random timing information, meaning the JSI is built up randomly over the total acquisition time. This has the added advantage of having no active tuning or moving parts, however also means that more complicated filtering functions are not available to reduce the total integrated time required to characterise the JSI.

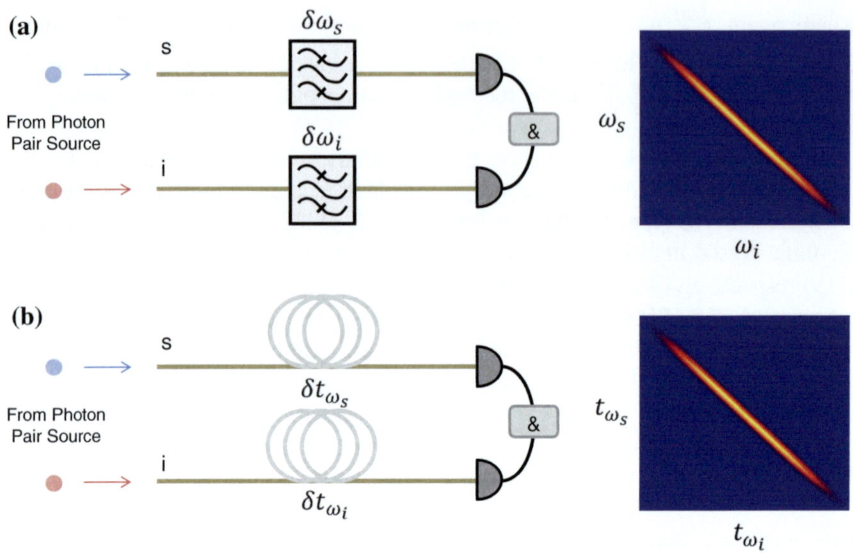

Fig. 12.20 Schematic of JSI measurement methods using **a** tunable optical filtering and **b** dispersive media and photon arrival time measurements

Fig. 12.21 The measured JSI from a chalcogenide fibre photon pair source measured with a high resolution automated method [123]

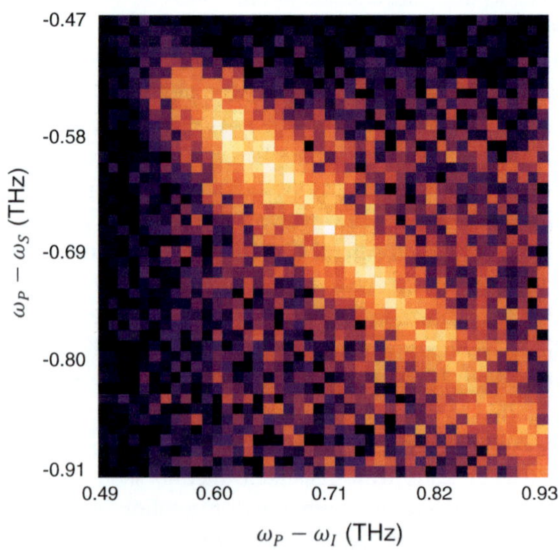

More recently there have been theoretical proposals for reconstructing the JSI using only classical measurements in a process known as stimulated emission tomography (SET) [126]. This method uses the same pumping characteristics as are used in the standard coincidence measurements above, however here a weak probe is also injected to perform the stimulated analogue of the spontaneous process, such as difference frequency generation for SPDC and stimulated four-wave mixing for SFWM. By scanning the probe and measuring the generated signal using an optical spectrum analyser it is possible to reconstruct the JSI with a high resolution and over a much shorter integration time as no coincidence measurements are required, shown very recently for SPDC [127].

12.2.5.5 Second-Order Correlation Functions

There are a number of other useful means of characterising the state that provide complementary information to direct measurement of the JSI or the full phase-sensitive JSA. In particular, indirect methods exist to measure its properties based on correlation functions [128].

Measurements of normalised correlation functions serve to quantify the coherence of the electromagnetic field produced by a particular source. While a first introduction to coherence generally deals with intensity fringes from the superposition of fields such as in a Michelson interferometer, this captures only first-order coherence. In fact important information can be extracted from the second order correlation function and higher order coherences introduced by Glauber [129, 130], following the invention of intensity interferometry by Hanbury Brown and Twiss [131].

The classical second-order correlation function

$$g^{(2)}(\mathbf{r}_1, t_1; \mathbf{r}_2, t_2) = \frac{\langle E^*(\mathbf{r}_1, t_1) E^*(\mathbf{r}_2, t_2) E(\mathbf{r}_1, t_1) E(\mathbf{r}_2, t_2) \rangle}{\langle E^*(\mathbf{r}_1, t_1) E(\mathbf{r}_1, t_1) \rangle \langle E^*(\mathbf{r}_2, t_2) E(\mathbf{r}_2, t_2) \rangle}, \tag{12.38}$$

is the intensity analogue of the first order function, and in the limit of stationary fields it depends only the difference in time $\tau = t_2 - t_1 - |\mathbf{r}_2 - \mathbf{r}_1|/c$. Switching to a simple quantum mechanical picture and replacing classical fields by corresponding Heisenberg field operators, $g^{(2)}(\tau)$ can be written in terms of the number of counts on a photon detector, which is proportional to the intensity, as

$$g^{(2)}(\tau) = \frac{\langle \hat{n}_{\mathrm{A}}(t) \hat{n}_{\mathrm{B}}(t+\tau) \rangle}{\langle \hat{n}_{\mathrm{A}}(t) \rangle \langle \hat{n}_{\mathrm{B}}(t+\tau) \rangle}, \tag{12.39}$$

where $\langle \cdots \rangle$ now denotes a temporal average. Here $\langle \hat{n}_{\mathrm{A,B}}(t) \rangle = \left\langle \hat{a}^{\dagger}_{\mathrm{A,B}}(t) \hat{a}_{\mathrm{A,B}}(t) \right\rangle$ is the average number of clicks on photo detector A or B over some integration time T. While derived for stationary fields, it turns out that (12.39) also holds for pulsed sources provided T encompasses a single pulse width and τ is an integer number of temporal pulse spacings. Written this way, $g^{(2)}(\tau)$ can be interpreted as simply the likelihood of detecting a photon at detector A in coincidence with the detection of a second photon delayed by time τ at detector B. As such, the second-order correlation function is usually measured with a Hanbury Brown and Twiss (HBT) setup of the form sketched in Fig. 12.22a.

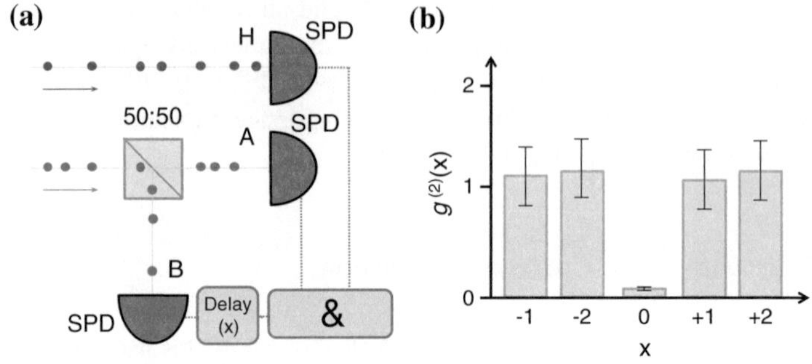

Fig. 12.22 **a** Illustration of a single photon HBT experiment for measuring $g^{(2)}(x)$. Idler photons are detected to herald signal photons which are incident on a 50:50 beam splitter and are either transmitted for reflected in two arms A and B. Single photon detectors (SPD) detect the photons and a coincidence counter correlates photon arrival times. An electronic delay can shift the signal from detector B by a pulse period, x. **b** Experimental $g^{(2)}(x)$ for a heralded single photon source. For $x = 0$, $g^{(2)}(x) \to 0$ showing non-classical statistics, while for $x \neq 0$, $g^{(2)}(x) \sim 1$, showing the recovery of uncorrelated Poissonian statistics

Figure 12.22b shows a typical $g^{(2)}(\tau)$ trace where τ, the optical delay in one arm, is varied while counting coincidences from two single photon detectors (SPDs) A and B. When the **path** length is equal, $g^{(2)}(0) \to 0$ for a single photon source. For a coherent source, the spacing between each photon in time is random and follows Poissonian statistics, resulting in $g^{(2)}(0) = 1$. For a chaotic/thermal light source $g^{(2)}(0) = 2$, which can be interpreted as a signature of photon bunching. The HBT experiment in a photon picture gives a direct measurement of the $g^{(2)}(\tau)$ of a light source.

As mentioned above $g^{(2)}(\tau)$ is just one of many possible correlation functions. To write them in a convenient form for calculations involving multimode squeezed states, we first note that (12.30) allows us to write a generated multimode squeezed state as

$$\begin{aligned}\left|\psi_{\text{gen}}\right\rangle &= \exp\left(\frac{v}{\sqrt{2}}\sum_{\alpha,\beta}\int_0^\infty \mathrm{d}\omega_1 \int_0^\infty \mathrm{d}\omega_2 \phi_{\alpha\beta}(\omega_1,\omega_2)\hat{a}^\dagger_{\alpha\omega_1}\hat{a}^\dagger_{\beta\omega_2} - \text{H.c}\right)|\text{vac}\rangle \\ &= \exp\left(\frac{v}{\sqrt{2}}\sum_{\alpha,\beta,j}\sqrt{p_j}\hat{A}^\dagger_{\alpha j}\hat{A}^\dagger_{\beta j} - \text{H.c}\right)|\text{vac}\rangle\end{aligned} \tag{12.40}$$

where

$$\hat{A}^\dagger_{\alpha(\beta)j} = \int_0^\infty \mathrm{d}\omega \Phi_{\alpha(\beta)j}(\omega)\hat{a}^\dagger_{\alpha(\beta)\omega}. \tag{12.41}$$

While the analysis proceeds similarly for many different states, we focus here on the state produced by a type-II process

$$\left|\psi_{\text{gen}}\right\rangle = \exp\left(v\sum_j \sqrt{p_j}\hat{A}^\dagger_{\alpha j}\hat{A}^\dagger_{\beta j} - \text{H.c}\right)|\text{vac}\rangle. \tag{12.42}$$

We note that this state contains an average number of

$$\mu = \left\langle \psi_{\text{gen}}\middle|\hat{N}_\theta\middle|\psi_{\text{gen}}\right\rangle = \sum_j \sinh^2\left(|v|\sqrt{p_j}\right), \tag{12.43}$$

photon pairs, where $\hat{N}_\theta = \sum_j \hat{A}^\dagger_{\theta j}\hat{A}_{\theta j}$ and $\theta = \alpha$ or β. Furthermore, for factorable bi-photon wave functions with $K = 1$, (12.42) represents a two-mode squeezed vacuum with a probability of n pairs expressed by (12.15). On the other hand, when K is so large that there are a very large number of Schmidt coefficients, all equal, and not more than one pair is produced in each Schmidt mode, the probability of n pairs is instead expressed by (12.13). Using (12.42) we may write zero-delay, $\tau = 0$, unheralded auto-correlation functions as [128]

$$g^{(n)}_{\text{Unheralded}} = \frac{\left\langle \psi_{\text{gen}}\right| : \hat{N}^n_\theta : \left|\psi_{\text{gen}}\right\rangle}{\left\langle \psi_{\text{gen}}\middle|\hat{N}_\theta\middle|\psi_{\text{gen}}\right\rangle^n}, \tag{12.44}$$

where the operators between the colons are normal ordered. Similarly, we may write zero-delay cross-correlation functions as [128]

$$g^{(m,n)}_{\text{Cross}} = \frac{\left\langle \psi_{\text{gen}}\right| : \hat{N}^m_\alpha :: \hat{N}^n_\beta : \left|\psi_{\text{gen}}\right\rangle}{\left\langle \psi_{\text{gen}}\middle|\hat{N}_\alpha\middle|\psi_{\text{gen}}\right\rangle^m \left\langle \psi_{\text{gen}}\middle|\hat{N}_\beta\middle|\psi_{\text{gen}}\right\rangle^n}. \tag{12.45}$$

In the low probability of pair production limit, we will see that $g^{(2)}_{\text{Unheralded}}$ can give information regarding the Schmidt number, K, and that $g^{(1,1)}_{\text{Cross}}$ can give information regarding the average number of generated photons per pump pulse. Following a bit of algebra, it can be shown that

$$g^{(2)}_{\text{Unheralded}} = 1 + \frac{\sum_j \sinh^4\left(|v|\sqrt{p_j}\right)}{\left[\sum_j \sinh^2\left(|v|\sqrt{p_j}\right)\right]^2}, \tag{12.46}$$

and

$$g_{\text{Cross}}^{(1,1)} = 1 + \frac{\sum_j \sinh^4\left(|v|\sqrt{p_j}\right)}{\left[\sum_j \sinh^2\left(|v|\sqrt{p_j}\right)\right]^2} + \frac{1}{\sum_j \sinh^2\left(|v|\sqrt{p_j}\right)}. \tag{12.47}$$

The low pair production probability limit implies that the number of photons produced, here $\left\langle \psi_{\text{gen}} \middle| \hat{N}_\alpha + \hat{N}_\beta \middle| \psi_{\text{gen}} \right\rangle \ll 1$, and thus $|v|^2 \ll 1$. In this limit

$$g_{\text{Unheralded},}^{(2)} \approx 1 + \sum_j p_j^2 = 1 + \frac{1}{K}, \tag{12.48}$$

and

$$g_{\text{Cross}}^{(1,1)} \approx 1 + \sum_j p_j^2 + \frac{1}{|v|^2} = g_{\text{Unheralded}}^{(2)} + \frac{1}{\left\langle \psi_{\text{gen}} \middle| \hat{N}_\theta \middle| \psi_{\text{gen}} \right\rangle}. \tag{12.49}$$

as initially suggested. Finally, the heralded auto-correlation function

$$g_{\text{Heralded}}^{(2)} = \frac{\int_0^\infty \mathrm{d}\omega \langle \omega_\theta | : \hat{N}_\theta^2 : \hat{\rho}_\theta | \omega_\theta \rangle}{\left(\int_0^\infty \mathrm{d}\omega \langle \omega_\theta | \hat{N}_\theta \hat{\rho}_\theta | \omega_\theta \rangle \right)^2}, \tag{12.50}$$

can easily be shown to be zero, if a number-resolving scheme is used to ensure that only a single α or β photon has been detected to herald the presence of its partner and ρ_θ takes the form of (12.33). Otherwise, it will be larger, as we seen in the experiments discussed throughout this chapter.

12.2.6 Multiplexing: A Route to On-demand Photons

In Sect. 12.2.1.2 we saw that all existing approaches to producing single photons have technical issues, making it very difficult to obtain a single photon in a truly on-demand fashion. This is a fundamental challenge in quantum optics. The non-deterministic nature of photon generation limits the feasibility of next-generation quantum processors that will require multiple simultaneous single photon inputs. In particular we noted that both attenuated laser sources and heralded pair sources emit a portion of their light as multiple photons, with very similar photon statistics (see (12.13) and (12.15)). At first sight then, the heralded sources have no clear advantage over attenuated lasers, other than perhaps the ability to gate single photon detectors using the idler detection.

In fact however, the concept of *photon multiplexing* can break the intrinsic limits on the photon statistics expressed by the inverse relationship of the CAR on the mean photon rate (see (12.17)). In this section, we explain these approaches and discuss the demonstrations to date.

12.2.6.1 Spatial Multiplexing

Single photons generated by parametric processes have an advantage over attenuated laser sources—because a photon pair is generated, the detection of one photon heralds the existence of the other. Using this knowledge, it is possible to deterministically combine the output of several sources. This results in a higher single-photon rate without a corresponding increase in multiple-photons. This occurs because the rate of higher order pairs grows quadratically with the pair generation probability μ in each source. Distributing the mean rate across N channels means the number of higher order pairs scales as $N(\mu/N)^2 \propto 1/N$ for the same net single pair rate. In principle then, the signal to noise level is multiplied by N.

Practically this is done by using the herald detections to control an $N \times 1$ low loss optical switch, routing single-photons from multiple sources to a common single mode output. This is shown in Fig. 12.23, where 8 nonlinear photon pair sources (in this case silicon PhC waveguides) have their outputs combined by an optical switch, triggered by a herald detection, to produce an enhanced single photon output. The probability of a single photon reaching the output of the multiplexing system is then, in the limit of $\mu \ll 1$

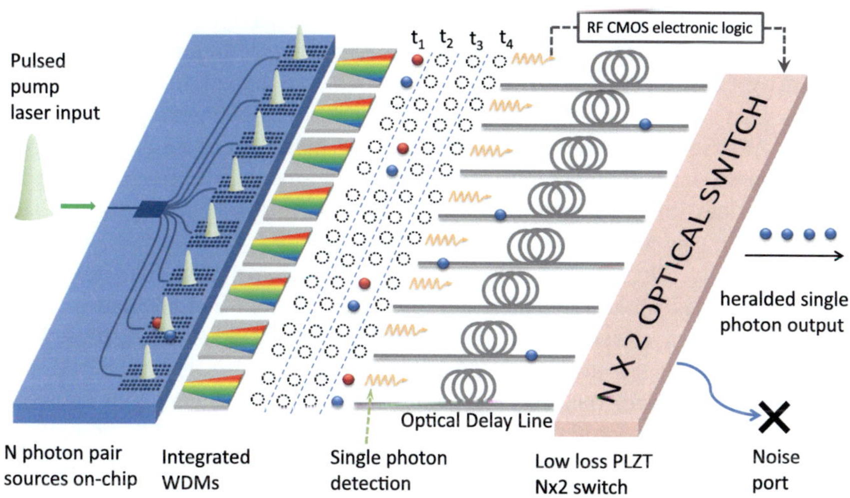

Fig. 12.23 Schematic illustration of integrated spatial multiplexing of 8 nonlinear photon pair sources. In this example, the photonic crystal waveguides monolithically fabricated on a single silicon chip are excited by a common pump [44]. PLZT refers to the switch material and is the optoceramic lead lanthanum zirconium titanate

$$P_{\mathrm{MUX_{Out}}} = \left(1 - (1 - \mu\eta_{\mathrm{h}})^{N}\right)\eta_{\mathrm{delay}}\eta_{\mathrm{couple}}\eta_{\mathrm{filter}}\eta_{\mathrm{switch}}^{\log_2(N)}, \tag{12.51}$$

where N (a power of 2) is the number of multiplexed sources, $\eta_{\mathrm{h}} = \eta_{\mathrm{couple}}\eta_{\mathrm{filter}}\eta_{\mathrm{detector}}$ is the efficiency with which a heralding click occurs, η_{delay} is the delay line efficiency, η_{couple} the efficiency of coupling the photons from the source into the fibre, η_{filter} the efficiency of the filtering and η_{switch} the efficiency of a single switch. For large N and zero losses and assuming perfect detector efficiency, the probability tends towards an on-demand single photon source. In reality there will always be loss and imperfect detection limiting the performance of any scheme. Also for a larger number of sources, the probability of two or more sources generating a single pair increases, so there are diminishing returns beyond a certain number of sources. The number of switches required to implement this scheme scales linearly, with $(N - 1)$ switches needed to multiplex N sources.

This idea of *spatial* multiplexing was proposed by Migdall et al. in 2002 [132]. An alternate, potentially lower-loss, method using electro-optic polarisation controllers and polarising beam splitters rather than 2×2 Mach-Zehnder interferometric switches, was suggested in 2007 [133]. This was the scheme used in the first demonstration by Ma et al. [134]. Using a double pumped BBO crystal and free space polarisation switches they achieved a four-fold increase in single photon rate for fixed $g^{(2)}(0)$ values, however the instability of bulk optics impeded scaling to more components. Other publications have examined the limits of multiplexing [43] finding that a deterministic source could be built from 17 ideal sources and photon number resolving detectors, and the practical requirements of using such a source for linear optical quantum computing [135], which found that some multiplexing schemes are more efficient than others and there should be a viable route to an on-demand multiplexed source which will be useful for some QIP protocols.

The first demonstration of partially integrated spatial multiplexing was performed by Collins et al. in 2013, using two photonic crystal waveguide photon sources combined on a single silicon chip [44, 136] implementing the Migdall scheme. In this paper the heralded photon rate was enhanced by 62.4 % for a fixed signal-to-noise ratio. Figure 12.24 shows a clear improvement to the count rate for all CAR values (blue triangles) when compared to the two separate sources (red squares and green circles).

Subsequently, Meany et al. demonstrated the multiplexing of four monolithic periodically-poled lithium niobate waveguides which were butt-coupled to direct laser-written silica wavelength division separation components [137]. This demonstrated not only the integration of four sources, but also several of the key components needed for multiplexing. In 2013 an alternate scheme with an asymmetric architecture for multiplexing was proposed, and it is predicted that for certain regimes this scheme will outperform the traditional symmetric switching architecture [138].

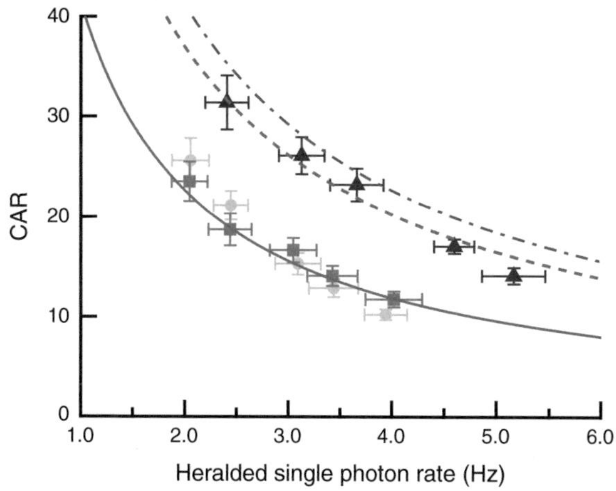

Fig. 12.24 Plot to show the enhancement of multiplexing two sources together. The single source heralded photon rates are plotted against the coincidence-to-accidental ratio for the two separate sources (*red squares* and *green circles*). After multiplexing the sources together the same metrics are measured (*blue triangles*) showing a clear enhancement of the heralded photon rate for all CAR values, corresponding to an enhancement of 62.4 % [44]

12.2.6.2 Temporal Multiplexing

In the strategy described in the previous section, multiple spatially-separated sources are combined using multiplexing. Each source has a single spatial mode, and these modes are combined deterministically using an $N \times 1$ optical switch. We now turn to temporal multiplexing where one or more photon pair sources have many temporal modes actively combined by routing the heralded photon to a different length of delay line. Depending on the temporal mode in which a photon pair is created, indicated by the timing of the herald detection, a different delay time is applied. This idea was proposed theoretically in 2011 by Mower and Englund [139], who envisaged the scheme being implemented with a sophisticated photonic integrated circuit (PIC).

Figure 12.25 shows a sketch of the concept. A pulse train with period T pumps a nonlinear device to produce photon pairs. The output is then divided, for this example, into four time-bins t_i, $i \in [1, 4]$. A herald detection, in for example time-bin t_1, triggers its partner single photon to be switched through $3T$ of relative optical delay line moving it into time bin t_4. On the other hand, if a heralded detection is made in time-bin t_4, zero relative delay is applied. The output has a period now of $4T$ with $4\times$ the probability of a single photon per clock cycle. A possible setup for temporal multiplexing is shown in Fig. 12.26 with Mach-Zehnder 2×2 switches used to route the heralded photon though the $3T$ of relative delay.

The number of switches required to implement this scheme scales logarithmically, with $(\log_2 N + 1)$ switches needed to multiplex N time-bins, equivalent to

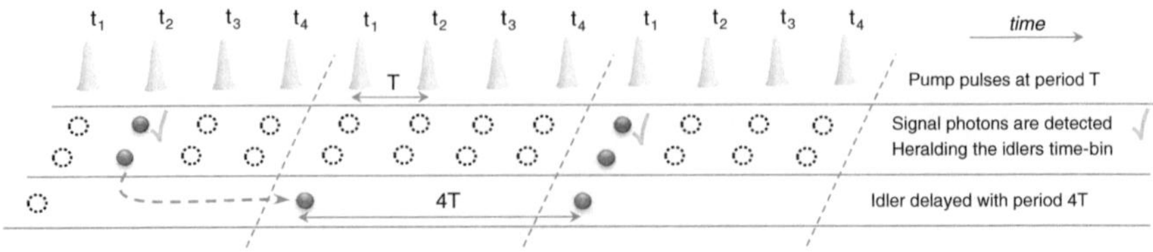

Fig. 12.25 Timing diagram of a four fold temporal multiplexing scheme

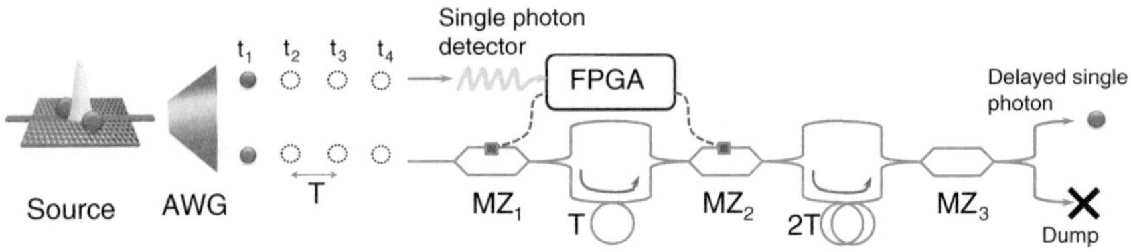

Fig. 12.26 Schematic illustration of temporal multiplexing of four temporal modes using Mach-Zehnder optical switches (MZi)

N sources. Temporal multiplexing is therefore much more resource efficient than spatial schemes for a high number of multiplexed temporal modes. It is however more challenging to implement, because switching triggered on temporal slots is difficult, and to date there have been no experimental demonstrations. In the future this idea will need to be extended to combine many more devices onto a single optical chip. Undoubtedly the hybrid integration of several material platforms will be required to realise all of the components necessary, including delay lines and fast switches. Multiplexing will allow many single photons to be generated nearly on-demand, enabling a new regime of complex quantum processing.

12.3 Nonlinear Optics for Quantum Frequency Conversion

As in classical communications, a range of controllable light sources is not the only requirement for effective transfer of quantum information. Using the fibre network effectively requires the ability to transfer information from one wavelength band to another, and this will increasingly be an important property of quantum information.

A working definition for quantum frequency conversion (QFC) is the translation of a quantum state of light from one frequency band to another. First discussed explicitly in the literature nearly 25 years ago by Kumar and colleagues [140, 141] in the context of making a tunable source of squeezed light, QFC may play an important role in photonic quantum information processing networks as an interface to link different components (operating at different wavelengths) together. One application would be to map the emission from a single quantum emitter (e.g., InAs/GaAs quantum dots, single neutral alkali atoms, or nitrogen vacancy centers in diamond) to the 1550 nm telecommunications band, for which single mode silica optical fibres show their lowest attenuation levels, and back to the original wavelength for interaction with another single quantum emitter [142]. A second application is in detection, by translating the wavelength of a quantum state for which detection is currently difficult (e.g., due to limitations in detector quantum efficiency, timing resolution, dark count rates, etc.) to one in which high performance detectors have already been developed [143, 144]. More generally, complete and efficient spectro-temporal control of quantum states of light would in principle allow arbitrary quantum photonic components, individually optimised to achieve a desired functionality, to be linked together as part of a high performance quantum internet [145].

In this section, we briefly discuss the fundamentals of QFC in second- and third-order nonlinear media, review the current status of frequency conversion devices, and highlight recent experiments demonstrating frequency conversion of single photon states of light. Finally, we discuss some perspective towards future directions with QFC, including recent progress using integrated device platforms and proposals for achieving temporal wavepacket shaping using QFC.

12.3.1 Second-Order Nonlinear Media

Nonlinear media governed by the second-order electronic susceptibility $\chi^{(2)}$ can be used for sum- and difference-frequency generation, in which an input signal at frequency ω_i is combined with a strong pump at frequency ω_p to generate light at the sum and difference frequencies $\omega_o = \omega_i \pm \omega_p$. Because optical nonlinearities are weak, efficient conversion only occurs over an interaction region that is many wavelengths in length. This places the requirement of phase-matching in a pre-eminent position; that is, for efficient sum- or difference-frequency generation, the wavevectors of the three fields must satisfy $\mathbf{k}_o = \mathbf{k}_i \pm \mathbf{k}_p$. Very much the same phase-matching issues apply for frequency conversion as were discussed in Sect. 12.2.3 for pair generation. Without phase matching, energy oscillates between the three waves involved in the mixing process rather than being monotonically transferred to the output field, and the overall conversion efficiency is low.

Improvements in conversion efficiency were initially gained using birefringent phase-matching, which takes advantage of the fact that orthogonally polarised light beams often travel at different velocities within a birefringent material. By tuning the angles of the input beams appropriately, phase matching can be achieved, although the interaction regions may be limited in size, particularly if tight focusing of the beams is used (which is often the case in order to achieve high intensities). As a result, efficient conversion (on the order of tens of percent) is typically not achieved in these systems without the application of extremely high pump powers.

A development that has increased the utility of frequency conversion devices and been particularly beneficial to subsequent QFC experiments is the use of quasi-phase-matching in a waveguide-based geometry [146] (see Sect. 12.2.3.2).

Conversion efficiencies (not including input/output coupling losses) in excess of 80 % have been achieved for both frequency upconversion and downconversion [147, 148]. Periodically-poled lithium niobate (PPLN) waveguides are perhaps the most common devices to have been used in QFC experiments to this point. Theoretically, sum- and difference-frequency generation can be noise-free, suitable for QFC all the way down to the level of single photon Fock states. In practice, one might expect that scattering and fluorescence processes associated with the interaction of a strong pump field with a solid material might lead to significant noise contributions. Raman scattering can be a dominant noise source if the pump and input signal are within a couple hundred nanometres of each other [146]. Keeping the pump on the long-wavelength side of the input signal [146] and increasing the pump-signal separation to several hundred nanometres [147, 149] can increase signal-to-noise levels to >100:1 (Fig. 12.27b). Spontaneous parametric downconversion can also be of concern, and can be exacerbated due to fabrication errors in the poling process [148].

The first experimental demonstration of QFC using a non-classical input signal was performed by Huang and Kumar [141] soon after Kumar introduced the concept [140], with non-classical intensity correlations between a pair of fields shown to be preserved after one of the fields was frequency converted. Over a decade later, Tanzilli and colleagues [142] showed that entangled fields remain so after one of the fields is frequency converted. This work using non-classical fields came at approximately the same time that several groups began pushing the

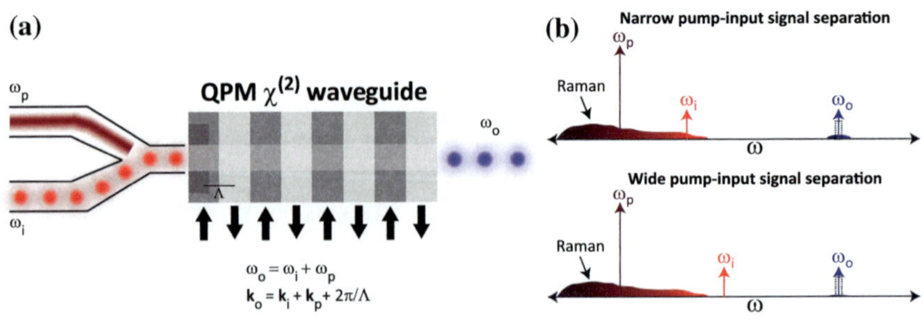

Fig. 12.27 **a** Schematic for quasi-phase-matching (QPM) in a $\chi^{(2)}$ waveguide (here, sum-frequency generation is shown). An input signal at frequency ω_i is combined with a strong pump at frequency ω_p to generate an output signal at ω_o. Phase matching is achieved through periodic poling, which generates a grating momentum $2\pi/\Lambda$ that compensates for wavevector mismatch. **b** The strong pump field can generate broadband Raman scattering in the QPM material. If this Raman scattered light reaches the input signal frequency, it will be frequency converted along with the signal and will thus be a limiting noise source. This source of noise can be limited by: (1) Placing the input signal on the anti-Stokes (higher frequency) side of the pump; (2) Increasing the pump-input signal separation as much as possible; (3) Narrowband spectral filtering around the target conversion frequency

conversion efficiency levels achievable in $\chi^{(2)}$ nonlinear crystals towards 100 %, for applications such as low light level detection of telecommunications-band photons using visible wavelength single photon counters [143, 144, 146].

In recent years, a number of experiments have looked at frequency upconversion of single photon Fock states using quasi-phase-matched $\chi^{(2)}$ nonlinear crystals (Fig. 12.28). For example, one experiment combined a 1300 nm triggered single photon source based on an InAs/GaAs quantum dot with a 1550 nm pump in a PPLN waveguide to generate 710 nm photons, with antibunched photon statistics measured to explicitly confirm the single photon nature of the upconverted light [150]. The utility of frequency conversion for studying the dynamics of telecommunications-band single quantum emitters was also highlighted, where conversion to the visible improved the dynamic range of time-correlated single photon counting measurements of the quantum dot excited state lifetime by a factor of 25, due to the superior characteristics of Si single photon counters (sensitive at 710 nm) relative to InGaAs devices (sensitive at 1300 nm). Subsequent work on upconversion of quantum dot single photon sources has included temporal shaping of the photon wavepackets using an amplitude modulated 1550 nm pump [151] and

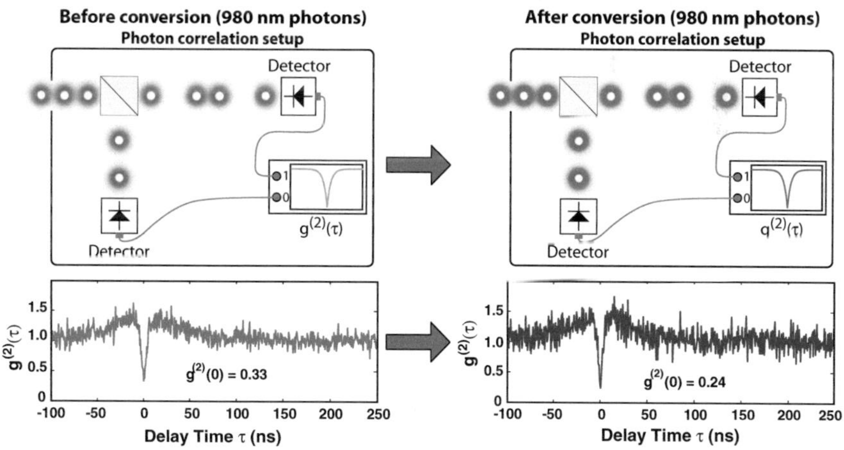

Fig. 12.28 Photon statistics measured before and after frequency upconversion, as in [149]. (*Top left*) 980 nm photons generated by a single InAs/GaAs quantum dot pumped with a continuous-wave 780 nm laser are sent into a beamsplitter, and photons in each output port detected by a single photon counter. A histogram of coincidence counts as a function of the difference in arrival times τ between the two paths (*bottom left*) is generated, showing antibunching with a value $g^{(2)}(0) = 0.33 \pm 0.03$ see (12.38). (*Top right*) After frequency conversion to 600 nm using a 1550 nm pump in a PPLN waveguide, photons are directed into a beamsplitter, and photons in each output port are detected by a single photon counter. The generated coincidence histogram (*Bottom right*) shows anti-bunching ($g^{(2)}(0) = 0.24 \pm 0.04 < 0.5$), indicating that the single photon nature of the input signal is preserved. The improved $g^{(2)}$ value is likely due to the extra spectral filtering provided by the quasi-phase-matching response of the PPLN waveguide. The uncertainty in the $g^{(2)}(0)$ values is due to fluctuations in the detected photon count rate and represents a one standard deviation value

experiments demonstrating that two-photon interference is preserved during frequency conversion [149]. The latter work also considered the use of frequency conversion to erase spectral distinguishability by converting the light from spectrally distinct sources to a common frequency. This work was quite analogous to an experiment that preceded it by a few years [152], which converted signal and idler photons from a correlated photon pair source to a common wavelength using sum frequency generation in a pair of PPLN waveguides, and then demonstrated two-photon interference of the now spectrally indistinguishable photons.

Frequency downconversion through difference frequency generation in a $\chi^{(2)}$ medium has also been experimentally explored in the context of single photon input fields. One experiment [153] used an entangled photon pair source and demonstrated that entanglement between signal and idler photons was preserved after downconversion of one of the two fields (essentially the downconversion equivalent of the upconversion experiment of [142]), as well as explicitly showing that the downconverted field exhibited antibunched photon statistics (through a heralded $g^{(2)}$ measurement). Frequency downconversion has also been demonstrated with triggered single photon sources based on a single semiconductor quantum dot, with photon anti-bunching and coherence shown to be preserved [154] and downconversion to the 1550 nm band used as an integral part of a spin-photon entanglement experiment [155].

12.3.2 Third-Order Nonlinear Media

Quantum frequency conversion using the third-order nonlinear susceptibility $\chi^{(3)}$ is achieved by the process of four-wave-mixing Bragg scattering [156], sometimes referred to in the literature as wavelength exchange [157]. This is a non-degenerate four-wave-mixing process in which two pump fields create an effective modulation in the $\chi^{(3)}$ nonlinearity, enabling the (in principle) noise-free translation of an input signal to an output that is shifted by an amount equal to the difference in the two pump frequencies (Fig. 12.29). The importance of using four-wave-mixing Bragg scattering for QFC was highlighted by McKinstrie and colleagues [156]. They outlined the noise considerations that arise when considering different four-wave-mixing processes, including degenerately pumped processes, which can provide wavelength conversion but will also be accompanied by spontaneous emission noise (i.e., pump photons converted to the output wavelength band).

The dual pumps in four-wave-mixing Bragg scattering provide a great deal of flexibility in the span over which frequency conversion can take place, with both narrow and wide ranges for both upconversion and downconversion possible. Figure 12.29b schematically shows some of the different configurations of pumps, input signal, and output converted signal that can be utilised to achieve these different functionalities. Interestingly, it should be pointed out that for each configuration, two possible idlers can be generated, as both are energy conserving.

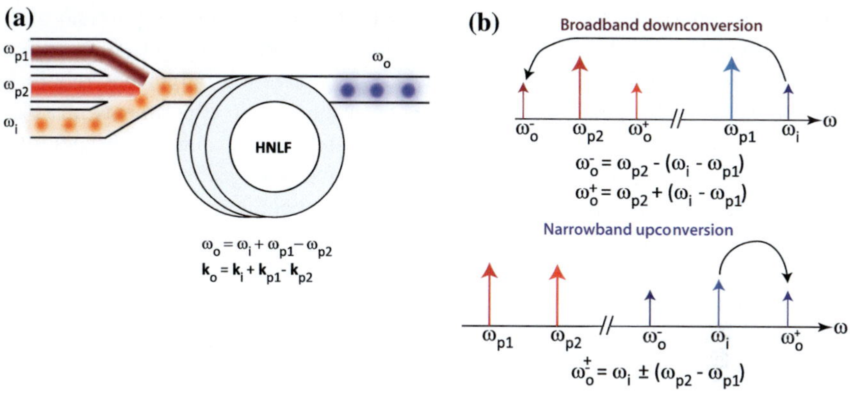

Fig. 12.29 Four-wave-mixing Bragg scattering. **a** Two strong pumps at frequencies $\omega_{\rm p1}$ and $\omega_{\rm p2}$ are combined with an input signal at $\omega_{\rm i}$ to generate an output converted field at $\omega_{\rm o}$ which, in this case, is shifted from the input by an amount $\omega_{\rm p2} - \omega_{\rm p1}$. **b** Different configurations for four-wave-mixing Bragg scattering. (*Top*) Broadband downconversion can be achieved by placing pumps $\omega_{\rm p1}$ and $\omega_{\rm p2}$ ($\omega_{\rm p1} > \omega_{\rm p2}$) in widely separated bands, with input signal $\omega_{\rm i}$ at a wavelength closer to $\omega_{\rm p1}$ than $\omega_{\rm p2}$; broadband upconversion (not shown) can similarly be achieved, with input signal closer to $\omega_{\rm p2}$. (*Bottom*) Narrowband conversion can be achieved by placing pumps $\omega_{\rm p1}$ and $\omega_{\rm p2}$ relatively close to each other in frequency. In all of these cases, two possible idlers can be generated, consistent with energy conservation, and phase matching determines the efficiency with which each idler is produced

Phase-matching then determines which idler is more efficiently generated (noting that for a long interaction length, even a small phase mismatch can be enough to select for one idler).

The most common medium for four-wave-mixing Bragg scattering has been some form of an optical fibre, including photonic crystal fibre [158] and highly nonlinear dispersion shifted fibre [157, 159, 160]. Though the $\chi^{(3)}$ nonlinearity is generally much weaker than the $\chi^{(2)}$ nonlinearity (e.g., comparing the values for materials like photonic crystal or highly nonlinear fibre with lithium niobate), the combination of relatively localised field confinement, the low propagation losses, and ability to fabricate optical fibres uniformly over length scales of kilometres has enabled four-wave-mixing Bragg scattering frequency converters to achieve a conversion efficiency of about 30 % for experiments at visible wavelength [158] and nearing unity in the telecommunications band [156–160] (typical length scales are $\sim$30 m for photonic crystal fibre and $\sim$1 km for highly nonlinear fibre in comparison to a few cm for lithium niobate). Phase-matching is achieved in these optical fibres by careful dispersion engineering, where waveguide dispersion (the variation in lights phase velocity with wavelength, due to the wavelength-dependent distribution of the optical field within the fibre core and cladding) can compensate for material dispersion.

While noise-free in principle, just as in the case of sum-frequency-generation and difference-frequency-generation in $\chi^{(2)}$ media, four-wave-mixing Bragg scattering in optical fibres can be limited by a number of different possible noise sources. Spontaneous Raman scattering, a known phenomenon in silica optical fibres, can be a dominant noise source when the converted output lies relatively close in wavelength to the pumps. This has been the case in experiments in which all four fields are near the telecommunications band, as explicitly discussed in [160], where the Raman noise signal was of similar strength to the weak (single photon level) coherent state input signal. For pumps that are more widely separated from the input signal and frequency converted output, noise due to spontaneous Raman scattering is more limited [158]. Other possible noise sources can include undesired four-wave-mixing processes, which in some circumstances, could be phase-matched (or close to phase matched). For example, this could include spontaneous four-wave-mixing from one of the pumps (degenerate) or both of the pumps (non-degenerate).

Quantum frequency conversion experiments in $\chi^{(3)}$ media utilising non-classical states of light have not been as widespread as they have in $\chi^{(2)}$ materials, with the first experiment being performed a few years ago [158]. Here, the output of a heralded single photon source produced by a photonic crystal fibre was upconverted by 24 nm (to 659 nm) in a second photonic crystal fibre. With input and frequency converted signals >150 nm on the anti-Stokes (higher frequency) side of the pumps, spontaneous Raman scattering was limited and $g^{(2)}(0)$ was essentially unchanged after frequency conversion.

12.3.3 Future Directions with Quantum Frequency Conversion

New media and devices for quantum frequency conversion are an active area of research, spurred in part by recent work in demonstrating nonlinear optical phenomena in integrated and scalable nanophotonic geometries, including some of the systems discussed earlier in this chapter. Within $\chi^{(2)}$ media, efforts at producing higher levels of device integration and more compact form factors within the lithium niobate platform have shown promise, including recent work on developing thin-film lithium niobate devices [161] and integrating them with silicon-based materials. Other $\chi^{(2)}$ media have included GaAs, GaP, and GaN, for which there has been work to demonstrate second harmonic generation and sum and difference frequency generation in waveguide geometries [162, 163], as well as recent efforts to establish second harmonic generation in nanophotonic cavity geometries [164–167]. Though these works are currently focused on the classical domain, improvements in conversion efficiency and an understanding of potential noise mechanisms could enable QFC experiments, with future prospects including direct integration with single photon emitters based on InAs/GaAs quantum dots, for example.

Recently, four-wave-mixing Bragg scattering based on the $\chi^{(3)}$ nonlinearity in silicon nitride (Si_3N_4) waveguides fabricated on a Si substrate has been demonstrated [168, 169], with both narrowband (few nanometres) and wideband (few hundreds of nanometres) upconversion and downconversion ranges shown in the classical regime. Along with its broadband optical transparency and lack of two-photon absorption for pump wavelengths >700 nm, Si_3N_4 does not appear to show significant amounts of spontaneous Raman scattering, so early indications suggest that high signal-to-noise levels can be achieved in this system. Furthermore, the combination of strong field confinement and the relatively high $\chi^{(3)}$ in Si_3N_4 create an effective nonlinearity that can be two orders of magnitude greater than that of highly nonlinear fibre. However, the relatively short Si_3N_4 waveguide length (1 cm) is in stark contrast to the typical hundreds of metres used for QFC in fibres, and conversion efficiencies have been limited to 5 %, with all experiments done strictly in the classical regime. Chalcogenide materials such as As_2S_3, which also has a broad optical transparency and limited two-photon absorption for pump wavelengths >800 nm, might also be a strong candidate for four-wave-mixing Bragg scattering, particularly due to its high nonlinear refractive index. Silicon is also a possibility for certain applications (e.g., telecom), particularly if pump wavelengths >2.0 μm can be used to avoid two-photon absorption, as has recently been demonstrated for degenerate four-wave-mixing experiments [170, 171].

Over the past few years, there have been a number of works looking to extend QFC from only shifting the color of quantum states of light to also providing temporal wavepacket shaping [145], a necessary resource for many applications in photonic quantum information science. For example, optimal storage of a single photon in a quantum memory will generally require control of not only the wavelength but also the spectro-temporal profile of the photon. Frequency conversion provides a versatile route to wavepacket shaping in large part because the nature of the process is such that the frequency converted field has characteristics that depend both on the input signal (a quantum field) and the input pump field(s) (strong classical field(s)). An example of this was demonstrated experimentally in [151], where classical pump pulses were used to produce upconverted single photons of a controllable duration (in the regime where the pump pulses were shorter than the input single photon pulses). That work essentially applied a nonlinear time gating operation to the input single photon states, and was therefore not lossless (even in principle). Fast nonlinear time gating of single photon pulses was also recently used to eliminate which-path information in the photon energy in a recent quantum dot spin-photon entanglement experiment [155].

There has been much progress on the theoretical front in developing lossless protocols for full spectro-temporal shaping of quantum light fields. In [172], a two-step process was proposed. The first step consists of spectral broadening and shaping of a single photon wavepacket through nonlinear mixing with a classical pump that has been imprinted with a specified temporal phase. The output single photon wavepacket has the desired frequency spectrum, and the spectral phase can then be corrected (using classical pulse shaping approaches) to allow for full temporal shaping. This approach could be used to strongly compress single photon pulses produced by single quantum emitters, which often have nanosecond (or longer) time constants. In [173], engineering of the group velocities of the pump and input signal in a $\chi^{(2)}$ medium was considered as a means to either compress or stretch a single photon wavepacket, depending on whether the pump moves faster or slower than the input signal. Single photon wavepacket shaping using four-wave-mixing Bragg scattering has also been considered [174, 175], with a recent result [175] showing an arbitrary input field temporal profile can be mapped to an arbitrary output field profile with unity efficiency. In comparison to the previous work [174], in which there was a tradeoff between selectivity (ability to map the input to only one desired output) and efficiency, [175] provides a prescription for providing perfect (in principle) selectivity and efficiency.

12.4 Nonlinear Optics for Quantum Communication

The first real impact of quantum information research on wider society is in the area of quantum communication and quantum cryptography: the field of using quantum laws to enhance privacy in optical communications. In this section we introduce one of the key concepts of this field—quantum key distribution—and describe how it can be used to provide completely secure quantum communication. We also provide state-of-the-art examples from the literature.

Fast and secure information transfer is vital to economic and social stability in the modern internet enabled world. The current standard cryptographic algorithms are not based on information-theoretic proofs, but on the assumption that some mathematical calculations are hard to compute. If a fast classical algorithm for these problems is discovered or a modestly sized practical quantum computer is developed, the integrity of these encryption protocols will be compromised. Quantum key distribution (QKD) is one avenue that offers practical information-theoretic security and is ideally realised using photonics platforms.

Typically photons are used, as their quantum properties remain intact over the large distances required for useful communication. fibre optics provides the ideal channel for transmitting QKD photons, making use of the existing global fibre network. QKD differs fundamentally from other forms of encryption in the way it provides security. If a photon is intercepted in flight by a malicious third party, and the result is re-sent in the measured basis then the eavesdropper can be readily detected and communication ceased [176]. Note that in most, though not all, cases a key is distributed by quantum means rather than the whole message, simply because the bandwidth of quantum information transfer is very much lower than current classical optical communications.

However, in reality there are many challenges associated with the practical implementation of QKD schemes and in some areas nonlinear photonics can play a key role. Reliable photon sources such as those described earlier in this chapter can be used in these networks, and the inherent entanglement in photon pair sources can be exploited for communication. Quantum frequency conversion will allow disparate users of a quantum network to be able to swap information and entanglement, and for long distance quantum communication to become a reality resonant and cavity based sources of photons may be able to reach the stringent requirements to interface with quantum memories.

12.4.1 Single Photon Schemes

The most common QKD schemes [20, 177, 178] encode the key information using one of the degrees of freedom of a single photon. Photons must be generated that are completely indistinguishable in every degree of freedom except one, typically the polarisation, but arrival time and orbital angular momentum are also common choices. The polarisation state of a photon can take a continuum of values, so two orthogonal values are selected to represent the states $|0\rangle$ and $|1\rangle$. If one measures in that orthogonal basis and and receives a photon in state $|0\rangle$, one is guaranteed to measure $|0\rangle$. However, detection of quantum particles such as a single photon is probabilistic and if an eavesdropper makes a measurement in an arbitrary polarisation basis, there is a nonzero probability that they will get the wrong answer. Most importantly, when the eavesdropper makes a measurement in the incorrect basis the photon state is modified increasing the number of errors and revealing the presence of an attacker. Figure 12.30 shows a simple schematic of the BB84 protocol [20], the first and best-known representative of this class of schemes.

The single photon source required is imperfectly realised with an attenuated laser, having the advantages of a high clock speed, low timing jitter, robustness and cheap manufacturing. The major drawback is that the photon emission from a laser is described by a Poisson distribution and in this case the output is attenuated to the point where the probability of emitting one photon is larger than the probability of emitting two or more. A source that emits two or more photons is potentially vulnerable to attacks such as the "beam splitter attack" [176]. This multi-photon rate can be reduced by multiplexing many nonlinear sources according to the strategies discussed in Sect. 12.2.6, or by using true single photon emitters such as diamond N-V centres and quantum dots, but the technical challenges of these sources makes their use impractical at present.

Current commercial products offered allow links over 100 km with key distribution rates of several Mbit/s [179] using optical fibres, with each link costing at least USD $100,000. State of the art research has demonstrated real QKD channels running in parallel down the same fibre as secured 10 GBit/s bright classical signals. All channels were located in the telecommunication C-band of wavelengths, with standard channel spacing of 100 and a 200 GHz spacing between the quantum and classical signals [180, 181]. For ultimate security, the gold standard is a one-time pad which does not allow repeated use of any random key, with the key size equal to the message size. 100 kbps one-time pad video conferencing over 45 km

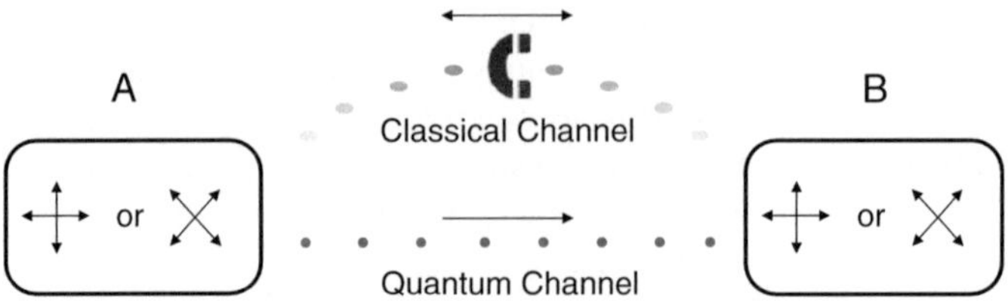

Fig. 12.30 BB84 QKD scheme. Alice, at location *A*, sends a stream of photons to Bob, at location *B*, with the photon polarisation randomly orientated between the orthogonal states of vertical (**0**) and horizontal (**1**). Alice also randomly changes the basis to a diagonal basis while sending the stream. Bob measures the incoming photons and records a **0** or **1** for vertical and horizontally polarised photons respectively. Both users then broadcast which photon they sent/measured in which basis, the normal or diagonal. When the bases are the same, they keep those **0**, **1** values and use them for their key. This is called key sifting. They can then encrypt messages and send them securely across public classical channels

was demonstrated at the Tokyo QKD network in a 2011 joint effort by NICT, NEC, Mitsubishi Electric and NTT [182].

In any optical fibre communication link photons are lost when they scatter off imperfections in the fibre. In the case of conventional communication, the information can be preserved by regenerating or amplifying the signal. In the quantum case, the "no-cloning theorem", which forbids the reliable copying of a quantum state, prevents a quantum signal from being amplified [176], thus the quantum key distribution rate drops with distance, eventually falling below a usable threshold.

In a recently developed technique called differential-phase-shift QKD [183] the key is encoded onto the phase of a coherent pulse, which is attenuated as it travels through the fibre, ensuring only 1 or fewer photons remain for detection. This has enabled impressive demonstrations of QKD over 100 km at 24 kbps by Japan NTT corporation [184] and more recently 260 km QKD over standard optical fibre was shown by China's Anhui Asky Quantum Technology Corporation [185].

12.4.2 Entanglement-Based Schemes

In 1992 the Eckert scheme for QKD was published [9], which exploits another quantum mechanical resource called entanglement. A pair of photons can be generated in such a way that the observable properties of one photon are inseparably linked to the other, and describing the state of one such photon independently of the other is both impossible and meaningless. Pairs of entangled photons can be generated using a nonlinear optical crystal, or from an atomic relaxation. If these photons are generated at the mid-point of a QKD link, they can be sent down an optical fibre to Alice and Bob, potentially doubling the maximum distance of a QKD link. When Alice and Bob happen to measure their photon in the same basis, entanglement ideally guarantees that their results will be perfectly correlated. Using

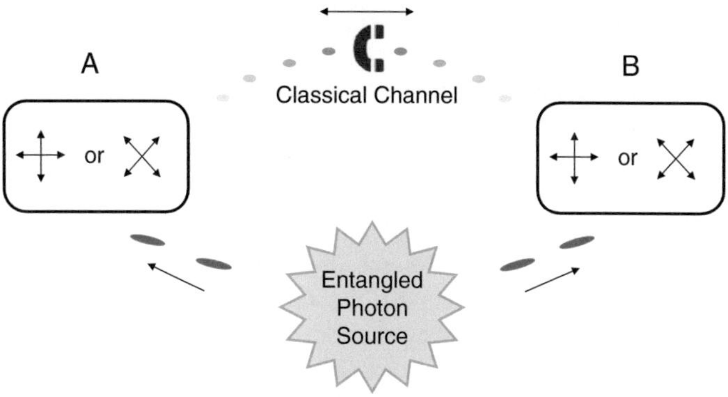

Fig. 12.31 Schematic of entanglement based QKD

this to generate identical keys allows secure communication over a public channel, as shown in Fig. 12.31. Observing either photon in the pair destroys the entanglement and the key correlations immediately drop, flagging the presence of an eavesdropper.

NuCrypt, [186] a spin-out from Northwestern University in the USA in 2003, offers the only entanglement based QKD product with a rack mountable 2.5 Gbit/s unit that sends the encrypted signal over the same fibre as the quantum keys. The entanglement is generated using highly nonlinear silica fibre, with the option of cooling in liquid nitrogen for increased performance. The European project SEC-OQC [187] has used such entangled photon pairs in combination with several other schemes, to connect 6 nodes securely. Toshiba demonstrated a 200 km fibre optic link using practical avalanche photo diode (APD) detectors [188]. Japan's NTT demonstrated a 100 km link using polarisation entangled photons generated in a CMOS compatible silicon chip.

12.4.3 Long Distance Quantum Communication

Long distance QKD is currently limited by losses in real world communication systems, whether they are based on free-space or optical fibre links, limiting communication to roughly less than 200 km [185, 189, 190]. As already mentioned, when dealing with single photons the no-cloning theorem disallows deterministic copying of quantum states, meaning that classical optical amplifiers are rendered useless. There are however a number of proposals to extend this distance. In free space links the use of a satellite based relay, or a simple reflector, could link two distant points on Earth, whereas in optical fibre links one would require a "quantum repeater" or "quantum relay" to forward on the photons carrying the quantum information.

12.4.3.1 Space-Based Systems

There have been a number of feasibility studies undertaken to discern whether satellite based QKD is a viable method to securely transfer information on a global scale [191, 192]. There are several possible methods for communication via this method—whether the receiver is placed on the ground or on the satellite, or if the satellite is a simple retro-reflector—each with its own inherent advantages and problems. One problem recently addressed was that of having a shared fixed reference frame to make basis choices for the sender and receiver, a hard problem when dealing with a satellite moving through space. Through the use of circular polarisation and integrated optics, researchers were able to devise a method of forming reference frame independent QKD protocols [193].

A seminal experiment in 2007 [189] demonstrated a 144 km free space link between two of the Canary Islands. On accounting for the degree of atmospheric scattering one would encounter between two points of equal height from sea level, this is comparable to sending a beam into space as the atmosphere becomes thinner with increasing height above ground level. Recently more advanced proposals have been put forward by a large number of universities to collaborate with the European Space Agency to implement satellite-based entanglement experiments and experiments to form a quantum link between the International Space Station and a telescope site on Earth [194]. One such experiment has been completed where a satellite is used as a weak reflector for a classical laser resulting in a channel containing single photons which is then experimentally verified through photon counting statistics, however no QKD has been performed using this method to date.

12.4.3.2 Quantum Repeaters and Relays

When sending classical communication through long distance optical fibre links one has the ability to amplify the channel at various locations to ensure that the signal-to-noise ratio is not degraded between users wishing to transfer data. Once again, this is impossible for quantum communication due to the impossibility of cloning states. Instead, proposed architectures for long distance quantum communication employ either a quantum repeater [195] shown in Fig. 12.32, which allows two-way communication, or a quantum relay [196] shown in Fig. 12.33, for one-way communication.

The aim of a quantum repeater is to supply one photon from an entangled pair to the next node in the communication chain whilst storing the other until all links have been set up. An entanglement swapping routine is then performed at each repeater on the stored photons until the two photons at the very end of the communication chain are themselves entangled. One can then perform a standard Eckert style QKD protocol. To perform the entanglement swapping routine one requires a small scale quantum information processing element, such as a controlled-NOT

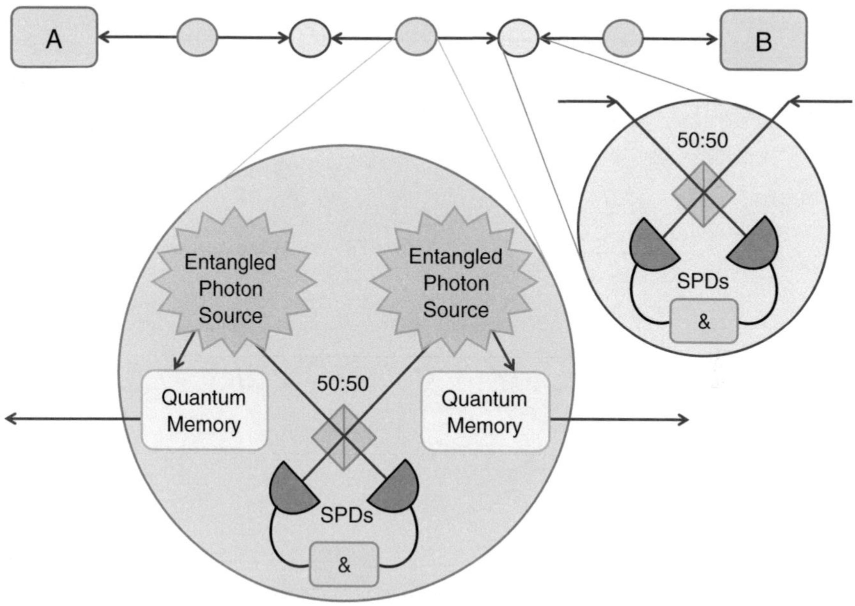

Fig. 12.32 A schematic of a quantum repeater. For Alice to communicate with Bob they must set up some entanglement between them. This is done through the use of a quantum repeater, shown in the *purple circle*, and joint quantum measurement, shown in the *red circle*. The quantum repeater contains entangled sources of photon pairs which are further entangled through measurement until entanglement swapping means that the photons at *A* and *B* are entangled. The quantum memory allows the initial entanglement to be set up before the other swapping occurs

gate [197], and single photon detection. The most challenging component in a quantum repeater is the quantum memory used to store the photons to ensure a successful swapping of entanglement along the chain. Current memory proposals [198] are either based on bulky and fragile atomic ensemble setups which are not easily used outside the lab, low temperature devices based on quantum dots, or diamond N-V centres which require significant further engineering of cavities and out-coupling before they can be viable as memories.

On the other hand, a quantum relay is a teleportation assisted device where an incoming photon is jointly measured with one photon of a correlated pair, shown schematically in Fig. 12.33. After the joint measurement the state of the input photon is teleported to the state of the other photon of the correlated pair which can then be transmitted. The joint measurement requires a form of quantum interference [18] as well as single photon detection. A preliminary experiment was recently published [196] demonstrating such a quantum relay device in a lithium niobate integrated architecture, however imperfect quantum interference and losses limit the usefulness of the device and no teleportation has been demonstrated thus far.

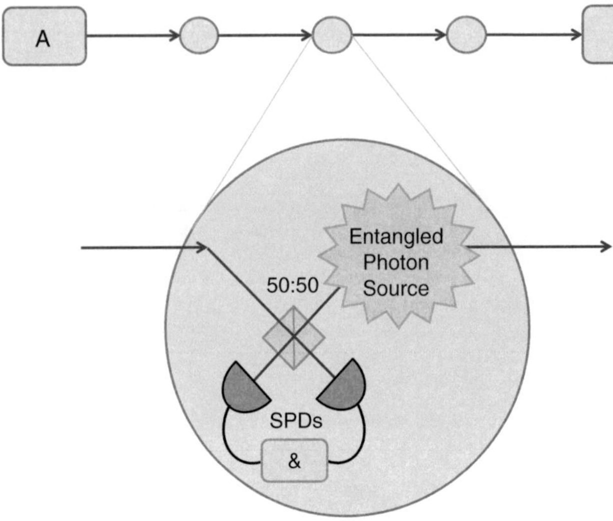

Fig. 12.33 A schematic of a quantum relay. A photon enters the relay device and is interfered with a photon from an entangled pair in a quantum measurement, which teleports the quantum state to the output photon for transmission. These can then be concatenated to extend the distance

12.5 Conclusion

It is clear that nonlinear photonics can provide enhanced functionality to a number of quantum technologies including in a quantum network, where quantum information can be transported securely when encoded to single photon states. There are many challenges to be overcome before we see this quantum photonic hardware in everyday use, however research into this area is fast paced and we expect to see quantum enhanced technologies and networks become prolific in the coming decades.

Acknowledgments This work was supported in part by the Centre of Excellence (CUDOS, project number CE110001018), Laureate Fellowship (FL120100029) and Discovery Early Career Researcher Award (DE130101148 and DE120100226) programs of the Australian Research Council (ARC). L.G.H. acknowledges support from a Macquarie University Research Fellowship (MQRF). K.S. would like to thank Matthew Rakher, Serkan Ates, and Imad Agha for helpful comments and contributions to work presented in Sect. 12.3.

References

1. W.K. Wooters, W.H. Zurek, A single quantum cannot be cloned. Nature **299**, 802–803 (1982)
2. D. Dieks, Communication by EPR devices. Phys. Lett. A **92**, 271–272 (1982)
3. W.J. Munro, K. Nemoto, T.P. Spiller, Weak nonlinearities: a new route to optical quantum computation. New J. Phys. **7**, 137 (2005)

4. N. Matsuda, R. Shimizu, Y. Mitsumori, H. Kosaka, K. Edamatsu, Observation of optical-fibre Kerr nonlinearity at the single-photon level. Nature Photon. **3**, 95–98 (2009)
5. V. Venkataraman, K. Saha, A.L. Gaeta, Phase modulation at the few-photon level for weak-nonlinearity-based quantum computing. Nature Photon. **7**, 138–141 (2013)
6. J.H. Shapiro, Single-photon Kerr nonlinearities do not help quantum computation. Phys. Rev. A **73**, 062305 (2006)
7. N.K. Langford, S. Ramelow, R. Prevedel, W.J. Munro, G.J. Milburn, A. Zeilinger, Efficient quantum computing using coherent photon conversion. Nature **478**, 360–363 (2011)
8. J.S. Bell, *Speakable and Unspeakable in Quantum Mechanics* (Cambridge University Press, Cambridge, 1987)
9. A. Ekert, Quantum cryptography based on Bell's theorem. Phys. Rev. Lett. **67**, 661–663 (1991)
10. A.K. Ekert, J.G. Rarity, P.R. Tapster, G.M. Palma, Practical quantum cryptography based on two-photon entanglement. Phys. Rev. Lett. **69**, 1293–1295 (1992)
11. J. Ye, D.W. Vernooy, H.J. Kimble, Trapping of single atoms in cavity QED. Phys. Rev. Lett. **83**, 4987–4990 (1999)
12. J.I. Cirac, P. Zoller, Quantum computation with cold trapped ions. Phys. Rev. Lett. **74**, 4091–4094 (1995)
13. S. Prawer, A.D. Greentree, Diamond for quantum computing. Science **320**, 1601–1602 (2008)
14. A. Politi, M. Cryan, J. Rarity, S. Yu, J. O'Brien, Silica-on-silicon waveguide quantum circuits. Science **320**, 646–649 (2008)
15. J. Brendel, N. Gisin, W. Tittel, H. Zbinden, Pulsed energy-time entangled twin-photon source for quantum communication. Phys. Rev. Lett. **82**, 2594–2597 (1999)
16. P.C. Humphreys, B.J. Metcalf, J.B. Spring, M. Moore, X.-M. Jin, M. Barbieri, S. Kolthammer, I.A. Walmsley, Linear optical quantum computing in a single spatial mode. Phys. Rev. Lett. **111**, 150501 (2013)
17. E. Knill, R. Laflamme, G.J. Milburn, A scheme for efficient quantum computation with linear optics. Nature **409**, 46–52 (2001)
18. C.K. Hong, Z.Y. Ou, L. Mandel, Measurement of subpicosecond time intervals between two photons by interference. Phys. Rev. Lett. **59**, 2044 (1987)
19. J.G. Rarity, P.R. Tapster, Fourth order interference in parametric down conversion. J. Opt. Soc. Am. B **6**, 1221 (1989)
20. C. Bennett, G. Brassard, Quantum cryptography: public key distribution and coin tossing, in *Proceedings of the International Conference on Computers, Systems & Signal Processing* (1984), pp. 175–179
21. M.A. Nielsen, I.L. Chuang, *Quantum Computation and Quantum Information* (Cambridge University Press, New York, 2004)
22. C. Santori, D. Fattal, J. Vucković, G.S. Solomon, Y. Yamamoto, Indistinguishable photons from a single-photon device. Nature **419**, 594–597 (2002)
23. S.V. Polyakov, A. Muller, E.B. Flagg, A. Ling, N. Borjemscaia, E. Van Keuren, A. Migdall, G.S. Solomon, Coalescence of single photons emitted by disparate single-photon sources: the example of InAs quantum dots and parametric down-conversion sources. Phys. Rev. Lett. **107**, 157402 (2011)
24. A. Huck, S. Kumar, A. Shakoor, U.L. Andersen, Controlled coupling of a single nitrogen-vacancy center to a silver nanowire. Phys. Rev. Lett. **106**, 096801 (2011)
25. F. Jelezko, J. Wrachtrup, Single defect centres in diamond: a review. Phys. Status Solidi A **203**, 3207–3225 (2006)
26. P. Gold, A. Thoma, S. Maier, S. Reitzenstein, C. Schneider, S. Höfling, M. Kamp, Two-photon interference from remote quantum dots with inhomogeneously broadened linewidths. Phys. Rev. B **89**, 035313 (2014)
27. R. Loudon, *The Quantum Theory of Light*, 3rd edn. (Oxford University Press, New York, 2000)
28. S. Scheel, Single-photon sources-an introduction. J. Mod. Opt. **56**, 141–160 (2009)

29. S. Fasel, O. Alibart, S. Tanzilli, P. Baldi, A. Beveratos, N. Gisin, H. Zbinden, High-quality asynchronous heralded single-photon source at telecom wavelength. New J. Phys. **6**, 163–164 (2004)
30. T. Pittman, B. Jacobs, J. Franson, Heralding single photons from pulsed parametric down-conversion. Opt. Commun. **246**, 545–550 (2005)
31. O. Alibart, D.B. Ostrowsky, P. Baldi, S. Tanzilli, High-performance guided-wave asynchronous heralded single-photon source. Opt. Lett. **30**, 1539–1541 (2005)
32. A. Soujaeff, S. Takeuchi, K. Sasaki, T. Hasegawa, M. Matsui, Heralded single photon source at 1550 nm from pulsed parametric down conversion. J. Mod. Opt. **54**, 467–471 (2007)
33. P. Mosley, J. Lundeen, B. Smith, P. Wasylczyk, A. U'Ren, C. Silberhorn, I. Walmsley, Heralded generation of ultrafast single photons in pure quantum states. Phys. Rev. Lett. **100**, 133601 (2008)
34. Z.H. Levine, J. Fan, J. Chen, A. Ling, A. Migdall, Heralded, pure-state single-photon source based on a potassium titanyl phosphate waveguide. Opt. Express **18**, 3708–3718 (2010)
35. M.D. Cunha Pereira, F.E. Becerra, B.L. Glebov, J. Fan, S.W. Nam, A. Migdall, Demonstrating highly symmetric single-mode, single-photon heralding efficiency in spontaneous parametric downconversion. Opt. Lett. **38**, 1609–11 (2013)
36. S. Ramelow, A. Mech, M. Giustina, Highly efficient heralding of entangled single photons. Opt. Express **21**, 6707–6717 (2013)
37. E. Goldschmidt, M. Eisaman, J. Fan, S. Polyakov, A. Migdall, Spectrally bright and broad fiber-based heralded single-photon source. Phys. Rev. A **78**, 013844 (2008)
38. A.R. McMillan, J. Fulconis, M. Halder, C. Xiong, J.G. Rarity, W.J. Wadsworth, Narrowband high-fidelity all-fibre source of heralded single photons at 1570 nm. Opt. Express **17**, 6156–6165 (2009)
39. C. Söller, O. Cohen, B.J. Smith, I.A. Walmsley, C. Silberhorn, High-performance single-photon generation with commercial-grade optical fiber. Phys. Rev. A **83**, 031806 (2011)
40. L. Yang, X. Ma, X. Guo, L. Cui, X. Li, Characterization of a fiber-based source of heralded single photons. Phys. Rev. A **83**, 053843 (2011)
41. M. Davanço, J.R. Ong, A.B. Shehata, A. Tosi, I. Agha, S. Assefa, F. Xia, W.M.J. Green, S. Mookherjea, K. Srinivasan, Telecommunications-band heralded single photons from a silicon nanophotonic chip. Appl. Phys. Lett. **100**, 261104 (2012)
42. J. Spring, P. Salter, B. Metcalf, P. Humphreys, M. Moore, N. Thomas-Peter, M. Barbieri, X.-M. Jin, N. Langford, W. Kolthammer, M. Booth, I. Walmsley, On-chip low loss heralded source of pure single photons. Opt. Express **21**, 13522–13532 (2013)
43. A. Christ, C. Silberhorn, Limits on the deterministic creation of pure single-photon states using parametric down-conversion. Phys. Rev. A **85**, 023829 (2012)
44. M.J. Collins, C. Xiong, I.H. Rey, T.D. Vo, J. He, S. Shahnia, C. Reardon, T.F. Krauss, M. J. Steel, A.S. Clark, B.J. Eggleton, Integrated spatial multiplexing of heralded single-photon sources. Nat. Commun. **4**, 2582 (2013)
45. A. Yariv, *Quantum Electronics,* 2nd edn. (Wiley, New York, 1975)
46. G.P. Agrawal, *Nonlinear Fiber Optics*, 4th edn. (Academic Press, Burlington, 2007)
47. L.G. Helt, M. Liscidini, J.E. Sipe, How does it scale? comparing quantum and classical nonlinear optical processes in integrated devices. J. Opt. Soc. Am. B **29**, 2199–2212 (2012)
48. H. Takesue, K. Shimizu, Effects of multiple pairs on visibility measurements of entangled photons generated by spontaneous parametric processes. Opt. Commun. **283**, 276–287 (2010)
49. M.J. Stevens, S. Nam, in *Single-Photon Generation and Detection—Physics and Applications*, ed. by J.F.A. Migdall, S. Polyakov, J.C. Bienfang. Photon Statistics, Measurements, and Measurement Tools (Academic Press, Burlington, 2013)
50. J.G. Rarity, J. Fulconis, J. Duligall, W.J. Wadsworth, P.S.J. Russell, Photonic crystal fiber source of correlated photon pairs. Opt. Express **13**, 534 (2005)
51. E. Engin, D. Bonneau, C.M. Natarajan, A.S. Clark, M.G. Tanner, R.H. Hadfield, S.N. Dorenbos, V. Zwiller, K. Ohira, N. Suzuki, H. Yoshida, N. Iizuka, M. Ezaki, J.L. O'Brien,

M.G. Thompson, Photon pair generation in a silicon micro-ring resonator with reverse bias enhancement. Opt. Express **21**, 27826–27834 (2013)

52. A. Christ, A. Fedrizzi, H. Hübel, T. Jennewein, C. Silberhorn, in *Single-Photon Generation and Detection—Physics and Applications,* ed. by J.F.A. Migdall, S. Polyakov, J.C. Bienfang. Parametric Down-Conversion (Academic Press, Burlington, 2013)
53. P. Kwiat, K. Mattle, H. Weinfurter, A. Zeilinger, A. Sergienko, Y. Shih, New high-intensity source of polarization-entangled photon pairs. Phys. Rev. Lett. **75**, 4337–4341 (1995)
54. A. Peruzzo, M. Lobino, J. Matthews, N. Matsuda, A. Politi, K. Poulios, X.-Q. Zhou, Y. Lahini, N. Ismail, K. Wörhoff, Y. Bromberg, Y. Silberberg, M. Thompson, J. O'Brien, Quantum walks of correlated photons. Science **329**, 1500–1503 (2010)
55. A. Aspuru-Guzik, P. Walther, Photonic quantum simulators. Nature Phys. **8**, 285–291 (2012)
56. P. Kwiat, Nonclassical Effects from Spontaneous Parametric Down-conversion: Adventures in Quantum Wonderland, Ph.D thesis, http://research.physics.illinois.edu/QI/Photonics/theses.html, 1993
57. A. Halevy, E. Megidish, L. Dovrat, H. Eisenberg, P. Becker, L. Bohatý, The biaxial nonlinear crystal BiB_3O_6 as a polarization entangled photon source using non-collinear type-II parametric down-conversion. Opt. Express **19**, 20420–20434 (2011)
58. R. Rangarajan, M. Goggin, P. Kwiat, Optimizing type-I polarization-entangled photons. Opt. Express **17**, 18920–18933 (2009)
59. M. Fiorentino, S. Spillane, R. Beausoleil, T. Roberts, P. Battle, M. Munro, Spontaneous parametric down-conversion in periodically poled KTP waveguides and bulk crystals. Opt. Express **15**, 7479–7488 (2007)
60. T. Zhong, F. Wong, T. Roberts, P. Battle, High performance photon-pair source based on a fiber-coupled periodically poled $KTiOPO_4$ waveguide. Opt. Express **17**, 12019–12030 (2009)
61. S.V. Zhukovsky, L.G. Helt, P. Abolghasem, D. Kang, J.E. Sipe, A.S. Helmy, Bragg reflection waveguides as integrated sources of entangled photon pairs. J. Opt. Soc. Am. B **29**, 2516–2523 (2012)
62. R.T. Horn, P. Kolenderski, D. Kang, P. Abolghasem, C. Scarcella, A.D. Frera, A. Tosi, L.G. Helt, S.V. Zhukovsky, J.E. Sipe, G. Weihs, A.S. Helmy, T. Jennewein, Inherent polarization entanglement generated from a monolithic semiconductor chip. Sci. Rep. **3**, 2314 (2013)
63. A. Vallés, M. Hendrych, J. Svozilík, R. Machulka, P. Abolghasem, D. Kang, B.J. Bijlani, A. S. Helmy, J.P. Torres, Generation of polarization-entangled photon pairs in a Bragg reflection waveguide. Opt. Express **21**, 10841–10849 (2013)
64. Z. Yang, P. Chak, A.D. Bristow, H.M. van Driel, R. Iyer, J.S. Aitchison, A.L. Smirl, J.E. Sipe, Enhanced second-harmonic generation in AlGaAs microring resonators. Opt. Lett. **32**, 826–828 (2007)
65. R. Boyd, *Nonlinear Optics*, 2nd edn. (Academic Press, Burlington, 2003)
66. M. Hunault, H. Takesue, O. Tadanaga, Y. Nishida, M. Asobe, Generation of time-bin entangled photon pairs by cascaded second-order nonlinearity in a single periodically poled $linbO_3$ waveguide. Opt. Lett. **35**, 1239–1241 (2010)
67. A. Martin, A. Issautier, H. Herrmann, W. Sohler, D. Ostrowsky, O. Alibart, S. Tanzilli, A polarization entangled photon-pair source based on a type-II PPLN waveguide emitting at a telecom wavelength. New J. Phys. **12**, 103005 (2010)
68. M. Lobino, G. Marshall, C. Xiong, A. Clark, D. Bonneau, C. Natarajan, M. Tanner, R. Hadfield, S. Dorenbos, T. Zijlstra, V. Zwiller, M. Marangoni, R. Ramponi, M. Thompson, B. Eggleton, J. O'Brien, Correlated photon-pair generation in a periodically poled MgO doped stoichiometric lithium tantalate reverse proton exchanged waveguide. Appl. Phys. Lett. **99**, 081110 (2011)
69. A.M. Brańczyk, A. Fedrizzi, T.M. Stace, T.C. Ralph, A.G. White, Engineered optical nonlinearity for quantum light sources. Opt. Express **19**, 55–65 (2011)
70. H. Zhang, X.-M. Jin, J. Yang, H.-N. Dai, S.-J. Yang, T.-M. Zhao, J. Rui, Y. He, X. Jiang, F. Yang, G.-S. Pan, Z.-S. Yuan, Y. Deng, Z.-B. Chen, X.-H. Bao, S. Chen, B. Zhao, J.-W. Pan,

Preparation and storage of frequency-uncorrelated entangled photons from cavity-enhanced spontaneous parametric downconversion. Nature Photon. **5**, 628–632 (2011)
71. C. Clausen, I. Usmani, F. Bussières, N. Sangouard, M. Afzelius, H. Riedmatten, N. Gisin, Quantum storage of photonic entanglement in a crystal. Nature **469**, 508–511 (2011)
72. X.-H. Bao, Y. Qian, J. Yang, H. Zhang, Z.-B. Chen, T. Yang, J.-W. Pan, Generation of narrow-band polarization-entangled photon pairs for atomic quantum memories. Phys. Rev. Lett. **101**, 190501 (2008)
73. E. Pomarico, B. Sanguinetti, C. Osorio, H. Herrmann, R. Thew, Engineering integrated pure narrow-band photon sources. New J. Phys **14**, 033008 (2012)
74. A. McMillan, Y. Huang, B. Bell, A.S. Clark, P. Kumar, J. Rarity, in *Single-Photon Generation and Detection—Physics and Applications, chap. 12*, ed. by A. Migdall, S. Polyakov, J. Fan. Four-Wave Mixing in Single-Mode Optical Fibers (Academic Press, Burlington, 2013), pp. 411–465
75. C. Xiong, Nonlinearity in Photonic Crystal Fibres. Ph.D thesis, University of Bath, 2008
76. J. Sharping, M. Fiorentino, P. Kumar, Observation of twin-beam-type quantum correlation in optical fiber. Opt. Lett. **26**, 367–369 (2001)
77. M. Fiorentino, P. Voss, J. Sharping, P. Kumar, All-fiber photon pair source for quantum communications. IEEE Photon. Tech. Lett. **14**, 983–985 (2002)
78. J. Sharping, J. Chen, X. Li, P. Kumar, Quantum correlated twin photons from microstructured fiber. Opt. Express **12**, 3086–3094 (2004)
79. X. Li, P. Voss, J. Sharping, P. Kumar, Optical-fiber source of polarization-entangled photons in the 1550 nm telecom band. Phys. Rev. Lett. **94**, 053601 (2005)
80. H. Takesue, K. Inoue, 1.5-μm band quantum-correlated photon pair generation in dispersion-shifted fiber suppression of noise photons by cooling fiber. Opt. Express **13**, 7832–7839 (2005)
81. S. Dyer, M. Stevens, B. Baek, S. Nam, High-efficiency, ultra low-noise all-fiber photon-pair source. Opt. Express **16**, 9966–9977 (2008)
82. J. Fulconis, O. Alibart, J. O'Brien, W. Wadsworth, J. Rarity, Nonclassical interference and entanglement generation using a photonic crystal fiber pair photon source. Phys. Rev. Lett. **99**, 120501 (2007)
83. J. Fulconis, O. Alibart, W. Wadsworth, J. Rarity, Quantum interference with photon pairs using two micro-structured fibers. New J. Phys. **9**, 276 (2007)
84. M. Halder, J. Fulconis, B. Cemlyn, A. Clark, C. Xiong, W. Wadsworth, J. Rarity, Nonclassical 2-photon interference with separate intrinsically narrowband fibre sources. Opt. Express **17**, 4670–4676 (2009)
85. A. Clark, B. Bell, J. Fulconis, M. Halder, B. Cemlyn, A. Olivier, C. Xiong, W. Wadsworth, J. Rarity, Intrinsically narrowband pair photon generation in microstructured fibres. New J. Phys. **13**, 065009 (2011)
86. J. Fan, A. Dogariu, L. Wang, Generation of correlated photon pairs in a microstructure fiber. Opt. Lett. **30**, 1530–1532 (2005)
87. J. Chen, K. Lee, C. Liang, P. Kumar, Fiber-based telecom-band degenerate-frequency source of entangled photon pairs. Opt. Lett. **31**, 2798–2799 (2006)
88. C. Xiong, G. Marshall, A. Peruzzo, M. Lobino, A. Clark, D.-Y. Choi, S. Madden, C. Natarajan, M. Tanner, R. Hadfield, S. Dorenbos, T. Zijlstra, V. Zwiller, M. Thompson, J. Rarity, M. Steel, B. Luther-Davies, B. Eggleton, J. O'Brien, Generation of correlated photon pairs in a chalcogenide As_2S_3 waveguide. Appl. Phys. Lett. **98**, 051101 (2011)
89. M. Collins, A. Clark, J. He, D.-Y. Choi, R. Williams, A. Judge, S. Madden, M. Withford, M. Steel, B. Luther-Davies, C. Xiong, B. Eggleton, Low raman-noise correlated photon-pair generation in a dispersion engineered chalcogenide As_2S_3 planar waveguide. Opt. Lett. **37**, 3393–3395 (2012)
90. J. He, C. Xiong, A. Clark, M. Collins, X. Gai, D.-Y. Choi, S. Madden, B. Luther-Davies, B. Eggleton, Effect of low-Raman window position on correlated photon-pair generation in a chalcogenide $Ge_{11.5}$ As_{24} $Se_{64.5}$ nanowire. J. Appl. Phys. **112**, 123101 (2012)

91. J. Sharping, K. Lee, M. Foster, A. Turner, B. Schmidt, M. Lipson, A. Gaeta, P. Kumar, Generation of correlated photons in nanoscale silicon waveguides. Opt. Express **14**, 12388–12393 (2006)
92. K. Harada, H. Takesue, H. Fukuda, T. Tsuchizawa, T. Watanabe, K. Yamada, Y. Tokura, S. Itabashi, Generation of high-purity entangled photon pairs using silicon wire waveguide. Opt. Express **16**, 20368–20373 (2008)
93. S. Clemmen, K. Huy, W. Bogaerts, R. Baets, P. Emplit, S. Massar, Continuous wave photon pair generation in silicon-on-insulator waveguides and ring resonators. Opt. Express **17**, 16558–16570 (2009)
94. K. Harada, H. Takesue, H. Fukuda, T. Tsuchizawa, T. Watanabe, K. Yamada, Y. Tokura, S. Itabashi, Indistinguishable photon pair generation using two independent silicon wire waveguides. New J. Phys. **13**, 065005 (2011)
95. S. Clemmen, A. Perret, S. Selvaraja, W. Bogaerts, D. van, Thourhout, R. Baets, P. Emplit, S. Massar, Generation of correlated photons in hydrogenated amorphous-silicon waveguides. Opt. Lett. **35**, 3483–3485 (2010)
96. B. Eggleton, B. Luther-Davies, K. Richardson, Chalcogenide photonics. Nature Photon. **5**, 141–148 (2011)
97. S. Azzini, D. Grassani, M. Strain, M. Sorel, L. Helt, J. Sipe, M. Liscidini, M. Galli, D. Bajoni, Ultra-low power generation of twin photons in a compact silicon ring resonator. Opt. Express **20**, 23100–23107 (2012)
98. C. Xiong, C. Monat, M. Collins, L. Tranchant, D. Petiteau, A. Clark, C. Grillet, G. Marshall, M. Steel, J. Li, L. O'Faolain, T. Krauss, B. Eggleton, Characteristics of correlated photon pairs generated in ultra-compact silicon slow-light photonic crystal waveguides. IEEE J. Sel. Topics Quantum Electron. **18**, 1676–1683 (2012)
99. N. Matsuda, H. Takesue, K. Shimizu, Y. Tokura, E. Kuramochi, M. Notomi, Slow light enhanced correlated photon pair generation in photonic-crystal coupled-resonator optical waveguides. Opt. Express **21**, 8596–8604 (2013)
100. W. Jiang, X. Lu, J. Zhang, O. Painter, Q. Lin, A silicon-chip source of bright photon-pair comb. arXiv:1210.4455 (2012)
101. A. Clark, C. Husko, M. Collins, G. Lehoucq, S. Xavier, A. De, Rossi, S. Combrié, C. Xiong, B. Eggleton, Heralded single-photon source in a III–V photonic crystal. Opt. Lett. **38**, 649–651 (2013)
102. C. Husko, A. Clark, M. Collins, A. De, Rossi, S. Combrié, G. Lehoucq, I. Rey, T. Krauss, C. Xiong, B. Eggleton, Multi-photon absorption limits to heralded single photon sources. Sci. Rep. **3**, 3087 (2013)
103. J. Li, T.P. White, L. O'Faolain, A. Gomez-Iglesias, T.F. Krauss, Systematic design of flat band slowlight in photonic crystal waveguides. Opt. Express **16**, 6227–6232 (2008)
104. T.F. Krauss, Slow light in photonic crystal waveguides. J. Phys. D: Appl. Phys. **40**, 2666–2670 (2007)
105. J. Li, L. O'Faolain, I.H. Rey, T.F. Krauss, Four-wave mixing in photonic crystal waveguides: slow light enhancements and limitations. Opt. Express **19**, 4458–4463 (2011)
106. A. Yariv, Y. Xu, R.K. Lee, A. Scherer, Coupled-resonator optical waveguide, a proposal and analysis. Opt. Lett. **24**, 711–713 (1999)
107. R. Kumar, J. Ong, J. Recchio, S. Mookherjea, K. Srinivasan, Wide-span and tunable-wavelength photon pairs around 1.55 μm from a silicon nanophotonic chip, 1304:4989v1 (2013)
108. Z. Yang, M. Liscidini, J. Sipe, Spontaneous parametric down-conversion in waveguides: a backward Heisenberg picture approach. Phys. Rev. A **77**, 033808 (2008)
109. L.G. Helt, E.Y. Zhu, M. Liscidini, L. Qian, J.E. Sipe, Proposal for in-fiber generation of telecom-band polarization-entangled photon pairs using a periodically poled fiber. Opt. Lett. **34**, 2138–2140 (2009)
110. M. Liscidini, L.G. Helt, J.E. Sipe, Asymptotic fields for a Hamiltonian treatment of nonlinear electromagnetic phenomena. Phys. Rev. A **85**, 013833 (2012)

111. P. Kwiat, E. Waks, A. White, Ultrabright source of polarization-entangled photons. Phys. Rev. A **60**, 773–776 (1999)
112. S.V. Zhukovsky, L.G. Helt, D. Kang, P. Abolghasem, A.S. Helmy, J.E. Sipe, Generation of maximally-polarization-entangled photons on a chip. Phys. Rev. A **85**, 013838 (2012)
113. J. Bell, On the Einstein Podolsky Rosen paradox. Physics **1**, 195–200 (1964)
114. J. Clauser, M. Horne, Proposed experiment to test local hidden-variable theories. Phys. Rev. Lett. **23**, 880–884 (1969)
115. T. Humble, W. Grice, Effects of spectral entanglement in polarization-entanglement swapping and type-I fusion gates. Phys. Rev. A **77**, 022312 (2008)
116. C. Law, J. Eberly, Analysis and interpretation of high transverse entanglement in optical parametric down conversion. Phys. Rev. Lett. **92**, 127903 (2004)
117. W. Grice, A. URen, I. Walmsley, Eliminating frequency and space-time correlations in multiphoton states. Phys. Rev. A **64**, 063815 (2001)
118. K. Garay-Palmett, H. McGuinness, Photon pair-state preparation with tailored spectral properties by spontaneous four-wave mixing in photonic-crystal fiber. Opt. Express **15**, 2915–2920 (2007)
119. O. Cohen, J. Lundeen, B. Smith, G. Puentes, P. Mosley, I. Walmsley, Tailored photon-pair generation in optical fibers. Phys. Rev. Lett. **102**, 123603 (2009)
120. Z. Walton, A. Sergienko, B. Saleh, M. Teich, Generation of polarization-entangled photon pairs with arbitrary joint spectrum. Phys. Rev. A **70**, 052317 (2004)
121. L.G. Helt, Z. Yang, M. Liscidini, J.E. Sipe, Spontaneous four-wave mixing in microring resonators. Opt. Lett. **35**, 3006–3008 (2010)
122. B.J. Smith, P. Mahou, O. Cohen, J.S. Lundeen, I.A. Walmsley, Photon pair generation in birefringent optical fibers. Opt. Express **17**, 23589–23602 (2009)
123. I. Jizan, A. Clark, L. Helt, M. Collins, E. Mägi, C. Xiong, M. Steel, B. Eggleton, High-resolution measurement of spectral quantum correlations in the telecommunication band. Opt. Commun. **327**, 45–48 (2014)
124. J.M. Lukens, A. Dezfooliyan, C. Langrock, M.M. Fejer, D.E. Leaird, A.M. Weiner, Biphoton manipulation with a fiber-based pulse shaper. Opt. Lett. **38**, 4652–4655 (2013)
125. M. Avenhaus, A. Eckstein, P.J. Mosley, C. Silberhorn, Fiber-assisted single-photon spectrograph. Opt. Lett. **34**, 2873–2875 (2009)
126. M. Liscidini, J.E. Sipe, Stimulated emission tomography. Phys. Rev. Lett. **111**, 193602 (2013)
127. A. Eckstein, G. Boucher, A. Lemaître, P. Filloux, I. Favero, G. Leo, J.E. Sipe, M. Liscidini, S. Ducci, High-resolution spectral characterization of two photon states via classical measurements. Laser Photonics Rev. **8**, L76–L80 (2014)
128. A. Christ, K. Laiho, A. Eckstein, K.N. Cassemiro, C. Silberhorn, Probing multimode squeezing with correlation functions. New J. Phys. **13**, 033027 (2011)
129. R.J. Glauber, The quantum theory of optical coherence. Phys. Rev. **130**, 2529–2539 (1963)
130. R.J. Glauber, Coherent and incoherent states of the radiation field. Phys. Rev. **131**, 2766–2788 (1963)
131. R.H. Brown, R.Q. Twiss, Correlation between photons in two coherent beams of light. Nature **177**, 27–29 (1956)
132. A. Migdall, D. Branning, S. Castelletto, Tailoring single-photon and multiphoton probabilities of a single-photon on-demand source. Phys. Rev. A **66**, 1–4 (2002)
133. J.H. Shapiro, F.N. Wong, On-demand single-photon generation using a modular array of parametric downconverters with electro-optic polarization controls. Opt. Lett. **32**, 2698–2700 (2007)
134. X.-S. Ma, S. Zotter, J. Kofler, T. Jennewein, A. Zeilinger, Experimental generation of single photons via active multiplexing. Phys. Rev. A **83**, 043814 (2011)
135. T. Jennewein, M. Barbieri, A.G. White, Single-photon device requirements for operating linear optics quantum computing outside the post-selection basis. J. Mod. Opt. **58**, 276–287 (2011)

136. C. Xiong, T.D. Vo, M.J. Collins, J. Li, T.F. Krauss, M.J. Steel, A.S. Clark, B.J. Eggleton, Bidirectional multiplexing of heralded single photons from a silicon chip. Opt. Lett. **38**, 5176–5179 (2013)
137. T. Meany, L.A. Ngah, M.J. Collins, A.S. Clark, R.J. Williams, B.J. Eggleton, M.J. Steel, M. J. Withford, O. Alibart, S. Tanzilli, Hybrid photonic circuit for multiplexed heralded single photons. Laser Photonics Rev. **8**, L42–L46 (2014)
138. L. Mazzarella, F. Ticozzi, A.V. Sergienko, G. Vallone, P. Villoresi, Asymmetric architecture for heralded single-photon sources. Phys. Rev. A **88**, 023848 (2013)
139. J. Mower, D. Englund, Efficient generation of single and entangled photons on a silicon photonic integrated chip. Phys. Rev. A **84**, 052326 (2011)
140. P. Kumar, Quantum frequency conversion. Opt. Lett. **15**, 1476–1478 (1990)
141. J. Huang, P. Kumar, Observation of quantum frequency conversion. Phys. Rev. Lett. **68**, 2153–2156 (1992)
142. S. Tanzilli, W. Tittel, M. Halder, O. Alibart, P. Baldi, N. Gisin, H. Zbinden, A photonic quantum information interface. Nature **437**, 116–120 (2005)
143. A.P. Vandevender, P.G. Kwiat, High efficiency single photon detection via frequency up-conversion. J. Mod. Opt. **51**, 1433–1445 (2004)
144. M.A. Albota, F.N.C. Wong, Efficient single-photon counting at 1.55 microm by means of frequency upconversion. Opt. Lett. **29**, 1449–1451 (2004)
145. M.G. Raymer, K. Srinivasan, Manipulating the color and shape of single photons. Phys. Today **65**, 32 (2012)
146. C. Langrock, E. Diamanti, R.V. Roussev, Y. Yamamoto, M.M. Fejer, H. Takesue, Highly efficient single-photon detection at communication wavelengths by use of upconversion in reverse-proton-exchanged periodically poled $LiNbO_3$ waveguides. Opt. Lett. **30**, 1725–1727 (2005)
147. J. Pelc, L. Ma, C. Phillips, Q. Zhang, Long-wavelength-pumped upconversion single-photon detector at 1550 nm: performance and noise analysis. Opt. Express **19**, 21445–21456 (2011)
148. J.S. Pelc, C. Langrock, Q. Zhang, M.M. Fejer, Influence of domain disorder on parametric noise in quasi-phase-matched quantum frequency converters. Opt. Lett. **35**, 2804–2806 (2010)
149. S. Ates, I. Agha, A. Gulinatti, I. Rech, M.T. Rakher, A. Badolato, K. Srinivasan, Two-photon interference using background-free quantum frequency conversion of single photons emitted by an InAs quantum dot. Phys. Rev. Lett. **109**, 147405 (2012)
150. M. Rakher, L. Ma, O. Slattery, X. Tang, K. Srinivasan, Quantum transduction of telecommunications-band single photons from a quantum dot by frequency upconversion. Nature Photon. **4**, 786–791 (2010)
151. M.T. Rakher, L. Ma, M. Davanço, O. Slattery, X. Tang, K. Srinivasan, Simultaneous wavelength translation and amplitude modulation of single photons from a quantum dot. Phys. Rev. Lett. **107**, 083602 (2011)
152. H. Takesue, Erasing distinguishability using quantum frequency up-conversion. Phys. Rev. Lett. **101**, 173901 (2008)
153. R. Ikuta, Y. Kusaka, T. Kitano, H. Kato, T. Yamamoto, M. Koashi, N. Imoto, Wide-band quantum interface for visible-to-telecommunication wavelength conversion. Nat. Commun. **2**, 1544 (2011)
154. S. Zaske, A. Lenhard, C.A. Keßler, J. Kettler, C. Hepp, C. Arend, R. Albrecht, W.-M. Schulz, M. Jetter, P. Michler, C. Becher, Visible-to-telecom quantum frequency conversion of light from a single quantum emitter. Phys. Rev. Lett. **109**, 147404 (2012)
155. K. De Greve, L. Yu, P.L. McMahon, J.S. Pelc, C.M. Natarajan, N.Y. Kim, E. Abe, S. Maier, C. Schneider, M. Kamp, S. Höfling, R.H. Hadfield, A. Forchel, M.M. Fejer, Y. Yamamoto, Quantum-dot spin-photon entanglement via frequency downconversion to telecom wavelength. Nature **491**, 421–425 (2012)
156. C. McKinstrie, J. Harvey, S. Radic, M. Raymer, Translation of quantum states by four-wave mixing in fibers. Opt. Express **13**, 9131–9142 (2005)

157. K. Uesaka, K.K.-Y. Wong, Wavelength exchange in a highly nonlinear dispersion-shifted fibre: theory and experiments. IEEE J. Sel. Top. Quantum Electron. **8**, 560–568 (2002)
158. H.J. McGuinness, M.G. Raymer, C.J. McKinstrie, S. Radic, Quantum frequency translation of single-photon states in a photonic crystal fiber. Phys. Rev. Lett. **105**, 093604 (2010)
159. A.H. Gnauck, R.M. Jopson, C.J. McKinstrie, J.C. Centanni, S. Radic, Demonstration of low-noise frequency conversion by bragg scattering in a fiber. Opt. Express **14**, 8989–8994 (2006)
160. A.S. Clark, S. Shahnia, M.J. Collins, C. Xiong, B.J. Eggleton, High-efficiency frequency conversion in the single-photon regime. Opt. Lett. **38**, 947–949 (2013)
161. A. Guarino, G. Poberaj, D. Rezzonico, R. Degl'Innocenti, P. Günter, Electro-optically tunable microring resonators in lithium niobate. Nature Photon. **1**, 407–410 (2007)
162. L. Lanco, S. Ducci, J.P. Likforman, M. Ravaro, P. Filloux, X. Marcadet, G. Leo, V. Berger, Backward difference frequency generation in an AlGaAs waveguide. Appl. Phys. Lett. **89**, 031106 (2006)
163. J. Han, P. Abolghasem, B.J. Bijlani, A. Helmy, Continuous-wave sum-frequency generation in AlGaAs Bragg reflection waveguides. Opt. Lett. **34**, 3656–3658 (2009)
164. K. Rivoire, Z. Lin, F. Hatami, W.T. Masselink, J. Vukovi, Second harmonic generation in gallium phosphide photonic crystal nanocavities with ultralow continuous wave pump power. Opt. Express **17**, 22609–22615 (2009)
165. C. Xiong, W. Pernice, K.K. Ryu, C. Schuck, K.Y. Fong, T. Palacios, H.X. Tang, Integrated GaN photonic circuits on silicon (100) for second harmonic generation. Opt. Express **19**, 10462–10470 (2011)
166. Y. Ota, K. Watanabe, S. Iwamoto, Y. Arakawa, Nanocavity-based self-frequency conversion laser. Opt. Express **21**, 19778–19789 (2013)
167. P.S. Kuo, J. Bravo-Abad, G.S. Solomon, Second-harmonic generation using quasi-phasematching in a GaAs whispering-gallery-mode microcavity. Nat. Commun. **5**, 3109 (2014)
168. I. Agha, M. Davanço, B. Thurston, K. Srinivasan, Low-noise chip-based frequency conversion by four-wave-mixing Bragg scattering in SiN(x) waveguides. Opt. Lett. **37**, 2997–2999 (2012)
169. I. Agha, S. Ates, M. Davanço, K. Srinivasan, A chip-scale, telecommunications-band frequency conversion interface for quantum emitters. Opt. Express **21**, 1476–1478 (2013)
170. X. Liu, R. Osgood, Y. Vlasov, W. Green, Mid-infrared optical parametric amplifier using silicon nanophotonic waveguides. Nature Photon. **4**, 2–5 (2010)
171. S. Zlatanovic, J. Park, S. Moro, Mid-infrared wavelength conversion in silicon waveguides using ultracompact telecom-band-derived pump source. Nat. Photonics **4**, 561–564 (2010)
172. D. Kielpinski, J.F. Corney, H.M. Wiseman, Quantum optical waveform conversion. Phys. Rev. Lett. **106**, 130501 (2011)
173. B. Brecht, A. Eckstein, A. Christ, H. Suche, C. Silberhorn, From quantum pulse gate to quantum pulse shaper—engineered frequency conversion in nonlinear optical waveguides. New J. Phys. **13**, 065029 (2011)
174. C.J. McKinstrie, L. Mejling, M.G. Raymer, K. Rottwitt, Quantum-state-preserving optical frequency conversion and pulse reshaping by four-wave mixing. Phys. Rev. A **85**, 053829 (2012)
175. D. Reddy, M. Raymer, C. McKinstrie, Efficient sorting of quantum-optical wave packets by temporal-mode interferometry. Opt. Lett. **39**, 2924–2927 (2014)
176. N. Gisin, G. Ribordy, W. Tittel, Quantum cryptography. Rev. Mod. Phys. **74**, 145–195 (2002)
177. C. Bennet, Quantum cryptography using any two nonorthogonal states. Phys. Rev. Lett. **68**, 3121–3124 (1992)
178. V. Scarani, A. Acin, G. Ribordy, N. Gisin, Quantum cryptography protocol robust against photon number splitting attacks for weak laser pulse implementations. Phys. Rev. Lett. **92**, 057901 (2004)
179. ID-Quantique website, retrieved on 6. Aug 2012

180. P. Eraerds, N. Walenta, M. Legre, N. Gisin, H. Zbinden, Quantum key distribution and 1gbps data encryption over a single fiber. New J. Phys. **12**, 063027 (2010)
181. I. Choi, R.J. Young, D. Townsend, Quantum key distribution on a 10Gb/s WDM-PON. Opt. Express **18**, 9600–9612 (2010)
182. NICT News, No. 401 Feb 2011 (2011)
183. K. Inoue, E. Wake, Y. Yamamoto, Differential-phase-shift quantum key distribution using coherent light. Phys. Rev. A **68**, 022317 (2003)
184. N. Namekate, H. Takesue, Y. Honjo, T. Tokura, S. Inoue, High-rate quantum key distribution over 100 km using ultra-low-noise, 2-GHz sinusoidally gated InGaAs/InP avalanche photodiodes. Opt. Express **19**, 10632–10639 (2011)
185. S. Wang, W. Chen, Z.-Q. Guo, Z.-Q. Yin, W.-H. Li, Z. Zhou, G.-C. Guo, Z.-F. Han, 2 GHz clock quantum key distribution over 260 km of standard telecom fiber. Opt. Lett. **37**, 1008–1011 (2012)
186. NuCrypt website, retrieved on 6. Aug 2012
187. SECOQC, Business White Paper: Quantum Cryptography: An Innovation in the Domain of Secure Information Transmission (2008)
188. J.F. Dynes, H. Takesue, Z.L. Yuan, A.W. Sharpe, K. Harada, K. Honjo, H. Kamada, O. Tadanaga, Y. Nishida, M. Asobe, A.J. Shields, Efficient entanglement distribution over 200 kilometers. Opt. Express **17**, 11440–11449 (2009)
189. R. Ursin, F. Tiefenbacher, T. Schmitt-Manderbach, H. Weier, T. Scheidl, M. Lindenthal, B. Blauensteiner, T. Jennewein, J. Perdigues, P. Trojek, B. Omer, M. Furst, M. Meyenburg, J. Rarity, Z. Sodnik, C. Barbieri, H. Weinfurter, A. Zeilinger, Entanglement-based quantum communication over 144 km. Nature Phys. **3**, 481 (2007)
190. N. Namekate, H. Takesue, Y. Honjo, T. Tokura, S. Inoue, High rate, long-distance quantum key distribution over 250 km of ultra low loss fibres. Opt. Express **19**, 10632–10639 (2011)
191. J.G. Rarity, P.R. Tapster, P.M. Gorman, P. Knight, Ground to satellite secure key exchange using quantum cryptography. New J. Phys. **4**, 82 (2002)
192. J.E. Nordholt, R.J. Hughes, G.L. Morgan, C.G. Peterson, C.C. Wipf, Present and future free-space quantum key distribution. Proc. SPIE **4635**, 116 (2002)
193. A. Laing, V. Scarani, J.G. Rarity, J.L. O'Brien, Reference-frame-independent quantum key distribution. Phys. Rev. A **82**, 012304 (2010)
194. R. Ursin, T. Jennewein, J. Kofler, J.M. Perdigues, L. Cacciapuoti, C.J. de Matos, M. Aspelmeyer, A. Valencia, T. Scheidl, A. Fedrizzi, A. Acin, C. Barbieri, G. Bianco, C. Brukner, J. Capmany, S. Cova, D. Giggenbach, W. Leeb, R.H. Hadfield, R. Laflamme, N. Lutkenhaus, G. Milburn, M. Peev, T. Ralph, J. Rarity, R. Renner, E. Samain, N. Solomos, W. Tittel, J. P. Torres, M. Toyoshima, A. Ortigosa-Blanch, V. Pruneri, P. Villoresi, I. Walmsley, G. Weihs, H. Weinfurter, M. Zukowski, A. Zeilinger, Space quest: Experiments with quantum entanglement in space. IAC Proceedings **08**, A2.1.3 (2008)
195. H.-J. Briegel, W. Dür, J.I. Cirac, P. Zoller, Quantum repeaters: the role of imperfect local operations in quantum communication. Phys. Rev. Lett. **81**, 5932–5935 (1998)
196. A. Martin, O. Alibart, M.P. De Micheli, D.B. Ostrowsky, S. Tanzilli, A quantum relay chip based on telecommunication integrated optics technology. New J. Phys. **14**, 025002 (2012)
197. A.S. Clark, J. Fulconis, J.G. Rarity, W.J. Wadsworth, J.L. O'Brien, All-optical-fiber polarization-based quantum logic gate. Phys. Rev. A **79**, 030303 (2009)
198. C. Simon et al., Quantum memories. a review based on the European integrated project Qubit Applications (QAP). Eur. Phys. J. D **58**, 1–22 (2010)

Chapter 13
Biphoton Pulse Shaping

Joseph M. Lukens and Andrew M. Weiner

Abstract Recent research has demonstrated how classical optical signal-processing techniques can be extended to nonclassical entangled-photon states, permitting unprecedented control of the time-frequency correlations shared by these light quanta. In this chapter, we introduce the basic theory behind such biphoton pulse shaping and highlight several key experiments conducted in this field. Our emphasis is not only on how entangled photons benefit from these traditionally classical techniques, but also how their specifically quantum properties produce interesting effects not observable with classical fields. We consider both Fourier-transform pulse shaping, which relies on programmable spectral filtering, and electro-optic modulation, in which the temporal phase or amplitude of the biphoton is manipulated by an electrical signal. Both avenues of research have facilitated new insights into biphoton correlations and look to play an important role in the future of quantum state manipulation and the next generation of quantum communication networks.

13.1 Introduction

The groundbreaking experiments of Hanbury Brown and Twiss in the 1950s [1, 2] highlighted a grave need in the optical community to better understand the correlations shared by multiple photons. Virtually all optical interference experiments prior to that time had consisted of first-order intensity measurements, which quantum mechanically could be understood as the interference of a single photon with itself. Yet Hanbury Brown and Twiss instead examined the correlations between *two* photoelectric detectors, finding that a pair of photons from a thermal light source are twice as likely to arrive together than at a specific nonzero delay,

J.M. Lukens (✉) · A.M. Weiner
Purdue University, West Lafayette, IN, USA
e-mail: jlukens@purdue.edu

A.M. Weiner
e-mail: amw@purdue.edu

S. Wabnitz and B.J. Eggleton (eds.), *All-Optical Signal Processing*,
Springer Series in Optical Sciences 194, DOI 10.1007/978-3-319-14992-9_13

an effect now termed bunching [3]. The surrounding debate [4–7], coupled with the advent of the laser, prompted Glauber to formulate the first fully quantized theory of optical coherence in 1963 [8, 9], and the modern era of quantum optics was born.

Since then, a particularly influential nonclassical state of light has proven to be the entangled photon pair, or biphoton [10]. Although such photons can be entangled in many possible degrees of freedom, in this paper our focus is on time-frequency entanglement. This is frequently produced through spontaneous parametric downconversion (SPDC) of a narrowband pump field, in which a single pump photon decays into two daughter photons traditionally denoted "signal" and "idler." Whereas individually these photons possess an essentially random creation time, collectively their detection times are highly correlated, since they share the same birth; likewise, their frequencies can assume any value from a potentially wide bandwidth, but once the energy of one photon is measured, the value of its partner is known with high precision. Figure 13.1 provides a sketch of the generation process and joint spectro-temporal distributions of such biphoton states. The simultaneously narrow correlations in both time and frequency variables violate the seemingly reasonable tenet of classical locality; although critiqued by even the likes of Einstein [11], such entanglement has proven essential in foundational tests of quantum mechanics [12] and even some forms of quantum cryptography [13].

As might be expected, a key tool for characterizing biphoton entanglement derives from Glauber's coherence theory, particularly the second-order correlation function in either temporal or spectral form. Such an entangled state displays "super-bunching," for the probability of simultaneous photon detections can be radically higher than the 2:1 contrast observed for thermal light [3]. And while biphoton *characterization* is certainly of great importance in itself, the last decade has also witnessed the emergence of pulse-shaping techniques for biphoton *manipulation*, which exploit methods previously developed in classical ultrafast optics [14]. The purpose of this chapter is to review these recent advances in biphoton pulse shaping, emphasizing both the impact they have already had on the field of quantum optics and their promise in future applications in quantum

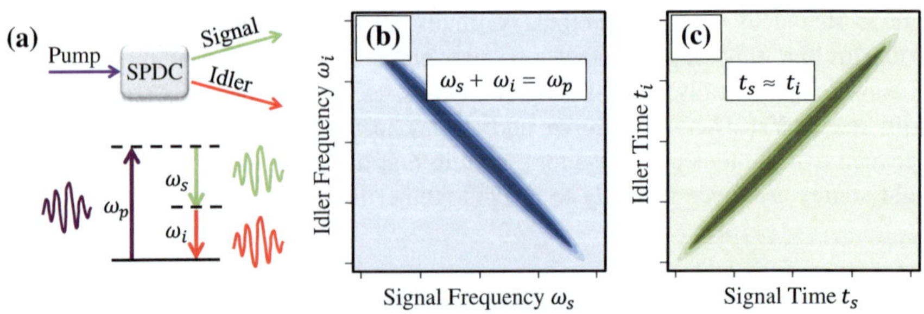

Fig. 13.1 **a** Biphoton generation through downconversion of a narrowband pump field. **b** Resulting biphoton spectral correlations. **c** Joint temporal correlations. Even though individual photons have a wide spread of spectro-temporal content, they are highly correlated with their partner

information. Additionally, although pulse shaping is typically associated with programmable filtering in the spectral domain to achieve shaping in time, for the purposes of this work we will also generalize the term to include its Fourier dual: modulation in the time domain to transform the spectrum. Both specializations demonstrate how classical communication techniques can assume new roles in quantum optics.

In Sect. 13.2 we introduce the basics of classical Fourier-transform pulse shaping and temporal modulation, showing thereafter in Sect. 13.3 how these techniques can be applied to biphotons. Many of the key experiments are summarized in Sect. 13.4, with two specific examples taken up in detail in Sects. 13.5 and 13.6. Finally, in Sect. 13.7 we provide an outlook for future work, with an emphasis on applications in quantum cryptography.

13.2 Classical Pulse Shaping

Biphoton pulse shaping represents the union of two optical disciplines which historically have remained largely independent of one another: the study of quantum states of light and the development of pulse-shaping techniques for manipulation of classical light sources. Since most of the biphoton shaping examples covered in this chapter can be viewed as outgrowths of similar experiments on classical pulses, we focus in this section on the classical foundations of both Fourier-transform pulse shaping and electro-optic modulation, introducing the basic theory behind these technologies which have proven so valuable in a variety of optical applications. This discussion furnishes a springboard for their extension to entangled photons, a task assumed in Sect. 13.3.

Following its development in the 1980s [15], Fourier-transform pulse shaping has enabled major advances in an assortment of fields, ranging from optical arbitrary waveform generation to coherent control, single-cycle pulse generation to microwave photonics [16]. And whereas synthesizing user-defined optical waveforms on the femtosecond timescale is impossible with the fastest available electronics, Fourier-transform pulse shaping permits the generation such arbitrary fields through parallel manipulation of the frequency content. An archetypal pulse shaper is shown in Fig. 13.2a. The frequency components of a broadband input field are spatially dispersed via a diffraction grating and focused to small spots at a plane where a programmably controlled spatial light modulator (SLM) is placed. Due to the physical separation of distinct frequencies, it is then possible to control the amplitude and phase at each wavelength independently; after recombining the shaped frequency components with a second lens and grating, the desired waveform is produced at the output, whose complexity is limited only by the spectral resolution and total bandwidth of the pulse shaper [16]. Mathematically, if we take the complex input and output time-domain fields as $E_{\text{in}}(t)$ and $E_{\text{out}}(t)$—with associated spectra $\mathscr{E}_{\text{in}}(\omega)$ and $\mathscr{E}_{\text{out}}(\omega)$—the effect of the pulse-shaper mask $H(\omega)$ is to produce the output

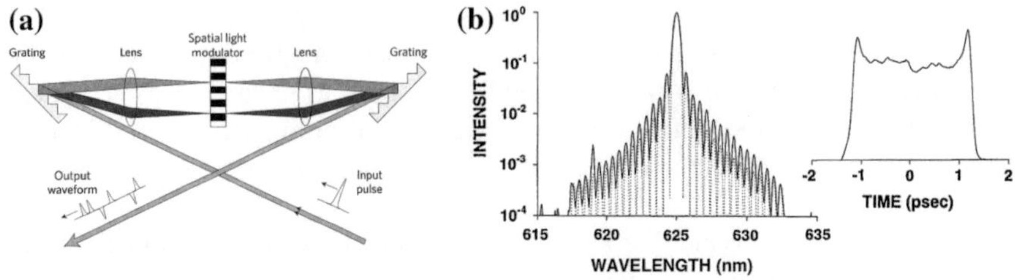

Fig. 13.2 **a** Schematic of a prototypical pulse shaper. **b** Example shaped power spectrum (*left*) and temporal intensity (*right*) for a square pulse (Image in **a** from [18] and that in **b** from [15])

$$E_{\text{out}}(t) = \int d\omega \mathscr{E}_{\text{in}}(\omega) H(\omega) e^{-i\omega t}, \tag{13.1}$$

so that Fourier-transform pulse shaping can be viewed as a general method to achieve nearly arbitrary and ultrafast linear transformations on optical fields. Figure 13.2b shows just one example from the early days of pulse shaping, in which a fixed spatial mask was used to generate a square pulse [15]. The measured spectro-temporal properties are in good agreement with theory, even displaying the expected overshoot resulting from truncation of the ideal sinc spectrum. The basic model for pulse-shaper operation embodied in (13.1) proves sufficient for the purposes of this chapter; for more detailed discussion of pulse-shaper theory and technology, the reader is directed to the reviews in [14, 16, 17]. The key point here is that Fourier-transform pulse shaping manipulates an optical field through frequency-domain multiplication by a user-defined filter.

The Fourier dual shaping process is effected through direct modulation of the temporal phase or intensity of an incident optical field. Although a range of modulator technologies have been considered in practice, we specialize our discussion to devices based on the electro-optic effect, as these have been employed exclusively in the biphoton modulation experiments described later in this chapter. In electro-optic materials, the refractive index changes with an applied external voltage, in turn permitting electrical control of the phase acquired by an optical field propagating through it. Figure 13.3a shows a cross-section of a modulator based on lithium niobate [19], which possesses a particularly strong linear electro-optic response along its z-axis. In the arrangement depicted here, an electric voltage applied across the signal and ground electrodes induces a refractive index change for light polarized along the z-direction and propagating along y in the embedded waveguide. Furthermore, by splitting the input optical field into two arms—each of the form of Fig. 13.3a—and recombining them after different phase shifts, it is possible to convert this phase modulation to intensity modulation, via the Mach-Zehnder interferometer configuration of Fig. 13.3b [19]. And by utilizing traveling-wave electrodes, in which the driving electrical signal propagates parallel to the optical waveguide, extremely high modulation bandwidths exceeding tens of gigahertz are possible, permitting rapid switching of either intensity or phase in

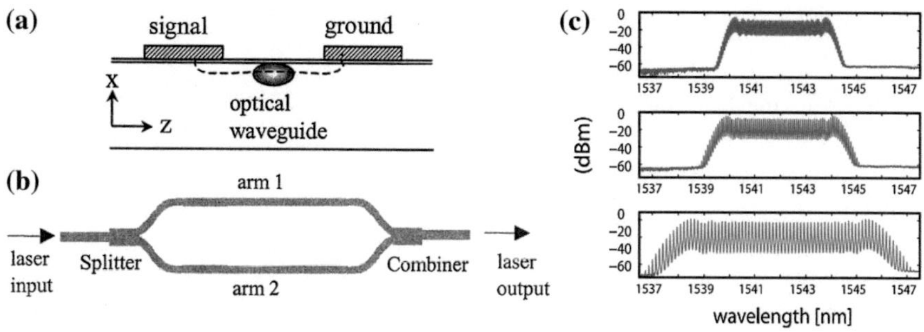

Fig. 13.3 **a** Cross-section of an x-cut lithium niobate phase modulator. **b** Arrangement for combining two phase-modulation arms to produce an intensity modulator. **c** Example comb spectra generated by cascading three phase modulators and one intensity modulator. The spectral line spacing, from *top* to *bottom*, is 7, 10, and 17 GHz (Images in **a** and **b** from [19] and **c** from [20])

high-speed optical communication networks. Formally, we can cover both phase and intensity cases through a general modulation function $m(t)$, so that the input-output complex field relationship can be expressed as

$$E_{\text{out}}(t) = E_{\text{in}}(t)m(t), \tag{13.2}$$

where, for example, sinusoidal phase modulation at frequency ω_m is represented by a function of the form $m(t) = \exp[i\delta \sin(\omega_m t + \phi)]$, with δ signifying the modulation index, a quantity which is proportional to the applied voltage amplitude.

In addition to their extensive use in transmission of data [21], electro-optic modulators have also found use as time lenses—devices which can compress and spread waveforms in time just as lenses do in space [22]—and frequency-comb generators. In the latter case, cascaded phase and intensity modulators driven by a single microwave source can be used to convert a monochromatic input spectrum into a series of discrete spectral lines; Fig. 13.3c provides an example in which one intensity modulator and three phase modulators generate broad and flat frequency combs with tunable line spacing [20]. Such optical sources are well-suited for applications including line-by-line pulse shaping [23] and radio-frequency photonic filtering [24].

In light of the maturity of these manipulation techniques for both spectral and temporal control of classical fields, the motivating question behind biphoton pulse shaping is: how can we apply such classical approaches to entangled photons? As highlighted in the experiments discussed below, pulse shaping indeed reveals interesting physics, with potential applications in quantum information. Yet the quantum extension involves a more refined understanding of what is entailed by the term "pulse shaping" when nonclassical sources are considered. Accordingly, we tackle the fundamentals of biphoton shaping theory in the next section.

13.3 Biphoton Pulse Shaping: Theory

As noted in the introduction, SPDC of a monochromatic pump can be used to generate entangled photons that share both high temporal and high spectral correlation. In order to derive an explicit form for the resultant state, we invoke the formalism of the interaction picture of quantum mechanics [25], in which the quantum state evolves in time according to an interaction Hamiltonian (in our case a second-order nonlinear interaction), and the field operators according to the free Hamiltonian (simply related to linear propagation of the corresponding classical fields). For a pump frequency $\omega_p = 2\omega_0$, a perturbative calculation [25] yields the entangled state

$$|\Psi\rangle = M|\text{vac}\rangle_s|\text{vac}\rangle_i + \int d\Omega\phi(\Omega)|\omega_0 + \Omega\rangle_s|\omega_0 - \Omega\rangle_i, \qquad (13.3)$$

where M $\sim$ 1, "vac" denotes the vacuum state, s and i signify signal and idler modes, and $\phi(\Omega)$ is a complex weight function determined by phase-matching conditions. In this form we have neglected any spatial or polarization dependencies, which prove unimportant for the experiments considered in this review. The vacuum term, with a near-unity weighting coefficient, signifies the relative improbability of this spontaneous process; downconversion efficiencies of only up to about 10^{-5} have been reported in the experiments discussed below [26]. Of course the more interesting term in (13.3) is the second one, which consists of a superposition of signal and idler frequency combinations, subject to the constraint that they must sum to $2\omega_0$. Since $\phi(\Omega)$ can be extremely broadband, a wide range of frequency pairs are possible, thereby producing strong entanglement. The degree of such entanglement can be operationally defined as the ratio of the single-photon bandwidth to that conditioned on a frequency measurement of its partner [3]; for a separable (non-entangled) state, this value is one, whereas for an entangled state with terahertz-scale bandwidth and perhaps megahertz pump linewidth—typical for the experiments considered here—the ratio can be well over a million. Such massive entanglement lies at the heart of the nonclassical behavior which these states can display.

For the Fourier-transform pulse-shaping experiments, manipulation and measurement of the time-frequency biphoton state follows the basic setup in Fig. 13.4a. Signal and idler photons are sent through programmable spectral filters, $H_s(\omega)$ and $H_i(\omega)$, and coincidences between two detectors are measured as the signal delay is varied. Assuming the detectors are ideal—i.e., possess infinite timing resolution and respond to all frequencies of the incoming field—the coincidence rate is directly proportional to the second-order Glauber correlation function [8]

$$G^{(2)}(\tau) = \langle\Psi|\hat{E}_i^{(-)}(t)\hat{E}_s^{(-)}(t+\tau)\hat{E}_s^{(+)}(t+\tau)\hat{E}_i^{(+)}(t)|\Psi\rangle. \qquad (13.4)$$

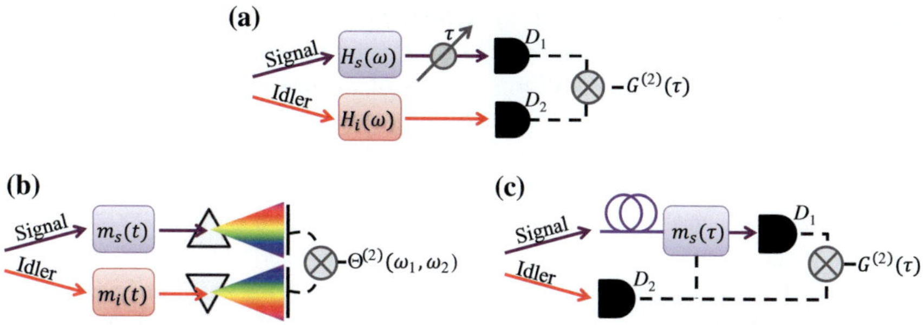

Fig. 13.4 Basic experimental setups for biphoton manipulation. **a** Pulse shaping. Each photon of an entangled pair is sent through a programmable spectral filter and detected on a photon counter. The coincidences between these (ideal) detectors map out the second-order Glauber correlation function $G^{(2)}(\tau)$. **b** Free-running electro-optic modulation. The photons are modulated in the time domain, with spectral correlations determined through coincidences between monochromators. This experiment represents the Fourier dual to that in **a**. **c** Modulation of a heralded single photon. Now the detection of the idler photon is used to trigger modulation of the signal, with $G^{(2)}(\tau)$ representing the measured quantity, as in **a**

The positive-frequency electric field operators $\hat{E}_s^{(+)}(t+\tau)$ and $\hat{E}_i^{(+)}(t)$ are associated with annihilation of a signal or idler photon at time $t+\tau$ or t, respectively; the corresponding negative-frequency operators create photons at these times. Since the monochromatic pump is statistically stationary, the correlation function depends only on the delay τ, making the choice of t arbitrary. Moreover, because the positive- and negative-frequency operators are related through Hermitian conjugation, the classical expression for $G^{(2)}(\tau)$, in which the electric field operators become complex numbers, is of the form of an intensity correlation between the signal and idler waveforms, so we can view $G^{(2)}(\tau)$ as the general quantum mechanical version of an intensity cross-correlation measurement.

When the quantum state contains no more than two photons, $G^{(2)}(\tau)$ can be conveniently expressed as the modulus squared of a biphoton wavepacket $\psi(t+\tau, t)$, defined as [10]

$$\psi(t+\tau, t) = \langle \text{vac} | \hat{E}_s^{(+)}(t+\tau) \hat{E}_i^{(+)}(t) | \Psi \rangle. \tag{13.5}$$

For the quantum state in (13.3) and the setup in Fig. 13.4a, the wavepacket assumes the form [26]

$$\psi(\tau) = \int d\Omega \, \phi(\Omega) H_s(\omega_0 + \Omega) H_i(\omega_0 - \Omega) e^{-i\Omega\tau}, \tag{13.6}$$

where we have neglected an unimportant overall constant and a t-dependence which vanishes on taking the modulus $G^{(2)}(\tau) = |\psi(\tau)|^2$. This expression represents the main relation for biphoton pulse shaping and well describes the spectral filtering experiments in this review. Therefore, analogous to a classical pulse [16], the signal-idler temporal correlations can be shaped through programmable filtering, permitting synthesis of waveforms orders of magnitude too fast for even the speediest electronics. Equation (13.6) also indicates why extremely broadband biphotons are so desirable in Fourier-transform pulse shaping; for a given pulse-shaper spectral resolution, the wider the bandwidth of $\phi(\Omega)$, the larger the number of independent spectral features which can be programmed on the biphoton spectrum. As in classical pulse shaping, the ratio of total bandwidth to spectral resolution provides a measure of the maximum complexity of the generated waveforms [16], so high-bandwidth biphotons prove most amenable to spectral shaping. However, unlike the classical case summarized in (13.1), the net transfer function is the *product* of the complex filters applied to the signal and idler photons, evaluated at each entangled frequency combination. Thus the filtering applied to the signal can actually be undone by filtering the idler with the inverse; for example, even when both photons are dispersively broadened *individually*, it is possible that *collectively* their correlations remain unaltered, an effect known as nonlocal dispersion cancellation [27] and considered in more detail in Sect. 13.5.

The second class of experiments we examine here is essentially the Fourier dual to the pulse-shaping arrangement above. A schematic of this general setup is provided in Fig. 13.4b. Now the signal and idler photons are multiplied in the time domain by the modulation functions $m_s(t)$ and $m_i(t)$, and then the spectral correlations are measured—i.e., the coincidences between two narrow frequency bins. In the absence of modulation, the entangled state in (13.3) indicates that only frequency pairs (ω_s, ω_i) satisfying $\omega_s + \omega_i = 2\omega_0$ will register coincidences; however, the presence of the modulators can introduce new frequency correlations, quantified by the second-order cross-spectral density:

$$\Theta^{(2)}(\omega_1, \omega_2) = \left\langle \Psi \middle| \hat{\mathscr{E}}_i(-\omega_2)\hat{\mathscr{E}}_s(-\omega_1)\hat{\mathscr{E}}_s(\omega_1)\hat{\mathscr{E}}_i(\omega_2) \middle| \Psi \right\rangle. \tag{13.7}$$

These operators are the Fourier transforms of those in (13.4), with negative- and positive-frequency arguments signifying photon creation and annihilation, respectively. Therefore $\Theta^{(2)}(\omega_1, \omega_2)$ is proportional to the probability of detecting a signal photon at frequency ω_1 and an idler at ω_2. Additionally, in the same fashion as the temporal case, this correlation function can be decomposed into the modulus squared of a biphoton spectral amplitude, denoted by $\theta(\omega_1, \omega_2)$:

$$\theta(\omega_1, \omega_2) = \langle \text{vac} | \hat{\mathscr{E}}_s(\omega_1)\hat{\mathscr{E}}_i(\omega_2) | \Psi \rangle. \tag{13.8}$$

The effect of temporal modulation is to produce convolution in the frequency domain, and in general the full expression for $\Theta^{(2)}(\omega_1, \omega_2)$ in the arrangement of Fig. 13.4b can become quite complicated. Yet if we restrict ourselves to the case in which the biphoton correlation time is much shorter than the scale over which the modulators vary, we can treat the inverse Fourier transform of $\phi(\Omega)$ as an impulse, giving for the biphoton spectrum

$$\theta(\Delta) = \int dt\, m_s(t) m_i(t) e^{i\Delta t}, \tag{13.9}$$

where we have neglected irrelevant overall constants and defined $\Delta = \omega_1 + \omega_2 - 2\omega_0$, the deviation of the sum frequency from that of the narrowband pump. In analogy with the spectral filtering case in (13.6), the effect of signal and idler modulation is cumulative, with the potential for one modulator to nonlocally cancel the other in the measured spectral correlation [28].

In the above arrangement, the signal and idler modulators are free-running in the sense that they are not synchronized with photon emission times. As an alternative for biphoton manipulation, also useful for temporal shaping, the detection of one photon can be used to trigger modulation of the other. Figure 13.4c shows how an experiment like this could work; detection of the idler photon prompts a voltage signal to be applied to the signal modulator, and then the temporal correlation function is measured. If $G_0^{(2)}(\tau)$ represents the correlation function with no modulation, the modified function measured is [29]

$$G^{(2)}(\tau) = |m_s(\tau)|^2 G_0^{(2)}(\tau), \tag{13.10}$$

where, as before, $m_s(\tau)$ is the complex modulation applied to the signal photon, but now it is a function of the *relative* signal-idler delay τ instead of some global time variable, since modulation is triggered only by detection of the idler photon. While causality limits modulator operation to delays satisfying $\tau > 0$, this restriction can be effectively mitigated by delaying the signal photon prior to modulation so that its entire duration falls in positive delays. A similar capability is often tacitly exploited in Fourier-transform pulse shaping as well; due to the physical delay of the pulse shaper itself, superficially non-causal spectral filters can be implemented by simply subtracting off the overall delay in defining the output time origin.

In the form of (13.10), one can view such manipulation as the shaping of a heralded single photon—that is, a photon whose presence is announced through detection of its entangled partner. We note that while the free-running modulator setup of Fig. 13.4b requires biphotons with correlation times much shorter than the modulation period for (13.9) to hold, this alternative detection setup relies on biphotons with the opposite property—namely, correlation times greatly exceeding the modulator period. Under this condition, the modulator can significantly alter the shape of $G^{(2)}(\tau)$. Thus, just as broadband biphotons are optimal for Fourier-transform pulse shaping in the setup of Fig. 13.4a, temporally long biphotons prove advantageous for electro-optic modulation, as they maximize the number of independent features which can be imprinted by the modulator. And so the theory here, typified in (13.6), (13.9), and (13.10), has motivated several experimental tests of Fourier-transform pulse shaping and high-speed electro-optic modulation of biphoton states—both well-established techniques in ultrafast photonics and optical communications, but only considered in the quantum optics community within the last decade.

13.4 Biphoton Pulse Shaping: Important Experiments

13.4.1 Ultrafast Coincidence Detection

In the conceptual pictures of Fig. 13.4, we have envisioned correlating the photoelectric pulses generated from two single-photon counters in order to determine coincident photon arrivals. In the spectral measurement of Fig. 13.4b, such a scheme poses no problems, for high frequency resolution is attainable with well-chosen spectral dispersers; the timing resolution of the detector is not a limiting factor. However, since most single-photon detectors suffer from timing jitters of tens of picoseconds—or more—a realistic implementation of Fig. 13.4a is unable to resolve biphoton correlations on the few- and sub-picosecond time scales. And although there are several interesting examples of shaping experiments with such slow detectors [30–32], the full potential of programmable pulse shaping is only realized on the ultrafast time scale. This timing resolution problem permeates classical optics as well, where correlation techniques based on nonlinear optical interactions have facilitated high-resolution temporal measurements of ultrashort pulses [14]. Borrowing from these ideas, one naturally hopes for a nonlinear process at the single-photon level which would likewise provide sufficient temporal resolution to characterize sub-picosecond biphoton correlations.

Such a technique has indeed been implemented: biphoton sum-frequency generation (SFG). In this process, signal and idler photons are mixed on a second nonlinear crystal; with some small probability, temporally coincident photons can then recombine and generate a new photon at the sum frequency [33]. Since only photons that overlap within a femtosecond timescale can mix, the rate of upconversion in a properly designed crystal is proportional to the Glauber correlation function, and so by sweeping the delay of the signal with respect to the idler, $G^{(2)}(\tau)$ is directly obtained. In this way SFG enforces temporal gating optically, instead of electronically, thereby enabling huge improvements in timing resolution.

Unfortunately, since the interacting fields have extremely low intensities, the efficiency of biphoton SFG is quite low, with a record upconversion rate of only 10^{-5} [26]. Yet such low efficiencies can be somewhat compensated by generating an ultrahigh flux of entangled photons [33]. As a general rule, in order to observe quantum mechanical effects, one must ensure that only a single biphoton is located within a given detection window; otherwise, there will result spurious coincidences in which the two detections actually correspond to different photon pairs. Since photons from separate pairs are not entangled, the desired quantum correlations are degraded. This limiting timescale for separate detectors—as in Fig. 13.4a—is electronic in origin, which for 1-ns resolution, say, fixes the maximum biphoton generation rate at $\sim 10^9\ \mathrm{s}^{-1}$. On the other hand, with ultrafast SFG which measures the biphoton directly, we need only ensure that successive biphotons *themselves* do not temporally overlap. With correlation times below 1 ps quite common, the maximum allowable quantum flux can exceed $10^{12}\ \mathrm{s}^{-1}$, permitting upconversion rates that are easily detectable even at such low SFG efficiencies [33]. For more details regarding the conditions under which biphoton SFG accurately reproduces the Glauber correlation function, the reader is directed to the theoretical development in [34].

Accordingly, the first demonstration of successful biphoton SFG [33] provided the necessary basis for biphoton pulse shaping, which itself was realized shortly thereafter [35]. The original experimental setup is reproduced in Fig. 13.5a. Signal and idler photons are emitted collinearly, with a prism sequence removing remaining pump light and spectrally decomposing the biphoton; by placing a spatial light modulator (SLM) in the path of the dispersed photons, the spectral phase of the signal and idler can be programmably controlled. (Since signal and idler are distinguished by frequency, both filters can be implemented on a single pulse shaper.) The shaped biphotons are then recombined on a second nonlinear crystal, and the SFG photons are measured by a single-photon counting module (SPCM). Additionally, because a shift in time is equivalent to linear phase in the frequency domain, the pulse shaper can also impose the relative photon delay, omitting the need for a dedicated delay stage. This general Fourier relationship is readily derived by considering the linear spectral phase filter $H(\omega) = \exp[i\omega\tau]$ in (13.1), which gives

$$E_{\text{out}}(t) = \int d\omega\, \mathscr{E}_{\text{in}}(\omega) e^{-i\omega(t-\tau)} = E_{\text{in}}(t-\tau), \tag{13.11}$$

so that, indeed, the output field is simply a time-shifted version of the input. Thus by applying similar linear phase terms to the pulse shaper, it is possible to directly control the delay of one photon with respect to the other.

Figure 13.5b shows some results from this experiment [35]. With linear phase swept on the pulse shaper, but no filtering otherwise, a narrow correlation peak is found; by adding a π phase jump halfway through the signal spectrum, this single correlation peak is transformed into a doublet. These seminal results represent the foundation of biphoton pulse shaping, upon which subsequent researchers have expanded with more sophisticated filtering or efficient detection.

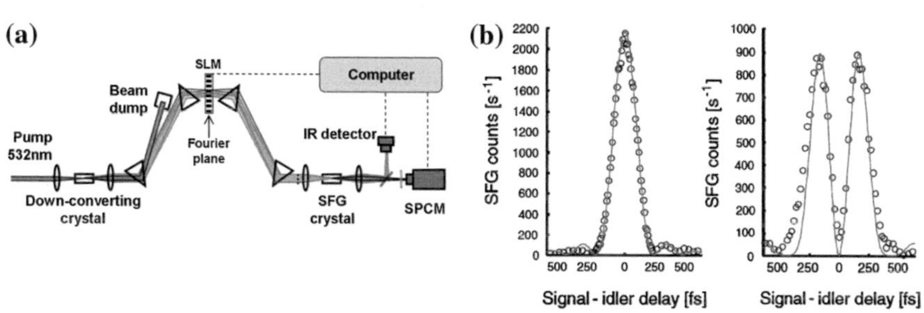

Fig. 13.5 First demonstration of biphoton pulse shaping. **a** Experimental setup. **b** Shaping results. In the *left* case, the pulse shaper sweeps the relative signal-idler delay, allowing measurement of a narrow temporal correlation function. On the *right*, an additional π phase shift is added halfway through the signal spectrum, turning this peak into a doublet (Images from [35])

13.4.2 Additional Biphoton Shaping Experiments

The first demonstration of biphoton pulse shaping [35] utilized a relatively low-resolution, phase-only pulse shaper, and succeeding work has focused at least in part on enhancing these capabilities. In 2008, the first biphoton pulse shaper which controlled both amplitude and phase was implemented [36]. One interesting consequence of independent phase and amplitude control is the ability to simulate a Franson interferometer [37]. In the typical Franson arrangement, signal and idler photons are sent through separate Mach-Zehnder interferometers (MZIs), identical except for phase shifts in their long arms, with coincidences recorded between one output port of each MZI. If the relative delay between the long and short arms of each MZI exceeds the single-photon coherence time, but not that of the original pump, the single detector rates show no interference while the coincidence rate does, due to the interference of indistinguishable quantum paths [38]. Yet because an MZI represents a linear, time-invariant system, it can instead be implemented in the frequency domain, by programming the associated amplitude and phase on a pulse shaper, as first demonstrated in [36], indicating a new application of biphoton shaping—namely, stable interferometer emulation. This basic setup was recently employed for creating and characterizing frequency-entangled biphoton states with programmable dimensions, through partitioning of the biphoton spectrum into discrete bins with the pulse shaper [39].

Pulse-shaping capabilities were further broadened by our experiments utilizing a high-resolution, commercial telecom pulse shaper based on liquid-crystal-on-silicon technology [40]. In this work, we were able to demonstrate complex amplitude coding for efficient verification of spectral entanglement, as well as additional pulse-shaper impersonations of various interferometers [32]. Due to the ultrahigh spectral resolution, much more complicated binning can be achieved; whereas signal and idler were divided into up to four bins each in [39], length-40 amplitude codes were readily demonstrated with the arrangement in [32]. In subsequent experiments with this pulse shaper [41], we have also been able to demonstrate the temporal Talbot effect with biphotons, in which a periodic correlation function is replicated on propagating through discrete amounts of optical dispersion [42]; in this example, the pulse shaper simultaneously creates and manipulates a periodic correlation function through spectral amplitude and phase filtering. Although the correlation trains generated in this way are limited in maximum duration by the pulse-shaper time aperture (around 50 ps in this experiment), such an approach nonetheless offers a means to yield periodic biphotons with extremely high and tunable repetition rates, without the need for a stable optical cavity.

In parallel with these developments in Fourier-transform pulse shaping, electro-optic modulation of biphoton sources was demonstrated first in 2008 [29], using a setup similar in form to Fig. 13.4c. The generation process—based on the phenomenon of electromagnetically induced transparency—produces biphotons with long correlation times approaching a microsecond, well-suited for temporal modulation and direct measurement with electronic detectors. Accordingly, a range of temporal shapes for the heralded single photon were demonstrated, including square, Gaussian, and exponential pulses. With slight modifications to this setup, electro-optic modulation was also exploited to perform a Fourier-based measurement of the temporal correlation function [43]. These examples, coupled with the developments in Fourier-transform pulse shaping, provide just a sampling of the unique potential of spectro-temporal biphoton shaping techniques. In the next sections we give a more detailed discussion of two particularly interesting classes of programmable shaping and modulation, which we hope will further highlight the fascinating aspects of biphoton pulse shaping. The first, cancellation of dispersion or modulation, emphasizes the truly nonclassical properties afforded by entangled states; the second, biphoton coding, shows how these properties may be exploited in quantum information. Both cases rely on a combination of classical techniques and quantum attributes.

13.5 Detailed Example I: Cancellation of Dispersion or Modulation

The possibility for nonlocal cancellation of dispersion—the spreading of broadband pulses on propagation through a medium—was mentioned briefly in Sect. 13.3, as a consequence of the combined effects of the signal and idler spectral filters. In general, dispersion results from the variation of group velocity with frequency, so that the spectral components of a short pulse travel through a dispersive medium at different speeds, thereby separating on propagation and causing the pulse to spread. Such spreading is closely related to the spectral phase imparted by the medium; if we assume a phase-only transfer function of the form $H(\omega) = \exp[i\Phi(\omega)]$, the group delay (under our time-harmonic convention) follows via [16]

$$\tau(\omega) = \frac{\partial \Phi}{\partial \omega}, \tag{13.12}$$

so that any variation of the spectral phase which is second order or higher in ω produces a frequency-dependent delay—i.e., the frequency components separate in time. This concept extends to broadband entangled photons as well, ordinarily causing the correlation function to spread. But as shown theoretically by Franson [27], if the signal and idler photons travel through separate media with second-order dispersion coefficients equal in magnitude but opposite in sign, the joint correlation function is unaffected, even when the photons are arbitrarily far apart. Although some aspects of this effect have been shown using electronic coincidence detection [44], it was not until utilization of ultrafast biphoton SFG that dispersion cancellation could be observed on a femtosecond scale [45]. With the variety of low-loss dispersive media available, programmable pulse shaping is in no way required to exhibit this effect; indeed, prism sequences alone proved sufficient in the first observation of ultrafast dispersion cancellation [45]. However, these initial experiments considered only cancellation of second-order spectral phase, whereas control of higher orders proves essential for compressing the correlations of ultrabroadband biphotons [46]. Cancellation of such arbitrary phase is virtually impossible with standard optical media, but it is achievable with Fourier-transform pulse shaping.

The value of such programmable pulse shaping for high-order dispersion compensation was first demonstrated on classical pulses. In one experiment [47], a pulse shaper was exploited to achieve distortion-free transmission of an approximately 500-fs pulse through ~50 km of single-mode fiber. Figure 13.6 shows the results from this test; after propagation through the 50-km fiber link followed by a matched dispersion-compensating fiber module, the pulse shows an additional tail due to uncompensated higher orders, since the dispersion-compensating fiber cannot simultaneously match all second- and third-order phase. But by inserting a pulse shaper to apply third-order phase, the original bandwidth-limited pulse is recovered. This example stresses both the importance of high-order dispersion management and the difficulty in achieving it with standard optical materials and components. Thus the experiments considering high-order dispersion cancellation for biphotons can be viewed as quantum extensions of these classical tests, with the fundamental difference that they can be achieved on separated photons.

To understand explicitly how arbitrary-order biphoton dispersion cancellation ensues, we specialize (13.6) to phase-only filters $H_s(\omega) = \exp[i\Phi_s(\omega)]$ and $H_i(\omega) = \exp[i\Phi_i(\omega)]$, expressing each phase function as a Taylor series about ω_0:

$$\Phi_{s,i}(\omega) = \sum_{n=0}^{\infty} \frac{\Phi_{s,i}^{(n)}}{n!} (\omega - \omega_0)^n. \tag{13.13}$$

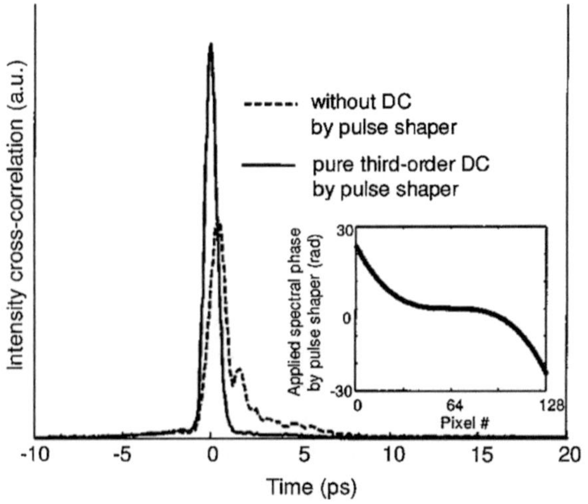

Fig. 13.6 Using a pulse shaper to compensate for high-order dispersion. After propagation through ~50 km of single-mode fiber and the matched length of dispersion-compensating fiber, the measured pulse is still distorted due to residual third-order spectral phase. Yet this is compensated by the pulse shaper, retrieving the original ~500-fs pulse (Result from [47])

Under these conditions, the measured biphoton wavepacket is

$$\psi(\tau) = \int d\Omega\phi(\Omega)\exp\left\{-i\Omega\tau + i\sum_{n=0}^{\infty}\frac{1}{n!}\left[\Phi_s^{(n)} + (-1)^n\Phi_i^{(n)}\right]\Omega\right\}, \tag{13.14}$$

so that high-order dispersion cancellation is achieved whenever the signal-idler expansion coefficients satisfy $\Phi_s^{(n)} + (-1)^n\Phi_i^{(n)} = 0$ for $n = 2, 3, \ldots$. [The $n = 0, 1$ cases do not impact the overall shape of $G^{(2)}(\tau)$.] Interestingly, opposite signs of dispersion give cancellation for even orders only, as odd orders instead require identical phase coefficients to achieve compression. We note that this cancellation behavior stands in stark contrast to dispersion compensation with classical pulses, where all orders must be flipped in the compensating medium. Yet nonetheless this biphoton cancellation condition can be understood intuitively as well. Because of the spectral correlation shared by these entangled photons (the fixed sum frequency of $2\omega_0$), each signal photon at $\omega_0 + \Omega$ is correlated with the idler at $\omega_0 - \Omega$, so that we would naturally expect the correlation function to remain unchanged when each of these signal-idler frequency pairs are delayed the same amount, that is, when

$$\tau_s(\omega_0 + \Omega) = \tau_i(\omega_0 - \Omega), \tag{13.15}$$

where here τ_s and τ_i are the frequency-dependent group delays for the signal and idler photons. And since these delays are related to the corresponding spectral phases through the derivative operation of (13.12) (which flips the parity), cancellation should be achieved whenever the signal-idler spectral phases are *anti-symmetric* with respect to each other:

$$\Phi_s(\omega_0 + \Omega) = -\Phi_i(\omega_0 - \Omega). \tag{13.16}$$

This equation leads immediately to the alternating-sign condition found previously through direct calculation of the biphoton wavepacket, justifying the validity of this more heuristic interpretation.

Utilizing the pulse shaper from [32], we have been able to verify this fascinating behavior for high-order spectral phase, making use of highly efficient periodically poled lithium niobate waveguides for both SPDC and biphoton SFG [26]. The basic setup mirrored that in Fig. 13.5a, although the prism sequence was replaced by optical fiber and a fiber-pigtailed pulse shaper. The results for high-order dispersion cancellation are reproduced in Fig. 13.7, showing examples for the orders $n = 3$, 4, 5. With the signal portion of the spectrum programmed with a particular order of spectral phase, and nothing applied to the idler, a lowered and stretched correlation function is measured ("signal dispersed" cases). Similarly, when the complementary phase—equal in the odd-order cases and with a sign flip in the even cases—is applied to the idler, but nothing on the signal, another dispersed wavepacket is observed ("idler dispersed" cases). These results can be interpreted as individually spreading one photon or the other. However, when both photons are

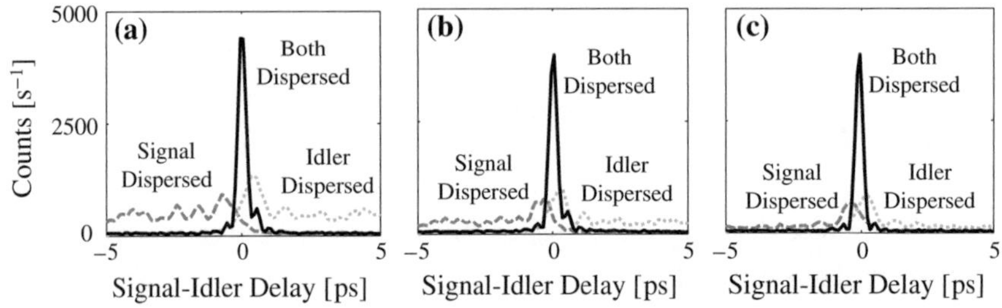

Fig. 13.7 High-order dispersion cancellation results for **a** third-order, **b** fourth-order, and **c** fifth-order spectral phase. In all cases, dispersing both photons recovers a narrow correlation peak, whereas classical correlations would only broaden further (Plots adapted from [26])

dispersed simultaneously, the original narrow correlation function is recovered, verifying complete cancellation ("both dispersed" cases). These pulse-shaping experiments therefore not only improve understanding of arbitrary biphoton phase, but also make significant strides in the quest for single-cycle biphotons—entangled photons with correlation times equal to a single optical period [46]—for likely only through improved control of high-order dispersion cancellation will single-cycle compression be possible.

As alluded to in Sect. 13.3, the electro-optic modulation setup of Fig. 13.4b permits realization of the Fourier dual to biphoton dispersion cancellation—namely, nonlocal cancellation of modulation [28]. Specifically, considering equi-amplitude sinusoidal phase modulation in (13.9), $m_s(t) = \exp[i\delta \sin \omega_m t]$ and $m_i(t) = \exp[i\delta \sin(\omega_m t + \alpha)]$, yields the complex joint spectrum

$$\theta(\Delta) = \int dt\; \exp\{i\Delta t + i\delta[\sin \omega_m t + \sin(\omega_m t + \alpha)]\}. \tag{13.17}$$

Thus the modulators completely cancel each other when driven out of phase ($\alpha = \pi$). As in the case of dispersion cancellation, this result can also be given a more intuitive footing. If we consider the general phase-modulation functions $m_{s,i}(t) = \exp\left[i\varphi_{s,i}(t)\right]$, analogously to the definition of group delay in (13.12), one can view the modulation as imparting a time-dependent instantaneous frequency shift to each of the photons, defined via the relation

$$\Delta\omega_{s,i}(t) = -\frac{\partial \varphi_{s,i}}{\partial t}. \tag{13.18}$$

Because the arrival time of a given signal or idler photon is random, the actual frequency shift is undefined prior to measurement. However, due to the strong temporal correlation (fixing joint photon arrivals to time windows much smaller than the modulator period), the frequency shift experienced by the signal photon can be exactly cancelled by that imposed on the idler when

$$\Delta\omega_s(t) = -\Delta\omega_i(t), \tag{13.19}$$

leaving a single spectral correlation peak at the original pump frequency. For the specific case of sinusoidal modulation considered here, this condition is satisfied by signal-idler modulator functions which are π radians out of phase, as found previously. Accordingly, one can view cancellation of electro-optic modulation as shifting signal and idler frequencies by equal and opposite amounts.

We note that, classically speaking, such modulation cancellation can only be achieved locally. In one recent example [48] (which actually appeared after the quantum experiment described below), this more limited form of cancellation was employed to achieve high-speed temporal cloaking, or hiding of events in time through manipulation of a probe beam. With a combination of sinusoidal electro-optic phase modulation and dispersion, a monochromatic probe was converted to a pulse train with intensity gaps; the event, intensity modulation, operated only in these gaps and therefore had no impact on the probing field. Thus, subsequently propagating the probe through matched dispersion and complementary modulation produced the original continuous-wave field at the output, thereby hiding the event from an observer. Importantly, in this classical case cancellation of phase modulation was obtainable only by sending the *same* photons through complementary modulators, whereas the version utilizing entangled photons can achieve cancellation on *separated* photons, an example of one of the uniquely nonclassical properties of such states of light.

An experiment demonstrating this quantum behavior was carried in the Harris group [49]; the key results are depicted in Fig. 13.8. Recalling that $\Delta = \omega_1 + \omega_2 - 2\omega_0$ represents the deviation of the sum frequency from the monochromatic pump, we see that with no modulators running, a sharp correlation at $\Delta = 0$ is observed, with two additional sidebands appearing when only one photon is phase-modulated. Switching on the second modulator in phase with the first widens the correlation bandwidth even further, whereas running the second one 180° out of phase recovers the original sharp spectral correlation, with no trace of modulation whatsoever. Taken together, these dispersion and modulation investigations provide a solid foundation for applying more sophisticated optical processing techniques to entangled photons, an example of which is discussed in the next section.

13.6 Detailed Example II: Encoding and Decoding of Biphoton Wavepackets

The previously described experiments with cancellation of polynomial spectral phase or sinusoidal temporal modulation can be viewed as building blocks for a wide range of complex coding schemes. For indeed, an arbitrary well-behaved function can be expanded as a linear superposition of polynomials (Taylor's theorem) or sinusoids (Fourier series), implying that many alternative phase

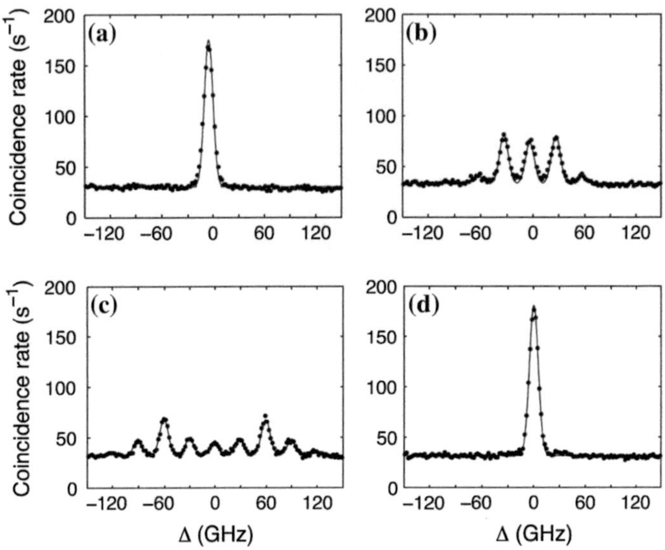

Fig. 13.8 Spectral correlations of temporally modulated biphotons, for **a** no modulation, **b** one photon modulated, **c** both photons modulated in phase, and **d** both photons modulated out of phase. Proper phasing allows the frequency spreading of one photon to be canceled by modulation of the other (Results from [49])

combinations could also find application in biphoton manipulation. In particular, we focus here on binary spectral or temporal phase sequences, common in optical code-division multiple-access communications (O-CDMA) [50]. In the classical spectral case, a code of 0 and π phases is imposed on the frequency spectrum of an ultrashort pulse, which spreads and lowers the waveform in time; only by applying the conjugate code is the original pulse recovered, thereby allowing users with different codes to communicate over the same time-frequency space. In the Fourier dual, the phase code is temporally modulated onto a narrowband optical signal, spreading the frequency content; as before, the original narrowband message is recovered by demodulating with the correct temporal code. This latter process is an example of spread-spectrum communication, which has proven itself adept at covert transmission and jamming resistance [51]. In this section, we describe two experiments which extend these classical communication techniques to entangled photons, showing how such quantum states can be "hidden" and "recovered" in either time or frequency.

First we consider frequency coding and time spreading, which is based on applying appropriate codes to a programmable pulse shaper. One particularly relevant approach, developed for classical fields, utilizes narrowband SFG to correlate phase codes applied on two halves of the optical spectrum. In this version of O-CDMA, the sender applies a phase code to one half of the optical spectrum, say all frequencies greater than ω_0; then the receiver applies another code to the lower half ($\omega < \omega_0$) and sends the field into a nonlinear crystal for SFG. If we denote the coded spectrum by $\mathscr{E}(\omega)$ and assume that phase-matching permits all frequency components to combine, the resultant SFG field at $2\omega_0$ is given by

$$\mathscr{E}_{\text{SFG}}(2\omega_0) = \int_0^{\omega_0} d\Omega\,\mathscr{E}(\omega_0 + \Omega)\mathscr{E}(\omega_0 - \Omega), \tag{13.20}$$

which for a sufficiently flat input spectrum represents the inner product of the codes applied to each half. Therefore, with codes chosen from an orthogonal family, matched pairs give a high SFG yield, whereas mismatched codes ideally produce nothing. Figure 13.9a provides a schematic of this coding process [52], with results from an early experiment in Fig. 13.9b [53]. When the two codes are identical, a strong upconverted signal is measured, with this dropping close to zero when the codes are different.

In the above classical coding example, narrowband SFG is required to correlate the two halves of the input spectrum; however, in the case of time-frequency entangled photons, the desired spectral correlation is available *automatically*. Examining the general biphoton shaping formula, (13.6), we observe that for $\phi(\Omega)$ roughly flat over signal and idler passbands, and at $\tau = 0$, the biphoton wavepacket represents nothing more than the inner product of the spectral filters $H_s(\omega)$ and $H_i(\omega)$, mathematically equivalent to the classical example of (13.20). Therefore, if these filters are codes chosen from some orthogonal family (such as Walsh-Hadamard codes), matched codes will give a correlation peak, whereas mismatched codes will generate a null. Thus while the classical version requires some form of spectral post-selection to ensure that only frequency components equally offset from ω_0 can recombine, no such post-selection is necessary in the quantum version, for the biphotons possess the requisite correlations intrinsically. This connection between classical narrowband SFG and time-frequency entangled photons has also been utilized to derive interesting (but limited) classical analogues of Hong-Ou-Mandel interference [54, 55] and dispersion cancellation [56]. And so in this light,

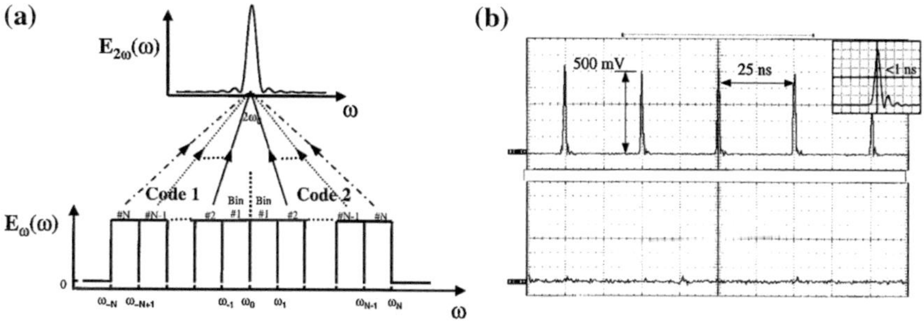

Fig. 13.9 Classical spectral coding. **a** Schematic of the coding process, in which phase codes are applied to each half of the optical spectrum, followed by narrowband SFG. **b** SFG output measured for matched codes (*top*) and mismatched ones (*bottom*) (Image in **a** from [52] and that for **b** from [53])

our work on coding of entangled photons can be viewed as exploiting this connection in the opposite direction; instead of looking for classical versions of quantum phenomena, we have developed a quantum coding approach based on ideas first developed classically.

An outline of how this biphoton coding process works is provided in Fig. 13.10a. The signal half-spectrum of a temporally narrow biphoton is encoded with a sequence of 0 and π phases, which spreads the correlation function and produces a null at $\tau = 0$. Then the idler is decoded by applying a second Hadamard code to the lower half of the biphoton spectrum. If the codes match in the sense that they are symmetric about ω_0, the original correlation peak is restored, but if they differ, Hadamard orthogonality ensures that the correlation function instead remains spread with a zero at $\tau = 0$. As an aside, we note that this symmetric cancellation condition in no way contradicts the general antisymmetric phase relation expressed in (13.16) regarding dispersion cancellation, but rather is a result of the binary codes; for truly antisymmetric phase, the π phase chips on the signal would need to be accompanied by $-\pi$ chips on the idler, but since $-\pi$ phase is physically indistinguishable from $+\pi$, symmetric and antisymmetric conditions are equivalent in this case.

With this coding approach, the biphoton can even be encoded and decoded nonlocally, if the signal and idler are spatially separated, although in the proof-

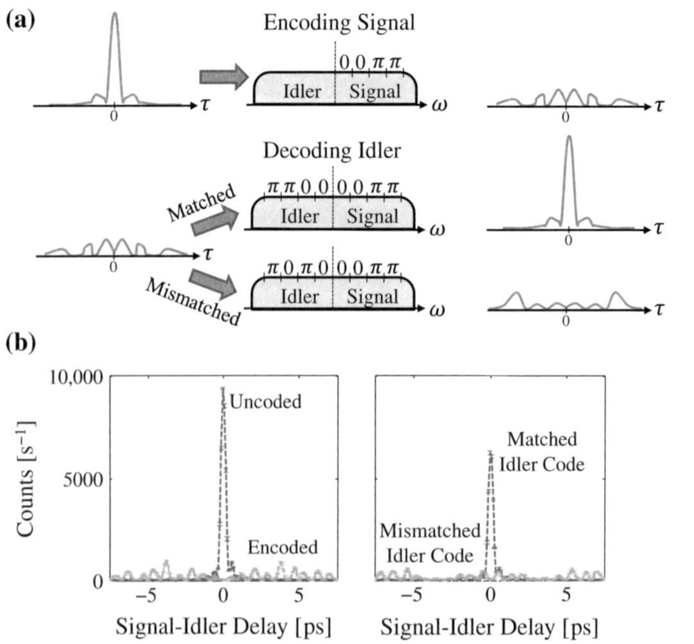

Fig. 13.10 Orthogonal spectral coding of entangled photons. **a** Schematic of the basic process, in which the signal photon is encoded and the idler decoded; only when the codes match is the ultrashort biphoton correlation recovered. **b** Experimental results for length-40 Hadamard codes, confirming the conceptual picture in **a** (Images adapted from [57])

of-principle demonstration discussed here, the photons are detected locally with biphoton SFG. An experimental example with length-40 Hadamard codes is furnished in Fig. 13.10b [57], using a setup identical to that in the previously discussed high-order dispersion cancellation experiments [26]. When the signal is encoded with a particular Hadamard sequence, the initially sharp correlation is temporally spread. Applying the correct idler code then retrieves the original biphoton, aside from a peak reduction due to scattering losses in the pulse shaper. On the other hand, the wrong idler code yields a new, but still spread, biphoton state. These results hint at the fascinating possibilities for complex spectral coding in quantum cryptography, although there remain several key hurdles, addressed in Sect. 13.7.

An example of the Fourier dual, temporal phase coding, was demonstrated in 2010 [58]. This experiment employed extremely narrowband photons in order to permit fractionally large spectral spreading with gigahertz modulators. Moreover, although coding and decoding could indeed be implemented nonlocally with the arrangement of Fig. 13.4b, the authors instead applied both modulators to the same photon, with the entangled partner serving as a trigger for coincidence counting; this can be viewed as a special case of Fig. 13.4c in which the net modulation $m(\tau)$ is the product of encoding and decoding modulators $m_1(\tau)$ and $m_2(\tau)$. In this way the signal photon is first temporally encoded with a pseudorandom sequence of $\pm\pi/2$ phases, which frequency-spreads the 3.5-MHz photon to around 10 GHz. Then a second modulator is applied, followed by a narrowband filter, before detection. To guarantee a high probability of transmission, the second modulator must undo the effects of the first one and restore the narrow photon spectrum; otherwise, the filter will most likely reject the frequency-spread photon. This provides an interesting way to hide single photons.

To demonstrate this capability, Belthangady et al. injected a laser field—at the same frequency and with a flux over 1000× that of the original photon—after the first modulator but before the second. As shown in Fig. 13.11a, when both

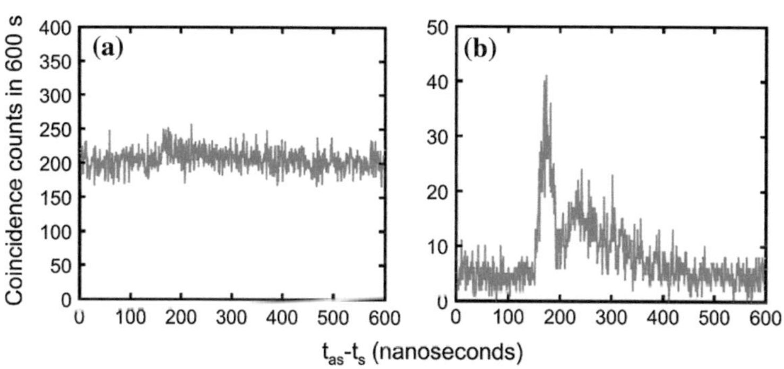

Fig. 13.11 Hiding single photons. **a** With no temporal modulation, the single-photon field is completely hidden by a much more powerful laser beam. **b** But when a temporal encoding and decoding process analogous to classical spread spectrum is applied, the laser is rejected and the characteristic shape of the single photon is recovered (Results from [58])

modulators are turned off, the laser field dominates the coincidence counts, with no evidence of time correlation. But when the modulators are turned on, the shape of the hidden single photon is observed (Fig. 13.11b). This ensues because the photon is frequency-spread by the first modulator and despread by the second, thereby passing through the narrowband detection filter with ease; on the other hand, the laser is transformed only by the second modulator, so its spectrum is spread right before the filter and is therefore highly attenuated. As this experiment highlights, even the thoroughly classical concept of spread spectrum can be applied to single-photon signals.

13.7 Outlook

The examples of biphoton pulse shaping and modulation considered in this chapter indicate the rapid development of this field, which over the last decade has succeeded in combining these classical technologies with quantum states of light. Nonetheless, there remain several important avenues of research to bring these proof-of-principle experiments to more applied specializations, such as secure quantum key distribution (QKD). In some ways SFG, which proved essential in the above demonstrations of biphoton pulse shaping, now sits as a hindrance to further development. For by requiring signal-idler recombination, any experiments utilizing biphoton SFG are intrinsically ill-suited for QKD implementations, in which the two communicating parties must obviously be spatially separated for any useful application. Moreover, the use of SFG in the previous demonstrations of *ultrafast* biphoton dispersion cancellation [26, 45] means that a fully nonlocal implementation—in which both photons are measured on separated detectors—has not been realized with such high timing resolution. It is therefore desirable to find alternative high-time-resolution detection methods which could observe these effects nonlocally. Pending further improvements in single-photon detector jitter to the picosecond level [59], methods based on single-photon mixing with ultrashort [60] or chirped [61] classical pulses seem particularly promising, and so nonlocal generalizations of ultrafast dispersion cancellation and orthogonal coding may be possible with such techniques, although they have yet to be demonstrated. Furthermore, high-dimensional time-frequency QKD protocols based on dispersion [62] or temporal modulation [63] have recently been proposed, and it would be interesting to investigate how sophisticated biphoton control could be exploited to realize these schemes in practice.

In fact, it is their potential as information carriers that has motivated much of the previous work on biphoton pulse shaping, and so here we briefly discuss some of the basic ideas behind time-frequency entanglement for QKD. Such correlations prove particularly promising for key distribution in fiber networks, as they are well-preserved through long propagation distances and can generate multiple bits of information per biphoton. A rough schematic of how this process could work is provided in Fig. 13.12. Signal and idler photons produced through SPDC are sent to

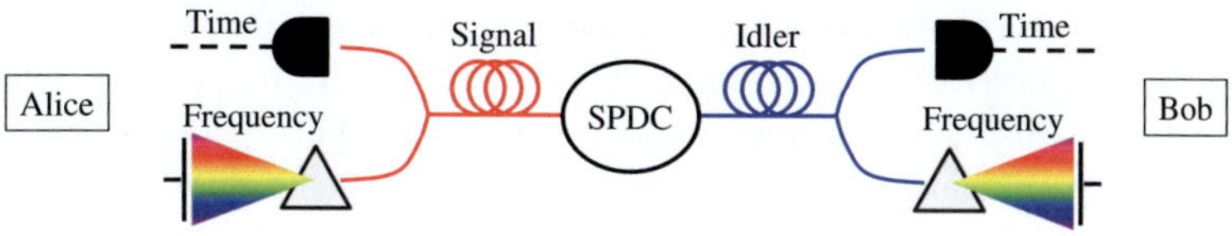

Fig. 13.12 Schematic of time-frequency QKD with entangled photons. Alice and Bob measure either the arrival time or frequency of their respective photons, using results with the same bases to construct a secure key

Alice and Bob, respectively, who measure randomly either the arrival time or frequency of their received photon. Due to the strong spectro-temporal entanglement, measurements in the same basis produce highly correlated results, so by dividing frequency and time into discrete bins, it is possible to construct a secret key through measurements on successive biphotons. And since results in two different bases are uncorrelated—as time and frequency are Fourier conjugates—the presence of an eavesdropper is revealed by publicly comparing some subset of the measured bits and looking for an increased error rate.

The expanded information potential resulting from high-dimensional entanglement can be quantified through the Schmidt decomposition [64, 65], in which the biphoton spectrum is expressed as a sum of factorable two-photon states. Roughly speaking, the number of modes with non-negligible coefficients can be estimated by the Schmidt number K [65], a slighly more rigorous measure of entanglement than the bandwidth ratio discussed briefly in Sect. 13.3. The larger K is, the greater the amount of information obtainable from the biphoton. As a reference, we note that the idealized biphoton state of (13.3) with a perfectly monochromatic pump has $K = \infty$; in reality, the finite linewidth of our pump drops our realistic estimate of K to several million or so. Yet actually extracting this information potential can prove extremely difficult. For example, in our experiments [57], the maximum number of frequency chips is limited by the 10-GHz spectral resolution of the pulse shaper—not the $\sim$200-kHz pump linewidth. Similar behavior has been shown for time binning [66] and orbital angular momentum [67], in which there were observed maximum dimensions beyond which the secure information capacity dropped. The ability to optimize both the intrinsic and practical information potential of biphoton states will prove essential in the experimental development of high-dimensional QKD.

Interestingly, electro-optic modulators have already been proposed for implementing a time-frequency QKD system [63]. Conceptually, the protocol matches Fig. 13.12; however, instead of directly measuring the time variable with a single-photon detector, the authors propose to use a second spectrometer preceded by a time-to-frequency converter. In this fashion, the arrival time of each photon is converted to a particular frequency, so that a wavelength measurement is equivalent to a temporal measurement. The time-to-frequency converter itself consists of a dispersive element and an electro-optic phase modulator. An advantage of this approach is the improved timing resolution over simple avalanche photodiodes; a disadvantage is the increased dark count rate due to utilization of a detector array in the spectrometer. As might be expected, the Fourier dual to this approach has the reverse advantage/disadvantage relationship; as proposed in [62], the spectrometers in the setup of Fig. 13.12 can be replaced by two frequency-to-time converters (highly dispersive elements) with opposite signs of dispersion, followed by single-photon detectors. Thus if both photons are measured in the dispersed-time basis, the dispersion cancellation effect described in Sect. 13.5 ensures that their arrival times are still synchronized. Intuitively, the connection with a frequency measurement can be understood by the fact that dispersion cancellation is itself a consequence of narrowband spectral entanglement. Although Fourier-transform pulse shaping was not explicitly considered in this proposal—and admittedly, pulse shapers cannot apply dispersions as large as other optical devices such as fiber Bragg gratings—the basic idea does lend itself to such pulse shaping. In fact, it would be interesting to consider how the spectral coding demonstrated in [57] could be used for similar time spreading and despreading in a form of code-based time-frequency QKD.

And so while experiments up to this point have already revealed new insights into the behavior of entangled photons, numerous opportunities remain unrealized, particularly in the context of quantum cryptography. Thus we expect the next decade to witness even more advances in biphoton pulse shaping; we have only begun to scratch the surface of the potential within such entangled quanta, and technologies developed in classical optics will no doubt continue to find unanticipated uses in the quantum regime.

References

1. R. Hanbury Brown, R.Q. Twiss, Nature **177**, 27 (1956)
2. R. Hanbury Brown, R.Q. Twiss, Nature **178**, 1046 (1956)
3. M. Chekhova, *Progress in Optics* (Elsevier, New York, 2011), vol. 56, chap. 4
4. E. Brannen, H.I.S. Fergusen, Nature **178**, 481 (1956)
5. R. Hanbury Brown, R.Q. Twiss, Nature **178**, 1447 (1956)
6. E.M. Purcell, Nature **178**, 1449 (1956)
7. P. Fellgett, Nature **179**, 956 (1957)
8. R.J. Glauber, Phys. Rev. **130**, 2529 (1963)
9. R.J. Glauber, Phys. Rev. **131**, 2766 (1963)
10. Y. Shih, Rep. Prog. Phys. **66**, 1009 (2003)
11. A. Einstein, B. Podolsky, N. Rosen, Phys. Rev. **47**, 777 (1935)
12. A. Aspect, Nature **398**, 189 (1999)
13. N. Gisin, G. Ribordy, W. Tittel, H. Zbinden, Rev. Mod. Phys. **74**, 145 (2002)
14. A.M. Weiner, *Ultrafast Optics* (Wiley, Hoboken, NJ, 2009)
15. A.M. Weiner, J.P. Heritage, E.M. Kirschner, J. Opt. Soc. Am. B **5**, 1563 (1988)
16. A.M. Weiner, Opt. Commun. **284**, 3669 (2011)
17. A.M. Weiner, Rev. Sci. Instr. **71**, 1929 (2000)
18. S.T. Cundiff, A.M. Weiner, Nat. Photon. **4**, 760 (2010)
19. G.L. Li, P.K.L. Yu, J. Lightw. Technol. **21**, 2010 (2003)
20. A.J. Metcalf, V. Torres-Company, D.E. Leaird, A.M. Weiner, IEEE J. Sel. Top. Quantum Electron. **19**, 3500306 (2013)

21. E.L. Wooten, K.M. Kissa, A. Yi-Yan, E.J. Murphy, D.A. Lafaw, P.F. Hallemeier, D. Maack, D.V. Attanasio, D.J. Fritz, G.J. McBrien, D.E. Bossi, IEEE J. Sel. Top. Quantum Electron. **6**, 69 (2000)
22. B.H. Kolner, IEEE J. Quantum Electron. **30**, 1951 (1994)
23. Z. Jiang, C.B. Huang, D.E. Leaird, A.M. Weiner, Nat. Photon. **1**, 463 (2007)
24. V.R. Supradeepa, C.M. Long, R. Wu, F. Ferdous, E. Hamidi, D.E. Leaird, A.M. Weiner, Nat. Photon. **6**, 186 (2012)
25. L. Mandel, E. Wolf, *Optical Coherence and Quantum Optics* (Cambridge University Press, Cambridge, UK, 1995)
26. J.M. Lukens, A. Dezfooliyan, C. Langrock, M.M. Fejer, D.E. Leaird, A.M. Weiner, Phys. Rev. Lett. **111**, 193603 (2013)
27. J.D. Franson, Phys. Rev. A **45**, 3126 (1992)
28. S.E. Harris, Phys. Rev. A **78**, 021807 (2008)
29. P. Kolchin, C. Belthangady, S. Du, G.Y. Yin, S.E. Harris, Phys. Rev. Lett. **101**, 103601 (2008)
30. B. Dayan, Y. Bromberg, I. Afek, Y. Silberberg, Phys. Rev. A **75**, 043804 (2007)
31. E. Poem, Y. Gilead, Y. Lahini, Y. Silberberg, Phys. Rev. A **86**, 023836 (2012)
32. J.M. Lukens, A. Dezfooliyan, C. Langrock, M.M. Fejer, D.E. Leaird, A.M. Weiner, Opt. Lett. **38**, 4652 (2013)
33. B. Dayan, A. Pe'er, A.A. Friesem, Y. Silberberg, Phys. Rev. Lett. **94**, 043602 (2005)
34. B. Dayan, Phys. Rev. A **76**, 043813 (2007)
35. A. Pe'er, B. Dayan, A.A. Friesem, Y. Silberberg, Phys. Rev. Lett. **94**, 073601 (2005)
36. F. Zäh, M. Halder, T. Feurer, Opt. Express **16**, 16452 (2008)
37. J.D. Franson, Phys. Rev. Lett. **62**, 2205 (1989)
38. L. Mandel, Rev. Mod. Phys. **71**, S274 (1999)
39. C. Bernhard, B. Bessire, T. Feurer, A. Stefanov, Phys. Rev. A **88**, 032322 (2013)
40. M.A.F. Roelens, S. Frisken, J.A. Bolger, D. Abakoumov, G. Baxter, S. Poole, B.J. Eggleton, J. Lightw. Technol. **26**, 73 (2008)
41. J.M. Lukens, O. Odele, C. Langrock, M.M. Fejer, D.E. Leaird, A.M. Weiner, Opt. Express **22**, 9585 (2014)
42. V. Torres-Company, J. Lancis, H. Lajunen, A.T. Friberg, Phys. Rev. A **84**, 033830 (2011)
43. C. Belthangady, S. Du, C.S. Chuu, G.Y. Yin, S.E. Harris, Phys. Rev. A **80**, 031803 (2009)
44. S.Y. Baek, Y.W. Cho, Y.H. Kim, Opt. Express **17**, 19241 (2009)
45. K.A. O'Donnell, Phys. Rev. Lett. **106**, 063601 (2011)
46. S.E. Harris, Phys. Rev. Lett. **98**, 063602 (2007)
47. Z. Jiang, S.D. Yang, D.E. Leaird, A.M. Weiner, Opt. Lett. **30**, 1449 (2005)
48. J.M. Lukens, D.E. Leaird, A.M. Weiner, Nature **498**, 205 (2013)
49. S. Sensarn, G.Y. Yin, S.E. Harris, Phys. Rev. Lett. **103**, 163601 (2009)
50. J.P. Heritage, A.M. Weiner, IEEE J. Sel. Top. Quantum Electron. **13**, 1351 (2007)
51. R. Pickholtz, D. Schilling, L. Milstein, IEEE Trans. Commun. **30**, 855 (1982)
52. Z. Zheng, A.M. Weiner, Opt. Lett. **25**, 984 (2000)
53. Z. Zheng, A.M. Weiner, K.R. Parameswaran, M.H. Chou, M.M. Fejer, IEEE Photon. Technol. Lett. **13**, 376 (2001)
54. R. Kaltenbaek, J. Lavoie, D.N. Biggerstaff, K.J. Resch, Nat. Phys. **4**, 864 (2008)
55. R. Kaltenbaek, J. Lavoie, K.J. Resch, Phys. Rev. Lett. **102**, 243601 (2009)
56. R. Prevedel, K.M. Schreiter, J. Lavoie, K.J. Resch, Phys. Rev. A **84**, 051803 (2011)
57. J.M. Lukens, A. Dezfooliyan, C. Langrock, M.M. Fejer, D.E. Leaird, A.M. Weiner, Phys. Rev. Lett. **112**, 133602 (2014)
58. C. Belthangady, C.S. Chuu, I.A. Yu, G.Y. Yin, J.M. Kahn, S.E. Harris, Phys. Rev. Lett. **104**, 223601 (2010)
59. R.H. Hadfield, Nat. Photon. **3**, 696 (2009)
60. O. Kuzucu, F.N.C. Wong, S. Kurimura, S. Tovstonog, Phys. Rev. Lett. **101**, 153602 (2008)
61. J.M. Donohue, M. Agnew, J. Lavoie, K.J. Resch, Phys. Rev. Lett. **111**, 153602 (2013)

62. J. Mower, Z. Zhang, P. Desjardins, C. Lee, J.H. Shapiro, D. Englund, Phys. Rev. A **87**, 062322 (2013)
63. J. Nunn, L.J. Wright, C. Söller, L. Zhang, I.A. Walmsley, B.J. Smith, Opt. Express **21**, 15959 (2013)
64. C.K. Law, I.A. Walmsley, J.H. Eberly, Phys. Rev. Lett. **84**, 5304 (2000)
65. C.K. Law, J.H. Eberly, Phys. Rev. Lett. **92**, 127903 (2004)
66. I. Ali-Khan, C.J. Broadbent, J.C. Howell, Phys. Rev. Lett. **98**, 060503 (2007)
67. J. Leach, E. Bolduc, D.J. Gauthier, R.W. Boyd, Phys. Rev. A **85**, 060304 (2012)

Chapter 14
Harnessing Nonlinear Optics for Microwave Signal Processing

David Marpaung, Ravi Pant and Benjamin J. Eggleton

Abstract Harnessing nonlinear optical effects in a photonic chip scale has been proven useful for a number of key applications in optical communications. Microwave photonics (MWP) can also benefit from the adoption of such a technology, creating a new concept of nonlinear integrated microwave photonics. Here, we look at the potential of using nonlinear optical effects in a chip scale to enable RF signal processing with enhanced performance. We review a number of recent results in this field, with particular focus on the creation of frequency agile and high suppression microwave bandstop filters using on-chip stimulated Brillouin scattering (SBS). We also discuss the future prospect of nonlinear integrated MWP to enable a general purpose, programmable analog signal processor, as well as compact, high performance active microwave filters with enhanced energy efficiency.

14.1 Microwave Photonics

Microwave photonics (MWP) is a field that deals with the use of optoelectronic devices and systems for enhanced radio frequency (RF) signal manipulation, including the generation, distribution, processing, and analysis of high-speed microwave signals [11, 44, 45, 53]. MWP technologies have also enabled various functionalities which are not feasible to achieve only by traditional microwave

D. Marpaung · R. Pant · B.J. Eggleton (✉)
Centre for Ultrahigh Bandwidth Devices for Optical Systems (CUDOS),
School of Physics, Institute of Photonics and Optical Science (IPOS),
University of Sydney, Camperdown, NSW, Australia
e-mail: egg@physics.usyd.edu.au

D. Marpaung
e-mail: david.marpaung@sydney.edu.au

R. Pant
e-mail: rpant@physics.usyd.edu.au

S. Wabnitz and B.J. Eggleton (eds.), *All-Optical Signal Processing*,
Springer Series in Optical Sciences 194, DOI 10.1007/978-3-319-14992-9_14

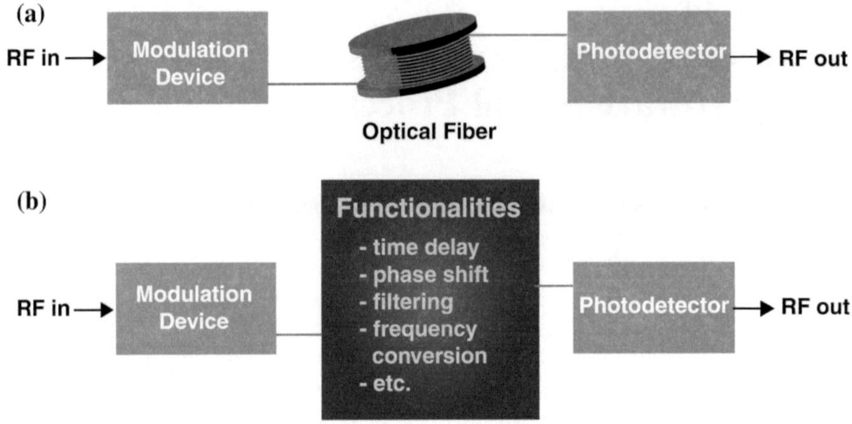

Fig. 14.1 Schematics of **a** an MWP link and **b** a simple MWP system. The MWP link basically consists of a modulation device for E/O conversion and a photodetector for O/E conversion. Such an MWP link with added functionalities between the conversion will make an MWP system. From [28]

techniques, such as the generation of ultra-wide bandwidth waveforms, or filtering with reconfigurable shape and central frequency.

The basis of any MWP system is a so-called MWP link [16–18]. This link consists of a modulation device to convert RF signals to modulated optical signals (E/O conversion) connected by an optical fiber to a photodetector that restores the RF signal back from the optical domain (O/E conversion), as shown in Fig. 14.1a The majority of MWP links employ intensity modulation-direct detection (IMDD) scheme, with a high speed electro-optic modulator (EOM) as the modulation device and a p-i-n photodiode as the detector.

Initially, the main use of MWP links was to replace long-length coaxial-cable links which were very lossy for high RF frequencies. Optical fibers offer low attenuation for virtually any microwave frequency, as well as massive weight reduction, increased flexibility, and immunity to electromagnetic interference (EMI). Subsequently, more efforts were directed to use these links to process the modulated RF signals beyond only signal transport. This was the birth of MWP systems, which established by means of adding functionalities between the two conversions, i.e. signal processing in the optical domain (Fig. 14.1b).

MWP has a wide range of applications in the processing of high speed RF signals. The generation of high purity microwave signals and high complexity ultra-broadband waveforms are the latest development in MWP. The main attraction for signal generation using MWP technique is the large frequency tunability and the potential of reaching very high frequencies (up to the THz region) using relatively simpler technique compared to the traditional microwave/electronic approach. Moreover, the distribution of such high frequency signals using extremely low-loss optical fibers is also attractive, which would have been very lossy using coaxial cables [43]. For waveform generations MWP techniques offer broad bandwidth and the full reconfigurabilities of the phase and amplitude of the RF waveforms [21, 54].

For microwave signal processing, MWP techniques have enabled filtering [12, 31], tunable true time delay [41, 55, 61] and wideband phase shifting of microwave signals [8, 15, 23]. The MWP concept added value are the operation bandwidth and the potential of fast and agile reconfigurabilities of these functionalities. Combining these basic functionalities lead to the realization of MWP processors for optical beamforming and phase array antenna systems [30, 34, 60].

14.1.1 MWP System Performance

To achieve optimum functionalities from MWP systems, the underlying MWP link needs to reach sufficient performance, which is limited by the losses, noise, and nonlinear distortions introduced by the E/O and O/E conversions [17]. Their performance can be described by a set of parameters, namely the link gain, the noise figure, the intercept points, and the spurious-free dynamic range (SFDR) [16, 18].

The link gain describes the RF to RF power signal transfer in the MWP link or system. Due to the limited conversion efficiencies in the modulation device and photodetector, it is common that the MWP link shows negative link gain in the decibel scale, i.e. a net loss. For externally modulated MWP link, the link gain is proportional to the square of the modulated optical power. Thus, a common technique to improve the link gain is to pump more optical power to the EOM. The technique has been effectively used to demonstrate MWP links with net gain (i.e. positive link gain) instead of loss, where the value as high as +44 dB has been demonstrated [49].

The noise figure (NF) is a useful metric that measures the signal-to-noise ratio (SNR) degradation in the system, expressed in decibels. The noise figure is proportional to the total noise power in the link, and is inversely proportional to the link gain. The total link noise is usually dominated by laser relative intensity noise (RIN) and the photodetector shot noise. State-of-the-art MWP links can exhibit NF below 10 dB [1, 29], which can be achieved using an ultra-low half-wave voltage EOM in the order of 1 V, and a very high input power.

The E/O and O/E conversions in the MWP link also add nonlinear distortions to the output RF signal. Due to the nonlinear response of the link (i.e. components like the EOM or the photodetector), input RF tones will generate new frequency components called the intermodulation distortions (IMDs). The second-order intermodulation (IMD2) is generated due to the quadratic nonlinearity in the link and the frequency components appear at the sum and the difference of the modulating frequencies, while the third-order intermodulation (IMD3) is generated by cubic nonlinearity in the link and appear at the sum and the difference of twice of one frequency with the other frequency.

The figure of merit that incorporates the effect of noise and nonlinearity in the MWP link is the spurious-free dynamic range (SFDR). The SFDR is defined as the ratio of input powers where, on the one hand, the fundamental signal power is equal to the noise power and, on the other hand, the *n*th-order intermodulation distortion (IMD*n*) power is equal to the noise power. In terms of output powers, this can be interpreted as the maximum output SNR that can be achieved while keeping the IMD*n* power below the noise floor. The SFDR generally governs the lowest and the strongest signal power that can be accommodated in the link.

14.1.2 Integrated Microwave Photonics

For the last 25 years, MWP systems and links have relied almost exclusively on discrete optoelectronic devices and standard optical fibers. These configurations are bulky, expensive and power-consuming while lacking in flexibility. A second generation of MWP systems, known as integrated microwave photonics (IMWP) [28], aims at the incorporation of MWP components/subsystems in photonic integrated circuits. With photonic integration, one can achieve a reduction in footprint, inter-element coupling losses, packaging cost as well as power dissipation.

IMWP clearly promises a number of advantages compared to conventional fiber-based approach. The potential of integrating a number of different signal processing functionalities in a single chip will lead to significant reduction in size, weight, power consumption, and cost. Moreover, the ability to integrate various functions like modulation, passive signal processing, and detection in a single chip will lead to reduction in insertion losses, which is the key for MWP signal processor to compete with RF signal processor in terms of performance.

14.1.3 Nonlinear Integrated Microwave Photonics

Thus far, IMWP has largely been limited to linear optical processing. It is interesting, however, to examine the potential of nonlinear optics in integrated platform for microwave signal processing. Optical nonlinearities (e.g., $\chi^{(2)}$ and $\chi^{(3)}$ [2], such as sum and difference-frequency generation (SFG/DFG), cross-phase modulation (XPM), degenerate/non-degenerate four-wave mixing (FWM), and stimulated Brillouin scattering (SBS), are suitable candidates for MWP signal processing. In the context of optical signal processing, these nonlinear optics effects have efficiently been harnessed, both in fibers and in integrated devices, for a number of functionalities such as all-optical delay [35, 47], all-optical switching and multiplexing [20], and wavelength multicasting [52].

Here, we look at the potential of harnessing nonlinear optical effects in a chip scale to enable RF signal processing with enhanced performance. As an example, we focus on stimulated Brillouin scattering (SBS) for MWP signal processing, namely for frequency agile and high suppression microwave bandstop filters, continuously tunable phase shifter, and tunable slow light delay lines.

14.2 Stimulated Brillouin Scattering

Stimulated Brillouin scattering (SBS) is a third-order optical nonlinearity, which results from three way interaction between light and acoustic vibrations in the medium. The process is characterized by the inelastic scattering of a pump beam from an acoustic vibration in the medium to generate a Stokes wave [2]. For the pump-probe configuration, the SBS process leads to a gain resonance centered at the Stokes frequency where the width of the resonance is determined by the phonon lifetime (τ_p). Typical phonon lifetime in commonly used optical materials is 10 ns, which results in narrow resonance bandwidth $(1/\tau_p)$ of tens-of-MHz. The bandwidth of the SBS resonance, however, can be controlled by tailoring the pump profile, enabling broadband (GHz) gain response. The Brillouin scattering process can therefore provide wide bandwidth tunability, ranging from MHz to GHz. Further, the centre frequency of the gain resonance can be tuned by tuning the pump frequency ω_p, which enables SBS gain realization at any wavelength.

The flexibility of tailoring the SBS process parameters has enabled a number of applications in MWP [39], including tunable delay lines [14, 35, 38, 47], phase shifters [15, 23], and filtering [9, 26, 32, 51, 56, 57]. In the SBS process, a signal centered at the Stokes frequency experiences gain whereas absorption resonance occurs for a signal centered at the anti-Stokes frequency. Both the gain and absorption resonance exhibit a phase response, obtained using the Kramers-Kronig relation, which can be used to realize slow light induced tunable delay and also to control the phase of a signal. The stimulated Brillouin scattering process is therefore emerging as a potential platform for microwave photonics, having been utilized in long length (≈km) optical fibers for a number microwave applications. Integrated MWP signal processing however, requires on-chip realization of SBS [40].

14.2.1 On-Chip SBS

Recently, on-chip SBS was demonstrated using a travelling wave geometry in a photonic chip fabricated using chalcogenide (As_2S_3) glass. Chalcogenide glasses are known to have large refractive index, which results in large nonlinearity [20]. The strong confinement of the optical and acoustic modes (see Fig. 14.2) in a chalcogenide photonic-chip provides a small effective area and a large overlap between acoustic and optical modes, which, when combined with large refractive index of chalcogenide glass, resulted in large Brillouin gain coefficient and thus

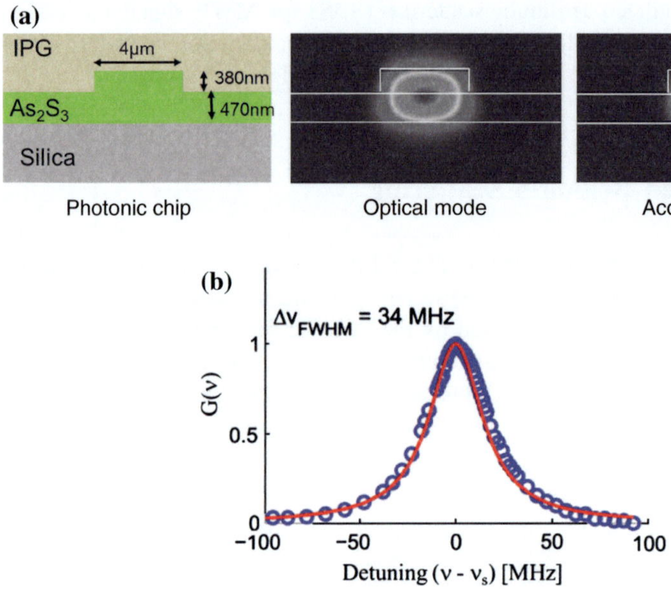

Fig. 14.2 **a** Schematic of the chalcogenide photonic chip showing strongly confined optical and acoustic modes, critical for realizing on-chip stimulated Brillouin scattering. **b** Measured gain profile for the Stokes signal showing a Lorentzian response with narrow linewidth of 34 MHz. From [40]

large Brillouin gain in a 7 cm long rib waveguide using moderate pump power (≈100 mW).

The measured frequency shift for the chip was $\nu_B = 7.7$ GHz, and the SBS gain profile, measured in a pump-probe experiment, is plotted in Fig. 14.4 as a function of frequency detuning from the Stokes frequency. It has a Lorentzian line shape, with a full-width at half-maximum (FWHM) of 34 MHz. The narrow bandwidth of the SBS resonance and the large dispersion associated with it is the key for enabling a number of applications in microwave photonics.

14.3 Reconfigurable Microwave Filters Using SBS

RF filters are fundamental signal processing components that play a critical role in determining the systems ultimate performance. The ultimate goal of a filter is to improve the signal-to-noise ratio of the signal of interest either by rejecting unwanted signals (i.e. band-stop filter) or by allowing only the desired signal to propagate to the receiver (i.e. band-pass filter). To operate in the increasingly crowded radio spectrum, RF receivers require frequency agility to operate in different frequency bands. A smart, reconfigurable filter that operates over a wide frequency range is a critical technology to achieve this agility.

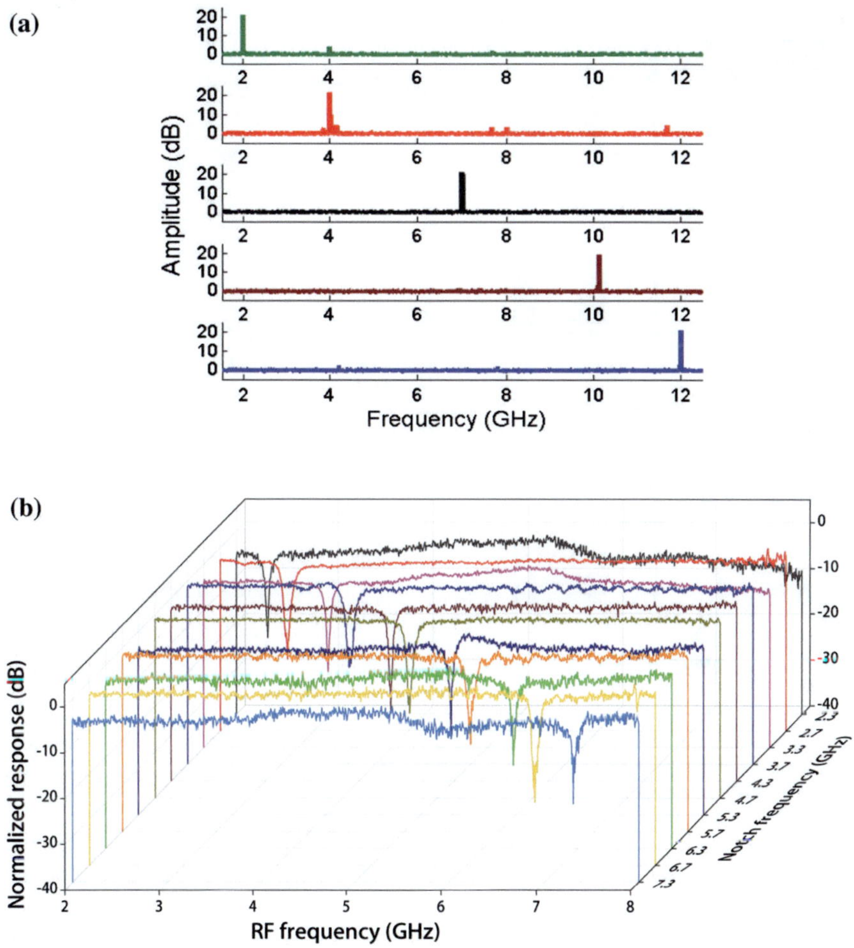

Fig. 14.3 **a** Measured response of a photonic-chip based MWP bandpass filter showing wide filter tunability of 2–12 GHz. From [9]. **b** Measured response of a photonic-chip based MWP notch filter based on single sideband modulation and SBS absorption. From [32]

The gain (absorption) responses of the SBS process can be used as a band-pass (band-reject) optical filter, where the filter extinction depends on the G parameter. Microwave photonic filters with shape reconfiguration and widely tunable pass-band and notch have recently been demonstrated exploiting SBS in several km long optical fibers [51, 56, 57]. While high-Q MWP filters ($\approx$1000) with wide tunability ($\approx$120 GHz) and large extinction $\approx$40 dB were reported, the use of long lengths in these demonstrations prohibits photonic integration.

Recently on-chip band-pass and band-reject MWP filters were demonstrated exploiting SBS in a chalcogenide photonic-chip discussed in Sect. 14.2.1. For the bandpass response, on-chip SBS gain was used to break the amplitude symmetry of

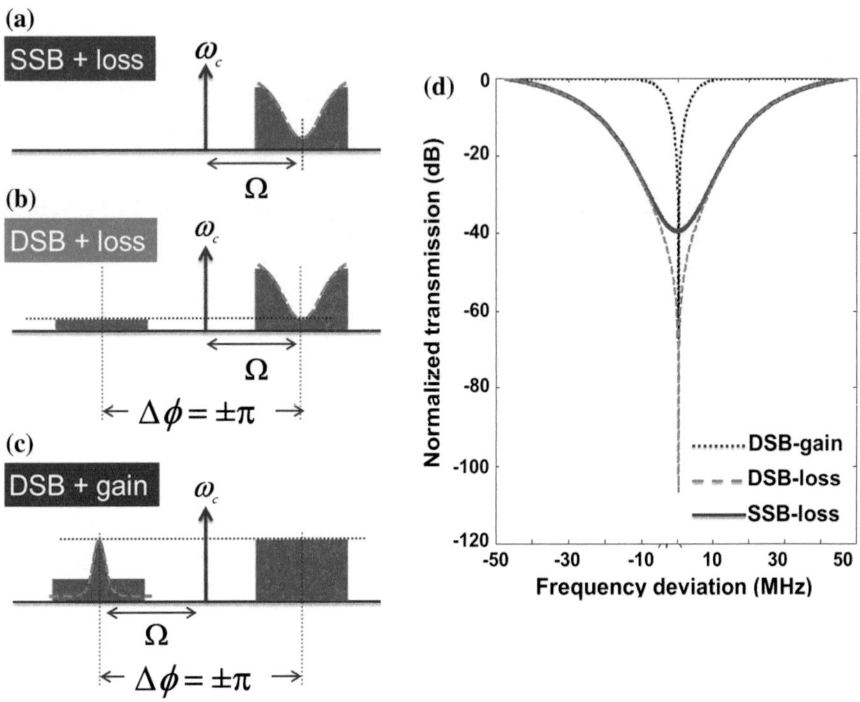

Fig. 14.4 Principle of a MWP notch filter using **a** single side band modulation and SBS absorption **b** double side band modulation and SBS absorption and **c** double side band modulation and SBS gain. From [26]

the upper and lower side bands, obtained by phase modulating an optical carrier with a radio-frequency (RF) signal (see Fig. 14.7). In the absence of SBS gain, equal amplitude and π phase shift between the two sidebands causes destructive interference at the microwave frequency on opto-electronic conversion, resulting in no RF signal. By applying SBS gain, the part of the microwave spectrum that sees the gain is filtered with the filtered spectrum having the shape and bandwidth of the SBS gain profile. In [9], the MWP filter frequency was tuned from 2 to 12 GHz by tuning the pump frequency (Fig. 14.3a) while maintaining stable amplitude (20 ± 2 dB) and 3 dB bandwidth ($f_{3dB} = 23 \pm 2$ MHz).

Additionally, in the demonstration of Byrnes et al. [9], the 3 dB bandwidth and filter shape factor S, defined as the ratio of the 20 to 3-dB bandwidth, were reconfigured using pump profile tailoring. The 3-dB bandwidth and shape factor S were improved from $f_{3dB} = 20$ MHz and $S = 3.5$ for the single-pump case to $f_{3dB} = 40$ MHz and $S = 2$ for the dual pump configuration. For a photonic-chip based MWP filter, an extremely large Q factor of 520 was obtained using the filter response at 12 GHz.

As in the case of bandpass filter, SBS absorption can be used as a band-stop filter to realize a notch response in microwave domain. Single-side band (SSB)

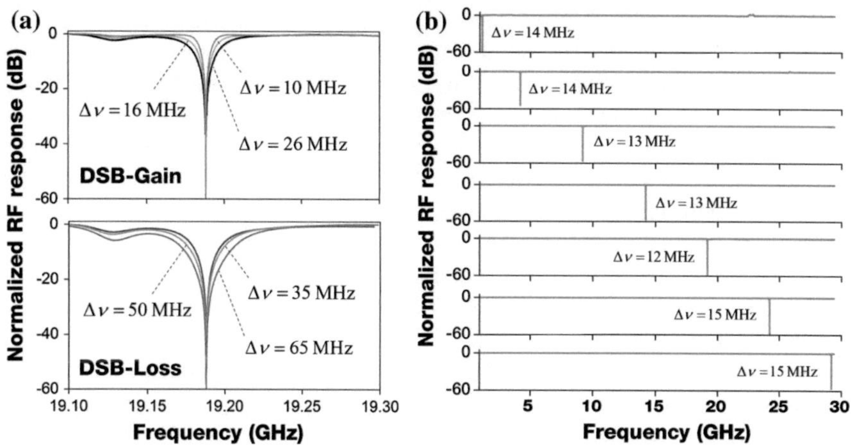

Fig. 14.5 **a** Experimental results of the MWP notch filter bandwidth tuning achieved by means of tuning the SBS pump power. **b** Experimental results of ultra-wide frequency tuning of the MWP notch filter, preserving a narrow isolation bandwidth of 13 ± 2 MHz and ultrahigh rejection of 60 ± 2 dB. From [26]

modulation was used to encode the microwave signal to the optical domain, where a SBS pump was used to create a band-stop response at the desired frequency. The frequency of the notch was tuned by tuning the pump frequency. For this configuration, the depth of the notch depends on the pump power via the Brillouin absorption parameter. A photonic-chip based MWP notch filter with a tuning range of 28 GHz (see Fig. 14.3b) was realized using this scheme [32]. The measured notch suppression of 19 ± 2 and 3 dB bandwidth (126 ± 7 MHz) in this demonstration showed good stability in the filter operation. Although MWP notch filter based on SSB modulation and SBS operation in absorption regime provides an easy to implement scheme, it is highly inefficient and it requires high loss for large suppression, resulting in large power consumption. Achieving large suppression (>60 dB) using SSB modulation scheme with on-chip SBS is therefore highly inefficient.

Recently, a remarkable progress has been made regarding MWP tunable notch filters based on SBS. A long standing problem in RF electronic filters is to achieve wide frequency tuning simultaneously with high resolution filtering. For example, state-of-the-art absorptive band-stop RF filters are capable of a high peak attenuation (>50 dB) and high resolution (<10 MHZ 3 dB isolation bandwidth measured from the passband) but have limited notch frequency tuning range in the order of 1.4 GHz [25]. MWP notch filters, on the other hand, are capable of multi-gigahertz tuning, but are often limited in peak attenuation (<40 dB) and resolution (gigahertz bandwidth instead of tens of megahertz). This limitation stems from the traditional way of implementing notch filters using SSB modulation and an optical filter. In this scheme, the resolution and peak suppression of the MWP filter are limited by those of the optical filter. Unfortunately, for some optical filters, there is a trade-off between resolution and peak suppression. Hence, the resolution and peak suppression of these MWP filters cannot be optimized simultaneously.

In [26], a solution to this limitation was proposed in the form of a new class of MWP notch filter that operates by a novel concept of optical sidebands coherent control using an electro-optic (EO) modulator and an optical resonant filter. This scheme allows the creation of an anomalously high rejection MWP notch filter from virtually any kind of optical resonance, irrespective of its type (gain or absorption), or its magnitude. Instead of SSB modulation, this scheme relies on double-side band (DSB) modulation to control the relative amplitude and phase response of both modulation side bands, as shown in Fig. 14.4. This amplitude and phase control was achieved using a dual-parallel Mach-Zehnder modulator (DPMZM) driven using a quadrature hybrid coupler. Voltage biases to the DPMZM modulator were used to tune the phase and amplitude of the two sidebands. The amplitude and phase response of an optical resonance, which can be passive e.g. ring resonator or active e.g. SBS, was then used to make the amplitude of two side bands equal and out-of-phase by π. The destructive interference between the signals obtained by beating of two sidebands with the carrier results in anomalously large suppression.

Microwave photonic notch filters with large suppression (>60 dB) were demonstrated by implementing this scheme using SBS resonance in optical fibers [26]. The filter provides large suppression (>60 dB) with wide frequency tuning ($\approx$30 GHz), and wide bandwidth tuning from 10–65 MHz, achieved using optical pump power tuning for both SBS gain and loss (see Fig. 14.5). For this scheme, large suppression is achieved irrespective of the SBS gain or loss. The bandwidth of the notch filter depends on the bandwidth of the SBS gain (loss). The SBS gain (loss) bandwidth reduces (increases) with increase in the pump power. Bandwidth tuning over a wide range can therefore be realized by varying the pump power. Although excellent performance was achieved by exploiting SBS gain and loss, the fiber-based demonstration is incompatible with photonic integration.

A widely tunable ($\approx$20 GHz) microwave photonic notch filter with large suppression (>55 dB) using on-chip SBS in combination with DSB modulation was recently reported [24]. The 3 dB bandwidth of the notch was tuned from 31 to 77 MHz by varying the SBS gain parameter from 12.5 to 7.6 dB. In contrast to the notch filter based on SSB modulation + band-stop response of the SBS process, where a suppression of 55 dB would require a Brillouin gain of 55 dB (with a pump power of 600 mW), this scheme realizes a notch depth >55 dB using only 0.8 dB of gain (pump power of tens of mW), which results in highly efficient operation.

14.4 SBS Tunable Delay Line and Phase Shifter

Reconfigurable optical delay lines and wideband tunable phase shifters have primary importance in a number of MWP signal processing applications like optical beamforming [30, 61], and MWP filtering. The simplest way for generating delay in the optical domain is through physical length of optical fibers. However, this can become rather bulky. For this reason, integrated photonic solutions are used.

14.4.1 On-Chip SBS Tunable Delay Lines

SBS can be harnessed as tunable delay lines with flexible operation bandwidth and wavelength. For the SBS process, the delay is given by:

$$T_{\mathrm{d}} = \frac{g_{\mathrm{B}} P_{\mathrm{p}} L}{A_{\mathrm{eff}} \Gamma_{\mathrm{B}}}, \tag{14.1}$$

where P_{p} is the pump power, A_{eff} is the effective mode area, Γ_{B} is the full-width at half-maximum (FWHM) linewidth of the Brillouin gain, L is the medium length, and g_{B} is the Brillouin gain coefficient. This suggests that the time delay can be tuned by varying the pump power. Further, for a device that provides large g_{B} and small A_{eff}, large delay can be obtained using small power.

Many of the earlier SBS delay line demonstrations [14, 15] used silica based optical fibers where small g_{B} and large mode area A_{eff}, necessitated the use of long length (≈km) to achieve large delay using small pump power (≈100 mW).

Recently, SBS based slow- and fast-light was demonstrated in a 7 cm long chalcogenide photonic chip with a delay of 23 ns which was obtained using a gain of 23 dB. This was obtained from a pulse delay measurement [38] for an input of 100 ns long nearly Gaussian pulse. For a 25 ns pulse, a pulse delay of nearly one pulse width was obtained with a gain of 23 dB. Figure 14.6 show a comparison of the measured delay with the theoretical delay calculated using (14.1), showing good agreement between theory and experiments.

On-chip SBS enabled optically controlled continuously tunable delay is a step towards realizing tap-delay based microwave photonic filters. The magnitude of the delay can be varied from ps to ns range by tailoring the pump profile to vary the SBS gain bandwidth [15].

14.4.2 On-Chip SBS Phase Shifter

Radio frequency (RF) phase shifters are a fundamental building block for analog signal processors [8], and are widely employed in phased array antenna systems. The characteristics of an ideal phase shifter include a continuously tunable, full-360° phase shift, constant over a wide, multi-GHz frequency band. In addition, it is desirable to have low insertion loss over the whole band of operation, with a flat amplitude response across the entire phase tuning range. These specifications however cannot be met by conventional electrical phase shifters, which can operate over only a narrow frequency range, and cannot achieve continuous phase tuning.

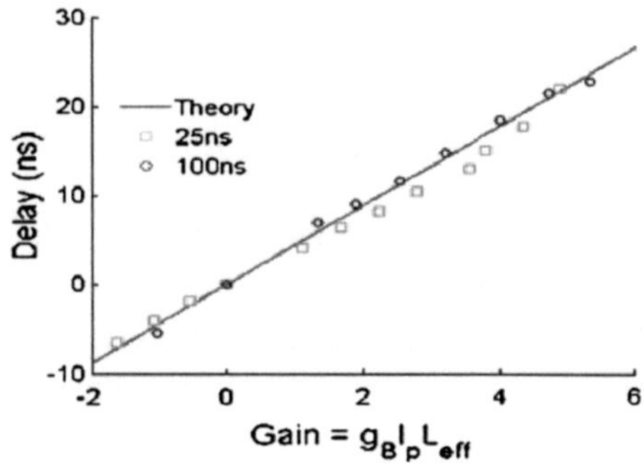

Fig. 14.6 Comparison of the measured and theoretical tunable time delay from SBS in a 7 cm chalcogenide waveguide. From [38]

The most promising approaches for ultra-wideband phase shifter operation rely on optical carrier processing [8, 15], in which a phase shift imparted on the optical carrier is converted by square-law photodetection to a phase shift in the electrical domain, with no theoretical electrical bandwidth limitations. In optical carrier processing, it is highly beneficial to use a narrowband optical effect to shift the phase of the optical carrier, since this does not affect the optical sidebands and as a consequence, does not restrict the phase shifter to high-frequency operation. Due to its low threshold power and narrow bandwidth, SBS is ideal for optical carrier processing.

In 2006, Loayssa and Lahoz [23] proposed a SBS phase shifter topology with two SBS pumps at frequencies up and down-shifted from the optical carrier, as shown in Fig. 14.7. The upper frequency pump will induce a gain response at the carrier frequency, which will be exactly cancelled by the loss response induced by the lower frequency pump. However, due to the profile of the SBS phase response, the phase shifts introduced by the two pumps will not cancel, but will in fact add up. In this way, the carrier amplitude will not be affected by the SBS process, while the carrier phase will be shifted by twice the SBS phase response. Tuning of the phase shift experienced by the carrier is achieved by changing the frequency of the SBS pumps relative to the carrier.

Recently, SBS MWP phase shifter using the dual-pump technique was demonstrated in a chalcogenide photonic chip [36]. The phase shifter exhibited a maximum of 2.5 dB amplitude fluctuations as the phase was continuously tuned over the range of 210° (Fig. 14.8). This results were achieved using 25 dBm of pump power. Extension to a full 360° range can be obtained by increasing the pump power by 3.6 dB, or by reducing the insertion loss of the photonic hip by the same amount.

14.5 FWM and XPM for MWP Signal Processing

On-chip ultrafast Kerr nonlinearity has also been exploited in the context MWP signal processing, for example for ultra-wideband (UWB) signal generation [48] and reconfigurable multi-tap filtering [13]. In [48], Tan et al. combine the effects of cross-phase modulation (XPM) and birefringence in an As_2S_3 rib waveguide to

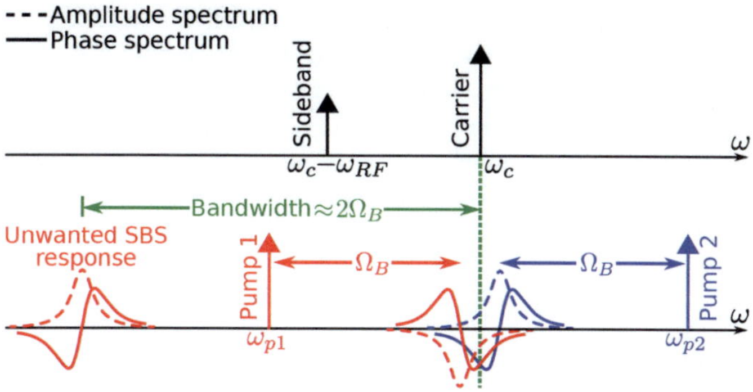

Fig. 14.7 SBS tunable MWP phase shifter using dual-pump SBS. The SBS magnitude responses (*dashed lines*) cancel out at the carrier frequency, while the phase responses add up. From [37]

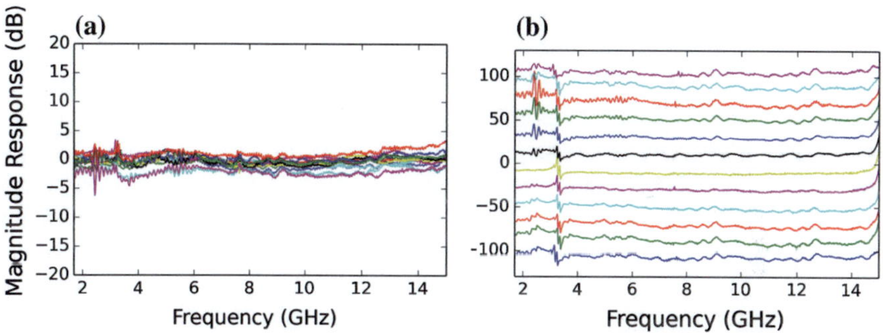

Fig. 14.8 Experimentally measured frequency response of the SBS phase shifter. **a** Magnitude response with ±2.5 dB amplitude fluctuations. **b** Phase response with 210° continuous tuning range. From [37]

generate polarity-inverted UWB monocycles with a single optical carrier. The high Kerr-nonlinearity of ChG in a chip platform enables efficient XPM in a short length of 7 cm. The combination of XPM and birefringence essentially acts as an all-optical differentiator that converts a train of Gaussian pulses into a train of monocycle electrical pulses after photo-detection process. Using this technique, Tan et al. demonstrated the generation of a wide variety of UWB pulses, including polarity-inverted monocycles and doublets. A key advantage in exploiting XPM was the chirp erasure of the input optical pulses, because XPM depends solely on the pulses intensity. Thus, shaped UWB optical pulses have a good tolerance to dispersion over fiber and are more suitable for long-distance transmission in UWB over fiber communication systems.

Recently, Chen et al. [13] have reported a step towards the integration of a reconfigurable MWP filter concept in [30], by exploiting an FWM process in a silicon nanowire, instead of using HLNF. The silicon waveguide ($h = 220$ nm, $w = 650$ nm, length = 12 mm) was dispersion engineered to yield broadband

FWM. They generated 2 idlers with measured conversion efficiency of −25 dB. These wavelengths taps were then modulated and amplitude controlled using a waveshaper. They used 4 km of single mode fiber (SMF) as the dispersive medium required to generate time delay between the taps. They measured the response of a 4-tap MWP filter and demonstrated that the 12 mm silicon nanowire can be used to replace 1 km of silica highly nonlinear fiber (HNLF) without compromising the quality of the MPF response.

14.6 Future Directions

In this section we discuss the potential of nonlinear IMWP and identify two key concepts that we believe will shape the field in the near future.

14.6.1 General Purpose Analog Processor

The exponential growth in bandwidth demand of radio communications has put unprecedented challenges in the RF signal processing chain. Traditionally, strong emphasis in such a chain is put on digital signal processing (DSP). However, for increasing operating frequencies and larger bandwidths, a reliable analog signal processor (ASP) that can match the flexibility of a DSP is desired [10, 25]. The aim here is to create a robust analog signal processing engine, which can be flexibly tailored in terms of group delay and amplitude responses, that works in millimeter wave or terahertz frequency range where DSP based systems are inefficient or inapplicable.

Due to its high bandwidth and potential reconfigurability, IMWP processor is an attractive candidate for such a flexible ASP. However, current state-of the-art IMWP processors were developed only for very specific tasks, for example beamforming and filtering, frequency discrimination, or pulse shaping. Thus, there is a lack of generality and multi-functionality of these processors. To emulate DSP, the IMWP processor has to be application agnostic, multi-purpose and programmable.

The incorporation of optical nonlinearities in an IMWP processor will open up the path towards a general purpose ASP. A large number of MWP signal processing functions can be generalized as processing of a number copies of RF modulated optical signals using reconfigurable optical filters. Nonlinear optical processes that create new optical wavelengths such as FWM, SFG, or DFG are attractive candidates for generating such signal copies. In fact, this concept of wavelength multicasting has recently been implemented for MWP signal processing techniques such as channelization [5, 6] and reconfigurable filtering [50]. In the view of the general processor, these wavelength copies will subsequently be processed by a reconfigurable optical filter. The typical signal processing tasks carried out by these filters include tunable time delay, carrier phase tuning, and complex (amplitude and

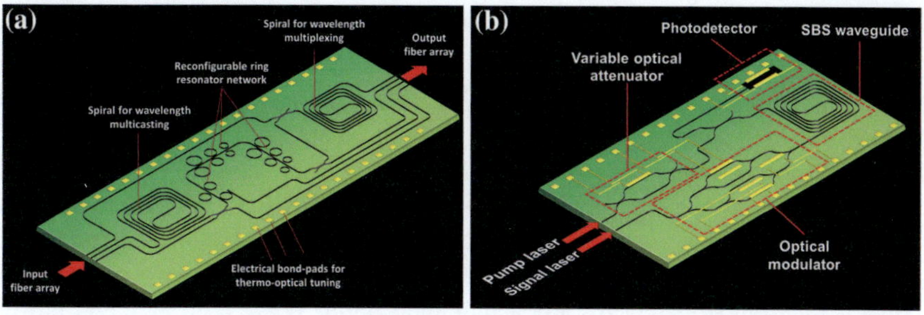

Fig. 14.9 **a** Illustration of the envisioned general purpose analog signal processor architecture combining ultrafast nonlinear optics for signal copies generation (for example wavelength multicasting via on chip FWM) and reconfigurable linear optical filtering (for example Vernier ring resonators tunable using thermo-optical tuning). **b** Possible implementations of high performance RF notch filter using sideband cancellation and SBS hybrid/monolithically integrated in a CMOS compatible photonic chip that consists of electro-optic modulator, tunable optical coupler for pump tuning, a nanophotonic waveguide with high SBS gain, and a photodetector. From [25]

phase) filtering. These processed optical signals will then be recombined or directed to separate dedicated outputs for a photodetection process.

The key challenge here is to create a monolithic photonic chip that hosts the nonlinear optics for wavelength multicasting and the reconfigurable linear optical filtering. Figure 14.9a depicts an illustration of a possible design of such a photonic chip in Si_3N_4 technology [42]. The chip consists of three main sections: two long spiral waveguides for wavelength multicasting and multiplexing based on cascaded FWM processes, and a reconfigurable optical filter based on a network of ring resonators that can be tuned using thermo-optics effect. This ring network might host a mixture of all pass ring filters, add-drop rings, as well as Vernier (cascade of non-identical) rings which have been recently exploited for tunable delay lines in a multi-wavelength beamformer [58]. The programmability of the ring network will allow the chip to be reconfigured in real time, to synthesize different responses according to the user demand.

Other material platforms such silicon can also be good candidates to construct this ASP. However, Si_3N_4 waveguides might be more suited for this purpose due to the following characteristics: ultra-low propagation losses ($\approx$0.1 dB/cm [59]) which is very important for MWP systems, moderately high optical nonlinearity (10 times of silica) [33] and free from nonlinear losses such as two-photon absorption (TPA) and free-carrier absorption (FCA). On top of this, Si_3N_4 waveguides have shown compatibility with efficient thermo-optics tuning and is a CMOS compatible material. While silicon also ticks some of the boxes and successful wavelength multicasting has been demonstrated in this platform [3], the propagation loss of a typical SOI nanowires is relatively high (2–3 dB/cm) [4] and they suffer from TPA and FCA. These losses might be prohibitive for creating an efficient nonlinear IMWP processor.

14.6.2 Highly Integrated Tunable RF Filter

Harnessing stimulated Brillouin scattering (SBS) in nanophotonic devices has recently attracted significant interest due to its great potential for realizing tunable slow light, narrow-linewidth laser, optical frequency comb, and RF photonic signal processor in a compact footprint. Observation of SBS in a CMOS-compatible platform has recently been reported in a hybrid Si-Si_3N_4 photonic/phononic waveguide [46], but the amount of realized SBS gain was only in the order of 1 dB, which is far from sufficient for any actual application. While high gain and power efficient SBS nanophotonic devices are ultimately desired, elegantly harnessing the low SBS gain from current on-chip low-power devices to function in real-life applications represents a technological breakthrough.

As discussed previously, using the novel notch filter scheme one can create a high performance RF notch with an ultra-high suppression using less than 1 dB of on-chip SBS and very low a pump power. The ultimate goal for such filter is, however, to integrate more functionalities in a single photonic chip, together with the optical waveguides where SBS is induced. This vision in integration is illustrated in Fig. 14.9b.

The realization of this integrated notch filter can go through two different paths: monolithic integration, or hybrid integration. Silicon for example, stands out as an ideal platform for monolithic integration of this filter. The fact that high speed modulators and photodetectors are readily available in silicon is, indeed, a big advantage. However, thus far there is still no observation of SBS in silicon, due to a number of effects such as acoustic mode leakage [19]. Thus, the realization of monolithic integration of this SBS notch filter will depend strongly on the efforts in efficiently inducing SBS in silicon, or other CMOS compatible materials. It is worth to mention that monolithic integration can still be achieved with OFs other than SBS, such as silicon ring resonators [27], or silicon integrated Bragg-gratings [7].

The hybrid integration route, on the other hand, has recently seen some encouraging progress. hybrid integration of chalcogenide and lithium niobate waveguides has recently been explored [22]. This is a very promising step to combine currently two best materials for SBS on chip (chalcogenide) and high speed electro-optic modulator (lithium niobate).

References

1. E.I. Ackerman, G.E. Betts, W.K. Burns, J.C. Campbell, C.H. Cox, N. Duan, J.L. Prince, M.D. Regan, H.V. Roussell, in *Proceedings of the IEEE MTT-S International Microwave Symposium* (*IMS 2007*) (Honolulu, HI, 2007) pp. 51–54
2. G. Agrawal, *Nonlinear Fiber Optics,* 3rd edn. (Academic Press, London, 2001)
3. A. Biberman, B.G. Lee, A.C. Turner-Foster, M.A. Foster, M. Lipson, A.L. Gaeta, K. Bergman, Wavelength multicasting in silicon photonic nanowires. Opt. Express **18**(17), 18047–18055 (2010)

4. W. Bogaerts, P. De Heyn, T. Van Vaerenbergh, K. De Vos, S.K. Selvaraja, T. Claes, P. Dumon, P. Bienstman, D. Van Thourhout, R. Baets, Silicon microring resonators. Laser Photonics Rev. **6**(1), 47–73 (2012)
5. C.S Bres, S. Zlatanovic, A.O.J. Wiberg, J.R. Adleman, C.K. Huynh, E.W. Jacobs, J.M. Kvavle, S. Radic, Parametric photonic channelized rf receiver. IEEE Photonics Technol. Lett. **23**(6), 344–346 (2011)
6. C.-S. Brès, S. Zlatanovic, A.O.J. Wiberg, S. Radic, Reconfigurable parametric channelized receiver for instantaneous spectral analysis. Opt. Express **19**(4), 3531–3541 (2011)
7. M. Burla, L.R. Cortés, M. Li, X. Wang, L. Chrostowski, J. Azaña, Integrated waveguide Bragg gratings for microwave photonics signal processing. Opt. Express **21**(21), 25120–25147 (2013)
8. M. Burla, D. Marpaung, L. Zhuang, C. Roeloffzen, M.R. Khan, A. Leinse, M. Hoekman, R. Heideman, On-chip cmos compatible reconfigurable optical delay line with separate carrier tuning for microwave photonic signal processing. Opt. Express **19**(22), 21475–21484 (2011)
9. A. Byrnes, R. Pant, E. Li, D.-Y. Choi, C.G. Poulton, S. Fan, S. Madden, B. Luther-Davies, B.J. Eggleton, Photonic chip based tunable and reconfigurable narrowband microwave photonic filter using stimulated Brillouin scattering. Opt. Express **20**(17), 18836–18845 (2012)
10. C. Caloz, S. Gupta, Q. Zhang, B. Nikfal, Analog signal processing: a possible alternative or complement to dominantly digital radio schemes. IEEE Microwave Mag. **14**(6), 87–103 (2013)
11. J. Capmany, D. Novak, Microwave photonics combines two worlds. Nat. Photonics **1**(6), 319–330 (2007)
12. J. Capmany, B. Ortega, D. Pastor, S. Sales, Discrete-time optical processing of microwave signals. J. Lightwave Technol. **23**(2), 702 (2005)
13. L.R. Chen, J. Li, M. Spasojevic, R. Adams, Nanowires and sidewall Bragg gratings in silicon as enabling technologies for microwave photonic filters. Opt. Express **21**(17), 19624–19633 (2013)
14. S. Chin, L. Thévenaz, Tunable photonic delay lines in optical fibers. Laser Photonics Rev. **6** (6), 724–738 (2012)
15. S. Chin, L. Thévenaz, J. Sancho, S. Sales, J. Capmany, P. Berger, J. Bourderionnet, D. Dolfi, Broadband true time delay for microwave signal processing, using slow light based on stimulated Brillouin scattering in optical fibers. Opt. Express **18**(21), 22599–22613 (2010)
16. C. Cox, E. Ackerman, R. Helkey, G.E. Betts, Techniques and performance of intensity-modulation direct-detection analog optical links. IEEE Trans. Microw. Theory Tech. **45**(8), 1375–1383 (1997)
17. C.H. Cox, *Analog Optical Links: Theory and Practice* (Cambridge University Press, Cambridge, 2004)
18. C.H. Cox, E.I. Ackerman, G.E. Betts, J.L. Prince, Limits on the performance of RF-over-fiber links and their impact on device design. IEEE Trans. Microw. Theory Tech. **54**(2), 906–920 (2006)
19. B.J. Eggleton, C.G. Poulton, R. Pant, Inducing and harnessing stimulated Brillouin scattering in photonic integrated circuits. Adv. Opt. Photon. **5**(4), 536–587 (2013)
20. B.J. Eggleton, T.D. Vo, R. Pant, J. Schr, M.D. Pelusi, D.Y. Choi, S.J. Madden, B. Luther-Davies, Photonic chip based ultrafast optical processing based on high nonlinearity dispersion engineered chalcogenide waveguides. Laser Photonics Rev. **6**(1), 97–114 (2012)
21. M.H. Khan, H. Shen, Y. Xuan, L. Zhao, S. Xiao, D.E. Leaird, A.M. Weiner, M. Qi, Ultrabroad-bandwidth arbitrary radiofrequency waveform generation with a silicon photonic chip-based spectral shaper. Nat. Photonics **4**(2), 117–122 (2010)
22. J. Kim, W.J. Sung, O. Eknoyan, C.K. Madsen, Linear photonic frequency discriminator on as2s3-ring-on-ti:linbo3 hybrid platform. Opt. Express **21**(21), 24566–24573 (2013)
23. A. Loayssa, F.J. Lahoz, Broad-band rf photonic phase shifter based on stimulated Brillouin scattering and single-sideband modulation. IEEE Photonics Technol. Lett. **18**(1), 208–210 (2006)

24. D. Marpaung, B. Morrison, R. Pant, D.Y. Choi, S. Madden, B. Luther-Davies, B.J. Eggleton, A tunable rf photonic notch filter with record 55 db suppression using sub-1 db on-chip Brillouin gain. in *Frontiers in Optics 2013 Postdeadline*. Opt. Soc. A., p. FW6B.9 (2013)
25. D. Marpaung, M. Pagani, B. Morrison, B.J. Eggleton, Nonlinear integrated microwave photonics. J. Lightwave Technol. **32**(20), 3421–3427 (2014)
26. D. Marpaung, B. Morrison, R. Pant, B.J. Eggleton, Frequency agile microwave photonic notch filter with anomalously high stopband rejection. Opt. Lett. **38**(21), 4300–4303 (2013)
27. D. Marpaung, B. Morrison, R. Pant, C. Roeloffzen, A. Leinse, M. Hoekman, R. Heideman, B.J. Eggleton, Si3n4 ring resonator-based microwave photonic notch filter with an ultrahigh peak rejection. Opt. Express **21**(20), 23286–23294 (2013)
28. D. Marpaung, C. Roeloffzen, R. Heideman, A. Leinse, S. Sales, J. Capmany, Integrated microwave photonics. Laser Photonics Rev. **7**(4), 506–538 (2013)
29. J.D. McKinney, M. Godinez, V.J. Urick, S. Thaniyavarn, W. Charczenko, K.J. Williams, Sub-10-dB noise figure in a multiple-GHz analog optical link. IEEE Photonics Technol. Lett. **19** (7), 465–467 (2007)
30. A. Meijerink, C.G.H. Roeloffzen, R. Meijerink, L. Zhuang, D.A.I. Marpaung, M.J. Bentum, M. Burla, J. Verpoorte, P. Jorna, A. Hulzinga, W. van Etten, Novel ring resonator-based integrated photonic beamformer for broadband phased array receive antennas—part i: design and performance analysis. J. Lightwave Technol. **28**(1), 3–18 (2010)
31. R.A. Minasian, Photonic signal processing of microwave signals. IEEE Trans. Microw. Theory Tech. **54**(2, Part 2), 832–846 (2006)
32. B. Morrison, D. Marpaung, R. Pant, E. Li, D.-Y. Choi, S. Madden, B. Luther-Davies, B.J. Eggleton, Tunable microwave photonic notch filter using on-chip stimulated Brillouin scattering. Opt. Commun. **313**, 85–89 (2014)
33. D.J. Moss, R. Morandotti, A.L. Gaeta, M. Lipson, New CMOS-compatible platforms based on silicon nitride and Hydex for nonlinear optics. Nat. Photonics **7**(8), 597–607 (2014)
34. W. Ng, A.A. Walston, G.L. Tangonan, J.J. Lee, I.L. Newberg, N. Bernstein, The first demonstration of an optically steered microwave phased array antenna using true-time-delay. IEEE Trans. Microw. Theory Tech. **9**(9), 1124–1131 (1991)
35. Y. Okawachi, M.S. Bigelow, J.E. Sharping, Z. Zhu, A. Schweinsberg, D.J. Gauthier, R.W. Boyd, A.L. Gaeta, Tunable all-optical delays via Brillouin slow light in an optical fiber. Phys. Rev. Lett. **94**, 153902 (2005)
36. M. Pagani, D. Marpaung, D.Y. Choi, S.J. Madden, B. Luther-Davies, B.J. Eggleton, On-chip wideband tunable rf photonic phase shifter based on stimulated Brillouin scattering. in *Proc. OptoElectronics and Communications Conference* (*OECC*), pp. 51–54 (2014)
37. M. Pagani, D Marpaung, B Morrison, B.J. Eggleton, Bandwidth tunable, high suppression rf photonic filter with improved insertion loss. in *CLEO: 2014*, p. STu2G.7. Opt. Soc. Am. (2014)
38. R. Pant, A. Byrnes, C.G. Poulton, E. Li, D.-Y. Choi, S. Madden, B. Luther-Davies, B.J. Eggleton, Photonic-chip-based tunable slow and fast light via stimulated Brillouin scattering. Opt. Lett. **37**(5), 969–971 (2012)
39. R. Pant, D. Marpaung, I.V. Kabakova, B. Morrison, C.G. Poulton, B.J. Eggleton, On-chip stimulated Brillouin scattering for microwave signal processing and generation. Laser Photonics Rev. **8**(5), 653–666 (2014)
40. R. Pant, C.G. Poulton, D.-Y. Choi, H. Mcfarlane, S. Hile, E. Li, L. Thevenaz, B. Luther-Davies, S.J. Madden, B.J. Eggleton, On-chip stimulated Brillouin scattering. Opt. Express **19** (9), 8285–8290 (2011)
41. M.S. Rasras, C.K. Madsen, M.A. Cappuzzo, E. Chen, L.T. Gomez, E.J. Laskowski, A. Griffin, A. Wong-Foy, A. Gasparyan, A. Kasper, J. Le Grange, S.S. Patel, Integrated resonance-enhanced variable optical delay lines. IEEE Photonics Technol. Lett. **17**(4), 834–836 (2005)
42. C.G.H. Roeloffzen, L. Zhuang, C. Taddei, A. Leinse, R.G. Heideman, P.W.L. van Dijk, R.M. Oldenbeuving, D.A.I. Marpaung, M. Burla, K.J. Boller, Silicon nitride microwave photonic circuits. Opt. Express **21**(19), 22937–22961 (2013)

43. J.E. Roman, L.T. Nichols, K.J. Wiliams, R.D. Esman, G.C. Tavik, M. Livingston, M.G. Parent, Fiber-optic remoting of an ultrahigh dynamic range radar. IEEE Trans. Microw. Theory Tech. **46** (12), 2317–2323 (1998)
44. A.J. Seeds, K.J. Williams, Microwave photonics. J. Lightwave Tech. **24**(12), 4628–4641 (2006)
45. A.J. Seeds, Microwave photonics. IEEE Trans. Microw. Theory Tech. **50**(3), 877–887 (2002)
46. H. Shin, W. Qiu, R. Jarecki, J.A. Cox, R.H. Olsson, A. Starbuck, Z. Wang, P.T. Rakich, Tailorable stimulated Brillouin scattering in nanoscale silicon waveguides. Nat. Commun. **4** (2013). doi:10.1038/ncomms2943
47. K.Y. Song, M.G. Herráez, L. Thévenaz, Observation of pulse delaying and advancement in optical fibers using stimulated Brillouin scattering. Opt. Express **13**(1), 82–88 (2005)
48. K. Tan, D. Marpaung, R. Pant, F. Gao, E. Li, J. Wang, D.-Y. Choi, S. Madden, B. Luther-Davies, J. Sun, B.J. Eggleton, Photonic-chip-based all-optical ultra-wideband pulse generation via xpm and birefringence in a chalcogenide waveguide. Opt. Express **21**(2), 2003–2011 (2013)
49. V.J. Urick, M.S. Rogge, F. Bucholtz, K.J. Williams, Wideband (0.045–6.25 GHz) 40 km analogue fibre-optic link with ultra-high (>40 dB) all-photonic gain. Electron. Lett. **42**(9), 552–553 (2006)
50. B. Vidal, J. Palaci, J. Capmany, Reconfigurable photonic microwave filter based on four-wave mixing. IEEE Photonics J. **4**(3), 759–764 (2012)
51. B. Vidal, M.A. Piqueras, J. Martí, Tunable and reconfigurable photonic microwave filter based on stimulated Brillouin scattering. Opt. Lett. **32**(1), 23–25 (2007)
52. A.E. Willner, O.F. Yilmaz, J. Wang, X. Wu, A. Bogoni, L. Zhang, S.R. Nuccio, Optically efficient nonlinear signal processing. IEEE J. Sel. Top. Quantum Electron. **17**(2), 320–332 (2011)
53. J. Yao, Microwave photonics. IEEE Photonics Technol. Lett. **27**(3), 314–335 (2009)
54. J. Yao, Microwave photonics arbitrary waveform generation. Nat. Photonics **4**(2), 79–80 (2010)
55. J. Zhang, A.N. Hone, T.E. Darcie, Limitation due to signal-clipping in linearized microwave-photonic links. IEEE Photonics Technol. Lett. **19**(14), 1033–1035 (2007)
56. W. Zhang, R.A Minasian, Widely tunable single-passband microwave photonic filter based on stimulated Brillouin scattering. IEEE Photonics Technol. Lett. **23**(23), 1775–1777 (2011)
57. W. Zhang, R.A Minasian, Ultrawide tunable microwave photonic notch filter based on stimulated Brillouin scattering. IEEE Photonics Technol. Lett. **24**(14), 1182–1184 (2012)
58. L. Zhuang, M. Hoekman, W. Beeker, A. Leinse, R. Heideman, P. van Dijk, C. Roeloffzen. Novel low-loss waveguide delay lines using Vernier ring resonators for on-chip multi-λ microwave photonic signal processors. Laser Photonics Rev. **7**(6), 994–1002 (2013)
59. L. Zhuang, D. Marpaung, M. Burla, W. Beeker, A. Leinse, C. Roeloffzen, Low-loss, high-index-contrast si3n4/sio2 optical waveguides for optical delay lines in microwave photonics signal processing. Opt. Express **19**(23), 23162–23170 (2011)
60. L. Zhuang, C.G.H. Roeloffzen, A. Meijerink, M. Burla, D.A.I. Marpaung, A. Leinse, M. Hoekman, R.G. Heideman, W. van Etten, Novel ring resonator-based integrated photonic beamformer for broadband phased array receive antennas—part ii: Experimental prototype. J. Lightwave Technol. **28**(1), 19–31 (2010)
61. H. Zmuda, R.A. Soref, P. Payson, S. Johns, E.N. Toughlian, Photonic beamformer for phased array antennas using a fiber grating prism. IEEE Photonics Technol. Lett. **9**(2), 241–243 (1997)

Chapter 15
Ultrafast Optical Techniques for Communication Networks and Signal Processing

Bhavin J. Shastri, John Chang, Alexander N. Tait, Matthew P. Chang, Ben Wu, Mitchell A. Nahmias and Paul R. Prucnal

Abstract Wireless communications, for data services in particular, have witnessed an exponential growth, and wireless spectrum shortages necessitate increasingly sophisticated methods to use spectrum efficiently. The backhaul of nearly all wireless data networks is fiber-optic. Analog optical signal processing techniques, or microwave photonics, provides an ideal platform for processing wireless information before it is transported to data aggregation centers by fibers. It is in this context that we present recent advances in optical signal processing techniques for wireless radio frequency (RF) signals. Specifically, this chapter is devoted to the discussion of photonic architectures for wideband analog signal processing, including RF beamforming, co-channel interference cancellation, and physical layer security. Photonics offers the advantages not only of broadband operation, but reduced size, weight, and power, in addition to low transmission loss, rapid reconfigurability, and immunity to electromagnetic interference.

15.1 Introduction

The use of wireless communication is growing exponentially. In June 2012, more than 5.6 billion subscribers had access to and were using a wireless device, nearly 80 % of the total world population of 7.02 billion [1]. By the end of 2017, more than 90 % of the world's population is expected to have access to mobile broadband 3G devices. This statistic demonstrates the importance of mobile wireless com-

---

B.J. Shastri and J. Chang—equal contribution.

B.J. Shastri · J. Chang · A.N. Tait · M.P. Chang · B. Wu · M.A. Nahmias · P.R. Prucnal (✉)
Princeton University, Princeton, NJ 08544, USA
e-mail: prucnal@princeton.edu

S. Wabnitz and B.J. Eggleton (eds.), *All-Optical Signal Processing*,
Springer Series in Optical Sciences 194, DOI 10.1007/978-3-319-14992-9_15

munication in the current decade and beyond. By the end of 2015, mobile traffic is expected to double. Nearly 11 exabytes will be transmitted *per month* in 2016, more than four times the quantity transmitted in the last 2 years. Mobile broadband, growing with a compound rate of 80 %, is shaping to become one of the most successful and fastest growing industries in history. This growth in the mobile market has been spurred by two key factors: fast, powerful, and ubiquitous mobile networking, and high-performance smartphones with a million-plus mobile applications [2]. The rapid shift from traditional headsets to smartphones has been attributed to increased performance and lower energy consumption.

As wireless communication continues to grow, the radio frequency spectrum will become increasingly scarce. The climb in the demand for smartphones and faster, large-coverage next-generation mobile networks will be paralleled by a demand for available bandwidth—but this demand comes with an equal increase in wireless spectrum use. There will be two technological barriers that will result from these increased demands: first, wireless spectrum is finitely limited and is constrained, and spectral efficiency use must be carefully managed as bandwidth begins to grow. Second, a heavily saturated spectrum means interference from neighboring devices and frequency bands could negatively impact data rates and signal bandwidth.

One way of addressing both these escalating problems is through the use of interference cancellation. Although noise-filtering and other post-processing methods could reduce interference, a more effective technique would involve cancelling interference directly in the radio frequency (RF) domain. This technique reduces interference and increases bandwidth availability. As a result, the application of RF interference cancellation could have a significant impact on the alleviation of overcrowding in the wireless domain.

Figure 15.1 shows a typical scenario seen by a wireless communication system (such as a radar, for instance). There is a clear target, whose information we wish to receive. Along with that information, however, is environmental clutter and other interfering signals. Clutter results from the local environment, which can be exacerbated by complex terrain or weather conditions. Noise and signals from other sources also contribute to interference. Our goal is to design a system to extract a target signal, or signal of interest (SOI), from this noise.

Adaptive arrays in electronics (Fig. 15.2) have been used to increase signal-to-noise ratio (SNR) in the presence of dynamic interference and noise in wireless communication systems since the 1960s and are a perfect solution for the problem outlined above [3]. Arrays, instead of singular antennas, give the advantages of higher gain (through multiple antennas), electronic beam forming (steering of the beam-pattern without physical movement), and adaptive cancellation (by taking advantage of beamforming capabilities). Because these adaptive arrays are both tunable and easily reconfigurable, they are more reliable than conventional antennas and can be used in multiple scenarios across different fields. However, electronic systems suffer from two major deficits: beam squinting and limited bandwidth. Beam squinting arises from the use of phase shifters, which degrades an antenna array's beam-pattern with frequency, resulting in different responses for signals of

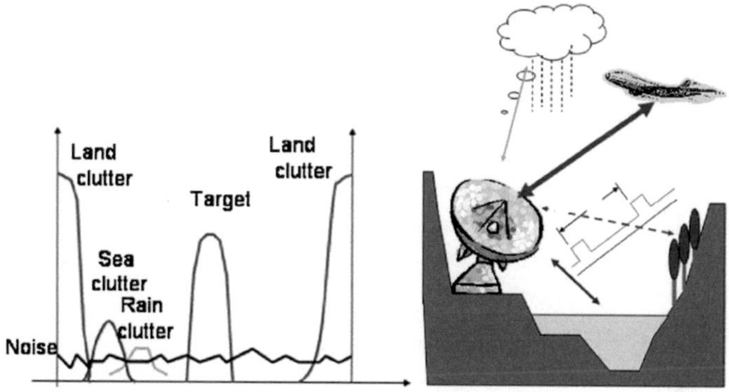

Fig. 15.1 Typical clutter in wireless environment [5]

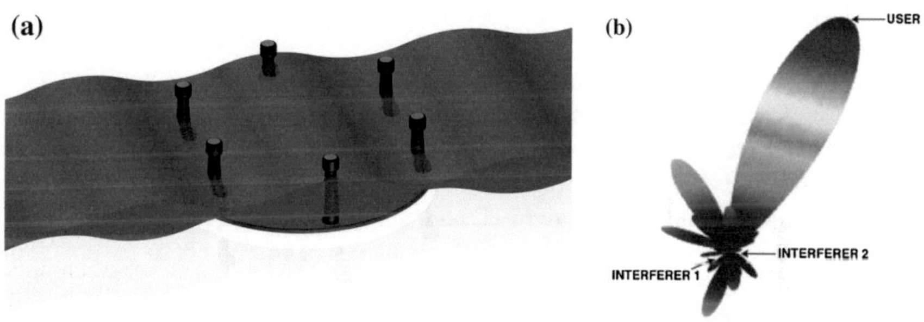

Fig. 15.2 **a** An antenna array uses phase information to detect the direction of arrival of an incident wave. **b** Adaptive array coverage: a representative depiction of a main lobe extending toward a user with null directed toward two co-channel interferers [3]

different frequencies. Bandwidth limitations result from the limited speed of electronic devices. This limits the use of electronic systems to narrow-band signals. Furthermore, losses increase for electronic systems past 100 MHz, limiting systems that wish to operate in the tens of GHz.

While wireless networks have seen a rapid increase their bandwidth capabilities, optical networks have been popular for a decade and their capabilities greatly overshadow even the fastest wireless standard. Figure 15.3 compares wireless and optical networks. Optical network growth follows that of wireless networks and is at least 10× faster at each step. Because of the inherent bandwidth superiority of optics, wireless data is always aggregated and backhauled to data centers by the way of optical networks. This serves as the perfect backdrop for an optical cancellation system sandwiched in-between the wireless base stations and the optical backhaul network. Wireless signals are modulated onto the optical domain and then directly processed using a photonic cancellation system before being sent over an optical network. This field is known as microwave photonics, which is an interdisciplinary area that studies the interaction between microwave and optical signals [4].

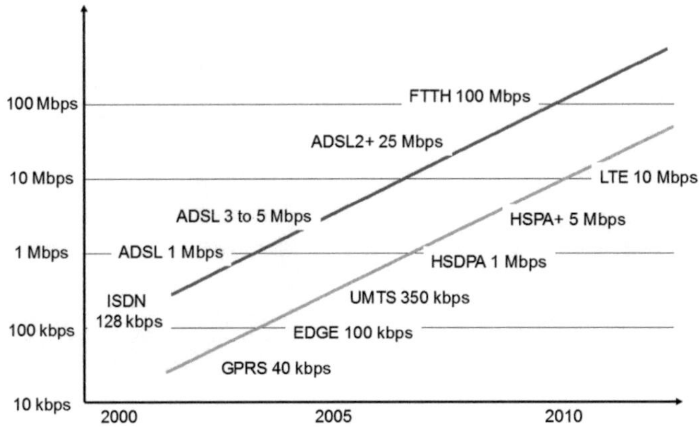

Fig. 15.3 Advances in wireless and optical capacity [1]

Optical systems bring advantages inherent to the physics of photonics, which include low loss, high bandwidth, immunity to electromagnetic interference (EMI), tunability, and reconfigurability [5]. Most importantly, optical systems solve both the beam squinting and bandwidth limitations of electronic systems. By using optical delay lines, true-time delay (TTD) antenna arrays eliminate beam squinting [4]. Electronic systems oftentimes use multiple pieces of equipment to cover a wide bandwidth. In contrast, because photonic systems keep the signal in the optical domain, they can utilize the tremendous bandwidth of optics (up to THz) to process fast-varying RF signals without degrading bandwidth using a single set of optical equipment. This gives optical systems savings in size, weight, and power (SWAP), especially for situations where compactness and light weight are important (i.e. aviation). Furthermore, optical interference cancellation in conjunction with optical backhaul networks makes it possible to centralize signal processing in one shared location. Centralization allows for equipment sharing, dynamic allocation of resources, simplified system operation and maintenance [5]. Different operators can share resources using the same optical backhaul network to minimize cost and energy.

Microwave photonics filters (MPFs) are an alternative photonic technology and have an enormous amount of bandwidth selectivity. They can be employed in optical systems for either channel rejection or channel selection to cancel interfering signals that are picked up by antennas [4, 6, 7]. Unique to photonics filters is the rejection of signals in the optical domain using optical signal processing techniques. MPFs can be tuned to cancel a selected frequency band across a large bandwidth from MHz to hundreds of GHz [5, 8].

This chapter investigates an optical method for wireless interference cancellation systems using *optical signal processing* techniques. We marry the field of adaptive arrays and microwave photonics. Beam-steering is done by adaptive antenna arrays with processing done all-optically by MPFs. By utilizing the broad bandwidth and high capacity of optics to address growing problems in mobile communication market, we hope to create a photonic interference cancellation system with wide-ranging impact, both economically and scientifically.

15.1.1 Scenario of Interest

This chapter is interested in a scenario with dynamically changing noise and interference in the presence of a SOI. We assume that we have a receiver either in a dense urban environment, such as in a mobile phone setting, or a crowded indoor environment, such as in a WIFI network. We wish to follow a mobile transmitter as it is broadcasting the SOI, so that its location, and possibly frequency, is constantly changing. To make the problem even more challenging, we assume that we cannot know anything about the SOI, other than its frequency, and have no information of the content of this signal.

We make several additional assumptions: we must cancel an interfering signal to receive the SOI. The SOI is much smaller in power than the interference. We also lack knowledge of both the location and frequency of the interfering signal. The interference is both dynamic and spontaneous, with multiple interferers appearing and disappearing with shifting frequencies. We also do not know the relative power or direction between the SOI and the interferers.

Since we are in a crowded environment (either urban or indoors), we can assume high levels of dynamically changing noise. Transmitters and other users may appear or disappear spontaneously and change locations, which is in addition to the background noise that permeates throughout the wireless spectrum. Interference in the form of multipath fading will be detrimental to signal quality. In the indoor scenario, near-field signals dominate, with wavefronts that may not be planar or uniform with respect to the receiver.

The problem requires a two-part solution. First, filters for both space and frequency are required to rid the interference and noise from the SOI. Secondly, an adaptive method is needed to quickly keep track of moving targets and rapidly changing interference. However, we do not have access to either the SOI or the interference and we cannot use a pilot signal. Unique to our scenario is that the signal incident on the receiver—the input signal—is inaccessible and only a signal processed by the filter is usable. This scenario nullifies the use of traditional adaptive algorithms, which typically require pre-steering, training, or an input signal. Multipath effects are also significant in a dynamically noisy environment, which fluctuate and are highly unpredictable, preventing a device from distinguishing between the loss of signal (LOS) or some *n*th order multipath. As a result, we cannot make traditional direction-of-arrival (DOA) estimates. Because broadband noise can fluctuate wildly in frequency and direction, stochastic gradient method are also unusable; even if signals remain the same, no two measurements will be the same as a result of noise. This chapter investigates solutions to the scenario introduced above.

We first discuss the basic theory of beamforming and the unique challenges of broadband interference cancellation. Next, we present the requirements and challenges for building optical finite impulse response (FIR) filters, and review several innovative MPF designs including optical tunable delays and weighting schemes. We also review the state of the art optical beamformers recently proposed and experimentally demonstrated in literature. Furthermore, we detail a highly scalable photonic beamforming architecture designed for a particularly non-stationary, interfering environment. Finally, we introduce a specific application for the photonic beamformer namely, physical layer security in optical backhaul networks.

15.2 Primer on Antenna Arrays and Beamforming

Antennas arranged in some physical geometry—array—exploit the spatial relationship between the antenna elements to either measure or manipulate the spatial properties of RF signals which is also called beam steering. Typically, several antennas are located with uniform spacing of scale similar to the frequency of interest. If the signal source is far away from the antenna array, the beam can be considered as a plane wave with a wavefront orthogonal to the direction of propagation. This wavefront arrives at the antenna array at some angle, and as it propagates across the array it impinges upon each antenna with a delay experiencing a phase shift. The delay is related to both the angle of incidence of the beam and the physical layout of the array. The antenna array is able to resolve the signal's DOA by calculating the time differences between the responses of each antenna. This information can then be used to steer the beampattern. A signal processor can follow each antenna. The signal processor can be used to manipulate the signal in a useful way; that is, to suppress interferers and enhance the SOI.

More formally, consider an array of M antenna elements arranged in an xy plane. The DOA denoted by a monochromatic plane wave incident on the array is denoted as (θ, ϕ) where θ is the elevation angle measured from the z axis and ϕ is the azimuth angle measured from the x axis. The received signal at antenna m is transformed by the steering vector $\mathbf{a}(\theta, \phi)$ given by

$$a_m(\theta, \phi) = e^{j\omega\tau_m(\theta,\phi)}, \quad m = 0, 1, \ldots, M - 1 \tag{15.1}$$

where ω is the signal carrier frequency, τ_m is the time delay associated with antenna m characterized by the array geometry. To implement beam steering, the array elements are weighted and summed. The incident signal $s(t)$ produces an output signal $y(t)$

$$y(t) = \mathbf{w} \cdot \mathbf{a}(\theta, \phi) \cdot s(t). \tag{15.2}$$

where $\mathbf{w}$ is the weight matrix vector which comprises of the complex coefficients w_m, and inner product $\mathbf{w} \cdot \mathbf{a}(\theta, \phi)$ is called the array factor $AF(\theta, \phi)$ [9].

To generalize, if the antenna array is receiving L signals $s_0(t), \ldots, s_{L-1}(t)$ with different DOAs $(\theta_0, \phi_0), \ldots, (\theta_{L-1}, \phi_{L-1})$, then the $M \times L$ steering matrix is

$$\mathbf{A}(\theta, \phi) = [\mathbf{a}(\theta_0, \phi_0) \ldots \mathbf{a}(\theta_{L-1}, \phi_{L-1})] \tag{15.3}$$

and the system output is given by

$$y(t) = \mathbf{w}^T \mathbf{A}(\theta, \phi) \cdot s(t). \tag{15.4}$$

If some of the signals are considered noise, the adaptive system becomes an optimization problem over the space of the M antenna weights. The weights can be chosen in such a way as to emphasize a signal coming from one direction while attenuating signals from other directions.

Consider the linear and circular antenna arrays shown in Fig. 15.4 which we use to illustrate the example of beam steering. The linear array consists of M antenna

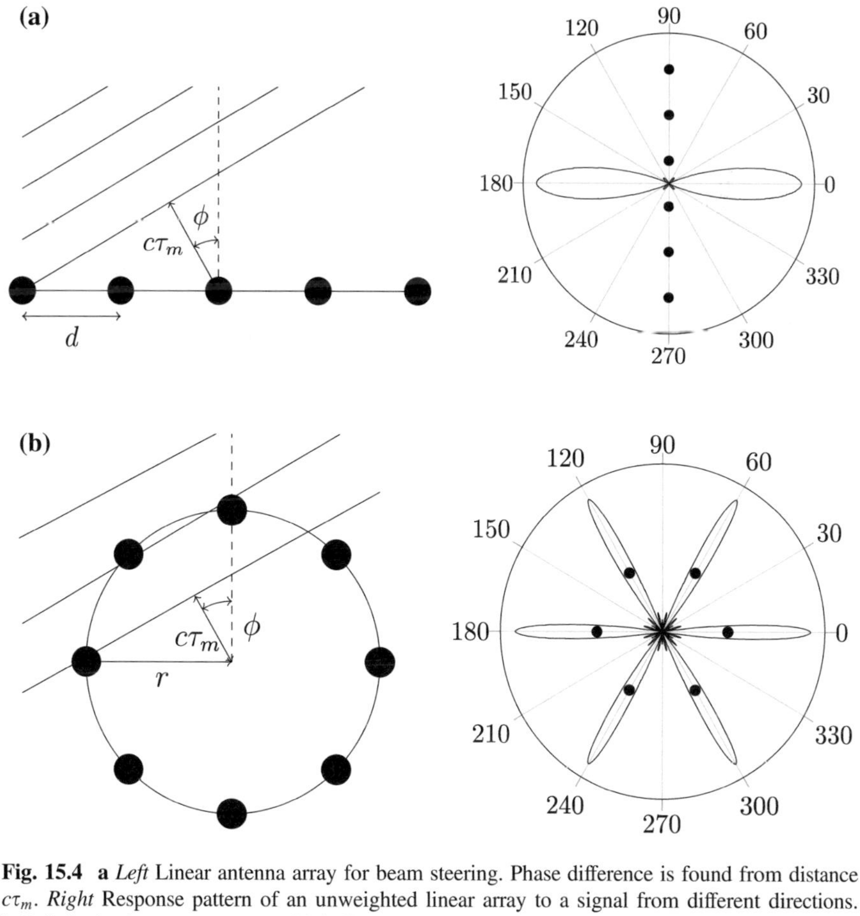

Fig. 15.4 **a** *Left* Linear antenna array for beam steering. Phase difference is found from distance $c\tau_m$. *Right* Response pattern of an unweighted linear array to a signal from different directions. **b** *Left* A circular antenna array. *Right* Response pattern of an unweighted circular array

elements equally spaced distance d apart arranged in a straight line along the y axis. If the first antenna element has phase 0, then the time delay to antenna m can be calculated by trigonometry to be

$$\tau_m(\theta,\phi) = m\frac{d}{c}\sin\phi\sin\theta \tag{15.5}$$

where c is the speed of light. The linear array with weights w_m then has an array factor

$$AF(\theta,\phi) = \sum_{m=0}^{M-1} w_m e^{j\omega m\frac{d}{c}\sin\phi\sin\theta} \tag{15.6}$$

which is comparable in form to the frequency response of an M-tap FIR filter: $F(\Omega) = \sum_{m=0}^{M} w_m e^{-jm\Omega}$. Hence, selecting weights for a particular spatial response is equivalent to selecting the weights of an FIR filter. The more popular circular antenna array can steer a beam in any direction. Here, the M antenna elements are arranged uniformly around a circle of radius r, with antenna 0 at $\phi = 0$. If the center of the circle has phase 0, then the time delay at each antenna element m is given by

$$\tau_m(\theta,\phi) = \frac{r}{c}\cos\left(\phi - \frac{2\pi m}{M}\right)\sin\theta \tag{15.7}$$

which gives a response of

$$AF(\theta,\phi) = \sum_{m=0}^{M-1} w_m e^{j\omega\frac{r}{c}\cos\left(\phi-\frac{2\pi m}{M}\right)\sin\theta} \tag{15.8}$$

15.2.1 *Narrowband Beamforming*

To use the antenna array to steer a signal, the complex weights at each antenna are adjusted to achieve the desired response. In the case of a narrowband signal, the weights can be simple phase shifts chosen to replicate a plane wave in the desired direction. These weights are pure phase delays, and can be realized with analog phase shifters attached to each antenna or with digital signal processing (DSP). Such a system is known as a phased array. For steering a beam with the linear and circular arrays, the weights are

$$w_m = e^{-j\omega m\frac{d}{c}\sin\phi} \text{ and } w_m = e^{-j\omega\frac{r}{c}\cos\left(\phi - \frac{2\pi m}{M}\right)}, \text{ respectively.} \tag{15.9}$$

Figure 15.5 illustrates an example of directional response of a steered circular phase array. Phase shift steering provides control over the direction of the main lobe which is centered on the look direction, but not over the relative placement of the smaller side lobes on other directions or nulls which are the minima of the response; signals from these directions are rejected by the array. To change the shape of the response, variable weights are applied to the phase shifted signals as shown in Fig. 15.6. An arbitrary phase shift is realized by adjusting the ratio between the original signals and a quarter-shifted copy of the signal [10]: $\Delta\varphi = -\arctan\frac{w_{m,1}}{w_{m,0}}$.

By controlling both the amplitude and phase of each antenna, this filter design implements weights as complex numbers. The complex weight vector $\mathbf{w}$ can be selected to solve the complex linear systems of (15.4) to achieve high gain of the SOI and suppression of the interference signals. The equation can be solved for up to $M-1$ interference signals which means the null steering array of M antennas can place up to $M-1$ nulls [3]. If there are fewer than $M-1$ interference signals, the additional degrees of freedom can be used to refine the gain levels and shapes of the side lobes. However, if there are more than $M-1$ interference signals, then the array cannot cancel them perfectly, but can still suppress them. In this case adaptive algorithms can be used to find the optimal weight vector for the incident signals. These standard optimization algorithms include gradient estimation, digital least mean square (LMS), and asynchronous LMS [11].

Filtering using phase shifts is only effective for narrowband signals, where the phase shift corresponds with the time delay between the antennas. For signals with different frequencies, the time delay between the antennas $\tau = d/c$ is constant, but the phase shift $\Delta\varphi_m = \omega\tau_m$ changes with frequency. A phased array therefore has a different response for each frequency and this is referred to as beam squint [4]. As a result, a broadband signal can get distorted after passing through the filter. To receive signals in a broad spectrum, the phase shifts can be replaced with TTDs

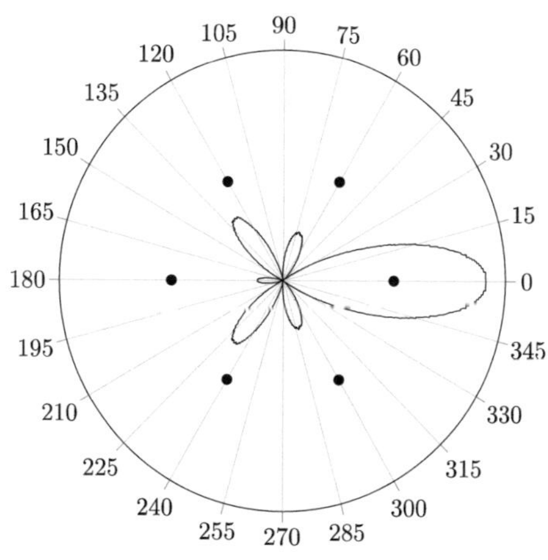

Fig. 15.5 Directional response of a circular phase array steered to angle 0. Power is depicted in linear scale

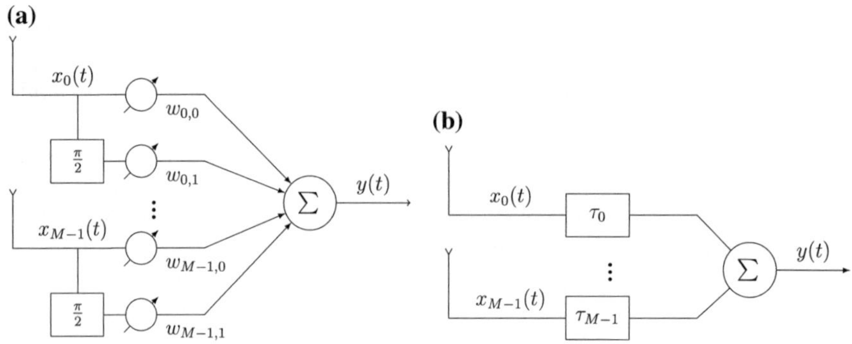

Fig. 15.6 **a** A weighted phased array beamformer applies both phase shifts and amplitude weights to alter the shape of the response pattern. **b** A delay-and-sum beamformer uses true time delays to receive a wideband signal

[12], as shown in Fig. 15.6b. A TTD can implement a phase shift greater than 2π for some frequencies, as required by extremely broadband signals. Then the spatial response is nearly the same for all frequencies.

Systems that use time delays for filtering are called delay-and-sum arrays. Digital receivers can trivially shift input signals using hardware shift registers or software signal processing. Optics is the perfect medium for creating TTD systems for broadband signals. Since optical beamforming is based on using fiber optics or other types of discrete optical components that already exist in current broadband optical networks, they are by nature broadband. TTD filters work well for receiving broadband signals, but they only filter in the spatial domain. For interference signals at different frequencies, it would be useful to control the frequency response as well —wideband beamforming. In this chapter, we will focus on photonic broadband beamformers.

15.2.2 Wideband Beamforming

For wideband arrays, it is desirable to control not only the spatial response but also the frequency response. To achieve an arbitrary frequency response, the weights w_m are replaced by linear filters $h_m(t)$ with frequency response $H_m(\omega)$. The combined spatial and frequency of the system is the sum of the filtered array factors from (15.8)

$$P(\omega, \theta, \phi) = \sum_{m=0}^{M-1} H_m(\omega) e^{j\omega\frac{r}{c}\cos\left(\phi - \frac{2\pi m}{M}\right)\sin\theta}. \tag{15.10}$$

A signal whose frequency and direction of arrival falls near the peak of (15.8) will be amplified whereas the one near null will be rejected by the filter. In the adaptive array proposed by Widrow et al. in 1967 [10], the filters were analog tapped delay lines (TDLs).

TDL filters, also known as transversal filters [13], are the continuous-time analogue of the well-known discrete-time FIR filter. As shown in Fig. 15.7, the input signal is delayed by successive time intervals, called taps. These delayed signals are then weighted and summed. For an N-tap filter with weights w_n and a constant delay T_d between each tap, the output is given by

$$y(t) = \sum_{n=0}^{N-1} w_n(t)x(t - nT_d) \tag{15.11}$$

and the frequency response is

$$H(\omega) = \sum_{n=0}^{N-1} w_n e^{-j\omega n T_d}. \tag{15.12}$$

Note that both the frequency response of the TDL filter (15.12) and the spatial response of the phase array (15.8) have a similar form to an FIR filter. Hence, in the most general terms, the physical formation of the antennas provides spatial filtering while the filters provide frequency filtering. By changing the weights of the FIR filters, one can manipulate a beamformer to act as a spatial and frequency filter and change the sensitivity of the beampattern to different frequencies and angles. Mathematically, if an N tap filter is used for each of the M antenna elements in the array, then there are MN total weights and delays in the system. The overall response of the array is found by substituting (15.12) into (15.10):

$$P(\omega, \theta, \phi) = \sum_{m=0}^{M-1} \sum_{n=0}^{N-1} w_{m,n} e^{-j\omega\left(nT_d - \frac{r}{c}\cos\left(\phi - \frac{2\pi m}{M}\right)\sin\theta\right)}. \tag{15.13}$$

The task of beamforming is to select the optimal set of MN weights to achieve the desired shape of $P(\omega, \theta, \phi)$. Figure 15.8 shows the simulation response of a

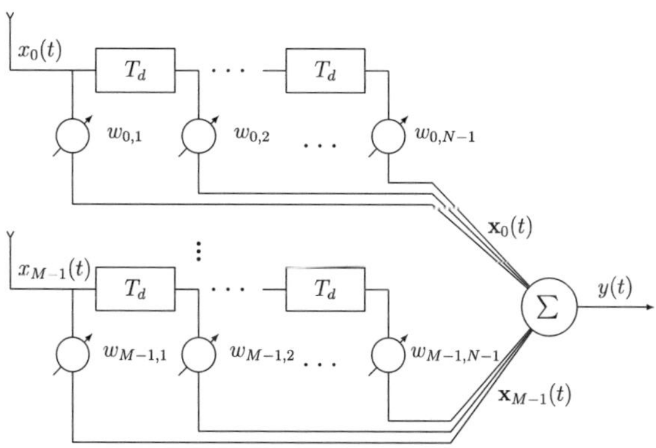

Fig. 15.7 Antenna array with tapped delay line filters

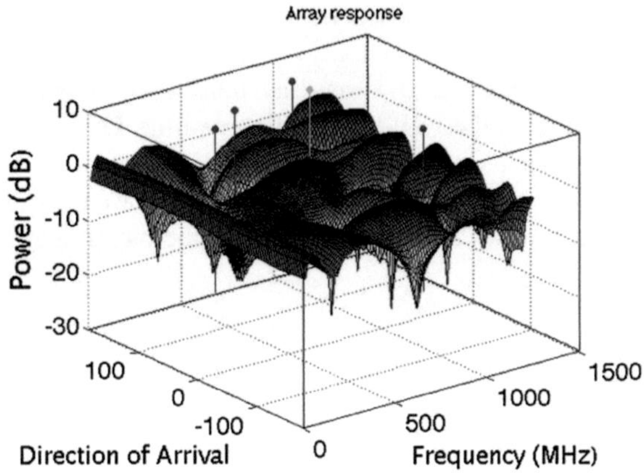

Fig. 15.8 Simulated spatial and frequency response of a tapped delay line filter using arbitrary weights

wideband 4-antenna, 8-tap circular beamformer. The beam pattern for a system with SOI of 500 MHz at 90° and interferences at 900, 1000, and 300 MHz at 120°, −70°, and 60° respectively demonstrates both the spatial filtering on the *x*-axis and the frequency filtering on the *y*-axis. We notice that nulls occur at the interferers while there is a peak at the SOI frequency and angle.

15.3 Microwave Photonic Filters

The previous section introduced the FIR filter as a TDL architecture. This section details the requirements and challenges for building *optical* FIR filters. We also review several innovative MPF designs including optical tunable delays and weighting schemes proposed and experimentally demonstrated in literature.

Optical filters or microwave photonic filters, are photonic subsystems specifically designed with the objective of carrying out the same functions as those of electronic or digital filters within the microwave range in an RF system or link [5]. The unique properties of MPFs offer many advantages, including high bandwidth, low loss across the entire bandwidth, reconfigurability, and immunity to EMI. Thus, there has been a considerable interest in the field of photonic signal processing for microwave filtering applications. Traditional RF and electronic approaches cannot practically handle wide bandwidths in the GHz range, whereas processing in the optical domain takes advantage of the broadband capabilities of optical delay schemes. Optical filters require only a single set of equipment to be able to cancel across a huge band of frequencies. Consequently, optical systems can potentially offer SWAP advantages, that is, savings through size, weight, and power. This is particularly critical for field-tested military or aviation applications where size and power are of key importance.

In this chapter, we are specifically interested in the possible application of MPFs in photonic phased array antennas, where they can provide the capability of steering without any physical movement, and offer the ability for precise spatial and frequency control of broadband signals. The key-motivating factor is that MPFs are easily tunable and reconfigurable in real time, allowing for filtering with adaptive capability. One can leverage these properties when designing such a system. Moreover, the field of photonic TTD units is essential for transversal, FIR filters (in phased array antenna systems) as it allows for wide bandwidth signal processing with little or no distortion or pulse broadening without being limited to a 'design frequency'. MPFs are the perfect optical complement for frequency filtering to the spatial filtering of RF antenna arrays.

15.3.1 Requirements for MPFs

From (15.11), an FIR filter is also a TDL filter. In order to replicate the equation, we need to be able to complete the same mathematical operations using optics. Each tap of an FIR filter is a weighted delay—methods for optical weighting and optical delay lines are necessary. The weighted delays are summed together to complete the FIR filter, so an optical method of adding the symbols together is also needed. All optical (and even analog) based filters require these three components: *optical weights*, *optical delay lines*, and a *summation method*. Traditional optical fused couplers are built specifically for the purpose of combining optical signals (and their powers), and since they are used widely in any optical communication system, we will focus on optical methods for delays and weights here.

Figure 15.9 shows a general reference layout of an MPF. In all optical filters, the RF signal must be converted to an optical signal. Generally, a continuous-wave (CW) laser source modulates the RF signal onto the optical domain using modulators such as electro-optic modulators (EOMs) like the Mach-Zehnder modulator (MZM), or electro-absorption modulators (EAMs). The optical signal is then fed

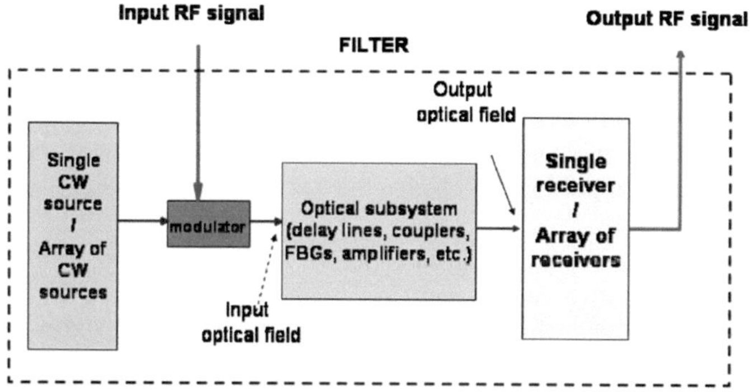

Fig. 15.9 General reference layout of a microwave photonic filter showing the relevant electrical and optical signals [5]

into a photonic circuit that delays and weights the signal on parallel taps and then combines them with a fused coupler (or other method). The optical signal can be converted back to an electrical signal using a photodetector.

15.3.2 Popular MPF Architectures

Many optical tunable delays and innovative weighting schemes have been proposed and experimentally demonstrated in literature. We review some of the popular techniques and present two unique MPF architectures.

Popular optical delaying techniques include filters that use fiber-Bragg gratings (FBGs) to delay filter taps [14]. Delays are created by forcing the optical signals to propagate along the FBG until reflected by a wavelength-specific grating. FBGs are popular since a whole array can be fabricated onto a single piece of optical fiber. Fiber delay lines have also been proposed for TTD in MPFs [15, 16], but are bulkier than FBG-based delays when filter taps become large.

Fully tunable and programmable weighting using free space methods such as spatial light modulators [17] and multi-port programmable wavelength processors [18] have been demonstrated. Unfortunately, typical optical systems are generally limited to positive tap coefficients, as light intensities are intrinsically "positive". Therefore, the shapes of their transfer functions are severely constrained and can only be used to implement low-pass filters (LPFs). The addition of negative coefficients in filter schemes allows for a practical filter with passband capabilities [5].

There are many ways to create negative optical weights. Recent advances in negative coefficient weighting have used cross-phase modulation (XPM) in an semiconductor optical amplifiers (SOA)-MZ interferometer [19] and self-wavelength cross gain modulation (XGM) in an SOA [20], but are suitable only for realizing a few taps. Another technique involves using multiple phase inversion in an SOA-based XGM wavelength converter [21]. An innovative method involving sinusoidal group delay gratings, which use FBGs to achieve positive/negative weighting and delays was proposed in [22].

Weight and delay are 'coupled', and it is difficult to have freely tunable attenuation without affecting the delays. Other optical weighting schemes involve polarization modulation and optical polarizers [19, 23]. While these approaches are simple, scalable, and tunable, the method relies heavily on mechanical polarization controllers, which cannot rapidly change the polarity of the taps (to sub millisecond-length time).

Simpler techniques entail using the negative bias slope of MZM to π-shift the output to create negative coefficients [24–27]. However, some techniques necessitate using two MZMs, which can be bulky when scaled [24]. Others using one MZM rely on the dependence on wavelength but positive and negative coefficient outputs are of different power. A novel technique using a specially designed integrated 2×1 MZM to achieve π-shifting was introduced in [26]. On the other hand, a simple technique using a 1 × 2 dual output MZM to achieve negative weighting by using phased-inversed dual outputs was proposed in [27].

Figure 15.10 shows an example of MPF using a spectrally-sliced free-space configuration [5]. The most distinctive and important part of the setup is the optical tapping through a sliced broadband source. A broadband laser is sliced by a diffraction grating into 10 different wavelengths. The slicing into individual wavelengths allows summing to be done incoherently. Use of spatial light modulators achieves the weighting and delaying of the samples. Fast tunability can be achieved as individual spatial light modulators (SLMs) can be turned ON and OFF, which are used to eliminate specific spectrum slices.

Figure 15.11 shows another realization of a MPF employing a discrete time optical transversal filtering scheme [28]. The unique characteristic of this system is the pulsed laser source being used. This means that the optical pulses sample the RF

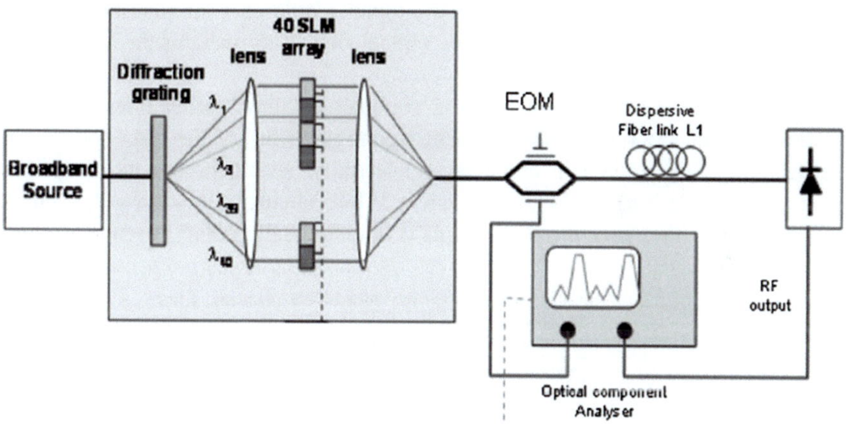

Fig. 15.10 Transversal MPF using spectrally-sliced free-space optics [5]

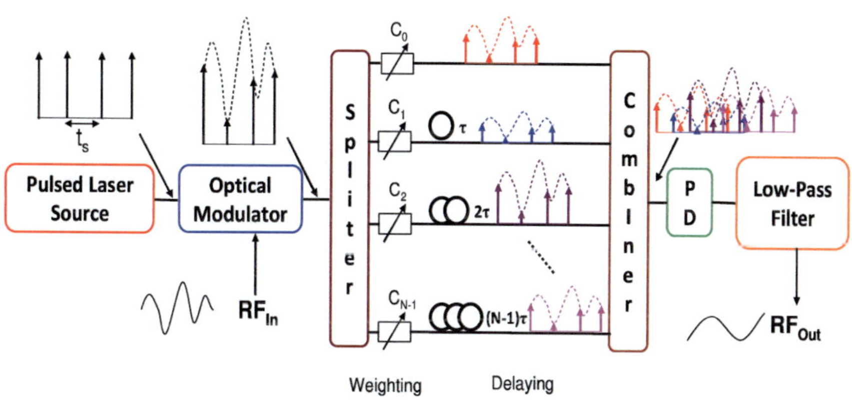

Fig. 15.11 Discrete time optical transversal MPF [28]

input data (discretely). There is no need to have a separate analog-to-digital converter (ADC) as the laser source serves that role. The delay is achieved through fiber delay lines. In order to avoid coherent summing, the delays are chosen so that no optical pulses from one tap overlaps with pulses from another tap. Optical attenuators provide weighting. There are no negative coefficients in this scheme.

15.3.3 Optical Technologies for MPFs

We now present the state of the art optical components and techniques that are employed in MPFs. As stated earlier, MPFs need an optical delay scheme and also an optical weighting scheme complete with negative coefficients. For an adaptable filter, tunability of (at least) the weights is required which need to be electronically programmable so that they can be quickly tuned. Table 15.1 summarizes some of the novel optical components used for optical delays or weights in MPFs from recent literature.

FBG arrays can be used for creating compact but fixed delay lines [29]. The drawback is that the fixed delays prevent reconfigurability of the filter bandwidth. Furthermore, this approach employs a spectrally sliced broadband amplified spontaneous emission (ASE) noise source. While slicing this source will create multiple optical filter taps, the ASE source is fundamentally noisy. An improvement

Table 15.1 State of the art optical technologies for microwave photonic filters

References	Functionality	Technology	Advantages	Limitations
[29]	Optical delays	FBG arrays	No need to use bulky and slowly tunable fiber delay lines	Spectrally sliced broadband ASE noise source; hard to tune and reconfigure
[30]	Optical delays	WDM filter with FBG arrays	WDM system, implementation of negative coefficients	No tunability/ programmability of delays
[31]	Optical delays	WDM filter with chirped FBG arrays	WDM system, reconfigurable bandwidth	No negative coefficients; lacks easy tunability/ programmability of weights
[17]	Optical weights	Spatial light modulators	Programmability of weights	Free space optical setup (not compact); spectrally slice a noisy ASE source
[18]	Optical weights	Liquid crystal on silicon	Programmability of weights	Free space optical setup (not compact); spectrally slice a noisy ASE source
[27]	Optical weights and delays	1×2 MZM	Simple implementation of negative weights, delay using FBG arrays	Spectrally slice a noisy ASE source; lack *easy* tunability/programmability of weights

can be achieved by using a wavelength-division multiplexing (WDM) scheme paired with FBG delay lines [30]. This technique uses an array of distributed feedback (DFB) lasers for the optical taps. While this gives the filter more dynamic range, it sacrifices simplicity and also lacks tunable delay lines. In contrast, using chirped FBGs with tunable DFB lasers allows controllably variable delay lines [31]. However, this technique does not incorporate negative coefficients.

Similarly, many technologies have been developed for optical weighting schemes. Free-space weighting schemes have recently garnered popularity including the weighting technique based on SLM [17] and liquid crystal on silicon (LCOS) [18]. Both these methods are extremely precise, can scale to filter orders of hundreds, and are easily programmable. However, they still have a couple of drawbacks. Firstly, they are free space optical techniques that are bulky and occupy a much larger footprint than discrete optics. Secondly, they lack the ability to create negative coefficients. A simple and compact method for creating negative coefficients can be integrated into the modulators that convert the electrical signal to the optical signal [27]. However, this negative weighting technique must be combined with another set of tunable optical weights for programmability.

15.4 Photonic Adaptive Beamformers

In this section, we first review the state of the art optical beamformers recently proposed and experimentally demonstrated in literature. We also detail a highly scalable photonic beamforming architecture designed for a particularly non-stationary, interfering environment. Finally, we introduce a specific application for the photonic beamformer namely, physical layer security in optical backhaul networks.

Beamformers have attracted significant interest because of their wide range of applications from radar, communication, and sensing. Beamforming is a technique that utilizes an array of antennas as a spatial and frequency filter to manipulate the beampattern to maximize the SNR of the SOI while cancelling/suppressing interferers and noise without a priori knowledge [32]. For instance, a beamformer can spatially separate two signals that have overlapping frequencies but originate from two different spatial locations [33]. Beamformers use optical FIR filters to process the signals temporally with thermo-optic optical attenuators—controlling the signal amplitudes and delays from each antenna element—to adaptively and rapidly adjust the beampattern; that is, the whole array can act in unison to steer the beam pattern. Meanwhile, the geometric configuration of the antennas allows the signals to be processed spatially. This makes beamforming attractive for overcoming the directivity problems of a single antenna while offering a higher gain [3]. Adaptive beamformers are particularly crucial in applications involving a highly non-stationary target environment. Compared to conventional RF beamformers which have limited narrowband performance due to their reliance on electrical phase shifters, the wideband nature of optics offers a clear advantage. The architecture also offers the distinct advantage of scalability to hundreds of antennas, as needed for practical systems.

The first adaptive beamformers were used as self-phased antennas that reradiated signals in the direction in which they were received without prior signal knowledge. In the 1960s, Howell and Applebaum introduced their versions of adaptive antennas for interference nulling [3]. Widrow followed closely with self-optimizing adaptive algorithms [10]. Applebaum and Widrow focused on beamforming specifically for applications in the field of sonar and radar signal reception. Seismic array development was especially popular in the 1960s. Recently, the major area of interest is radar and communication systems for interference suppression [3]. Direction finding in severe interference environments has also been a popular topic, as is scanning for high angular resolution imaging. We will focus on interference cancellation techniques in this section.

15.4.1 State of the Art Photonic Beamformers

This section provides a review of the recently demonstrated state of the art optical beamformers in literature. As explained previously, a photonic beamformer is an array of antennas with a set of *optical signal processors* attached to each antenna. In the previous section we stipulated the requirements needed for an FIR filter: a method for creating both optical delays and optical weights. The same requirements are necessary for a photonic beamformer. Moreover, optical delays with TTD capability is essential for a beamformer, as explained in Sect. 15.2. However, since a beamformer consists of multiple FIR filters (one for each antenna), scalability with optical components is very important as well. We focus the literature review on methods for TTD optical beamformers with an emphasis on scalability.

Figure 15.12 shows a TTD beamformer architecture based on FBG prisms. FBG arrays are created in a prism format in which different sets of delays (five as depicted in the figure) are selected for a specific wavelength of a laser. The FBG prism is built so that the delays create a pre-steered beamformer [34]. However, this architecture cannot be fully tunable, as the number of pre-steered directions is limited by the number of FBGs etched in the array.

A similar approach but that is tunable is shown in Fig. 15.13. In this scheme, instead of FBGs, the beamformer uses dispersive fibers that can be precisely tuned by a finely tunable laser [35]. However, this method, while more precise, is inherently slowed by the tuning speed of the laser.

Figure 15.14 depicts a scheme for a TTD beamformer based on WDM demultiplexer (demux) delays [36]. The architecture consists of a single antenna, a tunable laser with 8 λ's, and a corresponding WDM demux which selects one of eight possible delays. The limitation in its resolution depends on the tunable laser and the amount of wavelengths supported by the arrayed-waveguide (AWG) demux.

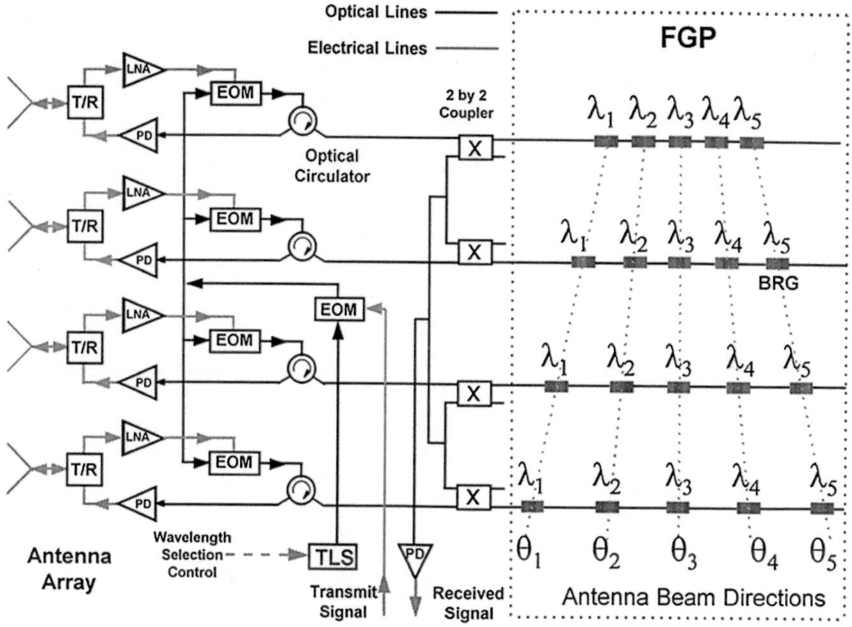

Fig. 15.12 Beamforming architecture based on FBG prism [34]

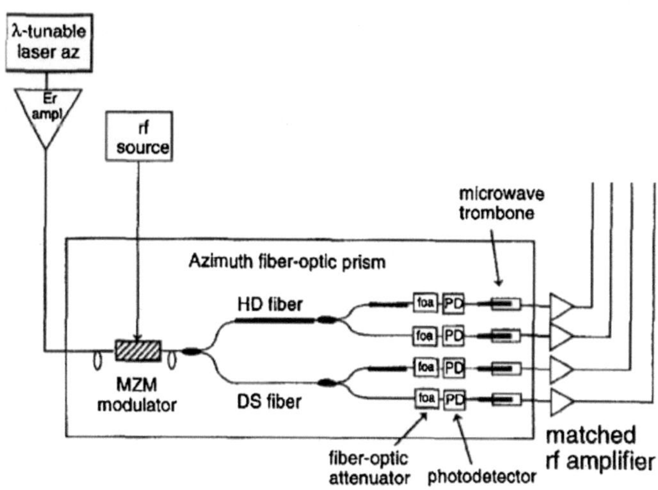

Fig. 15.13 Beamforming architecture based on dispersive fiber prism [35]

Furthermore, the number of lasers required scales linearly with the number of antennas, which can become bulky with large beamforming systems that necessitate hundreds of antennas.

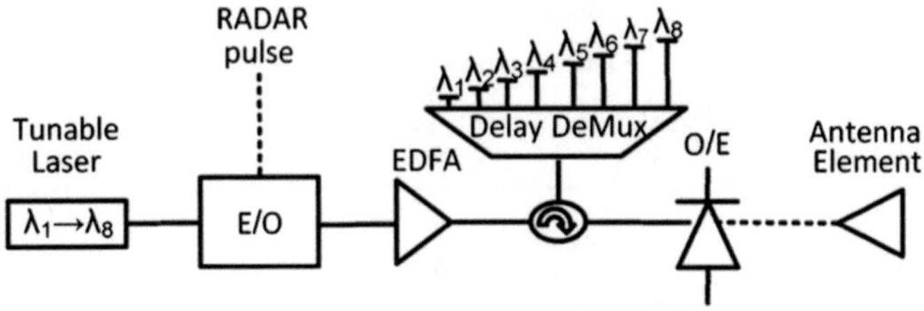

Fig. 15.14 Beamforming architecture based on WDM demux delay lines [38]

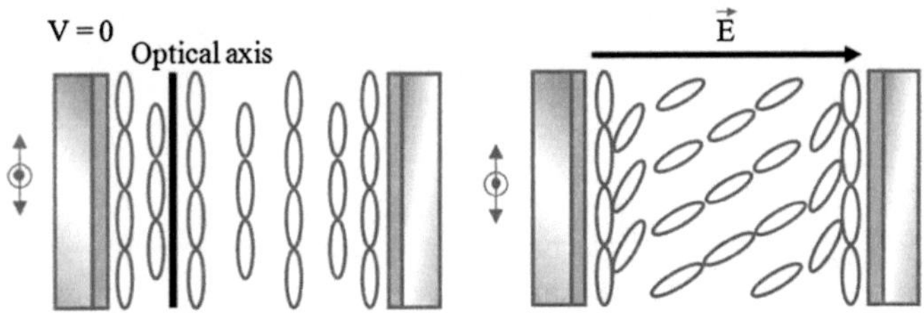

Fig. 15.15 SLM delay lines [37]

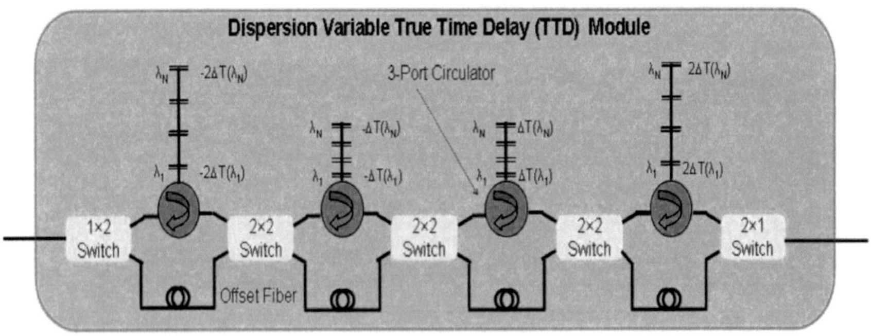

Fig. 15.16 Bit-switched delay lines [41]

Antenna arrays that use SLMs to perform optical delays have also been proposed [37]. By apply an external electrical field as shown in Fig. 15.15, dielectric molecules in the SLM form different polarizations. The polarization changes the index of refraction of the delay line which affects the length of time the light travels through the device.

Photonic crystal fiber and XGM in SOAs have also been used to create TTD lines in photonic beamformers for multi-transmit systems [38, 39]. On the other hand, bit-switched delay lines are also popular [40, 41]. Figure 15.16 shows an example of bit-switched delay line that selects a particular delay from a chain of FBG arrays using a tunable laser.

A beamforming architecture based on opto-VLSI is shown in Fig. 15.17 [42]. The opto-VLSI processor is an array of liquid crystal (LC) cells driven by a Very-Large-Scale-Integrated (VLSI) circuit. Delays can be created for each antenna by slicing a broad-band optical source (ASE-based source) and routing specific sliced wavebands through the Opto-VLSI processor to a high-dispersion fiber [42].

A beamforming architecture based on high dispersion fiber has also been reported [43]. Here the beamformer uses an array of lasers that each experiences a different delay based on passing through a wavelength-dependent high dispersion fiber. Figure 15.18 depicts the architecture of a single antenna. By using two π-shifted EOMS, negative coefficients can also be implemented. However, this technique requires a laser array for *each* antenna array.

Recently, the integration of the photonic components into an integrated *beamformer-on-chip* has been explored [44, 45]. Next generation systems investigate a scalable integrated photonic beamformer that can be electronically controlled for adaptive interference cancellation. The possibility of creating a beamformer-on-chip has become a question of not "if" but "when". Figure 15.19a illustrates an integrated beamformer. Figure 15.19b shows a 16-antenna beamformer integrated on a chip, roughly the size of a 20-cents of an euro coin. These systems show the potential for extreme savings in SWAP along with full broadband processing ability.

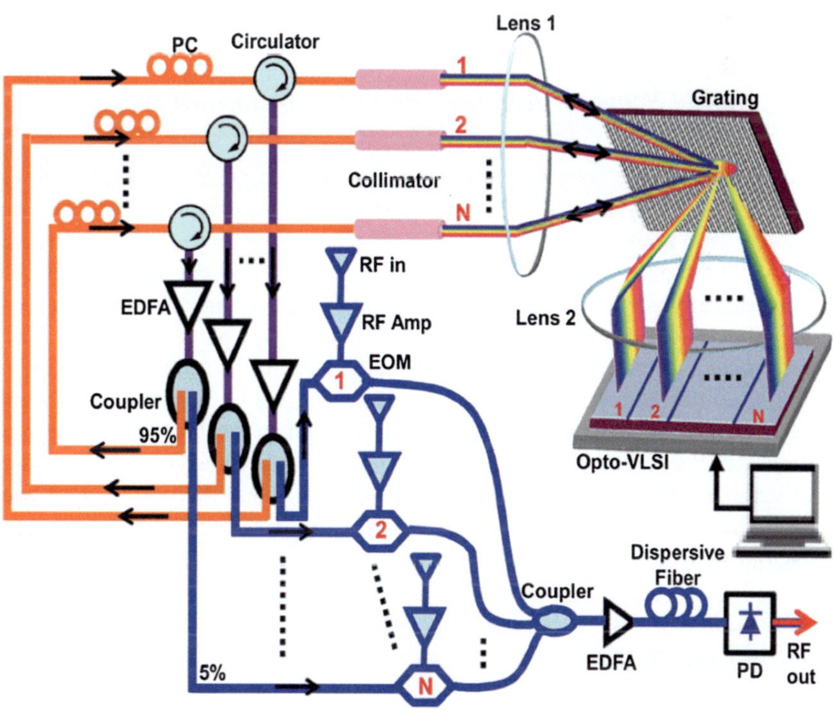

Fig. 15.17 Beamforming architecture based on Opto-VLSI [42]

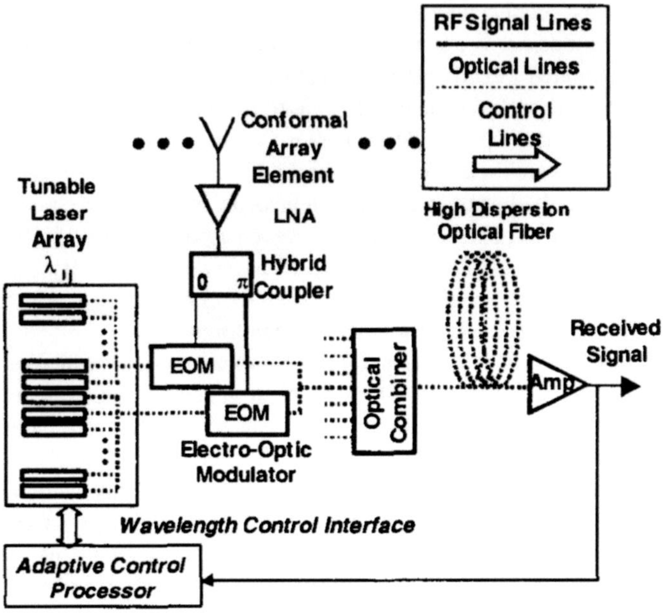

Fig. 15.18 Beamforming architecture based on high dispersion fiber [43]

15.4.2 Highly Scalable Adaptive Photonic Beamformer

The key characteristic of a reconfigurable and tunable photonic beamformer is scalability. Thus far, most of the beamformer architectures reviewed, are not scalable to large arrays—the number of lasers required by these systems increases linearly with the number of antennas. In this section, we present a highly scalable photonic beamformer [46], unique to our lab, which is specifically designed for a particularly non-stationary, interfering environment. Using optical transversal filters for each antenna element and thermo-optic optical attenuators, the array is capable of both spatial beamforming and frequency-domain filtering while adaptively and rapidly adjusting the beampattern. Our architecture offers the distinct advantage of scalability to hundreds of antennas, as needed for practical systems, by using a novel single-mode to multimode (SM-MM) combiner, and the same set of laser wavelengths can be used for every antenna in the system. By eliminating coherent effects, our system uses the same fixed set of optical wavelengths for each antenna in the system, resulting in a simple and compact architecture. We present experimental results to show proof-of-concept and demonstrate the proposed adaptive beamformer performance.

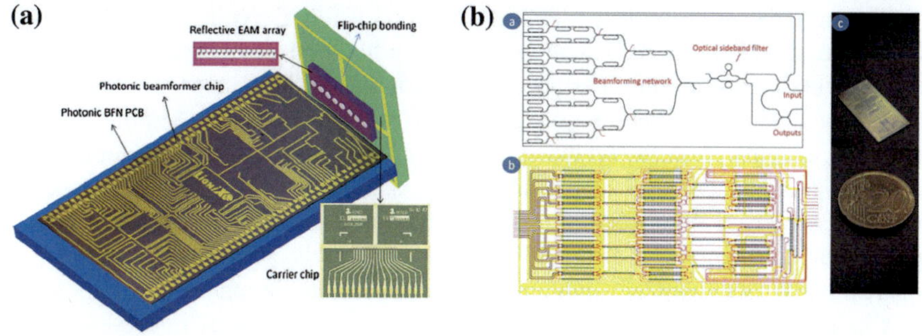

Fig. 15.19 **a** Potentially integrated beamforming network with EAM array [44]. **b** Schematic of 16 × 1 beamforming network with chip mask layout and photograph of a fabricated chip [45]

15.4.2.1 System Overview

Figure 15.20 shows the architecture of a wideband photonic beamformer that we recently demonstrated experimentally. The key component of the beamformer is the optical transversal filter which is driven by two eight-channel DFB laser arrays. The first array of wavelengths λ_{1-8} corresponds to the positive coefficients whereas the second array of wavelengths λ_{9-16} corresponds to the negative coefficients. The optical sources are inserted into a compact 16-channel thermal-optic attenuator for easy adaptive control of the weights through a computer or voltage source. The attenuators have a response time of 10 μs per 0.1 and a 20 dB range.

The weighted taps are then combined using an AWG multiplexer (mux). The RF signal to be processed is modulated onto the optical carrier using a dual output electro-optic MZM. The modulated signals of the outputs are biased at the inverse, π-shifted, parts of the modulator transfer function. We use this complimentary output to implement negative coefficients. Both outputs have equal insertion losses of 3.7 dB. The weighted signals exit from both the positive and negative outputs of the MZMs. The complementary outputs are launched in FBG arrays that only reflect and delay the wavelengths assigned to the respective coefficients, via an optical circulator (OC).

The coefficients encounter FBGs with the same delays, but at different wavelengths. As a result, each delay has both a positive and negative tap, and the attenuators are used to switch on the tap and weight each tap by enabling/disabling a certain wavelength. In this way, our 16-wavelength filter provides eight positive/negative taps. Since time delays and filter bandwidth are inversely proportional, fabricating FBGs with closer spacing and shorter delays can increase bandwidth.

There is a total optical insertion loss of ~19.5 dB for each filter. If each DFB laser has 13 dBm of power, each filter receives 9.4 dBm of power after splitting to four antennas and outputs −10.1 dBm into the combiner. Assuming a four-antenna beamformer, −4.6 dBm of power finally reaches the high speed MM photodetector, well above the −25 dBm limit.

The outputs of the optical filters are summed using a special SM-MM optical combiner, as in Fig. 15.20. The architecture is a blind adaptive approach, in which the adaptive algorithm only has access to the output of the system. This requires a single conversion to RF at the output, whereas traditional systems require an ADC for each antenna element, which is impractical for large antenna systems. As a result, conventional criteria such as minimum mean square error (MMSE) cannot be used and analytical Wiener solutions cannot be found. Instead, blind algorithms rely on correlating the processed signal with some known characteristic of the signal of interest (such as frequency), and applying a gradient-based algorithm.

Our main advantage is that the same set of 16 wavelengths is used for each antenna, reducing complexity and increasing scalability. Typically, when signals of the same optical wavelength are combined, beat noise from coherent summing will occur and severely degrade the performance. Therefore, without the use of a SM-MM combiner the architecture would require 16 lasers for each antenna. To re-use the same wavelengths, the SM-MM combiner is used. The combiner couples signals from several individual single-mode fibers to distinct modes inside a multimode fiber. The combiner offers the advantage of phase-insensitivity and coupling without optical interference. In-depth information and experimental data demonstrating operation can be found in [47].

The architecture scales by simply adding optical splitters and amplifiers up to the limit imposed by the ASE of the amplifiers. The optical weights, which are integrated sixteen per chip and electrically controlled, do not limit the scalability of this

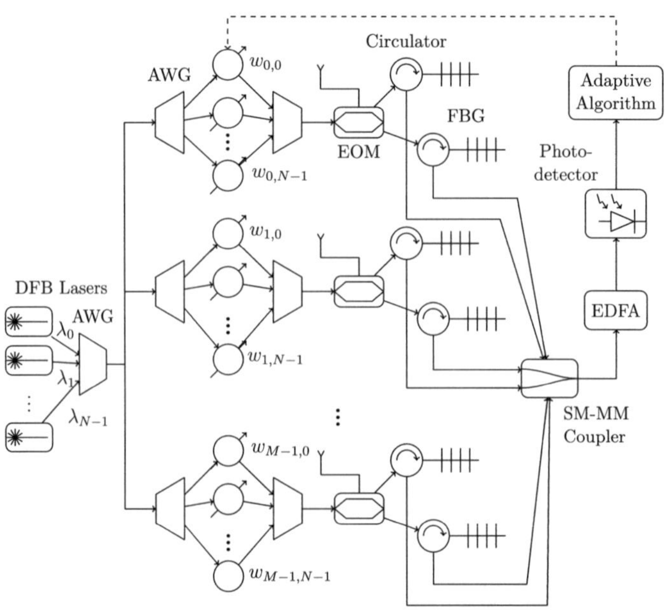

Fig. 15.20 Highly scalable photonic beamformer architecture (N = 16, M = 4)

architecture, nor does the addition of FBGs. A 100 μm multimode fiber can accept up to 113 inputs (or antennas).

15.4.2.2 Proof-of-Concept Experimental Results

We experimentally demonstrate an 8-tap filter with tap weights [1.0914 1.0617 −1.0715 −1.0814 −1 −0.9683 0.9795 0.0594] and delays incremented by 4 ns. The dark blue dotted curve in Fig. 15.21 shows the theoretical values and the thick light green curve shows the measured magnitude response. We are able to achieve a maximum extinction ratio of ~40 dB. The bandwidth and the depth and placement of the notches depend on the precision of both the delays and weights. Moreover, our optical system adds no additional noise to the processed signal, as seen in Fig. 15.21.

The data shows a 4.5 dB loss associated with the architecture. RF systems in the low GHz range exhibit typical losses around 1 dB. RF systems in the high GHz range (60 GHz), an area in which photonic systems are expected to excel, exhibit ~35 dB loss with one filter and a ~17 dB loss with a four-filter beamformer [48]. RF signal losses in optical systems in general originate from the electrical-to-optical conversion efficiency of the modulators and the modulation depth of the signal during this conversion. To reduce loss further, removal of the optical dc level of the processed signal would allow the signal to be optically amplified further, reducing the system insertion loss.

We experimentally demonstrate an adaptive single-antenna beamformer consisting of 8 fully tunable taps with fixed delays of 400 ps and a bandwidth of 2.5 GHz. We employ a modified version of the LMS algorithm called block LMS with a block size of 256 and stepsize of 512. The adaptive results are preliminary. The authors only had access to a single 5 Gs/s ADC card, instead of the two required for the LMS algorithm.

A signal generator is fed into two narrowband tones at 200 MHz and 1 GHz to the filter. The algorithm is programmed to cancel the 1 GHz interference and pass the 200 MHz signal. Our workaround involves only digitizing the filtered output signal. The input signals are replicated digitally and are used to calculate the error signal. The adaptive algorithm then calculates optical weights that are sent to the attenuators. We observe an SIR (signal-to-interference ratio) improvement of ~20 dB after ~60 iterations at the output of the filter. We are able to drop the interference to just 5 dB above the noise floor. The results are good and can be easily improved with a second digitizer. In the future, we plan to apply a blind adaptive technique as described previously. Limited resolution associated with the optical attenuators degrades performance.

Future work will include the construction of multiple transversal filters to complete true spatial beamforming capability.

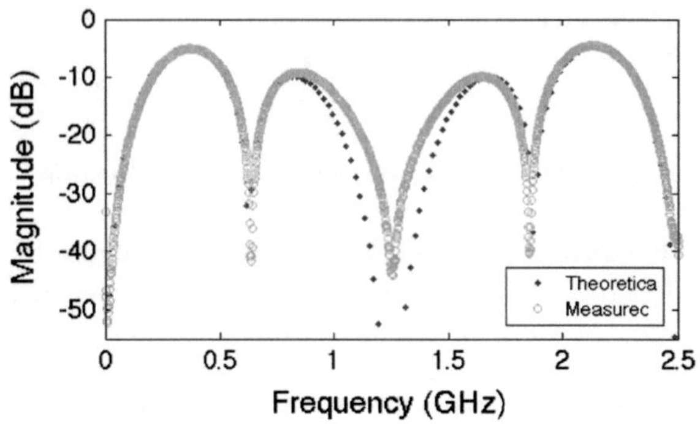

Fig. 15.21 Measured and predicted magnitude response of eight-tap FIR filter

15.4.3 Photonic Beamforming for Physical Layer Security

We introduce a specific application for using the photonic beamformer—physical layer security in optical backhaul networks. The purpose of this section is to demonstrate the applicability of the beamformer in commercial systems and to test the capabilities of the beamformer.

Emerging mobile services, such as mobile banking and desktop-to-mobile applications, require both increasingly high data rates and high data security. To address the rising bandwidth demand, strong emphasis is being placed on fiber-optic back- and fronthaul of mobile data [49–51]. While optical architectures that can satisfy the required data rate and latency requirements have matured [52–58], enhancing data security via optical-layer techniques is still in its infancy [59–66]. The potential of such physical-layer security is tremendous. For example, while higher-layer approaches, such as encryption or steganography, can be cracked by malicious eavesdropping, optical-layer techniques can prevent eavesdroppers from physically receiving the signal in the first place. The highest form of security thus resides in the physical layer, rendering it attractive for data-sensitive mobile fronthaul applications.

To implement security on the physical-layer, photonic beamforming can be used in the optical fronthaul network to cancel the signal (i.e. create a signal null) in the physical direction of an eavesdropper. For the downlink, this can enable sensitive data to only be received by an intended user within a meter-scale radius, based on the intended user's spatial location and signal carrier frequency. Given the user's spatial/ frequency coordinates, a photonic beamformer can create the desired signal null by effectively manipulating a beampattern that propagates through the optical fronthaul network. This is achieved by the beamformer's array of RF antennas and a series of adaptive optical FIR filters. The physical formation of the antennas provides spatial filtering while the FIR filters provide frequency filtering. By changing the weights of the FIR filters, an adaptive beamformer that can respond to dynamically-changing/ noisy environments can be implemented. Similarly, physical-layer security can also be provided on the uplink side, with potential application in a shared, multi-operator environment. In this case, various signals picked up by remote antennas can jointly propagate through a shared fronthaul infrastructure and be securely separated on a per-operator basis via adaptive photonic beamforming at a centralized processing site. It is noted that the proposed optical-layer security approach is also transparent to the modulation format of the underlying wireless signal.

To the best of our knowledge, we have proposed and demonstrated the first experimental adaptive photonic beamforming technique for physical-layer security in optical fronthaul of mobile traffic, achieving up to 43 dB signal cancellation of an undesired eavesdropper and 30 dB power budget over 8 km standard single mode fiber (SSMF) fronthaul distance. The proposed approach is attractive for secure optical fronthaul of advanced mobile services and is transparent across heterogeneous mobile technologies and standards.

15.4.3.1 Experimental Setup, Results, and Discussion

The proposed fronthaul architecture using photonic beamforming in the downlink and the uplink is shown Fig. 15.22. In both cases, the beamformer acts as an optical spatial and frequency filter that preserves optical signals arising from desired physical orientations and carrier frequencies, while canceling signals from undesired physical directions and frequencies. In the downlink, mobile data signals from a centralized site are applied to parallel optical adaptive FIR filters, which optically modulate and filter the input. The processed optical signals are then transmitted over parallel SSMF optical fronthaul links to an array of remote antennas. The optical FIR filters and remote antenna array together form the beamformer. Specifically, by proper assignment of the beamformer's filter coefficients, the remote antennas are turned into a distributed spatial filter that directs the output only to intended users. The configuration of Fig. 15.22a can thus be used to securely deliver data to users 1 and 3 while producing a null in the direction of user 2, or vice versa. In the uplink in Fig. 15.22b, wireless signals from multiple users are first detected by the remote antenna array, and optically processed by the FIR filters such that, after SSMF transmission and photodetection, only the signal from intended users (e.g. user 1 and user 3) is delivered to the centralized site, while undesired interferers are suppressed (e.g. user 2).

For downlink experiments, 2.5 and 5 GHz sinusoids are used as the SOI to be directed by the photonic beamformer to an intended user. After SSMF transmission and photodetection, the received SOI power is measured at both intended user and eavesdropper locations, mutually separated by 0.5 m. Downlink beamformer performance is thus measured in terms of the signal strength ratio (SSR) at different user locations. The beamformer is adjusted for maximum cancellation. Figure 15.23 shows the SSR for the 2.5 and 5 GHz signal. Similarly, for uplink measurements, 2.5 or 5 GHz sinusoids are used as the SOI, while a sinusoid at 900 MHz emulates an interferer to be suppressed. The two wireless RF signals are generated at different locations (0.5 m separation), picked up by the antenna array, processed by optical FIR filters, propagated over parallel SSMFs, combined using a SM-MM coupler, and detected by a multimode PD. In the uplink, the SSR of the SOI versus the

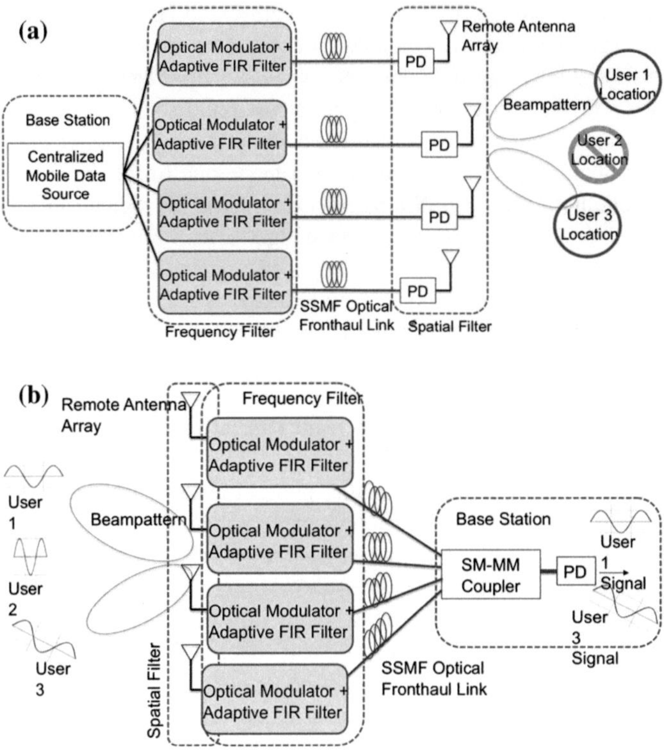

Fig. 15.22 Proposed architectures **a** Downlink fronthaul. **b** Uplink fronthaul

interfering signal is measured at the same location. Figure 15.24 shows the SSR for the 2.5 and 5 GHz SOI.

Maximum SSR of 33 and 43 dB is achieved in the downlink and uplink, respectively, over 8 km SSMF fronthaul distance. The 10 dB SSR gain for the uplink is attributed to the fact that physical-layer signal cancellation physically occurs in stable passive optical components, while in the downlink, it occurs in the wireless multi-path channel, resulting in additional undesired interference. Downlink SSR results in Fig. 15.23 are achieved for intended user versus eavesdropper separations from 0.1 to 0.5 m, confirming efficacy for small-cell scenarios, with similar results also measured for signals at 5 GHz, verifying performance for various frequency bands of prominent wireless standards.

The SSR curves of Figs. 15.23 and 15.24 feature a flat plateau in the 0–10 dBm optical power range, followed by a linear decrease versus optical power governed by PD sensitivity. At SSR = 10 dB, power budgets of 28 and 30 dB are supported by the downlink and uplink, respectively. Finally, from the experiments it may be observed that there is no SSR penalty for increasing fronthaul distance compared to optical back-to-back results. Consequently, so long as minimum PD sensitivity requirements are met, increasing fronthaul reach will only result in optical FIR coefficient scaling that is proportional to aggregate received power, without degrading SSR integrity.

To assess the bit error rate (BER) of the fronthaul link, downlink and uplink experiments are repeated with a baseband 6 Gbit/s non-return-to-zero (NRZ) pseudorandom binary sequence (PRBS) signal as the SOI. Results for the downlink and uplink architectures shown in Fig. 15.25, reveal a maximum transmission distance of 8 and 18 km SSMF for the downlink and uplink, respectively. The longer uplink transmission distance is attributed to the higher sensitivity of a multimode PD compared to a single mode PD, confirming that both SSR and BER performance are primarily governed by received optical power.

In summary, this investigation experimentally verifies adaptive photonic beamforming for advanced physical-layer security in optical fronthaul of mobile traffic. Up to 33 and 43 dB signal cancellation to undesired eavesdroppers are experimentally demonstrated over 8 km SSMF, with 28 and 30 dB power budgets achieved in downlink and uplink transmission scenarios, respectively. No power penalty is observed by our system. By enhancing physical-layer security with modulation format transparency, the proposed approach is promising for secure optical fronthaul of emerging mobile services.

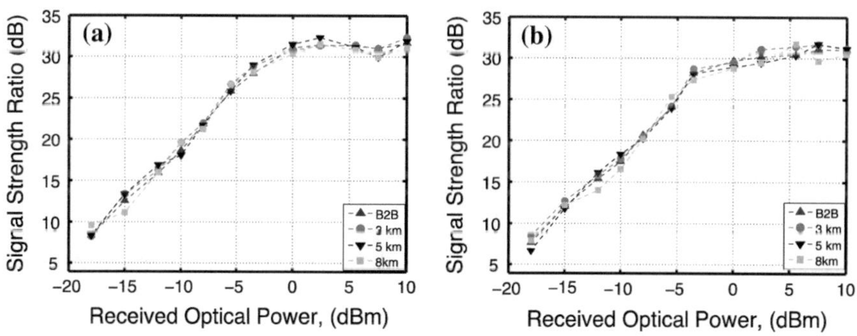

Fig. 15.23 Downlink signal strength ratio. **a** 2.5 GHz SOI. **b** 5 GHz SOI

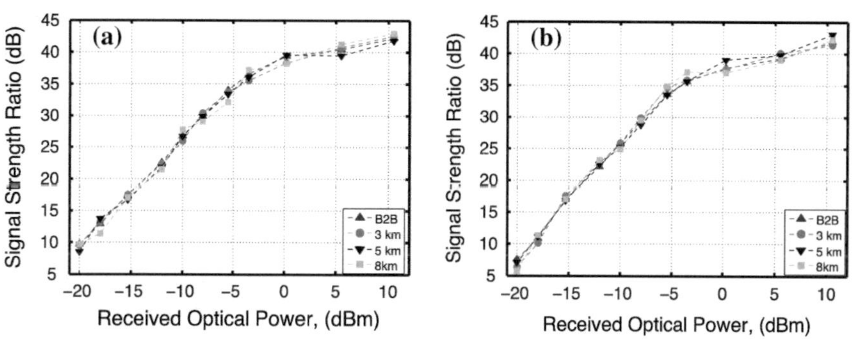

Fig. 15.24 Uplink signal strength ratio. **a** 2.5 GHz SOI. **b** 5 GHz SOI

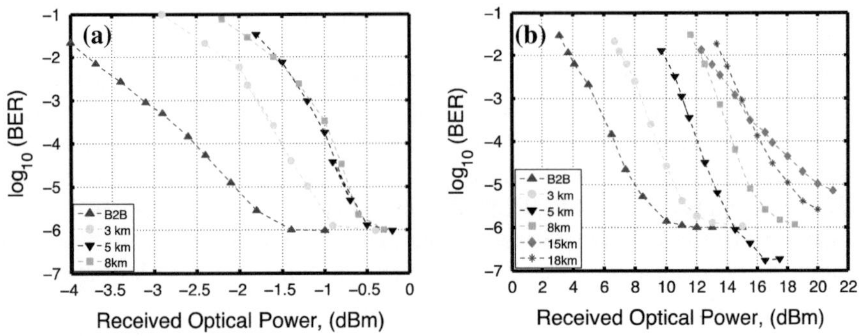

Fig. 15.25 BER versus optical signal power as a function of fronthaul distance. **a** Downlink architecture. **b** Uplink architecture

15.5 Summary and Concluding Remarks

In the past half-decade, wireless communication for data services has witnessed an exponential growth, with the advent of smart phones, WIFI, and wireless bluetooth devices. Mobile apps and computing power has spurred the use of iPhones and Android-powered devices, and it seems a wireless device occupies every pocket. People crave the ability to access anything anywhere at anytime, but, unfortunately, this popularity has caused both significant network burdens and a wireless spectrum crunch. As the growth of mobile Internet traffic continues, the need for an easily implementable method for processing all of this data is essential. Efficient use of the wireless spectrum also creates its own unique requirements.

The growth of wireless bandwidth mirrors the proliferation of optical network bandwidth a decade before, and the backhaul of nearly all wireless data networks is fiber-optic. Analog optical signal processing techniques, or microwave photonics, provides an ideal platform for processing wireless information before it is transported to data aggregation centers by fibers. Photonics offer the advantages not only of broadband operation, but reduced SWAP, in addition to low transmission loss, rapid reconfigurability, and immunity to electromagnetic interference. It is in this context that we presented recent advances in optical signal processing techniques for wireless RF signals. This chapter was devoted to the discussion of photonic architectures for wideband analog signal processing, including RF beamforming and physical layer security.

Beamforming is a technique that utilizes an array of antennas as a spatial and frequency filter to manipulate the beampattern for cancelling unwanted signals. The beamformer uses optical FIR filters to process the signals temporally with thermo-optic optical attenuators to adaptively and rapidly adjust the beampattern. Meanwhile, the geometric configuration of the antennas allows the signals to be processed spatially. Compared to conventional RF beamformers which have limited narrowband performance due to their reliance on electrical phase shifters, the wideband nature of optics offers a clear advantage. The architecture also offers the distinct advantage of scalability to hundreds of antennas, as needed for practical systems. Such an integrated system could potentially form a frontend of an emerging all-optical signal processing techniques such as photonic neuromorphic processing allowing massive network integration [67–76].

Interference cancellation plays a crucial part in increasing capacity as it enables full duplexing. The ability to reject interference, while receiving an SOI at the same frequency, is referred to as co-channel interference cancellation. Further, a system needs to be adaptive to react to a highly non-stationary environment. Photonic implementations of interference cancellation systems exhibit significant capabilities beyond what is possible with electronic processing.

An emergence of markets relating to mobile applications, such as mobile banking, mobile-to-desktop, and mobile-to-cell towers, will require a high degree of security. Data should only be received by their intended users, and eavesdroppers should not have access to this data. The ultimate form of security is in the physical layer, and it is easiest to place a null in the direction of an eavesdropper. Unlike the software layer, such as encryption or steganography, which can be cracked, physical layer security prevents eavesdroppers from even detecting the presence of data transmission in the first place. This creates a unique application of photonic beamformers for physical layer security in optical fronthaul of mobile traffic.

In summary, this chapter discussed the basic theory of beamforming and the unique challenges of broadband interference cancellation. Next, we presented the requirements and challenges for building optical FIR filters, and reviewed several innovative MPF designs including optical tunable delays and weighting schemes. We also reviewed the state of the art optical beamformers recently proposed and experimentally demonstrated in literature. Furthermore, we detailed a highly scalable photonic beamforming architecture designed for a particularly non-stationary, interfering environment. Finally, we introduced a specific application for the photonic beamformer namely, physical layer security in optical backhaul networks.

We hope that this chapter will contribute to stimulating research in the cross-disciplinary areas of photonics and optics, microwave engineering, wireless communications, and signal processing, from fundamental principles to practical applications.

References

1. Mobile broadband explosion, White Paper, Rysavy Research/4G Americas, Aug 2012
2. H. Bauer, F. Grawert, S. Schink. Semiconductors for wireless communications: growth engine of the industry, McKinsey & Company on Semiconductors, Technical Report (2012)
3. R.A. Monzingo, R.L. Haupt, T.W. Miller, *Introduction to Adaptive Antennas* (SciTech Publishing, Raleigh, 2011)
4. J. Yao, Microwave photonics. J. Lightwave Technol. **27**(3), 314–335 (2009)
5. J. Capmany, B. Ortega, D. Pastor, A tutorial on microwave photonic filters. J. Lightwave Technol. **24**(1), 201–229 (2006)

6. J. Chang, M.P. Fok, J. Meister, P.R. Prucnal, A single source microwave photonic filter using a novel single-mode fiber to multimode fiber coupling technique. Opt. Express **21**(5), 5585–5593 (2013)
7. J. Chang, Y. Deng, M.P. Fok, J. Meister, P.R. Prucnal, A photonic microwave FIR filter using a spectrally sliced supercontinuum source. Appl. Opt. **51**(19), 4265–4268 (2012)
8. J. Chang, P.R. Prucnal, A novel analog photonic method for broadband multipath interference cancellation. IEEE Microwave Wirel. Compon. Lett. **23**(7), 377–379 (2013)
9. R. Haupt, *Antenna Arrays: A Computational Approach* (Wiley, Hoboken, 2010)
10. B. Widrow, P.E. Mantey, L.J. Griffiths, B.B. Goode, Adaptive antenna systems. Proc. IEEE **55**, 2143–2159 (1967)
11. B. Widrow, S.D. Stearns, *Adaptive Signal Processing* (Prentice-Hall, Englewood Cliffs, 1985)
12. L.V. Blake, M.W. Long, *Antennas: Fundamentals, Design, Measurement* (SciTech Publishing, Daryaganj, 2009)
13. S.O. Haykin, *Adaptive Filter Theory* (Prentice-Hall, Englewood Cliffs, 2002)
14. D.B. Hunter, R.A. Minasian, Microwave optical filters using in-fiber Bragg grating arrays. IEEE Microwave Guided Waveguide Lett. **6**(2), 103–105 (1996)
15. X. Yi, R.A. Minasian, New spectrum-sliced microwave photonic filter for high- frequency signal processing. IEEE Photonics Technol. Lett. **21**(4), 230–232 (2009)
16. T. Chen, X. Yi, T. Huang, R.A. Minasian, Spectrum sliced microwave photonic signal processor with tunablity and reconfigurablity, in *Proceeding of the OptoElectronics and Communications Conference* (*OECC*), pp. 1–2, July 2009
17. J. Capmany, J. Mora, D. Pastor, B. Ortega, High-quality online-reconfigurable microwave photonic transversal filter with positive and negative coefficients. IEEE Photonic Technol. Lett. **17**(12), 2730–2732 (2005)
18. X. Yi, T.X.H. Huang, R.A. Minasian, Microwave photonic filter with tunability, reconfigurability and bipolar taps. Electron. Lett. **45**(16), 840–841 (2009)
19. M.D. Manzanedo, J. Mora, J. Capmany, Continuously tunable microwave photonic filter with negative coefficients using cross-phase modulation in an SOA-MZ interferometer. IEEE Photonics Technol. Lett. **20**(7), 526–528 (2008)
20. E.H.W. Chan, R.A. Minasian, Coherence-free equivalent negative tap microwave photonic notch filter based on delayed self-wavelength conversion. IEEE Trans. Microw. Theory Tech. **5**(11), 3199–3205 (2010)
21. J. Mora, A. Martinez, M.D. Manzanedo, J. Capmany, B. Ortega, D. Pastor, Microwave photonic filters with arbitrary positive and negative coefficients using multiple phase inversion in SOA based XGM wavelength converter. Electron. Lett. **41**(16), 53–54 (2005)
22. X. Yi, R.A. Minasian, Novel multitap, flat-top microwave photonic filter based on sinusoidal group delay gratings. J. Lightwave Technol. **26**(15), 2578–2583 (2008)
23. C.-K. Oh, T.-Y. Kim, C.-S. Park, Reconfigurable photonic microwave band-pass filter with negative coefficients based on polarisation modulation. Electron. Lett. **43**(11), 639–641 (2007)
24. S. Mansoori, A. Mitchell, K. Ghorbani, Photonic reconfigurable microwave filter with negative coefficients. Electron. Lett. **40**(9), 541–543 (2004)
25. B. Vidal, J.L. Corral, and J. Marti, Multi-tap all-optical microwave filter with negative coefficients based on multiple optical carriers and dispersive media, in *Proceeding of the International Topical Meeting on Microwave Photonics* (*MWP*), pp. 201–204, Oct 2005
26. D. Pastor, J. Capmany, B. Ortega, A. Martinez, L. Pierno, M. Varasi, Reconfigurable RF photonic filter with negative coefficients and flat-top resonances using phase inversion in a newly designed 2 × 1 integrated Mach-Zehnder modulator. IEEE Photonics Technol. Lett. **16** (9), 2126–2128 (2004)
27. D.B. Hunter, Incoherent bipolar tap microwave photonic filter based on balanced bridge electro-optic modulator. Electron. Lett. **40**(12), 856–858 (2004)
28. L.A. Bui, K.S. Dayaratne, A. Mitchell, Discrete time microwave photonic transversal filter, in *Proceeding of the International Topical Meeting on Microwave Photonics* (*MWP*), pp. 1–3, Oct 2009

29. X. Yi, R.A. Minasian, Dispersion induced RF distortion of spectrum-sliced microwave-photonic filters. IEEE Trans. Microw. Theory Tech. **54**(2), 880–886 (2006)
30. J. Mora, S. Sales, M.D. Manzanedo, R. Garcia-Olcina, J. Capmany, B. Ortega, D. Pastor, Continuous tuning of photonic transversal filter based on the modification of tapped weights. IEEE Photonic Technol. Lett. **18**(15), 1594–1596 (2006)
31. J. Capmany, D. Pastor, B. Ortega, New and flexible fiber-optic delay-line filters using chirped Bragg gratings and laser arrays. IEEE Trans. Microw. Theory Tech. **47**(7), 1321–1326 (1999)
32. W. Liu and S. Weiss. *Wideband Beamforming: Concepts and Techniques* (Wiley, Chichester, 2010)
33. B.D. Van Veen, K.M. Buckley, Beamforming: a versatile approach to spatial filtering. IEEE ASSP Mag. **5**(2), 4–24 (1988)
34. H. Zmuda, R.A. Soref, P. Payson, S. Johns, E.N. Toughlian, Photonic beamformer for phased array antennas using a fiber grating prism, IEEE Photonics Technol. Lett. **9**(2), 241–243 (Feb 1997)
35. M.Y. Frankel, P.J. Matthews, R.D. Esman, Two-dimensional fiber-optic control of a true time-steered array transmitter. IEEE Trans. Microw. Theory Tech. **44**(12), 2696–2702 (1996)
36. L. Yaron, R. Rotman, S. Zach, M. Tur, Photonic beamformer receiver with multiple beam capabilities. IEEE Photonics Technol. Lett. **22**(23), 1723–1725 (2010)
37. L. Jofre, C. Stoltidou, S. Blanch, T. Mengual, B. Vidal, J. Marti, I. Mckenzie, J.M. del Cura, Optically beamformed wideband array performance. IEEE Trans. Antennas Propag. **56**(6), 1594–1604 (2008)
38. H. Subbaraman, M.Y. Chen, R.T. Chen, Simultaneous dual RF beam reception of an X-band phased array antenna utilizing highly dispersive photonic crystal fiber based true-time-delay, in *Proceedings of the Asia Optical Fiber Communication and Optoelectronic Exposition and Conference* (*AOE*), pp 1–3, 2008, paper SaJ2
39. W. Xue, J. Mork, Microwave photonic true time delay based on cross gain modulation in semiconductor optical amplifiers, in *Proceeding of the OptoElectronics and Communications Conference* (*OECC*), pp. 202–203, July 2010
40. M.A. Piqueras, G. Grosskopf, B. Vidal, J. Herrera, J.M. Martinez, P. Sanchis, V. Polo, J.L. Corral, A. Marceaux, J. Galiere, J. Lopez, A. Enard, J.-L. Valard, O. Parillaud, E. Estebe, N. Vodjdani, M.-S. Choi, J.H. den Besten, F.M. Soares, M.K. Smit, J. Marti, Optically beamformed beam-switched adaptive antennas for fixed and mobile broad-band wireless access networks. IEEE Trans. Microw. Theory Tech. **54**(2), 887–899 (2006)
41. Y. Liu, G. Burnham, G. Jin, J. Zhao, Wideband multi-beam photonics-based RF beamformer, in *Proceedings of the IEEE International Symposium on Phased Array Systems and Technology* (*ARRAY*), pp. 581–585, Oct 2010
42. B. Juswardy, F. Xiao, K.E. Alameh, Opto-VLSI-based RF beamformer for space division multiple access network, in *Proceedings of the High-Capacity Optical Networks and Enabling Technologies* (*HONET*), pp. 222–226, Dec 2010
43. H. Zmuda, E. N. Toughlian, Broadband nulling for conformal phased array antennas using photonic processing, in *Proceedings of the International Topical Meeting on Microwave Photonics* (*MWP*), pp. 17–19, Sept 2000
44. D. Marpaung, L. Zhuang, M. Burla, C. Roeloffzen, B. Noharet, Q. Wang, W.P. Beeker, A. Leinse, R. Heideman, Photonic integration and components development for a Ku-band phased array antenna system, in *Proceedings of the International Topical Meeting on Microwave Photonics* (*MWP*), pp. 458–461, Oct 2011
45. L. Zhuang, D. Marpaung, M. Burla, C. Roeloffzen, W. Beeker, A. Leinse, P. van Dijk, Low-loss and programmable integrated photonic beamformer for electronically-steered broadband phased array antennas, in *Proceedings of the IEEE Photonics Conference* (*PHO*), pp. 137–138, 2011
46. J. Chang, M.P. Fok, R.M. Corey, J. Meister, P.R. Prucnal, Highly scalable adaptive photonic beamformer using a single mode to multimode optical combiner. IEEE Microwave Wirel. Compon. Lett. **23**(10), 563–565 (2013)
47. M.P. Fok, Y. Deng, K. Kravtsov, P.R. Prucnal, Signal beating elimination using single-mode fiber to multimode fiber coupling. Opt. Lett. **36**(23), 4578–4580 (2011)

48. S. Lin, K.B. Ng, K.M. Luk, S.S.Wong, A. Poon, A 60 GHz digitally controlled RF beamforming array in 65 nm CMOS with off-chip antennas, in *Proceedings of the Radio Frequency Integrated Circuits Symposium* (*RFIC*), pp. 1–4, June 2011
49. D. Bojic, E. Sasaki, N. Cvijetic, T. Wang, J. Kuno, J. Lessmann, S. Schmid, H. Ishii, S. Nakamura, Advanced wireless and optical technologies for small-cell mobile backhaul with dynamic software-defined management. IEEE Commun. Mag. **51**(9), 86–93 (2013)
50. N. Cvijetic, A. Tanaka, Y.-K. Huang, M. Cvijetic, E. Ip, Y. Shao, T. Wang, 4-G mobile backhaul over OFDMA/TDMA-PON to 200 cell sites per fiber with 10 Gb/s upstream burst-mode operation enabling 1 ms transmission latency, in *Proceedings of the Optical Fiber Communication Conference and Exposition* (*OFC/NFOEC*), 2012, paper PDP5B.7
51. C. Liu, K. Sundaresan, M. Jiang, S. Rangarajan, G.-K. Chang, The case for re-configurable backhaul in cloud-ran based small cell networks, in *Proceedings of the IEEE INFOCOM*, pp. 1124–1132, 2013
52. B.J. Shastri, D.V. Plant, Scaling technologies for terabit fiber optic transmission systems, in *Proceedings of the SPIE*, vol. 7942, (San Francisco, CA, Feb 2011), paper 794206
53. B.J. Shastri, P.R. Prucnal, D.V. Plant, 20-GSample/s (10 GHz × 2 clocks) burst-mode CDR based on injection locking and space sampling for multiaccess networks. IEEE Photonics J. **4** (5), 1783–1793 (2012)
54. B.J. Shastri, D.V. Plant, Truly modular burst-mode CDR with instantaneous phase acquisition for multiaccess networks. IEEE Photonics Technol. Lett. **24**(2), 134–136 (2012)
55. B.J. Shastri, D.V. Plant, 5/10-Gb/s burst-mode clock and data recovery based on semiblind oversampling for PONs: theoretical and experimental. IEEE J. Sel. Top. Quantum Electron. **16** (5), 1298–1320 (2010)
56. B.J. Shastri, Y.B. M'Sallem, N. Zicha, L.A. Rusch, S. LaRochelle, D.V. Plant, Experimental study of burst-mode reception in a 1300 km deployed fiber link. J. Opt. Commun. Networking **2**(1), 1–9 (2010)
57. B.J. Shastri, Z.A. El-Sahn, M. Zeng, N. Kheder, L.A. Rusch, D.V. Plant, A standalone burst-mode receiver with clock and data recovery, clock phase alignment, and RS(255, 239) codes for SAC-OCDMA applications. IEEE Photonics Technol. Lett. **20**(5), 363–365 (2008)
58. Z.A. El-Sahn, B.J. Shastri, M. Zeng, N. Kheder, D.V. Plant, L.A. Rusch, Experimental demonstration of a SAC-OCDMA PON with burst-mode reception: local versus centralized sources. J. Lightwave Technol. **26**(10), 1192–1203 (2008)
59. B. Wu, B.J. Shastri, P.R. Prucnal, Secure communication in fiber-optic networks, in *Emerging Trends in Information and Communication Technologies Security*, ed. by B. Akhgar, H. Arabnia (Elsevier, Waltham, 2013), ch. 11, pp. 173–183
60. B. Wu, Z. Wang, Y. Tian, M.P. Fok, B.J. Shastri, D.R. Kanoff, P.R. Prucnal, Optical steganography based on amplified spontaneous emission noise. Opt. Express **21**(2), 2065–2071 (2013)
61. B. Wu, Z. Wang, B.J. Shastri, M.P. Chang, N.A. Frost, P.R. Prucnal, Temporal phase mask encrypted optical steganography carried by amplified spontaneous emission noise. Opt. Express **22**(1), 954–961 (2014)
62. B. Wu, M.P. Chang, B.J. Shastri, Z. Wang, P.R. Prucnal, Analog noise protected optical encryption with two-dimensional key space. Opt. Express **22**(12), 14568–14574 (2014)
63. B. Wu, B.J. Shastri, P.R. Prucnal, System performance measurement and analysis of optical steganography based on amplifier noise. IEEE Photonics Technol. Lett. **26**(19), 1920–1923 (2014)
64. B. Wu, M.P. Chang, Z. Wang, B.J. Shastri, P.R. Prucnal, Optical encryption based on cancellation of analog noise, in *Proceedings Conference on Lasers and Electro-Optics* (*CLEO*), San Jose, CA, June 2014, paper AW3P.5
65. B. Wu, Z. Wang, B.J. Shastri, Y. Tian, P.R. Prucnal, Phase mask encrypted optical steganography based on amplified spontaneous emission noise, in *Proceedings of the IEEE Photonics Conference* (*IPC*), Seattle, Sept 2013, paper MG3.3, pp. 137–138

66. B. Wu, Z. Wang, B.J. Shastri, Y. Tian, P.R. Prucnal, Two dimensional encrypted optical steganography based on amplified spontaneous emission noise, in *Proceedings of the Conference on Lasers and Electro-Optics* (*CLEO*), San Jose, June 2013, paper AF1H.5
67. A. N. Tait, M.A. Nahmias, Y. Tian, B.J. Shastri, P.R. Prucnal, Photonic neuromorphic signal processing and computing, in *Nanophotonic Information Physics*, ser. Nano-Optics and Nanophotonics, ed. by M. Naruse, Springer, Heidelberg, 2014, pp. 183–222. http://dx.doi.org/10.1007/978-3-642-40224-1_8
68. B.J. Shastri, A.N. Tait, M.A. Nahmias, P.R. Prucnal, Photonic spike processing: ultrafast laser neurons and an integrated photonic network. IEEE Photonics Soc. Newsl. **28**(3), 4–11 (2014)
69. M.A. Nahmias, B.J. Shastri, A.N. Tait, P.R. Prucnal, A leaky integrate-and-fire laser neuron for ultrafast cognitive computing, IEEE J. Sel. Top. Quantum Electron. **19**(5), 1800212 (Sept–Oct 2013)
70. A.N. Tait, M.A. Nahmias, B.J. Shastri, P.R. Prucnal, Broadcast and weight: an integrated network for scalable photonic spike processing. J. Lightwave Technol. **32**(21), 3427–3439 (2014)
71. B.J. Shastri, M.A. Nahmias, A.N. Tait, P.R. Prucnal, Simulations of a graphene excitable laser for spike processing. Opt. Quant. Electron. **46**(10), 1353–1358 (2014)
72. A.N. Tait, B.J. Shastri, M.A. Nahmias, M.P. Fok, P.R. Prucnal, The DREAM: an integrated photonic thresholder. J. Lightwave Technol. **31**(8), 1263–1272 (2013)
73. A.N. Tait, M.A. Nahmias, B.J. Shastri, P.R. Prucnal, Broadcast-and-weight interconnects for integrated distributed processing systems, in *Proceedings of the IEEE Optical Interconnects Conference* (*OI*), Coronado Bay, CA, May 2014, paper WA3, pp. 108–109
74. B.J. Shastri, M.A. Nahmias, A.N. Tait, Y. Tian, B. Wu, P.R. Prucnal, Graphene excitable laser for photonic spike processing, *IEEE Photonics Conf.* (*IPC*), Sept 2013, pp. 1–2. http://dx.doi.org/10.1109/IPCon.2013.6656424
75. M.A. Nahmias, A.N. Tait, B.J. Shastri, P.R. Prucnal, An evanescent hybrid silicon laser neuron, in *Proceedings of the IEEE Photonics Conference* (*IPC*), Seattle, WA, Sept 2013, paper ME3.4, pp. 93–94
76. B.J. Shastri, M.A. Nahmias, A.N. Tait, Y. Tian, M.P. Fok, M.P. Chang, B. Wu, P.R. Prucnal, Exploring excitability in graphene for spike processing networks, in *Proceedings of the IEEE Numerical Simulation of Optoelectronic Devices* (*NUSOD*), Vancouver, Canada, Aug 2013, paper TuC5, pp. 83–84

Index

S. Wabnitz and B.J. Eggleton (eds.), *All-Optical Signal Processing*,
Springer Series in Optical Sciences 194, DOI 10.1007/978-3-319-14992-9

MIX
Papier aus verantwortungsvollen Quellen
Paper from responsible sources
FSC® C105338

If you have any concerns about our products,
you can contact us on
ProductSafety@springernature.com

In case Publisher is established outside the EU,
the EU authorized representative is:
Springer Nature Customer Service Center GmbH
Europaplatz 3, 69115 Heidelberg, Germany

Printed by Libri Plureos GmbH
in Hamburg, Germany